Network Routing

The Morgan Kaufmann Series in Networking

Series Editor, David Clark, M.I.T.

*For further information on these books and for a list of
forthcoming titles,
please visit our Web site at http://www.mkp.com.*

Network Routing

Algorithms, Protocols, and Architectures

Deepankar Medhi

Karthikeyan Ramasamy

AMSTERDAM • BOSTON • HEIDELBERG • LONDON
NEW YORK • OXFORD • PARIS • SAN DIEGO
SAN FRANCISCO • SINGAPORE • SYDNEY • TOKYO

Morgan Kaufmann is an imprint of Elsevier

ELSEVIER

Senior Acquisitions Editor	Rick Adams
Acquisitions Editor	Rachel Roumeliotis
Publishing Services Manager	George Morrison
Senior Production Editor	Dawnmarie Simpson
Cover Design	Eric DeCicco/Yvo Riezebos Design
Cover Image	Getty Images
Composition	VTEX
Copyeditor	SPi
Proofreader	SPi
Indexer	SPi
Interior printer	The Maple-Vail Book Manufacturing Group
Cover printer	Phoenix Color, Inc.

Morgan Kaufmann Publishers is an imprint of Elsevier.
500 Sansome Street, Suite 400, San Francisco, CA 94111

This book is printed on acid-free paper.

Library of Congress Cataloging-in-Publication Data
Medhi, Deepankar.
    Network routing : algorithms, protocols, and architectures / Deepankar Medhi, Karthikeyan Ramasamy.
       p. cm.
    Includes bibliographical references and index.
    ISBN-13: 978-0-12-088588-6 (hardcover : alk. paper)
    ISBN-10: 0-12-088588-3 (hardcover : alk. paper) 1. Computer networks. 2. Routers (Computer networks) 3. Computer network architectures. I. Ramasamy, Karthikeyan, 1967- II. Title.
    TK5105.5.M425 2007
    004.6–dc22

                                    2006028700

ISBN 13: 978-0-12-088588-6
ISBN 10: 0-12-088588-3

Printed in the United States of America
07 08 09 10   5 4 3 2 1

To Karen, Neiloy, and Robby: the core routers in our dynamic network where the distance cost varies instantaneously and over time, and where alternate routing and looping occur ...

Love,

Deep/Dad

To my parents, R. Chellammal and N. Ramasamy—backplane of my life.

Love and regards,

Karthik

Contents

16 IP Packet Filtering and Classification

Foreword

My involvement with computer networking started with TheoryNet (1977), an e-mail system for theoretical computer scientists. Later (1981) I helped lead the computer science network (CSNET) project, which eventually connected most academic and many industrial computer research groups. In the early days, our efforts were primarily focused on providing connectivity and being able to use applications such as e-mail, ftp, and telnet. However, even in the simple (by today's standards) environment of the 1970s and early 1980s (Arpanet, CSNET, and other experimental Internet networks), getting routing "right" turned out to be quite challenging.

I was fortunate to be part of the NSFNET regional/backbone model development. This is when I began to fully understand the significance of routing in a large-scale multi-domain network and, in particular, the central role of policy issues in such a decentralized environment. Over the past decade, as the Internet became ubiquitous and global in scale, routing has become ever more important. Packets must be forwarded efficiently from one end of the world to the other with minimal perception of delay. This has required tremendous efforts on many fronts: how to evolve routing protocols for large-scale loosely-coupled networking environments, how to engineer a network for efficient routing from an operational point of view, how to do efficient packet processing at routers, and how to effectively take into account the complexity of policy issues in the determination of routes. And while there have been many exciting advances over the past two decades, much work remains to be done.

In parallel, we have seen tremendous advances in traditional telephony. The underlying telecommunication system has changed from analog to digital and has incorporated the latest advances in optical technologies and, more recently, voice over IP. Throughout these revolutionary changes, routing has continued to play a critical role.

We are now at a crossroad. Various efforts are underway to determine a framework for next generation networks that allow seamless convergence of services and a platform to more easily create new services. Among other things, this requires a fresh look at routing. To be successful, it is important that we understand what has worked to date. To better understand the issues and complexities, we should look at this broadly, considering a variety of *different network architectures*, not just for the Internet. For each such network architecture, we can benefit from understanding its principles, protocols, algorithms, and functions, with a focus on routing. This will help give us perspective as we consider how to design routing for the next-generation network.

In this regard, Deepankar Medhi and Karthikeyan Ramasamy's book, *Network Routing: Algorithms, Protocols, and Architectures*, is very timely. Departing from most other works, it is unique in providing an in-depth understanding of routing in a wide variety of types of networks. It includes extensive coverage of the evolution of routing over time. Particularly appealing is its in-depth coverage across a spectrum of algorithmic, technical, experiential, and practical issues. In addition, the detailed coverage of routers and switches is particularly valuable, as it helps the reader gain an understanding of why different approaches and components are needed to address packet processing, especially for scalability. In this regard, it is uniquely successful in drawing an important connection between routing and routers.

Medhi and Ramasamy's presentation is clear and approachable, allowing a wide audience to understand and gain an appreciation of network routing. I believe that it will become a core reference book on routing for router developers, network providers, students, and researchers for both today's practitioners and those who are interested in next-generation routing.

LAWRENCE LANDWEBER
Past John P. Morgridge Chair and Past Department Chairman
Computer Science Department, University of Wisconsin–Madison
Fellow, Association for Computing Machinery and
Recipient of IEEE Award on International Communication
Former President and Chair of the Board of Trustees, Internet Society

Preface

In the span of a quarter-century, network routing in communication networks has evolved tremendously. Just a quarter-century ago, the public switched telephone network (PSTN) was running hierarchical routing, ARPANET routing was operational, and the telecommunication infrastructure had fixed static transport routes. In the 1980s, we saw the first tremendous growth in routing: Internet routing was deployed under the TCP/IP stack starting, first with the RIP protocol; the telephone network started deploying dynamic call routing schemes; and the telecommunication transport network deployed SONET transport mechanisms, which could reroute in a ring topology in 40 millisec in the event of a failure. In the past fifteen years, we have seen the need for policy routing because of multiprovider settings, and the need to develop fast lookup algorithms for packet processing that enables efficient routing. We have also seen interdependency between addressing and routing as first addressed through classless interdomain routing (CIDR) and more recently, because of number portability in the PSTN. More importantly, we saw how the way an addressing scheme is deployed can impact routing and lookup algorithms.

Network routing can be broadly divided into three basic fundamental categories: packet routing, circuit-switched routing, and transport routing; certainly, a combination is possible. The evolution over the past quarter-century has brought to the foreground the need to understand and examine where and how different dimensions of routing, from algorithms to protocols to architectures, can differ for different types of networks, and where they intersect. Certainly, the goal is to learn from our past experiences and prepare ourselves for next generation networks and routing.

While numerous papers have been written on the subject of network routing, and several books are now available on routing for specific networks, the field still lacks a comprehensive or systematic guide that encompasses various routing paradigms. Second, even in discussions of a single routing type (for example, either the Internet or PSTN), the focus often appears to be either on protocols or algorithms without tying them together with analysis and implementation; or, the work delves more into router command-line for router configuration; or, being informational without explaining the whys. Furthermore, how the addressing mechanism can affect routing decisions is yet another important topic that is rarely discussed. For efficient routing, how routers are architectured—and why—is yet another mystery. Finally, the relation between traffic engineering and efficient routing is also another topic. In the end, one needs to be somewhat of an "expert" in different routing paradigms to get a well-rounded view.

Last, after investigating routing in different networks for a number of years, we have come to the conclusion that network routing is like an economy. Similar to macroeconomics and microeconomics, network routing also has macro- and micro-centric issues. In addition, seemingly different and conflicting systems can and do co-exist. Not all of the issues are purely technical; business relations and regulatory issues are also important to recognize and consider. This book is an attempt to paint a broad picture that encompasses various aspects of network routing in one place.

AUDIENCE

Our goal has been to create a book that can be used by a diverse set of audiences, with varied levels of background. Specifically, we set out to create a book that can be used by professionals, as well as students and researchers. In general, this is intended as a self-study. We assume that the reader already has some basic knowledge of networking. Among professionals, the intent has been to cover two broad groups: router developers, including protocol designers and router architects, and network designers and operators, with the overall goal to bring out issues that one group might want to understand that the other group faces. For students, this book is intended to help learn about routing in depth, along with the big picture and lessons from operational and implementation experience. For researchers who want to know what has been done so far and what critical issues to address for next-generation routing, this is intended as a helpful reference. In general, this book has been intended as a one-stop treat for all interested in network routing in different networks.

ORGANIZATION AND APPROACH

The book is organized into six parts. Each part starts with a chapter-level summary. We present below a brief overview of each part:

- Part I (four chapters): We cover the basic foundations of routing from algorithms to protocols, along with network flow modeling.

- Part II (five chapters): This part is about IP network routing, from standardized protocols for both intra- and inter-domain routing, to IP traffic engineering and Internet routing architectures.

- Part III (four chapters): This part covers PSTN routing, from hierarchical routing to dynamic routing, and from addressing to traffic engineering, including the role of signaling in routing, along with the impact of number portability in routing.

- Part IV (three chapters): In this part, we cover router architectures for different scale routers for efficient packet processing, along with address lookup algorithms and packet filtering and classification mechanisms.

- Part V (four chapters): As impetuses for next generation routing, we present quality-of-service routing, multiprotocol label switching, generalized multiprotocol label switching, and routing at the intersection of IP-PSTN for voice over IP.

- Part VI (five chapters): This bonus material (available on the CD-ROM) is made up of two sub-parts: the first three chapters continue beyond Part IV by delving more into routers by presenting efficient switching, packet queueing and scheduling, and traffic conditioning; the remaining two chapters extend Part V by covering transport network routing, optical network routing, and multi-layer routing.

At the beginning of each chapter, a reading guideline is provided. This gives a brief description on the background needed to read the chapter; it also discusses which other chapters this chapter is connected to or has dependency on. In general, it is not necessary to read the chapters in sequential order. Furthermore, the chapters are organized in a way so that the reader who has familiarity with a particular topic can move on and read other chapters of interest. Similarly, there are a few chapters on traffic engineering that require a certain level of mathematical background. They can be read independently if the reader has the background, or can be skipped for later reading, without missing the broad picture. Regardless, each chapter contains a Further Lookup section, which includes a brief discussion on additional reading; followed by a set of exercises that is meant for a wide audience. Notations, conventions, and symbols used in the book are summarized in Appendix A. Miscellaneous refresher topics that are helpful in understanding the material presented in this book are included in Appendix B.

In general, we have given special attention to being concise about describing each topic, while ensuring that the material is approachable for a wider audience. The book is still hefty in size in order to cover routing in different networks. Despite our keen interest, we needed to make the decision to leave out certain important topics instead of cutting corners on the topics presented. The topics *not* covered in the book (except for cursory remarks) are: multicast routing, routing in ATM networks, routing in cellular/wireless networks, routing in sensor networks, and security vulnerabilities in routing. The router command-line–based configuration of protocols is not included in this book, because there are many detailed books available on this aspect for various Internet routing protocols. Finally, there is a direct connection between routing and capacity design and planning. For an in-depth treatment of capacity design and planning, the reader is referred to the companion book [564].

BONUS MATERIALS AND ONLINE RESOURCES

The book, in its printed form, has 20 chapters. A CD-ROM is provided with the book that contains an additional five chapters labeled "Advanced Topics." Of these five chapters, three chapters are related to router architectures: switching packets (Chapter 21), packet queueing and scheduling (Chapter 22), and traffic conditioning (Chapter 23). The remaining two chapters are related to transport and next-generation routing: transport network routing (Chapter 24), and optical network routing and multilayer routing (Chapter 25).

Additional support materials (for example, instructional materials and additional exercises) will be available at http://www.mkp.com/?isbn=9780120885886 and http://www.NetworkRouting.net. The latter site will also serve as a resource site and will provide links to materials available on the web on network routing.

ACKNOWLEDGMENTS

To quote Jeff Doyle, "An author of a technical book is just a front man for a small army of brilliant, dedicated people." We could not have said it better.

Our official technical reviewers did a tremendous job of reading carefully and providing detailed comments. We thank Jennifer Rexford (Princeton University), Ibrahim Matta (Boston University), K. R. Krishnan (Telcordia Technologies), and Kannan Varadhan (Juniper Networks) for lending their expertise, time, and effort.

In addition, many afforded their expertise by reading one or more chapters and by providing valuable feedback; we gratefully acknowledge Amit Shukla (Microsoft), Arthi Ayyangar (Nuova Systems), Caterina Scoglio (Kansas State University), Chelian Pandian (Juniper Networks), Dana Blair (Cisco Systems), David Walden (BBN, retired), Debashis Talukdar (Embarq), Dock Williams (Juniper Networks), Driss Benhaddou (University of Houston), Hua Qin (Beijing University of Technology), Hui Zhang (Carnegie Mellon University), Jeff Naughton (University of Wisconsin–Madison), Jignesh M. Patel (University of Michigan), Johannes Gehrke (Cornell University), John Strand (AT&T Labs), Mario Baldi (Politecnico di Torino), Prasad Deshpande (IBM), Prosper Chemouil (France Telecom R&D), Rahul Agrawal (Juniper Networks), Ravi Chandra (Sonoa Systems), Raymond Reeves (Sprint), Saad Siddiqi (Sprint), Shachi Sharma (Alcatel), Srinivas Seshadri (Kosmix), Steve Dispensa (Positive Networks), Vamsi Valluri (Cisco Systems), Venkatesh Iyengar (Sun Microsystems), and Vijay Talati (Juniper Networks).

The first author's colleagues in the Networking group at the University of Missouri–Kansas City, Appie van de Liefvoort, Baek-Young Choi, Cory Beard, Jerry Place, Ken Mitchell, and Khosrow Sohraby, served as great resources. They read one or more chapters, were around to have a quick discussion and to provide their theoretical as well as practical expertise when needed. Appie van de Liefvoort and Khosrow Sohraby, in their roles as administrators, provided a much-needed environment for the first author to carry out a project of this magnitude without too many distractions. More than a decade ago, a former colleague, Adrian Tang, was instrumental and believed in the importance of creating a separate course on network routing; with his interest and the nod from Richard Hetherington (the then director), the first author developed and taught a course on network routing encompassing different networks for the first time in fall 1995; he also benefited from the publication of [667] in 1995 that helped jump-start this course. Since then, he has been teaching this course every fall (except when he was on a sabbatical leave). The content has changed significantly in this short span of time to keep up with what has been happening in the field, providing an exciting challenge and opportunity. He gratefully acknowledges having a sabbatical in 2004 to plan for the initial preparation for this book.

The current and recent PhD students of the first author also read many chapters and provided valuable feedback. Many thanks to Amit Sinha, Balaji Krithikaivasan, Dijiang Huang, Gaurav Agrawal, Haiyang Qian, Plarent Tirana, and Shekhar Srivastava.

Several students who took the course, Network Routing, from the first author, in the fall of 2005, read the initial version of the first few chapters. When he taught it again in the fall 2006 semester, the entire manuscript was ready in its draft form; the entire class helped debug it by carefully reading various chapters and providing detailed feedback. For their help, we would like to thank Aditya Walavalkar, Ajay Karanam, Amol Rege, Dong Yoo, Francisco Jose Landeras, Hafeez Razzaq, Haiyang Qian, Hui Chang, Jignesh K. Patel, Jin-Ho Lee,

Jorge Rodriguez, Palani Ramalingam, Phaneesh Gururaj, Ravi Aute, Rehan Ishrat, Ron Mc-Manaman, Satoru Yamashita, Sreeram Gudipudi, Swapnil Deshmukh, Shamanth Kengeri, Shashank Manchireddy, Sundeep Udutha, Tongan Zhao, and Venkat Pagadala. Needless to say, the first author greatly benefited from many questions and discussions from teaching this course over the past decade that altogether attracted more than 300 students. The second author also benefited from his many interactions with colleagues while working at Juniper Networks. As a result, a range of interrelated topics is included in the book to give a broader perspective of network routing.

Over the years, we have both benefited from informative and enlightening discussions on routing in different domains and related topics from many individuals; many also answered queries during the preparation of this book. We like to thank the following: Aekkachai Rattanadilokochai (Cisco Systems), Åke Arvidsson (Ericsson), Amarnath Mukherjee (Clarifyre), Ananth Nagarajan (Juniper Networks), André Girard (INRS-EMT), Bharani Chadalavada (Juniper Networks), Brion Feinberg (Sereniti), Brunilde Sansò (University of Montréal), David DeWitt (University of Wisconsin–Madison), David Tipper (University of Pittsburgh), David Mills (University of Delaware), David Walden (BBN, retired), Debasis Mitra (Bell Labs), Di Yuan (Linköping Institute of Technology), Fu Chang (Academia Sinica), Gerald Ash (AT&T Labs), Gerald Combs (CACE Technologies, creator of Ethereal/Wireshark), Geoff Huston (APNIC), Götz Gräfe (HP Labs), Hadriel Kaplan (Acme Packet), Indrajanti (Yanti) Sukiman (Cisco Systems), Iraj Saniee (Bell Labs), Jean-François Labourdette (Verizon), Jeff Naughton (University of Wisconsin–Madison), Jim Pearce (Sprint), John Strand (AT&T Labs), Keith Ross (Polytechnic University), Larry Landweber (University of Wisconsin–Madison), Lindsay Hiebert (Cisco Systems), Lorne Mason (McGill University), Michał Pióro (Warsaw University of Technology and Lund University), Mikkel Thorup (AT&T Labs–Research), Mostafa Ammar (Georgia Tech), Mukesh Kacker (NetApp), Nitin Bahadur (Juniper Networks), Oscar González-Soto (ITU), Philip Smith (Cisco Systems), Pramod Srinivasan (Juniper Networks), Prosper Chemouil (France Telecom R&D), Rajat Monga (Attributor), Ravi Chandra (Sonoa Systems), Richard Harris (Massey University), Robert Doverspike (AT&T Labs–Research), Ron Skoog (Telcordia Technologies), Saad Siddiqi (Sprint), Samir Shah (Cisco Systems), Saravan Rajendran (Cisco Systems), Sergio Beker (France Telecom R&D), Shankar Satyanarayanan (Cisco Systems), Srinivasa Thirumalasetty (Ciena Corporation), Steve Dispensa (Positive Networks), Steve Robinson (University of Wisconsin–Madison), Toshikane Oda (Nippon Ericsson), Ulka Ranadive (Cisco Systems), Vamsi Valluri (Cisco Systems), Villy Bæk Iversen (Technical University of Denmark), Wayne Grover (TR-Labs & University of Alberta), Wen-Jung Hsin (Park University), Wesam Alanqar (Sprint), Yufei Wang (VPI Systems), and Zhi-Li Zhang (University of Minnesota). Furthermore, the first author benefited from Karen Medhi's insight and expertise in transport network routing and design.

Folks at AS3390 often provided their perspective from the viewpoint of running a stub AS by answering our questions. Our sincere thanks to the following individuals at AS3390: David Johnston, George Koffler, Jim Schonemann, II, and Justin Malyn.

We thank David Clark (M.I.T.), Series Editor for the Morgan Kaufmann series in Networking, for recognizing the importance of having a book that spans network routing in different networks, and for greenlighting our book proposal. We are honored that Larry Landweber

(University of Wisconsin–Madison) gladly accepted our request to write the Foreword for this book.

The first author would like to thank the Defense Advanced Research Project Agency, the National Science Foundation, the University of Missouri Research Board, and Sprint Corporation for supporting his networking research.

Two individuals deserve special note: (1) Jane Zupan took Network Routing from the first author in 2000. She typed his scattered notes, which gave the initial idea for planning this book. Then, at a conference in 2003, the three of us casually joked about doing this book, and the plan finally started. Jane was an initial partner in this project but could not continue to work on it because of time constraints. She, however, continued to provide much help, even at the last stage, by reading and editing, despite her busy schedule. We sincerely thank her for her great help. (2) Balaji Krithikaivasan, who completed a Ph.D. under the first author, passionately read many chapters of the book and provided feedback despite his deteriorating health. Unfortunately, he did not live long enough to see the final publication of the book. Thank you, Bala, wherever you are.

It has been a pleasure to work with Rick Adams and Rachel Roumeliotis of Morgan Kaufmann Publishers/Elsevier. From the initial proposal of the book to final production, they provided guidance in many ways, not to mention the occasional reminder. We appreciate their patience with us during the final stages of the manuscript preparation. Arline Keithe did a nice job of copyediting. In the pre-production phase, we received help from Project Managers Dawnmarie Simpson and Tracy Grace. Folks at VTEX did an excellent job of taking our LATEX files and transforming them to production quality, and Judy Ahlers did great work on the final proofread. We thank them all.

Our immediate family members suffered the most during our long hours of being glued to our laptops. Throughout the entire duration, they provided all sorts of support, entertainments, and distractions. And often they queried "Are you ever going to get this done?" Deep would like to thank his wife, Karen, and their sons, Neiloy and Robby, for love and patience, and for enduring this route (for the second time). He would also like to thank cc76batch—you know who you are—for their friendship. Karthik would like to thank his wife, Monika, for her love and patience. He would also like to acknowledge his family members, Sudarshan Kumar and Swarn Durgia, Sonu and Rajat Monga, and Tina and Amit Babel for keeping him motivated. In addition, he would like to thank his many friends for their support. Finally, we like to thank our parents and our siblings for their support and encouragement.

DEEPANKAR (DEEP) MEDHI   KARTHIKEYAN (KARTHIK) RAMASAMY
Kansas City, Missouri, USA   Santa Clara, California, USA
dmedhi@umkc.edu   karthik@cs.wisc.edu

About the Authors

Deepankar Medhi is Professor of Computer Networking in the Computer Science & Electrical Engineering Department at the University of Missouri–Kansas City, USA. Prior to joining UMKC in 1989, he was a member of the technical staff in the traffic network routing and design department at the AT&T Bell Laboratories. He was an invited visiting professor at Technical University of Denmark and a visiting research fellow at the Lund University, Sweden. He is currently a Fulbright Senior Specialist. He serves as a senior technical editor of the *Journal of Network & Systems Management*, and is on the editorial board of *Computer Networks*, *Telecommunication Systems*, and *IEEE Communications Magazine*. He has served on the technical program committees of numerous conferences including IEEE INFOCOM, IEEE NOMS, IEEE IM, ITC, and DRCN. He received B.Sc. (Hons.) in Mathematics from Cotton College, Gauhati University, India, an M.Sc. in Mathematics from the University of Delhi, India, and an M.S. and a Ph.D. in Computer Sciences from the University of Wisconsin–Madison, USA. He has published more than 70 papers, and is co-author of the book *Routing, Flow, and Capacity Design in Communication and Computer Networks*, also published by Morgan Kaufmann (July 2004).

Karthikeyan Ramasamy has 15 years of software development experience, including working with companies such as Juniper Networks, Desana Systems, and NCR. His primary areas of technical expertise are networking and database management. As a member of the technical staff at Juniper, he developed and delivered a multitude of features spanning a diverse set of technologies including protocols, platforms, databases, and high availability solutions for the JUNOS routing operating system. As a principal engineer at Desana Systems, he was instrumental in developing and delivering many fundamental components of an L7 switch for managing data centers. While pursuing his doctorate, he worked on a parallel object relational database system, which was spun off as a company and later acquired by NCR. Subsequently, he initiated a project in data warehousing which was adapted by NCR. As a consulting software engineer at NCR, he assisted in the commercialization of these technologies. Currently, he serves as an independent consultant. He received a B.E. in Computer Science and Engineering with distinction from Anna University, India, an M.S. in Computer Science from the University of Missouri–Kansas City, USA, and a Ph.D. in Computer Sciences from the University of Wisconsin–Madison, USA. He has published papers in premier conferences and holds 7 patents.

Part I: Network Routing: Basics and Foundations

We start with basics and foundations for network routing. It has four chapters.

In Chapter 1, we present a general overview of networking. In addition, we present a broad view of how addressing and routing are tied together, and how different architectural components are related to routing.

A critical basis for routing is routing algorithms. There are many routing algorithms applicable to a variety of networking paradigms. In Chapter 2, we present shortest and widest path routing algorithms, without referring to any specific networking technology. The intent here is to understand the fundamental basis of various routing algorithms, both from a centralized as well as a distributed point of view.

In Chapter 3, we consider routing protocols that play complementary roles to routing algorithms. The important point to understand about routing protocols is the nature of its operation in a distributed environment in which information are to be exchanged, and when and what information to be exchanged. Fundamentally, there are two routing protocol concepts: distance vector and link state. The path vector routing protocol extends the distance vector approach by including path information; however, this results in an operational behavior that can be drastically different than a distance vector protocol. Thus, the scope of this chapter is to present routing protocols in details, along with illustrations, however, without any reference to a particular networking technology.

This part concludes by presenting background material that is important for traffic engineering of networks. It may be noted that routing and traffic engineering are inter-twined. Thus, a good understanding of the fundamentals of how network flow modeling and optimization can be helpful in traffic engineering is important. Chapter 4 covers network flow modeling and introduces a number of objectives that can be application for network traffic engineering, and describes how different objectives can lead to different solutions.

1

Networking and Network Routing: An Introduction

Not all those who wander are lost.

J. R. R. Tolkien

It is often said that if anyone were to send a postcard with minimal address information such as "Mahatma Gandhi, India" or "Albert Einstein, USA," it would be routed to them due to their fame; no listing of the street address or the city name would be necessary. The postal system can do such routing to famous personalities usually on a case-by-case basis, relying on the name alone.

In an electronic communication network, a similar phenomenon is possible to reach *any* website or to contact *any* person by telephone anywhere in the world without knowing where the site or the person is currently located. Not only that, it is possible to do so very efficiently, within a matter of a few seconds.

How is this possible in a communication network, and how can it be done so quickly? At the heart of the answer to this question lies *network routing*. Network routing refers to the ability of an electronic communication network to send a unit of information from point A to point B by determining a path through the network, and by doing so efficiently and quickly. The determination of an efficient path depends on a number of factors, as we will be discussing in detail throughout this book.

First, we start with a key and necessary factor, known as *addressing*. In a communication network, addressing and how it is structured and used plays a critical role. In many ways, addressing in a communication network has similarities to postal addressing in the postal system. Thus, we will start with a brief discussion of the postal addressing system to provide an analogy.

A typical postal address that we write on a postcard has several components—the name of the person, followed by the street address with the house number ("house address"), followed by the city, the state name, and the postal code. If we, on the other hand, take the processing view to route the postcard to the right person, we essentially need to consider this address in the reverse order of listing, i.e., start with the postal code, then the city or the state name, then the house address, and finally the name of the person. You may notice that we can reduce this information somewhat; that is, you can just use the postal code and leave out the name of the city or the name of the state, since this is redundant information. This means that the information needed in a postal address consists of three main parts: the postal code, the street address (with the house number), and the name.

A basic routing problem in the postal network, then, is as follows: the postcard is first routed to the city or the geographical region where the postal code is located. Once the card reaches the postal code, the appropriate delivery post office for the address specified is identified and delivered to. Next, the postman or postwoman delivers the postcard at the address, without giving much consideration to the name listed on the card. Rather, once the card arrives at the destination address, the residents at this address take the responsibility of handing it to the person addressed.

You may note that at a very high-level view, the routing process in the postal system is broken down to three components: how to get the card to the specific postal code (and subsequently the post office), how the card is delivered to the destination address, and finally, how it is delivered to the actual person at the address. If we look at it in another way, the place where the postcard originated in fact does not need to know the detailed information of the street or the name to start with; the postal code is sufficient to determine to which geographical area or city to send the card. Thus, we can see that postal routing uses *address hierarchy* for routing decisions. An advantage of this approach is the decoupling of the routing decision to multiple levels such as the postal code at the top, then the street address, and so on. An important requirement of this hierarchical view is that there must be a way to divide the complete address into multiple distinguishable parts to help with the routing decision.

Now consider an electronic communication network; for example, a critical communication network of the modern age is the Internet. Naturally, the first question that arises is: how does addressing work for routing a unit of information from one point to another, and is there any relation to the postal addressing hierarchy that we have just discussed? Second, how is service delivery provided? In the next section, we address these questions.

1.1 Addressing and Internet Service: An Overview

In many ways, Internet addressing has similarities to the postal addressing system. The addressing in the Internet is referred to as *Internet Protocol (IP) addressing*. An IP address defines two parts: one part that is similar to the postal code and the other part that is similar to the house address; in Internet terminology, they are known as the *netid* and the *hostid*, to identify a network and a host address, respectively. Thus, a host is the end point of communication in the Internet and where a communication starts. A host is a generic term used for indicating many different entities; the most common ones are a web-server, an email server, and certainly the desktop, laptop, or any computer we use for accessing t he Internet. A netid identifies a contiguous block of addresses; more about IP Addressing later in Section 1.3.

Like any service delivery system, we also need a delivery model for the Internet. For example, in the postal system, one can request guaranteed delivery for an additional fee. The Internet's conceptual framework, known as *TCP/IP (Transmission Control Protocol/Internet Protocol)*, relies on a delivery model in which TCP is in charge of the reliable delivery of information, while IP is in charge of routing, using the IP addressing mechanism. IP, however, does not worry about whether the information is reliably delivered to the address or is lost during transit. This is somewhat similar to saying that the postal system will route a postcard to the house address, while residents at this address (not the postal authority) are responsible for ensuring that the person named on the card receives it. While this may seem odd at first, this paradigm has been found to work well in practice, as the success of the Internet shows.

A key difference in the Internet as opposed to the postal system is that the sending host first sends a beacon to the destination address (host) to see if it is reachable, and waits for an acknowledgment *before* sending the actual message. Since the beacon also uses the same transmission mechanism, i.e., IP, it is possible that it may not reach the destination. In order to allow for this uncertainty to be factored in, another mechanism known as a *timer* is used. That is, the sending host sends the beacon, then waits for a certain amount of time to see if it receives any response. If it does not hear back, it tries to send the beacon a few more times, waiting for a certain amount of time before each attempt, until it stops trying after reaching the limit on the maximum number of attempts. The basic idea, then, requires that the receiving host should *also* know the address of the sender so that it can acknowledge the receipt of the beacon. As you can see, this means that when the sending host sends its beacon, it must also include its source IP address.

Once the connectivity is established through the beacon process, the actual transmission of the content transpires. This is where a good analogy is not available in the postal system;

rather, the road transportation network is a better fit to describe an analogy. If we imagine a group of 100 friends wanting to go to a game, then we can easily see that not all can fit in one car. If we consider that a car can hold five people, we will need twenty cars to transport this entire group. The Internet transfer model also operates in this fashion. Suppose that a document that we want to download from a host (web-server) is 2 MB. Actually, it cannot be accommodated entirely into a single fundamental unit of IP, known as *packet* or *datagram*, due to a limitation imposed by the underlying transmission system. This limitation is known as the *Maximum Transmission Unit* (MTU). MTU is similar to the limitation on how many people can fit into a single car. Thus, the document would need to be broken down into smaller units that fit into packets. Each packet is then labeled with both the destination and the source address, which is then routed through the Internet toward the destination. Since the IP delivery mechanism is assumed to be unreliable, any such packet can possibly get lost during transit, and thus would need to be retransmitted if the timer associated with this packet expires. Thus another important component is that content that has been broken down into smaller packets, once it arrives at the destination, needs to be reassembled in the proper order before delivering the document.

We conclude this section by pointing out that the acknowledgment and retransmission mechanism is used for most well-known applications on the Internet such as web or email. A slightly different model is used for applications that do not require reliable delivery; this will be discussed later in the chapter.

1.2 Network Routing: An Overview

In the previous section, we provided a broad overview of addressing and transfer mechanisms for data in Internet communication services. Briefly, we can see that eventually packets are to be routed from a source to a destination. Such packets may need to traverse many cross-points, similar to traffic intersections in a road transportation network. Cross-points in the Internet are known as *routers*. A router's functions are to read the destination address marked in an incoming IP packet, to consult its internal information to identify an outgoing link to which the packet is to be forwarded, and then to forward the packet. Similar to the number of lanes and the speed limit on a road, a network link that connects two routers is limited by how much data it can transfer per unit of time, commonly referred to as the *bandwidth* or *capacity* of a link; it is generally represented by a data rate, such as 1.54 megabits per second (Mbps). A network then carries *traffic* on its links and through its routers to the eventual destination; traffic in a network refers to packets generated by different applications, such as web or email.

Suppose that traffic suddenly increases, for example, because of many users trying to download from the same website; then, packets that are generated can possibly be queued at routers or even dropped. Since a router maintains a finite amount of space, known as a *buffer*, to temporarily store backlogged packets, it is possible to reach the buffer limit. Since the basic principle of TCP/IP allows the possibility of an IP packet not being delivered or being dropped enroute, the finite buffer at a router is not a problem. On the other hand, from an efficient delivery point of view, it is desirable not to have any packet loss (or at least, minimize it) during transit. This is because the reliable delivery notion works on the principle of retransmission and acknowledgment and any drop would mean an increase in delay due to the need for retransmission. In addition, during transit, it is also possible that the content enclosed in a data packet is possibly corrupted due to, for example, an electrical signaling problem on a communication link. This then results in garbling of a packet. From an end-to-end communication point of view, a garbled packet is the same as a lost packet.

Thus, for efficient delivery of packets, there are several key factors to consider: (1) routers with a reasonable amount of buffer space, (2) links with adequate bandwidth, (3) actual transmission with minimal error (to minimize packets being garbled), and (4) the routers' efficiency in switching a packet to the appropriate outgoing link. We have already briefly discussed why the first two factors are important. The third factor, an important issue, is outside the scope of this book since encoding or development of an error-free transmission system is an enormous subject by itself; interested readers may consult books such as [666]. Thus, we next move to the fourth factor.

Why is the fourth factor important? A packet is to be routed based on the IP address of the destination host; however, much like street address information in a postal address, there are far too many possible hosts; it is impossible and impractical to store *all* host addresses at any router. For example, for a 32-bit address, theoretically a maximum of 2^{32} hosts are possible—a very large number (more about IP addressing in the next section). Rather, a router needs to consider a coarser level of address information, i.e., the netid associated with a host, so that an outgoing link can be identified quickly just by looking up the netid. Recall that a netid is very much like a postal code. There is, however, a key difference—netids do not have any geographical proximity association as with postal codes. For example, postal codes in the United States are five digits long and are known as ZIP (Zonal Improvement Plan) codes. Consider now Kansas City, Missouri, where a ZIP code starts with 64 such as 64101, 64102, and so on. Thus, a postcard can be routed to Kansas City, MO ("64") which in turn then can take care of routing to the specific ZIP code. This idea is not possible with IP addressing since netids do not have any geographical proximity. In fact, an IP netid address such 134.193.0.0 can be geographically far away from the immediately preceding IP netid address 134.192.0.0. Thus, at the netid level, IP addressing is flat; there is no hierarchy.

You might be wondering why IP address numbering is not geographic. To give a short answer, an advantage of a nongeographic address is that an organization that has been assigned an IP address block can keep its address block even if it moves to a different location or if it wants to use a different provider for connectivity to the Internet. A geographically based address system usually has limitations in regard to providing location-independent flexibility.

In order to provide the flexibility that two netids that appear close in terms of their actual numbering can be geographically far away, core routers in the Internet need to maintain an explicit list of all valid netids along with an identified outgoing link so that when a packet arrives the router knows which way to direct the packet. The list of valid netids is quite large, currently at 196,000 entries. Thus, to minimize switching time at a router, efficient mechanisms are needed that can look up an address, identify the appropriate outgoing link (direction), and process the packet quickly so that the processing delay can be as minimal as possible.

There is, however, another important phase that works in tandem with the lookup process at a router. This is the updating of a table in the router, known as the *routing table*, that contains the identifier for the next router, known as the *next hop*, for a given destination netid. The routing table is in fact updated ahead of time. In order to update such a table, the router would need to store all netids it has learned about so far; second, if a link downstream is down or congested or a netid is not reachable for some reason, it needs to know so that an alternate path can be determined as soon as possible. This means that a mechanism is required for *communicating* congestion or a failure of a link or nonreachability of a netid. This mechanism is known as the *routing protocol* mechanism. The information learned through a routing protocol is used for generating the routing table ahead of time.

If new information is learned about the status of links or nodes, or the reachability of a netid through a routing protocol, a *routing algorithm* is then invoked at a router to determine the best possible next hop for each destination netid in order to update the routing table. For efficient packet processing, another table, known as the *forwarding table*, is derived from the routing table that identifies the outgoing link interfaces. The forwarding table is also known as the Forwarding Information Base (FIB). We will use the terms forwarding table and FIB interchangeably.

It should be noted that a routing algorithm may need to take into account one or more factors about a link, such as the delay incurred to traverse the link, or its available bandwidth, in order to determine the best possible path among a number of possible paths. If a link along a path does not have adequate bandwidth, congestion or delay might occur. To minimize delay, an important function, called *traffic engineering*, is performed. Traffic engineering is concerned with ways to improve the operational performance of a network and identifies procedures or controls to be put in place ahead of time to obtain good network performance.

Finally, there is another important term associated with networking in general and network routing in particular, labeled as *architecture*. There are two broad ways the term architecture from the architecture of a building is applicable here: (1) a floor inside a building may be organized so that it can be partitioned efficiently for creating office spaces of different sizes by putting in flexible partitions without having to tear down any concrete walls, (2) it provides standardized interfaces, such as electrical sockets, so that equipment that requires power can be easily connected using a standardized socket without requiring modification to the building or the floor or the equipment. Similarly, there are several ways we use the term *architecting a network*: for example, from the protocol point of view, various functions are divided so that each function can be done separately, and one function can depend on another through a well-defined relationship. From a router's perspective, architecting a network refers to how it is organized internally for a variety of functions, from routing protocol handling to packet processing. From a network perspective, this means how the network topology architecture should be organized, where routers are to be located and bandwidth of links determined for efficient traffic engineering, and so on. Later, we will elaborate more on architectures.

To summarize, we can say that the broad scope of network routing is to address routing algorithms, routing protocols, and architectures, with architectures encompassing several different aspects for efficient routing. In this book, we will delve into these aspects in depth. With the above overview, we now present IP addressing in detail.

1.3 IP Addressing

If one has to send data to any host in the Internet, there is a need to uniquely identify all the hosts in the Internet. Thus, there is a need for a global addressing scheme in which no two hosts have the same address. Global uniqueness is the first property that should be provided in an addressing scheme.

1.3.1 Classful Addressing Scheme

An IP address assigned to a host is 32 bits long and should be unique. This addressing, known as IPv4 addressing, is written in the bit format, from left to right, where the left-most bit is considered the most significant bit. The hierarchy in IP addressing, similar to the postal code and the street address, is reflected through two parts, a network part and a host part referred as the pair *(netid, hostid)*. Thus, we can think of the Internet as the *interconnection* of networks identified through netids where each netid has a collection of hosts. The network part (netid) identifies the network to which the host is attached, and the host part (hostid) identifies a host on that network. The network part is also referred as the *IP prefix*. All hosts attached to the same network share the network part of their IP addresses but must have a unique host part.

To support different sizes for the (netid, hostid) part, a good rule on how to partition the total IP address space of 2^{32} addresses was needed, i.e., how many network addresses will be allowed and how many hosts each of them will support. Thus, the IP address space was originally divided into three different classes, Class A, Class B, and Class C, as shown in Figure 1.1 for networks and hosts. Each class was distinguished by the first few initial bits of a 32-bit address.

For readability, IP addresses are expressed as four decimal numbers, with a dot between them. This format is called the *dotted decimal notation*. The notation divides the 32-bit IP address into 4 groups of 8 bits and specifies the value of each group independently as a decimal number separated by dots. Because of 8-bit breakpoints, there can be at most 256 $(= 2^8)$ decimal values in each part. Since 0 is an assignable value, no decimal values can be more than 255. Thus, an example of an IP address is 10.5.21.90 consisting of the four decimal values, separated by a dot or period.

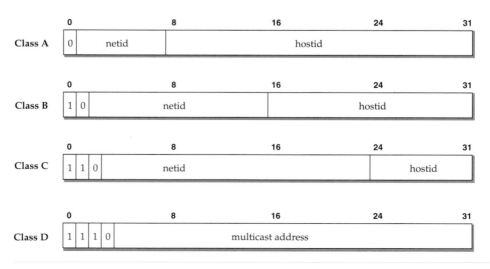

FIGURE 1.1   Classful IP addressing scheme.

Each Class A address has the first bit set to 0 and is followed by 7 bits for the network part, resulting in a maximum of 128 ($= 2^7$) networks; this is then followed by a 24-bit host part. Thus, Class A supports a maximum of $2^{24} - 2$ hosts per network. This calculation subtracts 2 because 0s and 1s in the host part of a Class A address may not be assigned to individual hosts; rather, all 0s that follows a netid such as 10.0.0.0 identify the network, while all 1s that follow a netid such as 10.255.255.255 are used as the broadcast address for this network. Each Class B network address has the first two bits set to "10," followed by a 14-bit network part, which is then followed by a 16-bit host part. A maximum of 2^{14} networks can be defined with up to $2^{16} - 2$ hosts per network. Finally, a Class C network address has the first three bits set as "110" and followed by a 21-bit network part, with the last 8 bits to identify the host part. Class C provides support for a maximum of $2^{21} (= 2,097,152)$ networks with up to 254 ($2^8 - 2$) hosts. In each class, a set of network addresses is reserved for a variety of purposes; see [319].

Three address classes discussed so far are used for unicasting in the Internet, that is, for a host-to-host communication. There is another class of IP addresses, known as Class D addressing, that is used for *multicasting* in the Internet; in this case, the first four bits of the 32-bit address are set to "1110" to indicate that it is a multicast address. A host can use a multicast address as the destination address for a packet generated to indicate that the packet is meant for any hosts on the Internet; in order for any hosts to avail this feature, they must use another mechanism to tune into this address. Multicast address on the Internet can be thought of as similar to a radio station frequency; a radio station transmits on a particular frequency—any listener who wants to listen to this radio station must tune the radio dial to this frequency.

The original rationale behind classes of different sizes was to provide the flexibility to support different sized networks, with each network containing a different number of hosts. Thus, the total address length can still be kept fixed at 32 bits, an advantage from the point of view of efficient address processing at a router or a host. As the popularity of the Internet grew, several disadvantages of the addressing scheme came to light. The major concerns were the rate at which the IP address blocks that identify netids were being exhausted, especially when it was necessary to start assigning Class C level netids. Recall from our earlier discussion that IP netids are nongeographic; thus, all valid netids are required to be listed at the core routers of the Internet along with the outgoing link, so that packets can be forwarded properly. If we now imagine all Class C level netids being assigned, then there are over 2 million entries that would need to be listed at a core router; no current routers can handle this number of entries without severely slowing packet processing. This issue, first recognized in the early 1990s, led to the development of the concept of *classless* addressing. In order to understand this concept, we first need to understand subnetting/netmask.

1.3.2 Subnetting/Netmask

Consider the IP address 192.168.40.3 that is part of Class C network 192.168.40.0. A subnet or sub-network is defined through a network mask boundary using the specified number of significant bits as 1s. Since Class C defines networks with a 24-bit boundary, we can then consider that the most significant 24 bits are 1s, and the lower 8 bits are 0s. This translates to the dotted decimal notation 255.255.255.0, which is also compactly written as "/24" to indicate how many most significant bits are 1s. We can then do a bit-wise logical "AND" operation between the host address and the netmask to obtain the Class C network address as shown below:

```
        11000000 10101000 00101000 00000011    → 192.168.40.3
AND     11111111 11111111 11111111 00000000    → netmask (/24)
        11000000 10101000 00101000 00000000    → 192.168.40.0
```

As you can see, both the host address and the netmask have 1s in the first two positions from the left; thus, the "AND" operation results in 1s for these two positions. For the third position from left, the host has 0 while the netmask has 1; thus, the result of the "AND" operation is zero; and so on. Note that for network addresses such as Class C address, the netmask is implicit and it is on a /24 subnet boundary. Now consider that we want to change the netmask *explicitly* to /21 to identify a network larger than a 24-bit subnet boundary. If we now do the bit-wise operation

```
        11000000 10101000 00101000 00000011    → 192.168.40.3
AND     11111111 11111111 11111000 00000000    → netmask (/21)
        11000000 10101000 00101000 00000000    → 192.168.40.0
```

we note that the network address is again 192.168.40.0. However, in the latter case, the network boundary is 21 bits. Thus, to be able to clearly distinguish between the first and the second one, it is necessary to explicitly mention the netmask. This is commonly written for the second example as 192.168.40.0/21, where the first part is the netid and the second part is the mask boundary indicator. In this notation, we could write the original Class C address as 192.168.40.0/24 and thus, there is no ambiguity with 192.168.40.0/21.

1.3.3 Classless Interdomain Routing

Classless Interdomain Routing (CIDR) uses an explicit netmask with an IPv4 address block to identify a network, such as 192.168.40.0/21. An advantage of explicit masking is that an address block can be assigned at any bit boundaries, be it /15 or /20; most important, the assignment of Class C level addresses for networks that can show up in the global routing table can be avoided or minimized. For example, a contiguous address block can be assigned at the /21 boundary which can be thought of as an aggregation of subnets at the /24 boundary. Because of this, the term *supernetting* or *variable-length subnet masking* (VLSM) is also used in reference to the explicit announcement of the netmask.

   Through such a process, and because of address block assignment at boundaries such as /21, the routing table growth at core routers can be delayed. In the above example, only the netid 192.168.40.0/21 needs to be listed in the routing table entry, instead of listing *eight* entries from 192.168.40.0/24 to 192.168.47.0/24. Thus, you can see how the routing table growth can be curtailed. CIDR was introduced around the mid-1990s; the current global routing table size, as of this writing, is about 196,000 entries. The routing table growth over time, along with projection, is shown later in Figure 9.10. In order for CIDR to take effect, any network address reachability announcement that is communicated with a routing protocol such as the *Border Gateway Protocol* must also carry the mask information explicitly. Its usage and applicability will be discussed in more detail in Chapter 8 and Chapter 9. In Table 1.1, we show a set of IP addresses reserved for a variety of purposes; see [319] for the complete list.

TABLE 1.1 Examples of reserved IP address blocks.

Address Block	Current Usage
0.0.0.0/8	Identifies source hosts in the current network
10.0.0.0/8	Private-use IP networks
127.0.0.0/8	Host loopback address
169.254.0.0/16	Link local for communication between links on a single link
172.16.0.0/12	Private-use IP networks
192.168.0.0/16	Private-use IP networks
240.0.0.0/4	Reserved for future use

1.4 On Architectures

Architectures cover many different aspects of networking environments. Network routing must account for each of the following architectural components. Some aspects of the architectures listed below are critical to routing issues:

- *Service Architecture:* A service model gives the basic framework for the type of services a network offers.

- *Protocol Stack Architecture:* A protocol stack architecture defines how service delivery may require different functions to be divided along well-defined boundaries so that responsibilities can be decoupled. It does not describe how actual resources might be used or needed.

- *Router Architecture:* A router is a specialized computer that is equipped with hardware/software for packet processing. It is also equipped for processing of routing protocols and can handle configuration requirements. A router is architected differently depending on its role in a network, such as a core router or an edge router, although all routers have a common set of requirements.

- *Network Topology Architecture:* For efficient operation as well as to provide acceptable service to its users, a network is required to be organized based on a network topology architecture that is scalable and allows growth. In order to address efficient services, there is also a direct connection among the topology architecture, traffic engineering, and routing.

- *Network Management Architecture:* A network needs to provide several additional functions in addition to carrying the user traffic from point A to point B; for clarity, the user data traffic forwarding is considered as the *data plane*. For example, from an operational point of view, a *management plane* handles the configuration responsibility of a network, and a *control plane* addresses routing information exchanges.

In the following sections, we elaborate on the above architectural facets of networking. To simplify matters, most of the following discussions will center around IP networks. Keep in mind that these architectures are applicable to most communication networking environments as well.

1.5 Service Architecture

An important aspect of a networking architecture is its service architecture. The service architecture depends partly also on the communication paradigm of its information units. Every networking environment has a service architecture, much like the postal delivery system. In the following, we focus on discussing three service models associated with IP networks.

BEST-EFFORT SERVICE ARCHITECTURE

Consider an IP network. The basic information unit of an IP network is a packet or a datagram which is forwarded from one router to another towards the destination. To do that, the IP network uses a switching concept, referred to as *packet switching*. This means that a router makes decisions by identifying an outgoing link on a packet-by-packet basis instantaneously after the packet arrives. At the conceptual level, it is assumed that no two packets are related, even though they might arrive one after another and possibly for the same web-page downloaded. Also, recall that at the IP level, the packet forwarding function is provided without worrying about reliable delivery; in a sense, IP makes its best effort to deliver packets. Because of this, the IP service paradigm is referred to as the *best-effort service*.

INTEGRATED SERVICES ARCHITECTURE

Initially, the best-effort service model was developed for the reliable delivery of data services, since it was envisioned that services would be data-oriented services that can tolerate delay, but not loss of packets. This model worked because the data rate provided during a session can be adaptive.

The concept for integrated services ("int-serv") architecture was developed in the early 1990s to allow functionalities for services that are real-time, interactive, and that can tolerate some loss, but require a bound on the delay. Furthermore, each session or connection requires a well-defined bandwidth guarantee and a dedicated path. For example, interactive voice and multimedia applications fall into this category. Note that the basic best-effort IP framework works on the notion of statelessness; that is, two consecutive packets that belong to the same connection are to be treated independently by a router. Yet, for services in the integrated services architecture that require a connection or a session for a certain duration of time, it became necessary to provide a mechanism to indicate the longevity of the session, and the ability for routers to know that resources are to be reserved for the entire duration.

Since the basic IP architecture works on the notion of statelessness, and it was infeasible to completely change the basic IP service architecture, a soft-state concept was introduced to handle integrated-services. To do that, a session setup and maintenance protocol was also developed that can be used by each service—this protocol is known as the resource ReSerVation Protocol (RSVP). The basic idea was that once a session is established, RSVP messages are periodically generated to indicate that the session is alive. The idea of integrated services was a novel concept that relies on the soft-state approach. A basic problem is the scalability of handling the number of RSVP messages generated for all sessions that might be simultaneously active at a router or a link.

DIFFERENTIATED SERVICES ARCHITECTURE

The differentiated services ("diff-serv") architecture was developed to provide prioritized service mechanisms without requiring connection-level information to be maintained at routers. Specifically, this approach gives priority to services by marking IP packets with diff-serv code points located in the IP header. Routers along the way then check the diff-serv code point and prioritize packet processing and forwarding for different classes of services. Second, this model does not require the soft-state concept and thus avoids the connection-level scalability issue faced with RSVP. Diff-serv code points are identified through a 6-bit field in the IPv4 packet header; in the IPv6 packet header, the traffic class field is used for the same purpose.

SUPPLEMENTING A SERVICE ARCHITECTURE

Earlier in this section, we introduced the best-effort service model. In a realistic sense, and to provide acceptable quality of service performance, the basic concept can be supplemented with additional mechanisms to provide an acceptable service architecture, while functionally it may still remain as the best-effort service architecture. For example, although the basic conceptual framework does not require it, a router can be designed to do efficient packet processing for packets that belong to the same web-page requested by a user since they are going to the same destination. That is, a sequence of packets that belongs to the same pair of origination and destination IP addresses, to the same pair of source and destination port numbers, and to the same transport protocol (either TCP or UDP) can be thought of as a single entity and is identified as a *microflow*. Thus, packets belonging to a particular microflow can be treated in the same manner by a router once a decision on forwarding is determined based on the first packet for this microflow.

Another way to fine-tune the best-effort service architecture is through traffic engineering. That is, a network must have enough bandwidth so that delay or backlog can be minimal, routers must have adequate buffer space, and so on, so that traffic moves efficiently through the network. In fact, both packet processing at a router and traffic engineering work in tandem for providing efficient services.

Similarly, for both integrated-services and differentiated-service architecture, packet handling can be optimized at a router. Furthermore, traffic engineering can be geared for integrated services or differentiated services architectures.

1.6 Protocol Stack Architecture

Another important facet of a networking environment is the protocol stack architecture. We start with the OSI (Open Systems Interconnections) reference model and then discuss the IP protocol stack architecture and its relation to the OSI reference model.

1.6.1 OSI Reference Model

The OSI reference model was developed in the 1980s to present a general reference model for how a computer network architecture should be functionally divided. As part of OSI, many protocols have also been developed. Here, we will present the basic reference model.

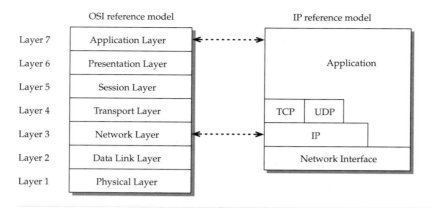

F I G U R E 1.2   The OSI reference model and the IP reference model.

The OSI reference model uses a layered hierarchy to separate functions, where the layering is strictly enforced. That is to say that an N-layer uses services provided by layer $N - 1$; it cannot receive services directly from layer $N - 2$. In the OSI model, a seven-layer architecture is defined; this is shown in Figure 1.2. The seven layers are also referenced by layer numbering counting from bottom up. From a functional point of view, layer 1 provides physical layer functions. Layer 2 provides the data link function between two directly connected entities. Layer 3 is the network layer, where addressing and routing occurs. Layer 4 is the transport layer that can provide either reliable or unreliable transport services, with or without defining a connection ("connection-oriented" or "connection-less"). Layer 5 is the session layer, addressing communication that may transcend multiple connections. Layer 6 is the presentation layer that addresses structured information and data representation. Layer 7 is where the application layer protocols are defined.

While not every computer networking environment strictly adheres to the OSI reference model, it does provide an easy and simple way to check and compare what a particular networking environment might have to consider. Thus, this reference model is often quoted; in fact, you will hear use of terms such as "layer 2" device or "layer 3" device in the technical community quite often, assuming you know what they mean.

1.6.2 IP Protocol Stack Architecture

The IP architectural model can be classified into the following layers: the network interface, the IP layer, the transport layer, and the application layer (see Figure 1.2). We can easily see that it does not exactly map into the seven-layer OSI reference model. Actual applications are considered on the top of the application layer, although the IP model does not strictly follow layering boundaries as in the OSI reference model. For example, it allows an application to be built without using a transport layer; *ping* is such an example. We have discussed earlier that IP includes both the destination and the source address—this is accomplished through a header part in the IP packet that also contains additional information. The IP model does not explicitly declare how the layer below the IP layer needs to be; this part is simply referred to as the network interface that can support IP and will be discussed later in the chapter.

NETWORK AND TRANSPORT LAYER

The IP addressing is defined at the IP layer, where the delivery mode is assumed to be unreliable. The transport layer that is above the IP layer provides transport services, which can be either reliable or unreliable. More important, the transport layer provides another form of addressing, commonly known as the *port number*. Port numbers are 16 bits long. Thus, the unreliable transport layer protocol, known as the User Datagram Protocol (UDP), can be thought of as allowing the extension of the address space by tagging a 16-bit port number to the 32-bit IP address. However, the role of the port number is solely at the host while routing is still done using the IP address. This is similar to the decoupling of the postal code and the house address in the postal addressing system. The reliable transport counterpart of UDP is known as the Transmission Control Protocol (TCP) which also uses a 16-bit port number, but provides reliable transport layer service by using a retransmission and acknowledgment mechanism. To be able to include the port number and other information, both TCP and UDP have well-defined headers. Because of two-way communication, similar to an IP packet including both the source and the destination address, TCP and UDP also include port numbers both for the source and the destination side. Since both TCP and UDP are above IP, a field in the IP header, known as the protocol type field, is used to be able to distinguish them. That is, through five pieces of information consisting of the source and the destination IP addresses, the source and the destination port numbers, and the transport protocol type, a connection in the Internet can be uniquely defined. This is also known as a microflow.

There are two IP packet formats: IPv4 and IPv6 (see Figure 1.3). IPv4 uses the 32-bit IP address and is the most widely deployed addressing scheme. IPv6 uses a longer 128-bit address that was developed in the mid-1990s; initially, it was designed anticipating that IPv4 addresses would be running out soon. This did not happen as initially thought, partly because of the proliferation of private IP address usage (see Table 1.1) that has been made possible by mechanisms known as network address translation (NAT) devices, which can map and track multiple private IP addresses to a single IP address. Packet formats for TCP and UDP are shown in Figure 1.4. So far, we have already discussed several well-known fields in these packets, such as IP source and destination addresses, source and destination port numbers, the protocol type field, and the diff-serv code point; other key fields shown in packets formats will be discussed later in Appendix B.14.

APPLICATION LAYER AND APPLICATIONS

Information structure at the transport layer is still at the byte level; there is no structured, semantic information considered at this level. However, structural information is needed for a particular application. For example, an email requires fields such as "From," "To" before the body of a message is added; this then helps the receiving end know how to process the structured information. In order to provide the structured information for different applications, the IP architectural model allows the ability to define application layer protocols on the top of the transport layer protocols. Application layer protocols use unique listening port numbers from the transport layer level to distinguish one application from another. In other words, the IP architectural model cleverly uses the transport layer port number to streamline different application layer protocols, instead of defining yet another set of addresses at the application layer protocol level. Examples of application layer protocols are Simple Mail Transfer Protocol (SMTP), and HyperText Transport Protocol (HTTP), which are used by email and web

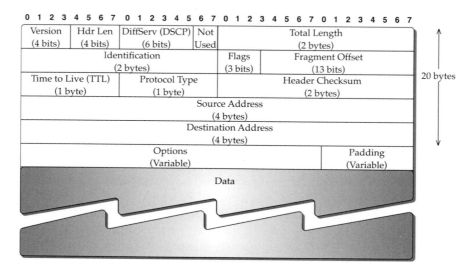

(a) IPv4 packet

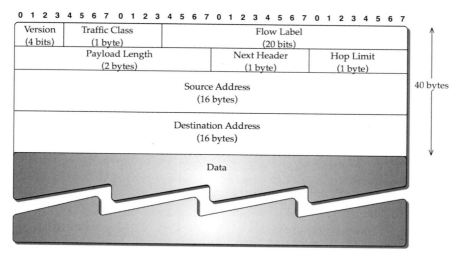

(b) IPv6 packet

FIGURE 1.3   Packet formats: IPv4 and IPv6.

applications, respectively. However, the terminology is a bit confusing with some of the older application layer protocols, since both the application layer protocol and its associated application are described by the same name; for example, File Transfer Protocol (FTP), and telnet. It may be noted that this set of application layer protocols (SMTP, HTTP, FTP, telnet) requires reliable data delivery and, thus, uses TCP as the transport layer protocol.

There are other applications that do not require reliable data delivery. Voice over IP protocol, commonly referred to as VoIP, is one such application that can tolerate some packet loss and thus, retransmission of lost packets is not necessary. Such an application can then use

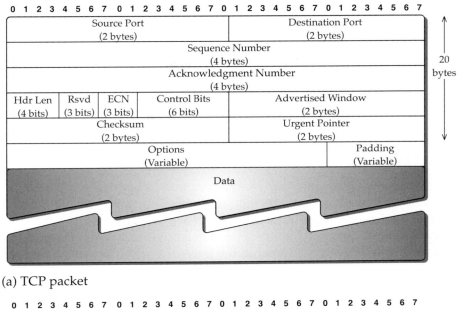

(a) TCP packet

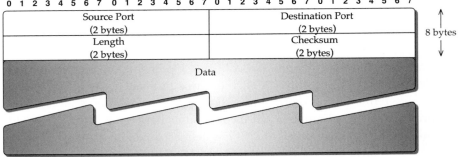

(b) UDP packet

FIGURE 1.4   Packet formats: TCP and UDP.

UDP. Since UDP does not provide any structural boundaries, and because many real-time communications, such as voice and video, require similar structural formats with the ability to distinguish different encoding mechanisms, Real-time Transport Protocol (RTP) has been defined above UDP. For example, a voice stream, with its coding based on G.711 PCM coding standards, can use RTP, while a motion JPEG video can also use RTP; they are distinguished through a payload-type field in RTP.

ROLE OF HEADERS

By now, it might have become apparent that each layer needs to add a *header* to provide its functionality; and it then encapsulates the content received from the layer above. For example, RTP adds a header so that the payload time, among other things, can be indicated. How is then a message or a web page generated at an application level related to the layered data units, along with a header? To see this, consider transferring a web page. First, the

HTTP protocol adds its header to the entire content of the page. Suppose that the combination of this header and the content of the page comes to 50 kbytes. This content is then broken into smaller units. If such a unit is to be of 1000 bytes each, for example, due to a limitation imposed by the maximum transmission unit of a link, then we have to create 50 units of information. First, TCP will include its header which is 20 bytes in the default case, to make each unit, commonly referred to as a *segment*, to be 1020 bytes. Then, IP will include its own header, which is 20 bytes in the default mode. Thus, altogether each unit becomes a packet of size 1040 bytes at the IP level.

WHERE DO ROUTING PROTOCOLS FIT IN?

We next discuss the exchange of information required for routing protocols. It is important to note that such exchanges of information for routing protocols also use the same protocol architectural framework. The information content of a routing protocol exchange has specific semantic meaning so that two routers can exchange and understand this information using these semantics. Interestingly, a router in the Internet is a also a host and is assigned an IP address. Thus, any communication between two adjacent routers is similar to any communication between any two hosts. Since IP is the core network layer, this means that IP is also used for this information exchange, much like using IP for communications related to the web or email. This is where the protocol-type field in the IP header, and the port numbering at the transport layer, can be used for distinguishing information exchanges related to different routing protocols. Three well-known routing protocols that we will be discussing later in the book are: Routing Information Protocol (RIP), Open Shortest Path First protocol (OSPF), and Border Gateway Protocol (BGP). Each of these protocols uses a different approach and exchanges different types of information. RIP is a protocol defined on top of UDP through a well-known listening port number and the unreliable delivery provided by UDP is used. Although not a transport layer protocol, OSPF is defined directly on top of IP by being assigned a protocol-type field at the IP level. It has its own retransmission and acknowledgment mechanism since it requires reliable delivery mechanisms. BGP is defined on top of TCP through a well-known listening port number, and BGP relies on TCP's reliable service to transfer its structured contents. An application, usually a command-line interface, is available with each router so that specific commands can be issued for each of these routing protocols, which are then translated into respective routing protocol exchange messages for communication with its adjacent routers.

AUXILIARY APPLICATIONS

Besides applications for actual user data traffic and applications for providing routing information exchanges, the IP architecture also supports auxiliary applications needed for a variety of functions. A well-known application is the name-to-address translation function provided through the Domain Name System (DNS), such that a domain name like www.NetworkRouting.net can be mapped into a valid IP address. This function can be either invoked indirectly when a user accesses a website or can be invoked directly by using the command, *nslookup*. DNS is an application layer protocol that typically uses UDP for the transport layer function, but it can use TCP if needed. This example also shows that it is possible to define end applications that may depend on more than one transport layer protocol.

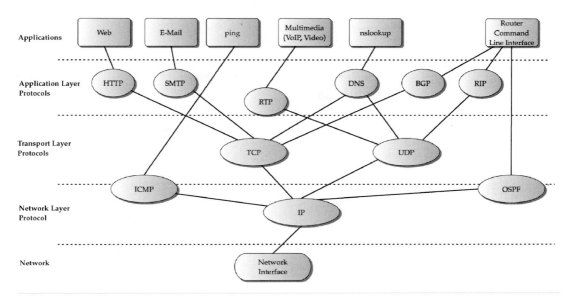

FIGURE 1.5   Protocol layering in IP architecture.

Another well-known utility application is *ping*, which is written on top of Internet Control Message Protocol (ICMP), that is directly over IP.

In Figure 1.5, we summarize the protocol dependency of different applications in terms of the application, transport, and network layer in the IP architecture.

1.7 Router Architecture

A router provides several important functions in order to ensure proper packet forwarding, and to do so in an efficient manner. A router is a specialized computer that handles three primary functions:

- *Packet Forwarding:* On receiving an incoming packet, a router checks whether the packet is error free. After inspecting the header of a packet for destination address, it performs a table lookup function to determine how to find the appropriate outgoing link.

- *Routing Protocol Message Processing:* A router also needs to handle routing protocol packets and determine if any changes are needed in the routing table by invoking a routing algorithm, when and if needed.

- *Specialized Services:* In addition, a router is required to handle specialized services that can aid in monitoring and managing a network.

A high-level functional view of a router is shown in Figure 1.6; it also shows how the routing table and the forwarding table fit in the overall process. In Part IV of this book, we will examine in detail router architectures, address lookup, packet processing, and so on.

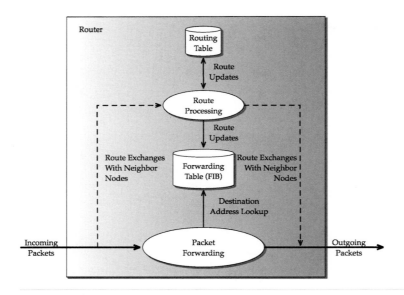

FIGURE 1.6   Router architecture: a functional view.

1.8 Network Topology Architecture

The network topology architecture encompasses how a network is to be architected in an operational environment while accounting for future growth. What does topology mean? It refers to the form a network will adopt, such as a star, ring, manhattan-street network, or a fully mesh topology, or a combination of them. The topological architecture then covers architecting a network topology that factors in economic issues, different technological capabilities, and limitations of devices to carry a certain volume of expected traffic and types of traffic, for an operational environment. Certainly, a network topology architecture also needs to take into account routing capability, including any limitation or flexibility provided by a routing protocol. It is up to a network provider, also referred to as a network operator or a service provider, to determine the best topological architecture for the network.

It is important to note that the operational experience of an existing network can contribute to the identification of additional features required from a routing protocol, or the development of a new routing protocol, or the development of a new routing algorithm or modification of an existing algorithm. We briefly discuss two examples: (1) when it was recognized in the late 1980s that the Internet needed to move from being under one network administrative domain to more flexible loosely connected networks managed by different administrative domains, BGP was developed, (2) when it was felt in the late 1970s that the telephone network needed to move away from a hierarchical architecture that provided limited routing capability to a more efficient network, dynamic call routing was developed and deployed. This also required changes in the topological architecture.

It may be noted that the term *network architecture* is also fairly commonly used in place of network topology architecture. One difficulty with the term *network architecture* is that it is also used to refer to a protocol architecture. It is not hard to guess that network providers are the ones who usually use the term network architecture to refer to a topological architecture.

1.9 Network Management Architecture

From the discussion in the previous sections, we can see that the routing information exchange uses the same framework as the user data traffic in the Internet. For an operational network, it is important to have a network management architecture where various functions can be divided into "planes." Specifically, we consider three different planes: the management plane, the control plane, and the data plane.

The management plane addresses router configuration and collection of various statistics, such as packet throughput, on a link. Router configuration refers to configuration of a router in a network by assigning an IP address, identifying links to its adjacent routers, invoking one or more routing protocols for operational usage, and so on. Statistics collection may be done, for example, through a protocol known as Simple Network Management Protocol (SNMP). The management plane of a router is closely associated with network operations.

The control plane exchanges control information between routers for management of a variety of functions, such as setting up a virtual link. The control plane is also involved in identifying the path to be taken between the endpoints of this virtual link, which relies on the routing information exchange.

Another clarification is important to point out. Since these functions are different, the routing-related functions are in the *control plane*, and the data transfers, such as the web or email, are in the *data plane*. These two planes, as well as the management plane, use IP for communication, so at the IP layer, there is no distinction between these functional planes. As we go through additional networking environments in this book, you will find that there are environments in which the control plane and the management plane are completely partitioned from the data plane.

It may be noted that for efficient traffic engineering of a network, certain information is also required from different routers. Such information exchanges can be conducted either through the control plane or through the management plane. In certain networking environments, some functions can overlap across different planes. Thus, the three planes can be thought of as interdependent. A schematic view is presented in Figure 1.7.

1.10 Public Switched Telephone Network

So far, our discussions have been primarily related to the Internet. In this section, we present a brief overview of Public Switched Telephone Network (PSTN), another important communication network.

An information unit in the PSTN is a call. Many of the architectural aspects discussed so far apply to the PSTN as well. The PSTN has a global addressing scheme to uniquely identify an end device; an end device is commonly referred to as a telephone, while a more generic term is customer premise equipment (CPE). The global addressing scheme is known as E.164 addressing. It is a hierarchical addressing scheme that identifies the country code at the top level followed by the city or area code, and finally the number assigned to a subscriber. Nodes in the PSTN are called *switches*, which are connected by intermachine trunks (IMTs), also known as *trunkgroups*.

From a protocol architecture point of view, and using the OSI reference model, PSTN can be simply summed up as consisting of application layer, network layer, and physical layer. The application layer enables the telephone service, the network layer handles addressing

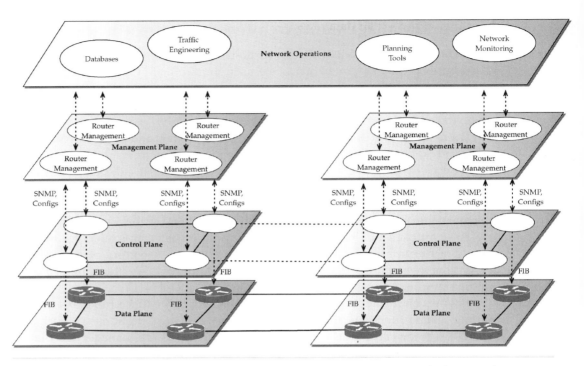

FIGURE 1.7   Network management architecture: data plane, control plane, and management plane.

and routing, while the physical transmission system carries the actual signal for voice communication. From a service architecture perspective, it provides the service model of *blocked-calls-cleared* mode using circuit switching. Circuit switching means that for a call requesting to be connected, a dedicated path is to be established instantaneously on demand from the source to the destination. The dedicated path is in fact a dedicated circuit with a specific bandwidth allocated—this value is 4 kilohertz (kHz) in an analog circuit and 64 kbps in a wireline digital circuit. The bandwidth of the circuit cannot be used by any other calls as long as this call is actively using it. Blocked-calls-cleared mode means that if the sequence of trunkgroups on all possible paths attempted from the source to destination does not have a circuit available for this call, then the call request is blocked and cleared from the system (not queued). Typically, a blocked call is indicated through a fast busy tone. Certainly, a user may retry.

More detail about routing in PSTN and its evolution will be covered later in Part III of this book. Routing in the IP-PSTN interworking environment will be presented in Chapter 20.

1.11 Communication Technologies

Communication technologies are used for carrying network layer services, whether for the Internet or PSTN. In this sense, communication technologies provide *transport* services for both the Internet and PSTN. Note that the use of the term *transport services* is not to be confused with the term *transport layer* of the OSI reference model. Unfortunately, the term transport is used in several ways in networking; these are two such examples. To provide transport

TABLE 1.2   Modular data rates.

Signal/data rate name	Bit rate (Mbps)
DS0 (voice circuit)	0.064
T1 (DS-1)	1.54
E1	2.04
Ethernet	10.00
T3 (DS-3)	45.00
E3	34.36
STS-1	51.84
Fast Ethernet	100.00
OC-3/STS-3/STM-1	155.52
OC-12/STS-12/STM-4	622.08
Gigabit Ethernet	1,000.00
OC-48/STS-48/STM-16	2,488.32
OTU1 (Optical Transport Unit-1)	2,666.06
OC-192/STS-192/STM-64	9,953.28
OTU2 (Optical Transport Unit-2)	10,709.22
OC-768/STS-768/STM-256	39,813.12
OTU3 (Optical Transport Unit-3)	43,018.41

services, transport networks are deployed that may be based on one or more communication technologies. At the real physical (duct) level though, fibers or coaxial cables are used for wired transport services. Such cables are either buried underground or carried overground on poles; submarine cabling is used for connecting different continents. Nowadays, submarine cables are almost exclusively based on fiber cables; for a recent map of global submarine cabling, see [693].

On top of cabling, a set of digital communication technologies can be provided; for example, SONET, T1/E1, T3/E3, and so on with well-defined data rates. A summary of different technologies and data rates is listed in Table 1.2, with all data rates listed using Mbps. A network is formed at any technological level, for example, SONET can use different rates such as OC-3 or OC-12. Similarly, a network can be formed at the T1 level or the T3 level. In particular, data rate multiplexing is also possible to go from one rate to another, such as from T1 to T3. The telecommunication infrastructure uses a mix of technologies, and transport services are provided either through networks at different levels, such as a network of T1s, a network of T3s, a network of SONET rings, or a combination of them. Each such transport network also needs to handle routing. For example, if a customer wants a T1 dedicated permanent circuit from Los Angeles to New York, the routing path needs to be mapped out. Certainly, the customer who wants the T1 transport service does not care how the T1 is routed in the transport network. However, for the T1 provider, it is an important problem since for all its T1 customers it needs to find efficient routing between different places.

In reference to the OSI terminology, the communication technologies reside mostly at layer 1 and sometimes in layer 2. Thus, instead of thinking about routing "purely" at the network layer (layer 3), routing problems also arise below layer 3 for transport network providers. In recent years, virtual private networking has become immensely popular. It requires another form of routing that is above layer 2, but below layer 3, often dubbed as layer 2.5. For example, MultiProtocol Label Switching (MPLS) and Asynchronous Transfer Mode (ATM) fall into this category.

Essentially, to provide transport services using communication technologies, a variety of transport network routing problems arises that need to take into account the capability of a particular communication technology and the "routing" device. Second, multilayered networking and multilayered routing can also be envisioned going from layer 3 down to layer 1 due to transport network routing. Third, new technologies for transport networking are being continually developed with new capabilities, creating new opportunities in transport network routing. Finally, traditionally, different transport networks had very little capability to communicate with each other and thus relied on manual configurations. We are now starting to see development of new capabilities that allow dynamic configuration and the ability to exchange information between networks at different layers so that dynamically reconfigurable multilayer routing will be possible in the coming years. However, such multilayer routing brings new challenges. In Part V and Part VI of this book, we will cover transport network routing and multilayered routing, and the evolution of next-generation routing.

1.12 Standards Committees

It is important to note that for all technologies developed, standards play important roles. In fact, standards have been defined from a specific technology, such as T1, to packet formats, such as an IP packet. Standardization allows different vendors to develop products that can talk to each other so that customers can choose products from multiple vendors; this helps bring the price down. Furthermore, vendors look for innovative ways to implement specific standards to reduce their costs and be competitive with other vendors, who are offering similar products.

There are two types of standards: *de jure* and *de facto*. De jure standards are arrived at through consensus by national or international standards bodies; for example, ITU-T and IETF. De facto standards are usually the result of an effort by one or more vendors to standardize a technology by forming a consortium. Sometimes, an early effort for de facto standards eventually transitions to de jure standards. There are many standards bodies that address issues related to networking and networking technologies. We briefly discuss some of them below.

1.12.1 International Telecommunication Union

ITU (http://www.itu.int/) plays the role of standardizing international telecommunications; it is a United Nations specialized agency. One of the key sections of ITU is known as ITU Telecommunication Standardization Sector (ITU-T). ITU-T brings both the public and private sectors together in an international forum. ITU-T is in charge of standardization of the international telephone numbering system, such as E.164 addressing. It also defines signaling protocol standards, and so on. Standards generated by ITU-T are called *Recommendations*.

You will see in the bibliography at the end of the book a number of ITU-T recommendations that we have referenced.

1.12.2 Internet Engineering Task Force

IETF (http://www.ietf.org/), as its web site says, "is a large, open international community of network designers, operators, vendors, and researchers concerned with the evolution of the Internet architecture and the smooth operation of the Internet. It is open to any interested individual." The IETF is structured around working groups, which then are grouped into areas. Areas have Area Directors (ADs). The ADs are members of the Internet Engineering Steering Group (IESG).

Standards generated by IETF are published as *Requests for Comments* (RFCs). This name stuck since its original use. The intent was to request for comments from the networking community; over time, it has become the avenue for IETF to publish standards. It may be noted that IETF also publishes informational documents as RFCs. Thus, each RFC is marked with a category such as standards track or informational. RFCs are available online from many web sites, for example, http://www.rfc-editor.org/. In the bibliography, we have referenced many routing-related RFCs.

In relation to IETF, there are several associated bodies. For example, the Internet Advisory Board (IAB) is chartered as a committee of IETF; it is also an advisory body of the Internet Society (ISOC). IAB handles architectural oversight of IETF activities, Internet Standards Process oversight and appeal. The IAB is also responsible for the management of the IETF protocol parameter registries.

Another important organization, the Internet Corporation for Assigned Names and Numbers (ICANN) (http://www.icann.org/), is an internationally organized, nonprofit corporation that now has responsibility for IP address space allocation, protocol identifier assignment, generic and country code top-level domain name system management, and root server system management functions. These services were originally performed by the Internet Assigned Numbers Authority (IANA) (http://www.iana.org/) and other entities. ICANN now performs the IANA function. Any new protocol parameter values identified by the IETF in a standard must be coordinated with the IANA to avoid any ambiguity.

1.12.3 MFA Forum

The MPLS and Frame Relay Alliance (MFA) Forum (http://www.mfaforum.org/) is an international, industry-wide forum consisting primarily of telecommunications and networking companies. It is focused on the creation of specifications on how to build and deliver MPLS, Frame Relay and ATM networks, and services. MFA also handles interoperability testing of different vendors' products.

1.13 Last Two Bits

In this section, we present two topics. The first, TLV, is a concept used in many protocols. The second topic is the protocol analyzer.

1.13.1 Type-Length-Value

An important concept used in protocol messages is *Type-Length-Value* (TLV). This concept is used in headers as well as the body of a packet, and by different layers of a networking architecture. For simplicity, consider that the IP header includes 32-bit IP addresses, one for the source and the other for the destination. First, for each receiving end to interpret properly, the source and the destination address must be listed in the same order in the header. Second, such information has a well-defined structure: it is of a certain type (IP address, in this case), it is of certain length (32 bits in this case), and it contains a value (the actual IP address). When such information is well-structured within a packet header and because of the well-known nature of such information, it is not often necessary to explicitly indicate the type and the length; just the allocation of the 32-bit space for an IP address in the header suffices. That is, for well-structured information that has a well-defined position in a packet header, the type and the length can be *implicit*.

In many instances, the length may vary, or the type is preferred to be left open for future extensions of a protocol. To do that, the type and the length need to be explicitly declared along with the value—this notion is what is known as TLV. As you go through this book, you will see many examples of how the TLV notion is used. Briefly, when the type and the length are to be explicit, then the length for each of these must be clearly defined, so that the value can be allowed to be of variable length. For example, a byte may be assigned to indicate the type (so that up to 256 different types can be defined), followed by two bytes for the length (to indicate through its 16 bits the length of value, that is counted in bytes), such that the value field can be up to 65,536 ($=2^{16}$) bytes. Because of the well-defined structure of TLV, the information content can be processed and another TLV can follow. Furthermore, a nested notion of TLV is also possible where the "V" part may include one or more TLV encoded sets of data.

1.13.2 Network Protocol Analyzer

Packet formats for networking protocols are described in standards documents by respective standards bodies. Many details about what a protocol can do lie in the header of a packet. Yet, just by looking at a packet format and reading a standards document, it is still difficult to grasp. Network protocol analyzers are used to capture packets from live networks. By studying headers captured through such analyzers, it is often easier to understand a packet header, and more important, a protocol.

In this book, we have presented sample headers (or relevant parts of headers) associated with a few protocols to illustrate them. Sample header captures for many routing protocols are available from the website of public-domain network protocol analyzers such as WIRE-SHARK [743]. Additionally, packet headers of both request-and-response messages of a protocol can be studied from such captures—this is sometimes very helpful in understanding a protocol. Sample captures using WIRESHARK for many protocols are found at [744]. We strongly recommend studying sample captures from this site or similar sites for helping you to understand protocols better.

1.14 Summary

In this introductory chapter, we have presented a brief overview of networking, and the scope and goal of network routing. We have also presented architectural aspects of communication networks that are useful in network routing.

All of these components have a history and a set of issues to address. The state of network routing today is the result of theoretical progress, technological advances, and operational experience. It is also impacted by economic and policy issues. From which angle should these interconnected facets of network routing be viewed? In an email to the authors, Ibrahim Matta wrote:

> "To me, it would be invaluable to highlight concepts and techniques in routing that survived the various instances in different networks; for example, the concepts of scalability-performance tradeoff (scalability techniques include area hierarchy, virtual paths, periodic updates . . .), routing information propagation vs. forwarding, etc."

The rest of the book will explore each aspect of network routing, with a nod toward the historical part, due respect for the scalability-performance tradeoff, and lessons learned from operational experience.

Further Lookup

Early works in the development of ARPANET have been instrumental in understanding today's computer communication network. ARPANET design decisions are discussed in [464]. Cerf and Kahn's seminal paper [112] discusses the TCP/IP protocol communication. The design philosophy of the Internet is discussed, for example, in [143]. A comprehensive discussion on architecture can be found in [142].

A comprehensive summary of the telecommunication network can be found in Bell System's Engineering and Operations handbook, last published in 1984 [596]. While this book is almost a quarter century old and out of print, it still serves as a good resource book on basic telecommunication networking.

Naming, addressing, and routing are interrelated topics for a communication network. In 1978, Shoch [639] wrote "The name of a resource indicates what we seek, an address indicates where it is, a route tells how to get there." Shoch's original work has a lot to do with how we think about naming, addressing, and routing in the Internet, even today. Certainly we can no longer say that an address is where it is. Also, the naming and addressing are now blurry. For additional discussions on naming, addressing, and routing, see [285], [366], [497], [618].

Finally, the focus of this book, as the title says, is network routing. You may consult books such as [152], [386], [562], [668], [683], to improve your understanding of computer networking in general; in fact, it might be handy to have one of them with you as you read through this book. If you are interested in understanding in depth the OSI architecture and protocols that were developed for OSI, you may consult books such as [567], [684]. For a comprehensive discussion of protocols developed by IETF for the Internet, you may consult [211]. For a summary of technology-specific standards, see [560].

Exercises

1.1 Review questions:

    (a) Given the IP address of a host and the netmask, explain how the network address is determined.

(b) Identify the key differences between the differentiated services architecture and the integrated services architecture.

(c) What is TLV?

1.2   Consider IP address 10.22.8.92 that is given to be part of a /14 address block. Determine the IP prefix it belongs to in the CIDR notation.

1.3   Consider IP address 10.21.5.90 that is given to be part of a /17 address block. Determine the IP prefix it belongs to in the CIDR notation.

1.4   From the TCP packet format, you will notice that it does not have a field that indicates the length of a TCP packet. How can you determine the TCP payload length, i.e., the length of the data carried in a TCP packet?

1.5   Why is it necessary to reserve some addresses from an address space rather than making all of them available?

1.6   Consider an IPv4 packet going through a router.

(a) Determine which fields in the header are minimally changed before the packet is forwarded.

(b) Which fields are also possibly changed at a router?

1.7   Are there any fields from the header of an IPv4 packet that are no longer maintained in the header of an IPv6 packet?

1.8   Investigate why the header fields in an IPv6 packet are significantly different than the header fields in an IPv4 packet.

1.9   Visit the IETF web-site (http://www.ietf.org/), and identify routing related working groups. Familiarize yourself with the type of routing protocols issues currently being addressed by these working groups.

1.10 Find out about other standards bodies, such as Institute of Electrical and Electronics Engineers (IEEE), American National Standards Institute (ANSI), Optical Internetworking Forum (OIF), especially regarding networking standards they are actively involved in.

2

Routing Algorithms: Shortest Path and Widest Path

"If everybody minded their own business," the Duchess said in a hoarse growl, "the world would go round a deal faster than it does."

Lewis Carroll in *Alice in Wonderland*

Reading Guideline

Shortest path algorithms are applicable to IP networks and widest path algorithms are useful for telephone network dynamic call routing and quality-of-service-based routing. If you are primarily interested in learning about routing in IP networks, you may read material on shortest path routing algorithms, and then come back to read about widest path algorithms later. If you are interested in understanding routing in a voice over IP (VoIP) environment or a Multiprotocol Label Switching (MPLS) network, researching widest path routing is also recommended.

In this chapter, we will describe two classes of routing algorithms: shortest path routing and widest path routing. They appear in network routing in many ways and have played critical roles in the development of routing protocols. The primary focus of this chapter is to describe how they work, without discussing how they are used by a specific communication network, or in the context of routing protocols. These aspects will be addressed throughout this book.

2.1 Background

In general, a communication network is made up of nodes and links. Depending on the type of the network, nodes have different names. For example, in an IP network, a node is called a *router* while in the telephone network a node is either an *end (central) office* or a *toll switch*. In an optical network, a node is an *optical or electro-optical switch*. A link connects two nodes; a link connecting two routers in an IP network is sometimes called an *IP trunk* or simply an IP link, while the end of a link outgoing from a router is called an *interface*. A link in a telephone network is called a *trunkgroup*, or an *intermachine trunk (IMT)*, and sometimes simply a *trunk*.

We first briefly discuss a few general terms. A communication network carries traffic where traffic flows from a *start* node to an *end* node; typically, we refer to the start node as the *source* node (where traffic originates) and the end node as the *destination* node. Consider now the network shown in Figure 2.1. Suppose that we have traffic that enters node 1 destined for node 6; in this case, node 1 is the source node and node 6 is the destination node. We may also have traffic from node 2 to node 5; for this case, the source node will be node 2 and the destination node will be node 5; and so on.

An important requirement of a communication network is to flow or *route* traffic from a source node to a destination node. To do that we need to determine a route, which is a path from the source node to the destination node. A route can certainly be set up manually; such a route is known as a *static route*. In general, however, it is desirable to use a routing *algorithm* to determine a route. The goal of a routing algorithm is in general dictated by the requirement of the communication network and the service it provides as well as any additional or specific goals a service provider wants to impose on itself (so that it can provide a

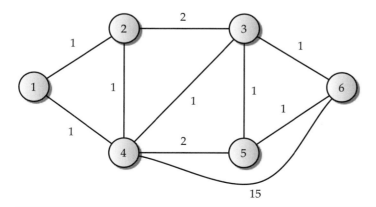

FIGURE 2.1   A six-node network (the entry shown next to a link is the cost of the link).

better service compared to another provider). While goals can be different for different types of communication networks, they can usually be classified into two broad categories: user-oriented and network-oriented. User-oriented means that a network needs to provide good service to each user so that traffic can move from the source to the destination quickly for this user. However, this should not be for a specific user at the expense of other users between other source–destination nodes in the network. Thus, a network's goal ("network-oriented") generally is to address how to provide an efficient and fair routing so that *most* users receive good and acceptable service, instead of providing the "best" service to a specific user. Such a view is partly required because there are a finite amount of resources in a network, e.g., network capacity.

We next consider two very important algorithms that have profound impact on data networks, in particular on Internet routing. These two algorithms, known as the *Bellman–Ford algorithm* and *Dijkstra's algorithm*, can be classified as falling under user-oriented in terms of the above broad categories. They are both called shortest path routing algorithms, i.e., an algorithm where the goal is to find the shortest path from a source node to a destination node. A simple way to understand a shortest path is from road networks where shortest can be defined in terms of distance, for example, as in what is the shortest distance between two cities, which consists of the link distance between appropriate intermediate places between the end cities. However, it is possible that notions other than the usual distance-based measure may be applicable as well, for instance, time taken to travel between two cities. In other words, an equally interesting question concerns the shortest route between two cities in terms of *time*. This means that the notion of distance need not always be in terms of physical distance; it can be in other measures such as time.

Instead of worrying about the unit of measurement, it is better to have an algorithm that works *independent* of the measuring unit and considers a generic measure for distance for each link in a network. In communication networks, a generic term to refer to a distance measure without assigning any measure units is called *cost, link cost, distance cost,* or *link metric*. Consider again Figure 2.1. We have assigned a value with each link, e.g., link 4-6 has the value 15; we will say that the link cost, or distance cost, or link metric of link 4-6 is 15. No measuring units are used; for example, in road networks, it could be in miles, kilometers, or minutes. By simple inspection, it is not hard to see that the shortest path between nodes 1 and 6 is the path 1-4-3-6 with a total minimum cost of 3. It may be noted that the shortest path in this case did not include the link 4-6, although from the viewpoint of the number of nodes visited, it would look like the path 1-4-6 is the shortest path between nodes 1 and 6. In fact, this would be the case if the link cost was measured in terms of nodes visited, or *hops*. In other words, if the number of hops is important for measuring distance for a certain network, we can then think about the network in Figure 2.1 by considering the link cost for each link to be 1 instead of the number shown on each link in the figure. Regardless, having an algorithm that works without needing to worry about how cost is assigned to each link is helpful; this is where the Bellman–Ford and Dijkstra's algorithms both fit in.

At this point, it is important to point out that in computing the shortest path, the *additive* property is generally used for constructing the overall distance of a path by adding a cost of a link to the cost of the next link along a path until all links for the path are considered, as we have illustrated above. Thus, we will first start with this property for shortest path routing in describing the Bellman–Ford and Dijkstra's algorithms, and their variants. You will see later that it is possible to define distance cost between two nodes in terms of *nonadditive* concave properties to determine the widest path (for example, refer to Section 2.7, 2.8, 10.9.2, 17.3, or 19.2.2). To avoid confusion, algorithms that use a non-additive concave property will be generally referred to as *widest* path routing algorithms.

We conclude this section by discussing the relation between a network and a *graph*. A network can be expressed as a graph by mapping each node to a unique vertex in the graph where links between network nodes are represented by edges connecting the corresponding vertices. Each edge can carry one or more weights; such weights may depict cost, delay, bandwidth, and so on. Figure 2.1 depicts a network consisting of a graph of six nodes and ten links where each link is assigned a link cost/weight.

2.2 Bellman–Ford Algorithm and the Distance Vector Approach

The Bellman–Ford algorithm uses a simple idea to compute the shortest path between two nodes in a centralized fashion. In a distributed environment, a distance vector approach is taken to compute shortest paths. In this section, we will discuss both the centralized and the distributed approaches.

2.2.1 Centralized View: Bellman–Ford Algorithm

To discuss the centralized version of the Bellman–Ford algorithm, we will use two generic nodes, labeled as node i and node j, in a network of N nodes. They may be directly connected as a link such as link 4-6 with end nodes 4 and 6 (see Figure 2.1). As can be seen from Figure 2.1, many nodes are not directly connected, for example, nodes 1 and 6; in this case, to find the distance between these two nodes, we need to resort to using other nodes and links. This brings us to an important point; we may have the notion of cost between two nodes, irrespective of whether they are directly connected or not. Thus, we introduce two important notations:

d_{ij} = Link cost between nodes i and j

$\overline{D}_{ij}$ = Cost of the computed minimum cost path from node i to node j.

Since we are dealing with different algorithms, we will use overbars, underscores, and hats in our notations to help distinguish the computation for different classes of algorithms. For example, overbars are used for all distance computation related to the Bellman–Ford algorithm and its variants. Note that these and other notations used throughout this chapter are summarized later in Table 2.5.

If two nodes are directly connected, then the link cost d_{ij} takes a finite value. Consider again Figure 2.1. Here, nodes 4 and 6 are directly connected with link cost 15; thus, we can write $d_{46} = 15$. On the other hand, nodes 1 and 6 are not directly connected; thus, $d_{16} = \infty$. What then is the difference between d_{ij} and the minimum cost $\overline{D}_{ij}$? From nodes 4 to 6, we see that the minimum cost is actually 2, which takes path 4-3-6; that is, $\overline{D}_{46} = 2$ while $d_{46} = 15$. For nodes 1 and 6, we find that $\overline{D}_{16} = 3$ while $d_{16} = \infty$. As can be seen, a minimum cost path can be obtained between two nodes in a network regardless of whether they are directly

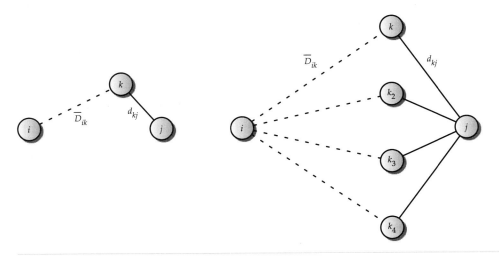

FIGURE 2.2   Centralized Bellman–Ford Algorithm (solid line denotes a direct link; dashed line denotes distance).

connected or not, as long as one of the end nodes is not completely isolated from the rest of the network.

The question now is how to compute the minimum cost between two nodes in a network. This is where shortest path algorithms come in. To discuss such an algorithm, it is clear from the six-node example that we also need to rely on intermediate nodes. For that, consider a generic node k in the network that is directly connected to either of the end nodes; we assume that k is directly connected to the destination node j, meaning d_{kj} has a finite value. The following equations, known as Bellman's equations, must be satisfied by the shortest path from node i to node j:

$$\overline{D}_{ii} = 0, \quad \text{for all } i, \tag{2.2.1a}$$

$$\overline{D}_{ij} = \min_{k \neq j} \{\overline{D}_{ik} + d_{kj}\}, \quad \text{for } i \neq j. \tag{2.2.1b}$$

Simply put, Eq. (2.2.1b) states that for a pair of nodes i and j, the minimum cost is dependent on knowing the minimum cost from i to k and the direct link cost d_{kj} for link k-j. A visual is shown in Figure 2.2. Note that there can be multiple nodes k that can be directly connected to the end node j (say they are marked k, k_2, and so on; note that $k = i$ is not ruled out either); thus, we need to consider all such ks to ensure that we can compute the minimum cost. It is important to note that technically, a node k that is not directly connected to j is also considered; since for such k, we have $d_{kj} = \infty$, the resulting minimum computation is not impacted. On close examination, note that Eq. (2.2.1b) assumes that we know the minimum cost, $\overline{D}_{ik}$, from node i to k first somehow! Thus, in an actual algorithmic operation, a slight variation of Eq. (2.2.1b) is used where the minimum cost is accounted for by iterating through the number of hops. Specifically, we define the term for the minimum cost in terms of number of hops h as follows:

$$\overline{D}_{ij}^{(h)} = \quad \text{cost of the minimum cost path from node } i \text{ to node } j \text{ when up to } h \\ \text{number of hops are considered.}$$

The Bellman–Ford algorithm that iterates in terms of number of hops is given in Algorithm 2.1. Note the use of (h) in the superscript in Eq. (2.2.2c); while the expression on the right side is up to h hops, with the consideration of one more hop, the expression on the left hand side is now given in $h + 1$ hops.

ALGORITHM 2.1   Bellman–Ford centralized algorithm.

Initialize for nodes i and j in the network:

$$\overline{D}_{ii}^{(0)} = 0, \quad \text{for all } i; \qquad \overline{D}_{ij}^{(0)} = \infty, \quad \text{for } i \neq j. \tag{2.2.2a}$$

For $h = 0$ to $N - 1$ do

$$\overline{D}_{ii}^{(h+1)} = 0, \quad \text{for all } i \tag{2.2.2b}$$

$$\overline{D}_{ij}^{(h+1)} = \min_{k \neq j}\left\{\overline{D}_{ik}^{(h)} + d_{kj}\right\}, \quad \text{for } i \neq j. \tag{2.2.2c}$$

For the six-node network (Figure 2.1), the Bellman–Ford algorithm is illustrated in Table 2.1. A nice way to understand the hop-iterated Bellman–Ford approach is to visualize through an example. Consider finding the shortest path from node 1 to node 6 as the number of hops increases. When $h = 1$, it means considering a direct link path between 1 and 6; since there is none, $D_{16}^{(1)} = \infty$. With $h = 2$, the path 1-4-6 is the only one possible since this is a two-link path, i.e., it uses two hops, consisting of the links 1-4 and 4-6; in this case, the hop-iterated minimum cost is 16 ($= D_{16}^{(2)}$). At $h = 3$, we can write the Bellman–Ford step as follows (shown only for k for which $d_{k6} < \infty$) since there are three possible paths that need to be considered:

$$k = 3: \quad \overline{D}_{13}^{(2)} + d_{36} = 2 + 1 = 3$$
$$k = 5: \quad \overline{D}_{15}^{(2)} + d_{56} = 3 + 1 = 4$$
$$k = 4: \quad \overline{D}_{14}^{(2)} + d_{46} = 1 + 15 = 16.$$

In this case, we pick the first one since the minimum cost is 3, i.e., $\overline{D}_{16}^{(3)} = 3$ with the shortest path 1-4-3-6. It is important to note that the Bellman–Ford algorithm computes only the minimum cost; it does not track the actual shortest path. We have included the shortest path in Table 2.1 for ease of understanding how the algorithm works. For many networking environments, it is not necessary to know the entire path; just knowing the next node k for which the cost is minimum is sufficient—this can be easily tracked with the min operation in Eq. (2.2.2c).

TABLE 2.1   Minimum cost from node 1 to other nodes using Algorithm 2.1.

h	$\overline{D}_{12}^{(h)}$	Path	$\overline{D}_{13}^{(h)}$	Path	$\overline{D}_{14}^{(h)}$	Path	$\overline{D}_{15}^{(h)}$	Path	$\overline{D}_{16}^{(h)}$	Path
0	∞	–	∞	–	∞	–	∞	–	∞	–
1	1	1-2	∞	–	1	1-4	∞	–	∞	–
2	1	1-2	2	1-4-3	1	1-4	3	1-4-5	16	1-4-6
3	1	1-2	2	1-4-3	1	1-4	3	1-4-5	3	1-4-3-6
4	1	1-2	2	1-4-3	1	1-4	3	1-4-5	3	1-4-3-6
5	1	1-2	2	1-4-3	1	1-4	3	1-4-5	3	1-4-3-6

2.2.2 Distributed View: A Distance Vector Approach

In a computer network, nodes need to work in a distributed fashion in determining the shortest paths to a destination. If we look closely at the centralized version discussed above, we note that a source node needs to know the cost of the shortest path to all nodes immediately prior to the destination, i.e., $\overline{D}_{ik}^{(h)}$, so that the minimum cost to the destination can be computed; this is the essence of Eq. (2.2.2c), communicated through Figure 2.2. This view of the centralized Bellman–Ford algorithm is not directly suitable for a distributed environment. On the other hand, we can consider an important rearrangement in the minimum cost computation to change the view. That is, what if we change the *order* of consideration in Eq. (2.2.1b) and instead use the minimum cost computation step as follows:

$$\overline{D}_{ij} = \min_{k \neq i}\{d_{ik} + \overline{D}_{kj}\}, \quad \text{for } i \neq j. \tag{2.2.3}$$

Note the subtle, yet distinctive difference between Eq. (2.2.1b) and Eq. (2.2.3); here, we first look at the outgoing link out of node i to a directly connected node k with link cost d_{ik}, and then consider the minimum cost $\overline{D}_{kj}$ from k to j without knowing how k determined this value. The list of directly connected nodes of i, i.e., neighbors of i, will be denoted by $\mathcal{N}_i$. In essence, what we are saying is that if node i finds out from its neighbor the cost of the minimum cost path to a destination, it can then use this information to determine cost to the destination by adding the outgoing link cost d_{ik}; this notion is known as the *distance vector approach*, first applied to the original ARPANET routing. With this approach, the computational step Eq. (2.2.3) has a nice advantage in that it helps in building a computational model for a distributed environment.

We illustrate the change of order and its advantage for the distributed environment using Figure 2.3. Suppose that node i periodically receives the minimum cost information $\overline{D}_{kj}$ from its neighboring node k for node k's minimum cost to node j; this variation can be addressed by introducing the dependency on the time domain, t, using $\overline{D}_{kj}(t)$ for node k's cost to j—this will then be available *to* node i (compare this expression to hop-based $\overline{D}_{kj}^{(h)}$). Now, imagine for whatever reason that node k recomputes its cost to j and makes it available to another source node, say i_2, but not to node i as shown in Figure 2.3. In other words, from the view of the source node i, the best we can say is that the minimum cost value from node k to node j that is available to node i is as node i has been able to receive; that is, it is more appropriate to use the term $\overline{D}_{kj}^{i}(t)$ than $\overline{D}_{kj}(t)$ to indicate that the minimum cost from node k to node j, as

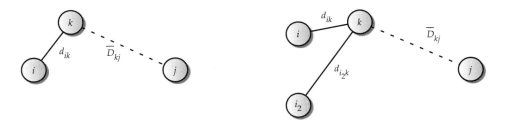

FIGURE 2.3   Distance vector view for computing the shortest path.

available to source node i at time t; to distinguish, the cost availability to i_2 can be written as $\overline{D}_{kj}^{i_2}(t)$ since i_2 may receive at a different time instant t. Furthermore, in a dynamic network, the direct link cost d_{ik} from node i to k can also change with time t, for example, to reflect change in network load/traffic. Thus, we can generalize the direct link cost to write as $d_{ik}(t)$ to indicate dependency on time t. With these changes, we present the distributed distance vector approach in Algorithm 2.2.

ALGORITHM 2.2   **Distance vector algorithm (computed at node i).**

Initialize

$$\overline{D}_{ii}(t) = 0; \quad \overline{D}_{ij}(t) = \infty, \quad \text{(for node j that node i is aware of).} \tag{2.2.4a}$$

For (nodes j that node i is aware of) do

$$\overline{D}_{ij}(t) = \min_{k \text{ directly connected to } i} \left\{ d_{ik}(t) + \overline{D}_{kj}^{i}(t) \right\}, \quad \text{for } j \neq i. \tag{2.2.4b}$$

We will now illustrate the distributed variation. For simplicity, we will assume that node k, which is directly connected to node i, is sending $\overline{D}_{kj}^{i}(t)$ at the same instant to other directly connected nodes like i. Furthermore, we will assume that the direct link cost does not change with time, i.e., $d_{ik}(t)$ does not change with time.

Our illustration will consider the same six-node example by considering computing the shortest path cost from node 1 to node 6 (see Table 2.2). This time, we will interpret the hop-based cost computation in terms of discrete time windows; for example, $t = 0$ means what node 4 sees about cost to node 6 when zero hops away, $t = 1$ means what node 4 sees about cost to node 6 when information from one hop away is received, and so on. Note that node 1 is directly connected to node 2 and node 4. Thus, node 1 will have $\overline{D}_{26}^{1}(t)$, the cost between node 2 and node 6 from node 2, and $\overline{D}_{46}^{1}(t)$, the cost between node 4 and node 6 from node 4.

To summarize, in the distance vector approach, a node relies on its neighboring nodes' known cost to a destination to determine its best path. To do so, it does periodic computation as and when it receives information from its neighbor. For this entire process to work,

TABLE 2.2 Distance vector based distributed computation at time t from node 1 to node 6.

Time, t	$\overline{D}^1_{46}(t)$	$\overline{D}^1_{26}(t)$	Computation at node 1 $\min\{d_{14}(t) + \overline{D}^1_{46}(t),\, d_{12}(t) + \overline{D}^1_{26}(t)\}$	$\overline{D}_{16}(t)$
0	∞	∞	$\min\{1 + \infty, 1 + \infty\}$	∞
1	15	∞	$\min\{1 + 15, 1 + \infty\}$	16
2	2	3	$\min\{1 + 2, 1 + 3\}$	3

the key idea is that a node k needs to distribute its cost to j given by $\overline{D}_{kj}(t)$ to all its directly connected neighbor i—the dependency on i and t means that each node i may get such information potentially at a different time instant t. The difference between this idea and the centralized Bellman–Ford algorithm is subtle in that the order of computation along with the link considered in computation leads to different views to computing the shortest path.

2.3 Dijkstra's Algorithm

Dijkstra's algorithm is another well-known shortest path routing algorithm. The basic idea behind Dijkstra's algorithm is quite different from the Bellman–Ford algorithm or the distance vector approach. It works on the notion of a candidate neighboring node set as well as the source's own computation to identify the shortest path to a destination. Another interesting property about Dijkstra's algorithm is that it computes shortest paths to all destinations from a source, instead of just for a specific pair of source and destination nodes at a time—which is very useful, especially in a communication network, since a node wants to compute the shortest path to all destinations.

2.3.1 Centralized Approach

Consider a generic node i in a network of N nodes from where we want to compute shortest paths to all other nodes in the network. The list of N nodes will be denoted by $\mathcal{N} = \{1, 2, \ldots, N\}$. A generic destination node will be denoted by j ($j \neq i$). We will use the following two terms:

d_{ij} = link cost between node i and node j

$\underline{D}_{ij}$ = cost of the minimum cost path between node i and node j.

Note that to avoid confusing this with the computation related to the Bellman–Ford algorithm or the distance vector approach, we will be using *underscores* with uppercase D, as in $\underline{D}_{ij}$, for the cost of the path between nodes i and j in Dijkstra's algorithm.

Dijkstra's algorithm divides the list $\mathcal{N}$ of nodes into two lists: it starts with permanent list S, which represents nodes already considered, and tentative list S', for nodes not considered yet. As the algorithm progresses, list S expands with new nodes included while list S' shrinks when nodes newly included in S are deleted from this list; the algorithm stops when

---

ALGORITHM 2.3   **Dijkstra's shortest path first algorithm (centralized approach).**

---

1. Start with source node i in the permanent list of nodes considered, i.e., $S = \{i\}$; all the rest of the nodes are put in the tentative list labeled as S'. Initialize

$$\underline{D}_{ij} = d_{ij}, \quad \text{for all } j \in S'.$$

2. Identify a neighboring node (intermediary) k not in the current list S with the minimum cost path from node i, i.e., find $k \in S'$ such that $\underline{D}_{ik} = \min_{m \in S'} \underline{D}_{im}$.

    Add k to the permanent list S, i.e., $S = S \cup \{k\}$,

    Drop k from the tentative list S', i.e., $S' = S' \setminus \{k\}$.

    If S' is empty, stop.

3. Consider the list of neighboring nodes, $\mathcal{N}_k$, of the intermediary k (but do not consider nodes already in S) to check for improvement in the minimum cost path, i.e., for $j \in \mathcal{N}_k \cap S'$

$$\underline{D}_{ij} = \min\{\underline{D}_{ij}, \underline{D}_{ik} + d_{kj}\}. \tag{2.3.1}$$

    Go to Step 2.

---

list S' becomes empty. Initially, we have $S = \{i\}$ and $S' = \mathcal{N} \setminus \{i\}$ (i.e., all nodes in $\mathcal{N}$ except node i).

The core of the algorithm has two parts: (1) how to expand the list S, and (2) how to compute the shortest path to nodes that are neighbors of nodes of list S (but nodes not in this list yet). List S is expanded at each iteration by considering a neighboring node k of node i with the least cost path from node i. At each iteration, the algorithm then considers the neighboring nodes of k, which are not already in S, to see if the minimum cost changes from the last iteration.

We will illustrate Dijkstra's algorithm using the network given in Figure 2.1. Suppose that node 1 wants to find shortest paths to all other nodes in the network. Then, initially, $S = \{1\}$, and $S' = \{2, 3, 4, 5, 6\}$, and the shortest paths to all nodes that are direct neighbors of node 1 can be readily found while for the rest, the cost remains at ∞, i.e.,

$$\underline{D}_{12} = 1, \underline{D}_{14} = 1, \quad \underline{D}_{13} = \underline{D}_{15} = \underline{D}_{16} = \infty.$$

For the next iteration, we note that node 1 has two directly connected neighbors: node 2 and node 4 with $d_{12} = 1$ and $d_{14} = 1$, respectively; all the other nodes are not directly connected to node 1, and thus, the "direct" cost to these nodes remains at ∞. Since both nodes 2 and 4 are neighbors with the same minimum cost, we can pick either of them to break the tie. For our illustration, we pick node 2, and this node becomes intermediary, k. Thus, we now have $S = \{1, 2\}$, and S' becomes the list $\{3, 4, 5, 6\}$. Then, we ask node 2 for cost to its direct

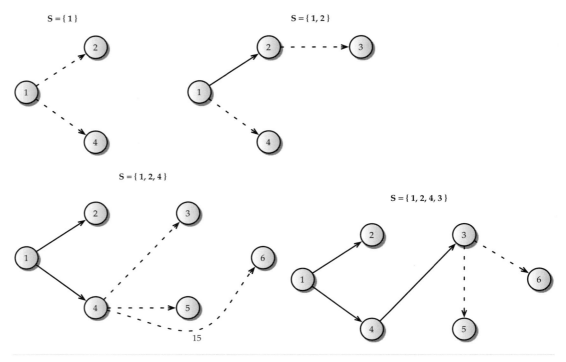

FIGURE 2.4   Iterative view of Dijkstra's algorithm.

neighbors not already in set S. We can see from Figure 2.1 that node 2's neighbors are node 3 and node 4. Thus, we compare and compute cost from node 1 for these two nodes, and see if there is any improvement:

$$\underline{D}_{13} = \min\{\underline{D}_{13}, \underline{D}_{12} + d_{23}\} = \min\{\infty, 1 + 2\} = 3$$
$$\underline{D}_{14} = \min\{\underline{D}_{14}, \underline{D}_{12} + d_{24}\} = \min\{1, 1 + 1\} = 1.$$

Note that there is no improvement in cost to node 4; thus, we keep the original shortest path. For node 3, we now have a shortest path, 1-2-3. For the rest of the nodes, the cost remains at ∞. This completes this iteration. We then move to the next iteration and find that the next intermediary is $k = 4$, and the process is continued as before. In Table 2.3, we summarize all the steps until all nodes are considered in list S, and in Figure 2.4, we give a visual illustration to how the algorithm adds a new intermediary k to list S. The centralized version of Dijkstra's algorithm is formally presented in Algorithm 2.3.

2.3.2 Distributed Approach

The distributed variant of Dijkstra's algorithm is very similar to the centralized version. The main difference is that the link cost of a link received by one node could be different from another node since this information is disseminated in an asynchronous manner. Thus, we

TABLE 2.3   Iterative steps in Dijkstra's algorithm.

Iteration	List, S	D_{12}	Path	D_{13}	Path	D_{14}	Path	D_{15}	Path	D_{16}	Path
1	$\{1\}$	1	1-2	∞	–	1	1-4	∞	–	∞	–
2	$\{1, 2\}$	1	1-2	3	1-2-3	1	1-4	∞	–	∞	–
3	$\{1, 2, 4\}$	1	1-2	2	1-4-3	1	1-4	3	1-4-5	16	1-4-6
4	$\{1, 2, 4, 3\}$	1	1-2	2	1-4-3	1	1-4	3	1-4-5	3	1-4-3-6
5	$\{1, 2, 4, 3, 5\}$	1	1-2	2	1-4-3	1	1-4	3	1-4-5	3	1-4-3-6
6	$\{1, 2, 4, 3, 5, 6\}$	1	1-2	2	1-4-3	1	1-4	3	1-4-5	3	1-4-3-6

denote the cost of link k-m as received at node i at time t by $d^i_{km}(t)$. Similarly, the minimum distance from i to j is time-dependent and is denoted by $\underline{D}_{ij}(t)$.

Dijkstra's algorithm for the distributed environment is presented in Algorithm 2.4. The steps are similar to the centralized version. Thus, in the distributed version, it is really the communication of the link cost information in a distributed manner that is taken into account by the algorithm. The steps are the same as in Table 2.3—this time we can think of the iterative

---

ALGORITHM 2.4   Dijkstra's shortest path first algorithm (a distributed approach).

1. Discover nodes in the network, $\mathcal{N}$, and cost of link k-m, $d^i_{km}(t)$, as known to node i at the time of computation, t.

2. Start with source node i in the permanent list of nodes considered, i.e., $S = \{i\}$; all the rest of the nodes are put in the tentative list labeled as S'. Initialize

$$\underline{D}_{ij}(t) = d^i_{ij}(t), \quad \text{for all } j \in S'.$$

3. Identify a neighboring node (intermediary) k not in the current list S with the minimum cost path from node i, i.e., find $k \in S'$ such that $\underline{D}_{ik}(t) = \min_{m \in S'} \underline{D}_{im}(t)$.

    Add k to the permanent list S, i.e., $S = S \cup \{k\}$,

    Drop k from the tentative list S', i.e., $S' = S' \backslash \{k\}$.

    If S' is empty, stop.

4. Consider neighboring nodes $\mathcal{N}_k$ of the intermediary k (but do not consider nodes already in S) to check for improvement in the minimum cost path, i.e.,
    for $j \in \mathcal{N}_k \cap S'$

$$\underline{D}_{ij}(t) = \min\{\underline{D}_{ij}(t), \underline{D}_{ik}(t) + d^i_{kj}(t)\}. \tag{2.3.2}$$

    Go to Step 3.

ALGORITHM 2.5   Dijkstra's shortest path first algorithm (with tracking of next hop).

0	// Computation at time t
1	$S = \{i\}$ // permanent list; start with source node i
2	$S' = \mathcal{N} \setminus \{i\}$ // tentative list (of the rest of the nodes)
3	for (j in S') do
4	if ($d^i_{ij}(t) < \infty$) then // if i is directly connected to j
5	$\underline{D}_{ij}(t) = d^i_{ij}(t)$
6	$H_{ij} = j$ // set i's next hop to be j
7	else
8	$\underline{D}_{ij}(t) = \infty$
9	$H_{ij} = -1$ // next hop not set
10	endif
11	endfor
12	while (S' is not empty) do // while tentative list is not empty
13	$Dtemp = \infty$ // find minimum cost neighbor k
14	for (m in S') do
15	if ($\underline{D}_{im}(t) < Dtemp$) then
16	$Dtemp = \underline{D}_{im}(t)$
17	$k = m$
18	endif
19	endfor
20	$S = S \cup \{k\}$ // add to permanent list
21	$S' = S' \setminus \{k\}$ // delete from tentative list
22	for (j in $\mathcal{N}_k \cap S'$) do
23	if ($\underline{D}_{ij}(t) > \underline{D}_{ik}(t) + d^i_{kj}(t)$) then // if cost improvement via k
24	$\underline{D}_{ij}(t) = \underline{D}_{ik}(t) + d^i_{kj}(t)$
25	$H_{ij} = H_{ik}$ // next hop for destination j; inherit from k
26	endif
27	endfor
28	endwhile

---

steps as the increment in time in terms of learning about different links and link costs in the network.

Determination of the next hop is important in many networking environments; next hop refers to the next directly connected node that the source node i should go to for reaching a destination j; ideally, the next hop should be on the optimal path. In Algorithm 2.5, we present a somewhat formal version of Dijkstra's algorithm—the purpose is to highlight the logic conditions for the benefit of the interested reader. In this algorithm, we have also included another identifier H_{ij} to track the next hop from i for destination j. Finally, in many situations, the shortest path to a specific destination j, instead of being to all destinations, is sufficient to compute. This can be easily incorporated in Algorithm 2.5 by inserting the following operation between line 19 and line 20: "if (k is same as destination j), then *exit* the while loop;" this means that we have found the shortest path to destination j.

2.4 Comparison of the Bellman–Ford Algorithm and Dijkstra's Algorithm

This is a good time to do a quick comparison between Dijksta's algorithm (Algorithm 2.3) and the Bellman–Ford algorithm (Algorithm 2.1). First, the Bellman–Ford algorithm computes the shortest path to one destination at a time while Dijksta's algorithm computes the shortest paths to all destinations (sometimes called the shortest path tree). When we compare the minimum cost computation for each algorithm, i.e., between Eq. (2.2.1b) and Eq. (2.3.1), it may seem that they are similar. However, there are actually very important subtle differences. In both of them, there is an intermediary node k; in the case of the Bellman–Ford algorithm, node k is over *all* nodes to find the best next hop to node j, while in the case of Dijkstra's algorithm, node k is an intermediary first determined and fixed, and then the shortest path computation is done to all j, not already covered. Table 2.1 and Table 2.3 are helpful in understanding this difference.

Since there are various operations in an algorithm, it is helpful to know the computational complexity (see Appendix B.3) so that a comparison of two or more algorithms can be done in terms of computational complexity using the "big-O" notation. Given N as the total number of nodes and L as the total number of links, the computational complexity of the Bellman–Ford algorithm is $O(LN)$. The complexity of Dijkstra's algorithm is $O(N^2)$ but can be improved to $O(L + N \log N)$ using a good data structure. Note that if a network is fully connected, the number of bidirectional links is $N(N-1)/2$; thus, for a fully connected network, the complexity of the Bellman–Ford algorithm is $O(N^3)$ while for Dijkstra's algorithm, it is $O(N^2)$.

Two key routing protocol concepts, the distance vector protocol concept and the link-state protocol concept, have fairly direct relation to the Bellman–Ford algorithm (or the distance vector-based shortest path computation approach) and Dijkstra's algorithm, respectively. These two key routing protocol concepts will be discussed later in Sections 3.3 and 3.4.

2.5 Shortest Path Computation with Candidate Path Caching

We will next deviate somewhat from the Bellman–Ford algorithm and Dijkstra's algorithm. There are certain networking environments where a list of possible paths is known or determined ahead of time; such a path list will be referred to as the *candidate* path list. *Path caching* refers to storing of a candidate path list at a node ahead of time. If through a distributed protocol mechanism the link cost is periodically updated, then the shortest path computation at a node becomes very simple when the candidate path list is already known.

Consider again the six-node network shown in Figure 2.1. Suppose that node 1 somehow *knows* that there are four paths available to node 6 as follows: 1-2-3-6, 1-4-3-6, 1-4-5-6, and 1-4-6; they are marked in Figure 2.5.

Using the link cost, we can then compute path cost for each path as shown in the table in Figure 2.5. Now, if we look for the least cost path, we will find that path 1-4-3-6 is the most preferred path due to its lowest end-to-end cost. Suppose now that in the next time period, the link cost for link 4-3 changes from 1 to 5. If we know the list of candidate paths, we can then recompute the path cost and find that path 1-4-3-6 is no longer the least cost; instead, both 1-2-3-6 and 1-4-5-6 are now the shortest paths—either can be chosen based on a tie-breaker rule.

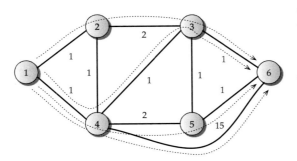

Path	Cost
1-2-3-6	$d_{12} + d_{23} + d_{36} = 4$
1-4-3-6	$d_{14} + d_{43} + d_{36} = 3$
1-4-5-6	$d_{14} + d_{45} + d_{56} = 4$
1-4-6	$d_{14} + d_{46} = 16$

FIGURE 2.5   Paths identified from node 1 to node 6, along with associated path cost.

We will now write the shortest path calculation in the presence of path caching in a generic way, considering that this calculation is done at time t. We consider a candidate path p between nodes i and node j, and its cost at time t as

$$\widehat{D}_{ij/p}(t) = \sum_{\text{link } l\text{-}m \text{ in path } p} d^i_{lm}(t), \tag{2.5.1}$$

where $d^i_{lm}(t)$ is the cost of link l-m at time t as known to node i, and the summation is over all such links that are part of path p. The list of candidate paths for the pair i and j will be denoted by $\mathcal{P}_{ij}$; the best path will be identified by $\widehat{p}$. The procedure to compute the shortest path is given in Algorithm 2.6.

ALGORITHM 2.6   **Shortest path computation when candidate paths are known.**

At source node i, a list of candidate paths $\mathcal{P}_{ij}$ to destination node j is available,
  and link cost, $d^i_{lm}(t)$, of link l-m at time t is known:
// Initialize the least cost:
$\widehat{D}_{ij}(t) = \infty$
// Consider each candidate path in the list
for (p in $\mathcal{P}_{ij}$) do
    $\widehat{D}_{ij/p}(t) = 0$
    for (link l-m in path p) do   // add up cost of links for this path
        $\widehat{D}_{ij/p}(t) = \widehat{D}_{ij/p}(t) + d^i_{lm}(t)$ (2.5.2)
    end for
    if ($\widehat{D}_{ij/p}(t) < \widehat{D}_{ij}(t)$) then   // if this is cheaper, note it
        $\widehat{D}_{ij}(t) = \widehat{D}_{ij/p}(t)$
        $\widehat{p} = p$
    end if
end do

It is important to note that the candidate path list is not required to include all possible paths between node i and j, only a sublist of paths that are, for some reason, preferable to consider for a particular networking environment. The way to think about this is to think of a road network in a city where to go from your home to school/office, you are likely to use only a selected set of paths. In a communication network, this approach of computing the shortest path involves a trade-off between storage and time complexity. That is, by storing multiple candidate paths ahead of time, the actual computation is simple when new link costs are received. The set of candidate paths can be determined using, for example, the K-shortest path algorithm (see Algorithm 2.10); since the interest in the case of path caching is obtain a good working set, any reasonable link cost can be assumed; for example, we can set all link costs to 1 (known also as hop count) and use Algorithm 2.10 to obtain a set of K candidate paths.

It is worth noting that such a candidate path-based approach can potentially miss a good path. For example, if a node is configured to keep only three candidate paths, it can potentially miss including 1-4-3-6; thus, in the first cycle of computation before the link cost d_{43} for link 4-3 was updated, this path would not be chosen at all.

2.6 Widest Path Computation with Candidate Path Caching

So far, we have assumed that the shortest path is determined based on the additive cost property. There are many networking environments in which the additive cost property is not applicable; for example, dynamic call routing in the voice telephone network (refer to Chapter 10) and quality of service based routing (refer to Chapter 17). Thus, determining paths when the cost is nonadditive is also an important problem in network routing; an important class among the nonadditive cost properties is *concave* cost property that leads to widest path routing. We will first start with the case in which path caching is used, so that it is easy to transition and compare where and how the nonadditive concave case is different from the additive case described in the previous section.

Suppose a network link has a certain bandwidth available, sometimes referred to as *residual capacity*; to avoid any confusion, we will denote the available bandwidth by b_{lm} for link l-m, as opposed to d_{lm} for the additive case. Note that $b_{lm} = 0$ then means that the link is not feasible since there is no bandwidth; we can also set $b_{lm} = 0$ if there is no link between nodes l and m (compare this with $d_{lm} = \infty$ for the additive case). We start with a simple illustration. Consider a path between node 1 and node 2 consisting of three links: the first link has 10 units of bandwidth available, the second link has 5 units of bandwidth available, and the third link has 7 units of bandwidth available. Now, if we say the cost of this path is additive, i.e., $22(= 10 + 5 + 7)$, it is unlikely to make any sense. There is another way to think about it. Suppose that we have new requests coming in, each requiring a unit of *dedicated* bandwidth for a certain duration. What is the maximum number of requests this path can handle? It is easy to see that this path would be able to handle a maximum of five additional requests simultaneously since if it were more than five, the link in the middle in this case would not be able to handle more than five requests. That is, we arrive at the availability of the path by doing $\min\{10, 5, 7\} = 5$. Thus the path "cost" is 5; certainly, this is a strange definition of a path cost; it is easier to see this as the *width* of a path (see Figure 2.6). Formally, similar to Eq. (2.5.1), for all links l-m that make up a path p, we can write the width of the path as

$$\widehat{B}_{ij/p}(t) = \min_{\text{link } l\text{-}m \text{ in path } p} \left\{ b^i_{lm}(t) \right\}. \tag{2.6.1}$$

FIGURE 2.6 Width of a path—a visual depiction.

Regardless, the important point to note is that this path cost is computed using a non-additive cost property, in this case the minimum function. It may be noted that the minimum function is not the only nonadditive cost property possible for defining cost of a path; there are certainly other possible measures, such as the nonadditive *multiplicative* property given by Eq. (B.8.1) discussed in Appendix B.8.

Now consider a list of candidate paths; how do we define the most preferable path? One way to define it is to find the path with the largest amount of available bandwidth. This is actually easy to do once the path "cost" for each path is determined since we can then take the maximum of all such paths. Consider the topology shown in Figure 2.7 with available bandwidth on each link as marked. Now consider three possible paths between node 1 and node 5:

Path	Cost
1-2-3-5	$\min\{b_{12}, b_{23}, b_{35}\} = 10$
1-4-3-5	$\min\{b_{14}, b_{43}, b_{35}\} = 15$
1-4-5	$\min\{b_{14}, b_{45}\} = 20$

ALGORITHM 2.7   **Widest path computation (non-additive, concave) when candidate paths are known.**

At source node i, a list of candidate paths $\mathcal{P}_{ij}$ to destination node j is available,
  and link bandwidth, $b^i_{lm}(t)$, of link l-m at time t is known:
// Initialize the least bandwidth:
$\widehat{B}_{ij}(t) = 0$
for p in $\mathcal{P}_{ij}$ do
  $\widehat{B}_{ij/p}(t) = \infty$
  for (link l-m in path p) do   // find bandwidth of the bottleneck link
  $\widehat{B}_{ij/p}(t) = \min\left\{\widehat{B}_{ij/p}(t), b^i_{lm}(t)\right\}$ (2.6.2)
  end for
  if $\left(\widehat{B}_{ij/p}(t) > \widehat{B}_{ij}(t)\right)$ then   // if this has more bandwidth, note it
    $\widehat{B}_{ij}(t) = \widehat{B}_{ij/p}(t)$
    $\widehat{p} = p$
  end if
end do

It is easy to see that the third path, 1-4-5, has the most bandwidth and is thus the preferred path. This means that we need to do a maximum over all paths in the case of the nonadditive property to find the widest path as opposed to the minimum over all paths when additive cost property is used. A widest path so selected is sometimes referred to as the *maximal residual capacity* path. The procedure is presented in detail in Algorithm 2.7. It is helpful to contrast this algorithm with its counterpart, Algorithm 2.6, where the additive cost property was used; for example, you can compare Eq. (2.6.2) with Eq. (2.5.2), especially the logical "if" condition statements.

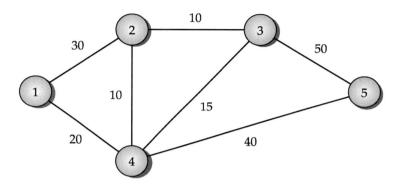

FIGURE 2.7   Network example for widest path routing.

Remark 2.1. *Relation between shortest path routing and widest path routing.*
    While the cost of a path is determined differently, by using the additive property in Eq. (2.5.1), and by the nonadditive property in Eq. (2.6.1), there is a direct relation between shortest and widest. If we imagine the path cost to be the *negative* of the quantity given in Eq. (2.6.1), then widest translates to being the minimum of this negative cost. Thus, the widest path is the cheapest path in the sense of this negative representation. In other words, the widest path can be thought of as the *nonadditive (concave) shortest path*.   ♦

2.7 Widest Path Algorithm

We are coming back full circle to *no* path caching for widest path routing algorithms. We present two approaches: first we show an extension of Dijkstra's shortest path first algorithm; next, we extend the Bellman–Ford algorithm.

2.7.1 Dijkstra-Based Approach

When there is no path caching, the algorithm is very similar to Dijkstra's algorithm that is adapted from [731], and is listed in Algorithm 2.8.
    Consider the network topology shown in Figure 2.7 where each link is marked with an available bandwidth. The algorithmic steps with Algorithm 2.8 are detailed in Table 2.4 for

---

ALGORITHM 2.8   Widest path algorithm, computed at node i (Dijkstra-based).

---

1. Discover list of nodes in the network, $\mathcal{N}$ and available bandwidth of link k-m, $b^i_{km}(t)$, as known to node i at the time of computation, t.

2. Initially, consider only source node i in the set of nodes considered, i.e., $S = \{i\}$; mark the set with all the rest of the nodes as S'. Initialize

$$\underline{B}_{ij}(t) = b^i_{ij}(t).$$

3. Identify a neighboring node (intermediary) k not in the current list S with the maximum bandwidth from node i, i.e., find $k \in S'$ such that $\underline{B}_{ik}(t) = \max_{m \in S'} \underline{B}_{im}(t)$

   Add k to the list S, i.e., $S = S \cup \{k\}$

   Drop k from S', i.e., $S' = S' \backslash \{k\}$.

   If S' is empty, stop.

4. Consider nodes in S' to update maximum bandwidth path, i.e., for $j \in S'$

$$\underline{B}_{ij}(t) = \max\{\underline{B}_{ij}(t), \min\{\underline{B}_{ik}, b^i_{kj}(t)\}\}. \tag{2.7.1}$$

   Go to Step 3.

TABLE 2.4   Iterative steps based on Algorithm 2.8.

Iteration	List, S	$\underline{B}_{12}$	Path	$\underline{B}_{13}$	Path	$\underline{B}_{14}$	Path	$\underline{B}_{15}$	Path
1	$\{1\}$	30	1-2	0	–	20	1-4	0	–
2	$\{1, 2\}$	30	1-2	10	1-2-3	20	1-4	0	–
3	$\{1, 2, 4\}$	30	1-2	15	1-4-3	20	1-4	20	1-4-5
4	$\{1, 2, 4, 3\}$	30	1-2	15	1-4-3	20	1-4	20	1-4-5
5	$\{1, 2, 4, 3, 5\}$	30	1-2	15	1-4-3	20	1-4	20	1-4-5

the distributed time-dependent case from the point of view of node 1, i.e., suppose that node 1 wants to find the path with most available bandwidth to all other nodes in the network. Then, initially, $S = \{1\}$ and $S' = \{2, 3, 4, 5\}$, and the widest paths to all nodes that are direct neighbors of node 1 can be readily found while for the rest, the "cost" remains at 0, i.e.,

$$\underline{B}_{12} = 30, \quad \underline{B}_{14} = 20, \quad \underline{B}_{13} = \underline{B}_{15} = 0.$$

Since $\max_{j \in S'} \overline{B}_{1j} = 30$ is attained for $j = 2$, we add node 2 to list S. Thus, we have the updated lists: $S = \{1, 2\}$ and $S' = \{3, 4, 5\}$.

Now for j not in $\mathcal{S}$, we update the available bandwidth to see if it is better than going via node 2, as follows:

$$\underline{B}_{13} = \max\{\underline{B}_{13}, \min\{B_{12}, b_{23}\}\} = \max\{0, \min\{30, 10\}\} = 10 \quad // \text{ use 1-2-3}$$

$$\underline{B}_{14} = \max\{\underline{B}_{14}, \min\{B_{12}, b_{24}\}\} = \max\{20, \min\{30, 10\}\} = 20 \quad // \text{ stay on 1-2}$$

$$\underline{B}_{15} = \max\{\underline{B}_{15}, \min\{B_{12}, b_{25}\}\} = \max\{0, \min\{30, 0\}\} = 0 \quad // \text{ no change}$$

Now we are in the second pass of the algorithm. This time, $\max_{j \in \mathcal{S}'} \overline{B}_{1j} = \max\{\underline{B}_{13}, \underline{B}_{14}, \underline{B}_{15}\} = 20$. This is attained for $j = 4$. Thus, $\mathcal{S}$ becomes $\{1, 2, 4\}$. Updating the available bandwidth to check via node 4 will be as follows:

$$\underline{B}_{13} = \max\{\underline{B}_{13}, \min\{B_{14}, b_{43}\}\} = \max\{10, \min\{20, 15\}\} = 15 \quad // \text{ use 1-4-3}$$

$$\underline{B}_{15} = \max\{\underline{B}_{15}, \min\{B_{14}, b_{45}\}\} = \max\{0, \min\{20, 40\}\} = 20 \quad // \text{ use 1-4-5}$$

This time, $j = 3$ will be included in $\mathcal{S}$. Thus, $\mathcal{S}$ becomes $\{1, 2, 4, 3\}$. There is no further improvement in the final pass of the algorithm.

2.7.2 Bellman–Ford-Based Approach

The widest path algorithm that uses the Bellman–Ford-based approach is strikingly similar to the Bellman–Ford shortest path routing algorithm given in Algorithm 2.2. For completeness, this is listed in Algorithm 2.9.

ALGORITHM 2.9   Widest path algorithm, computed at node i (Bellman–Ford-based).

Initialize

$$\overline{B}_{ii}(t) = 0; \quad \overline{B}_{ij}(t) = 0, \quad (\text{for node } j \text{ that node } i \text{ is aware of}). \tag{2.7.2a}$$

For (nodes j that node i is aware of) do

$$\overline{B}_{ij}(t) = \max_{k \text{ directly connected to } i} \min\left\{b_{ik}(t), \overline{B}^i_{kj}(t)\right\}, \quad \text{for } j \neq i. \tag{2.7.2b}$$

2.8 *k*-Shortest Paths Algorithm

We now go back to the class of shortest path algorithms to consider an additional case. In many networking situations, it is desirable to determine the second shortest path, the third shortest path, and so on, up to the k-th shortest path between a source and a destination. Algorithms used for determining paths beyond just the shortest paths are generally referred to as k-shortest paths algorithms.

A simple way to generate additional paths would be to start with, say Dijkstra's shortest path first algorithm, to determine the shortest path; then, by temporarily deleting each link on the shortest path one at a time, we can consider the reduced graph where we can apply again

---

ALGORITHM 2.10   *k*-**shortest paths algorithm.**

---

1. Initialize $k := 1$.

2. Find the shortest path $\mathcal{P}$ between source (i) and destination (j) in graph $\mathcal{G}$, using Dijkstra's Algorithm.

   Add $\mathcal{P}$ to permanent list $\mathcal{K}$, i.e., $\mathcal{K} := \{\mathcal{P}\}$.

   If $K = 1$, stop.

   Add $\mathcal{P}$ to set $\mathcal{X}$ and pair $(\mathcal{P}, i)$ to set $\mathcal{S}$, i.e., $\mathcal{X} := \{\mathcal{P}\}$ and $\mathcal{S} := \{(\mathcal{P}, i)\}$.

3. Remove $\mathcal{P}$ from $\mathcal{X}$, i.e., $\mathcal{X} := \mathcal{X} \backslash \{\mathcal{P}\}$.

4. Find the unique pair $(\mathcal{P}, w) \in \mathcal{S}$, and corresponding deviation node w associated with $\mathcal{P}$.

5. For each node v, except j, on subpath of $\mathcal{P}$ from w to j ($sub_\mathcal{P}(w, j)$):

   Construct graph $\mathcal{G}'$ by removing the following from graph $\mathcal{G}$:

   (a) All the vertices on subpath of $\mathcal{P}$ from i to v, except v.

   (b) All the links incident on these deleted vertices.

   (c) Links outgoing from v toward j for each $\mathcal{P}' \in \mathcal{K} \cup \{\mathcal{P}\}$, such that $sub_\mathcal{P}(i, v) = sub_{\mathcal{P}'}(i, v)$.

   Find the shortest path $\mathcal{Q}'$ from v to j in graph $\mathcal{G}'$ using Dijkstra's Algorithm.

   Concatenate subpath of $\mathcal{P}$ from i to v and path $\mathcal{Q}'$, i.e., $\mathcal{Q} = sub_\mathcal{P}(i, v) \oplus \mathcal{Q}'$.

   Add $\mathcal{Q}$ to $\mathcal{X}$ and pair $(\mathcal{Q}, v)$ to $\mathcal{S}$, i.e., $\mathcal{X} := \mathcal{X} \cup \{\mathcal{Q}\}$ and $\mathcal{S} := \mathcal{S} \cup \{(\mathcal{Q}, v)\}$.

6. Find the shortest path $\mathcal{P}$ among the paths in $\mathcal{X}$ and add $\mathcal{P}$ to $\mathcal{K}$, i.e., $\mathcal{K} := \mathcal{K} \cup \mathcal{P}$.

7. Increment k by 1.

8. If $k < K$ and $\mathcal{X}$ is not empty, go to Step 4, else stop.

Dijkstra's shortest path first algorithm. This will then give us paths that are longer than the shortest path. By identifying the cost of each of these paths, we can sort them in order of successively longer paths. For example, consider finding k-shortest paths from node 1 to node 6 in Figure 2.1. Here, the shortest path is 1-4-3-6 with path cost 3. Through this procedure, we can find longer paths such as 1-2-3-6 (path cost 4), 1-4-5-6 (path cost 4), and 1-4-3-5-6 (path cost 4). It is easy to see that paths so determined may have one or more links in common.

Suppose that we want to find k-shortest link disjoint paths. In this case, we need to temporarily delete all the links on the shortest path and run Dijkstra's algorithm again on the reduced graph—this will then give the next shortest link disjoint path; we can continue this process until we find k-shortest link disjoint paths. Sometimes it might not be possible to find two or more link disjoint paths, if the reduced graph is isolated into more than one network. Consider again Figure 2.1. Here, the shortest path from node 1 to node 6 is 1-4-3-6 with path cost 3. If we temporarily delete the links in this path, we find the next link-disjoint shortest path to be 1-2-4-5-6 of path cost 5. If we now delete links in this path, node 1 becomes isolated in the newly obtained reduced graph.

In Algorithm 2.10, we present a k-shortest path algorithm that is based on an idea, originally outlined in [756]; see also [454], [549] for additional references for this method. In this algorithm, a fairly complicated process is applied beyond finding the shortest path. For example, it uses an auxiliary list S in order to track/determine longer paths. This is meant for die-hard readers, though. A description with each step is included in Algorithm 2.10 to convey the basic idea behind this algorithm.

Finally, recall that we discussed widest path computations with candidate path caching; such candidate paths to be cached can also be determined using a k-shortest paths algorithm. Typically, in such situations, the link cost for all links can be set to 1 because usually hop-length-based k-shortest paths are sufficient to determine candidate paths to cache.

2.9 Summary

We first start with notations. In discussing different shortest path algorithms, we have used a set of notations. While the notations might look confusing at first, there is some structure to the notations used here.

First, for a link i-k connecting node i and node k, the link cost for the additive case has been denoted by d_{ik}, while the link cost for the nonadditive case was denoted by b_{ik}. From a computational results point of view, we needed to track the minimum path cost between node i and j for various algorithms that can be distinctly identifiable—they can be classified as follows:

Algorithm	Indicator	Additive	Nonadditive (Widest)
Bellman–Ford	"overbar"	$\overline{D}_{ij}$	$\overline{B}_{ij}$
Dijkstra	"underscore"	$\underline{D}_{ij}$	$\underline{B}_{ij}$
Path caching	"hat"	$\widehat{D}_{ij}$	$\widehat{B}_{ij}$

A superscript is used when we discuss information as known to a node, especially node i where the algorithmic computation is viewed from in the distributed environment. Thus, we have used $\overline{D}^i_{kj}$ to denote the minimum additive path cost from node k to node j as known to node i. Finally, the temporal aspect is incorporated by making an expression a function of time t. Thus, we use $\overline{D}^i_{kj}(t)$ to indicate dependency on time t. While there are a few more notations, such as path list, these notations basically capture the essence and distinction of various algorithms. In any case, all notations are summarized in Table 2.5.

We have presented several shortest path and widest path routing algorithms that are useful in communication network routing. We started with the centralized version of the Bellman–Ford algorithm and then presented the distance vector approach, first used in the ARPANET distributed environment. Similarly, we presented Dikstra's algorithm, both the centralized and its distributed variant. We then considered routing algorithms when a non additive cost property (based on minimum function) is applicable; such algorithms can be classified as widest path routing when nonadditive cost property is concave. It may be noted that there may be several widest paths between two nodes, each with a different number of

TABLE 2.5   Summary of notations used in this chapter

Notation	Remark
i	Source node
j	Destination node
k	Intermediate node
$\mathcal{N}$	List of nodes in a network
$\mathcal{S}$	Permanent list of nodes in the Dijkstra's algorithm (considered so far in the calculation)
$\mathcal{S}'$	Tentative list of nodes in the Dijkstra's algorithm (yet to be considered in calculation)
$\mathcal{N}_k$	List of neighboring nodes of node k
d_{ij}	Link cost between nodes i and j
$d_{ij}(t)$	Link cost between nodes i and j at time t
$\overline{D}_{ij}$	Cost of the minimum cost path from node i to node j (Bellman–Ford)
$\overline{D}_{ij}^{(h)}$	Cost of the minimum cost path from node i to node j when h hops have been considered
$\overline{D}_{ij}(t)$	Cost of the minimum cost path from node i to node j at time t
$d_{kj}^{i}(t)$	Link cost between nodes k and j at time t as known to node i
$\overline{D}_{kj}^{i}(t)$	Cost of the minimum cost path from node k to node j at time t as known to node i
$\underline{D}_{ij}$	Cost of the minimum cost path from node i to node j (Dijkstra)
$\widehat{D}_{ij/p}(t)$	Cost of path p from node i to node j (path caching)
$\underline{B}_{ij}(t)$	Nonadditive cost (width) of the best path from node i to node j at time t (Dijkstra)
$\widehat{B}_{ij}(t)$	Nonadditive cost (width) of the best path from node i to node j at time t (path caching)
$\mathcal{P}_{ij}$	The list of cached path at node i for destination j
H_{ij}	Next hop for source i for destination j

hops. It is sometimes useful to identify the *shortest-widest* path; if "shortest" is meant in terms of the number of hops, then it can be more appropriately referred to as the *least-hop-widest* path, i.e., the widest path that uses the least number of hops between two nodes. Another consideration is the determination of the *widest-shortest path*, i.e., a feasible path with the minimum cost, for example, in terms of hop count; if there are several such paths, one with the maximum bandwidth is used. These will be discussed later in Chapter 17.

Note that in this chapter, we have presented our discussion using origin and destination nodes. When a network structure is somewhat complicated, that is, when we have a backbone network that is the carrier of traffic between access networks, we also use the term *ingress* node to refer to a entry point in the core network and the term *egress* node to refer to an exit point in the core network. It is important to keep this in mind.

Finally, it is worth noting that the Bellman–Ford algorithm can operate with negative link cost while Dijkstra's algorithm requires the link costs to be nonnegative; on the other hand, Dijkstra's algorithm can be modified to work with negative link cost as well. Communication network routing protocols such as Open Shortest Path First (OSPF) and Intermediate System-to-Intermediate System (IS-IS) (refer to Chapter 6) that are based on the link state protocol concept do not allow negative weights. Thus, for all practical purposes, negative link cost rarely plays a role in communication networking protocols. Thus, in this book, we primarily consider the case when link costs are nonnegative. Certainly, from a graph theory point of view, it is important to know whether a particular algorithm works with negative link cost; interested readers may consult books such as [624].

Further Lookup

The Bellman–Ford shortest path algorithm for computing the shortest path in a centralized manner was proposed by Ford [231] in 1956. Bellman [68] described a version independently in 1958, by using a system of equations that has become known as Bellman's equations. Moore also independently presented an algorithm in 1957 that was published in 1959 [499]. Thus, what is often known as the Belllman–Ford algorithm, especially in communications networking, is also known as the *Bellman–Ford–Moore* algorithm in many circles.

The distance vector approach for the shortest path computation in a distributed environment is subtly as well as uniquely different from the centralized Bellman–Ford approach. The distance vector approach is also known as the original or "old" ARPANET routing algorithm, yet is sometimes attributed as the "distributed Bellman–Ford" algorithm. For a discussion on how these different naming and attributions came to be known, refer to [725]. For a comprehensive summary of ARPANET design decisions in the early years, see [464].

In 1959, Dijkstra presented his shortest path first algorithm for a centralized environment [178]. The "new" ARPANET routing took the distributed view of Dijkstra's algorithm along with considerations for numerous practical issues in a distributed environment; for example, see [368], [462], [463], [599].

Widest path routing with at most two links for a path has been known since the advent of dynamic call routing in telephone networks in the early 1980s. In this context, it is often known as *maximum residual capacity routing* or *maximum available trunk routing*; for example, see [680]. The widest path algorithm based on Dijkstra's framework given in Algorithm 2.8 and the distance vector framework given in Algorithm 2.9 are adapted from [731]. Widest path routing and its variations are applicable in a quality-of-service routing framework.

The *k*-shortest paths algorithm and its many variants have been studied by numerous researchers; see [202] for an extensive bibliography.

Exercises

2.1. Review questions:

(a) In what ways, are the Bellman–Ford algorithm (Algorithm 2.1) and the distance vector algorithm (Algorithm 2.2) different?

(b) What are the main differences between shortest path routing and widest path routing?

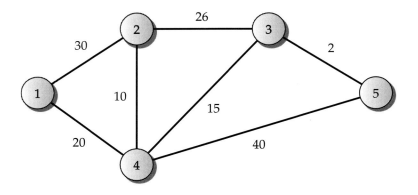

FIGURE 2.8   A 5-node example.

    (c) What is the difference between minimum hop routing and shortest path routing?

2.2. For the network example presented in Figure 2.1, compute the shortest paths from node 2 to all other nodes using the centralized Bellman–Ford algorithm (Algorithm 2.1).

2.3. For the network example presented in Figure 2.1, use Dijkstra's algorithm (Algorithm 2.3) to compute the shortest paths from node 6 to the other nodes. Next, consider that link 3-6 fails; recompute the shortest paths from node 6 to the other nodes.

2.4. Consider the network topology in Figure 2.1. Assume now that the links have the following bandwidth: 1-2: 1, 1-4: 1, 2-3: 2, 2-4: 2, 3-4: 1, 3-5: 1, 3-6: 4-5: 2; 4-6: 3; 5-6: 2. Determine the widest paths from node 6 to all the other nodes.

2.5. Consider the network topology in Figure 2.8. The number listed next to the links are link costs.

    (a) Determine the shortest path from node 1 to node 5 using Algorithm 2.2 and also using Algorithm 2.3.

    (b) Now suppose that the link cost for link 1-4 is changed to 45. Determine again the shortest path from node 1 to node 5 using Algorithm 2.2 and also using Algorithm 2.3. Also, generate an iterative view similar to Figure 2.4.

2.6. Consider the network topology in Figure 2.8. The number listed next to the links are assumed to be bandwidth. Determine the widest path from node 2 to node 5 using Algorithm 2.8.

2.7. Identify networking environments where path caching might be helpful that require either the shortest path or the widest path computation.

2.8. Develop a specialized *k*-shortest paths algorithm, given that a path cannot consist of more than two links.

2.9. Implement the *k*-shortest paths algorithm described in Algorithm 2.10.

3

Routing Protocols: Framework and Principles

There is nothing so annoying as to have two people talking when you're busy interrupting.

Mark Twain

Reading Guideline

This chapter is organized by topics for each routing protocol family. You may choose to read each one separately while we do encourage reading both Section 3.1 and Section 3.2 before doing so. For each routing protocol family, we include a general discussion followed by a descriptive discussion. Specific details of a protocol are listed as templates; thus, if you are interested in a general understanding, you may skip the templates. However, you can study the specifics of a protocol by directly going into its details in the template format. Certainly, you can get the most out of a protocol by reading both the descriptive discussion as well as studying the specifics given in the template format. Notations used in this chapter are closely related to routing algorithms described in Chapter 2; in particular, you will be able to see the connection between a protocol message and the mathematical representation of information.

In this chapter, we consider three classes of routing protocols: distance vector, link state, and path vector. We present the basic working of these classes of protocols along with how special cases are handled, and types of problems a protocol might face in a distributed operating environment. We also show how routing protocols are related to routing algorithms.

3.1 Routing Protocol, Routing Algorithm, and Routing Table

In Chapter 2, we presented two important classes of shortest path routing algorithms, the Bellman–Ford algorithm and Dijkstra's algorithm, including their distributed variants. The role of a routing protocol arises only when we consider a distributed environment. From the discussion of the distributed algorithms, we can see that for a node to compute the shortest paths, certain information must be available at that node; the node needs to somehow get this information from its neighbor or use the neighbor as a conduit to obtain information from other nodes. At the same time, the node might need to let its neighbors know the information it has. A major role of a routing protocol is to facilitate this exchange of information in a standardized manner, sometimes also coupled with routing computation.

From the discussion of the distributed variant of both the Bellman–Ford algorithm and Dijkstra's algorithm, we can see that what a node needs to know is different from one algorithm to another to compute the shortest paths. Thus, when we say standardization, we mean the standardization of information that is applicable within an algorithmic framework where all nodes conform to this framework. For example, all nodes that conform to computing the shortest paths using the Bellman–Ford algorithm would need the same type of information; thus, the exchange of information needs to be standardized around this requirement. Similarly, this is the case for Dijkstra's algorithm.

When the exchange of information is coupled to a particular algorithm such as the Bellman–Ford algorithm, it can give the impression that a routing protocol is closely tied to a specific shortest path route computation algorithm. While the concept of routing protocols has direct connection to routing algorithms such as the Bellman–Ford algorithm or Dijkstra's algorithm, it is also important to distinguish and decouple the basic routing protocol concept from a specific routing algorithm. Fundamentally, a routing protocol addresses what type of information a node may need to receive from its neighbors and also may need to pass information to its neighbor. In other words, a routing protocol need not necessary worry about how a node uses this information to compute the shortest paths (or multiple paths), how often it computes the shortest paths.

At the same time, historically, there has been a close tie between a shortest path algorithm and a routing protocol, for example, between the Bellman–Ford algorithm and a distance vector routing protocol, and between Dijkstra's algorithm and a link state routing protocol. That is, it is hard to sometimes distinguish and often this leads to confusion. Regardless, it is best to think about a routing protocol separately from a routing computation algorithm. As you go through the book, you will find out that there are link state routing protocols that do *not* use Dijkstra's shortest path first algorithm. Similarly, there are distance vector protocols that do not use just the distributed Bellman–Ford algorithm for determining the shortest paths; additional functionalities are needed. On the other hand, when a specific standardization or implementation of a particular protocol concept is considered, an algorithm is often closely tied to the protocol; for example, consider the *Routing Information Protocol* (RIP) (see

Section 5.3), which is based on a distance vector protocol concept and uses the distance vector ("distributed Bellman–Ford") *algorithm* for computing the shortest paths; similarly, consider the *Open Shortest Path First* (OSPF) protocol (refer to Section 6.2), which is based on the link state routing protocol concept and uses Dijkstra's shortest path first routing algorithm. However, a networking environment enabled to use MPLS or GMPLS (refer to Chapter 18) also uses OSPF/IS-IS as routing protocols, yet use of Dijkstra's shortest path first algorithm is not required. Consider another example, real-time network routing (RTNR), used for dynamic call routing in the telephone network, RTNR uses a link state protocol while the actual routing computation is quite different from Dijkstra's algorithm (refer to Section 10.6). Finally, there are routing environments where no information exchange is done in order to select/compute routes; while this may seem odd, there are adaptive algorithms that work (and work well) without the paradigm of information exchange (see Section 10.5).

ROUTING INFORMATION: PUSH OR PULL

We stated earlier that a routing protocol addresses what type of information a node needs to receive from its neighbors and to pass information to its neighbors. This brings up another important issue that can best be stated as information *push* and information *pull*. *Push* refers to a node pushing information to other nodes (usually on a periodic basis), while *pull* refers to a node contacting/requesting other nodes to obtain information needed usually for routing computation, but information can be for other network controls as well (refer to Section 11.6). Some routing protocols use the push mode while others use the pull mode, and yet others use a hybrid push-pull mode. For various routing protocols discussed in this book, we will identify which mode is used. A routing protocol is also required to handle special cases, such as when a link or a node fails; in such cases, special messages might need to be exchanged.

MODES OF COMMUNICATING ROUTING INFORMATION

We next discuss *how* the routing information exchange is accomplished in a particular network. Essentially, there are two communication modes for exchanging routing information: *in-band* and *out-of-band*. While these two terms are used in a variety of ways in communication networking, we first clarify the use of these terms as they relate to this book. Specifically, in-band, in our context, means that the communication network that carries user traffic also carries routing information exchange; that is, there are mechanisms that allow carrying these two types of traffic in the *same* communication network. For example, an in-band mechanism is used in the Internet since eventually all are IP packets; how this is done in specific instances of routing protocols has been already discussed briefly in Chapter 1 and will be discussed later in Part II of this book. Out-of-band, as we use here, means that a completely separate network or mechanism is used for exchanging routing information from the communication network where the user traffic is carried. For example, for the circuit-switched telephone network, routing information exchanges are accomplished through an out-of-band mechanism. A simple analogy might be helpful here. Consider the road network in a large metropolitan area; here user traffic is motorists driving various automobiles. To indicate congestion, out-of-band mechanisms are used such as a helicopter to monitor from the sky, reporting to radio stations, which in turn use their radio frequency to transmit to motorists; thus, motorists, on listening to a radio station, can make a decision to choose an alternate path. That is, from a networking point of view, the control plane and data plane are completely separate in the out-of-band mode.

ROUTING TABLE AND FORWARDING TABLE

Finally, there is another important component that goes with any routing environment. It is called the *routing table*. A routing table at a node provides a lookup entry for each destination by identifying an outgoing link/interface or a path. A routing table can be set up with an entry that remains static forever—this is the case when static routing is used. When dynamic routing is used, and based on exchange of information, a node performs route computation to determine paths to each destination and then creates/updates a routing table; this can be on a periodic basis or due to an extraordinary event such as a link failure. Thus, it is important to decouple the need for the routing table from whether the network employs static or dynamic routing—a routing table is needed in a communication network whether the mechanism is accomplished statically or dynamically. Typically, two forms of routing table entries are possible: (1) *next hop–based* or *hop-by-hop* routing based, and (2) *explicit route* based or *source routing* based. In the case of next hop–based, a node stores the pointer only to the next node(s) or hop(s) for destinations it is aware of, while in the case of explicit route based, a node stores the entire path to a destination; the second one is sometimes referred to as *route pinning*. The originating node then tags the pinned route to a packet for traversal through the nodes listed in the pinned route; however, this does not rule out the possibility of a node downstream replacing a pinned route with its own source route to destination. Next hop–based routing toward a destination is commonly used in the Internet, although source routing is an option possible within an intradomain routing protocol such as OSPF or IS-IS. Routing in the telephone network with progressive call control is also based on next hop routing, while routing is source-based if originating call control is used (see Section 10.1).

There is another terminology, *forwarding table*, often used in place of a routing table. While at the conceptual level there is no difference between these two terms, there are important differences when it comes to implementation at a routing device such as a router; this will be discussed later in Section 14.1.4. In this chapter, we will consistently use the term *routing table*.

3.2 Routing Information Representation and Protocol Messages

Several routing information entities will be covered starting in the next section. For consistency, we will follow notations that we have used earlier in Chapter 2. For example, d_{ik} refers to the distance cost (link-cost) on the direct link between node i and node k. $\overline{D}_{kj}$ refers to the computed cost *between* node k and j, whether on a direct link or through other nodes in the network, and $\overline{D}^i_{kj}$ refers to the cost *between* node k and j, as known to node i. Recall from Chapter 2 that an overbar over uppercase D, as in $\overline{D}$, is used in the the Bellman–Ford algorithm and the distance vector algorithm approach; for consistency, it will be used here in the discussion of a distance vector routing protocol. A lowercase d with subscript, such as d_{ik}, is used for link-cost from node i to node k in a link state protocol. For a path found from node i to node j, the immediate next hop of node i will be denoted by H_{ij}. The list of neighbors of node i will be denoted by $\mathcal{N}_i$. *All protocols will be discussed from the point of view of node i.*

There are two types of link representation, bidirectional links and unidirectional links; their usage depends on a routing protocol. A bidirectional link between nodes 1 and 2 will be denoted by 1-2, while the unidirectional link *from* node 1 *to* node 2 will be denoted by 1→2. Similarly, a bidirectional path between node 1 and node 3 where a link directly connects node 1 and node 2 and another connects node 2 and node 3 will be denoted by 1-2-3; its unidirectional counterpart from node 1 to node 3 will be denoted by 1→2→3. Finally, we will use another notation for a path when referring to nodes that are in series without referring to links connecting any two nodes. For example, path $(1, 2, 3)$ connects node 1 to node 2 to node 3 without saying anything about the links in between; in other words, this path notation does not rule out the possibility that there may be multiple links between between node 1 and 2, or node 2 and 3. In general, we denote a list of paths between nodes i and j by $\mathcal{P}_{ij}$, and between nodes k and j as known to node i by $\mathcal{P}_{kj}^i$; in the second case, we typically assume that k is a neighbor of node i.

We will use a box around texts to indicate a protocol message such as this one: $\boxed{\text{Protocol Message}}$. A protocol message will be partitioned by "|" between two different message entities such as $\boxed{\text{Message-1 | Message-2 | ...}}$. A general delimiter of information pieces within a message will be marked by "," while ";" will be used for special markers of separate information within a message, for example, $\boxed{2, 3; 3, 2, 1 \mid 3, 1; 1}$.

These are the main representations to keep in mind. We will introduce other representations as we progress through the chapter.

3.3 Distance Vector Routing Protocol

In some ways, a distance vector protocol is the oldest routing protocol in practice. Thus, much has been learned from and about distance vector protocols. In the following, we will start with the basic concept of a distance vector protocol, followed by issues faced and how they are addressed, leading to identifying factors in the design of recent distance vector protocols. It is important to read the following with a general view of what a generic distance vector protocol is and should/can be, rather than to tie in to a specific instance of a distance vector protocol such as RIP (see Section 5.3). In other words, limitations of a specific instance of a distance vector protocol should not be confused with limitations of the *concept* of a distance vector protocol in general.

3.3.1 Conceptual Framework and Illustration

In Chapter 2, we presented the distance vector routing algorithm (see Algorithm 2.2). It tells us that a node, say node i, needs to know the distance or cost from its neighbors to a destination, say node j, to determine the shortest path to node j. Since node i can have multiple neighbors, it is preferable to know the distance from all its neighbors to node j so that node i can compare and determine the shortest path. The basic information exchange aspect about a distance vector protocol is that a node needs the distance cost information from each of its neighbors for *all* destinations; this way, it can compare and determine the shortest paths to all destinations. This means that our discussion about a distance vector protocol is strongly tied to the distance vector routing algorithm (Algorithm 2.2), but keep in mind that it is not necessary to do so in general.

The main operation of a distance vector protocol needs to address dissemination and reception of information. Thus, we start with the basic operation of a distance vector protocol from the point of view of a node as shown in Figure 3.1. We clarify a few aspects about the protocol description:

- The protocol does not need to know ahead of time how many nodes are in the network; in other words, through the information received periodically that may contain a new node information, the receiving node can update the list of destination nodes.

- Actual numbering of a node can be done through some addressing scheme outside the protocol (for example, IP addressing with RIP).

- For each destination j (from node i), the protocol maintains/updates the next hop, to be denoted as H_{ij}.

- With the arrival of a distance vector message from a neighbor k, the protocol updates the cost to a destination if the currently stored next hop for this destination is also k.

- Steps indicated are not necessarily in any specific order (except for initialization).

- There is possibly a time wait between steps and within substeps of a step.

We will now illustrate the basic working of a distance vector protocol through the six-node network discussed earlier in Chapter 2. We reproduce the topology (along with link cost information) in Figure 3.2 for ease of reference. Since we do not consider time-dependent distance cost, we will ignore time parameter t, i.e., we will use d_{ik} and $\overline{D}_{ij}$, instead of $d_{ik}(t)$ and $\overline{D}_{ij}(t)$, respectively. Recall that d_{ik} refers to the link cost on the link i-k connecting node i to node k, while $\overline{D}_{ij}$ refers to the cost *between* node i and j. Consider that node 1 wants to compute the shortest paths to all its destinations. It needs to receive cost information from its neighbor nodes 2 and 4. Assuming for now that node 2 knows the best information (somehow), it is easy to see that node 2 needs to transmit the following protocol message to node 1:

$\overline{D}_{21} = 1$	$\overline{D}_{22} = 0$	$\overline{D}_{23} = 2$	$\overline{D}_{24} = 1$	$\overline{D}_{25} = 3$	$\overline{D}_{26} = 3$

Note that the first subscript with $\overline{D}$ is the node generating this distance vector information (2 in this case), while the second subscript indicates the destination for which this distance cost is provided. Since the first subscript is common, it can be ignored as long as the receiving node knows who it is coming from. The second subscript is the most relevant identifier that indicates the destination. Obviously, a routing protocol exchange message cannot understand or indicate a subscript as we can do with a notation in describing the protocol! Thus, the message format, shown above, looks more like the following where the destination j is identified first with a node number followed by the distance cost $\overline{D}$, which is repeated for every j:

$j = 1, \overline{D} = 1$	$j = 2, \overline{D} = 0$	$j = 3, \overline{D} = 2$	$j = 4, \overline{D} = 1$	$j = 5, \overline{D} = 3$	$j = 6, \overline{D} = 3$

Initialize:
-   Node is configured with a unique node ID, say i
-   Node i's distance vector to itself is set to zero, i.e., $\overline{D}_{ii} = 0$
-   Initialize module that determines the link cost to its directly connected neighbor (either manually or based on measurements), i.e., d_{ik} for all neighbors k, and set the routing table entry to indicate the next hop for k as k itself, i.e., $H_{ik} = k$

Transmit mode:
-   Transmit the most recently computed cost (distance) for all known destination nodes to all its neighbors k on a periodic basis

Receive mode:
-   Node i receives a distance vector from its neighbor k
    a. If the distance vector contains a new destination node j', then a new node entry in the routing table is created and set $\overline{D}_{ij'} = \infty$

    b. The distance vector $\overline{D}_{kj}^{i}$ for each destination node j received from neighbor k at node i is temporarily stored

    c. If the currently stored next hop for a destination j is the same as k itself, then update the distance cost for this destination, i.e.,
    $$\text{If } (H_{ij} = k) \text{ then} \quad // \text{ if next hop is } k$$
    $$\overline{D}_{ij} = d_{ik} + \overline{D}_{kj}^{i}$$
    Endif

-   Route Computation
    For each destination j: // shortest path computation
    For all neighbors k (or, the last received neighbor k)
    $$\text{Compute } temp = d_{ik} + \overline{D}_{kj}^{i}$$
    $$\text{If } (temp < \overline{D}_{ij}) \text{ then}$$
    $$\overline{D}_{ij} = temp \quad // \text{ update the new cost}$$
    $$H_{ij} = k \quad // \text{ update next hop in the routing table}$$
    Endif

Special Cases:
-   If for any neighbor k, link i-k goes down, then
    Set $d_{ik} = \infty$
    If $H_{ij} = k$, then $\overline{D}_{ij} = \infty$
    Broadcast a distance vector to each neighbor
    Endif
-   If for any neighbor k, link i-k is up again, then
    Update d_{ik} (fixed or dynamic)
    Broadcast a distance vector to each neighbor
    Endif

FIGURE 3.1   Distance vector protocol (node i's view): basic framework.

Based on the above message information, we can also see that the term distance vector then refers to the vector of distance or *direction*. This is communicated through the above message to a neighbor so that the neighbor can determine its best distance to a destination.

While it looks odd to include information about cost to node 1 itself in such a message, this is indeed the case in the basic or naïve distance vector protocol since node 2 does not differentiate who its neighbors are when disseminating such information. Upon receiving this information, node 1 will add its cost to this neighbor (1 in this case) for each destination separately to compare and determine the best cost to every destination. Assuming that until this instant, node 1 does not know the cost to any destination (except itself), it will then compare and update the cost (see Eq. (2.2.4b) in Chapter 2) really for itself resulting in no improvement; it also computes cost to all other destinations based on the information received and creates entries in its routing table, and tracks the outgoing link identifier. Thus, the routing table at node 1 is as given in Table 3.1.

Now compare Table 3.1 and the topology given in Figure 3.2; it is clear that for node 4 as destination, the entry is not optimal. This means that at this point, to route to node 4, node 1 will send to node 2 hoping that node 2 knows how to get to node 4. In fact, the routing table will stay in the above form as long as node 1 does not hear updated distance vector information from node 4 directly (to which it is also connected). In other words, a node actually never knows if it has the optimal cost as well as the best outgoing link identified for each destination (i.e., it is only the network administrator who can do optimal route analysis based on measurements). A node assumes that its neighbor is giving the correct distance vector information all the time. Furthermore, a node may or may not know if it has the view of the entire list of active nodes.

Now suppose that sometime after the above update, node 1 receives a distance vector from node 4 as given below:

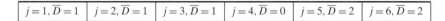

| $j=1, \overline{D}=1$ | $j=2, \overline{D}=1$ | $j=3, \overline{D}=1$ | $j=4, \overline{D}=0$ | $j=5, \overline{D}=2$ | $j=6, \overline{D}=2$ |

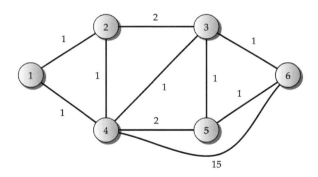

FIGURE 3.2   Six-node, ten-link network example (the entry shown next to a link is the cost of the link).

TABLE 3.1   Routing table information at node 1 (after receiving distance vector from node 2).

Destination Node	Cost	Outgoing Link
1	0	local
2	1	1-2
3	3	1-2
4	2	1-2
5	4	1-2
6	4	1-2

Upon receiving this information, node 1 performs an updated computation as shown in Table 3.2. You may note that this computation, marked as action-6 in Figure 3.1, is the same as Eq. (2.2.4b) in Algorithm 2.2; here, $d_{14}(t) = 1$, and node 1 receives $\overline{D}_{4j}(t), j = 1, 2, \ldots, 6$ as the distance vector message described above. The basic difference in the computation is that while the Bellman–Ford algorithm does not explicitly state that the outgoing link should be tracked, a distance vector routing protocol usually tracks this information for the purpose of creating/updating the routing table.

While it may not be apparent from the discussion so far, we learn the following lessons in regard to how timing (and timers) influences a routing protocol (see also Figure 3.3):

- *The order of information as received matters:* In the example discussed so far, we started by assuming that node 1 receives a distance vector from node 2 first *before* receiving a distance vector from node 4. Had node 1 received a distance vector from node 4 first, the entire routing table in the first round would have been different.

- *How often the distance vector information is disseminated matters:* Assume for the sake of argument that a distance vector is disseminated by node 2 every minute while node 4 does it every 10 min. Clearly, this would make a difference to node 1 in terms of how quickly it would be able to arrive at the best routing table.

- *The instant when a node broadcasts the distance vector (after an event) matters:* It is important to distinguish this item from the previous item. While the previous item discusses periodicity of the update interval, this one refers to whether a node should trigger an immediate update after a major event such as a failure of a link connected to it.

- *The instant when a routing computation is performed matters:* Suppose a node receives a distance vector from a neighbor every 2 min while it performs the route computation every 3 min. Certainly, these update cycles have an impact on obtaining the best route in a timely manner.

TABLE 3.2   Cost and routing able updating at node 1 (after receiving distance vector from node 4).

Destination Node	Current Cost	New Possible Cost	Updated Cost (Update?)	Update Outgoing Link (If Any)
1	0	1 + 1	0 (No)	local
2	1	1 + 1	1 (No)	1-2
3	3	1 + 1	2 (Yes)	1-4
4	2	1 + 0	1 (Yes)	1-4
5	4	1 + 2	3 (Yes)	1-4
6	4	1 + 2	3 (Yes)	1-4

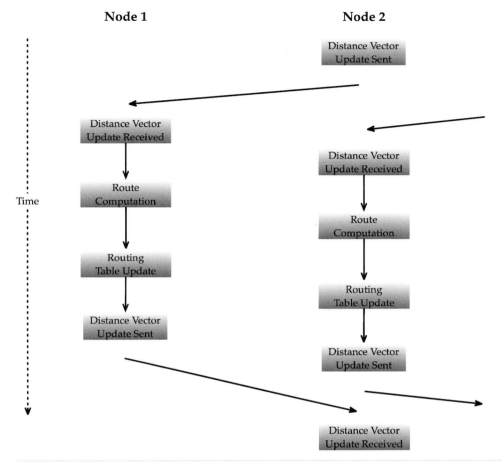

FIGURE 3.3   Time line of different activities at two different nodes.

- *The instant when the routing table is updated matters:* Suppose a node waits another 30 sec after performing a route computation before updating the routing table. This would impact the flow of user data.

An important corollary of the above discussion is that time (and timers) matter when it comes to a routing protocol, and that a routing environment encounters a transient period during which different nodes may have different views of the network; this is the root cause of many undesirable problems, which are discussed next. While common sense indicates that extended gaps between events as highlighted in Figure 3.3 are probably not necessary, it is important to understand what can happen if they do exist. This is explored in the next section.

3.3.2 Why Timers Matter

An important issue with any routing protocol is *convergence*; convergence refers to the same view arrived at by all nodes in a network from an inconsistent view, which may have resulted due to a failure or a cost change. Depending on how often the timers are activated, the convergence of a distance vector routing protocol can be delayed. We will discuss several aspects of this. There are some undesirable behaviors we will now highlight.

SLOW CONVERGENCE

Consider again Figure 3.2, but this time only with a partial view where we consider four nodes being activated at the same time: node 1, node 2, node 3, and node 6 (see Figure 3.4). We assume that nodes 4 and 5 are not activated yet; thus, nodes 1, 2, 3, and 6 form a linear network with a link connecting the adjacent nodes in the given order; also note that the link cost is 1 for every link in this linear network except for link 2-3 which has cost 2.

Suppose that at time $t = 0$ sec, when all four routers come alive simultaneously, they broadcast their presence to their neighbors, and then wait for 60 sec before doing another distance vector message. We assume that immediately after receiving a distance vector message, a node invokes the shortest path computation step and updates the routing table—for simplicity, we assume that this step takes 1 sec. We show below routing tables at various nodes as time passes (the destination node is labeled as D-node and the outgoing link is labeled as O-link):

Time: t= 0 sec: Nodes 1, 2, 3, and 6 are activated and the initial distance vector broadcast is sent.
Time: t= 1 sec: Routing tables at different nodes:

Node 1:

D-node	Cost	O-link
1	0	local
2	1	1-2

Node 2:

D-node	Cost	O-link
1	1	2-1
2	0	local
3	2	2-3

Node 3:

D-node	Cost	O-link
2	2	3-2
3	0	local
6	1	3-6

Node 6:

D-node	Cost	O-link
3	1	6-3
6	0	local

Time: t= 60 sec: Distance vector broadcast.
Time: t= 61 sec: Routing tables at different nodes:

Node 1:

D-node	Cost	O-link
1	0	local
2	1	1-2
3	3	1-2

Node 2:

D-node	Cost	O-link
1	1	2-1
2	0	local
3	2	2-3
6	3	2-3

Node 3:

D-node	Cost	O-link
1	3	3-2
2	2	3-2
3	0	local
6	1	3-6

Node 6:

D-node	Cost	O-link
2	3	6-3
3	1	6-3
6	0	local

Time: t= 120 sec: Distance vector broadcast.
Time: t= 121 sec: Routing tables at different nodes:

Node 1:

D-node	Cost	O-link
1	0	local
2	1	1-2
3	3	1-2
6	4	1-2

Node 2:

D-node	Cost	O-link
1	1	2-1
2	0	local
3	2	2-3
6	3	2-3

Node 3:

D-node	Cost	O-link
1	3	3-2
2	2	3-2
3	0	local
6	1	3-6

Node 6:

D-node	Cost	O-link
1	4	6-3
2	3	6-3
3	1	6-3
6	0	local

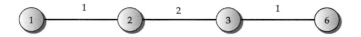

FIGURE 3.4   A four-node linear network.

From the above sequence of tables, we see that at different point of time, different nodes have different views of the network, including partial views. It is easy to see that all routing tables do not converge until $t = 121$ sec.

ROUTING LOOPS

A major problem with a distance vector protocol is that it can cause *routing loops*; this refers to a scenario in which a node's routing table points to the next hop, which in turn points to another hop and so on, and eventually the last node points back to the original node. In this situation, user traffic will go in a circular manner unless some mechanism is introduced to halt it. Looping can occur, for example, when a link fails. To illustrate, we first assume that the routing given in Figure 3.2 has fully converged. We first list below the sequence of events that occurs as viewed at node 2 and node 3:

time, t_0 —converged state; a routing computation performed for all destinations.
time, t_1 —nodes 2 and 3 update their respective routing tables (based on the result of routing computation at time t_0).
time, t_2 —link 3-6 fails.
time, t_3 —node 3 updates its routing table entry for destination node 6 by listing cost as ∞.
time, t_4 —node 2 sends distance vector to node 3.
time, t_5 —nodes 2 and 3 both perform a routing computation for all destinations.
time, t_6 —nodes 2 and 3 update their respective routing tables.

We now elaborate what is happening at some of these time instances. At time t_1, the routing table entries at node 2 and node 3 for destination node 6 are as follows:

At node 2:

Destination Node	Cost	Outgoing Link
6	3	2-3

At node 3:

Destination Node	Cost	Outgoing Link
6	1	3-6

Then, at time t_3, the routing table entry at node 3 for destination node 6 becomes:

Destination Node	Cost	Outgoing Link
6	∞	3-6

At time t_4, node 3 receives the following distance vector message from node 2:

$j=1, \overline{D}=1$	$j=2, \overline{D}=0$	$j=3, \overline{D}=3$	$j=4, \overline{D}=1$	$j=5, \overline{D}=3$	$j=6, \overline{D}=3$

In particular, note that node 3 receives node 2's cost to destination node 6 as 3. At time t_5, both node 2 and node 3 perform a routing computation. Since node 2 has not received any distance vector, there is no change at node 2; thus, there is no change in the routing table entry either. For clarity, we show the routing table at node 2 for destination nodes 3 and 6:

Destination Node	Cost	Outgoing Link
3	2	2-3
6	3	2-3

On the other hand, for node 3 the following update occurs:

Destination Node	Current Cost	New Possible Cost	Updated Cost (update?)	Update Outgoing Link (if any)
6	∞	$1 + 3$	4 (yes)	3-2

This results in node 3 pointing to node 2 for destination node 6. Thus, if user traffic now arrives at node 2 destined for node 6, node 2's routing table will send the packet on the outgoing link 2-3; on receiving this packet, node 3 looks up the routing table entry for destination node 6, and ships the user traffic on 3-2 to node 2! Thus, these data will keep looping back and forth between nodes 2 and node 3 forever unless the network provides a way to stop this through some other mechanism such as a time-to-live parameter. This looping effect is also called the bouncing effect.

COUNTING TO INFINITY

Counting to infinity is another problem faced with a distance vector protocol. Consider again Figure 3.2 and assume that routing tables have converged at all nodes. We will now consider the consequence of two-link failures in term of the following time line:

time, t_1 —node 4 sends distance vector to node 5.
time, t_2 —links 3-6 and 5-6 both fail.
time, t_3 —node 5 informs node 4 (and node 3) that its cost to node 6 is ∞.
              —node 3 informs node 4 (and node 5) that its cost to node 6 is ∞.
time, t_4 —node 4 performs a shortest path computation.
time, t_5 —node 4 receives a distance vector from node 1 indicating that its cost to node 6 is 3 (i.e., node 1 has not learned about the failures yet).
time, t_6 —node 4 performs a shortest path computation again.
time, t_7 —node 4 sends its distance vector to node 5 and node 3 (and others).
time, t_8 —node 3 updates their routing tables based on information from node 4.

Now, we walk through the details at each time event. At time t_1, node 4 has the following routing table:

Destination Node	Cost	Outgoing Link
1	1	4-1
2	1	4-2
3	1	4-3
4	0	local
5	2	4-5
6	2	4-3

and broadcasts the following distance vector:

$j=1, \overline{D}=1$	$j=2, \overline{D}=1$	$j=3, \overline{D}=1$	$j=4, \overline{D}=0$	$j=5, \overline{D}=2$	$j=6, \overline{D}=2$

The link failures occur at time t_2; thus, node 5 sets the cost to node 6 as ∞ since link 5-6 is the outgoing link for this destination. At time t_3, node 5 broadcasts the following distance vector to node 4 (and node 3):

$j=1, \overline{D}=3$	$j=2, \overline{D}=3$	$j=3, \overline{D}=1$	$j=4, \overline{D}=2$	$j=5, \overline{D}=0$	$j=6, \overline{D}=\infty$

while node 3 broadcasts the following distance vector to node 4 (and node 5):

$j=1, \overline{D}=2$	$j=2, \overline{D}=2$	$j=3, \overline{D}=0$	$j=4, \overline{D}=1$	$j=5, \overline{D}=1$	$j=6, \overline{D}=\infty$

At time t_4 and after receiving the above distance vectors from node 3 and node 5, node 4 performs a shortest path computation. In particular, consider the case of destination node 6; since node 3 reports the cost to node 6 to be ∞ and node 3 happens to be the next hop for node 4 in regard to destination node 6, node 4 updates its cost to ∞ for destination node 6 as required by the protocol.

At node 4, a new distance vector from node 1 is received at time t_5 that reports that the best cost to destination node 6 is 3. At time t_6, node 4 performs the shortest path computation again; for destination 6, node 4 notices that

Destination Node	Current Cost	New Possible Cost	Updated Cost (update?)	Update Outgoing Link (if any)
6	∞	$1+3$	4 (Yes)	4-1

and, thus, updates the routing table entry for destination node 6. This is, in fact, the start of the main trouble. Next at time t_7, node 4 send the following distance vector message to node 3 and node 5:

$j=1, \overline{D}=1$	$j=2, \overline{D}=1$	$j=3, \overline{D}=3$	$j=4, \overline{D}=0$	$j=5, \overline{D}=2$	$j=6, \overline{D}=4$

On receiving this message, node 3 notices node 6 is reachable via node 4 and the cost to be 4; thus, node 3 updates its cost to $5(=1+4)$. It then sends the distance vector to node 4 with this cost. Thus, it will continue on in the following cycle (assuming cost to be x, to start with):

- Node 4 sends a distance vector to node 3 with cost for destination node 6 as x

- Node 3 computes the distance to node 6 as $x + 1$

- Node 3 sends a distance vector to node 4 with cost for destination node 6 as $x + 1$

- Node 4 updates its cost to node 6 as $x + 2 = (1 + x + 1)$

- Node 4 sends a distance vector to node 3 with cost for destination node 6 as $x + 2$

until at node 4, the cost to node 6 is more than 15 (which is the direct cost to node 6 on link 4-6). Incidentally, all this time, node 4 knew about the direct link with cost 15 but ignored it, mistakenly assuming that there is a path through node 3 that is cheaper.

As you can see, due to periodic update, it will take several rounds before recognizing about the high cost path. This phenomenon is known as *counting to infinity*.

3.3.3 Solutions

In this section, we discuss a set of solutions to the issues discussed above.

SPLIT HORIZON AND SPLIT HORIZON WITH POISONED REVERSE

We have seen from previous illustrations that when a node transmits a distance vector update, it sends information about all nodes of which it is aware; it also includes information about itself as well as about the node to which this distance vector is sent. We also saw that this actually led to providing a false indication when a failure occurs, thus causing behavior such as the count to infinity. *Split horizon* is a technique that can help speed up convergence and can also solve the count to infinity problem in some instances. The basic idea of split horizon is quite simple: when transmitting a distance vector update on an outgoing link, send updates only for nodes for which this link is *not* on its routing table as the outgoing link.

To illustrate split horizon, consider node 4 in Figure 3.2. When fully converged, it will have the following values about cost to various destinations and the entry of the outgoing link in the routing table:

Destination Node	Cost	Outgoing Link
1	1	4-1
2	1	4-2
3	1	4-3
4	0	local
5	2	4-5
6	2	4-3

The basic distance vector protocol tells us to advertise the following:

$j = 1, \overline{D} = 1$	$j = 2, \overline{D} = 1$	$j = 3, \overline{D} = 1$	$j = 4, \overline{D} = 0$	$j = 5, \overline{D} = 2$	$j = 6, \overline{D} = 2$

With split horizon, node 4 will send the following distance vector to node 3

$j = 1, \overline{D} = 1$	$j = 2, \overline{D} = 1$	$j = 4, \overline{D} = 0$	$j = 5, \overline{D} = 2$

since nodes 3 and 6 have link 4-3 as the outgoing link for these nodes. However, node 4 will send the following distance vector to node 5 instead:

$j=1,\overline{D}=1$	$j=2,\overline{D}=1$	$j=3,\overline{D}=1$	$j=4,\overline{D}=0$	$j=6,\overline{D}=2$

From the above example, we note that a node may generate *different* distance vector updates depending on the outgoing link.

It can also be argued that *no news* is not necessarily always good news. There is, however, another variation of split horizon, called the *split horizon with poisoned reverse*, where news about all nodes is provided. In this case, the ones accessible on the outgoing link are marked as ∞. Thus, node 4 will send the following distance vector to node 3:

$j=1,\overline{D}=1$	$j=2,\overline{D}=1$	$j=3,\overline{D}=\infty$	$j=4,\overline{D}=0$	$j=5,\overline{D}=2$	$j=6,\overline{D}=\infty$

This essentially says that it is good to transmit bad news as well. *Split horizon with poisoned reverse* is more aggressive than split horizon and can stop looping between two adjacent nodes. Consider Figure 3.2 again. Suppose that node 4 incorrectly believes that it can reach node 1 via node 2—this can happen, for example, due to corrupted information. For the case of split horizon, node 3 would *not* indicate that it cannot reach node 1. On the other hand, for the case of split horizon with poisoned inverse, node 3 would indicate to node 4 that node 1 is unreachable—this then lets node 4 correct its misconception that there is a path to node 1 via node 2, and can avoid looping back and forth.

TIMER ADJUSTMENT (JITTER)

In discussing various examples, we have deliberately injected a time gap between different events in a distance vector protocol. Certainly, common sense would tell us that such time gaps are not necessary; moreover, such gaps can cause unnecessary problems during a transient period. Thus, the following are good steps to take in a distance vector protocol:

- Once the shortest path is computed by a node, it immediately updates the routing table—there is no reason to inject a time gap.

- When an outgoing link is found to be down, the routing table is immediately updated with an infinite cost for the destinations that would have used this link as the outgoing link, *and* a distance vector is generated to other outgoing links to communicate explicitly about nodes that are not reachable.

- As part of normal operations, it is a good idea to periodically send a distance vector to neighbors, even if the cost has not changed. This helps the neighbor to recognize/realize that its neighbor is not down.

- If a routing environment uses an *unreliable* delivery mechanism for dissemination of the distance vector information, then, besides the periodic update timer ("Keep-alive" timer), an additional timer called a *holddown timer* is also used. Typically, the holddown timer has a value several times the value of the periodic update timer. This way, even if a periodic update is sent and a neighboring node does not hear it, for example, due to packet corruption or packet loss, the node would not immediately assume that the node is unreachable;

it would instead wait till the holddown timer expires before updating the routing table (for more discussion, see Section 5.3 about Routing Information Protocol (RIP)—a protocol that uses unreliable delivery mechanism for routing information).

- If a routing environment uses a *reliable* delivery mechanism for dissemination of the distance vector information, the holddown timer does not appear to be necessary (in addition to the periodic update timer). However, the holddown timer can still play a critical role, for example, when a node's CPU is busy and cannot generate the periodic update messages within its timer window. Instead of assuming that it did not receive the periodic update because its neighbor is down, it can wait till the holddown timer expires. Thus, for such situations, the holddown timer helps to avoid unnecessary destabilization.

- The count to infinity situation was aggravated partly because one of the critical links had a much higher cost than the other links. Thus, in a routing environment running a distance vector protocol, it is often recommended that link costs be of comparable value and certainly should not be different in orders of magnitude.

- From the illustrations, it is clear that while the periodic update is a good idea, certain updates should be communicated as soon as possible; for example, when a node is activated, when a link goes down, or when a link comes up. In general, if the cost of a link changes significantly, it is a good idea to generate a distance vector update immediately, often referred to as the *triggered update*. This would then lead to faster convergence; furthermore, the count to infinity problem can be minimized (although it cannot be completely ruled out).

- If the cost on a link changes and then it changes back again very quickly, this would result in two triggered updates that can lead to other nodes updating their routing tables and then reverting back to the old tables. This type of oscillatory behavior is not desirable. Thus, to avoid such frequent oscillations, it is often recommended that there be a minimum time interval (*holddown timer*) between two consecutive updates. This certainly stops it from updating new information as quickly as possible and dampens convergence; but, at the same time, this also helps in stopping the spread of bad information too quickly.

- There is another effect possible with a distance vector protocol. Nodes are set up to send distance vector updates on a periodic basis, as mentioned earlier. Now, consider a node that is directly connected to 10 other nodes. Then, this node will be sending a distance vector on 10 outgoing links and at the same time it will be receiving from all of them. This situation can lead to congestion at the node including CPU overload. Thus, it is preferable that periodic updates from different nodes in a network be asynchronous. To avoid synchronous operations, instead of updating on expiration of an exact update time, a node computes the update time as a random amount around a mean value. For example, suppose that the average update time is 1 min; thus, an update time can be chosen randomly from a uniform distribution with 1 min as the average $\pm$ 10 sec. This way, the likelihood of advertising at the same time by different routers can be reduced.

The last item requires a bit of clarification. It is not hard to see that even if all the routers are set to start at a random start time and are independent events, all nodes eventually can synchronize in terms of update time, especially due to triggered update. This phenomenon is known as the *pendulum effect* [228]—the name stems from physics where you can start two independent pendulums physically close to each other with different swing cycles that eventually have synchronized swing cycles. Injection of random timer adjustment on the update time helps avoid the pendulum effect; however, the variation should be set to a *large* value to avoid synchronization. Note that this randomization is used for any subsequent update; certainly, if there is a failure, the triggered update still be generated.

From the illustrations and the above comments, it is also important to recognize that while a routing protocol needs a variety of timers, the actual value of the timers should not be rigidly defined as a part of the protocol description. In certain older routing protocols, values of timers were rigidly defined (see Section 5.3). Since then, we have learned enough to know that it is important to leave the rigid values out of the protocol; instead, include threshold values and range, and let the operational environment determine what are good values to use in practice. Needless to say, the timer values need to be chosen very carefully. A similar comment is also applicable to the link cost update timers; we will discuss this later in Section 3.6 and in subsequent chapters.

DISTANCE VECTOR MESSAGE CONTENT

The distance vector message discussed so far includes a distance cost for each destination node; there are some variations as to the value of the cost and/or which node should or should not be included in a distance vector broadcast, for example, to address for split horizon. It is, however, important to realize that perhaps additional attributes should be included with a distance vector message.

A simple first extension is to include next hop with a distance vector update for each destination as follows:

> Destination Node,
> Next Hop,
> Distance $\overline{D}$

It may be noted that if the next hop information is included, a node on receiving a distance vector from a neighbor has the ability to determine if the next hop for its neighbor goes through itself. Thus, what we were trying to accomplish by split horizon can be essentially accomplished by having this additional information. Note that this does not help solve the looping problem; it only helps to identify a possible loop and to stop doing a mistaken shortest path computation and avoid forwarding user traffic.

Another important aspect to consider is the type of event. If it is a link failure, or if the link cost has changed significantly, this is an important event compared to a normal periodic update. Thus, for example, for each distance, we may identify whether this is a normal periodic update, or a special update due to an event; this will then allow the receiving node a differential, based on which it may take different actions. Furthermore, a sequence number may be included that is incremented each time a new update is generated. Thus, a possible format for each destination may be as follows:

> Destination Node,
>
> Distance Vector Sequence Number,
> Normal Periodic Update or Special Event Indicator,
> Next Hop,
>
> Distance $\overline{D}$

Note that we have added a new field: Normal periodic update or special event indicator. It may be noted that if a distance cost is infinite, this may implicitly mean that an unusual event has occurred. On the other hand, a node may explicitly request information from its neighbor; for this case, it would be necessary to indicate this. Thus, the above format has some additional benefit. As an example, a distance vector message will have the following format:

> $j = 1$, Sequence Number $= 1$, Update=normal, Next Hop $= 7$, $\overline{D} = 3$
>
> $j = 2$, Sequence Number $= 1$, Update=normal, Next Hop $= 7$, $\overline{D} = 2$
>
> $\ldots$

Alternately, the sequence number may be done at a level of the message boundary (instead of for each distance direction), especially if the distance vector message is done for all nodes instead of a partial list of nodes.

3.3.4   Can We Avoid Loops?

So far, we have discussed a distance vector protocol and ways to circumvent a variety of issues faced in its operation. There is, however, one critical aspect that none of the mechanisms discussed so far can address adequately—the looping problem. In some ways, we could say that looping is the most serious problem since user packets will bounce back and forth between two (or more) nodes. Thus, an important question is: can looping be completely avoided in a distance vector protocol? To answer this question, we first need to understand the source of looping. On close scrutiny, you will note that the looping is induced by the Bellman–Ford computation in a distributed environment and it occurs when a link fails. In fact, looping can occur when the link cost increases also; incidentally, link failure can be thought of as a special case of increases in link cost (when link cost is set to ∞). You might wonder: what about a link cost decrease? This case is actually not a problem since Bellman–Ford can be applied as before.

To address loop-free routing for the case in which a link fails or its link cost increases, some extra work is required. Briefly, if the distance vector broadcast contains certain additional information beyond just the distance, and additionally, route computation is performed through inter-nodal coordination between a node and its neighbors, then looping can be avoided; this is considered in the next section.

3.3.5   Distance Vector Protocol Based on Diffusing Computation with Coordinated Update

The most well-known scheme that accomplishes loop-free routing in a distance vector protocol framework is the diffusing computation with coordinated update approach [244], [245], that incorporates the concept of diffusing computation [179] and the coordinated update approach [336]; this approach has been implemented in Enhanced Interior Gateway Routing Protocol (EIGRP) (see Section 5.6) and is known as the Diffusing Update Algorithm (DUAL). To identify that this approach is still built on the notion of a distance vector protocol framework, we will refer to this approach simply as the *loop-free distance vector protocol*, while we will refer to the original distance vector protocol discussed earlier as the *basic distance vector protocol*.

We start with three important assumptions for the loop-free approach: (1) within a finite time, a node must be able to determine the existence of a new neighbor or if the connectivity to a neighbor is lost, (2) distance vector updates are performed reliably, and (3) message events are processed one at a time, be it an update or a link failure message or a message about new neighbors being discovered.

It may be noted that the basic distance vector protocol does not explicitly address the first assumption. This assumption helps to build adjacency with neighbors in a finite time and is now a common feature in other, more recent routing protocols as well. The second assumption was also not addressed in the basic distance vector protocol—we have noted earlier that the lack of this functionality heavily affects convergence time and in the determination of whether a link has gone down. The third assumption is specifically about workings of diffusing computation with coordinated update.

The basic distance vector protocol has only a single message type, which is the distance vector update. However, the loop-free approach has multiple different message types: (1) *hello*—used in neighbor discovery, (2) *ack*—acknowledgment of a hello message, (3) *update*—for distance vector update, (4) *query*—for querying a neighbor for specific information, and (5) *reply*—for response to a query. Given the first assumption, you can clearly see the need for *hello/ack* messages. In the case of the loop-free distance vector (DV) protocol, update messages are not periodic, unlike the basic distance vector protocol, and can contain partial information. Query reply messages help in accomplishing loop-free routing and are used for coordination. This is also a good time to bring up information *push* and information *pull* in regard to protocol exchanges, as discussed earlier in Section 3.1. While the basic distance vector protocol operates in an information push mode, the loop-free distance vector protocol employs both the push mode for updates and the pull mode when *hello* or *query* are generated.

To execute the loop-free distance vector protocol, each node i maintains the following information:

- A list of neighboring/adjacent nodes, represented by $\mathcal{N}_i$.

- A network node table that includes *every* node j in the network along with the following information:

  - Lowest feasible distance to node j (represented as $\overline{D}_{ij}$).

  - A list of feasible next hops k—this means that a sublist ($\overline{\mathcal{N}}_{ij}$) of neighbors, $\mathcal{N}_i$, for which the distance from such a neighbor (as known to i) is smaller than its own distance to j; this means that $\overline{D}^i_{kj} < \overline{D}_{ij}$.

  - Feasible next hop's advertised distance (i.e., $\overline{D}^i_{kj}$ for $k \in \overline{\mathcal{N}}_{ij}$).

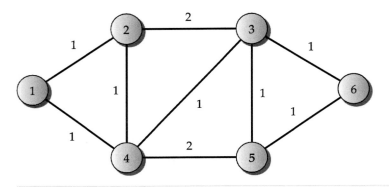

FIGURE 3.5   Six-node, nine-link network example.

- Distance through all feasible neighbors ("working distance") determined as $\tilde{D}_{ij} = d_{ik} + \overline{D}^{i}_{kj}$ for $k \in \overline{\mathcal{N}}_{ij}$.
- Active or passive states (more discussion later).

- A routing table that contains the next hop, H_{ij}, for which i has the lowest feasible distance for node j. If there are multiple next hops that have the same lowest feasible distance, then they are all entered in the routing table. There are two benefits to having suchentries in the routing table: (1) user traffic can be distributed among equal cost paths, and (2) if one link goes down, then there is a second one readily available to carry user traffic.

Given the information maintained by node i, it can generate a distance vector message that has the following components:

Destination Node, Message Type, Next Hop, Distance

It is important to note that in loop-free distance vector protocol, distance vector updates are not periodic and also need not contain the distance for all destinations; furthermore, through message type, it can be indicated whether a message is an update message or otherwise.

Example 3.1 *Illustration of a network node table and message type.*
Consider the six-node network that we have discussed earlier; this time with link 4-6 deleted; the new topology is shown in Figure 3.5.

Here, we will consider node 5's view; thus, $i = 5$. It has three neighboring nodes 3, 4, and 6. Thus, $\mathcal{N}_5 = \{3, 4, 6\}$. We will illustrate the node table and routing table for destination node 6. The shortest distance for node 5 to reach node 6 is the direct link 5-6 with cost 1. Thus, the lowest feasible distance is $\overline{D}_{56} = 1$. Consider the distance cost of its other neighbors to node 6; we see that $\overline{D}^{5}_{36} = 1$ and $\overline{D}^{5}_{46} = 2$, and that none is lower than its distance ($\overline{D}_{56} = 1$) to node 6, except node 6 itself. Thus, the only feasible next hop is node 6 itself; thus, $\overline{\mathcal{N}}_{56} = \{6\}$.

It also stores feasible next hop's advertised distance, i.e., $\overline{D}_{66}^{5} = 0$, and its current distance through them, i.e., $\tilde{D}_{56} = d_{56} + \overline{D}_{66}^{5} = 1$. Certainly, there is no difference here between this and the lowest feasible distance since there is only one feasible next hop for destination node 6. The next hop in the routing table for destination node 6 has one entry: $H_{56} = 6$.

The node table and routing table for all destinations are summarized below:

Network node table at node $i = 5$

Destination j	Distance $\overline{D}_{5j}$	Feasible Next Hop $k \in \overline{\mathcal{N}}_{5j}$	Advertised Distance, $\overline{D}_{kj}^{5}$	Working Distance $\tilde{D}_{5j} = d_{5k} + \overline{D}_{kj}^{5}$	State: Active (1)/Passive (0)
1	3	4	1	3	0
2	3	3, 4	2, 1	3, 3	0
3	1	3	0	1	0
4	2	4, 3	0, 1	2, 2	0
5	0	5	0	0	0
6	1	6	0	1	0

Routing table at Node $i = 5$:

Destination, j	Next Hop, H_{ij}
1	3
2	3, 4
3	3
4	4, 3
5	0
6	6

A distance vector update message generated at node 5 in regard to destination node 6 will be in the following form $\boxed{j = 6, \text{Update}, 6, 1}$. For brevity, we will write this as $\boxed{6, \text{U}, 6, 1}$, where U stands for update (similarly, Q stands for query, R for reply, and so on). ▲

Now consider the operation of the loop-free distance vector protocol. Recall that in the case of the basic distance vector protocol, a node does route computation using distributed Bellman–Ford and then updates the routing table. In the case of the loop-free distance vector protocol, route computation is a bit different depending on the situation; furthermore, there are three additional aspects that need to be addressed: building of the neighbor table, node discovery and creating entry in the network node table, and coordination activity when link cost changes or link fails. Building of the neighbor table typically occurs when a node is first activated; this is where *hello* messages are sent to neighbor—once an *ack* message is received in response, the neighbor relationship is set up. It is important to note that *hello* is also periodically transmitted to determine/maintain availability and connectivity to a neighboring node. Typically, node discovery occurs immediately after the initial *hello* message when the neighboring node sends a distance vector update to the newly activated node. Since the newly activated node may be connected to multiple nodes, it will do such an exchange with each of the neighbors; furthermore, it will do its own route computation and send an update message to its neighbors. For message exchange once a node is activated, there is a series of exchanges involved (see Figure 3.6).

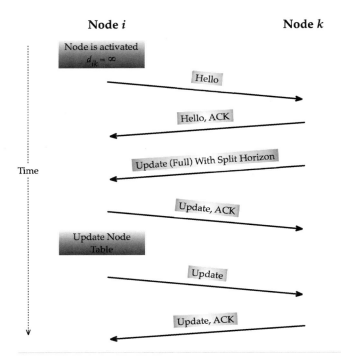

FIGURE 3.6 Protocol message exchanges when a node is activated.

A node performs various tasks depending on an event; primarily, a node provides updates under normal situations and coordinates activities when an event such as a link going down occurs. To do that, a node is required to maintain two states: passive (0) and active (1). When it is passive, it can receive or send normal distance vector updates. A node moves to an active state for a specific destination when, for example, a link failure occurs. When it is in an active state, node table entries are frozen. Note that when it is in an active state, a node generates the request message instead of the update message and keeps track of which requests are outstanding; it moves to the passive state when responses to all outstanding requests are received. It may be noted that a request may be forwarded if the receiving node does not have an answer (in this case, a feasible next hop). In Figure 3.7, we show two different instances of how request response messages are handled. Finally, in Figure 3.8, we present the loop-free distance vector protocol in a compact manner.

It is possible that one or more events occur before the completion of exchanges during an active state. For example, a link goes down; before the coordination phase is completely done, the link comes back up again. While it is possible to handle multiple events and computations through diffusing computation (for example, see [245]), it can be essentially restricted to just one event handling at a time by enforcing a holddown timer; we have discussed the notion of a holddown timer earlier in this chapter. That is, when a link cost changes or link failure event occurs, a holddown timer is started; another link status change event for the same link is not permitted to be communicated through the network until the first holddown timer expires. If the duration of the holddown timer is chosen properly, the coordination phase for a single event can be completed. In general, the holddown timer makes the state maintenance

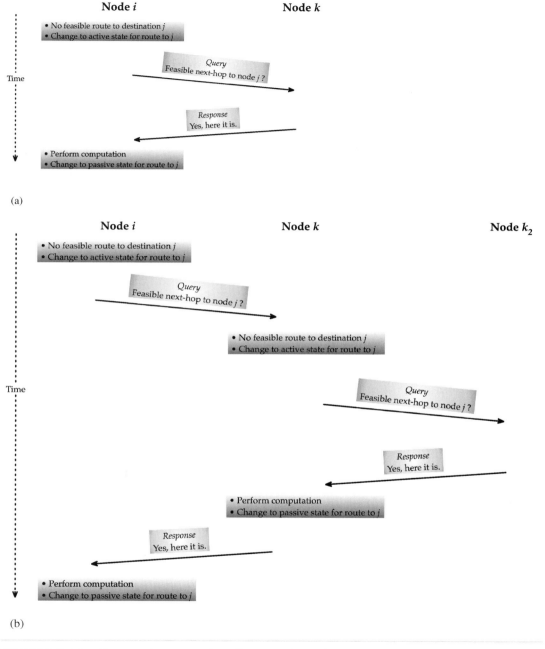

FIGURE 3.7   Request/response handling in a loop-free distance vector protocol.

mechanism much simpler compared to when multiple events are needed to be taken care of simultaneously.

In the following, we illustrate how a link failure is handled by the loop-free distance vector protocol.

Initialization:
- Node i initializes itself in a passive state with an infinite distance for all its known neighbors and zero distance to itself
- Initiate hello protocol with neighbor k and update value d_{ik}
- Receive update message from all neighbors k, and update node and routing table

Node i in passive mode:
- Node i detects change in a link cost/status to neighbor k that changes $\overline{D}ij$ for some j
    If (this is due to link failure) then $d_{ik} = \infty$, $\overline{D}_{kj}^i = \infty$
- Check for a feasible next hop already in the node table (i.e., a neighbor k in the node table that satisfied $\overline{D}_{kj} < \overline{D}_{ij}$ prior to the link status change)
- If (Node i finds one or more feasible next hops, $k \in \overline{\mathcal{N}}_{ij}$) then
    // initiate local computation (much like D-BF)
    If (there is a feasible k such that $d_{ik} + \overline{D}_{kj} < d_{i,H_{ij}} + \overline{D}_{H_{i,j},j}^i$, that is, cost through
        this k is better than the current next hop H_{ij}) then
            Set k as the new next hop, i.e., $H_{ij} = k$
    If ($\overline{D}_{H_{i,j},j}^i < \overline{D}_{ij}$) then $\overline{D}_{ij} = \overline{D}_{H_{i,j},j}^i$
    Send update, $\widetilde{D}_{ij}$, to all its neighbors
    Endif
- If (Node i cannot find a feasible next hop) then
    Set $\overline{D}_{ij} = \widetilde{D}_{ij} = d_{ih} + \overline{D}_{hj}^i$ where h is the current next hop for node j
    Set feasible distance to D_{hj}^i
    // initiate diffusing computation
    Change state to node j to active state, freeze any change for destination j, and set action flag to 1
    Send query to all neighbors $k \in \mathcal{N}_i$
    Endif

Node i in receiving mode:
    Node i in passive mode:
    Receive update message:
        Update node table, determine new next hop, routing table, and update message
        Receive Query:
        If (feasible next hop exists) respond with $\widetilde{D}_{ij}$
        If ($\overline{D}_{ij}$ changes) send update to neighbors
        If (no feasible next hop exists)
            change to active mode, $R = 1$ and send query to other neighbors
    Node i in active mode:
        If (response received from neighbors k to all queries) then
            Change to passive mode ($R = 0$)
            Reset $\widetilde{D}_{ij}$
            Update node table, determine new next hop, update routing table
        Endif

FIGURE 3.8   Loop-free distance vector protocol based on diffusing computation with coordinated update (node i's view).

Example 3.2 *Link failure handling in the loop-free distance vector protocol.*

We continue with Example 3.1 to illustrate the link failure case. Consider that the network has converged as illustrated earlier in Example 3.1 and all nodes are in passive states. We will consider only part of the information from the node table just for destination node 6: specifically $(\widetilde{D}, \overline{D},$ passive/active status), i.e., distance, working distance, and state. This in-

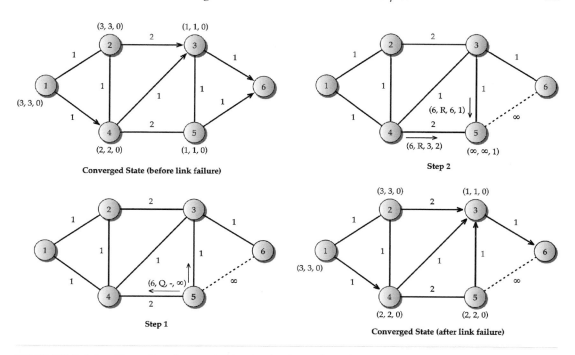

FIGURE 3.9   Coordination activities in the loop-free distance vector protocol.

formation is shown in Figure 3.9 at each node for destination node 6. For example, at node 5, for destination node 6, we have $(\widetilde{D}, \overline{D}, \text{passive/active status}) = (1, 1, 0)$.

Now suppose that link 5-6 goes down; it will be marked as the active state and node 5 will change the entry in the node table for node 6 from $(1, 1, 0)$ to $(\infty, \infty, 1)$. It will then generate the query message $\boxed{6, Q, -, \infty}$ to be sent to the other two outgoing links, 5-3 and 5-4. To keep track of outstanding requests, the following will be set for requested tracking parameters: $R^5_{36} = 1$ (i.e., node 5 querying node 3 in regard to node 6) and $R^5_{46} = 1$ (i.e., node 5 querying node 4 in regard to node 6). On receiving the query message from node 5 at node 3, it will check to see if it has a feasible next hop to node 6. In this case, it does. Thus, it will generate the following response message $\boxed{6, R, 6, 1}$ to node 5. Similarly, node 4 will generate a response message to node 5 as follows: $\boxed{6, R, 3, 2}$. Note that node 5 has to receive response messages to both its outstanding requests to move from the active state to the passive state. Once it does, it performs a computation for the node table and generates a new shortest path to destination node 6.   ▲

As you can see, the loop-free distance vector protocol resolves the looping problem faced due to a link-failure in the basic distance vector protocol. A general question (with any protocol): are there any limitations? Below, we discuss two possible problematic scenarios for the loop-free distance vector approach:

- In some situations, the loop-free distance vector protocol requires quite a bit of coordination between neighboring nodes, thus creating a *chatty* mode. A general problem with such a chatty protocol is that it can consume a significant percentage of the bandwidth if the link between two neighboring nodes is a low-speed link, thus affecting performance for user traffic.

- Recall that a holddown timer is started in this protocol once a node moves to the active state; before the holddown timer expires, the node is supposed to hear responses back about a query. However, under an unrelated multi-event situation, it is possible that the time expires before the situation is resolved; this is known as the *stuck in active* (SIA) condition. For example, a node, say A, loses a link to a neighboring node B that happens to isolate the network into two networks. Node A would not realize the isolation of the overall network; it will query its other neighbors, say C and D, about determining a path to node B. In turn, nodes C and D will inquire of its neighbors about any path to node B and change to the active state. Now, assume that there is a congestion problem between node D and its neighboring node E that delays getting a response back. In the meantime, the timer at node A expires, thus resulting in the SIA condition at node A. In general, this is a difficult problem to resolve.

To summarize, it is possible to use a distance vector protocol framework and extend it for loop-free routing by using diffusing computation with coordinated update. Like any protocols, it has limitations under certain conditions.

3.4 Link State Routing Protocol

The idea of a link state routing protocol has its roots in Dijkstra's shortest path first algorithm (see Section 2.3). An important aspect to understand about Dijkstra's algorithm is that it requires a node to have topological information to compute the shortest paths. By topological information, we mean links in the network and the nodes to which they are connected, along with the cost of each link; that is, just a node table as in the loop-free distance vector protocol is not sufficient. A node in the network needs to store the cost of a link and record whether this link is up or down—generally referred to as the *state* of the link. This then gives rise to the name *link state* and the information about links a node needs to store as the *link state database*. Thus, a link state protocol is a way to communicate information among the nodes so that all nodes have the consistent link state database. While this basic idea is very simple, it is not so easy to make this work in a distributed environment.

There are several important issues to consider. First, how does each node find out the link state information of all links? Second, do all nodes have the exact same link state information? What may cause one node to have different link state information than others? Finally, how can we disseminate link state information? How can inconsistency be avoided or minimized? And last, but not least, how does link state information affect the shortest path computation?

IN-BAND VERSUS OUT-OF-BAND

First, recall our brief discussion about in-band and out-of-band in Section 3.1. To address issues discussed there, we need to first know whether a particular network uses *in-band* or *out-of-band* mechanisms to communicate link state information, and whether this is accomplished through information push or information pull or a hybrid mechanism.

If a communication network uses an out-of-band mechanism for communicating the link state information, then there are two possibilities: (1) any pair of nodes talk to each other through this mechanism irrespective of their location, (2) all nodes communicate to a central system through a dedicated channel, which then communicate back to all nodes. Both these options have been applied for dynamic routing in the telephone network where either a *signaling network* or dedicated circuits are used to accomplish the communication of link state information. In the second case, it also typically means that the central system is used for doing route computation and disseminating the computed result back to the nodes. For example, dynamically controlled routing (see Section 10.4) uses dedicated circuits and a central system for routing computation, while real-time network routing (see Section 10.6) uses a signaling network with distributed route computation. In DCR, link state information is pushed to the central system, yet the central system may pull information from the nodes if needed. In the case of RTNR, no information push is done; a node pulls link state information from another node when needed. In Chapter 10, we will cover these aspects in detail.

In-band communication about routing information can be divided into two categories: in-band on a hop-by-hop basis and in-band on a connection/session basis. Why do we make these distinctions? In a data network, a simple distinction on a packet type can be used for communicating routing information on a hop-by-hop basis. On the other hand, a data network also provides the functionality of a virtual reliable connection (such as a TCP-based connection); thus, routing information can be exchanged using such a virtual connection between any two nodes. The rest of the discussion in this section mostly centers around exchange of routing information using in-band communication on a hop-by-hop basis. At the end of this section, we will also discuss in-band communication on a session basis.

3.4.1 Link State Protocol: In-Band Hop-by-Hop Disseminations

First and foremost, in-band hop-by-hop basis is possible for link state information exchange since packets can be marked either as user data packets or *routing* packets to communicate link state information. How this is specifically done will be covered in detail for protocols such as OSPF in later chapters. For now, our discussion will be limited to the basic idea of link state protocol when in-band communication on a hop-by-hop basis is used for exchanging link state routing information.

We start with two important points:

- The link state information about a particular link in one part of a network to another part can traverse on a hop-by-hop communication basis to eventually spread it throughout the network; this is often referred to as *flooding*.

- On receiving link state information that is forwarded through the hop-by-hop basis, a node can do its own route computation in a distributed manner.

The second component is really related to performing route computation and can be decoupled from the protocol itself. The first part is an essential part of the link state routing protocol.

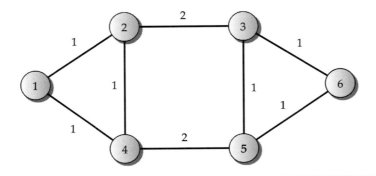

FIGURE 3.10 Six-node, eight-link network example.

LINK STATE ADVERTISEMENT AND FLOODING

A link state message, often referred to as a *link state advertisement* (LSA), is generated by a node for each of its outgoing links, and each LSA needs to contain at least

| Source node, Link ID, Link Cost | (3.4.1)

which is then forwarded throughout the network. Certainly, we need to ask the question: is the flooding reliable or unreliable? That is, is the information flooded received intact by each node in the network, or is it corrupted? From the discussion about a distance vector protocol, we know that routing information exchange using unreliable delivery mechanisms causes additional problems. Thus, since the early days of a distance vector protocol, we have learned one important thing: reliable delivery of routing information is important. We already saw its use in the loop-free distance vector protocol. You will find out that in fact almost all routing protocols since the early days of the basic distance vector protocol use reliable delivery of routing information. Henceforth, we will assume reliable flooding with the link state protocol.

We first examine the LSA format as given in protocol message (3.4.1). Consider the link that connects from node 1 to node 2 in Figure 3.10: this LSA will be generated by node 1; however, the reverse direction, LSA for the same link from node 2 to node 1, will be generated by node 2. In other words, links in a link state protocol are *directional* (while directionality is not an issue for a distance vector protocol). To avoid any confusion between a bidirectional and a unidirectional link, we will use 1-2 to denote the bidirectional link that connects node 1 and node 2 while 1→2 to denote the directional link from node 1 to node 2. In addition to the directional aspect, there is a critical issue we need to understand in regard to hop-by-hop traversal. Consider Figure 3.10, and the link cost $d_{12} = 1$ from node 1 to node 2, which needs to be disseminated. Thus, the link state information about the link that originates at node 1 and ends at node 2, that is for 1→2, would be generated at node 1 as the message $i = 1$, Link=1→2, $d_{12} = 1$, which can be written as 1, 1→2, 1 in short; this message is forwarded to both nodes 2 and 4. These nodes can, in turn, forward ("flood") on their outgoing links; for example, from node 2 to both node 4 and node 3. We can immediately see that node 4 would receive the *same* information in two different ways!

If the cost value of both the LSAs for the same link is the same, then it is not difficult to resolve. However, if the value is different, then a receiving node needs to worry about which LSA for a particular link was generated more recently. Consider the following illustration in terms of times event:

time t_0:   LSA $\boxed{1, 1 \rightarrow 2, 1}$ is generated at node 1 and is sent to node 2 and node 4.

time t_1:   LSA $\boxed{1, 1 \rightarrow 2, 1}$ is forwarded by node 2 to node 4.

time t_2: $1 \rightarrow 2$ fails; node 1 generates the new LSA $\boxed{1, 1 \rightarrow 2, \infty}$ to node 4.

time t_3:   LSA $\boxed{1, 1 \rightarrow 2, 1}$ is received at node 4 from node 2.

From the above illustration, node 4 would receive LSA for the same link with two different cost values: ∞ first and then 1 next; however, the failure occurred afterward! We can see that the LSA needs to carry at least another piece of information that helps to identify LSA at a receiving node based on *when* it was generated. Thus, some way to time-stamp an LSA would then avoid any ambiguity. Thus, instead of using (3.4.1), LSA should contain a time stamp resulting in the format:

$$\boxed{\text{Source Node, Link ID, Link Cost, Time stamp}} \hspace{4cm} (3.4.2)$$

The question is how to indicate a time stamp that works in a distributed networked environment. There are two possibilities: either all nodes are clock-synchronized through some geosynchronous timing system, or a clock-independent mechanism is used. While a geosynchronous timing system is a good idea, until recently this was not feasible; furthermore, a separate mechanism *independent* of the protocol would be required. Most link state routing protocols use a clock-independent mechanism called *sequence number* to indicate the notion of a time stamp that can be defined within the context of the protocol. That is, a node, when it generates an LSA for an outgoing link, stamps it with a sequence number and the LSA then has the following modified format:

$$\boxed{\text{Source Node, Link ID, Link Cost, Sequence Number}} \hspace{3cm} (3.4.3)$$

When the *same* node needs to generate a new LSA for the *same* outgoing link, it increments the sequence number counter, inserts this new value in the LSA message, and sends out the LSA packet. Going back to the previous example, if the sequence number for link $1 \rightarrow 2$ is 1 before failure, then the first LSA announcement would be $\boxed{1, 1 \rightarrow 2, 1, 1}$. After failure at time t_2, the sequence number counter would be incremented to 2, and the new LSA would be $\boxed{1, 1 \rightarrow 2, \infty, 2}$. Thus, when at time t_3, node 4 receives LSAs for the same link from two different directions, it can check the sequence number and discard the one with the older sequence number, in this case, the one received from node 2 with sequence number 1.

It is important that each node maintains a *different* sequence number counter for each outgoing link, and that other nodes maintain their own sequence number counters for their outgoing links; in other words, there is no dependency among nodes, which is an advantage of using the concept of a source-initiated, link-based sequence number. There is, however, a key issue to consider: the size of the sequence number space. In any data network environment, usually a fixed length field is used for the sequence number space. Suppose that the sequence number space is only 3 bits long; this would mean that it can take values 1 to 8, and after it reaches 7, it would need to wrap around and start at 1 again. Here is the first problem we encounter due to wrapping of the sequence number. When a node receives two LSAs for the same link ID from two different neighbors, one with sequence number 7 and the other with sequence number 2, the receiving node has no way of knowing if the sequence number 2 is *after* the number is wrapped or *before*; in other words, the receiving node has no way of knowing which is more recent. This tells us that the size of the sequence number space should not be small. Typically, the sequence number space is a 32-bit field; in most cases, this would solve the problem. However, there is still some ambiguity, for example, when a node goes down and then comes back up with the sequence number set to one, or when a network is isolated into two separate networks. Essentially, what this means is that some additional safeguard is required to ensure that a receiving node is not confused. A possible way to provide this safeguard is to use an additional field in LSA that tells the *age* of the LSA. Taking this information into account, the LSA takes the form:

$$\boxed{\text{Source Node, Link ID, Link Cost, Sequence Number, Age}} \qquad (3.4.4)$$

Now we describe how to handle the age field at different nodes. The originating node sets the starting age field at a maximum value; the receiving node decrements this counter periodically while storing the link state information in its memory. When the age field reaches zero for a particular link, the link state information for this link is considered to be too old or stale. The following is a classical example of what can happen if sequence number and age are not addressed properly.

Example 3.3 *ARPANET operational problem due to sequence number and age.*
     From an operational environment, we can learn a lot about what does or does not work in practice. A case in point is the sequence and age field, as used and as observed through its early use in ARPANET. This example is very nicely described in [559], and is reproduced here.
     ARPANET at that time used a 3-bit-long age field with 8 sec as the time unit. This means that the starting maximum age was 56 sec ($= 7 \times 8$), which was decremented every 8 sec. To avoid the age becoming stale by the time an LSA reaches a downstream node, each node needed to generate a new LSA for an outgoing link within 60 sec. When a node starts up (either initial activation, or if rebooted), it needed to wait for 90 sec before generating the first LSA. The idea was that this would allow any old LSA in the memory of the node to decrement the age counter to 0; at the same time, it can receive new LSAs from neighboring nodes.
     ARPANET was found to be nonfunctional one night (these things always happen at night!) with the queue at a router filled with multiple LSAs from a specific router, say Z, where each of these LSAs had different sequence numbers a_1, a_2, a_3 with $a_1 < a_2 < a_3$ and then wrap around to a_1. Now, consider a router that has a stored LSA from Z with sequence number a_1, and it receives an LSA with sequence number a_2; it would overwrite the one in memory since $a_2 > a_1$ and, in addition, it will flood this "new" LSA to its neighbors who in turn will update accordingly. This pattern of updating the sequence number was repeated.

It was found that LSAs did not age out. The problem was in the inherent assumption that the age counter will be decremented at a node every 8 sec. If a received LSA leaves a particular node within this 8 sec, its age field would not get decremented. However, it was originally envisioned that if a node receives an LSA and immediately sends it out, the age counter *would* get decremented. This simple logic problem caused the network to become nonfunctioning.   ▲

In recent protocols, the sequence number space is large enough to avoid any such problems; for example, a 32-bit signed sequence number space is used. Furthermore, in many protocol implementations, the sequence number space is considered as a *lollypop sequence number space*; in this scheme, from the entire range of possible numbers, two are not used. For example, consider a 32-bit signed sequence number space. The sequence number is varied from the negative number $-2^{31} + 1$ to the positive number $2^{31} - 2$ while the ends -2^{31} and $2^{31} - 1$ are not used. The sequence number begins in the negative space and continues to increment; once it reaches the positive space, it continues to the maximum value, but cycles back to 0 instead of going back to negative; that is, it is linear in the negative space and circular in the positive space giving the shape of a lollypop and thus the name. The lollypop sequence number is helpful when a router, say R1, restarts after a failure. R1 announces the sequence number $-2^{31} + 1$ to its neighbor R2. The neighbor R2 immediately knows that R1 must have restarted and sends a message to R1 announcing where R1 left off as the last sequence number before the failure. On hearing this sequence number, R1 now increments the counter and starts from the next sequence number in the next LSA. Note that not all protocols use lollypop sequence numbering—the complete linear sequence number space that starts at negative and continues to positive in a linear fashion is also used; if the maximum value is reached, other mechanisms such as flushing of LSA are used when the maximum positive value is reached.

LSA AND LSU

Along with LSA, there is another terminology commonly used: link state update (LSU). It is important to understand and distinguish LSA from LSU. An LSA is an announcement generated by a node for its outgoing links; a node receiving LSAs from multiple nodes may combine them in an LSU message to forward to other nodes.

Example 3.4 *LSA and LSU.*
    Consider Figure 3.10. Here, node 1 generates the link state for $1{\rightarrow}4$ as $\boxed{1, 1{\rightarrow}4, 1, 1, 60}$ using the originating age counter as 60 and sends to node 4. Similarly, node 2 generates the link state for $2{\rightarrow}4$: $\boxed{2, 2{\rightarrow}4, 1, 1, 60}$ and sends to node 4. Node 4 can combine these two LSAs along with the link state for link $4{\rightarrow}5$, and assuming it takes one time unit to process, it decrements the age counter by one for the received LSAs and sends out the link state update to node 5 as $\boxed{1, 1{\rightarrow}4, 1, 1, 59 \mid 2, 2{\rightarrow}4, 1, 1, 59 \mid 4, 4{\rightarrow}5, 2, 1, 60}$.

                                                                                      ▲

SPECIAL CASES

How does a link state protocol handle special cases? There are two scenarios we consider here: a node going down and coming back up again, and a link going down and coming back up again. The node failure has an immediate impact on the sequence number and the age field since nodes are, after all, specialized computing devices. When a node is restarted, the sequence number space may be reinitialized to 1 again; this again leaves a receiving node wondering whether it has received a new or old LSA generated from the node that just recovered from a failure. While in some cases such an exception can be handled through additional attributes in an LSA, it is usually done through additional mini-protocol mechanisms along with the proper logic control within the framework of the link state routing protocol. For example, there are several aspects to address: (1) the clock rate for aging needs to be about the same at all nodes, (2) receiving, storing, and forwarding rules for an LSA need to take into account the age information, (3) the maximum-age field should be large enough (for example, an hour), and (4) if the sequence number is the same for a specific link that is received from two incoming links at a receiving node, then the age field should be checked to determine any anomaly. Thus, typically a link state routing protocol consists of three subprotocol mechanisms:

- Hello protocol

- Resynchronization protocol

- Link state advertisement (normal).

The *hello* protocol is used for initialization when a node is first activated; this is somewhat similar to the hello protocol used in the loop-free distance vector protocol. In this case, the hello protocol is useful in letting neighbors know its presence as well as the links or neighbors to which it is connected and to learn about the rest of the network from its neighbor so that it can perform route computation and build a routing table. The hello protocol is also periodically invoked to see if the link is operational to a neighbor. Thus, the hello protocol has both information push and information pull. The *resynchronization* protocol is used after recovery from a link or a node failure. Since the link state may have been updated several cycles during the failure, resynchronization is merely a robust mechanism to bring the network to the most up-to-date state at the nodes involved so that LSA can be triggered. The resynchronization step includes a link state database exchange between neighboring nodes, and thus involves both information pull and push. The normal LSA by an originating node is information push. The entire logic control for a link state protocol is shown in Figure 3.11.

We will illustrate the need for the resynchronization step in the following example. Note that this step is also called "bringing up adjacencies."

Example 3.5 *Need for resynchronization.*

We will use the same network as the one shown in Figure 3.10. We will start by assuming that the network has converged with all links having sequence number 1. We will also consider that the failure of link 4-5 occurred resulting in a new sequence number for each

Initialization:
- Use hello protocol to establish neighbor relation and learn about links from neighbors

Link State Advertisement (normal):
- Generate LSA periodically (before the expiration of the age counter), incrementing the sequence number from the last time, and set the age field to the maximum value for this LSA; start the timer on the age field

Receive (new):
- Receive LSA about a new link $l \rightarrow m$ from neighboring node k
- Update the link state database and send LSA for this link to all neighbors (except k)
- Start the timer for decrementing the age counter

Receive (normal):
- Receive LSA about known link $l \rightarrow m$ from neighboring node k
- Compare the sequence number to check if I have the most recent LSA for this link
- If yes, send LSA for this link back to neighboring node k
- If not, decrement the age counter, store the recent LSA for link $l \rightarrow m$ in the link state database, and forward it on all other outgoing links to the rest of the neighbors (except for k)
- If it is the same sequence number and the age difference is small, do nothing; if the age difference is large, store the most recent record
- Start the timer for decrementing the age counter

Compute:
- Perform route computation using the local copy of the link state database
- Update the routing table

Special Cases:
- Link failure: set the link cost to ∞ and send LSA immediately to neighbors
- Link recovery: perform link state database resynchronization
- Node recovery: perform link state database resynchronization; alternately, flush records and perform hello protocol
- (Action mode when the age counter reaches 0 for a link ID):
    (1) Do not consider this link in route computation
    (2) Inform neighbors through advertisement stating that the age field is zero
    (3) Delete this link entry from the database
- (Receive mode with age 0): accept this LSA. This would set the age to zero, and thus, perform 'action mode when age counter reaches 0.' If the record is already deleted, ignore this advertisement.

FIGURE 3.11   Link state protocol from the point of view of node i (with in-band hop-by-hop communication for flooding).

direction with link cost set to ∞; this information has also converged. Our interest is in the second failure, i.e., the failure of the second link, 2-3. We show the two network states, before and after failure of link 2-3, in Figure 3.12.

Note that when 4-5 fails, both its end nodes (node 4 and node 5) will increment the sequence number count to 2 and generate the directional LSAs with cost set to ∞ to advertise to their neighbors. This information will be flooded throughout the network, and all nodes will eventually converge to having the link state database as shown in Table 3.3(a). When the second link 2-3 fails, we can see that the network will be isolated into two separate smaller networks. Node 2, on recognizing this failure, will increment the sequence number counter to 2 and generate an LSA for the directional link $2 \rightarrow 3$ with cost set to ∞; this will be disseminated which can reach only nodes 1 and 4. Similarly, node 3, on recognizing the same failure, will increment the sequence number counter to 2 and generate LSA for the directional link $3 \rightarrow 2$ with cost set to ∞; this will be disseminated, which can reach only nodes 5 and 6. Thus, in the network on the left side consisting of the nodes 1, 2, and 4, the link state database

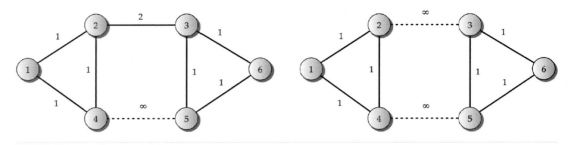

FIGURE 3.12 Six-node network: before and after failure of link 2-3 (assuming link 4-5 has already failed).

TABLE 3.3 Link state database as viewed before and after failure of link 2-3.

(a) Before Failure of Link 2-3 (as Seen by Every Node)			(b) After Failure of Link 2-3 (as Seen by Nodes 1, 2, 4)			(c) After Failure of Link 2-3 (as Seen by 3, 5, 6)		
Link-ID	Cost	Seq #	Link-ID	Cost	Seq #	Link-ID	Cost	Seq #
1→2	1	1	1→2	1	1	1→2	1	1
2→1	1	1	2→1	1	1	2→1	1	1
1→4	1	1	1→4	1	1	1→4	1	1
4→1	1	1	4→1	1	1	4→1	1	1
2→4	1	1	2→4	1	1	2→4	1	1
4→2	1	1	4→2	1	1	4→2	1	1
2→3	2	1	2→3	∞	2	2→3	2	1
3→2	2	1	3→2	2	1	3→2	∞	2
4→5	∞	2	4→5	∞	2	4→5	∞	2
5→4	∞	2	5→4	∞	2	5→4	∞	2
3→6	1	1	3→6	1	1	3→6	1	1
6→3	1	1	6→3	1	1	6→3	1	1
3→5	1	1	3→5	1	1	3→5	1	1
5→3	1	1	5→3	1	1	5→3	1	1
5→6	1	1	5→6	1	1	5→6	1	1
6→5	1	1	6→5	1	1	6→5	1	1

will become as shown in Table 3.3(b), while for the network on the right side consisting of nodes 3, 5, and 6, the link state database will become as shown in Table 3.3(c). Notice the subtle difference in regard to the entry for directional links 2→3 and 3→2 since either side would not find out about the directional entry after the failure.

So far we have not discussed the issue of age of the LSA. In fact, for now, we will ignore it and will come back to it soon. Due to possible changes in cost values of links, each part of the network will evolve over time, which means an increase in the sequence number counter value for other links in the network, cost change, and so on. Now consider that the link 4-5 has recovered. On recognizing that the link has recovered, node 4 will generate LSA for 4→5 and node 5 will generate for 5→4 and flood to the rest of the network. However, with normal flooding, node 4 or node 5 would not do anything in regard to link 2-3, although both have different views! This is partly why resynchronization is important. When a failed link or a node is recovered, the best thing to do is to exchange the entire list of link numbers along with the current sequence number between the neighbor nodes through a *database description message*. This allows the node on each side to determine where they differ, and then request the cost information for the ones where they differ in terms of sequence number. It may seem odd to request the database description first followed by the link cost update. There are two reasons why this approach is beneficial: (1) usually a link state message contains much more than just the link cost (especially if multiple metrics are allowed); thus, the message size stays smaller if only the database description is exchanged; and (2) a different message type for database description exchange avoids needing to invoke a full link state update between these two neighbors. Certainly, once the end nodes (nodes 4 and 5 in this case) have done resynchronization, each will generate a standard/normal link state update message to let the rest of the nodes know about the changed entries.

Now we will bring age into the picture. If two parts have been isolated for a long time, the age field of LSAs received from the other side will decrement all the way to zero. This will then trigger exception advertisement on both sides for the appropriate set of links. Through this process, links will be deleted from the local copy at the nodes. For example, nodes 1, 2, and 4 from the left side will not have any information about links on the right side. In this case, when link 4-5 recovers, nodes 4 and node 5 would do database description exchange and find out about the existence of new links will be synchronized and then flooded to the rest of the network through a normal link state update.   ▲

3.4.2 Link State Protocol: In-Band Based on End-to-End Session

From the above discussion, it is clear that many of the problems, such as sequence number and age, and safeguards needed to address them are primarily due to the flooding mechanism, which is based on in-band communication on a hop-by-hop basis. The flooding mechanism is used so that each node can build the link state database to perform route computation. Traditionally, the flooding mechanism is considered an inherent part of the link state protocol; as a result, it has become tightly coupled with a link state protocol.

We can, however, ask an opposing question: if the link state database is primarily what a node needs, is in-band hop-by-hop flooding the only mechanism that can accomplish it? The answer is clearly no. In fact, in a data network, since virtual connection functionality is available, the link state information can be exchanged by setting up virtual connections between any two nodes that are kept alive as long as both nodes are operational. This would mean that in an N node network, each node would need to set up $N - 1$ virtual connection sessions with the rest of the nodes. Through such a connection, a node can query another node about the link state status of all its outgoing links so that it can update its link state database, and vice versa. If a link fails, the node to which this link is connected can inform the other nodes through the virtual connection about the change of status.

With the *classical* link state protocol (based on flooding and LSA), there is no way for a node to know if the node to which it is directly connected has failed; at most, it knows that the *link* that connects these two nodes is not operational. It might be able to *infer* later through much flooding and learning from other neighboring nodes that perhaps the node has actually failed. However, for a link state protocol with an end-to-end session for link state exchange, a node can immediately determine if another node has failed since the virtual connection will die down.

There are, however, scalability issues with the notion of a virtual connection between any two nodes. For example, if N is large, there are too many sessions a node is required to keep activated simultaneously; this may not be desirable since a node is a specialized device whose primary job is to route/forward user traffic by looking up the routing table. A possible alternative is to have a pair of primary and secondary specialized link state servers in a network; this way, each node would need to have just two connections to this pair of servers to retrieve all the link state information. This would mean that such servers would have many connections to all the nodes—this is not too many when you consider a typical high-end web server that handles a large number of TCP connections to clients at a time.

You may discover that the idea of virtual connection is not much different from the idea of out-of-band communication used in circuit-switched routing, whether it is through a dedicated channel or a signaling network. Furthermore, there are examples in the Internet as well where the routing information exchange is done through virtual connections (through a TCP session); Border Gateway Protocol (BGP) discussed later in Chapter 8 is such an example. Certainly, BGP does not use a link state information exchange; however, reliable routing information communication is similar in need. In addition, in recent years, providers of large intradomain networks have started using servers similar to the notion link state servers to upload link state metrics to routers (see Chapter 7); thus, the entire concept of servers that provide link state information is quite realistic, and are being currently considered for networks known as overlay networks.

3.4.3 Route Computation

When a node has constructed the link state database, even if information for some links is not recent or not all links in the network are known to the node yet, it is ready to compute routes to known destinations. If the goal is to compute the shortest path, Dijkstra's algorithm can be used as described earlier in Section 2.3.2; thus, we will not repeat illustrations described there. In this section, we briefly discuss route computation as related to the protocol.

It is important to note that each node *independently* invokes Dijkstra's shortest path computation based on the most recent link state database available to the node; no centralized processor is involved in this computation. Due to the availability of the link state database, a node actually knows the entire route to any destination after the route computation, unlike a distance vector protocol. Thus, a node can use source route with route pinning, if this function is used in the network. Alternately, it can still create a routing table entry by identifying the next hop from the computed route. Because of the way Dijkstra's shortest path computation works, looping can be completely avoided when all nodes have the same copy of the link state database, even though the routing table entry at each node stores just the next hop for each of the destination nodes. To ensure that all nodes have the same link state database entries, flooding plays an important role along with link state database resynchronization to quickly reach a consistent link state database throughout the network.

Finally, there are networking environments where the basic essence of the link state protocol is used, yet the routing computation is not based on Dijkstra's algorithm; we will discuss this later in regard to dynamic call routing in telephone networks (refer to Chapter 10) and quality-of-service routing (refer to Chapter 17).

3.5 Path Vector Routing Protocol

A path vector routing protocol is a more recent concept compared to both a distance vector protocol and the link state routing protocol. In fact, the entire idea about the path vector protocol is often tightly coupled to BGP. In this section, we will present the general idea behind the path vector protocol mechanism; thus, it is important to distinguish and decouple the generic path vector protocol presented here from BGP4, a specific *instance* of a path vector protocol; we will discuss BGP in detail later in Chapter 8.

First and foremost, a path vector protocol has its roots in a distance vector protocol. We already know that there are several problems with the basic distance vector protocol; for example, looping, count to infinity, unreliable information exchange, and so on. We have already discussed that both the loop-free distance vector protocol and the link state protocol use reliable delivery for exchange of routing information; this itself takes care of certain undesirable behavior. In fact, you will find that all modern routing protocols now use a reliable delivery mechanism for routing information exchange along with a hello protocol for initialization. Thus, we will start with the built-in assumption that a path vector protocol uses reliable delivery mechanism for information exchange related to routing and a hello protocol is used for initialization.

In a path vector protocol, a node does not just receive the distance vector for a particular destination from it neighbor; instead, a node receives the distance *as well* as the entire path to the destination from its neighbor. The path information is then helpful in detecting loops. For example, consider the basic distance vector protocol: at time t, node i receives the distance cost $\overline{D}^i_{kj}(t)$ from its neighbor k for all known destinations j, as known to node i. In the case of a path vector protocol, node i receives both the distance *and* the entire path list from its neighbor k for each of the known destination nodes, that is, both $\overline{D}^i_{kj}(t)$ and the list of nodes, denoted by $\mathcal{P}^i_{kj}(t)$, from k to j as known to node i. A node is thus required to maintain two tables: the path table for storing the current path to a destination, and the routing table to identify the next hop for a destination for user traffic forwarding.

3.5.1 Basic Principle

We first explain the basic principle behind a path vector protocol through a simple illustration.

A SIMPLE ILLUSTRATION

We start with a simple illustration using the topology and cost given in Figure 3.13 and will ignore the time dependency on t. Consider node 2 (that is, $i = 2$) that is receiving a path vector

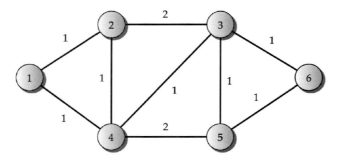

FIGURE 3.13   Network topology for illustration of the path vector protocol.

from node 3 (that is, $k = 3$), which is providing distance and path to the nodes of which it is aware. Thus, node 3 will need to send the following information for destination nodes 3, 4, 5, and 6 where a path is identified using the representation (3, 4), which means that the path contains the nodes 3 and 4 in that order:

$$\overline{D}_{33} = 0, \quad P^2_{33} \equiv (3)$$
$$\overline{D}_{34} = 1, \quad P^2_{34} \equiv (3, 4)$$
$$\overline{D}_{35} = 1, \quad P^2_{35} \equiv (3, 5)$$
$$\overline{D}_{36} = 1, \quad P^2_{36} \equiv (3, 6).$$

In a protocol message, the above information will need to be sent using the following format:

> Destination Node, Cost, Number of Nodes in the Path; Node List of the Path | ...

In particular, the information for the first entry (destination 3) will need to be embedded in the protocol message format as

> $j = 3, \overline{D}_{33} = 0$, Number of Nodes = 1; $P^2_{33} \equiv (3)$

Without writing the identifiers explicitly, this and the rest can be written in the following path vector message that node 3 will send to node 2:

> 3, 0, 1; 3 | 4, 1, 2; 3, 4 | 5, 1, 2; 3, 5 | 6, 1, 2; 3, 6

Assume that node 2 has received the above path vector for the first time from node 3. Furthermore, we assume that node 2, through the hello protocol, has learned about neighbors node 1 and node 4 and has already initialized paths to these neighbors. Thus, the initial path table at node 2 is

Destination	Cost	Path
1	1	(2, 1)
4	1	(2, 4)

On receiving the path vector from node 3, node 2 will compare and find out that it already has a path to node 4 that is cheaper than the new path, and thus, does not need to be changed, and that it has learned about new nodes. Thus, the path table at node 2 will be updated to

Destination	Cost	Path
1	1	$(2, 1)$
3	2	$(2, 3)$
4	1	$(2, 4)$
5	3	$(2, 3, 5)$
6	3	$(2, 3, 6)$

We can see from the above table that it is possible to have parts of paths being common for different destinations. Based on the above information, node 2 sends the following path vector message to nodes 4, 3, and 1:

$$1, 1, 2; 2, 1 \mid 2, 0, 1; 2 \mid 3, 2, 2; 2, 3 \mid 4, 1, 2; 2, 4 \mid 5, 3, 3; 2, 3, 5 \mid 6, 3, 3; 2, 3, 6 \qquad (3.5.1)$$

It is important to realize from the above message that the path vector message can include information for *all* known destinations since the entire path is included for each destination; inclusion of the path vector allows a receiving node to catch any looping immediately; thus, rules such as split horizon used in a distance vector protocol to resolve similar issues are not always necessary.

On receiving the path vector message from node 2, the receiving nodes (nodes 1, 2, and 4, in this case) can check for any looping problem based on the path list and discard any path that can cause looping, and update their path table that has the least cost. It is important to note that the path vector protocol inherently uses Bellman–Ford for computing the shortest path.

Next, we consider how a link failure is handled by the nodes. To show this, we first assume that the network has converged, and all nodes have built their path table for all destination nodes. For our illustration, we show the path table entry at each source node, just for the single destination node 6 (see Figure 3.14):

From Node	To Destination	Cost	Path Table Entry
1	6	3	$(1, 4, 3, 6)$
2	6	3	$(2, 3, 6)$
3	6	1	$(3, 6)$
4	6	2	$(4, 3, 6)$
5	6	1	$(5, 6)$

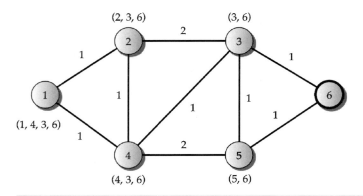

FIGURE 3.14 Path vector to destination node 6.

LINK FAILURE CASE

We will now consider the failure of the link between nodes 3 and 6. We know from the information presented above that this failure will affect routing for nodes 1, 2, and 4. However, node 3 has no way of knowing who will be affected. Thus, node 3 will send an "unreachable" message to its neighbors 2, 4, and 5 stating that path (3, 6) is not available. On receiving this message, node 2 will realize that its path to node 6 was through node 3, and will mark it unavailable, and send an "unreachable" message upstream to nodes 4 and 1, in case they are also using node 2 as the next hop for destination node 6. At about the same time, on receiving the "unreachable" message from node 3, node 4 will take the same action for destination node 6. However, node 5, on receiving the "unreachable" message from node 3, will realize that it can reach node 6, and thus will inform node 3 of its path vector to node 6. In turn, node 3, on learning about the new path vector, will send a follow-up path vector message to node 2 and node 4: 6, 2, 3; 3, 5, 6 . On receiving this message, node 2 will update its path table to node 6 as follows:

Destination	Cost	Path Table Entry
6	4	(2, 3, 5, 6)

and node 4 will update its path table to node 6 as follows:

Destination	Cost	Path Table Entry
6	3	(4, 3, 5, 6)

If after the above updates, a path vector message is received at node 4 from node 5, the path table will remain the same since the path cost does not decrease.

We can see that a path vector protocol requires more than the exchange of just path vector messages; specifically, a path vector protocol needs to provide the ability to coordinate with neighbors which is helpful in discovering a new route, especially after a failure.

3.5.2 Path Vector with Path Caching

There are additional aspects that may impact the actual working of a path vector protocol. To see this, observe that an advantage of the path vector announcement is that a node may learn about multiple nonlooping paths to a destination from its different neighbors. It may choose to *cache* multiple path entries in its path table for each destination that it has learned from all or some of its neighboring nodes, instead of caching just the best one; this is an advantageous feature of a path vector protocol. Note, however, that while the path table may have multiple entries for each destination, the routing table points only to the next hop for the best or single preferred path for each destination unless there is a tie; in case of a tie, the routing table will enter the next hop for both paths.

Thus, we will now illustrate a generalization of the basic concept of a path vector protocol where a node is allowed to cache multiple paths for every destination in the path table. For simplicity, we assume that each node stores two best paths per destination that it has learned from different neighbors. Thus, for destination node 6, each source node will have the following path entries:

From Node	To Destination	Cost	Path Table Entry
1	6	3	$(1, 4, 3, 6)$
1	6	4	$(1, 2, 3, 6)$
2	6	3	$(2, 3, 6)$
2	6	3	$(2, 4, 3, 6)$
3	6	1	$(3, 6)$
3	6	2	$(3, 5, 6)$
4	6	2	$(4, 3, 6)$
4	6	3	$(4, 5, 6)$
5	6	1	$(5, 6)$
5	6	2	$(5, 3, 6)$

Consider again the failure of the link between node 3 and node 6. When node 3 recognizes this link failure, it will realize that it has a second path to node 6, i.e., $(3, 5, 6)$. Thus, it will delete path $(5, 6)$, make route $(3, 5, 6)$ the preferred route in the path table, make appropriate adjustments in the routing table, and advertise path $(3, 5, 6)$ with its cost to its neighbors 2, 4, and 5 as $\boxed{6, 2, 3; 3, 5, 6}$.

We will first consider the situation at node 5. On receiving the above message, node 5 will immediately detect that this will create a loop, and thus it will ignore this message. Node 2, on receiving the new path vector from node 3, will notice that it will need to update the first path. This will make the second path cheaper, although the second path has an unreachable path.

Node 4, on receiving the new path vector from node 3, will change its path via node 3 to $(4, 3, 5, 6)$ and the cost to 3. Now, both paths for destination node 6 will be of cost 3. It will then transmit a path vector message to nodes 2 and 1 about the new path vector to destination 6.

Node 1 will receive an updated path vector from node 2 in regard to destination node 6, and update both its paths via node 2. Node 2 will receive another one from node 4 and will update that path as well.

Eventually, different nodes will have new path entries to destination node 6 as follows (invalid paths are marked with a strikethrough):

From Node	To Destination	Cost	Path Table Entry
1	6	4	$(1, 4, 3, 5, 6)$
1	6	5	$(1, 2, 3, 5, 6)$
2	6	4	$(2, 3, 5, 6)$
2	6	4	$(2, 4, 3, 5, 6)$
3	6	1	$\cancel{(3, 6)}$
3	6	2	$(3, 5, 6)$
4	6	3	$(4, 3, 5, 6)$
4	6	3	$(4, 5, 6)$
5	6	1	$(5, 6)$
5	6	2	$\cancel{(5, 3, 6)}$

It may be noted that path caching can be helpful as it provides a node with additional paths through its neighbors if the first one does not work because of a link failure. The basic idea behind a path vector protocol with path caching is outlined in Figure 3.15. It usually converges to a new solution quickly. Thus, it can give the impression that path caching is a good idea. However, path caching is not always helpful; in fact, it can result in poor convergence in case of a node failure, especially when the node is connected to multiple nodes. We will describe this next.

NODE FAILURE CASE

To illustrate the node failure case, we consider a four-node fully connected network example [594] shown in Figure 3.16; we assume that the distance cost between any two nodes is 1. Preferred direct path and alternate cached paths are shown in the following table (top of the next page) from nodes 1, 2, and 3 to destination node 0.

From Node	To Destination	Cost	Path Table Entry
1	0	1	$(1, 0)$
1	0	2	$(1, 2, 0)$
1	0	2	$(1, 3, 0)$
2	0	1	$(2, 0)$
2	0	2	$(2, 1, 0)$
2	0	2	$(2, 3, 0)$
3	0	1	$(3, 0)$
3	0	2	$(3, 1, 0)$
3	0	2	$(3, 2, 0)$

Initialization:
 – Node i is activated and exchanges "hello" message with neighbors; table exchange updates are performed to find paths for all known destination nodes j

Node i in the receiving mode:
 Announcement/Advertisement Message:
 – Receive a path vector $\mathcal{P}^i_{kj}$ from neighbor k regarding destination node j
 – If (destination node j is not previously in routing and path table)
     create a new entry and continue
 – If (for destination node j there is already an entry for the previously announced path from node k in the path table)
     Replace the old path by the new path $\mathcal{P}^i_{kj}$ in the path table
 – Update candidate path cost: $\widetilde{D}_{ij} = d_{ik} + \overline{D}^i_{kj}$
     If ($\widetilde{D}_{ij} < \overline{D}_{ij}$) then
     Mark best path as $i \to \mathcal{P}^i_{kj}$
     Update the routing table entry for destination j
     else
     For destination j, identify neighbor $\overline{k}$ that results in minimum cost
       over all neighbors, i.e., $\overline{D}_{ij} = d_{i\overline{k}} + \overline{D}^i_{\overline{k}j} = \min_{k \in \mathcal{N}_i}\{d_{ik} + \overline{D}^i_{kj}\}$
     Mark the best path through this new neighbor as $i \to \mathcal{P}^i_{\overline{k}j}$
     Update the routing table entry for destination j
     Node i in *sending* mode: send the new best path to all its neighbors
     Endif
 Withdrawal Message:
 – If (a withdrawal message is received from a neighbor for a particular destination j) then
     Mark the corresponding entry as "unreachable"
     If (there is a next best path in the path table for this destination)
       Advertise this to the rest of the neighbors
     Endif
Special Cases: (node i lost communication with neighbor k ["link failure"]):
 – For those destinations j for which the best path was through k, mark the path as "unreachable" in the path table (and also routing table)
 – If (another path is available in cache for the same destination through another neighbor k')
     Advertise/announce this to all the remaining neighbors
 – If (there is no other path available in cache for destination j)
     Send a message to all neighbors that path to node j is "withdrawn"

FIGURE 3.15 Path vector protocol with path caching (node i's view).

Consider now the failure of node 0. Immediately, all the other nodes lose their direct path to node 0. All of them switch to their second path in the path table entry. For example, node 1 will switch from $(1, 0)$ to $(1, 2, 0)$; furthermore, node 1 will send the following path vector announcement to node 3 and node 2: $\boxed{0, 2, 3; 1, 2, 0}$. This announcement will be understood by the receiving nodes as an *implicit* withdrawal, meaning its previous one does not work and it is replaced by a new one. This would lead to a cascading of path changes; for example, at node 2, first path $(2, 0)$ will be dropped and will be replaced by $(2, 1, 0)$. On hearing from node 1 that $(1, 0)$ is not reachable, it will then switch to $(2, 3, 0)$. It will go through another step to $(2, 1, 3, 0)$ before recognizing that no more paths are available, and noting then that a path to destination node 0 is unavailable.

In this illustration, we assume that node 2 finds out about the connectivity being down to node 0 *before* node 1 and node 3 have recognized that their connectivity to node 0 is down as well; in other words, node 2 does not receive the first path withdrawal messages from node 1 and node 3 before it sends out to them. The following sequence of steps will then take place:

- Node 2 sees that path $(2, 0)$ is no longer available and crosses it off. By inspecting paths cached in its routing table, it then switches to $(2, 1, 0)$, which is least as the next preferred path.
  Node 2 informs both node 1 and node 3 about withdrawal of path $(2, 0)$.

- Node 1 recognizes that path $(1, 0)$ is no longer available and crosses it off.
  Node 1 switches to its next preferred path $(1, 2, 0)$.
  Node 1 receives withdrawal of path $(2, 0)$ from node 2, before it has time to inform others about $(1, 0)$.
  Node 1 switches to $(1, 3, 0)$ and advertises this path to node 2.

- Node 2 receives the advertisement $(1, 3, 0)$ from node 1 and recognizes that the preferred path of node 1 to node 0 is *no* longer $(1, 0)$; thus, node 1 strikes off $(2, 1, 0)$.
  Node 2 compares the last path in its preferred list $(2, 3, 0)$ to the newly advertised path $(1, 3, 0)$ received from node 1, i.e., compare $(2, 3, 0)$ with $(2, 1, 3, 0)$, and switches to $(2, 3, 0)$ since this is preferred over $(2, 1, 3, 0)$.

- Node 3 recognizes that path $(3, 0)$ is no longer available and crosses it off.
  Node 3 receives withdrawal of path $(1, 0)$ from node 1 and withdrawal of path $(2, 0)$ from node 2.
  Node 3 thus realizes that $(3, 1, 0)$ and $(3, 2, 0)$ are no longer available and crosses them off.
  Node 3 informs node 1 and node 2 that path $(3, 0)$ is no longer available.

- Upon receiving withdrawal of path $(3, 0)$ from node 3, node 2 realizes that path $(2, 3, 0)$ is no longer available, thus, it switches to $(2, 1, 3, 0)$, since this is the only path remaining in its table.

- Upon receiving withdrawal of path $(3, 0)$ from node 3, node 1 realizes that $(1, 3, 0)$ is no longer available and thus inform node 2 that path $(1, 3, 0)$ is no longer available.

- Upon receiving withdrawal of path $(1, 3, 0)$ from node 3, node 2 finally realizes that it no longer has any path to node 0.

The key point about this illustration is that due to path exploration, convergence can take quite a bit of time in case of a node failure when the failing node has multiple connectivity. You may wonder why the node failure case is so important since the nodes are built to be robust to start with. There are two reasons for this:

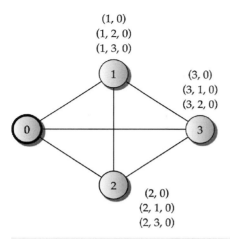

(1, 0)
(1, 2, 0)
(1, 3, 0)

(3, 0)
(3, 1, 0)
(3, 2, 0)

(2, 0)
(2, 1, 0)
(2, 3, 0)

FIGURE 3.16   Four-node fully connected network.

- From a theoretical point of view, it is important to understand how the entire protocol works so that fail-safe mechanisms can be added, if needed.

- While the node failure case discussed above seems like an unlikely case, this is a fairly common case in BGP since a node such as node 0 is a *specialized* node. Actually, such a node is not a real node and in fact represents an address block (IP prefix) that is conceptually represented as a specialized node in the above illustration; we will discuss this in detail in Chapter 8. In this sense, the above illustration is really about how quickly the overall network may learn about losing connectivity to an address block or a route and helps us see the implication of a node failure.

   Finally, an important comment here. A path vector protocol tries to avoid looping by not accepting a path from a neighboring node if this path already contains itself. However, in the presence of path caching, a node can switch to a secondary path, which can in fact lead to an unusual oscillatory problem during the transient period; thus, an important problem in the operation of a path vector protocol is the *stable paths* problem [265]. This says that instead of determining the shortest path, a solution that reaches an equilibrium point is desirable where each node has only a local minimum; note that it is possible to have multiple such solutions at equilibrium.

IMPLICIT COST AND RELATION TO BGP

In our basic description of the path vector protocol with path caching, we have assumed that the cost to destination is included; this is helpful when the cost between two neighboring nodes is different. When the cost between two neighboring nodes is implicitly set to 1, it means that the link cost is hop-based. In this case, it is not necessary to include the cost to destination in the path vector message. Thus, a message format would be:

| Destination Node, Number of Nodes in the Path; Node List of the Path \| ... |

As for example, instead of using the protocol message given in (3.5.1), the following will be transmitted:

```
1, 2; 2, 1 | 2, 1; 2 | 3, 2; 2, 3 | 4, 2; 2, 4 | 5, 3; 2, 3, 5 | 6, 3; 2, 3, 6
```

Certainly, this implicitly assumes that all costs are hop-based instead of link costs shown in Figure 3.2. In fact, BGP uses the notion of implicit hop-based cost. BGP has another difference; the node as described here is a model of a *supernode* this is called an *autonomous system* in the Internet. This also results in a difference, instead of the simplifying assumption we have made that the node ID is what is contained in a path vector message. Furthermore, two supernodes may be connected by multiple links where there may be a preference in regard to selecting one link over another. Thus, BGP is quite a bit more complicated than what we have described and illustrated here in regard to a path vector protocol; BGP will be discussed in detail in Chapter 8.

3.6 Link Cost

In this chapter, you may note that the term *link cost*, sometimes referred to also as the *distance* or the *distance cost*, is generically used to indicate the cost of a link in a network. Because of the generic nature, the term *metric* is also used instead. In this chapter, we do not give much indication on how the link cost is obtained or what factors need to be considered in determining the link cost.

As a matter of fact, determination of the link cost is itself an interesting and important problem. Common sense would tell us that since we want user traffic to move from one node to another as quickly as possible in a packet-switched network, we need to ensure that delay, and probably link speed, are somehow taken into account. In this regard, we briefly discuss below metrics considered in ARPANET routing and the lessons learned.

3.6.1 ARPANET Routing Metrics

During its life time, ARPANET routing used both the distance vector protocol framework ("old" ARPANET routing) and the link state protocol framework ("new" ARPANET routing). The metrics considered also were very much along the line of the protocol framework.

In the "old" ARPANET routing from early 1970s, the link metric was determined based on the queue length plus a fixed constant where the queue length was computed at the instant the link cost is to be updated to its neighboring nodes. The intent was to capture the delay cost based on the queue length, meaning preference was to be given in considering links with shorter queues over longer queues in computing the shortest path in the distance vector framework. However, the queue length can change significantly from one instant to the other that led to (1) packets going from one short queue to another, instead of moving toward the destination, and (2) routing oscillations; certainly, the fixed constant that was added helped in relieving the oscillation problem, but not completely. There were also looping problems, as you would guess from our discussions earlier with a distance vector protocol. Also, the link cost did not consider link speed.

When ARPANET moved to the link state protocol framework in the late 1970s, the link metric computation was also changed. Two link metrics were considered; in the first one, the delay was measured directly by first timestamping each incoming packet when it arrives and then recording the time when it leaves a node; thus, it captured both the queueing and processing delay for every packet. Then, for each outgoing link, delay is first averaged over the delay measurements for packets that arrived in the past 10 sec; this value is compared to the last value similarly computed to see if there was any significant change. If the change was significant, the link cost would take the new value and it is flooded. Note that if the link cost has not changed significantly over five such time windows, i.e., 50 sec, the link cost is reported anyway. For stability, the link cost was bounded below with a fixed value that was determined based on the link speed; this means that an idle link can never report its cost to be zero. This link metric was found to be more stable than the old metric. However, the difficulty with this new metric was in the assumption that the measured link delay is a good indicator, rather predictor, for efficient routing of traffic that will be arriving in the next 10 sec. It was found that it was indeed a good indicator for lightly and moderately loaded links, but not for highly loaded links. Note that the goal of each node was to find the *best* shortest path to all destinations. While it is good to be able to give the best path to everyone, it was problematic during heavy loads. Instead, during heavy loads, a better goal is to determine good paths to most destinations rather than trying to give everyone the best path.

Thus, a revision was made to ARPANET routing metric in the late 1980s. The goal was to resolve the problem faced with the new metric from the early 1980s so that the revised metric works in a heavy load situation; this was accomplished by taking into account link speed along with dampening any oscillation from one time window to another. First, the delay was measured as before. But, this time, it was transformed to a link utilization value using the simple $M/M/1$ queueing delay formula (refer to Appendix B.12.2) assuming the average packet size to be 600 bits; that is, the average delay, τ, for utilization ρ, link speed C, and average packet size of κ is given by:

$$\tau = \frac{\kappa}{C(1-\rho)}. \tag{3.6.1}$$

Rearranging, the average utilization, ρ, can be written as

$$\rho = 1 - \frac{\kappa}{\tau C}. \tag{3.6.2}$$

Since link speed C is known for each link type, thus, using the average packet size ($\kappa = 600$ bits) and the measured delay τ, the average utilization can be computed. The utilization value was then smoothed (see Appendix B.6) by averaging with the previous value to dampen any oscillatory behavior from one time window to the next. The link metric is then chosen as a linear function of the smoothed utilization, with limits put on how little and how much it can change from the last reported value. This way of computing the link metric was found to work well in heavy loads for ARPANET routing.

3.6.2 Other Metrics

We can see from the above discussion that ARPANET pursued dynamic link cost determination, starting from its early days. In the early 1980s, the first round of development of Routing

Information Protocol (RIP) for use in a TCP/IP environment took place; RIP is a distance vector protocol that we will cover in detail in Chapter 5. It may be noted that RIP uses just hop count as the link cost. In Chapter 5, we will discuss how link cost is calculated in other distance vector routing protocols such as Interior Gateway Routing Protocol (IGRP). Later we will discuss in the chapter on IP traffic engineering (see Chapter 7) how link cost can be determined through centralized coordination, especially for large networks, by taking into account network goals.

3.7 Summary

In this chapter, we have presented three important classes of routing protocols: distance vector protocol, link state protocol, and path vector protocol. In presenting these classes of protocols, we focus primarily on the basic principles behind them and their strengths and weaknesses.

In this chapter, we have purposefully stayed with the general principles rather than how a specific *instance* of a class of protocols works. There are two reasons for this: (1) to communicate that the basic idea is not always complicated, and (2) to be able to see how routing protocols may evolve or a new routing protocol may be developed knowing the strength and weakness from the basic framework and principles.

There are several important aspects about any routing protocol: (1) initialization of a protocol, for example, through a hello message, (2) ability to avoid looping, (3) what information to communicate, (4) transient behavior and the rate of convergence, (5) how an exception is handled, for example, when a link failure occurs, and (6) scalability. We have also commented that exchange of routing information through some reliable means is important as well.

It is also important to understand that for most routing protocols, what information is to be exchanged can depend on what action might have taken place. In fact, all nodes in a network seem to live in a symbiotic relationship. If certain aspects do not work properly, a large segment of a network can go into a tailspin (remember the ARPANET example). This brings up another important aspect about routing protocols: routing protocols other than the basic distance vector protocol are stateful protocols where the nodes have peer-to-peer relationships. It is important to not confuse this statefulness ("in" the network) with whether a node needs to maintain states for user traffic on a packet-by-packet basis. As an example, IP in the TCP/IP protocol stack is stateless; however, OSPF, a link state routing protocol used for routing information exchange, is a stateful protocol.

From the practical side, there is another important point to note that arises from protocol information exchanges. Some routing protocols require a significant number of message exchanges, especially in the event of some problem. Thus, the volume of message exchanges can be high enough to consume sizable bandwidth on a link; this then impacts how much is left for the user traffic to use on a link. Typically, routing protocols are not designed to consider how to take this bandwidth factor into account. Sometimes, an indirect mechanism is used such as a node that is not supposed to generate a message to a neighboring node more than once every x sec. The most sophisticated mechanism in practice is to introduce a dampening factor (refer to Section 8.9). Nowadays, a mechanism outside the protocol is also possible in which the data rate for different streams can be capped at specific values through a node's packet scheduling mechanism.

It is also important to note that a routing protocol is not just limited to serving network layer traffic routing; a routing protocol can be developed for overlay networks—an overlay network is conceived at a layer above the network layer where specialized hosts serve as routing nodes for the overlay network; these hosts form a peering relation and exchange routing information pertinent to the overlay network. Thus, basic principles discussed here can be applied in overlay network routing with appropriate adjustments.

Finally, we have given several examples of vulnerabilities with protocols presented in this chapter, mostly related to the issue of correct operation of a protocol. Besides the ones discussed, there is another type of vulnerabilities that can affect a routing protocol: this type can be referred to as security-related vulnerabilities. The question is: is a routing protocol secure and robust to defray any attacks? This is an important question, and there are many efforts currently being pursued to address this question for all routing protocols. In this book, we make only cursory remarks on security-related vulnerabilities; this is not to say that this problem is not important. Detailed discussion would require almost another book in its own right.

Further Lookup

There have been extensive work on routing protocols in the past three decades. Much has been learned from the early days of ARPANET routing protocols that started with a distance vector framework and moved to a link state framework; for example, [461], [462], [463], [599], [724]. A good summary of ARPANET routing metrics, including the revised metric and its performance, can be found in [368]. Also, see [610] for an early work on routing performance in packet-switched networks.

Topics such as how to avoid looping and how to provide stability in shortest path routing have received considerable attention starting in the late 1970s; for example, see [78], [79], [244], [336], [484], [627], [638].

Path vector protocols have received considerable attention in recent years in regard to convergence and stability issues due to wide-scale deployment of BGP; for example, see [95], [116], [265], [390], [434], [553], [554], [555], [728].

Exercises

3.1 Review questions:

    (a) How is split horizon with poisoned reverse different from split horizon?

    (b) What are the sub-protocols of a link state protocol?

    (c) List three differences between a distance vector protocol and a link state protocol.

    (d) Compare and contrast announcements used by a basic distance vector protocol and the enhanced distance vector protocol based on the diffusing computation with co-ordinated update.

3.2 Identify issues faced in a distance vector protocol that are addressed by a path vector protocol.

3.3 Consider a link state protocol. Now, consider the following scenario: a node must not accept an LSA with age 0 if no LSA from the same node is already stored. Why is this condition needed?

3.4 Study the ARPANET vulnerability discussed in RFC 789 [605].

3.5 Consider the network given in Figure 3.12. Write the link state database at different nodes (similar to Table 3.3) before and after failure of link 4-5.

3.6 Consider a seven-node ring network.

(a) If a distance vector protocol is used, determine how long it will take for all nodes to have the same routing information if updates are done every 10 sec.

(b) If a link state protocol is used, how long will it take before every node has the identical link-state database if flooding is done every 5 sec. Determine how many link-state messages in total are flooded till the time when all nodes have the identical database.

3.7 Solve Exercises 3.6, now for a fully-connected 7-node network.

3.8 Investigate how to resolve a stuck in active (SIA) situation that can occur in a distance vector protocol that is based on the diffusing computation with coordinated update.

3.9 Consider a fully-mesh N node network that is running a link state protocol. Suppose one of the nodes goes down. Estimate how many total link state messages will be generated.

3.10 Implement a distance vector protocol using socket programming where the different "nodes" may be identified using port numbers. For this implementation project, define your own protocol format and the fields it must constitute.

3.11 Implement a link state protocol using socket programming. You may do this implementation over TCP using different port numbers to identify different nodes. For this implementation project, define your own protocol format and the fields it must constitute.

4

Network Flow Modeling

If the weak were to
Tide across the rapids of life
With your help,
What do you stand to lose?

Bhupen Hazarika (Based on a translation by Pradip Acharya)

Reading Guideline

This chapter is useful for understanding traffic engineering approaches. The chapter is organized by considering first what traffic means for different communication networks. We then consider a single demand to show how the optimization plays a role depending on a goal. We then discuss a multiple flow model, and then complete the chapter with the general formalism to address a traffic engineering problem. We also comment on solution space, which can be very helpful in obtaining insights about what you might want to do in engineering a network. The background provided here is useful in understanding material presented in several subsequent chapters such as Chapters 7, 19, and 24.

A critical function of a communication network is to carry or flow the *volume* of user traffic. The traffic volume or demand volumes can impact routing and routing decisions, which are also influenced by the goal or objective of the network. So far in Chapter 2, we discussed routing algorithms for determining paths for flowing user traffic, and in Chapter 3, we discussed mechanisms needed in a distributed network environment to accomplish routing.

In this chapter, we present the basic foundation of network flow models along with a variety of objective functions that are applicable in different communication networks. Network flow models are used for traffic engineering of networks and can help in determining routing decisions. In later chapters, we will discuss how models presented in this chapter are applicable to different networking environments.

4.1 Terminologies

To discuss network flow models, we start with a few key terminologies.

The volume of traffic or demand, to be referred to as *traffic volume* or *demand volume*, is an important entity in a communication network that can impact routing. In general, traffic volume will be associated with traffic networks while demand volume will be associated with transport networks; for example, in regard to IP networks or the telephone network, we will use the term traffic volume; However, for transport networks such as DS3-cross-connect, SONET, or WDM networks where circuits are deployed on a longer term basis, we will use the term demand volume. Similarly, routing in a traffic network is sometimes referred to as *traffic routing* while in a transport network it is referred to as *transport routing*, *circuit routing*, or *demand routing*.

The measurement units can also vary depending on the communication network of interest. For example, in IP networks, traffic volume is measured often in terms of *Megabits per sec (Mbps)* or *Gigabits per sec (Gbps)*, while in the telephone network, it is measured in *Erlangs*. When we consider telecommunications transport networks, the demand volume is measured in terms of number of digital signals such as DS3, OC-3, and so on.

In this chapter, we will *uniformly* use the term *demand volume*, instead of switching between traffic volume and demand volume, without attaching a particular measurement unit or a network type since our goal here is to present the basic concepts of network flow models. For any demand volume between two nodes in a network, one or more paths may need to be used to carry it. Any amount of demand volume that uses or is carried on a path is referred to as *flow*; this is also referred to as *path flow*, or *flowing demand volume on a path*, or even *routing demand volume on a path*. A path is one of the routes possible between two end nodes with or without positive flows. Since a network consists of nodes and links, we will also use the term *link flow* to refer to the amount of flow on a link regardless of which end nodes the demand volume is for.

This is also a good time to point out that the term *flow* is used in many different ways in communication networking. For example, as we discussed earlier in Chapter 1, a connection in the TCP/IP environment is uniquely defined by a source/destination IP address pair, a source/destination port number pair, and the transport protocol used—this is also referred to as a flow in the networking literature; in this book, we use the term *microflow* to refer to such a connection identifier, a term used in many Internet RFCs; this helps us in distinguishing microflows from the use of the term flow in general for network flow modeling, and so on.

A given network may not always be able to carry all its demand volume; this can be due to limits on network capacity but also can be dictated by the stochastic nature of traffic. If the network capacity is given, then we call such a network a *capacitated network*. Typically, traffic engineering refers to the best way to flow the demand volume in a capacitated network—this is where network flow models are helpful in determining routing or flow decisions.

A communication network can be represented as a *directed network*, or an *undirected network*. A directed network is one in which the flow is directional from one node to another and the links are considered as directional links. An undirected network is a network in which there is no differentiation between the direction of flow; thus, it is common to refer to such flows as bidirectional and links as bidirectional links. For example, an IP network that uses OSPF protocol (refer to Chapter 6) is modeled as a directed network with directional links. However, a telephone network is an undirected network in which a link is bidirectional and where calls from either end can use the undirected link. In this chapter, we present network flow models assuming networks to be undirected since this allows small models to be explained in a fairly simple way. For instance, for a three-node network we need to consider only three links in an undirected network while six links are required to be considered in a directed network. A pair of demand nodes will be referred to as a *node pair* or a *demand pair*. A node pair in a network will be denoted by $i{:}j$ where i and j are the end nodes for this pair; if it is a directed network, the first node i should be understood as the *origin* or *source* while the second node should be understood as the *destination* or *sink*. For an undirected network, i and j are interchangeable while we will typically write the smaller numbered node first; for example, the demand pair with end nodes 2 and 5 will be written as 2:5. A link directly connecting two nodes i and j in general will be denoted as $i\text{-}j$; in case we need to illustrate a point about a directional link, we will specifically use $i \rightarrow j$ to denote the directional link from node i to node j. In fact, we have already used the notion of directed links and undirected links in earlier chapters. This summary is presented here for the purpose of understanding network flow models. Finally, we use the term *unit cost of flow* on a link or *unit link cost* of flow in regard to carrying demand volume; this term should not be confused with *link cost* or *distance cost* of a link used earlier in Chapter 2.

4.2 Single-Commodity Network Flow

We start with the single-commodity network flow problem. This means that only a node pair has positive demand volume, thus, the name *single-commodity* where the term *commodity* refers to a demand. For illustration of network flow models, we will use a three-node network in this section.

4.2.1 A Three-Node Illustration

Consider a three-node network where 5 units of demand volume need to be carried between node 1 and node 2 (see Figure 4.1); we assume that the demand volume is a deterministic number. We are given that all links in the network are bidirectional and have a capacity of 10 units each. It is easy to see that the direct link 1-2 can easily accommodate the 5 units of demand volume since there the direct link can handle up to 10 units of capacity; this remains the case as long as the demand volume between node 1 and node 2 is 10 units or less. As soon as the demand volume becomes more than 10 units, it is clear that the direct link path cannot

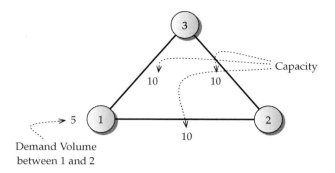

FIGURE 4.1   Three-node network with single demand between node 1 and node 2.

carry all of the demand volume between node 1 and node 2. In other words, any demand in *excess* of 10 units would need to be carried on the second path 1-3-2 .

This simple illustration illustrates that not all demand volume can always be carried on a single path or the shortest, hop-based path; the capacity limit on a link along a path matters. In addition, we have made an *implicit* assumption up to this point that the direct link path 1-2 is less costly per unit of demand flow than the two-link alternate path 1-3-2. However, in many networks, this may not always be true. If we instead suppose that the per-unit cost of the two-link path 1-3-2 is 1 while the per-unit cost on the direct link 1-2 is 2, then it would be more natural or optimal to route demand volume first on the alternate path 1-3-2 for up to the first 10 units of demand volume, and then route any demand volume above the first 10 units on the direct link path 1-2.

The above illustration helps us to see that the actual routing decision should depend on the goal of routing, irrespective of the hop count. This means that we need a generic way to represent the problem so that various situations can be addressed in a capacitated network in order to find the best solution.

4.2.2 Formal Description and Minimum Cost Routing Objective

We are now ready to present the above discussion in a formal manner using *unknowns* or *variables*. We assume here that the capacity of each link is the same and is given by c. Let the demand volume for node pair 1:2 be denoted by h. For example, in the above illustration capacity c was set to 10.

Since the demand volume for the node pair 1:2 can possibly be divided between the direct link path 1-2 and the two-link path 1-3-2 , we can use two unknowns or variables to represent this aspect. Let x_{12} be the amount of the total demand volume h to be routed on direct link path 1-2 , and let x_{132} be any amount of the demand volume to be routed on the alternate path 1-3-2 (see Figure 4.2). Note the use of subscripts so that it is easy to track a route with flows. Since the total demand volume is required to be carried over these two paths, we can write

$$x_{12} + x_{132} = h. \tag{4.2.1a}$$

This requirement is known as the *demand flow constraint,* or simply the *demand constraint.* It is clear that the variables cannot take negative values since a path may not carry any negative

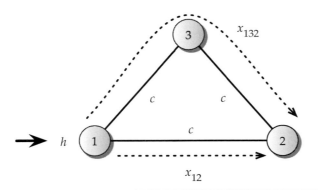

FIGURE 4.2   Single-commodity network flow modeling: three-node network.

demand; this means the lowest value that can be taken is zero. Thus, we include the following additional conditions on the variables:

$$x_{12} \geq 0, \qquad x_{132} \geq 0. \tag{4.2.1b}$$

In addition, we need to address the capacity limit on each link. Certainly, any flow on a path due to routing cannot exceed the capacity on any of the links that this path uses. An implicit assumption here is that the flow and the capacity are using the same measurement units; we will discuss deviations from this assumption in later chapters. Since we assume that the capacity limit is the same on all links in this three-node network, we can write

$$x_{12} \leq c, \qquad x_{132} \leq c. \tag{4.2.1c}$$

The first one addresses the flow on the direct link 1-2 being less than its capacity; flow x_{132} uses two links 1-3 and 2-3, and we can use only a single condition here since the capacity is assumed to be the same on each link. Constraints (4.2.1c) are called *capacity constraints*.

From the above discussion, we can see that we need conditions (4.2.1a), (4.2.1b), and (4.2.1c) to define the basic system. It is important to note that it is not a system of equations; while the first one, i.e., (4.2.1a), is an equation, the second and the third ones, i.e., (4.2.1b) and (4.2.1c), are inequalities. Together, the system of equations and inequalities given by Eq. (4.2.1), which consists of conditions (4.2.1a), (4.2.1b), and (4.2.1c), is referred to as *constraints* of the problem. Even when all the constraints are known, our entire problem is not complete since we have not yet identified the *goal* of the problem. In fact, without defining a goal, system (4.2.1) has infinite numbers of solutions since an infinite combination of values can be assigned to x_{12} and x_{132} that satisfies constraints (4.2.1a), (4.2.1b), and (4.2.1c).

As the first goal, we consider the cost of routing flows. To do that, we introduce a generic nonnegative cost *per unit* of flow on each path: ξ_{12} (≥ 0) for direct path 1-2 and ξ_{132} (≥ 0) for alternate path 1-3-2 . Thus, the total cost of the demand flow can be written as

$$\text{Total cost} = \xi_{12}x_{12} + \xi_{132}x_{132}. \tag{4.2.2}$$

The total cost is referred to as the *objective function*. In general, the objective function will be denoted by F. If the goal is to minimize the total cost of routing, we can write the complete problem as follows:

$$\begin{aligned}
&\textbf{minimize}_{\{x_{12}, x_{132}\}} \quad && F = \xi_{12}x_{12} + \xi_{132}x_{132} \\
&\textbf{subject to} \quad && x_{12} + x_{132} = h \\
& && x_{12} \leq c, \; x_{132} \leq c \\
& && x_{12} \geq 0, \; x_{132} \geq 0.
\end{aligned} \tag{4.2.3}$$

The problem presented in Eq. (4.2.3) is a *single-commodity network flow* problem; it is also referred to as a *linear programming problem* since the requirements given by Eq. (4.2.1) are all linear, which are either equations or inequalities, and the goal given by Eq. (4.2.2) is also linear. In general, a representation as given in Eq. (4.2.3) is referred to as the *formulation* of an optimization problem. The system given by Eq. (4.2.1) is referred to as *constraints*. To avoid any confusion, we will identify the variables in any formulation by marking them as subscripts with **minimize**. Thus, in the above problem, we have noted that x_{12} and x_{132} are variables by indicating so as subscripts with **minimize**. Often, the list of variables can become long; thus, we will also use a short notation such as x in the subscript with **minimize** to indicate that all xs are variables.

Because of the way the goal is described in Eq. (4.2.3), the problem is also known as the *minimum cost routing* or *minimum cost network flow* problem. An optimal solution to an optimization problem is a solution that satisfies the constraints of the problem, i.e., it is a *feasible solution* and the objective function value attained is the lowest (if it is a minimization problem) possible for any feasible solution. For clarity, the optimal solution to a problem such as Eq. (4.2.3) will be denoted with asterisks in the superscript, for example, x_{12}^* and x_{132}^*.

INSTANCE 1:

We now consider some specific cases discussed earlier in Section 4.2.1 to obtain solutions to problem (4.2.1). First, we consider the capacity to be 10, i.e., $c = 10$.

If the unit cost is based on a unit flow per link, then we can clearly write cost components as $\xi_{12} = 1$ (since it is a direct link path) and $\xi_{132} = 2$ (due to two links making a path). This will then correspond to the first case discussed in Section 4.2.1. In this case, optimal flows can be written as:

$$\begin{aligned}
x_{12}^* = 10, \quad & x_{132}^* = 0 && \text{when } 0 \leq h \leq 10 \\
x_{12}^* = 10, \quad & x_{132}^* = h - 10 && \text{when } h \geq 10, \text{ and } h \leq 20.
\end{aligned} \tag{4.2.4}$$

If $h > 20$, it is clear that the network does not have enough capacity to carry all of the demand volume—this is referred to as an *infeasible* situation and the problem is considered to be infeasible.

INSTANCE 2:

Consider the alternate case where per unit cost on the alternate path is 1 while on the direct path it is 2, i.e., $\xi_{12} = 2$ and $\xi_{132} = 1$. In this case, optimal flows can be written as:

$$\begin{aligned}
x_{12}^* = 0, \quad & x_{132}^* = 10 && \text{when } 0 \leq h \leq 10 \\
x_{12}^* = h - 10, \quad & x_{132}^* = 10 && \text{when } h \geq 10, \text{ and } h \leq 20.
\end{aligned} \tag{4.2.5}$$

ON SOLVING PROBLEM (4.2.3)

We now consider the general solution to Problem (4.2.3) when the demand volume is less than the capacity of a link, i.e., $h \leq c$. With two unknowns, problem (4.2.3) can be solved by using substitutions, i.e., by setting $x_{132} = h - x_{12}$ and using it back in the objective. Then, the objective becomes

$$F = \xi_{12}x_{12} + \xi_{132}(h - x_{12}) = (\xi_{12} - \xi_{132})x_{12} + \xi_{132}h.$$

Note that the last term, $\xi_{132}h$, remains constant for a specific problem instance. Thus, we need to consider the minimization of the rest of the expression, i.e.,

minimize$_{\{x\}}$ $(\xi_{12} - \xi_{132})x_{12}$

subject to appropriate constraints. We can easily see that if $\xi_{12} < \xi_{132}$, then the problem is at minimum when $x_{12}^* = h$; however, if $\xi_{12} > \xi_{132}$, then the minimum is observed when $x_{12}^* = 0$. When $\xi_{12} = \xi_{132}$, then x_{12} can take any value in the range $[0, h]$, that is, the problem has *multiple* optimal solutions.

Consider now the case in which demand volume, h, is more then c but the problem is still feasible, i.e., $h > c$, but $h \leq 2c$. In this case, we need to take the bounds into account properly; thus, if $\xi_{12} < \xi_{132}$, then $x_{12}^* = \min\{h, c\}$; similarly, if $\xi_{12} > \xi_{132}$, then the minimum is observed when $x_{12}^* = \max\{0, h - c\}$.

Thus, for values of h ranging from 0 to $2c$, we can see that optimal flows are as we have already identified in (4.2.4) and (4.2.5), corresponding to $\xi_{12} < \xi_{132}$ and $\xi_{12} > \xi_{132}$, respectively.

4.2.3 Variation in Objective: Load Balancing

In model (4.2.3), we have considered the goal to be based on minimizing the routing cost by incorporating the unit cost of a path. While this is applicable in some communication networks, other goals or objectives are also applicable in other communication networks.

We now consider another goal—*minimization of maximum link utilization*. This goal is also referred to as *load balancing* flows in the network. To illustrate this, we will again use constraints (4.2.1) discussed above. The link utilization is defined as the amount of flow on a link divided by the capacity on that link. We know that the only flow using link 1-2 is x_{12} while x_{132} uses both links 1-3 and 3-2 . Thus, the utilization on link 1-2 can be written as

$$\frac{x_{12}}{c}$$

while utilization on either link 1-3 or 3-2 can be written as

$$\frac{x_{132}}{c}.$$

Then, the maximum utilization over all links means the maximum over these two expressions, i.e.,

$$\max\left\{\frac{x_{12}}{c}, \frac{x_{132}}{c}\right\}.$$

Note that x_{12} and x_{132} are variables that are constrained given by Eq. (4.2.1). Thus, for load balancing, we want to pick the values of the variables in such a way that the maximum link utilization is at a minimum. That is, the load balancing problem can be formally written as

$$
\begin{aligned}
\textbf{minimize}_{\{x\}} \quad & F = \max\left\{\frac{x_{12}}{c}, \ \frac{x_{132}}{c}\right\} \\
\textbf{subject to} \quad & x_{12} + x_{132} = h \\
& x_{12} \le c, \ x_{132} \le c \\
& x_{12} \ge 0, \ x_{132} \ge 0.
\end{aligned}
\tag{4.2.6}
$$

To illustrate the meaning of maximum link utilization, consider $c = 10$ and $h = 5$. If all of the demand volume is routed on the direct link path 1-2 , then $x_{12} = 5$ and $x_{132} = 0$; the maximum of the link utilization is then $\max\{5/10, 0/10\} = 1/2$. However, if we were to route one unit of demand volume on the alternate path, i.e., $x_{132} = 1$, while keeping the rest on the direct link, i.e., $x_{12} = 4$, then the maximum link utilization is $\max\{4/10, 1/10\} = 2/5$; this utilization value is lower than if all of the demand volume were routed on the direct link path. The question is: can we do even better? The answer is yes, leading to the optimal solution for the load balancing case.

In fact, we can discuss the optimal solution for the general formulation given by Eq. (4.2.6) without needing to consider specific values of c or h. First note that the maximum in the objective is over only two terms; thus, the minimum can be achieved if they are equal, i.e., at optimality, we must have

$$
\frac{x_{12}^*}{c} = \frac{x_{132}^*}{c}.
$$

Note that the unknowns are related by $x_{12}^* + x_{132}^* = h$. Thus, substituting x_{132}^*, we obtain

$$
\frac{x_{12}^*}{c} = \frac{h - x_{12}^*}{c}.
$$

Transposing and noting that the denominators are the same on both sides, we get

$$
x_{12}^* = h/2.
\tag{4.2.7}
$$

Thus, we see that when the load balancing of flows is the main goal, the optimal solution for Eq. (4.2.6) is to split the flows equally on both paths. Certainly, this result holds true as long as the demand volume h is up to and including $2c$; the problem becomes infeasible when $h > 2c$.

Variation in Capacity

We now consider a simple variation in which the link capacities in the network are not the same. To consider this case, we keep the capacity of link 1-2 at c but increase the capacity of the other two links to 10 times that of 1-2 , i.e., to $10c$. Note that the utilization on links 1-3 and 3-2 are now $x_{132}/(10c)$, and Formulation (4.2.6) changes to the following:

$$
\begin{aligned}
\textbf{minimize}_{\{x\}} \quad & F = \max\left\{\frac{x_{12}}{c}, \ \frac{x_{132}}{10c}\right\} \\
\textbf{subject to} \quad & x_{12} + x_{132} = h \\
& x_{12} \le c, \ x_{132} \le 10c \\
& x_{12} \ge 0, \ x_{132} \ge 0.
\end{aligned}
\tag{4.2.8}
$$

In this case, too, the optimal load balance is achieved when

$$\frac{x_{12}^*}{c} = \frac{x_{132}^*}{10c}.$$

On simplification, we obtain

$$x_{12}^* = h/11$$

and thus, $x_{132}^* = 10h/11$. This essentially says that load balancing on a network with non-uniform capacity results in utilization being balanced, but not necessarily flows. A simple way to visualize this is to consider the capacity of link 1-2 to be 10 Mbps, and the capacity of other links to be 100 Mbps; it would be preferable to send more traffic to the fatter/higher capacity link.

4.2.4 Variation in Objective: Average Delay

Another goal commonly defined, especially in data networks, is the minimization of the average packet delay. For this illustration, we consider again the three-node network with demand volume h between node 1 and node 2; the capacity of all links is set to c. The average delay in a network (see Appendix B.13 for details) with flow x_{12} on the direct link path and x_{132} on the alternate path can be captured through the expression

$$\frac{x_{12}}{c - x_{12}} + \frac{2x_{132}}{c - x_{132}}.$$

Here again, the capacity is normalized so that the measurement units for flow and capacity are the same. The goal of minimizing the average delay is to solve the following problem:

$$
\begin{aligned}
&\textbf{\textit{minimize}}_{\{x\}} \quad && F = \frac{x_{12}}{c-x_{12}} + \frac{2x_{132}}{c-x_{132}} \\
&\textbf{\textit{subject to}} \quad && x_{12} + x_{132} = h \\
& && x_{12} \le c, \; x_{132} \le c \\
& && x_{12} \ge 0, \; x_{132} \ge 0.
\end{aligned}
\tag{4.2.9}
$$

First, we want to point out that the objective function is not defined when $x_{12} = c$ or $x_{132} = c$. In fact, the problem is meaningful only when $x_{12} < c$ and $x_{132} < c$. Thus, in reality, we want to limit the bounds on the unknowns below c by a small positive amount, say, $\varepsilon(> 0)$, i.e., $x_{12} \le c - \varepsilon$ and $x_{132} \le c - \varepsilon$.

To solve Eq. (4.2.9), we have two nonnegative variables, x_{12} and x_{132}, which are related by $x_{12} + x_{132} = h$; thus, we can rewrite the objective function in terms of just a single unknown, say in x_{12}, as

$$F = \frac{x_{12}}{c - x_{12}} + \frac{2(h - x_{12})}{c - (h - x_{12})}.$$

This is a nonlinear function that we want to minimize. We can use calculus to solve this problem. That is, we differentiate expression F with respect to x_{12} (i.e., $\frac{dF}{dx_{12}}$) and set the result to zero, i.e., $\frac{dF}{dx_{12}} = 0$; then, we can solve this as an equation to find solution x_{12}. You can do the "second-derivative" test at this solution to verify that it is indeed minimum, not a maximum. In our case, $\frac{dF}{dx_{12}} = 0$ translates to solving a quadratic equation; this means that we obtain two solutions. However, only one solution, $x_{12} = -h + 3c - 2\sqrt{2}c + \sqrt{2}h$, is relevant since we must have $x_{12} \geq 0$; furthermore, we need to ensure that the resulting solution is never beyond the demand volume h, i.e., $x_{12} \leq h$. Thus, we can write the solution from differentiation by incorporating the necessary bounds as follows:

$$x_{12}^* = \min\{h, \ -h + 3c - 2\sqrt{2}c + \sqrt{2}h\}. \tag{4.2.10}$$

From the above result, we can see that if the demand volume is low, the optimal solution is route all flow on the direct link path; but as the demand volume grows, it is optimal to flow some of the value on of the second path.

4.2.5 Summary and Applicability

In this section, we have considered the single-commodity network flow problem for three different goals: minimum cost routing, load balancing, and minimization of the average delay. In Figure 4.3, we have plotted the optimal flow on the direct link path, x_{12}^*, given by (4.2.4), (4.2.5), (4.2.7), and (4.2.10) for the three different objectives (including two cases for minimum cost routing) for demand volume, h, ranging from 0 to 20. From the optimal solutions obtained for these cases, it is clear that although the *domain* of each problem is the same, that is, the constraint set is the same, the actual optimal solution is different depending on the goal; furthermore, the actual values of h and c also matter.

While we have illustrated here solutions for only a three-node single-commodity case, the general behavior is quite applicable in any size network. That is, it is important to note that

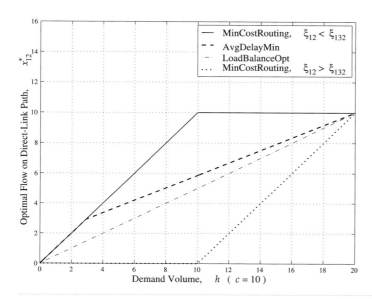

FIGURE 4.3 Optimal flow on the direct link path with different objectives.

the optimal solution with different objective functions usually has minimal difference when the demand volume is low compared to the capacity ("lowly-loaded case"), and also have surprisingly minimal difference when the demand volume is closer to the capacity ("highly-loaded case"); however, in the moderately-loaded region, the flow solution can vary significantly from one objective function to another. Thus, the important lesson here is that for a particular communication networking and/or operational environment, we need to be careful in choosing the primary goal and the load to capacity region considered matters. In other words, we cannot choose one goal and then wonder why the optimal solution for this goal is quite different from another goal.

4.3 Multicommodity Network Flow: Three-Node Example

In this section, we consider multiple commodities, that is, multiple demand pairs have positive demand volumes. As with the single-commodity case, we will consider again the three different objectives. We will again use a three-node network to explain the multicommodity network flow problem.

4.3.1 Minimum Cost Routing Case

For the multicommodity case in a three-node network, all three demand pairs can have positive demand volumes. For clarity, we will use a subscript with demand volume notation h to identify different demands; thus, the demand volume between nodes 1 and 2 will be identified as h_{12}, between 1 and 3 as h_{13}, and between 2 and 3 as h_{23}. For each demand pair, the volume of demand can be accommodated using two paths: one is the direct link path and the other is the alternate path through the third node. In Figure 4.4, we show all the possible paths for each demand pair. The amount of flow on each path is the unknown that is to be determined based on an objective; we denote the unknowns as x_{12} for path 1-2 for demand pair 1:2, and x_{132} for path 1-3-2 , and so on.

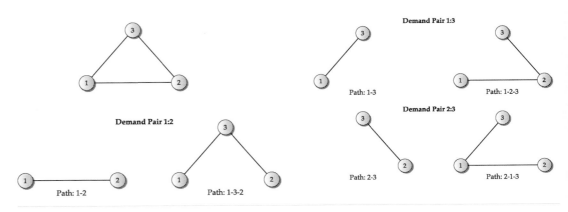

FIGURE 4.4   Three node example with all possible paths.

Much as shown earlier for the single-commodity flow problem, we can write that the demand volume for a node pair must be carried over the two paths. Thus, for demand pair 1:2, we can write

$$x_{12} + x_{132} = h_{12}. \tag{4.3.1a}$$

Similarly, for demand pairs 1:3 and 2:3, we can write the following:

$$x_{13} + x_{123} = h_{13} \tag{4.3.1b}$$

$$x_{23} + x_{213} = h_{23}. \tag{4.3.1c}$$

These unknown flow amounts, while satisfying the demand volume requirements, must also satisfy capacity limits on any link. We denote capacities of links 1-2 , 1-3 , and 2-3 by c_{12}, c_{13}, and c_{23}, respectively.

By following the paths listed in Figure 4.4, we note that three different paths from three different demand pairs use link 1-2; they are paths 1-2, 1-2-3, and 2-1-3 (see Figure 4.5). Since the sum of the flow over these three paths cannot exceed the capacity, c_{12}, of link 1-2 , we can write the following inequality (constraint):

$$x_{12} + x_{123} + x_{213} \leq c_{12}. \tag{4.3.2a}$$

Similarly, for the other two links 1-3 and 2-3, we can write

$$x_{13} + x_{132} + x_{213} \leq c_{13} \tag{4.3.2b}$$

$$x_{23} + x_{132} + x_{123} \leq c_{23}. \tag{4.3.2c}$$

We next consider the objective function for minimum cost routing. If the unit costs of routing on paths 1-2, 1-3-2, 1-3, 1-2-3, 2-3, and 2-1-3 are denoted by ξ_{12}, ξ_{132}, ξ_{13}, ξ_{123}, ξ_{23}, and ξ_{213}, respectively, then the total routing cost can be written as

$$\text{total cost} = \xi_{12}x_{12} + \xi_{132}x_{132} + \xi_{13}x_{13} + \xi_{123}x_{123} + \xi_{23}x_{23} + \xi_{213}x_{213}. \tag{4.3.3}$$

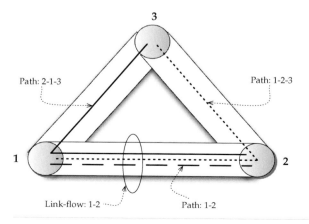

FIGURE 4.5   Link flow on link 1-2 for paths for different demand pairs.

Thus, the entire problem can be formulated as follows:

$$
\begin{aligned}
\textbf{\textit{minimize}}_{\{x\}} \quad & F = \xi_{12}x_{12} + \xi_{132}x_{132} + \xi_{13}x_{13} + \xi_{123}x_{123} + \xi_{23}x_{23} + \xi_{213}x_{213} \\
\textbf{\textit{subject to}} \quad & x_{12} + x_{132} = h_{12} \\
& x_{13} + x_{123} = h_{13} \\
& x_{23} + x_{213} = h_{23} \\
& x_{12} + x_{123} + x_{213} \leq c_{12} \\
& x_{13} + x_{132} + x_{213} \leq c_{13} \\
& x_{23} + x_{132} + x_{123} \leq c_{23} \\
& x_{12} \geq 0, \ x_{132} \geq 0, \ x_{13} \geq 0, \ x_{123} \geq 0, \ x_{23} \geq 0, \ x_{213} \geq 0.
\end{aligned}
\tag{4.3.4}
$$

The above problem has six nonnegative variables, and six constraints.

Example 4.1 *Illustration of solution for Eq. (4.3.4).*

Consider demand volumes to be $h_{12} = 5$, $h_{13} = 10$, and $h_{23} = 7$, and capacities to be $c_{12} = 10$, $c_{13} = 10$, $c_{23} = 15$ (see Figure 4.6). If the unit cost is based on the number of links a flow traverses, that is, 1 for a single-link path and 2 for a two-link path, then we can write $\xi_{12} = \xi_{13} = \xi_{23} = 1, \xi_{132} = \xi_{123} = \xi_{213} = 2$. Clearly, the optimal solution to Eq. (4.3.4) is to flow demand volume for each demand pair on the respective direct link path, i.e., we set $x_{12}^* = 5$, $x_{13}^* = 10$, $x_{23}^* = 7$ with the other variables taking the value zero since all constraints are satisfied; here, the total cost at optimality is 22.

However, if the unit costs are different, such as a single link path costing *twice* that of a two-link path, i.e., $\xi_{12} = \xi_{13} = \xi_{23} = 2, \xi_{132} = \xi_{123} = \xi_{213} = 1$, then the optimal solution that satisfies all the constraints would be: $x_{12}^* = 1, x_{132}^* = 4, x_{13}^* = 3.5, x_{123}^* = 6.5, x_{23}^* = 4.5, x_{213}^* = 2.5$, with the total cost being 31. ▲

Unlike the ease with which we were able to determine the optimal solution for the single-commodity network flow problem given by Eq. (4.2.3), it is not so easy to do so for the multi-commodity network flow problem given by Eq. (4.3.4) since the latter problem has six variables and six constraints. Thus, we need to resort to an algorithmic approach to solving this problem.

First, recall that problems such as Eq. (4.3.4) are classified as *linear programming (LP) problems* since all constraints as well as the objective function are linear. LP problems can be solved using the well-known simplex method, and other methods such as the interior point method; for example, see [164], [515], [711]. While these methods work well in practice, they are fairly complicated algorithms, and their description is beyond the scope of this book. Fortunately, there are many software packages for solving LP problems; for example, see [237] for a survey of LP solvers. Such a package allows a user to enter the problem almost in the way it is described in Eq. (4.3.4).

Example 4.2 *Solving Eq. (4.3.4) using CPLEX.*

We will illustrate here how to solve Eq. (4.3.4) when the alternate path is cheaper than the direct path, i.e., $\xi_{12} = 2, \xi_{132} = 1$, and so on. We will use CPLEX [158], a popular LP solver for this illustration. In CPLEX, you can enter the data for the second case of Example 4.1 as given below (see Appendix B.5 for additional information):

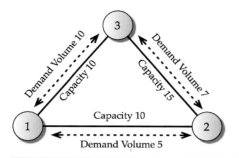

FIGURE 4.6 Demand volume and capacity data for three-node network.

```
Minimize 2 x12 + x132 + 2 x13 + x123 + 2 x23 + x213
subject to
      d12:  x12 + x132 = 5
      d13:  x13 + x123 = 10
      d23:  x23 + x213 = 7
      c12:  x12 + x123 + x213 <= 10
      c13:  x132 + x13 + x213 <= 10
      c23:  x132 + x123 + x23 <= 15
Bounds
      0 <= x12
      0 <= x132
      0 <= x13
      0 <= x123
      0 <= x23
      0 <= x213
End
```

The above representation is very similar to Formulation (4.3.4). Problem data in CPLEX are entered in the ASCII-based text format; thus, subscripts are directly tagged on to the variables; similarly, note the use of <= instead of $\leq$.

Using CPLEX, we can find the optimal solution to Eq. (4.3.4). Solutions can be displayed by giving the display command as follows:

```
CPLEX> display solution variables -
Variable Name      Solution Value
x12                1.000000
x132               4.000000
x13                3.500000
x123               6.500000
x23                4.500000
x213               2.500000
```

Thus, we have $x_{12}^* = 1, x_{132}^* = 4, x_{13}^* = 3.5, x_{123}^* = 6.5, x_{23}^* = 4.5, x_{213}^* = 2.5.$ ▲

It may be noted that the above solution gives fractional values, which is the case in general for an LP problem. That is, the multicommodity flow model as given by Eq. (4.3.4) is for variables taking values in the real number space. Sometimes we do have restrictions; for example, some or all variables are allowed to take only *integer* values. If some of the variables take integral values, then such problems are labeled as *mixed integer linear programming (MILP) problems*; if all variables take integral values, then such problems are referred to as *integer linear programming (ILP) problems*. Many real-world communication network problems are appropriately modeled as MILP or ILP problems; we will illustrate such real examples later in this book.

If variables take only integral values, then this is in fact a form of constraints; thus, they need to be explicitly stated as part of the problem definition. If we do in fact require Eq. (4.3.4) to include the requirement that all variables take integral values, then we can rewrite it as follows:

$$
\begin{aligned}
\textit{minimize}_{\{x\}} \quad & F = \xi_{12}x_{12} + \xi_{132}x_{132} + \xi_{13}x_{13} + \xi_{123}x_{123} + \xi_{23}x_{23} + \xi_{213}x_{213} \\
\textit{subject to} \quad & x_{12} + x_{132} = h_{12} \\
& x_{13} + x_{123} = h_{13} \\
& x_{23} + x_{213} = h_{23} \\
& x_{12} + x_{123} + x_{213} \leq c_{12} \\
& x_{13} + x_{132} + x_{213} \leq c_{13} \\
& x_{23} + x_{132} + x_{123} \leq c_{23} \\
& x_{12} \geq 0,\ x_{132} \geq 0,\ x_{13} \geq 0,\ x_{123} \geq 0,\ x_{23} \geq 0,\ x_{213} \geq 0 \\
& \text{all } xs \text{ integer.}
\end{aligned}
\tag{4.3.5}
$$

Example 4.3 *Multicommodity network flow with integer solution.*
Considering again demand volumes to be $h_{12} = 5$, $h_{13} = 10$, and $h_{23} = 7$, and capacities to be $c_{12} = 10$, $c_{13} = 10$, $c_{23} = 15$, in CPLEX, we can enter the above ILP problem in the following way where integrality of variables is explicitly listed:

```
Minimize 2 x12 +  x132 + 2 x13 +  x123 + 2 x23 +  x213
subject to
      d1:  x12 + x132 = 5
      d2:  x13 + x123 = 10
      d3:  x23 + x213 = 7
      c1:  x12 + x123 + x213 <= 10
      c2:  x132 + x13 + x213 <= 10
      c3:  x132 + x123 + x23 <= 15
Bounds
      0 <= x12 <= 10
      0 <= x132 <= 10
      0 <= x13 <= 10
      0 <= x123 <= 10
      0 <= x23 <= 10
      0 <= x213 <= 10
Integer
      x12 x132 x13 x123 x23 x213
End
```

An important point to note here is that when variables are explicitly declared as integers, an upper bound for these variables is required to be specified since CPLEX assumes that the default for the upper bound is 1. In the above case, we have artificially set the upper bound at 10 since from demand volume values we know that no variables will take more than 10 units of flow at optimality. The optimal solution with integrality requirement is obtained as

$$
x_{12}^* = 1, x_{132}^* = 4, x_{13}^* = 4, x_{123}^* = 6, x_{23}^* = 5, x_{213}^* = 2
$$

and the total cost at optimality is 32.   ▲

You may note that the optimal objective cost is higher when the variables take integral values compared to the counterpart when variables are allowed to take real values. It is indeed true that for a minimization problem, the optimal cost for the integral-valued problem is always higher than or equal to the counterpart problem when the variables take real values, i.e., when integrality is relaxed. Note that the integrality requirement can be thought of as additional constraints to a problem; any time additional constraints are added to a base problem, the optimal objective cost goes up as long as the objective is minimization based. Finally, note that problems with integrality constraints are in general harder to solve, i.e., more time-consuming in general. Furthermore, a problem with integrality constraints cannot be solved by the simplex method; instead, methods such as branch-and-bound and branch-and-cut are used. Tools such as CPLEX support these methods in addition to the simplex method; this is exactly what happened when we solved Eq. (4.3.5). For very large problems (with many variables and constraints), sometimes commercial solvers are not effective; thus, we resort to developing specialized algorithms by exploiting the structure of a problem; for various approaches to solving large communication network optimization problems, refer to [564].

4.3.2 Load Balancing

We now present the formulation for when maximum link utilization is minimized for the multicommodity case. We first introduce a set of *dependent* variables for flow on each link. If we denote the link flow on link 1-2 by y_{12}, then based on our discussion earlier, we can write

$$y_{12} = x_{12} + x_{123} + x_{213}. \tag{4.3.6a}$$

Similarly, for the other two links, we introduce y_{13} and y_{23}, and write

$$y_{13} = x_{13} + x_{132} + x_{213} \tag{4.3.6b}$$

$$y_{23} = x_{23} + x_{132} + x_{123}. \tag{4.3.6c}$$

Given y_{12}, y_{13}, y_{23}, we can write the link utilization as $y_{12}/c_{12}, y_{13}/c_{13}, y_{23}/c_{23}$ for links 1-2, 1-3, and 2-3, respectively. The maximum link utilization over these three links can be written as

$$\max\left\{\frac{y_{12}}{c_{12}}, \frac{y_{13}}{c_{13}}, \frac{y_{23}}{c_{23}}\right\}. \tag{4.3.7}$$

Note that the actual maximum value is influenced by link flow variables, which are, in turn, affected by how much flow is assigned to path flow variables. Recall that the goal of load balancing is to minimize the maximum link utilization. We can write the entire formulation as

$$\textbf{minimize}_{\{x,y\}} \quad F = \max \left\{ \frac{y_{12}}{c_{12}}, \frac{y_{13}}{c_{13}}, \frac{y_{23}}{c_{23}} \right\}$$

$$\textbf{subject to} \qquad x_{12} + x_{132} = h_{12}$$

$$x_{13} + x_{123} = h_{13}$$

$$x_{23} + x_{213} = h_{23}$$

$$x_{12} + x_{123} + x_{213} = y_{12}$$

$$x_{13} + x_{132} + x_{213} = y_{13} \qquad\qquad (4.3.8)$$

$$x_{23} + x_{132} + x_{123} = y_{23}$$

$$y_{12} \leq c_{12}, \ y_{13} \leq c_{13}, \ y_{23} \leq c_{23}$$

$$x_{12} \geq 0, \ x_{132} \geq 0, \ x_{13} \geq 0, \ x_{123} \geq 0, x_{23} \geq 0, \ x_{213} \geq 0$$

$$y_{12} \geq 0, \ y_{13} \geq 0, \ y_{23} \geq 0.$$

It is important to note that the maximum link utilization is a quantity that lies between 0 and 1 since link flow should be less then the capacity of the respective link for a feasible problem. Note that problem (4.3.8) is, however, *not* an LP problem since the objective function (4.3.7) is nonlinear; rather it is a piecewise linear function. That means that the standard LP approach that we discussed earlier cannot be directly used. Fortunately, the maximum link utilization function has a nice property. To discuss it, we first introduce the definition of a *convex function*. A function f is *convex* if for any two points z_1 and z_2 in its domain, and for a parameter α in $0 \leq \alpha \leq 1$, the following holds:

$$f(\alpha z_1 + (1 - \alpha)z_2) \leq \alpha f(z_1) + (1 - \alpha)f(z_2). \qquad\qquad (4.3.9)$$

This means that the convex combination of the line segment connecting the function values at $f(z_1)$ and $f(z_2)$ will be grater than or equal to the function between the points z_1 and z_2. A pictorial view is presented in Figure 4.7.

We now consider again the objective (4.3.7); it is easy to see that it is convex; also recall that it is piecewise linear. Thus, if we write (4.3.7) as

$$r = \max \left\{ \frac{y_{12}}{c_{12}}, \frac{y_{13}}{c_{13}}, \frac{y_{23}}{c_{23}} \right\}, \qquad\qquad (4.3.10)$$

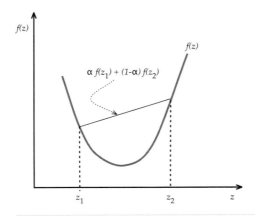

FIGURE 4.7 Convex function.

then clearly r is greater than or equal to each of its components, i.e.,

$$r \geq \frac{y_{12}}{c_{12}}, \quad r \geq \frac{y_{13}}{c_{13}}, \quad r \geq \frac{y_{23}}{c_{23}}. \tag{4.3.11}$$

Since the goal is to minimize Eq. (4.3.7), this is equivalent to minimizing r subject to constraints Eq. (4.3.11). Using this result and noting that $r \geq \frac{y_{12}}{c_{12}}$ is the same as $y_{12} \leq c_{12}\, r$, and similarly for the other two links, we can write Eq. (4.3.8) as the following equivalent problem:

$$
\begin{aligned}
&\textbf{\textit{minimize}}_{\{x,y,r\}} \quad && F = r \\
&\textbf{\textit{subject to}} \quad && x_{12} + x_{132} = h_{12} \\
& && x_{13} + x_{123} = h_{13} \\
& && x_{23} + x_{213} = h_{23} \\
& && x_{12} + x_{123} + x_{213} = y_{12} \\
& && x_{13} + x_{132} + x_{213} = y_{13} \\
& && x_{23} + x_{132} + x_{123} = y_{23} \\
& && y_{12} \leq c_{12}, \ y_{13} \leq c_{13}, \ y_{23} \leq c_{23} \\
& && y_{12} \leq c_{12}\, r, \ y_{13} \leq c_{13}\, r, \ y_{23} \leq c_{23}\, r \\
& && x_{12} \geq 0, \ x_{132} \geq 0, \ x_{13} \geq 0, \ x_{123} \geq 0, \ x_{23} \geq 0, \ x_{213} \geq 0 \\
& && y_{12} \geq 0, \ y_{13} \geq 0, y_{23} \geq 0.
\end{aligned} \tag{4.3.12}
$$

An advantage with the above formulation is that it is an LP problem and, thus, can be solved using an LP solver such as CPLEX. Note that in transforming Eq. (4.3.8) to Eq. (4.3.12), we have introduced an additional variable r and three additional constraints as given by Eq. (4.3.11).

Example 4.4 *Solution for the load balancing problem.*

We again considering demand volumes to be $h_{12} = 5$, $h_{13} = 10$, and $h_{23} = 7$, and capacities to be $c_{12} = 10$, $c_{13} = 10$, and $c_{23} = 15$. Using CPLEX on formulation (4.3.12), we obtain the optimal solution as

$$x^*_{12} = 5, x^*_{13} = 7.5, x^*_{123} = 2.5, x^*_{23} = 7$$

and the maximum link utilization at optimality is $r^* = 0.75$. ▲

Remark 4.1. *Dropping capacity constraints from Eq. (4.3.8) and Eq. (4.3.12).*

In formulations (4.3.8) and (4.3.12), constraints $y_{12} \leq c_{12}$, $y_{13} \leq c_{13}$, and $y_{23} \leq c_{23}$ can be dropped from the formulation without changing the problem. If for the optimal solution the maximum link utilization is found to be greater than 1, this automatically implies that the network does not have enough capacity to carry all the demand volume. ◆

4.3.3 Average Delay

We now consider the problem of minimizing the average delay. To do so, we will use link flow variables, y_{12}, y_{13}, and y_{23}, that we have introduced in the previous section. The average delay function (see Appendix B.13) is given by

$$f(y_1, y_2, y_3) = \frac{1}{h_{12} + h_{13} + h_{23}} \left(\frac{y_{12}}{c_{12} - y_{12}} + \frac{y_{13}}{c_{13} - y_{13}} + \frac{y_{23}}{c_{23} - y_{23}} \right). \tag{4.3.13}$$

Since the total external offered load, $h_{12} + h_{13} + h_{23}$, is a constant it can be ignored from the objective function. Similar to Eq. (4.3.8), this time with the average delay function, we can write the formulation as follows:

$$\textbf{minimize}_{\{x,y\}} \quad F = \left(\frac{y_{12}}{c_{12}-y_{12}} + \frac{y_{13}}{c_{13}-y_{13}} + \frac{y_{23}}{c_{23}-y_{23}}\right)$$

$$\textbf{subject to} \qquad x_{12} + x_{132} = h_{12}$$
$$x_{13} + x_{123} = h_{13}$$
$$x_{23} + x_{213} = h_{23}$$
$$x_{12} + x_{123} + x_{213} = y_{12} \tag{4.3.14}$$
$$x_{13} + x_{132} + x_{213} = y_{13}$$
$$x_{23} + x_{132} + x_{123} = y_{23}$$
$$y_{12} \le c_{12}, \; y_{13} \le c_{13}, \; y_{23} \le c_{23}$$
$$x_{12} \ge 0, \; x_{132} \ge 0, \; x_{13} \ge 0, \; x_{123} \ge 0, \; x_{23} \ge 0, \; x_{213} \ge 0$$
$$y_{12} \ge 0, \; y_{13} \ge 0, \; y_{23} \ge 0.$$

We note that the objective function is undefined whenever the link load for any of the links equals the capacity of that link, a situation that certainly is not desirable when solving the above problem. Thus, in developing an algorithm to solve this formulation, a slightly modified version is used where the link is restricted below the capacity by a small positive quantity, say, ε; that is, a capacity constraint such as $y_{12} \le c_{12}$ is replaced by $y_{12} \le c_{12} - \varepsilon$; this is similar for the other two links. Note that we discussed this adjustment earlier for the single-commodity network flow problem.

It is important to note that the objective function is convex when the link flow is less than capacity. However, unlike the maximum link utilization case, this function is *not* piecewise linear; rather, it is highly nonlinear as the load approaches the capacity. To solve this nonlinear optimization problem, approaches such as the flow deviation algorithm can be used; for example, see [564, Chapter 5]. We will now describe another approach that considers a piecewise linear approximation of the objective function.

First observe that the objective function includes the functional form

$$f(y) = \frac{y}{c-y}, \quad \text{for } 0 \le y/c < 1. \tag{4.3.15}$$

This function, scaled by c, i.e., $cf(y)$ can be approximated by the following piecewise linear function by matching the function at several points on the curve such as $\frac{y}{c} = 0, \frac{1}{3}, \frac{2}{3}$, and so on:

$$\tilde{f}(y) = \begin{cases} \frac{3}{2}y, & \text{for } 0 \le \frac{y}{c} \le \frac{1}{3} \\ \frac{9}{2}y - c, & \text{for } \frac{1}{3} \le \frac{y}{c} < \frac{2}{3} \\ 15y - 8c, & \text{for } \frac{2}{3} \le \frac{y}{c} < \frac{4}{5} \\ 50y - 36c, & \text{for } \frac{4}{5} \le \frac{y}{c} < \frac{9}{10} \\ 200y - 171c, & \text{for } \frac{9}{10} \le \frac{y}{c} < \frac{19}{20} \\ 4000y - 3781c, & \text{for } \frac{y}{c} \ge \frac{19}{20}. \end{cases} \tag{4.3.16}$$

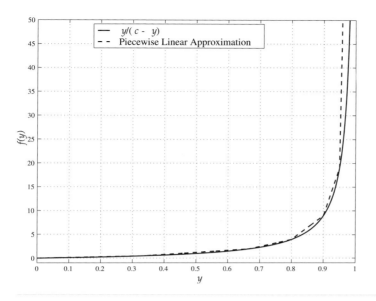

FIGURE 4.8   Piecewise linear approximation function (4.3.16) of $y/(c - y)$ (when $c = 1$).

In Figure 4.8, we have plotted both Eq. (4.3.15) and Eq. (4.3.16) when $c = 1$; it is easy to see that the match is very close. Thus, we can say that $cf(y) \approx \tilde{f}(y)$. The approximate function, (4.3.16), can also be written as the maximum of each of the linear pieces as follows:

$$\tilde{f}(y) = \max\left\{\tfrac{3}{2}y,\ \tfrac{9}{2}y - c,\ 15y - 8c,\ 50y - 36c,\ 200y - 171c,\right.$$
$$\left. 4000y - 3781c\right\}, \quad \text{for } y \geq 0. \tag{4.3.17}$$

Recall our approach to transform the minimization of max function given by Eq. (4.3.10) in the previous section to an equivalent problem with constraints. We can do the same transformation here, and write the minimization of Eq. (4.3.17) in the following form:

$$
\begin{aligned}
&\textbf{\textit{minimize}}_{\{r\}} && r \\
&\textbf{\textit{subject to}} && r \geq \tfrac{3}{2}y \\
& && r \geq \tfrac{9}{2}y - c \\
& && r \geq 15y - 8c \\
& && r \geq 50y - 36c \\
& && r \geq 200y - 171c \\
& && r \geq 4000y - 3781c \\
& && y \geq 0.
\end{aligned}
\tag{4.3.18}
$$

Now consider again Formulation (4.3.14); the objective function has three parts, one for each link, and each of the exact same form $y/(c - y)$. Thus, for each part, we can use the piecewise linear approximation as we have described above, and introduce variables r_{12}, r_{13}, and r_{23} and the related constraints. Using this approximation, and accounting for the scaling factor

c_{12}, c_{13}, and c_{23} in the approximation, we can write an approximate equivalent problem of Eq. (4.3.14) as follows:

$$\text{\textbf{\textit{minimize}}}_{\{x,y,r\}} \quad F = \frac{1}{c_{12}}r_{12} + \frac{1}{c_{13}}r_{13} + \frac{1}{c_{23}}r_{23}$$

$$\textbf{\textit{subject to}} \quad x_{12} + x_{132} = h_{12}$$

$$x_{13} + x_{123} = h_{13}$$

$$x_{23} + x_{213} = h_{23}$$

$$x_{12} + x_{123} + x_{213} = y_{12}$$

$$x_{13} + x_{132} + x_{213} = y_{13}$$

$$x_{23} + x_{132} + x_{123} = y_{23}$$

$$r_{ij} \geq \tfrac{3}{2} y_{ij}, \qquad\qquad\qquad (i,j) = (1,2), (1,3), (2,3)$$

$$r_{ij} \geq \tfrac{9}{2} y_{ij} - c_{ij}, \qquad\qquad (i,j) = (1,2), (1,3), (2,3) \qquad (4.3.19)$$

$$r_{ij} \geq 15 y_{ij} - 8c_{ij}, \qquad\qquad (i,j) = (1,2), (1,3), (2,3)$$

$$r_{ij} \geq 50 y_{ij} - 36c_{ij}, \qquad\qquad (i,j) = (1,2), (1,3), (2,3)$$

$$r_{ij} \geq 200 y_{ij} - 171c_{ij}, \qquad\quad (i,j) = (1,2), (1,3), (2,3)$$

$$r_{ij} \geq 4000 y_{ij} - 3781c_{ij}, \qquad (i,j) = (1,2), (1,3), (2,3)$$

$$x_{12} \geq 0, \ x_{132} \geq 0, \ x_{13} \geq 0$$

$$x_{123} \geq 0, x_{23} \geq 0, \ x_{213} \geq 0$$

$$y_{12} \geq 0, \ y_{13} \geq 0, \ y_{23} \geq 0$$

$$r_{12} \geq 0, \ r_{13} \geq 0, \ r_{23} \geq 0.$$

The good news about the above formulation is that it is an LP problem; as such, we can use tools such as CPLEX.

Example 4.5 *Minimization based on a piecewise linear approximation for the delay function.*

Again consider demand volumes to be $h_{12} = 5$, $h_{13} = 10$, and $h_{23} = 7$, and capacities to be $c_{12} = 10$, $c_{13} = 10$, and $c_{23} = 15$. For Formulation (4.3.19) using CPLEX, we obtain optimal flows as: $x_{12}^* = 5$, $x_{13}^* = 8$, $x_{123}^* = 2$, and $x_{23}^* = 7$.　▲

Remark 4.2. *On Formulation (4.3.19).*

First note that similar to the previous section, capacity constraints $y_{12} \leq c_{12}$, $y_{13} \leq c_{13}$, and $y_{23} \leq c_{23}$ can be ignored from this formulation since a high value of r_{ij} would indicate whether the capacity limit is exceeded. Second, due to piecewise linear approximation, we do not need to worry about issues such as the discontinuity of the nonlinear function, (4.3.15), at the capacity limit.　♦

4.4 Multicommodity Network Flow Problem: General Formulation

In previous sections, we considered a three-node network example, first for the single-commodity network flow problem and then for the multicommodity network flow problem. In this section, we generalize the multicommodity flow problem for an arbitrary size network. First we establish a set of notations that works well for representing a general problem.

4.4.1 Background on Notation

Consider a network with N nodes that has L number of links. With N nodes in a network, there are $N(N-1)/2$ demand pairs possible when we consider bidirectional demands, or $N(N-1)$ demand pairs possible when we consider unidirectional demands. In a practical network, not all nodes in the network are sources or sinks for external traffic; there are often transit nodes used solely for the purpose of routing; thus, it makes sense to say that we consider demand pairs that have *positive* demand volume. Thus, to avoid listing which node pairs have demand and which do not, a simple way to list them is by using a separate index, i.e., $k = 1, 2, \ldots, K$, where K is the total number of demand pairs having positive demand volumes. This means that from a pair of demand nodes i, j a mapping must be maintained to index k, i.e., $i : j \rightarrow k$.

We illustrate the new labeling/indexing though a three-node network example (see Figure 4.4). In this network (with nodes labeled 1, 2, 3) consider that there are positive demand volumes for node pairs 1 and 2, and 2 and 3, but not between 1 and 3. We can index the demand pairs as follows:

Pair	Index
1:2	1
2:3	2

with $K = 2$. However, if all demand pairs have positive demand volume we can index the demand pairs as follows:

Pair	Index
1:2	1
1:3	2
2:3	3

with $K = 3$. In other words, actual demand pair mapping to index can be different depending on the number of demands with positive demand volumes. Similarly, candidate paths can be indexed $p = 1, 2, \ldots, P_k$ for each demand identifier k, where P_k is the total number of candidate paths for demand k. Consider the three-node network with all demand pairs having positive demand volume. Each demand pair has two possible paths, one direct and the other via the third node; thus, $P_1 = 2$, $P_2 = 2$, and $P_3 = 2$. If we label the paths as $p = 1$ and $p = 2$ for each pair, we can define flow variable (unknown) for demand k and path p using the generic name x_{kp}. In the case of the three-node network, we can write

$$\begin{aligned}
x_{11} + x_{12} &= h_1 \\
x_{21} + x_{22} &= h_2 \\
x_{31} + x_{32} &= h_3.
\end{aligned}$$

$$(4.4.1)$$

Note that the above equations are the same as Eq. (4.3.1); only the subscripts are relabeled with the new indexing scheme for demand pairs and paths.

Now we consider indexing for the three links as follows:

Link	Index
1-2	1
1-3	2
2-3	3

With this indexing, the direct path for demand index 1 (i.e., for demand pair 1:2), which is the path index 1, uses link index 1; the alternate path for demand index 1, which is path index 2, uses links 2 and 3. This is similar for the other demand. Thus, we can write the capacity constraints as:

$$
\begin{aligned}
x_{11} + x_{22} + x_{32} &\le c_1 \\
x_{21} + x_{12} + x_{32} &\le c_2 \\
x_{31} + x_{12} + x_{22} &\le c_3.
\end{aligned}
\tag{4.4.2}
$$

Overall we can write the relationship between a link and a path by indicating which paths take a specific link—such a link-path match-up is marked with a 1 (and 0, otherwise). This is presented in Table 4.1.

Incidentally, the representation shown in Table 4.1 can be thought of as a matrix representation and if it is multiplied on the right with a vector of all path variables xs in the order listed, we arrive again at the left-hand side of Eq. (4.4.2); the matrix of information for the left side is shown in Table 4.1. As an example, the minimum cost routing problem described earlier in Eq. (4.3.4) can be rewritten with the new labels/indexes for demand pair, link, and paths, as follows:

$$
\begin{aligned}
&\textbf{minimize}_{\{x\}} \quad F = \xi_{11}x_{11} + \xi_{12}x_{12} + \xi_{21}x_{21} + \xi_{22}x_{22} + \xi_{31}x_{31} + \xi_{32}x_{32} \\
&\textbf{subject to} \\
&\quad x_{11} \; + x_{12} &&&&= h_1 \\
&\qquad\quad x_{21} \; + x_{22} &&&&= h_2 \\
&\qquad\qquad\qquad x_{31} \; + x_{32} &&&&= h_3 \\
&\quad x_{11} &&+ x_{22} \;\; + x_{32} &&\le c_1 \\
&\qquad x_{12} \; + x_{21} &&+ x_{32} &&\le c_2 \\
&\qquad x_{12} &&+ x_{22} \; + x_{31} &&\le c_3 \\
&\quad x_{11},\; x_{12},\; x_{21},\; x_{22},\; x_{31},\; x_{32} \ge 0.
\end{aligned}
\tag{4.4.3}
$$

Because of representation of the relationship between links and paths, the above representation is known as the *link-path* representation or formulation.

4.4.2 Link-Path Formulation

We are ready to consider the general case. Consider an N-node network with K demand pairs having positive demand volumes, each with $P_k (k = 1, 2, \ldots, K)$ candidate paths. Thus, extending from two paths for the three-node network given in Eq. (4.4.1) to the number of paths P_k, we can write

$$
x_{k1} + x_{k2} + \cdots + x_{kP_k} = h_k, \quad k = 1, 2, \ldots, K.
\tag{4.4.4}
$$

TABLE 4.1   Link-path incidence information.

Link\Path	$k=1, p=1$	$k=1, p=2$	$k=2, p=1$	$k=2, p=2$	$k=3, p=1$	$k=3, p=2$
1	1	0	0	1	0	1
2	0	1	1	0	0	1
3	0	1	0	1	1	0

Using summation notation, the above can be rewritten as

$$\sum_{p=1}^{P_k} x_{kp} = h_k, \quad k = 1, 2, \ldots, K. \tag{4.4.5}$$

For capacity constraints, we introduce a new notation, called the δ notation, which parallels the information presented in Table 4.1. Define

$\delta_{kp\ell} = 1$ if path p for demand pair k uses the link ℓ; 0, otherwise.

Now consider again Table 4.1; the information that is reflected in the table is in fact nothing but $\delta_{kp\ell}$. For example, when $k = 1$, $p = 1$ and we consider link $\ell = 1$, then we are referring to the direct link path for demand pair 1:2 and whether it uses the direct link 1-2 ; since it does, $\delta_{111} = 1$ as we can see from the table. On the other hand, this same path does not use 1-3 (link identifier 2), which means by definition of the δ notation that δ_{121} should be zero, which is reflected also in Table 4.1. If we work though a few more examples like these two, we can see that the δ notation allows us to reflect all capacity constraints, specifically the left-hand side. In general, consider first a specific demand pair k; for a specific link ℓ, the term

$$\sum_{p=1}^{P_k} \delta_{kp\ell} x_{kp}$$

indicates that we are to add flows for only the paths that use link ℓ. Now, if we do summation over all $k = 1, 2, \ldots, K$, i.e.,

$$\sum_{k=1}^{K} \sum_{p=1}^{P_k} \delta_{kp\ell} x_{kp},$$

then this term represents the summation of all paths for all demand pairs that use link ℓ, which is then the link flow for link ℓ; the link flow will be denoted by the dependent variables y_ℓ also, i.e.,

$$\sum_{k=1}^{K} \sum_{p=1}^{P_k} \delta_{kp\ell} x_{kp} = y_\ell, \quad \ell = 1, 2, \ldots, L.$$

Since the link flow must be less than or equal to the capacity of the link, then we can write

TABLE 4.2   Notation used in the link-path formulation.

Notation	Explanation
Given:	
K	Number of demand pairs with positive demand volume
L	Number of links
h_k	Demand volume of demand index $k = 1, 2, \ldots, K$
c_ℓ	Capacity of link $\ell = 1, 2, \ldots, L$
P_k	Number of candidate paths for demand k, $k = 1, 2, \ldots, K$
$\delta_{kp\ell}$	Link-path indicator, set to 1 if path p for demand pair k uses the link ℓ; 0, otherwise
ξ_{kp}	Nonnegative unit cost of flow on path p for demand k
Variables:	
x_{kp}	Flow amount on path p for demand k
y_ℓ	Link-flow variable for link ℓ

$$y_\ell = \sum_{k=1}^{K} \sum_{p=1}^{P_k} \delta_{kp\ell} x_{kp} \leq c_\ell, \quad \ell = 1, 2, \ldots, L. \tag{4.4.6}$$

It is easy to see that y_ℓ is really a dependent variable, and is used here for convenience; it can be dropped without affecting the overall model. Here we will discuss the general formulation for the minimum cost routing problem. Thus, if ξ_{kp} is the unit cost of path p for demand pair k, then we can write

$$\text{Total cost} = \sum_{k=1}^{K} \sum_{p=1}^{P_k} \xi_{kp} x_{kp}.$$

Thus, the general formulation for the minimum cost routing problem can be written as

$$\textbf{\textit{minimize}}_{\{x\}} \quad F = \sum_{k=1}^{K} \sum_{p=1}^{P_k} \xi_{kp} x_{kp}$$

subject to

$$\sum_{p=1}^{P_k} x_{kp} = h_k, \qquad k = 1, 2, \ldots, K \tag{4.4.7}$$

$$\sum_{k=1}^{K} \sum_{p=1}^{P_k} \delta_{kp\ell} x_{kp} \leq c_\ell, \quad \ell = 1, 2, \ldots, L$$

$$x_{kp} \geq 0, \qquad p = 1, 2, \ldots, P_k, \; k = 1, 2, \ldots, K.$$

We now discuss the size of the problem. For an undirected network with N nodes, there are $N(N-1)/2$ demand pairs, which is also the number of demand constraints if all demand pairs have positive demand volume; for L links, there are L capacity constraints. If on average $\overline{P}_k$ candidate paths are considered for each demand pair k, then the number of flow variables

TABLE 4.3   Size of minimum cost routing problem (undirected network).

N	L	$\overline{P}_k$ (average)	Variables	Constraints	
				Demand	Capacity
5	7	4	40	10	7
10	30	7	315	45	30
50	200	10	12,250	1225	200

is $N(N-1)\overline{P}_k/2$. In Table 4.3, we list the size of Problem (4.4.7) for several values of N, L, and $\overline{P}_k$. It is clear that the problem size grows significantly as the network size grows. Still such a formulation can be solved by tools such as CPLEX for problems of reasonable size. Certainly, specialized algorithms can be used as well; for example, see [564] for a survey of different specialized methods.

Remark 4.3. *Generating candidate paths for Eq. (4.4.7).*

It may be noted that any link-path formulation assumes that a set of candidate paths for each demand pair is available as an input. For many networks, the network administrators have a good sense about what candidate paths to consider–thus such paths are not too difficult to determine. Certainly, a *k-shortest path algorithm* (refer Section 2.8) can be used to generate a set of candidate paths, which can then be used as input to the above models. Since the purpose of candidate paths is to serve as a feeder to the link-path formulation, the *k*-shortest paths can be generated using such an algorithm by using just the hopcount to reflect cost on a link.   ◆

The above remark, however, does not indicate how many candidate paths to generate for each demand pair. The following important result gives us a very good clue, instead.

Result 4.1. *If Eq. (4.4.7) is feasible, then at most $K + L$ flow variables are required to be nonzero at optimality.*

To see the above result, consider just the demand and capacity constraints of Problem (4.4.7). Note that there are K demand constraints that are equations and L capacity constraints that are inequalities. For each capacity constraint, we add a nonnegative variable s_ℓ (called the *slack* variable) to convert it to an equation. Thus, we can write the following system of equations for demand and capacity constraints of Problem (4.4.7):

$$\sum_{p=1}^{P_k} x_{kp} = h_k, \qquad\qquad k = 1, 2, \ldots, K$$

$$\sum_{k=1}^{K}\sum_{p=1}^{P_k} \delta_{kp\ell}\, x_{kp} + s_\ell = c_\ell, \qquad \ell = 1, 2, \ldots, L. \tag{4.4.8}$$

Linear programming theory (for example, see Section 1.2 in [397]) says that the number of nonzero variables in any *basic feasible solution*, a requirement at optimality, is *at most* require to be the number of equations. Here, we can easily see that there are $K + L$ equations; thus, at most $K + L$ of the flow variables x_{kp} are required to be nonzero at optimality. It should be noted that Result 4.1 is true only when the objective function is linear.

Example 4.6 *Meaning of Result 4.1.*

Consider the case of $N = 50$ nodes from Table 4.3. Since this network has 200 links and all 1225 demand pairs are assumed to have positive demand volume, Result 4.1 implies that at most 1425 out of 12,250 flow variables will need to be nonzero at optimality. More importantly, even if we were to increase the number of candidate paths for each pair from 10 to 20, which increases the total number of flow variables to 24,450, the requirement for the number of nonzero flows at optimality remains at 1425. Note that every demand pair must have *at least* one nonzero flow due to demand constraints. This means that at most 200 demand pairs will need to have two or more paths with nonzero flows and one path with nonzero flow for at least the remaining 1025 ($= 1225 - 200$) demand pairs.   ▲

Remark 4.4. *Common sense on candidate path generation and relation to optimal solution.*

A general observation from the above discussion is that for a large network, only a very few paths for each demand pair will have nonzero flows at optimality, while most demand pairs will have only a single path with nonzero flow at optimality. The difficulty ahead of time is *not* knowing which demand pairs will have multiple paths with nonzero flows at optimality; in other words, we do not know exactly which candidate paths to consider as input to the model. Since for most communication networks in practice, network designers and administrators have a reasonable idea on what paths are likely to be on the optimal solution, the candidate path consideration can be tailored as needed. However, there is still a possibility that for a specific network scenario a path that could have been in the optimal solution was not even included in the set of candidate paths; thus, even after solving a model such as Eq. (4.4.7), we might not have arrived at the "ultimate" optimality. This is where we need to keep in mind another important aspect of real-world communication networks; the demand volume used in a multicommodity flow formulation is often an estimate based either on measurements and/or forecast; some error in the demand volume estimate cannot be completely ruled out. Thus, any attempt to find "ultimate" optimality based on demand volume information that has a certain error margin to begin with can be best construed as defying common sense. Thus, a candidate path set with a reasonable number of paths per demand pair suffices in practice. In our experience, we have found that for most moderate to large networks, considering/generating 5–10 candidate paths per demand pair is often sufficient.   ◆

We next consider the multicommodity network flow problem where the goal is to minimize the maximum link utilization, i.e., the load balancing case. We will write this using the additional variable r along with dependent link-flow variables y_ℓ. Thus, similar to Eq. (4.3.11) for the three-node case, we can write the constraints

$$y_\ell \le c_\ell r, \quad \ell = 1, 2, \ldots, L, \tag{4.4.9}$$

which captures the maximum link utilization factor. Overall, we can write the generation formulation for minimizing maximum link utilization as the following linear programming problem:

minimize$_{\{x,y,r\}}$ $F = r$
subject to

$$\sum_{p=1}^{P_k} x_{kp} = h_k, \qquad\qquad k = 1, 2, \ldots, K$$

$$\sum_{k=1}^{K}\sum_{p=1}^{P_k} \delta_{kp\ell}\, x_{kp} = y_\ell, \qquad \ell = 1, 2, \ldots, L \qquad\qquad\qquad (4.4.10)$$

$$y_\ell \leq c_\ell\, r, \qquad\qquad \ell = 1, 2, \ldots, L$$

$$x_{kp} \geq 0, \qquad\qquad p = 1, 2, \ldots, P_k, \ \ k = 1, 2, \ldots, K$$

$$y_\ell \geq 0, \qquad\qquad \ell = 1, 2, \ldots, L$$

$$r \geq 0.$$

Remark 4.5. *Revisiting Result 4.1 for Eq. (4.4.10).*

Note that Result 4.1 would come out somewhat different for Eq. (4.4.10). An important point about Result 4.1 is that it is driven by the constraints of a problem, other than nonnegative constraints. From Eq. (4.4.10), we can combine the first three sets of constraints into the following two sets of constraints by eliminating ys:

$$\sum_{p=1}^{P_k} x_{kp} = h_k, \qquad\qquad k = 1, 2, \ldots, K$$

$$\sum_{k=1}^{K}\sum_{p=1}^{P_k} \delta_{kp\ell}\, x_{kp} \leq c_\ell\, r, \qquad \ell = 1, 2, \ldots, L. \qquad\qquad\qquad (4.4.11)$$

Thus, we still have $K + L$ equations when we transform the second set to equality by adding slack variables. In this case, at most $K + L - 1$ flow variables need to be nonzero at optimality since variable r must be positive at optimality; thus, together, they total $K + L$ nonzero variables.   ♦

For the case of minimizing the average link delay, we can take a similar approach to formulate the general model, especially when using the piecewise linear approximation of the load-latency function as given earlier in Eq. (4.3.16). This formulation as well as how many flows need to be nonzero at optimality is left as an exercise.

4.4.3 Node-Link Formulation

In this section, we present another approach for representing a multicommodity network flow problem that is based on node-link representation. We will illustrate it for the minimization of maximum link utilization problem; the new notation is summarized in Table 4.4.

The idea here is that instead of taking the notion of paths, the point of view is taken from a node. For any demand volume for a demand pair, a node is either a source node (that is the demand starting point) or a destination node (that is the demand termination point), or an intermediary node where any flow that enters for this demand pair through one link must go out through another link to maintain conservation of flows. While it may not be apparent, the node-link formulation is inherently described for a directed network (with directed links). Somewhat similar to the δ notation for the link-path formulation, we introduce a pair of (a, b) notation for node-link formulation (see Table 4.4).

Regarding variables, they are considered to be flow variables in terms of links (*not* paths) for every demand pair. Thus, if we define $z_{\ell k}$ as the amount of flow on link ℓ for demand pair identifier k, then the flow conservation requirement for k ($k = 1, 2, \ldots, K$) can be written as follows:

$$\sum_{\ell=1}^{L} a_{v\ell} z_{\ell k} - \sum_{\ell=1}^{L} b_{v\ell} z_{\ell k} = \begin{cases} h_k, & \text{if } v = s_k \\ 0, & \text{if } v \neq s_k, t_k, \quad v = 1, 2, \ldots, N \\ -h_k, & \text{if } v = t_k. \end{cases}$$

The relation between $z_{\ell k}$ and y_ℓ can be stated as follows:

$$\sum_{k=1}^{K} z_{\ell k} = y_\ell, \quad \ell = 1, 2, \ldots, K.$$

The minimization of the maximum link utilization problem can be represented in the node-link representation by the following linear programming problem:

minimize$_{\{z,y,r\}}$　$F = r$
subject to

$$\sum_{\ell=1}^{L} a_{v\ell} z_{\ell k} - \sum_{\ell=1}^{L} b_{v\ell} z_{\ell k} = \begin{cases} h_k, & \text{if } v = s_k \\ 0, & \text{if } v \neq s_k, t_k, \quad v = 1, 2, \ldots, V \\ -h_k, & \text{if } v = t_k, \end{cases}$$
$$k = 1, 2, \ldots, K$$

$$\sum_{k=1}^{K} z_{\ell k} = y_\ell, \qquad \ell = 1, 2, \ldots, K$$

$$y_\ell \leq c_\ell r, \qquad \ell = 1, 2, \ldots, K$$
$$z_{\ell k} \geq 0, \qquad \ell = 1, 2, \ldots, L, \ k = 1, 2, \ldots, K$$
$$y_\ell \geq 0, \qquad \ell = 1, 2, \ldots, L$$
$$r \geq 0.$$

(4.4.12)

Thus, the above formulation is the counterpart of the link-path formulation presented in Eq. (4.4.10) for the minimization of maximum link utilization. An advantage of the node-link formulation is that it is not constrained by the candidate path set being an input. Thus, from a purely theoretical point of view, the node-link formulation is more general. However, once a node-link formulation is solved, it is not easy to construct the optimal paths as well as the flows (except for the single-commodity case). Knowing the paths that have positive flows is often the requirement of many network designers and administrators since this information allows them to see how the demand volume is flowing in the model and whether it is comparable to the observation from the actual network. Thus, the link-path formulation is more practical than the node-link formulation and will be primarily used in the rest of the book. We have presented here the node-link representation because this representation is often used in many scientific publications.

TABLE 4.4 Notation used in the node-link formulation.

Notation	Explanation
Given:	
N	Number of nodes (indexed by $v = 1, 2, \ldots, N$)
K	Number of demand pairs with positive demand volume
L	Number of links
h_k	Demand volume of demand identifier $k = 1, 2, \ldots, K$
s_k	Source node of demand identifier $k = 1, 2, \ldots, K$
t_k	Destination node of demand identifier $k = 1, 2, \ldots, K$
c_ℓ	Capacity of link $\ell = 1, 2, \ldots, L$
$a_{v\ell}$	Link-path indicator, set to 1 if path p for demand pair k uses the link ℓ; 0, otherwise
$b_{v\ell}$	Link-path indicator, set to 1 if path p for demand pair k uses the link ℓ; 0, otherwise
Variables:	
$z_{\ell k}$	Flow amount on link ℓ for demand k
y_ℓ	Link-flow variable for link ℓ
r	Maximum link utilization variable

4.5 Multicommodity Network Flow Problem: Nonsplittable Flow

In many instances, the demand volume between an origination-destination node pair is not allowed be split into multiple paths. In this section, we consider the case when the demand volume is nonsplittable. For ease of comparison and simplicity, we will consider the counterpart of the minimum cost routing case given by Eq. (4.4.7) in the link-path representation framework. A similar approach can be taken for other objectives.

From a modeling point of view, we need to pick only a single path out of a set of candidate paths for a demand pair. In other words, the decision to choose a path is a binary decision, however, with the additional requirement that only one of them per demand pair is to be selected. Thus, if we assign a 0/1 decision variable, u_{kp}, to path p for demand pair k, we must have the requirement that

$$\sum_{p=1}^{P_k} u_{kp} = 1, \quad k = 1, 2, \ldots, K. \tag{4.5.1}$$

You may compare this equation with Eq. (4.4.5) to note the differences. To determine the link flow on a link, we need to first identify which candidate paths are using a particular link for a particular demand pair k, i.e., $\sum_{p=1}^{P_k} \delta_{kp\ell} u_{kp}$ on link ℓ. To bring into account the demand volume h_k with this expression, we simply multiply it and obtain $\sum_{p=1}^{P_k} \delta_{kp\ell} h_k u_{kp}$. Since only one path is selected, h_k is counted only once for the path selected, although multiple candidate paths for a demand pair are potentially likely to use a particular link. Now, summing over all demand pair, and bringing the capacity constraint requirement, we can write

$$\sum_{k=1}^{K}\sum_{p=1}^{P_k}\delta_{kp\ell}h_k u_{kp} \le c_\ell, \quad \ell = 1, 2, \ldots, L \tag{4.5.2}$$

as the counterpart to Eq. (4.4.6). Then, the minimum cost routing problem with nonsplittable multicommodity flow can be written as:

$$\textbf{\textit{minimize}}_{\{u\}} \quad F = \sum_{k=1}^{K}\sum_{p=1}^{P_k}\xi_{kp}h_k u_{kp}$$

subject to

$$\sum_{p=1}^{P_k} u_{kp} = 1, \qquad\qquad k = 1, 2, \ldots, K \tag{4.5.3}$$

$$\sum_{k=1}^{K}\sum_{p=1}^{P_k}\delta_{kp\ell}\,h_k u_{kp} \le c_\ell, \qquad \ell = 1, 2, \ldots, L$$

$$u_{kp} = 0 \text{ or } 1, \qquad\qquad p = 1, 2, \ldots, P_k, \ k = 1, 2, \ldots, K.$$

It is easy to see the similarity between this formulation and Formulation (4.4.7). The main difference is that the nonsplittable flow problem is an *integer* multicommodity flow model, with the special requirement that only one path is be chosen for each demand pair. It is interesting to note that if we relax the binary requirement on u_{kp} in the above formulation and allow u_{kp}s to also take fractional values between 0 and 1 instead, the relaxed problem is equivalent to Formulation (4.4.7) since we can then write $x_{kp} = h_k u_{kp}$.

4.6 Summary

In this chapter, we introduce you to network flow modeling, especially to link-path formulation of the single and multicommodity network flow problems along with consideration of different objective functions. Such models are often used in traffic engineering of communication networks. Thus, the material here serves the purpose of introducing you to how to do abstract representations of flows and paths, especially when it comes to network modeling.

It may be noted that we have presented Formulations (4.4.7) and (4.4.10) assuming that flow variables take continuous values. In many communication networks problems, flow variables are integer-valued only or the demand volume for a demand pair is nonsplittable (refer to Formulation (4.5.3)). Similarly, objective functions other than the ones illustrated here can be developed for appropriate problems. These variations will be introduced later in the book as and when we discuss specific communication network routing problems.

Further Lookup

Kalaba and Juncosa [345] in 1956 were the first to address problems in communication networking using a multicommodity flow approach; incidently, their formulation can be considered the first node-link–based multicommodity flow representation in which the term *message* is used to generically describe communication demand. A quote from this paper is interesting: "In a system such as the Western Union System, which has some 15 regional switching centers all connected to each other, an optimal problem of this type would have about 450 conditions (constraints) and involve around 3000 variables." By referencing the Kalaba–Juncosa paper, Ford and Fulkerson [232] were perhaps the first to formulate the maximal flow multicommodity problem using the link-path representation; incidently, the origin of the "delta" notation (i.e., $\delta_{kp\ell}$) can be attributed to this work.

While we have briefly discussed solution approaches to models presented here, this chapter primarily focuses on problem formulation. We have, however, mentioned tools such as CPLEX to solve such formulations. While such tools work well, they might not be the best tools for *all* types and sizes of multicommodity network flow problems. A detailed discussion of various algorithms that might be applicable is outside the scope of this book. Thus, we direct you to the *companion* book [564] if you are interested in understanding the details of algorithmic approaches and implementation for a variety of network flow models.

Finally, for additional discussion on network flow modeling, see books such as [6], [80].

Exercises

4.1 Consider a three-node network where the nodes are denoted by 1, 2, and 3. You're given the following information:

Pair	Demand
1:2	5
1:3	9
2:3	–

Link	Capacity
1-2	10
1-3	10
2-3	5

Assume that only direct routing is allowed for pair 1:2 demand, while the other pair is allowed to split its demand volume.

(a) Formulate the minimum cost routing problem assuming that the cost of unit flow on any link is one, except 2-3 where it is zero. Determine the optimal solution.

(b) Formulate the problem of optimal load balancing (min-max) flows in the network. Determine the optimal solution.

4.2 Consider a four-node ring network where nodes are connected as follows: 1-2-3-4-1. Assume that demand volume between 1 and 3 is 25, between 2 and 4 is 30, and between 2 and 3 is 10. Capacity on each link is 50.

(a) Formulate an optimization problem in which the goal is to maximize free capacity availability. Determine the optimal flow for this objective.

(b) Formulate an optimization problem in which the goal is to load balance the network. Determine the optimal flow for this objective.

(c) What would happen when we consider either of the above objectives if the following additional requirement is added: the demand volume for each demand pair must not be split into two paths?

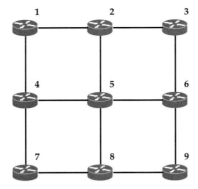

FIGURE 4.9   A nine-node Manhattan street network.

4.3   Consider a four-node network in which nodes are numbered 1, 2, 3, 4. All nodes are directly connected to each other except that there is no link between nodes 1 and 2.

Link capacities are given as follows: 30 on link 1-3, 5 on link 1-4, 15 on link 2-3, 10 on link 2-4, and 10 on link 3-4.

Demand volumes are given to be 15 for pair 1:2, 20 for pair 1:3, and 10 for pair 2:3.

(a)   Formulate the load balancing optimization problem, and determine the optimal solution.

(b)   Formulate the problem of minimizing average delay where the average delay is approximated using the piecewise linear function (4.3.16). Determine the optimal solution.

4.4   Consider the nine-node Manhattan street network in Figure 4.9.

(a)   Assume that all links have 100 units of capacity, and the goal is to load balance the network. Find optimal flows (1) if a single demand between node 1 and 9 with volume of 60 units is considered, (2) if two demands, one between 1 and 9 and another between 3 and 7, each of volume 60, considered.

(b)   Assume that all links have 100 units of capacity except for links 2-5, 4-5, 5-6, 5-8, which have 60 units of capacity. The goal is to load balance the network. Find optimal flows (1) if single demand between node 1 and 9 with volume of 60 units is considered, (2) if two demands, one each between 1 and 9 and another between 3 and 7, each of volume 60, are considered.

4.5   Consider the same demand/capacity scenarios described in Exercise 4.4. Find optimal flows if minimum cost routing is the objective used instead (assume unit cost of flow on each link).

4.6   Consider the same demand/capacity scenarios described in Exercise 4.4. Find optimal flows if a composite objective function that combines minimum cost routing with load balancing is used. Determine how the solution changes as the parameters associated with the cost components of the composite object are varied.

Part II: Routing in IP Networks

In this part, we focus on routing in IP networks. It is divided into five chapters.

In Chapter 5, we first present the basic background on IP routing in the presence of IP addressing, and how the routing table is organized and used by routers for packet forwarding. We then present protocols for Internet that falls into the distance vector protocol family. Specifically, we discuss three well-known protocols RIP, IGRP, and EIGRP. The connection is also drawn between a protocol specification such as RIP and and the basic concept of a distance vector protocol.

Chapter 6 covers OSPF and integrated IS-IS. In discussing OSPF, we point out why different types of link statement advertisements are required to cater to the needs in different operational configuration scenarios. For the integrated IS-IS protocol, we show its similarities and subtle differences with OSPF, although as of now there are no fundamental differences.

An important aspect of efficient routing in an operational network is proper traffic engineering. In Chapter 7, we show how network flow modeling can be applied to determine link weights for IP traffic engineering. In doing so, we also discuss how traffic demands are taken into account in the traffic engineering decision process.

Next, we present Border Gateway Protocol (BGP) in Chapter 8. The role of BGP in the Internet is critical as it allows exchange of reachable IP prefixes and in determinating AS-paths. There are, however, several attributes to consider in the path selection process; more importantly, policy constraints are also taken into account. Thus, many facets of BGP are covered in this chapter.

Finally, in Chapter 9, we present Internet routing architectures. This brings together how BGP is used, interaction between different domains either through public or private peering, and how points of presence are architected. Furthermore, we discuss growth in routing table entries.

5

IP Routing and Distance Vector Protocol Family

If I have seen further than others, it is by standing upon the shoulders of giants.

Isaac Newton

Reading Guideline

This chapter is geared to provide you with details of distance vector protocols RIP, IGRP, and EIGRP. Thus, each topic can be read separately. However, to understand the context of the protocols, some basics are included at the beginning about routers and networks, including addressing. A comparative summary of these protocols including a discussion on configuration complexity is provided at the end in the section titled Summary.

In this chapter, we start with the basics of IP routing. Thus, this builds on our overview discussion about IP addressing and routing, presented in Chapter 1. We then present IP distance vector routing protocol family: Routing Information Protocol (RIP), Interior Gateway Routing Protocol (IGRP), and Enhanced Interior Gateway Routing Protocol (EIGRP). In subsequent chapters, we will discuss Open Shortest Path First (OSPF) and Intermediate System to Intermediate System (IS-IS)—two link state routing protocols, followed by Border Gateway Protocol (BGP), the path vector protocol used in the Internet.

The primary goal here is to see how a specific protocol is described and its intrinsic features, including any limitations. We also discuss any specific issues that need addressing to go from the basic concept of a distance vector protocol to an actual protocol used in IP networks. To consider various protocols in an IP network, it is helpful to understand how IP addressing and routing tables are related to a router-based IP network. Furthermore, since communication of any routing information brings up the issue of whether this information is reliably delivered, we also discuss how the TCP/IP protocol stack plays a role.

The protocols described here were all originally intended for IP intradomain networks but are possible to use for interdomain routing interactions. In this regard, we also present a short discussion on route redistribution.

5.1 Routers, Networks, and Routing Information: Some Basics

In this section, we will discuss a few important points in regard to an IP network and communication of routing information. This is helpful in understanding and differentiating how a real protocol's applicability to a networking environment requires consideration of the addressing mechanism, and in considering unreliable or reliable delivery of routing information, the functionalities provided in the TCP/IP protocol stack. It is important that you are familiar with IPv4 addressing, subnetting, and CIDR, and the basics of the TCP/IP protocol stack, described earlier in Chapter 1.

5.1.1 Routing Table

A communication network connects a set of nodes through links so that traffic can move from an originating node to a destination node; for all the traffic to go to its destination, nodes in the network must provide *directions* so that the traffic goes toward the destination. To do that, each node in the network maintains a routing table so that user traffic can be forwarded by looking up the routing table to the next hop. In Chapter 3, we indicated that nodes need identifiers along with a link identifier so that those identified can be used in the routing table.

In an IP network, nodes are routers and links are often identified by interfaces at each end of routers. However, user traffic originates from host computers and goes to other host computers in the network; that is, the traffic does not originate or terminate at the router level (except routing information traffic between routers). Thus, we first need to understand what entries are listed in the routing table at an IP router if the traffic eventually terminates at a host.

To understand this, we need to refer to IP addressing and its relation to routing table entries. A routing table entry at a router can contain information at three levels: *addressable networks* (or IP prefixes, or network numbers), subnets, or directly at the host level, which is conceptually possible because the IP addressing structure allows all three levels to be specified without any change in the addressing scheme. These three levels are often referred to using the generic term *routes*. Furthermore, this also means that a router maintains entries for IP destinations, *not* to the router itself. We will now illustrate the relationship between an IP addressing and routing table through a three-node example shown in Figure 5.1. For simplicity, we consider routing table entries for addressable networks at Class C address boundaries, and thus, subnet masking is /24.

In Figure 5.1, the IP core network consists of three routers: "Alpha," "Bravo," and "Charlie;" they help movement of traffic between the following subnets: 192.168.4.0, 192.168.5.0, 192.168.6.0, and 192.168.7.0; as you can see, these are the networks attached to routers Alpha, Bravo, and Charlie, respectively. You will also notice that we use another set of IP addresses/subnets to provide interfacing between different routers; specifically, address block 192.168.1.0 between routers Alpha and Bravo, 192.168.2.0 between routers Alpha and Charlie, and 192.168.3.0 between routers Bravo and Charlie. Furthermore, each interface that connects to a router has a specific IP address; for example, IP address 192.168.1.2 is on an interface on router Bravo that router Alpha sees while IP address 192.168.3.1 is on another interface that router Charlie sees while 192.168.5.254 is on yet another interface that the addressable network 192.168.5.0 sees. We have shown the routing table at each router for all different address blocks in Table 5.1.

Now consider host "catch22" with IP address 192.168.4.22 in the network 192.168.4.0 that has an IP packet to send to host "49ers" with IP address 192.168.5.49 in network 192.168.5.0. This packet will arrive at router Alpha on the interface with IP address 192.168.4.254; through routing table lookup, router Alpha realizes that the next hop is 192.168.1.2 for network 192.168.5.0 and will forward the packet to router Bravo. On receiving this packet, router Bravo realizes that network 192.168.5.0 is directly connected and thus will send it out on interface 192.168.5.254. Now, consider an IP packet going in the reverse direction from 49ers to catch22. The packet will arrive at the interface with IP address 192.168.5.254 at router Bravo. Immediately, router Bravo realizes that for this packet, the next hop is 192.168.1.1 to forward to router Alpha. On receiving this packet, router Alpha will recognize that network 192.168.4.0 is

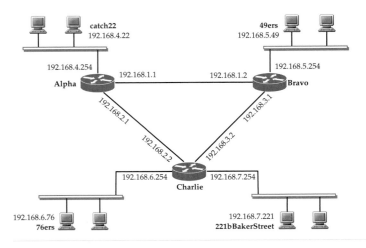

FIGURE 5.1   IP network illustration.

TABLE 5.1   Routing table at each router for the network shown in Figure 5.1.

Router: Alpha		Router: Bravo		Router: Charlie	
Network/Mask	Next Hop	Network/Mask	Next Hop	Network/Mask	Next Hop
192.168.1.0/24	direct	192.168.1.0/24	direct	192.168.1.0/24	192.168.2.1
192.168.2.0/24	direct	192.168.2.0/24	192.168.1.1	192.168.2.0/24	direct
192.168.3.0/24	192.168.1.2	192.168.3.0/24	direct	192.168.3.0/24	direct
192.168.4.0/24	direct	192.168.4.0/24	192.168.1.1	192.168.4.0/24	192.168.2.1
192.168.5.0/24	192.168.1.2	192.168.5.0/24	direct	192.168.5.0/24	192.168.3.1
192.168.6.0/24	192.168.2.2	192.168.6.0/24	192.168.3.2	192.168.6.0/24	direct
192.168.7.0/24	192.168.2.2	192.168.7.0/24	192.168.3.2	192.168.7.0/24	direct

directly connected and thus will forward it along the interface with IP address 192.168.4.254. Now let us consider what catch22 and 49ers might see based on interface addresses:

192.168.4.22 (catch22) $\longmapsto$ 192.168.4.254 (Alpha) $\longmapsto$ 192.168.1.2 (Bravo) $\longmapsto$ 192.168.5.49 (49ers)

192.168.5.49 (49ers) $\longmapsto$ 192.168.5.254 (Bravo) $\longmapsto$ 192.168.1.1 (Alpha) $\longmapsto$ 192.168.4.22 (catch22)

Thus, catch22 sees Alpha as 192.168.4.254, while Bravo sees the same router as 192.168.1.1. From an interface point of view, both are correct. How do we know that these two IP addresses "belong" to the same router? From a cursory look at IP interface addresses, there is no simple way to know this since there is going to be an address for each interface, and a router has to have at least two interfaces (otherwise, it is not routing/forwarding anything!). To avoid any confusion, a router is assigned a router ID, which is either one of the interface addresses or a different address altogether. For example, typically the interface address with the highest IP address is assigned as the address of the router. For ease of human tracking, a router with its different interfaces is typically associated with an easy to remember domain name, say Alpha.NetworkRouting.net; then, interface addresses are assigned relevant domain names such as 4net.Alpha.NetworkRouting.net and 1net.Alpha.NetworkRouting.net, so that the subnets can be easily identified and their association with a router is easy to follow.

In the above illustration, we have used a Class C address block for addressable networks. We can easily add a subnet in the routing table that is based on variable-length subnet masking (VLSM) where the subnet mask needs to be explicitly noted due to CIDR. Further more, a host can have an entry in the routing table as well. Suppose a host with IP address 192.168.8.88 is directly connected to router Charlie through a point-to-point link (not shown in Figure 5.1). If this is so, all routers will have an entry in the routing table for 192.168.8.88 (see Exercise 5.8). Usually, direct connection of hosts to a router is not advisable since this can lead to significant growth in the routing table, thus impacting packet processing and routing table lookup (see Chapter 15).

From the above illustration, you may also notice that the term *network* is used in multiple ways. Consider the following statement: user traffic moves from a network to another network that is routed through one or more routers in the IP network. Certainly, this is a confusing statement. To translate this, the first two uses of *network* refer to a network identified through an IP prefix where traffic originates or terminates at hosts, while the third use of *network* refers to a network in the topological sense where routers are nodes connected by links.

The first two uses of *network* are also referred to as *route*. Since a routing table can have an entry directly for a specific host (at least in theory), the term *route* is a good term without being explicit as to whether it is a network number or a host. For clarity and to avoid confusion, a *network* identified using an *IP prefix* will be referred to as *network number* or *addressable network*, or simply as *IP prefix*; we will also use the term *route* interchangeably. This then avoids any confusion with the generic term *network* used throughout the book.

5.1.2 Communication of Routing Information

An important aspect of the TCP/IP protocol stack is that all types of communications must take place within the same TCP/IP stack framework; that is, there are no separate networks or channels or dedicated circuits for communicating control or routing messages separately from user traffic. To accommodate different types of messages or information, the TCP/IP stack provides functionalities at both the IP layer and the transport layer; this is done differently for different routing protocols. For example, in the case of the RIP protocol, messages are communicated above the transport layer using a UDP-based port number; specifically, port number 520 is used with UDP as the transport protocol. How about other routing protocols? BGP is assigned port number 179 along with TCP as the transport protocol. However, for several routing protocols, identification is done directly above the IP layer using protocol number field; for example, protocol number 9 for IDRP protocol, 88 for EIGRP, and 89 for OSPF protocol. It may be noted that reliability of message transfer in BGP is inherently addressed since TCP is used; however, for OSPF and EIGRP, which require reliable delivery of routing information, reliability cannot be inherently guaranteed since they are directly above the IP layer; thus, for communication of routing information in OSPF, for example, through flooding, it is required that the protocol implementation ensures that communication is reliable by using acknowledgment and retransmission (if needed). In any case, while it may sound strange, all routing protocols act as applications in the TCP/IP framework where RIP and BGP are application layer protocols while OSPF and IS-IS are protocols that sit just above the IP layer. In other words, to get the routing information out for the IP layer to establish routing/forwarding of user traffic, the network relies on a higher layer protocol.

5.2 Static Routes

While routing protocols are useful to determine routes dynamically, it is sometimes desirable in an operational environment to indicate routes that remain static. These routes, referred to as *static routes*, are required to be configured manually at routers. For example, sometimes a network identified by an IP prefix is connected to only one router in another network; this happens to be the only route out to the Internet. Such a network is called a *stub network*; in such a case, a static route can be manually configured to a stub network. Static routes can also be defined when two autonomous systems must exchange routing information.

It is important to set up any static routes carefully. For example, if not careful, it is possible to get into *open jaw routes*. This terms means that there is a path defined from an origin to a destination; however, the destination takes a different path in return that does not make it back to the origin.

5.3 Routing Information Protocol, Version 1 (RIPv1)

RIP is the first routing protocol used in the TCP/IP-based network in an intradomain environment. While the RIP specification was first described in RFC 1058 in 1988 [290], it was available when RIP was packaged with 4.3 Berkeley Software Distribution (BSD) as part of *"routed"* daemon software in the early 1980s. The following passage from RFC 1058 is interesting to note: "The Routing Information Protocol (RIP) described here is loosely based on the program *routed*, distributed with the 4.3 Berkeley Software Distribution. However, there are several other implementations of what is supposed to be the same protocol. Unfortunately, these various implementations disagree in various details. The specifications here represent a combination of features taken from various implementations."

The name RIP can be deceiving since all routing protocols need to exchange "routing information." RIP should be understood as an *instance* of a distance vector protocol family, regardless of its name. It was one of the few protocols for which an implementation was available *before* a specification was officially complete. The original RIP is now referred to as RIP version 1, or RIPv1 in short. It has since evolved to RIPv2, which is standardized in RFC 2453 [442].

RIP remains one of the popular routing protocols for a small network environment. In fact, most DSL/cable modem routers such as the ones from Linksys come bundled with RIP. Thus, if you want to set up a private layer-3 IP network in your home or small office, you can do so by using multiple routers where you can invoke RIP.

5.3.1 Communication and Message Format

Since distance vector information is obtained from a neighboring router, the communication of routing information is always between two neighboring routers in the case of RIP. Furthermore, since RIP is UDP based, there is no guarantee that a routing information message is received by a receiving router. Also, no session is set up; a routing packet is just encapsulated and sent to the neighbor, normally through broadcast. Thus, we can envision a routing packet in the TCP/IP stack as shown in Figure 5.2.

Next we consider the format of a RIPv1 message; this is shown in Figure 5.3. As a commonly accepted convention in IP, the packet format for RIPv1 is shown in 32-bit (4-byte) boundaries. A RIPv1 message has a common header of 4 bytes, followed by a 20-byte message for each route for which the message is communicating, up to a maximum of 25 routes/addresses. Thus, the maximum size of a RIP message (including IP/UDP headers)

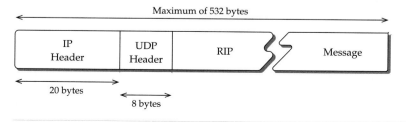

FIGURE 5.2   RIP message structure, with IP and UDP header.

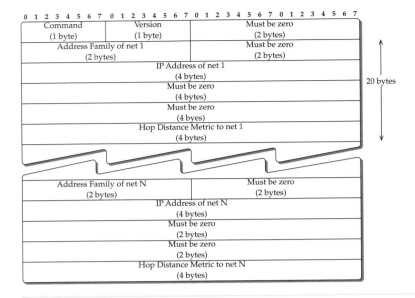

FIGURE 5.3   RIPv1 packet format.

is $20 + 8 + 4 + 25 \times 20 = 532$ bytes while the minimum is $20 + 8 + 4 + 20 = 52$ bytes. It is important to note that the message size does not limit the size of the network in terms of the number of routers; rather it is in terms of the number of addressable networks or routes. Consider again the three-router network shown in Figure 5.1 along with the routing table in Table 5.1. We can see that there are seven different addressable networks while there are three routers; thus, the routing table needs to have entries for all seven addressable networks, not in terms of routers.

It is important to note that the message size does not limit the size of the addressing networks to 25 networks (certainly not to routers); if an IP network has more than 25 addressable networks, say 40 of them, a neighbor can send distance vector information for 25 addressable networks in one message and the rest of the 15 addressable networks in another message.

Let us now look at the various fields. A common practice in many protocols is to have some spaces left out for future enhancement of the protocol; often, these spaces are marked with *Must Be Zero*. As can be seen, there are many places where this occurs in the RIPv1 message format; soon, we will see how some of them are utilized in the RIPv2 message format. Thus, a RIPv1 message has the following five fields: command, version, address family identifier, IP address, and metric. They are described below with command being discussed last:

- *Version* (1 byte): This field indicates the RIP protocol version. This is set to 1 for RIPv1. If this field happens to be zero, the message is to be ignored.

- *Address family identifier* (2 bytes): This field identifies the address family. This is set to 2 for the IP address family. Originally, the intent was to provide RIP for other address families, although in practice this RIP packet format has not been used for any other address family. There is a special use case when this field is set to zero; see command field below.

- *IP address* (4 bytes): This is the destination network, identified by a subnet or a host.

- *Metric* (4 bytes): This is based on hop count; it is a number between 1 and 16 where 16 means unreachable or infinity.

- *Command* (1 byte): This field is used for different command sets in a RIPv1 message. While there were five different commands originally defined, only two are used: *request* and *response*; the others are obsolete. The request command can be used by a router to request a neighboring router for distance vector information. If the entire routing table is desired, a request message (referred to as "request-full") is sent where the address family identifier is set to 0 and the metric to infinity; the response, however, follows a split horizon (see Section 3.3.3). However, if responses are sought for a set of address blocks (referred to as "request-partial"), the request flag is set, the address family identifier is set to IP, and the addresses are listed; the responding router sends a response to all addresses listed; no split horizon is done in this case. This is with the understanding that such a special request is not a normal request. It may be noted that the periodic distance vector update message is also sent with command set to response mode. Since there is no identification field in a RIPv1 message (unlike, say, a DNS message format), a receiving router has no direct way to determine whether the response was a periodic response or a response to its "request-full" or "request-partial."

Due to the availability of the request message type, RIP can do information pull, instead of completely relying on information push.

5.3.2 General Operation

The following are the primary operational considerations in regard to the RIP protocol:

- General packet handling: if any of the must-be-zero fields have nonzero values anywhere or if the version field is zero, the packet is discarded.

- Initialization: when a router is activated and it determines that all the interfaces are alive, and it broadcasts a request message that goes to all interfaces in the "request-full" mode. The neighboring routers handle responses following the split horizon rule. Once the responses are received, the routing table is updated with new routes the router has learned about.

- Normal routing updates: in the default case, this is done approximately every 30 sec ("*Autoupdate timer*") where updates are broadcasted with command fields set to the response mode; as discussed earlier about timer adjustment in Section 3.3.3, a large variation is added to avoid the *pendulum effect*.

- Normal response received: the routing table is updated by doing the distributed Bellman–Ford step; only a single best route is stored for each destination.

- Triggered updates: if the metric for an addressable network changes, an update message is generated containing only the affected networks.

- Route expiration: if an addressable network has not been updated for 3 min (*"expiration timer"*) in the default case, its metric is set to infinity and it is a candidate for deletion. However, it is kept in the routing table for another 60 sec; this extra time window is referred to as *garbage collection* or *flush* timer.

5.3.3 Is RIPv1 Good to Use?

In some sense, RIP has gone through the growing pains of being one of the oldest routing protocols in practice, coupled with the fact that it is a distance vector protocol that has various problems. Some key problems have been addressed through triggered update and avoiding the pendulum effect. However, it cannot avoid the looping problem and slow convergence.

In addition, RIP inherently imposes a few additional restrictions: the link cost is based only on hop count, a destination cannot be longer than 15 hops (since infinity is defined to be 16), and subnet masking is not provided. The last item deserves further elaboration. If you look at the RIPv1 message format, you will notice that it has a field for the addressable network, but no way to indicate anything specific about this network. This is partly because RIPv1 is an old protocol from the days of IP classful addressing; that is, RIPv1 *assumes* that an address included follows a Class A, Class B, Class C boundary implicitly. Subnet masking is an issue only for an address block that is not directly connected to a router. We illustrate this by considering the example network shown in Figure 5.1. Suppose that we want to connect subnet address block 172.16.1.0 to router Alpha and subnet 172.16.2.0 to router Charlie. RIPv1, however, implicitly assumes 172.16.0.0 to be a Class B address and thus cannot make the distinction; this means subnet allocation to different routers would not be routable, especially for traffic coming from a network attached to router Bravo.

From an actual operational point of view, RIPv1 is good to use in a small network environment where links are not likely to fail; this means looping is unlikely to occur. It is also good to use when link cost is not a factor, for example, a simple campus network or a small home network or a simple topology (e.g., hub-and-spoke) where the traffic may be low compared to the link speed. If a link or an interface card is likely to fail, RIPv1 faces serious transient issues including possibly creating *black hole routes*.

5.4 Routing Information Protocol, Version 2 (RIPv2)

RIPv2 [442] extends RIPv1 in several ways. Most importantly, it allows explicit masking; also, *authentication* is introduced. Authentication refers to using some mechanism to authenticate the message and/or its contents when a router receives it in such a way that it knows that the data can be trusted. To do that, changes were introduced in the RIP message format from v1 while keeping the overall format similar by taking advantage of fields previously marked as *must be zero*. This also shows why when designing a protocol, it is good to leave some room for future improvement. Thus, we start with the basic packet format as shown in Figure 5.4.

We can see from Figure 5.4 that the common header part, i.e., the first 4 bytes, is the same as in RIPv1; in this, case the version field is set to 2, and the must-be-zero field is labeled as

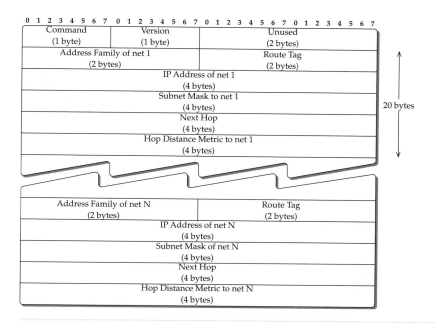

FIGURE 5.4   RIPv2 packet format.

unused while command can be either a request or a response. We now discuss the new ones beyond RIPv1:

- *Route Tag* (2 bytes): this field is provided to differentiate internal routes within a RIP routing domain from external routes. For internal routes, this field is set to zero. If a route is obtained from an external routing protocol, then an arbitrary value or preferably the autonomous system number of the external route is included here to differentiate it from internal routes.

- *Subnet mask* (4 bytes): this field allows routing based on subnet instead of doing classful routing, thus eliminating a major limitation of RIPv1. In particular, variable-length subnet masking (VLSM) may be used.

- *Next hop* (4 bytes): typically, an advertising router is the best next hop from its own view point when it lets its neighbors know about a route; at least, this is the basic assumption. However, in certain unusual circumstances, an advertising router might want to indicate a next hop that is different from itself, such as when two routing domains are connected on the same Ethernet network ([189], [441]).

Unlike RIPv1, RIPv2 allows a simple form of authentication. For the purpose of authentication, a first entry block of 20 bytes can be allocated for authentication instead of being a route entry. That is, when authentication is invoked, a RIPv2 message can contain only a maximum of 24 routes since one route table entry is used up for authentication. The address family identifier for the authentication part is tagged as 0xFFFF (i.e., all 1s, written in hexadec-

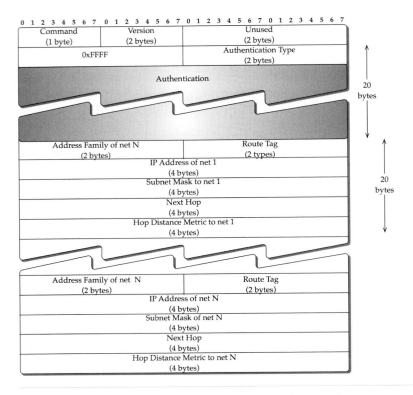

FIGURE 5.5   RIPv2 packet format with authentication.

imal notation), and the authentication type is set to 2 to indicate that it is a simple clear-text password; then the remaining 16 bytes contain the clear-text password. The packet format with authentication is shown in Figure 5.5. Certainly, a clear-text password is not a very good form of authentication. Thus, in practice, this is not used much.

From an operational consideration, RIPv2 messages are multicast on 224.0.0.9 instead of broadcast as was done in RIPv1. However, a network can be configured where routers can be on a *nonbroadcast network*; an example of a nonbroadcast network is an ATM network. Then, point-to-point unicast can be used for sending routing information. We also note that the address family identifier can now take three values: 2 for normal IP addressing, all 1s for authentication, which is done only in the first route entry after the common header, and 0 (coupled with metric set to 16) to a request message to obtain a full distance vector from a neighbor. In the common header, the unused field means that they do not need to be all zeros like RIPv1; that is, any information in this field will be ignored as opposed RIPv1's handling to discard the packet if this field contains nonzero entries.

RIPv2 has been extended for use with IPv6 addressing; this extension is known as *RIPng* [443]. They are very similar otherwise; see Table 5.2 later in the chapter for a quick comparison.

5.5 Interior Gateway Routing Protocol (IGRP)

IGRP was developed by Cisco primarily to overcome the hop count limit and hop count metric of RIPv1. In general, IGRP differs from RIPv1 in the following ways:

- IGRP runs directly over IP with protocol type field set to 9.

- Autonomous system is part of the message fields.

- Distance vector updates include five different metrics for each route, although one is not used in computing the composite metric.

- External routes can be advertised.

- It allows multiple paths for a route for the purpose of load balancing; this requires modification of the Bellman–Ford computation so that instead of a single best path to a destination, multiple "almost" equal cost paths can be stored.

   IGRP's normal routing update is sent every 90 sec on average with a variation of 10% to avoid synchronization. It has an invalid timer to indicate nonreachability of a route; this is set to three times the value of the update period. It is important to note that IGRP does not support variable length subnet masking, much like RIPv1; this is an instance in which IGRP differs from RIPv2.

5.5.1 Packet Format

IGRP packet is fairly compact consisting of 12-byte header fields followed by 14 bytes for each route entry (see Figure 5.6). The header field consists of the following fields:

- *Version* (4 bits): This field is set to 1.

- *Opcode* (4 bits): This field is equivalent to the command code in RIP. 1 is a Request and 2 is an Update. In case of a request, only the header is sent; there are no entries.

- *Edition* (1 byte): A counter that is incremented by the sender; this helps prevent a receiving router from using an old update; it essentially plays the role of a timestamp.

- *Autonomous system number* (2 bytes): ID number of an IGRP process.

- *Number of interior routes* (2 bytes): A field to indicate the number of routing entries in an update message that are subnets of a directly connected network.

- *Number of system routes* (2 bytes): This is a counterpart of the number of interior routes; this field is used to indicate the number of route entries that are not directly connected.

- *Number of exterior routes* (2 bytes): The number of route entries that are default networks. This and the previous two fields, the number of interior routes and the number of system routes, together constitute the total number of 14-byte route entries.

- *Checksum* (2 bytes): This value is calculated on the entire IGRP packet (header + entries).

   For each route entry, there are seven fields that occupy 14 bytes:

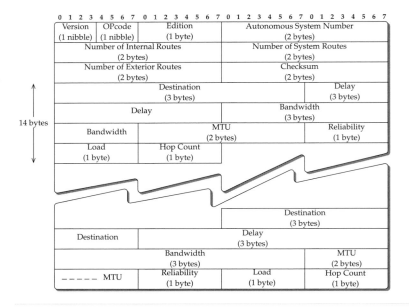

FIGURE 5.6   IGRP packet format.

- *Destination* (3 bytes): This is the destination network for which the distance vector is generated. It seems confusing to see that this field is only 3 bytes instead of the standard 4 bytes for IP addresses. However, for classful addresses, this is workable. If the update is for a system route, the first 3 bytes of the address are included; for example, if IP address of a route is 192.168.1.0, entry 192.168.1 is listed in this field. On the other hand, if it is an interior route, the last 3 bytes are listed; for example, if the field lists 16.2.0 for an interior route that is received on interface 172.16.1.254/24, it is meant for the subnet 172.16.2.0.

- *Delay* (3 bytes), *bandwidth* (3 bytes), *reliability* (1 byte), and *load* (1 byte): These fields are explained in Section 5.5.2 while discussing how the composite metric is computed.

- *Hop count* (1 byte): A number between 0 and 255 used to indicate the number of hops to the destination.

- *MTU* (2 bytes): The smallest MTU of any link along the route to the destination.

5.5.2 Computing Composite Metric

An interesting aspect of IGRP is the elaborate method it uses to compute the composite metric to represent the link cost; this was included to provide the flexibility to compute better or more accurate routes from a link cost rather than just using a hop count as a link cost as in RIPv1 or RIPv2. The composite metric in IGRP is based on four factors: bandwidth (B), delay (D), reliability (R), and load (L), along with five nonnegative coefficients (K_1, K_2, K_3, K_4, K_5) for weighing these factors. The composite metric, C ("cost of a link"), is given as follows:

$$C = \begin{cases} \left(K_1 \times B + K_2 \times \frac{B}{256-L} + K_3 \times D\right) \times \left(\frac{K_5}{R+K_4}\right), & \text{if } K_5 \neq 0 \\ K_1 \times B + K_2 \times \frac{B}{256-L} + K_3 \times D, & \text{if } K_5 = 0. \end{cases} \quad (5.5.1)$$

This composite cost metric is used in routing table computation. Here, the special case for $K_5 = 0$ means that the last part, $K_5/(R + K_4)$, which considers the reliability of a link, is not included; in other words, this means that if $K_5 = 0$, all links have the same level of reliability. In the default case, $K_1 = K_3 = 1$ and $K_2 = K_4 = K_5 = 0$. Thus, the composite metric reduces to

$$C_{\text{default}} = B + D. \quad (5.5.2)$$

This shows that the default composite metric is the summation of bandwidth and delay. Now, certainly this seems odd since bandwidth is typically given in mbps or kbps while delay is given in time unit such as sec or, millisec; that is, how do you add apples and oranges? IGRP uses a transformation process to map the raw parameters to a comparable level.

First, the raw bandwidth (B_{raw}) is expressed in kbps. Thus an Ethernet link with a data rate of 10 mbps is given the raw value 10,000. The calculated bandwidth, B, is the inverse of the raw bandwidth multiplied by the factor 10^7 to scale everything in terms of 10 gbps. That is,

$$B = \frac{10^7}{B_{\text{raw}}}. \quad (5.5.3)$$

Thus, in the case of an Ethernet link, $B = \frac{10^7}{10^4} = 1000$, and for a Fast-Ethernet link, $B = 100$. Essentially, this means that the faster the data rate of a link, the smaller the value of B is capping with $B = 1$ for a 10 Gbps link. Since 24 bits are provided for the bandwidth field, even for a 1 kbps link the value is within the range. Certainly, we do not expect any network to have a 1-kbps link anymore! In any case, the intent behind making a higher bandwidth data rate translate to a smaller value is that it takes less time to send a data packet—in other words, inverting the raw data rate allows us to think in terms of time. In some sense, this is no different than a road network in which you can drive to a place in much less time on a road with a speed limit of 120 Kmph compared to a road with a speed limit of 70 kmph.

Bandwidth or raw bandwidth assumes that the road is clear and your packet is the only packet traveling; it does not assume how much time the packet itself will take from the first bit of the packet to the last bit of the packet. Thus, the delay parameter is meant to capture the packet transmission delay on an interface, which is given in tens of μsec of raw delay, D_{raw}. That is,

$$D = D_{\text{raw}}/10. \quad (5.5.4)$$

Thus, if the raw delay is 1000 μsec, we have $D = 100$. Also, 24 bits are assigned for the delay field. Thus, for an interface running Ethernet and a delay computed to be 1000 μsec, the default composite metric value, C_{default}, is $1000 + 100 = 1100$. The default composite metric is computing a link cost that essentially reflects delay due to path delay and packet transmission delay.

Going beyond the default composite metric, consider the middle term with coefficient K_2 in the generic composite metric given in Eq. (5.5.1). This term incorporates delay "cost" due to load on a link, that is, due to traffic. For the load factor, an 8-bit field is used; thus, raw load, L_{raw}, which is a fraction between 0 and 1, can be written as

$$L_{\text{raw}} = \frac{L}{256} \tag{5.5.5}$$

so that L can take a value between 0 and 255 (inclusive) to represent link load.

The delay cost term in the middle term in Eq. (5.5.1) essentially follows the queueing delay formula modeled using an $M/M/1$ queue system (see Appendix B.12.2). If S is the average packet size and λ is the average arrival rate of packets, the average delay for an $M/M/1$ queueing system is given by

$$T = \frac{1}{\frac{B_{\text{raw}}}{S} - \lambda}. \tag{5.5.6}$$

By pulling B_{raw}/S out of the expression in the denominator, we can rewrite it as

$$T = \frac{S}{B_{\text{raw}}} \times \frac{1}{(1 - \frac{S\lambda}{B_{\text{raw}}})}.$$

However, $S\lambda/B_{\text{raw}} = L_{\text{raw}}$ is the raw load. Thus, we arrive at

$$T = \frac{S}{B_{\text{raw}}} \times \frac{1}{(1 - L_{\text{raw}})}.$$

Multiplying the numerator and denominator by 256, we get

$$T = \frac{S}{B_{\text{raw}}} \times \frac{256}{(256 - 256L_{\text{raw}})}.$$

Using relations for B and L given in Eq. (5.5.3) and Eq. (5.5.5), respectively, we can then finally write this expression as

$$T = \frac{S \times B}{10^7} \times \frac{256}{(256 - L)} = \frac{S \times 256}{10^7} \times \frac{B}{(256 - L)}. \tag{5.5.7}$$

Since the first term $S \times 256/10^7$ is a constant, we can assign a coefficient, K_2, including accounting for any proportion compared to other terms—this is then the middle term in Eq. (5.5.1).

We can see that IGRP provides an elaborate way to compute cost of a link. In practice, the default composite metric given in Eq. (5.5.2) is often used. You might notice that if the network has an interface of the same data rate, the value of the default composite metric will be the same for all links; this essentially means that network routing is working as if a hop count metric is in place much like RIP.

Next, we consider the reliability term $K_5/(R+K_4)$ in Eq. (5.5.1) and discuss possible ways to set K_4 and K_5. First observe that if a link is not fully reliable, we want the cost of the link to be higher than if it is reliable. It is given that the "base" reliability is 1 (when $K_5 = 0$); thus, for an unreliable link, we want

$$\frac{K_5}{R+K_4} > 1.$$

This means

$$K_5 > R + K_4.$$

Since the largest value of R is 255, this implies that K_5 and K_4 should be chosen such that

$$K_5 > K_4 + 255. \tag{5.5.8}$$

It is worth noting that the protocol message includes all the different metric components rather than the composite metric; in other words, the composite metric is left to a router to compute first, before doing the Bellman–Ford computation for the shortest path. This also means that it is extremely important to ensure that each router is configured with the *same* value of the coefficients K_1, K_2, K_3, K_4, K_5. For example, if in one router K_1 is set to 1 and the rest to zero, while in another router K_3 is set to 1 and the rest to zero, their view of the shortest path would be different, thus potentially causing yet another problem.

5.6 Enhanced Interior Gateway Routing Protocol (EIGRP)

EIGRP is another routing protocol from Cisco; it is, however, more than a simple enhancement of IGRP. The one thing in common between IGRP and EIGRP is the composite metric. Although EIGRP is also from the distance vector protocol family, in many ways it is completely different from protocols such as RIP and IGRP. A major difference is that EIGRP provides loop-free routing; this is accomplished through diffusing computation discussed earlier in Section 3.3.5; this also shows that not every distance vector protocol uses a straightforward Bellman–Ford computation for shortest path routing. There is an active coordination phase before routing computation when a link fails or link cost changes; to do that, additional information is sought for which the diffusing update algorithm (DUAL) needs to maintain states. DUAL allows EIGRP to attain faster convergence. In addition, EIGRP includes a hello protocol for neighbor discovery and recovery, and a reliable transfer mechanism for exchange of distance vector data.

EIGRP is provided directly over IP using protocol number 88. Furthermore, all EIGRP related message communication is multicast on the address 224.0.0.10; however, acknowledgments are unicasted. Since EIGRP requires reliable delivery, and given that the protocol is built directly over IP and multicast addressing is used, a reliable multicast mechanism is used.

5.6.1 Packet Format

The EIGRP packet is divided into two parts: an EIGRP header part, which is 20 bytes long, followed by various entities that are encoded using a variable-length TLV (Type-Length-Value)

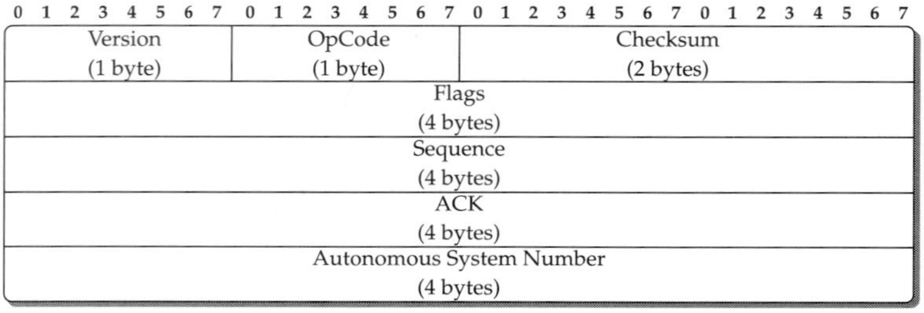

```
 0  1  2  3  4  5  6  7  0  1  2  3  4  5  6  7  0  1  2  3  4  5  6  7  0  1  2  3  4  5  6  7
```

Version (1 byte)	OpCode (1 byte)	Checksum (2 bytes)	
Flags (4 bytes)			
Sequence (4 bytes)			
ACK (4 bytes)			
Autonomous System Number (4 bytes)			

FIGURE 5.7　EIGRP packet header.

format (refer to Section 1.13.1). In the EIGRP header, there are seven fields (see Figure 5.7), which are described below:

- *Version* (1 byte): This field is set to 1.

- *OpCode* (1 byte): This field is used to specify the EIGRP packet type. There are four key types for IP networks: update, query, reply, and hello. Note that the need for these fields has been already discussed in Section 3.3.5.

- *Checksum* (2 bytes): Checksum is calculated over the entire EIGRP packet.

- *Flags:* If this value is 1, it indicates a new neighbor relationship. This value is set to 2 to indicate a conditional receive bit for a propriety multicast algorithm Cisco implements for reliable delivery using the multicast address 224.0.0.10.

- *Sequence:* This is a 32-bit sequence number used by the reliable delivery mechanism.

- *ACK:* This field lists the sequence number from the last heard from neighbor. For an initial hello packet, this field is set to zero. A hello packet type with a nonzero ACK value is an acknowledgment to an initial hello message. An important distinction is that acknowledgment is sent as a unicast message; this ACK field is nonzero only for unicast.

- *Autonomous system number:* This identifies the EIGRP domain.

Beyond the header, different entities are separated using the TLV format in an EIGRP packet (see Figure 5.8). Each TLV entity is of variable length where the type field is fixed at 1 byte, the length field is fixed at 1 byte, while the value field is of variable length; the length of the value field is indicated through the length field. Most importantly, through the type field, the packet type is identified; this field is not to be confused with the OpCode in the header field used for message type. Cisco has defined abilities to do different types such as general information, or network-specific information, such as whether the packet is for IP or other networks (e.g., IPX, developed by Novell NetWare, which many organizations deployed).

In our discussion, we specifically consider two types that are relevant and important: (1) EIGRP parameters and (2) IP internal routes. The type field is set with value 0x0001 for an EIGRP parameter description in which the information content includes coefficients K_1, K_2, K_3, K_4, and K_5, which are used in the calculation of the composite cost metric (see

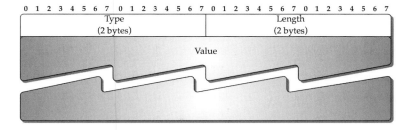

FIGURE 5.8 Data encoding in EIGRP packet: Generic TLV format.

Figure 5.9). Thus, unlike IGRP, EIGRP allows coefficients used by a router to be communicated to its neighboring routers. Despite that, a router has no way of knowing if the same or all coefficient values are used by *all* routers *internally* in their computation and in determining the shortest paths. Thus, inconsistency in computing the route, if different coefficient values are used by different routers, cannot be completely ruled out.

For the distance vector packet type for internal routes for IP networks, the type field is set to 0x0102; this type is for the route table entry having fields similar to the fields used in IGRP (compare Figure 5.10 with Figure 5.6). Thus, let us consider only the key differences between EIGRP and the other protocols. A next hop field is included in EIGRP much like RIPv2; this is not done in IGRP. Delay and bandwidth fields are 4 bytes long in EIGRP instead of 3 bytes in

```
0 1 2 3 4 5 6 7 0 1 2 3 4 5 6 7 0 1 2 3 4 5 6 7 0 1 2 3 4 5 6 7
```

Type = 0x0001		Length (2 bytes)	
K1 (1 byte)	K2 (1 byte)	K3 (1 byte)	K4 (1 byte)
K5 (1 byte)	Reserved (1 byte)	Hold Time (2 bytes)	

FIGURE 5.9 EIGRP: TLV type for EIGRP parameters.

```
0 1 2 3 4 5 6 7 0 1 2 3 4 5 6 7 0 1 2 3 4 5 6 7 0 1 2 3 4 5 6 7
```

Type = 0x0102		Length (2 bytes)	
Next Hop (4 bytes)			
Delay (4 bytes)			
Bandwidth (4 bytes)			
MTU (3 bytes)			Hop Count (1 byte)
Reliability (1 byte)	Load (1 byte)	Reserved (2 bytes)	
Prefix Length (1 byte)	Destination (3 or 4 bytes)		

FIGURE 5.10 EIGRP: TLV type for communicating distance vector of an internal route.

IGRP since EIGRP uses a 256 multiplier for a finer metric granularity than IGRP; thus, if the composite metric as given in Eq. (5.5.1) for IGRP is denoted by C_{IGRP}, the composite metric, C_{EIGRP}, for EIGRP can be written as follows:

$$C_{EIGRP} = 256 \times C_{IGRP}. \tag{5.6.1}$$

Through the combination of PrefixLength and Destination fields, variable-length subnet masking is communicated. For example, if an addressable network is 10.1.0.0/16, PrefixLength is 16 and Destination field will contain 10.1. If the addressable network is 167.168.1.128/25, PrefixLength will be 25 and the Destination field will be set to 167.168.1.128.

5.7 Route Redistribution

Often in practice, we face the situation of connecting two networks where each network speaks a different routing protocol. Then the question is: how does one network learn about the routes (IP prefixes) of the other network, and vice versa? The benefit is that when one network learns about IP prefixes in the other network, it can forward any user traffic to addresses in the other network. An important way to learn about IP prefixes in other networks is through Border Gateway Protocol (BGP)—this will be covered in detail in Chapter 8. However, BGP is not the only way to learn about routes. It is possible to learn about routes, for example, if one network uses RIPv2 and the other network uses IGRP, without relying on BGP.

To learn about routes (IP prefixes), a router at the boundary that is connected to both networks is required to perform *route redistribution*; this means that this router redistributes routes it has learned from the first network to the second network using the routing protocol used by the second network, and vice versa. Suppose that one network is running IGRP and the other is running RIPv2. Then the boundary router is configured to operate both IGRP and RIPv2. To let one network know about routes learned from the other network, protocols must provide functionalities to indicate routes so learned. Suppose that a boundary router has learned about an IP prefix from its IGRP side; it can use the RouteTag field in RIPv2 to tag that this route has been learned from another protocol/mechanism and let the routers in the RIPv2 side know. Similarly, if a boundary router learns a route from RIPv2 and wants to announce to the IGRP side, the number of external routes in the IGRP packet format must be positive and the route would be announced. In IGRP, internal and system routes are listed first; thus, it is easy to identify if a route is an external route. Note that route redistribution is often used for static routes learned.

Besides the capability of a protocol to advertise external routes, an important issue is metric compatibility. For example, RIPv2 uses a hop-based metric while IGRP uses a composite metric, while a static route has no metric value. Thus, the boundary router is required to somehow translate the metric value from one protocol to the other protocol. Since there is not really a direct translation, an externally learnt route is assigned an administrative distance instead; this is helpful if a route is learned from two different ways so that the most preferred route can be selected. Such administrative distances can be based on how much you can *trust* a routing protocol; for example, since EIGRP is a loop-free protocol, it is better to give a lower administrative distance for a route from EIGRP (e.g., say 90) than learned through IGRP (e.g., say 100); similarly, a route learned from IGRP can be given a lower distance than from RIPv2

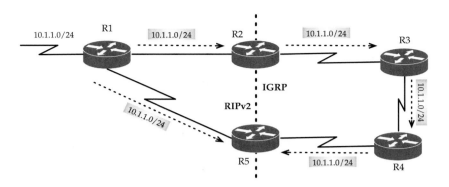

FIGURE 5.11   Route redistribution example with two routing protocols: RIPv2 and IGRP.

(e.g., say 120). Typically, static routes are given the lowest cost (e.g., 1) since it is assumed to be directly connected.

It is important to note that administrative distances do help in preferring path selections among different routing protocols; however, they cannot always solve problems that arise due to route redistribution. For example, looping can still occur, and convergence can be a problem. Consider Figure 5.11 in which we have two networks: one running RIPv2 among routers R1, R2, and R5, and the other running IGRP among R2, R3, R4, and R5. In this case, there are two boundary routers, R2 and R5, that are not directly connected to each other. Addressable network 10.1.1.0/24 attached to R1 is announced to R2 and R5 through the distance vector mechanism. Now for the IGRP side, R2, on learning route 10.1.1.0/24, announces to R3 about this external route, and so on. We can see that router R5 learns about 10.1.1.0/24 from R1 (RIPv2 side) and also from R4 (IGRP side). While from Figure 5.11, we can see that it is clearly better to forward traffic from R5 to 10.1.1.0/24 via R1, it would not do so *if* the administrative distance gives lower weight to a route learnt from IGRP over RIPv2; that is, packets arriving at R5 destined for 10.1.1.0/24 would instead be forwarded to R4 for further forwarding to R3, then to R2, and finally to R1. Note that if split horizon is implemented on the RIPv2 side, a routing loop can be avoided. However, convergence and looping can still occur if routers R3 and R4 fail and come back up again, i.e., during the transient time. Fortunately, such problems can be avoided by introducing a simple policy: do not announce routes originally received from a routing protocol back into the same routing protocol.

Finally, route redistribution is not just limited to RIPv2 and IGRP; this is possible between any combination of routing protocols, including protocols such as EIGRP and OSPF. An important point to note is that route redistribution requires careful handling, as we can see from the above example.

5.8 Summary

We have described different protocols in the distance vector family for IP networks. A summary comparison is listed in Table 5.2. It is important to recognize that a distance vector protocol can have a variety of manifestations. From a basic routing protocol point of view, fast convergence is important as well as whether a routing protocol is loop-free. From the

TABLE 5.2   Comparison of protocols in the distance vector protocol family.

Protocol	RIPv1	RIPv2	IGRP	EIGRP	RIPng
Address Family	IPv4	IPv4	IPv4	IPv4	IPv6
Metric	Hop	Hop	Composite	Composite	Hop
Information Communication	Unreliable, broadcast	unreliable, multicast	Unreliable, multicast	Reliable, multicast	Unreliable, multicast
Routing Computation	Bellman–Ford	Bellman–Ford	Bellman–Ford	Diffusing computation	Bellman–Ford
VLSM/CIDR	No	Yes	No	Yes	v6-based
Remark	Slow convergence; split horizon	Slow convergence; split horizon	Slow convergence; split horizon	Fast, loop-free convergence; chatty protocol	Slow convergence; split horizon

perspective of the addressing and the operational usage requirement of running a protocol in a network, some additional issues emerge: how the protocol handles the inherent need of the addressing scheme of the network, whether a protocol provides support for external protocols and routes, and whether the information is delivered reliably or not.

Often, we also learn much from operational experience. For example, triggered update and split horizon with poison reverse are important timer-based issues to implement in a distance vector family. Update synchronization, known as the *pendulum effect*, was a problem encountered in the early days of RIPv1 deployment; to overcome this problem, a large delay variation between updates was recommended. Finally, to overcome the looping problem that is inherent in the basic distance vector concept, a loop-free distance vector protocol based on diffusing computation emerged. While the loop-free distance vector protocol has certain added complexities in terms of state management, the routing convergence is fast and loop-free. From a network routing management point of view in an operational network, the complexity of the protocol is not always a critical issue.

In an operational environment, issues are more centered around whether a routing protocol and its features are easy to configure and manage; these may be labeled as *configuration complexities*. Here are a few to note:

- If at the command prompt level of a router, EIGRP can be configured as easily as RIPv1 or RIPv2, and knowing that EIGRP is loop-free, it will be natural for a network provider to opt for EIGRP. Furthermore, EIGRP is much easier to configure than link state protocols such as OSPF. Note that EIGRP is a chatty protocol; if routers are connected by low-bandwidth links, this factor can have an impact on inducing congestion in certain situations.

- Route management at routers for internal and external routes is an important feature.

- Scalability and growth of a network should be considered. For example, for a campus network, the number of routers may remain fixed at a small number for a number of years while for a regional or national ISP, the number of routers may rapidly increase over the years, sometimes in months. Thus, for a campus network, it is best to deploy EIGRP, especially since the link speed between routers now are at least at Fast-Ethernet (100 mbps) data rate; thus, the chattiness of EIGRP is not a concern.

- Sometimes, a rich feature available with a protocol is not always used since it is poorly understood; a case in point is the composite metric that can be used in IGRP and EIGRP. Given different coefficients and factors, the composite metric for IGRP/EIGRP can be confusing sometimes; thus, it is often found that in an operational environment, the simplest case of coefficients known as "defaults" is used. In many small or campus networking environments, the simplest form of the composite metric is even desirable, especially when the link speed between connecting routers is not a bottleneck.

In an operational environment, a common principle about deploying any routing protocol is to use routers from the same vendor with the same software release. Certainly, for business reasons, it is desirable to use multiple vendors so that a provider does not get locked in with one vendor. While RIP is a standardized specification and it should be possible to use products from multiple vendors in the same network, it is usually not advisable. A minor difference in implementation can cause unnecessary problem. Similarly, having the same software release on all routers from the same vendors in a network is also advisable. Furthermore, the command sets to configure routers can conceivably be different for different router vendors; thus, the network operational staff would need to be conversant with different command sets, an additional requirement that should be avoided if possible. Note that these are general guidelines, not cast in stone; a size of a network is also a critical factor in regard to consideration of products from multiple vendors, and proper training is required for operational personnel.

Finally, we note that configuration complexities are not the only issues in an operational network; there are other issues such as security and reliability that need to be addressed. To summarize, it is important to recognize that from a basic concept of a routing protocol, to its specification, to its vendor implementation, and finally to its operational deployment, the issues and needs can be quite different.

Further Lookup

RIPv1 is the oldest routing protocol in practice for intradomain routing in the Internet that was designed and implemented in the early 1980s, although an RFC was not officially available until 1988 [290]. IGRP was developed by Cisco in the mid-1980s. RIPv2 was first described in the early 1990s [442]. Cisco developed EIGRP at about the same time that implemented a loop-free distance vector algorithm called the diffusing coordination with coordinated update (see Section 3.3.5). Thus, it is not surprising that RIPv1 and IGRP are used for classful IP addressing while RIPv2 and EIGRP allow variable length subnet masking. RIPng was first described in 1997; for additional details, see RFC 2080 [443].

There are many books that cover the entire family of IP distance vector routing protocols; for example, see [301], [571]. For an excellent coverage of routing protocols along with
command line information on how to configure routers, see [189].

Exercises

5.1. Review questions:

   (a) What are the main differences between RIPv1 and RIPv2?

   (b) What are the three timers in RIPv1?

5.2. Under what conditions would RIPv2 and IGRP essentially behave the same way?

5.3. Consider adding a host with IP address 192.168.8.88 directly to router Charlie through a
     point-to-point link in the network shown in Figure 5.1. List the routing tables entries in
     each router for this route; include any additional consideration you need to address.

5.4. Consider the route redistribution example shown in Figure 5.11. Assume that networks
     have converged. Now consider that routers R3 and R4 went down and came back up
     again. Identify the sequence of steps during the transient time that will take place that
     would lead to a routing loop (assuming no policy rule is in place).

5.5. Why do some routing protocols message identification at above the transport layer while
     some other do so directly over the IP layer?

6

OSPF and Integrated IS-IS

In protocol design, perfection has been reached not when there is nothing left to add, but when there is nothing left to take away.

Ross Callon (RFC 1925)

Reading Guideline

This chapter provides specifics about OSPF, including its key features and protocol formats. We have also highlighted integrated IS-IS. The basic concept of a link state protocol discussed separately in Section 3.4 is recommended reading along with this material to see the distinction between the link state routing protocol family and instances of this protocol family. A basic knowledge of OSPF and/or IS-IS is also helpful in understanding IP traffic engineering, discussed later in Chapter 7.

In this chapter, we consider two important link state routing protocols: Open Shortest Path First (OSPF) and Intermediate System-to-Intermediate System (IS-IS). The currently used version of OSPF in IPv4 networks is known as OSPF, version 2 (OSPFv2); here, we will simply refer to it as OSPF. While OSPF is exclusively designed for IP networks, IS-IS was designed for the connection-less network protocol (CLNP) in the OSI reference model. For use in IP networks, an integrated IS-IS or dual IS-IS protocol has been used to support both CLNP and IP, thus allowing an OSI routing protocol in IP networks. Most of our discussion will focus on the OSPF protocol; at the same time, we will highlight a few key features of integrated IS-IS; however, as of now, there are no fundamental differences between OSPF and IS-IS. In any case, we will highlight certain similarities and differences between OSPF and Integrated IS-IS.

6.1 From a Protocol Family to an *Instance* of a Protocol

OSPF is an *instance* of a link state protocol based on hop-by-hop communication of routing information, specifically designed for intradomain routing in an IP network. Recall from our earlier discussion in Section 3.4.1 that such a routing protocol requires information about the state (e.g., cost) of a link, and the ability to advertise this link state reliably through in-band (in-network) communication. Furthermore, a link state protocol uses twosub protocols, one to establish a neighborhood relationship through a hello protocol, and another for database synchronization.

Going from a basic understanding of a protocol concept to an instance applicable in a specific networking environment requires certain customization, including provision for flexibility to handle various possible variations. Consider the following examples/scenarios:

- Flooding the link state advertisement (LSA) is not always necessary since a network may have different types of transmission media. For example, if N routers in a network are, say, in the same local area network (LAN), it unnecessarily creates $N(N-1)$ links while a single-link definition is sufficient; furthermore, it also results in unnecessary shortest path computation in each router without any gain. Thus, some summarization is desirable.

- Besides LAN, are there other types of networks for which any customization is needed?

- An intradomain network may consist of a large number of routers, possibly geographically spread out; thus, scalability is an important issue. Thus, from the point of view of manageability and scalability, it is desirable to have the ability to cluster the entire domain into several subdomains by introducing hierarchy. This, in turn, raises the possibility that an entire LSA from one subdomain to another may not need to be distributed, especially if two subdomains are connected by just a link; some form of summarization is sufficient since all traffic would need to use this link after all. A major benefit of this hierarchy is that the shortest path computation at a router needs to consider links only within its subdomain.

- How can flooding of a LSA be accomplished in an IP network?

From the above discussion, we can see that a protocol intended for use in practice is required to address many functionalities and features. In the following section, we describe primary key features of OSPF, a commonly deployed link state protocol in IP networks.

6.2 OSPF: Protocol Features

OSPF provides many features. We will highlight the key features below. The packet format for various OSPF packets and the key fields are described later in Section 6.3. For clarity, any packet that carries OSPF routing information or is used for an OSPF protocol will be referred to as an *OSPF packet*, to distinguish it from packets for user traffic.

6.2.1 Network Hierarchy

OSPF provides the functionality to divide an intradomain network (an autonomous system) into subdomains, commonly referred to as *areas*. Every intradomain must have a core area, referred to as a *backbone area*; this is identified with Area ID 0. Areas are identified through a 32-bit area field; thus Area ID 0 is the same as 0.0.0.0.

Usually, areas (other than the backbone) are sequentially numbered as Area 1 (i.e., 0.0.0.1), Area 2, and so on. OSPF allows a hierarchical setup with the backbone area as the top level while all other areas, connected to the backbone area, are referred to as low-level areas; this also means that the backbone area is in charge of summarizing the topology of one area to another area, and vice versa. In Figure 6.1, we illustrate network hierarchy using low-level areas.

6.2.2 Router Classification

With the functionality provided to divide an OSPF network into areas, the routers are classified into four different types (Figure 6.1):

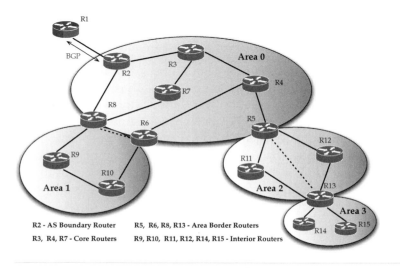

FIGURE 6.1   OSPF backbone and low-level areas.

- *Area-Border Routers:* These are the routers that sit on the border between the backbone and the low-level areas. Each area-border router must have at least one interface to the backbone; it also has at least one interface to each area to which it is connected.

- *Internal Routers:* These are the routers in each low-level area that have interfaces only to other internal routers in the same area.

- *Backbone Routers:* These are the routers located in Area 0 with at least one interface to other routers in the backbone. Area-border routers can also be considered as backbone routers.

- *AS Boundary Routers:* These routers are located in Area 0 with connectivity to other AS; they must be able to handle more than one routing protocol. For example, to exchange information with another AS, they must be able to speak BGP. These routers also have internal interfaces for connectivity to other backbone routers.

The above terminologies, as described, are OSPF specific; however, it is also common to use names such as backbone routers in general. You will see such usage throughout this book; such usage should not be confused with Backbone Routers as used in the context of OSPF.

6.2.3 Network Types

OSPF is designed to address five different types of networks: (1) point-to-point networks, (2) broadcast networks, (3) non–broadcast multiaccess (NBMA) networks, (4) point-to-multi-point networks, and (5) virtual links.

Point-to-point networks refer to connecting a pair of routers directly by an interface/link such as OC-3. A router may be connected to multiple different routers by such point-to-point interfaces. Point-to-point links are typically used when an OSPF domain is spread out in a geographically distributed region.

Broadcast networks refer to networks such as LANs connected by a technology such as Ethernet. Broadcast networks, by nature, are multiaccess where all routers in a broadcast network can receive a single transmitted packet. In such networks, a router is elected as a Designated Router (DR) and another as a Backup Designated Router (BDR).

Non–broadcast multiaccess networks use technologies such as ATM or frame relay where more than two routers may be connected without broadcast capability. Thus, an OSPF packet is required to be explicitly transmitted to each router in the network. Such networks require an extra configuration to emulate the operation of OSPF on a broadcast network. Like broad-cast networks, NBMA networks elect a DR and a BDR.

Point-to-multipoint networks are also non–broadcast networks much like NBMA networks; however, OSPF's mode of operation is different and is in fact similar to point-to-point links.

Virtual links are used to connect an area to the backbone using a nonbackbone (transit) area. Virtual links are configured between two area-border routers. Virtual links can also be used if a backbone is partitioned into two parts due to a link failure; in such a case, virtual links are tunneled through a nonbackbone area. Consider again Figure 6.1. Here Area 3 is con-nected to the backbone area using transit Area 2 through a virtual link that connects router 6 to router 7. Also note that if the link between router 2 and router 3 in the backbone area goes down, Area 0 becomes partitioned; to avoid that, a virtual link between Area-Border Routers 4 and 5 is established through Area 1.

Finally, an important point to understand about OSPF networks is that the neighborhood relation is not based on routers or networks connected by physical links, but is based on logical adjacencies established.

6.2.4 Flooding

OSPF uses *in-network* functionality to flood routing information such as LSAs. In-network means OSPF packets are carried in the same network as user traffic. From the discussion above, we note that there are different possible network types. Thus, transmission of OSPF packets requires some tailoring.

First note that multiple LSAs can be combined into an OSPF link state update packet. Flooding is required for link state update packets, as well as for LSA packets (for a discussion about different packet types, see Section 6.3); the protocol type field in an IP packet header is set to 89 for OSPF packets. Also note that flooding is selective in that a router forwards an update only if it is not stale; for this, it relies on checking the age and the sequence number field, discussed earlier in Section 3.4.1.

On point-to-point networks, updates use the IP multicast address 224.0.0.5, referred to as *AllSPFRouters*. A router on receiving an update forwards it to other routers, if needed (after checking the sequence number), again using the same multicast address.

On broadcast networks, all non-DR and non-BDR routers send link state update and LSA packets using the IP multicast address 224.0.0.6, referred to as AllDRouters. Any OSPF packets that originates from a DR or a BDR, however, use the IP multicast address 224.0.0.5.

In NBMA networks, LSAs are sent as unicast from non-DR/non-BDR routers to the DR and the BDR. DR, in turn, sends a copy of the LSA as unicast to all adjacent neighbors. On point-to-multipoint networks and virtual link networks, updates are sent as unicast using the interface's IP address of the adjacent neighbor.

Regardless of the network type, OSPF flooding must be reliable. Since OSPF sits directly on top of IP in the TCP/IP stack, OSPF is required to provide its own reliable mechanism, instead of being able to use a reliable transport protocol such as TCP. OSPF addresses reliable delivery of packets through use of either implicit or explicit acknowledgment. An implicit acknowledgment means that a duplicate of the LSA as an update is sent back to the router from which it has received the update. An explicit acknowledgment means that the receiving router sends a LSA packet on receiving a link state update. Since a router may not receive acknowledgment from its neighbor to whom it has sent a link state update message, a router is required to track a link state retransmission list of outstanding updates. An LSA is retransmitted, always as unicast, on a periodic basis (based on the value RxmtInterval) until an acknowledgment is received, or the adjacency is no longer available.

Finally, OSPF defines three global parameters in regard to flooding of LSAs. *LSRefreshTime* indicates the maximum acceptable time between generation of any particular LSA, regardless of whether the content of the LSA such as the metric value has changed; this time window is set to 30 min. *MinLSInterval* reflects the minimum time between generation of any particular LSA; this is set to 5 sec. Finally, *MinLSArrival* is the minimum time between reception of new LSAs during flooding, set to 1 sec; this parameter serves as the hold-down timer.

6.2.5 Link State Advertisement Types

From the discussion about network hierarchy and network types, it is clear that an OSPF network requires different LSA types. The five most commonly known LSA types are: Router LSA (type code = 1), Network LSA (type code = 2), Network Summary LSA (type code = 3), AS Border Router (ASBR) Summary LSA (type code = 4), and AS External LSA (type code = 5).

A *Router LSA* is the most basic or fundamental LSA that is generated for each interface. Such LSAs are generated for point-to-point links. Router LSAs are recorded in the link state database and are used by the routing computation module. Flooding of Router LSAs is restricted to the area where they originate.

Network LSAs are applicable in multiaccess networks where they are generated by the DR. All attached routers and the DR are listed in the Network LSA. Flooding of Network LSAs is also restricted to the area where they originate.

Area-Border Routers generate *Network Summary LSAs* that are used for advertising destinations outside an area. In other words, Network Summary LSAs allow advertising IP prefixes between areas. Area Border Routers also generate *ASBR Summary LSAs*; in this case, they advertise AS Border Routers external to an area.

AS External LSAs are generated by AS Border Routers. Destinations external to an OSPF AS are advertised using AS external LSAs.

There are six additional LSA types; they are described later in Section 6.2.8.

6.2.6 Subprotocols

In our discussion of a link state protocol earlier in Section 3.4.1, we mentioned that subprotocol mechanisms are also used for the operation of a link state protocol in addition to the main function of LSA through flooding. Two key subprotocols are the hello protocol and the database synchronization protocol. It should be noted that to accomplish these protocols, various packet types such as the hello packet, database description packet, link state request packet, and link state update packet have been defined as part of the OSPF protocol; these packet types are outlined in detail later in Section 6.3.

HELLO PROTOCOL

While its name seems to imply that the hello protocol is just for initialization, it is actually much more than that. Recall that the OSPF protocol is designed for several different types of networks as discussed earlier in Section 6.2.3. First, during initialization/activation, the hello protocol is used for neighbor discovery as well as to agree on several parameters before two routers become neighbors; this means that using the hello protocol, logical adjacencies are established; this is done for point-to-point, point-to-multipoint, and virtual link networks. For broadcast and NBMA networks, not all routers become logically adjacent; here, the hello protocol is used for electing DRs and BDRs. After initialization, for all network types, the hello protocol is used to keep alive connectivity, which ensures bidirectional communication between neighbors; this means, if the keep alive hello messages are not received within a certain time interval that was agreed upon during initialization, the link/connectivity between the routers is assumed to be not available.

To accomplish various functions described above, a separate hello packet is defined for OSPF; details about the field are described in Section 6.3.

DATABASE SYNCHRONIZATION PROCESS

Beyond basic initialization to discover neighbors, two adjacent routers need to build adjacencies. This is important more so after a failed link is recovered between two neighboring routers. Since the link state database maintained by these two routers may become out of sync during the time the link was down, it is necessary to synchronize them again. While a complete LSA of all links in the database of each router can be exchanged, a special database description process is used to optimize this step. For example, during database description, only *headers* of LSA are exchanged; headers serve as adequate information to check if one side has the latest LSA. Since such a synchronization process may require exchange of header information about many LSAs, the database synchronization process allows for such exchanges to be split into multiple *chunks*. These chunks are communicated using database description packets by indicating whether a chunk is an initial packet (using I-bit) or a continuation/more packet or last packet (using M-bit). Furthermore, one side needs to serve as a master (MS-bit) while the other side serves as a slave—this negotiation is allowed as well; typically, the neighbor with the lower router ID becomes the slave. It is not hard to see that the database synchronization process is a stateful process.

In Figure 6.2, we illustrate the database synchronization process, starting with initialization through the hello packet, adapted from [505]. After initialization, this process goes through several states: from exchange start to exchange using database description packets to synchronizing their respective databases by handling one outstanding database description packet at a time, followed by a loading state when the last step of synchronization is done. After that, for link state request and update for the entire LSA for which either side requires updated information, the communication takes place in the full state until there are no more link state requests.

6.2.7 Routing Computation and Equal-Cost Multipath

First note that LSAs are flooded throughout an area; this allows every router in an area to build link state databases with identical topological information. Shortest path computation based on Dijkstra's algorithm (see Section 2.3) is performed at each router for every known destination based on the directional graph determined from the link state database; the cost used for each link is the metric value advertised for the default type of service in the link LSA packet; see Figure 6.11 presented later for the metric field. Originally, it was envisioned that there be different types of services that might require different metrics. Thus, a type of service (TOS) field was created. The default TOS is indicated by setting field, Number of TOS, to 0. Metric field allows the value to be between 1 and 65,535, both inclusive. If additional types of services are defined and supported by routers in an area, then for each type of service the shortest path can be computed separately. While the default metric is dimensionless, additional types of services are identified based on attributes such as monetary cost, reliability, throughput, and delay. At the same time, the default metric being dimensionless provides the flexibility to *not* explicitly consider metrics for other types of services in an operational environment since through IP traffic engineering the link metric/cost can be determined and

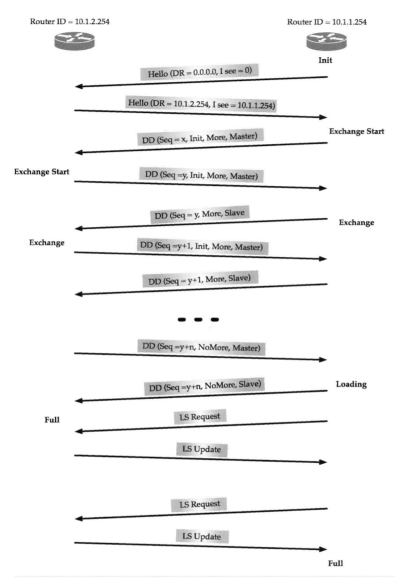

FIGURE 6.2   OSPF link state database synchronization process (based on [505]).

set just under the default TOS, which can still take into account diverse goals of a network and the network provider; we will discuss link cost determination for IP intradomain traffic engineering in Chapter 7.

A nice feature of Dijkstra's algorithm computed at each router is that the entire shortest path from a source to a destination (in fact, for all destinations) is available at the end. OSPF allows a source routing option that can be used by user traffic on the path determined by Dijkstra's algorithm. Certainly, OSPF allows the default next hop option commonly deployed in IP networks; thus, once the path is computed, the next hop is also extracted from the shortest path computation to update the routing table, and subsequently, the forwarding table. Note that routing table entries are for destinations identified through hosts or subnets or simply IP prefixes (with CIDR notation), not in terms of end routers. Thus, once the shortest path first computation is performed from a router to other reachable routers, reachable addresses from each destination router as learned through LSAs are identified, and the routing table entries are accordingly created for all such addresses. Because of CIDR, multiple similar route entries are possible. For example, there might be an entry for 10.1.64.0/24, and another for 10.1.64.0/18, where the difference is in the netmask. To select the route to be preferred by an arriving packet, OSPF uses a best route selection process. According to this process, the route(s) with the most specific match to the destination is to be selected, which is the one with the longest match. As an example, 10.1.64.0/24 would be preferred over 10.1.64.0/18. In case there are multiple paths available after this step, the second step selects the route where an intra-area path is given preference over an interarea path, which, in turn, gives preference over external paths for routers learned externally (refer to Section 6.2.8).

ECMP

An important feature of OSPF routing computation is the *equal-cost multipath* (ECMP) option; that is, if two paths have the same lowest cost, then the outgoing link (next hop) for both can be listed in the routing table so that traffic can be equally split. It may be noted the original Dijkstra's algorithm generates only one shortest path even if multiple shortest paths are available. To capture multiple shortest paths, where available, Dijkstra's algorithm is slightly modified. In line 23 of Algorithm 2.5 in Chapter 2, if the greater than sign ($>$) is changed to a greater than or equal to sign ($\geq$), it is possible to capture multiple shortest paths by identifying the next hops. In this case, line 25 is then updated to collect all next hops that meet the minimum, instead of just one when the strictly greater than sign is used. Thus, more than one outgoing link in the case of multiple shortest paths would need to be stored, instead of just one.

It is important to note that ECMP is based on the number of outgoing interfaces (links) involved on the shortest path at node level along the path, not at the source-destination path level. Consider the six-router network shown in Figure 6.3. Suppose the paths between router 1 and router 6 are of equal cost. Thus, for traffic from router 1 to router 6, it will be equally split at router 1 along the two directional links 1→2 and 1→5; traffic from router 1 that arrived at router 2 will be equally split further along its two outgoing links 2→3 and 2→4. Thus, of the traffic from router 1 destined for router 6, 25% each will arrive at router 6 from links 3→6 and 4→6, while one-half will arrive from link 5→6. This illustrates the meaning of equal-cost being node interface–based. Since OSPF routing is directional, the traffic splitting for this example will be different for traffic going the other direction from router 6 to router 1. Since router 6 has three outgoing links, traffic will be split equally among the links 6→3, 6→4, and 6→5. Note that the traffic sent on the first two links will be combined at router 2; thus, two-thirds of the traffic from router 6 destined for router 1 will arrive at router 6 from link 2→1, while one-third will arrive from link 5→1.

It may be noted that the OSPF specification does not say exactly how ECMP is to be accomplished from an implementation point of view. In concept, packets that arrive for the same destination router can be equally split among outgoing links of ECMP paths. However, this is not desirable for several reasons. For example, for a single TCP session (microflow), if

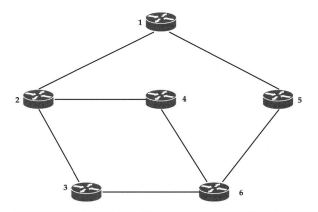

FIGURE 6.3   OSPF equal-cost multipath (ECMP) example.

the packets are sent on different ECMP paths that might consist of links with different link bandwidths, packets can arrive with different delays; this can then affect TCP throughput. If packets for this session are alternated across each link, then packets can arrive out of order. Thus, router software implementation handles ECMP path selection on a per-microflow basis. Yet, implementation at a per-microflow choice level at a router can have an effect as well. If every router in the network makes identical decisions because of the way flows are processed by the router software, for example, due to prefix matching, then microflows that arrive at router 1 destined for router 6 that are directed to link 1-2 would use only one of the paths 2-3-6 or 2-4-6 (see Figure 6.3). Thus, sophisticated, yet fast, router software implementation addresses randomization of microflows to different ECMP outgoing links so that such situations do not occur. In summary, ECMP is possible only as *approximate* split by randomizing at a per-microflow level (or destination address level) from a practical point of view.

The ECMP feature is helpful in load balancing traffic, but may not be helpful when troubleshooting a network. Thus, some providers try to avoid using ECMP; that is, they seek the single shortest paths, to the extent possible. This will be discussed further in relation to IP traffic engineering in Chapter 7.

INTERAREA ROUTING COMPUTATION

It is important to note that Dijkstra-based shortest path computation using link state information is applied only within an area. For routing update between areas, information from one area is summarized using Summary LSAs without providing detailed link information; thus, interarea routing computation in OSPF is similar to the distance vector flavor. Since OSPF employs only a two-level hierarchy, a looping problem typically known to occur with a distance vector approach is not conceptually possible. Yet, due to aggregation and hierarchy, in certain situations, it is possible to create a scenario where looping can occur [589].

6.2.8 Additional Features

OSPF has the capability to authenticate packet exchanges between two routers. Such authentication can be either simplex password-based or MD5-based. Furthermore, extensions to OSPF, to add digital signature authentication to LSA data and to provide a certification mechanism for router data, have been addressed in RFC 2154 [514]. We will highlight here a few additional features in OSPF.

STUB AREAS AND STUB NETWORKS

Recall that we discussed backbone and low-level areas earlier. OSPF provides additional ways to define low-level areas. A low-level area is considered to be a *regular* area if all types of LSAs are permitted into this area; thus, all routers in a regular area can determine the best path to a destination. A *stub area* is an area where information about external routes, communicated through AS external LSAs, is not sent; the area border router on this stub area creates a default external route for use by the stub area. Consider Figure 6.1 again; here, Area 3 is a stub area.

A *not-so-stubby area* (NSSA) is a stub area that can import AS external routes—this means that this stub area has an unusual connectivity to another AS. Since routes/networks from this AS would need to be known to the rest of the Internet, this information needs to be imported. To accomplish this, an additional LSA type called NSSA-LSA (type code = 7) is defined so that routes from an AS connected to an NSSA can be imported to an area border router where they can be converted to a type 5 LSA (AS-external-LSA) for flooding. For example, if you imagine such an area connected to Area 3 in Figure 6.1 (not shown in figure) that is *outside* the OSPF domain, then this area would be an NSSA. In addition, another area type called a *totally stubby area* is being used in practice; this type of area is useful for a large OSPF network since such an area can use default route for all destinations outside this area (in addition to external routes), thereby saving on memory requirement of routers in the area.

There is another term, *stub networks*, that should not to be confused with stub areas. A stub network is a network identified by an IP prefix that is connected to only one router within an area.

ADDITIONAL LSA TYPES

In addition to the LSA types described earlier in Section 6.2.5 and the one described above, there are five more LSA types have been defined so far. Group Membership LSA (type code = 6) is used in multicast OSPF. External Attributes (type code = 8) has been deprecated; in its place, three new LSA types, known as the Opaque LSA option, have been defined. The role of opaque LSA is to carry information that is not used by SPF calculation, but can be useful for other types of calculations. For example, traffic engineering extensions to OSPF utilize the opaque LSA option; see Section 18.3.4. Three types of opaque LSA options have been defined to indicate the scope of propagation of information, i.e., whether it is link-local, area-local, or AS scope.

ROUTE REDISTRIBUTION

In Section 5.7, we discussed route redistribution, especially for distance vector protocols. Route redistribution is similarly possible with OSPF (and IS-IS); for example, one side can be EIGRP while the other side is OSPF. For OSPF that learns a route from another protocol such as EIGRP, NSS External LSA (type = 7) can be used. To allow for route redistribution and metric compatibility, NSSS External LSA has an E-bit field to indicate whether to use cost that is the external cost plus the cost of the path to the AS border router ("E1 type external path"), or simply the external cost ("E2 type external path"), and External Route Tag field for help in external route management. For the purpose of path selection for external routes, an E1 type external path is given preference over an E2 type external path.

6.3 OSPF Packet Format

In this section, we describe packet formats for several key OSPF packet types.

Common Header

The common header has the following key fields (Figure 6.4):

- *Version:* This field represents the OSPF version number; the current version is 2.
- *Type:* This field specifies the type of packet to follow. OSPF has five packet types: hello (1), database description (2), link state request (3), link state update (4), and LSA (5).
- *Packet Length:* This indicates the length of the OSPF packet.
- *Router ID:* This field indicates the ID of the originating router. Since a router has multiple interfaces, there is no definitive way to determine which interface IP address should be the router ID. According to RFC 2328 [505], it could be either the largest or the smallest IP address among all the interfaces. It may be noted that if a router is brought up with no interface connected, then it has no ability to acquire a router ID. To avoid this scenario, a loopback interface, being a virtual interface, can be used to acquire a router ID. In general, a router ID that is based on a loopback interface provides much more flexibility to network operators in terms of management than a physical interface–based address.
- *Area ID:* This is the ID of the area where the OSPF packet originated. Value 0.0.0.0 is reserved for the backbone area.

0 1 2 3 4 5 6 7	0 1 2 3 4 5 6 7	0 1 2 3 4 5 6 7	0 1 2 3 4 5 6 7
Version (1 byte)	Type (1 byte)	Packet Length (2 bytes)	
Router ID (4 bytes)			
Area ID (4 bytes)			
Checksum (2 bytes)		Authentication Type (2 bytes)	
Authentication (4 bytes)			
Authentication (4 bytes)			

FIGURE 6.4 OSPF packet common header.

- *Checksum:* This is the IP checksum over the entire OSPF packet.

- *AuType* and *Authentication Field:* AuType works with the Authentication field for authentication. There are three authentication types:

AuType	Meaning	Authentication Field
0	No authentication	Can be anything
1	Simple, clear text password-based authentication	An 8-byte password
2	Cryptographic MD5 checksum authentication	8-byte is divided as shown in Figure 6.5

Note that when AuType is 2, it contains a KeyID, an Authentication Data Length, and a Cryptographic Sequence Number. MD5 checksum is used to produce a 16-byte message digest that is not part of the OSPF packet; rather, it is appended to the end of the OSPF packet.

FIGURE 6.5 OSPF AuType = 2.

HELLO PACKET

The primary purpose of the hello packet (Figure 6.6) is to establish and maintain adjacencies. This means that it maintains a link with a neighbor that is operational. The hello packet is also used in the election process of the DR and BDR in broadcast networks. The hello packet is also used for negotiating optional capabilities.

- *Network Mask:* This is the address mask of the router interface from which this packet is sent.

- *Hello Interval:* This field designates the time difference in seconds between any two hello packets. The sending and the receiving routers are required to maintain the same value; otherwise, a neighbor relationship between these two routers is not established. For point-to-point and broadcast networks, the default value is 10 sec, while for non–broadcast network the default value used is 30 sec.

- *Options:* Options field allows compatibility with a neighboring router to be checked.

- *Priority:* This field is used when electing the designated router and the backup designated router.

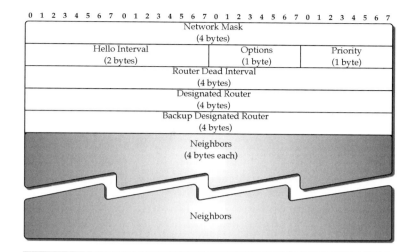

FIGURE 6.6   OSPF hello packet (OSPF packet type = 1).

- *Router Dead Interval:* This is the length of time in which a router will declare a neighbor to be dead if it does not receive a hello packet. This interval needs to be larger than the hello interval. Also note that the neighbors need to agree on the value of this parameter; this way, a routing packet that is received, which does not match this field on a receiving router's interface, is dropped. The default value is typically four times the default value for the hello interval; thus, in point-to-point networks and broadcast networks, the default value used is 40 sec while in non–broadcast networks, the default value used is 120 sec.

- *Designated Router (DR) (Backup Designated Router (BDR)):* DR (BDR) field lists the IP address of the interface of the DR (BDR) on the network, but not its router ID. If the DR (BDR) field is 0.0.0.0, then this means there is no DR (BDR).

- *Neighbor:* This field is repeated for each router from which the originating router has received a valid Hello recently, meaning in the past RouterDeadInterval.

DATABASE DESCRIPTION PACKET

The OSPF database description packet has the following key features (Figure 6.7):

- *Interface Maximum Transmission Unit (MTU):* This field indicates the size of the largest transmission unit the interface can handle without fragmentation.

- *Options:* Options fields consist of several bit-level fields. The most critical one is the E-bit, which is set when the attached area is capable of processing AS-external LSAs.

- *I/M/MS bits:* I-bit (initial-bit) is initialized to 1 for the initial packet that starts a database description session; for other packets for the same session, this field is set to 0. M-bit (more-bit) is used to indicate that this packet is not the last packet for the database description session by setting it to 1; the last packet for this session is set to 0. MS-bit (master-slave

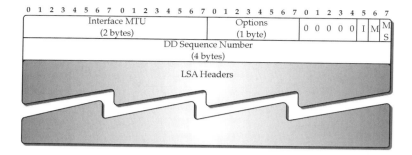

FIGURE 6.7   OSPF database description packet (OSPF packet type = 2).

bit) is used to indicate that the originator is the master by setting this field to 1, while the slave sets this field to 0. This was illustrated earlier in Figure 6.2.

- *DD Sequence number:* This field is used for incrementing the sequence numbers of packets from the side of the master during a database description session; the master sets the initial value for the sequence number.

- *LSA Header:* This field lists headers of the LSAs in the originator's link state database; it may list some or all of them.

LINK STATE REQUEST PACKET

The link state request packet is used for pulling information. For example, based on database description received from a neighbor, a router might want to know link state information from its neighbor. The link state request packet has the following fields, which are repeated for each unique entry (Figure 6.8):

- *Link State Type:* This field identifies a link state type such as a router or network.

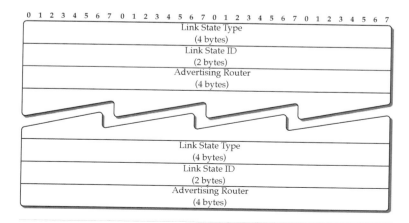

FIGURE 6.8   OSPF link state request packet (OSPF packet type = 3).

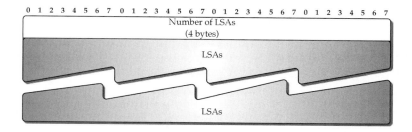

FIGURE 6.9 OSPF link state update packet (OSPF packet type = 4).

- *Link State ID:* This field is dictated by the link state type.

- *Advertising Router:* This is the address of the router that has generated this LSA.

LINK STATE UPDATE PACKET

This packet (Figure 6.9) contains the first field to be the number of LSAs followed by information on LSAs that follow the LSA packet format. Thus, a link state update packet can contain one or more LSAs.

LINK STATE ACKNOWLEDGMENT PACKET

The LSA packet is used in acknowledging each link state advertisement received from a neighboring router. This includes the LSA headers that follow the OSPF packet header where the type field is set to 5.

OSPF LINK STATE ADVERTISEMENT

In some sense, this is the heart of the link state protocol concept. A LSA packet consists of a header, followed by data for different link state types. Here, we will present packet formats for Router LSA and Network LSA. First, the common LSA header has the following fields:

- *Age:* This field reflects the time in seconds since the LSA was originated. The originating router sets this value to 0. Through a global parameter, MaxAge, the maximum life of an LSA is set to 1 hour. When the age field for an LSA reaches MaxAge, LSA is flooded again regardless of change in the link state of this LSA.

- *Options:* This is used to identify optional capabilities supported by the OSPF routing domain.

- *Type:* This field indicates the LSA type: 1 for Router LSA, 2 for Network LSA, and so on. This type field is not to be confused with the OSPF packet type discussed earlier.

- *Link State ID:* This field uniquely identifies an LSA.

- *Advertising router:* This is the OSPF router ID of the originating router.

- *Sequence number:* This field is incremented each time a new LSA is generated by the originating router.

- *Checksum* and *Length:* The checksum is over the entire packet except for the age field. Length is counted in bytes for the enter LSA including header.

Router LSA:

A Router LSA consists of the LSA header (Figure 6.10) followed by the content of the Router LSA (Figure 6.11). Every router generates a Router LSA that lists all the routers, outgoing interfaces (links); for each interface, the state and cost of the link are included. In addition to the LSA header, a Router LSA has the following fields:

- *V/E/B-bits:* V-bit indicates if it is a virtual link, E-bit indicates an AS boundary router, and B-bit indicates an area border router.

- *Number of Links:* This field indicates the total number of router interfaces.

- *Link ID, Link Data, and Link Type:* Link ID and Link Data are better understood in the context of Link Type ([189], [505]); this is summarized in Table 6.1.

- *Metric:* This is the cost of an interface/link. The value is in the range 1 to 65,535 ($=2^{16} - 1$). OSPF specification does not specify what values are to be used here. Rather, this is left to the network service provider to decide. In Chapter 7, we will discuss how values might be chosen for the purpose of traffic engineering of a network.

- *Number of TOS, TOS, and TOS Metric:* The Number of TOS field indicates the different number of Type of Service; if this field is zero, then TOS and TOS Metric fields are not applicable. If the Number of TOS is 2, then TOS and TOS Metric fields are repeated twice; here TOS would then refer to a particular type such as normal service, maximize reliability, or minimize delay, while the TOS Metric field would then include the cost for the associated TOS field.

Network LSA:

Network LSAs are generated by DRs. In addition to the LSA header, a Network LSA has the following fields (Figure 6.12):

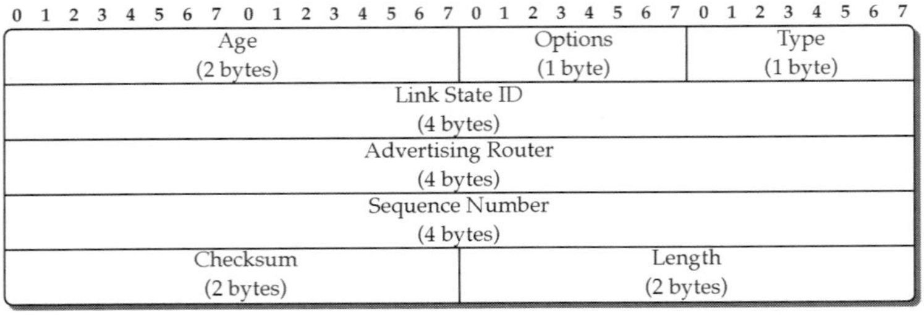

0 1 2 3 4 5 6 7	0 1 2 3 4 5 6 7	0 1 2 3 4 5 6 7	0 1 2 3 4 5 6 7
Age (2 bytes)		Options (1 byte)	Type (1 byte)
Link State ID (4 bytes)			
Advertising Router (4 bytes)			
Sequence Number (4 bytes)			
Checksum (2 bytes)		Length (2 bytes)	

FIGURE 6.10   OSPF link state advertisement header.

TABLE 6.1   Router LSA: Link Type, Link ID, and Link Data.

LinkType	Description	Link ID	Link Data
1	Point-to-point link	Neighboring router's Router ID	Interface IP address of originating router
2	Link to transit network	Interface IP address of Designated Router	Interface IP address of originating router
3	Link to stub network	IP network or subnet address	Network's IP address
4	Virtual link	Neighboring router's Router ID	Interface IP address

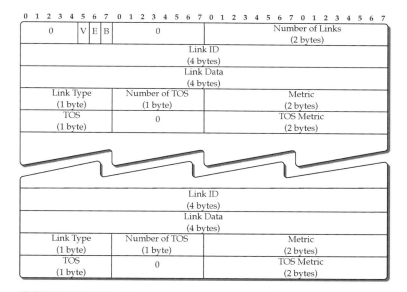

FIGURE 6.11   OSPF Router LSA content (LSA type = 1).

- *Network Mask:* This is the standard subnet mask information.

- *Attached Router:* This field is repeated once for each router that is fully adjacent to the DR.

6.4 Examples of Router LSAs and Network LSAs

In this section, we will illustrate router and network LSAs through an example, adapted from [505]. We will use private IP address space in this illustration. In Figure 6.13, we show two areas: Area 0 and Area 1. In Area 1, there are three stub networks: N1 identified by IP prefix 192.168.1.0, N2 by 192.168.2.0, and N3 by 192.168.3.0, which are off routers R1, R2, and R3, respectively. The transit network N4 is identified by 192.168.4.0 with R4 as the DR, while having IP interfaces to R1, R2, and R3, as noted. Both R3 and R4 area border routers are

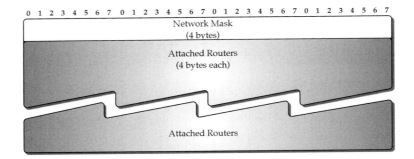

FIGURE 6.12 OSPF Network LSA content (LSA type = 2).

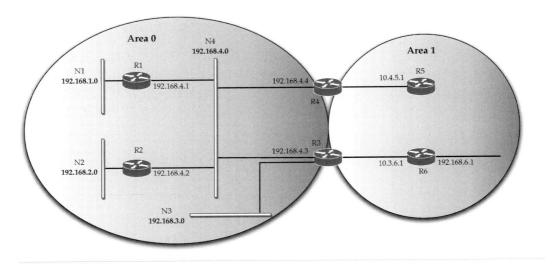

FIGURE 6.13 OSPF network example.

connected to Area 0. In our discussion, we will identify the router ID of a router by the highest IP address of all its interfaces. For example, router ID of R1 is 192.168.4.1 since this address is higher than its other interface to network N1 (192.168.1.0). For simplicity, we will assume all metric values to be 1.

Router R3 will generate two Router LSAs: one for Area 1 and the other for Area 0. For Area 1, it is necessary to identify two Link IDs: one for the transit network identified by DR 192.168.4.4 and the other for stub network 192.168.3.0. Now you can see how networks/IP prefixes are communicated to other routers in the network. Thus, when a router computes the shortest path tree, it first computes it for all the routers in its area. Then, any networks it learns about from any of the Router LSAs of various routers, it can add a leaf for a route to such networks. This means that once completed, a router's routing table contains entries for all other routers as well as for destination networks. The information content of Router LSA for R3 for Area 1 is shown in Figure 6.14.

```
// Router LSA of R3 for Area 1
LS age = 0                          //always true on origination
Options = (E-bit)
LS type = 1                         //indicates Router LSA
Link State ID = 192.168.4.3        //R3's Router ID
Advertising Router = 192.168.4.3   //R3's Router ID
bit E = 0                          //not an AS boundary router
bit B = 1                          //area border router
#links = 2
   Link ID = 192.168.4.4           //IP address of Designated Router
   Link Data = 192.168.4.3         //R3's IP interface to net
   Type = 2                        //connects to transit network
   # TOS metrics = 0
   metric = 1                      // End of first Link ID info
   Link ID = 192.168.3.0           //IP Network number
   Link Data = 255.255.255.0       //Network mask
   Type = 3                        //connects to stub network
   # TOS metrics = 0
   metric = 1                      // End of Second Link ID info
```

FIGURE 6.14   Router LSA of R3 in Area 1 for the network example in Figure 6.13.

```
// Network LSA for Network N4
LS age = 0                          //always true on origination
Options = (E-bit)
LS type = 2                         //indicates Network LSA
Link State ID = 192.168.4.4        //IP address of Designated Router
Advertising Router = 192.168.4.4   //R4's Router ID
Network Mask = 255.255.255.0
   Attached Router = 192.168.4.4    //Router ID
   Attached Router = 192.168.4.1    //Router ID
   Attached Router = 192.168.4.2    //Router ID
   Attached Router = 192.168.4.3    //Router ID
```

FIGURE 6.15   Network LSA for Network N4 for the network example in Figure 6.13.

Router LSAs within Area 1 generated by routers R1 and R2 will be similar. Since R4 is the DR in Area 1, it will generate a network LSA for transit network N4 (Figure 6.15), which also identifies all the attached routers.

Now consider Area 0. R3 is connected to router R6 by a point-to-point link in Area 0; Router LSA for R3 would be different than when it sends to Area 1, as shown in Figure 6.16. Router LSA for R4, which is connected to router R5, will be similar.

6.5 Integrated IS-IS

Integrated IS-IS for both CLNP and IP protocols is described in RFC 1195 [105] while the original IS-IS protocol was described in [321], [537]. IS-IS comes with its own terminology that is different from OSPF. For example, routers are referred to as *intermediate systems*; thus, the name intermediate systems-to-intermediate systems means router-to-router. For consistency, we will use the term routers instead of intermediate systems. LSAs are called *link state*

```
// R3's Router LSA for Area 0
LS age = 0                          //always true on origination
Options = (E-bit)
LS type = 1                         //indicates Router LSA
Link State ID = 192.168.4.3        //R3's router ID
Advertising Router = 192.168.4.3   //R3's router ID
bit E = 0                           //not an AS boundary router
bit B = 1                           //area border router
#links = 1
    Link ID = 192.168.6.1          //Neighbor's Router ID
    Link Data = 10.3.6.1           //MIB-II ifIndex of P-P link
    Type = 1                        //connects to router
    # TOS metrics = 0
    metric = 1
```

FIGURE 6.16   Router LSA for R3 in Area 0 for the network example in Figure 6.13.

protocol data units, or LSPs, in short. A broadcast network is referred to as a *pseudonode*; a *designated intermediate system* is elected from all the ISs to represent a broadcast network. An address to identify an intermediate system is called a *network service access point (NSAP)*. IS-IS runs directly over layer-2 protocols, unlike OSPF that runs over IP. Similar to OSPF, IS-IS has also been extended to provide traffic engineering capabilities; this will be discussed later in Section 18.3.4.

6.5.1 Key Features

We now highlight the main features of IS-IS protocols.

AREAS

IS-IS provides two-level network hierarchy using areas that are similar to OSPF. The routers in the backbone area are called L2 routers; the internal routers in low-level areas are called L1 routers. A network that has any low-level (L1) areas must also have at least one L1/L2 router that sits in the L1 area but is connected to the L2 (backbone) area by a link. Note that in IS-IS, a router is entirely within an area, unlike OSPF, where a router can sit on the border between two areas; connectivity between areas is only through a link.

ADDRESSING IN IS-IS

Addressing in IS-IS is based on OSI-NSAP addressing and is compatible with USA GOSIP version 2.0 NSAP address format. GOSIP stands for Government Open Systems Interconnection Profile, the federal standard for network systems procurement that was standardized in the early 1990s. The OSI-NSAP addressing has, key fields, as shown on the table on the next page.

Field	Size	Value
AFI (Authority and Format Identifier)	1 byte	"47"
ICD (International Code Designator)	2 bytes	"00 05"
DFI (Domain-Specific Path Format Identifier)	1 byte	"xx"
AAI (Administrative Authority Identifier)	3 bytes	"xx xx xx"
Reserved	2 bytes	Must be "00 00"
RDI (Routing Domain identifier)	2 bytes	Contains autonomous system number
Area	2 bytes	Assigned by the authorities responsible for the routing domain to uniquely identify areas
System ID	6 bytes	Use either (1) "02 00" prepended to the 4-byte IP address of the router, or (2) IEEE 802 48-bit MAC address
N-Selector (upper layer identifier)	1 byte	Set to zero

As you can see, several fields are used for setup in IP networks. The important ones are: RDI, Area, and System ID. When the last byte, N-selector, is set to zero, there is no upper-layer user, and the address is meant purely for routing; such routing-layer-only NSAP addresses are called *Network Entity Titles (NET)*. In effect, IS-IS for IP networks uses NET addressing. It is important to note that NET is a router identifier, not an interface identifier.

PSEUDONODES AND NONPSEUDONODES

IS-IS allows handling of different network types. For example, a broadcast network is treated as a pseudonode where one of the routers serves as the pseudonode, which is labeled the designated intermediate system (DIS), with links to each attached router.

For links that are not for broadcast networks but are for point-to-point networks and stub networks, a nonpseudonode is created. Essentially, a nonpseudonode is similar to a router LSA in OSPF.

SHORTEST PATH CALCULATION

Shortest path calculation is based on Dijkstra's algorithm. Once a router receives a new LSP, it waits for 5 sec before running the shortest path calculation. There is a 10 sec hold-down timer between two consecutive shortest-path calculations within the same area. However, L1/L2 routers that reside in L1 areas must run separate shortest path calculations, one for the L1 area and the other for the L2 area.

Link metric in IS-IS has been originally limited to 6 bits and, thus, the value ranges from 0 to 63 and the total path cost in an IS-IS domain can have a maximum value of 1023. This 6-bit metric is known as a *narrow metric*. A *wide metric* extension is now available through traffic engineering extensions to IS-IS that permits a 24-bit metric, thus allowing a range of 0 to 16,777,215 (=$2^{24} - 1$).

CATEGORIZATION OF PACKETS

IS-IS defines four categories of protocol packets, or protocol data units (PDUs): hello packet, link state PDUs (LSP), complete sequence number PDUs (CSNP), and partial sequence number PDUs (PSNP).

The purpose of the hello packet is similar to the hello packet for OSPF. IS-IS defines three types of hello packets; they are for (1) point-to-point interfaces, (2) L1 routers, and (3) L2 routers.

There are two types of LSPs—one for level 1 and the other for level 2. Each LSP contains three key pieces of information: LSP ID (8 bytes), the sequence number (4 bytes), and the remaining lifetime (2 bytes). LSP ID is system ID (6 bytes) followed by pseudonode ID; if the first byte of the pseudonode ID field is nonzero, then this LSP orginated from a DIS in a broadcast network; the last byte is used for identification in case the LSP needs to be fragmented because it exceeds the maximum transmission unit of an interface. The remaining lifetime field is the same as the age field in an OSPF LSA; the difference is that the lifetime field value is set at the maximum age of 1200 sec (20 min) at the beginning and is then decreased unlike OSPF where it is set to zero at the beginning and is then increased. In addition to this, an LSP uses a Type-Length-Value (TLV) format to include information such as a list of connected IP prefixes along with subnet masks. This makes it possible to determine destination networks to which the domain is connected so that the routing table can properly list such destinations.

CSNPs are like database description packets in OSPF and are used for link state database synchronization. A router creates CSNPs with all LSPs in its local link state database. A PSNP is created when upon receiving a CSNP from a neighbor, it realizes that some parts are missing; this means that this router has received certain other LSPs that are in its location link state database but the neighbor's CSNP did not include them. Thus, the receiving router generates a PSNP to request newer copy of the missing LSPs. In essence, PSNPs are similar to the link state request packet in OSPF.

PACKET FORMAT AND INFORMATION ENCODING THROUGH TLV

The first 8 bytes of IS-IS PDUs form the common header that includes fields such as version number, header length, and PDU types. After the common header, PDU-specific fields are included followed by variable-length fields. For example, PDU-specific fields in the hello packet include fields equivalent to Router Dead Interval in OSPF hello packet. IS-IS Link state PDUs are similar to OSPF LSAs. A subtle difference is that while OSPF LSAs start with the age field set to zero and the counter is incremented until MaxAge to indicate the expiry of time, IS-IS link state PDUs start with a remaining lifetime and the counter is decremented until zero to indicate that the lifetime has expired. Since there are two types of routers in IS-IS, L1 and L2, a field is included to indicate the originating router type.

The variable-length field that follows the header is encoded using TLV encoding where 1 byte is assigned for code type (T), 1 byte for length (L), and a variable-length value (V) field not to exceed 255 bytes since one byte is assigned for the length field. A representative set of well-known types is listed in Table 6.2; note that many types are as originally described in ISO 10589 [321] while for IP environments several additional types were added in RFC 1195 [105] and in recent RFCs such as RFC 3784 [649]. An updated list is maintained at [316].

TABLE 6.2   TLV codes for Integrated IS-IS protocol.

Type	TLV
1	Area Addresses (ISO 10589 [321])
2	IS Neighbors (LSPs) (ISO 10589 [321])
3	ES Neighbors (ISO 10589 [321])
4	Partition Designated level 2 IS (ISO 10589 [321])
5	Prefix Neighbors (ISO 10589 [321])
6	IS Neighbors (Hellos) (ISO 10589 [321])
8	Padding (ISO 10589 [321])
9	LSP Entries (ISO 10589 [321])
10	Authentication Information (ISO 10589 [321])
14	LSP Buffersize (ISO 10589 [321])
22	Extended IS reachability (RFC 3784 [649])
128	IP Internal Reachability Information (RFC 1195 [105])
129	Protocols Supported (RFC 1195 [105])
130	IP External Reachability Information (RFC 1195 [105])
131	Inter-Domain Routing Protocol Information (RFC 1195 [105])
132	IP Interface Address (RFC 1195 [105])
133	Authentication Information (RFC 1195 [105])
134	Traffic Engineering router ID TLV (RFC 3784 [649])
135	Extended IP reachability TLV (RFC 3784 [649])
138	Shared Risk Link Group (RFC 4205 [373])

6.6 Similarities and Differences Between IS-IS and OSPF

It is helpful to consider the similarities and differences between IS-IS and OSPF. First, it should be noted that fundamentally there is little difference between OSPF and IS-IS. Thus, the differences center more on how certain things are done, often stylistic differences.

SIMILARITIES

There are several similarities between IS-IS and OSPF:

- Both protocols provide network hierarchy through two-level areas.

- Both protocols use Hello packets to initially form adjacencies and then continue to maintain them.

- Both protocols have the ability to do address summarization between areas.

- Both protocols maintain a link state database, and shortest path computation performed using Dijkstra's algorithm.

- Both protocols have the provision to elect a designated router for representing a broadcast network.

DIFFERENCES

While there are similarities as noted above, there are several differences:

TABLE 6.3   IS-IS and OSPF development/deployment timeline (adapted from [354]).

Year	Note
1987	IS-IS (CLNP) chosen as the OSI intradomain protocol from DECnet proposal
1988	NSFnet deployed; routing protocol uses an early draft of IS-IS
	Work on OSPF started
	IP extensions to IS-IS defined
1989	OSPFv1 (RFC 1131) published
	Proteon ships OSPF implementation
	IS-IS becomes ISO proposed standard
1990	Integrated IS-IS (RFC 1195) published
1991	OSPF v2 (RFC 1247) published
	Cisco ships its OSPF implementation
	Cisco ships its OSI-only IS-IS implementation
1992	Cisco ships dual–IS-IS implementation
	Many deployment of OSPF
1993	Novell publishes NLSP
1994	Cisco ships NLSP, rewriting IS-IS as well
	IS-IS is recommended for large ISPs due to recent rewrite and OSPF field experience, and CLNP mandate by NSF
1995	ISPs begin deployment of IS-IS
1996–1998	IS-IS niche popularity continues to grow (some ISPs switch to it from OSPF)
	IS-IS becomes barrier to entry for router vendors targeting large ISPs
	Juniper and other vendors ship IS-IS–capable routers
1999–present	Extensions continue for both protocols in parallel (e.g., Traffic Engineering)

- With OSPF, an area border router can sit on the boundary between the backbone area and a low-level area with some interfaces in the area while other interfaces are in the other area. In IS-IS, routers are entirely within one or the other area—the area borders are on links, not on routers.

- While OSPF packets are encapsulated in IP datagrams, IS-IS packets are directly encapsulated in link layer frames.

- The OSPF dimension-less link metric value is in the range 1 to 65,535, while IS-IS allows the metric value to be in the range 0 to 63 (narrow metric), which has been extended to the range 0 to 16,777,215 (wide metric).

- IS-IS being run directly over layer 2 is relatively safer than OSPF from spoofs or attacks.

- IS-IS keepalives can be used for MTU detection since they are MTU-sized TLVs that are explicitly checksummed and need to be verified as such.

- IS-IS allows overload declaration through an overload bit by a router to other routers. This is used, for example, by other routers to not consider an overloaded router in path computation.

Along with similarities and differences, it is helpful to also consider a timeline of evolution of OSPF and IS-IS as outlined in Table 6.3.

6.7 Summary

In this chapter, we have presented the OSPF protocol, discussing its main features at length. We have also described packet formats for certain key packets in OSPF. Furthermore, we have presented examples of LSAs for OSPF networks. We also presented the integrated IS-IS protocol through a summary of its key features. It is important to note that OSPF and IS-IS are stateful protocols. Note that both OSPF and IS-IS allow route redistribution capability (refer to Section 5.7).

We also provided a brief summary on similarities and differences between the OSPF and the IS-IS protocol. It may be noted that as of now there are no fundamental differences between OSPF and IS-IS. In retrospect, it can be said that market competition made both protocols as robust as possible. Thus, the choice of routing protocol for deployment in an ISP's network is based on issues such as configuration management, maintainability of large networks, in-house expertise, and so on. Typically, medium- to large-scale ISPs use either OSPF or IS-IS protocol, while small providers or campus networks use routing protocols such as EIGRP. Finally, while OSPF defines the concept of areas, many providers deploy their networks configured simply with a single area (Area 0); in many instances, a single area is found to be easy to manage since all routers see the same view, which can be helpful in troubleshooting any routing problem.

Further Lookup

While the link state routing protocol goes back to ARPANET when the "new" ARPANET routing was introduced [463], this approach gained significance during the early days of OSI protocol development. IS-IS was introduced in 1987. Factors including NSFnet deployment were key drivers in creating the first version of the OSPF protocol [503]. It was also recognized in the late 1980s that the IS-IS protocol can be tweaked to work in an IP environment [105].

The current OSPF standard, known as version 2, is described in RFC 2328 [505]; also, see [504] for a detailed discussion of OSPF. For a comparative discussion on OSPF and IS-IS, see [354]. For additional discussions on the similarities and differences between OSPF and IS-IS, see [83], [189], [211], [558]. For details on command line level configuration of OSPF and IS-IS, there are several books available; for example, see [189] for an excellent coverage.

OSPF has been extended for use with IPv6 addressing; often, this version of OSPF is referred to as OSPFv3; for details, see RFC 2740 [150].

Exercises

6.1. Review questions:

    (a) What are the different OSPF packet types?

    (b) What is the range of allowable metric values in OSPF and IS-IS?

    (c) What is a database description packet?

(d)   What is a link state advertisement?

(e)   What is a designated router?

(f)   What is an network entity title?

6.2.   Describe an usage of a not-so-stubby area.

6.3.   Explore route redistribution between OSPF and EIGRP.

6.4.   Identify the functionality in OSPF that allows a static route to be injected into an OSPF domain.

6.5.   Consider a five-router OSPF network. How many entries will be in the routing table at each router?

6.6.   Consider a fully-connected *N*-router OSPF network. Suppose one of the routers goes down. Estimate how many total link state messages would be generated before the network converges.

6.7.   Why are different types of LSAs defined in OSPF?

6.8.   How is the router ID determined in OSPF? How about IS-IS?

6.9.   How is an OSPF area different from an IS-IS area?

6.10.   Can you redistribute route learned from OSPF to IS-IS and vice-versa?

6.11.   Refer to the discussion about the generic link state routing protocol framework in Section 3.4. Present a comparative assessment between the basic framework and OSPF/IS-IS.

7

IP Traffic Engineering

As late as 1842 a train was started only when sufficient traffic was waiting along the road to warrant the use of the engine.

John Moody

Reading Guideline

To get the most out of this chapter, we assume that you are already familiar with network flow modeling (discussed in Chapter 4); some familiarity with IP routing protocols such as OSPF and IS-IS (see Chapter 6) is necessary. By reading this chapter, you will know how to determine link weights for IP traffic engineering for an interior gateway protocol (IGP) such as OSPF or IS-IS.

In this chapter, we discuss traffic engineering for IP intradomain networks. The role of traffic engineering is to optimize an operational network so that performance requirements are met, yet network resources are well utilized. Traffic engineering is an essential component of IP intradomain operational networks, especially if the network is large. Traffic engineering addresses medium-term goals of a network and overall behavior of operational networks; typically, it does not cover adding new capacity, which falls under the *network dimensioning* problem. Furthermore, traffic engineering does not address issues such as traffic surge that last a few seconds to a few minutes, which may result in excessive delay for a very brief period; this is important to keep in mind in order to understand the context of traffic engineering.

7.1 Traffic, Stochasticity, Delay, and Utilization

7.1.1 What Is IP Network Traffic?

We start with a discussion about traffic in IP networks. To describe traffic, we first need to consider sources that generate IP traffic.

An IP network provides many services such as web and email; there are also interactive services such as telnet, ssh for terminal services. In current IP networks, the predominant traffic is due to applications that use TCP for transport layer; it has been reported that on a backbone link approximately 90% of traffic is TCP based [350]. A message content created by applications is broken into smaller TCP pieces, called *TCP segments*, by including TCP header information, which are then transmitted over the IP network after including IP header information; the data entity at the IP level is *IP datagrams*, while *packet* is also a commonly used term. Thus, traffic in an IP network is IP datagrams generated by various applications, without wondering which among the applications it is for.

Thus, when we talk about traffic volume on an IP network link, we are interested in knowing the number of IP packets flowing on a link in a certain unit of time. Usually, the time unit is considered in *seconds*. Thus, traffic volume can be specified in terms of IP packets offered per second, or *packets per sec (pps)*. On the other hand, there is another measure of traffic volume that is often used—raw data rate units such as Megabits per sec (Mbps) or Gigabits per sec (Gbps). Indeed, there is a relation between pps and Mbps (or Gbps). Suppose we consider the average packet size to be K Megabits. Then pps is related to Mbps as follows:

$$\text{Traffic data rate (Mbps)} = \text{Packets per sec} \times \text{Average packet size (Megabits)}. \qquad (7.1.1)$$

It is, however, not required that the average packet size be counted separately to obtain the data rate. With the sophisticated network monitoring system in current IP networks, the traffic data rate in Mbps (or Gbps) can be estimated based on measurements through either an active or passive monitoring system.

7.1.2 Traffic and Performance Measures

In an IP network environment, delay is a critical performance parameter since we are interesting in ensuring that a packet generated from one end reaches the other end as soon as possible. Interestingly, there is an analogy between road transportation networks and IP networks. In road transportation networks, delay depends on the volume of traffic as well as the number of street lanes (and speed limit) imposed by the system. Similarly, delay in an IP network can depend on the amount of traffic as well as the capacity of the system; thus, the following functional relation can be generally written:

$$\text{Delay} = \mathcal{F}(\text{Traffic volume data rate, Capacity}). \qquad (7.1.2)$$

To be specific, the above relation is true only in a single-link system. When we consider a network where routing is also a factor, then a more general functional relation is as follows:

$$\text{Delay} = \mathcal{F}(\text{Traffic volume data rate, Capacity, Routing}). \qquad (7.1.3)$$

7.1.3 Characterizing Traffic

So far, we have not said much about traffic volume except for Eq. (7.1.1); that is, traffic volume may be given through a single number, such as packets per second or Megabits per second. How do we obtain a number like this one?

Consider the arrival of packets to a network link. If we consider just a single request for a web page that is traversing the link, it may appear that the packets at the IP level are arriving in a deterministic fashion; the page is generated by the web server, which is broken into TCP segments that are wrapped with an IP header, and is then transmitted one after another; this is certainly from the point of view of a single web session. However, in a network link, many web sessions are active, each one being requested at a random start point by a user; this is similar for other traffic due to applications such as email, and so on. Thus, from the point of view of a network link, if we consider only the IP level, the link then sees *random* arrival of packets. Thus, a number that may represent pps cannot be a fixed, deterministic number; it is rather dictated by the randomness of traffic arrival. Thus, at most what we can say is *average* pps or *average* Mbps in regard to random traffic arrival. The primary question is: can we say anything about the characteristics of the random behavior?

Traditionally, it has been assumed that arrival of packets follows a well-known random process called the *Poisson process*, and the average arrival rate is the average rate for this Poisson process. However, in the early to mid-1990s, there were a series of studies based on actual measurements of packet traffic that reported that packet arrival behavior is not Poisson; rather traffic is *self-similar* in different time scales following heavy-tailed distributions, exhibiting long-range dependency; for example, see [160], [404], [545], [551], [740], [741]. Not to clutter the discussion here, definitions for Poisson process, self-similarity, long-range dependency, and heavy-tailed distributions are provided in Appendix B.10 and in Appendix B.11. The key point to note here is that self-similarity contradicts the Poisson assumption; furthermore, a self-similar process with a heavy-tailed distribution impacts the delay behavior much worse than for a Poisson process. In a recent illuminating study [350] based on measurements from a backbone network link at OC-48 speed, it was observed that it is indeed possible that *both* Poisson behavior and self-similarity *can* co-exist; it is a matter of the time frame being considered. Specifically, they reported that in a subsecond time scale, the behavior is Poisson while in the scale of seconds long-range dependency is observed. We thus start with the assumption of the Poisson model and discuss how self-similarity can be factored in indirectly for the purpose of traffic engineering.

7.1.4 Average Delay in a Single-Link System

First, we assume that packet arrival to a network link follows a Poisson process with the average arrival rate as λ packets per sec. The average service rate of packets by the link is assumed to be μ packets per sec. We consider here the case in which the average arrival rate is lower than the average service rate, i.e., $\lambda < \mu$; otherwise, we would have an overflow situation. If we assume that the service time is *exponentially distributed* (see Appendix B.10), in addition to packet arrival being Poissonian, then the average delay, τ, can be given by the following formula, which is based on the $M/M/1$ queueing model (see Appendix B.12.2):

$$\tau = \frac{1}{\mu - \lambda}. \tag{7.1.4}$$

Now consider that the average packet size is κ Megabits, and that the packet size is exponentially distributed. Then, there is a simple relation between the link speed c (in Mbps), the average packet size κ, and the packet service rate μ, which can be written as:

$$c = \kappa \mu. \tag{7.1.5}$$

This is then essentially the relation discussed earlier in Eq. (7.1.1). Combining κ with the packet arrival rate λ, we can consider the arrival rate, h, in Mbps as follows:

$$h = \kappa \lambda. \tag{7.1.6}$$

If we multiply the numerator and the denominator by κ, we can then transform Eq. (7.1.4) as follows:

$$\tau = \frac{\kappa}{\kappa(\mu - \lambda)} = \frac{\kappa}{c - h}. \tag{7.1.7}$$

This relation can be rewritten as:

$$\frac{\tau}{\kappa} = \frac{1}{c - h}. \tag{7.1.8}$$

If we now compare Eq. (7.1.4) and Eq. (7.1.8), we see that the average packet delay can be derived directly from the link speed and arrival rate given in a measure such as Mbps; the only difference is the factor κ, the average packet size. Second, although it may sound odd, the quantity, $\frac{\tau}{\kappa}$, can be thought of as the average "bit-level" delay on a network link where the average traffic arrival rate is assumed to be h Mbps. In other words, if we track the traffic volume in Mbps on a link and know the link data rate, we can get a pretty good idea about the average delay. There are a couple of advantages to this observation: first, we can use traffic volume, h, and link speed, c, in other units such as Gbps without changing the basic behavior on delay given by $1/(c - h)$; second, it is not always necessary to track the average packet size; third, if the delay is to be measured in millisec instead of sec, then $1/(c - h)$ must be multiplied by the constant, 1000, without changing the basic structure of the formula. Finally, whether we consider measures in packets per sec or Mbps (or Gbps), the link utilization parameter, ρ, that captures the ratio of traffic volume over the link rate, remains the same regardless of the average packet size since

$$\rho = \frac{h}{c} = \frac{\kappa\lambda}{\kappa\mu} = \frac{\lambda}{\mu}. \tag{7.1.9}$$

In essence, we can say that under the $M/M/1$ queueing assumption, the average delay, $t(= \tau/\kappa)$, can be given in terms of the link speed c and the traffic rate h where $h < c$ as

$$t = \frac{1}{c - h} \tag{7.1.10}$$

with utilization given by $\rho = h/c$. Incidentally, Eq. (7.1.10) then gives a functional relation mentioned earlier in Eq. (7.1.2). What happens if we were to consider self-similarity of traffic? Unfortunately, there is no simple formula like the above when traffic is self-similar. It has been reported that the delay behavior with heavy-tail traffic is worse than that of the $M/M/1$ delay. Thus, we will create a fictitious delay function for self-similar traffic and plot it along with the $M/M/1$ delay as shown in Figure 7.1; note that in this figure the link speed c is kept fixed while the traffic rate h is increased—this is why the x-axis is marked in terms of link utilization, ρ, given in percentage as ρ goes from 0 to 100%.

Figure 7.1 is, in fact, very helpful in letting us see a problem from the perspective of traffic engineering. For instance, suppose that to provide acceptable perception to users, we want to maintain the average delay at say 20 millisec. From the graph, we can see that with the $M/M/1$ average delay formula, the link can handle an arrival traffic rate up to about 80% of the link capacity while maintaining the acceptable average delay. However, if the traffic does *not* follow the Poisson process, then the delay would be much worse at the same utilization

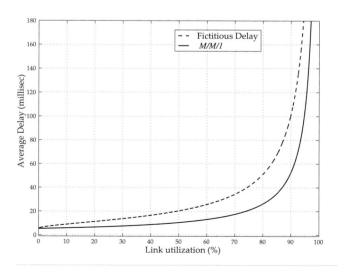

FIGURE 7.1 The $M/M/1$ average delay curve along with a fictitious delay curve.

value; for instance, in this fictitious graph of delay, we can see that the delay would be about 50 millisec instead. Certainly this is not desirable when the acceptable delay is required to be below 20 millisec. Thus, instead of taking a vertical view at a certain utilization, we take the horizontal view at an acceptable average delay. If we do so, we see that to maintain the average delay at or below 20 millisec, the non-Poisson traffic cannot go beyond 50% link utilization.

In regard to traffic engineering, there are two important points to note from the above discussion. First, there is a direct relation between delay and utilization; because of this, requiring a network to maintain a certain delay can be recast as requiring the utilization to be kept below an appropriate level. Second, since there is no simple formula to consider delay for self-similar traffic, being conservative on the requirement on utilization can often be sufficient for the purpose of traffic engineering. For example, in the above example, we observe that keeping utilization at 50% would be more appropriate than letting it grow to 80%. Due to the relation between traffic volume and capacity through utilization ($\rho = h/c$), this means that for a fixed link speed c, we need to keep the traffic volume at a lower level than would otherwise be indicated for Poisson traffic in order to address traffic engineering needs.

7.1.5 **Nonstationarity of Traffic**

The analysis/discussion above is based on stationary traffic assuming that an *average* traffic data rate is given. However, network traffic has been observed to be nonstationary and can be time dependent. For example, consider a 24-hour network traffic profile on a link as shown in Figure 7.2. We can see that the data rate is different depending on the time of the day; in this specific instance, the traffic volume data rate range is from below 8 Mbps to as high as 30 Mbps. If, for the purpose of traffic engineering, we were to use traffic volume to be the data rate, say at midnight (about 8 Mbps), and determine link capacity needed to be, say 15 Mbps (based on utilization being about 50%), then we will certainly be overlooking many

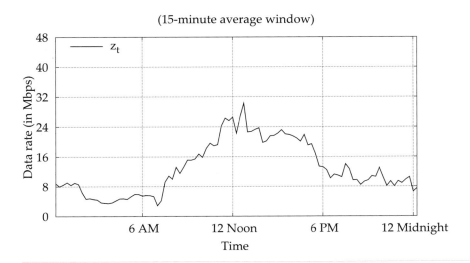

FIGURE 7.2   Traffic data rate over a 24-hour period.

time windows when the traffic volume will overflow this capacity. This tells us that it would make more sense to consider the *peak* of the traffic data rate (or, say 90% of the peak traffic) over the 24-hour window as the traffic volume needed for traffic engineering consideration. In this example, the peak traffic volume rate is about 30 Mbps; thus, for an acceptable delay or utilization, at least a 45 Mbps link would be desirable.

Remark 7.1. *Traffic engineering and network dimensioning.*

From the illustration above, it could be argued that a 45-Mbps link is not sufficient since the utilization at the peak traffic rate would be over 60%. For example, 60 Mbps would be minimally necessary considering 50% as acceptable utilization. However, the determination of actual link speed to put in place or lease, especially in a backbone network, also depends on the actual cost of establishing or leasing the link. The problem of determining the appropriately sized link, especially taking into consideration network cost minimization, is often considered under *network dimensioning* rather than under traffic engineering, while the distinction is sometimes blurry if you read the current literature. We will assume that the goal of traffic engineering is to see if the network can provide acceptable delay or utilization for offered traffic in a *capacitated* environment.   ◆

Remark 7.2. *Traffic engineering and traffic estimation.*

From Figure 7.2, it is clear that the offered traffic should be chosen wisely and may depend on the time of day. In fact, traffic estimation is itself a challenging problem; there has been much recent work on understanding how to do it and how to do it as accurately as possible, especially for a large network. We assume that through some process, the offered traffic is determined for use in traffic engineering.   ◆

7.2 Applications' View

In the previous section, we presented traffic as viewed from the network layer. Since applications are the ones that generate IP traffic, it is helpful to understand the requirements in regard to applications. Since most commonly used applications such as web, email are TCP-based, from an application point of view, not only should the delay perception be minimized, but the throughput of data rate transfer is also an important consideration; this is necessary since TCP uses an adaptive sliding window mechanism to regulate how much data to be pumped based on perception of congestion (see Section 22.2). Thus, we discuss two important aspects: TCP throughput and bandwidth-delay product and how they need to be accounted for in traffic engineering.

7.2.1 TCP Throughput and Possible Bottlenecks

It has been noted that TCP throughput depends primarily on three factors: the maximum segment size (S), the round-trip time ($\mathcal{RTT}$), and the average packet loss probability (q). A key result [224], [225], [450] on TCP throughput is the following:

$$\text{TCP throughput} = \frac{1.22\,S}{\mathcal{RTT} \times \sqrt{q}}. \tag{7.2.1}$$

An important question is: from the traffic engineering perspective, where and how does an IP network fit in the three factors and the relation shown in Eq. (7.2.1)? First, we see that the segment size should be as large as possible. However, note that the maximum segment size is not entirely within the control of the network since it is negotiated by the end hosts; at the same time, this tells us that the network link should be set for the maximum transmission unit possible so that the network link itself does not become the bottleneck in reducing the TCP throughput of end applications. Since end hosts are connected to Ethernet (where the maximum transmission unit that can be handled is 1500 bytes), it is imperative that the core network links have the ability to handle packets of at least this size to avoid any fragmentation of packets into multiple smaller packets.

The second factor that affects TCP throughput is the round-trip time. From Eq. (7.2.1), we see that the round-trip time should be minimized, which means that one-way delay must be minimized. While many factors, including processing at the end hosts can impact delay, from the point of view of the network, it is important that the delay on a network link be minimized. Since numerous TCP sessions traverse through a network for different source destinations, delay minimization in an IP network is an important goal. Recall our discussion earlier about the direct relationship between delay and utilization, which tells us that utilization should be kept below a desirable value in lieu of considering delay.

The third factor is the average packet loss probability. The average packet loss can depend on many points along a TCP connection; the end hosts may drop a packet, the edge network may drop a packet, there may be bit error rate, and so on. A core network can minimize its contribution to the packet loss probability by ensuring that the bit error is not a dominant factor, which is a fair assumption in fiber-based transmission networks now commonly deployed in core networks. However, there is another factor that can contribute to the increase in packet loss probability—that is, if the buffer size at a router is not sized properly. Since packets arrive at random time, it is quite possible that the queue builds up at a router. If there is not enough buffer space, a router is forced to drop packets. If this happens, the affected TCP sessions are forced to reduce the data rate since a drop packet is commonly understood by a TCP session to be an indication of congestion. That is, even if a network link has enough bandwidth, it is quite possible that if a router buffer is not sized properly, it may appear as congestion to TCP sessions; in other words, the router buffer size has the potential to be another bottleneck in reducing TCP throughput. Thus, the router buffer should be sized properly for the benefit of traffic engineering of a network. How do we estimate router buffer size? To determine this, it is helpful to consider the *bandwidth-delay product*.

7.2.2 Bandwidth-Delay Product

The term bandwidth-delay product means exactly what it says—that is, to take the product of the bandwidth and the delay. In case of a network link, the bandwidth then refers to the link speed and the delay refers to what the network would like to account for. For example, if the link speed is given in Mbps and the round-trip time delay in seconds, then the product will result in a quantity in Megabits. What does this quantity signify? This is none other than the amount of data the network link needs to handle *in-flights*, often referred to as the *window size*. To put it formally, if c is the data rate of a link ("bandwidth") and $\mathcal{RTT}$ is the round-trip time delay, then the bandwidth-delay product defines the window W given by

$$W = c \times \mathcal{RTT}. \tag{7.2.2}$$

Router buffer size has a strong relation to the bandwidth-delay product, which will be addressed next.

7.2.3 Router Buffer Size

From the network's point of view, the window determined by the bandwidth-delay product is an important factor to consider without this becoming a bottleneck for end applications, especially for synchronized TCP flows. In other words, this window allows the number of packets that can be generated by end applications that are still outstanding, without being acknowledged. Since such outstanding packets can arrive at a router in a short span of time (and for many different TCP sessions), the router buffer needs to be sized to account for the bandwidth-delay product so that it does not become a bottleneck. This rule of thumb for sizing the router buffer based on the bandwidth-delay product has been around for some time [335], [721].

For consideration of buffer size, we need to be careful about how we interpret delay. The delay here is *not* the propagation delay of the immediate outgoing link; the delay is rather an estimate of the round-trip delay for *most* applications that use this link. A commonly used value of round-trip delay for this purpose is 250 millisec. As an illustration, consider a T3 network link that has a data rate of 45 Mbps; if we assume the delay to be 250 millisec, the window size is 11.25 Megabits, or approximately 1.4 Megabytes. Certainly, this buffer size is a reasonable number for current hardware technology. Now, consider an ultra–high-speed link such as OC-768 that has a data rate of 40 Gbps; for 250 millisec estimate on round-trip time, the rule of thumb would result in 1.25 Gigabytes of buffer size— a number difficult to implement in hardware technology. Thus, a fresh look at buffer sizing is required.

Note that the rule of thumb is quite valid when bulk TCP microflows are synchronized. However, due to the random arrival of TCP sessions, such synchronization may be unlikely. In a series of recent studies, a number of new schemes for core router buffer sizing have been proposed; for example, one proposal [25] suggests that if there are n simultaneous TCP microflows, the buffer size can be set to the bandwidth-delay product divided by $\sqrt{n}$, while another suggests a different view in that it should be proportional to the number of TCP connections [177], and while another proposal [259] suggests that the buffer size in terms of number of IP packets can be set to two times the number of links.

Whether the old rule of thumb is used, or any new rule is used, it is important that router buffer sizing is done adequately for all types of applications that may traverse a network link. Note, however, that the router buffer size is set by the router vendor when a router is shipped. Typically, buffers are carved into different sizes based on the configured maximum transmission unit of each interface. Thus, a network provider does not have the option to change it, except to inquire about it.

The important lesson from the perspective of traffic engineering is that if the router buffer is not sized properly, a network router has the potential to be a bottleneck leading to dropping of packets, thereby reducing TCP throughput between end hosts. Thus, we will assume for the rest of the chapter that buffer sizing is adequately addressed.

7.3 Traffic Engineering: An Architectural Framework

So far our discussion has centered primarily on a single-link system. What are the issues in a network once we go beyond a single-link system? Since a network consists of a number of routers, it is important to estimate source-destination traffic volume rather than on a link basis to obtain a *traffic matrix* that can be used for traffic engineering. Given the traffic volume between different demand pairs and the capacity of network links, the primary traffic engineering goal is to optimize a suitable objective function to obtain the *optimal link weight system* while recognizing that the network uses shortest path routing for forwarding traffic.

The above description requires a bit more clarity in light of OSPF and IS-IS protocols, and where and how traffic engineering fits in. First and foremost, traffic engineering occurs *outside* the actual network. This can be illustrated through an architectural framework of the traffic engineering system as shown in Figure 7.3. From the actual network, traffic measurements are collected to estimate the traffic matrix; furthermore, topology and configuration are also obtained from the network. Based on topology and configuration, along with the traffic matrix, a link weight determination process determines link weights keeping in mind that OSPF/IS-IS uses shortest path routing. The computed link weight for each link is then injected into the network; that is, each router receives metrics for its outgoing links through this external process. Once a router receives these link metrics, it then disseminates through flooding of link-state advertisements (LSAs) to other routers through the normal OSPF/IS-IS flooding process. This would mean that if no new link weights are obtained from the traffic engineering system when the age field of an LSA expires, the router will generate a new LSA by continuing to use the link metric value it received last from the the traffic engineering system. An obvious question then is: how often does the traffic engineering system update the link weights? This is certainly up to each network provider. Currently, most network

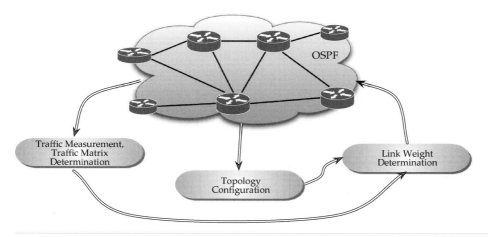

FIGURE 7.3   IP traffic engineering: an architectural framework.

providers use such an approach to update link weights either once a day or once a week, partly to avoid short-term traffic fluctuations by changing link weights too often, and partly since accurate traffic matrix determination from the measurements is a fairly complex and time-consuming process. For additional discussion, see Section 7.8.

7.4 Traffic Engineering: A Four-Node Illustration

We will first discuss the traffic engineering problem in a network by considering a four-node network. Assume that in this four-node network, there is traffic volume for only a single demand pair; this is then a single-commodity problem. We first briefly revisit the single commodity network flow problem described earlier in Chapter 4, assuming that the reader is already familiar with the material presented in Chapter 4. We will then indicate how this problem changes when link weights are introduced.

7.4.1 Network Flow Optimization

We will consider traffic volume to exist for the demand pair 1:2; this pair will be identified as demand identifier 1. We will denote the path from node 1 to 2 via 3 as path number 1, and the path from 1 to 2 via node 4 as path number 2, and denote the flow variables as x_{11} and x_{12}, respectively (see Figure 7.4). Thus, to carry the traffic volume h_1 for demand identifier 1, i.e., from node 1 to node 2, the following must be satisfied:

$$x_{11} + x_{12} = h_1. \tag{7.4.1}$$

Certainly, we require that flow on each path is non-negative, i.e., $x_{11} \geq 0, x_{12} \geq 0$. Let the link be identified as 1 for 1-3, 2 for 3-2, 3 for 1-4, and 4 for 4-2. Then, we can list the flows to satisfy the capacity constraints as follows:

$$x_{11} \leq c_1, \qquad x_{11} \leq c_2, \qquad x_{12} \leq c_3, \qquad x_{12} \leq c_4. \tag{7.4.2}$$

Note that we can combine constraints $x_{11} \leq c_1$, and $x_{11} \leq c_2$ to a single constraint by considering whichever capacity is more stringent, i.e., as $x_{11} \leq \min\{c_1, c_2\}$; this is similar for the other two constraints. However, we will list them all since this is the general representation, un-

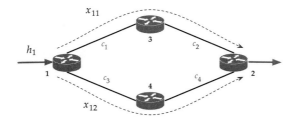

FIGURE 7.4 A four-node network example.

less we consider specific values of capacity. Suppose our goal is to minimize maximum link utilization (see Sections 4.2.3 and 4.3.2). Then, we can write the optimization problem as

$$
\begin{aligned}
minimize_{\{x\}} \quad & F = \max\left\{ \frac{x_{11}}{c_1}, \frac{x_{11}}{c_2}, \frac{x_{12}}{c_3}, \frac{x_{12}}{c_4}, \right\} \\
subject\ to \quad & x_{11} + x_{12} = h_1 \\
& x_{11} \le c_1, \qquad x_{11} \le c_2, \qquad x_{12} \le c_3, \qquad x_{12} \le c_4 \\
& x_{11} \ge 0, \qquad x_{12} \ge 0.
\end{aligned}
\tag{7.4.3}
$$

As discussed earlier in Section 4.3.2, the above problem can be written as the following equivalent linear programming (LP) problem:

$$
\begin{aligned}
minimize_{\{x,r\}} \quad & F = r \\
subject\ to \quad & x_{11} + x_{12} = h_1 \\
& x_{11} \le c_1\,r, \qquad x_{11} \le c_2\,r, \qquad x_{12} \le c_3\,r, \qquad x_{12} \le c_4\,r \\
& x_{11} \ge 0, \qquad x_{12} \ge 0.
\end{aligned}
\tag{7.4.4}
$$

We now consider two specific examples.

Example 7.1 *All links of the same capacity (Figure 7.5).*
Suppose all links are of same capacity, say 100 Mbps, and we are given that $h_1 = 60$ Mbps. Then, two constraints can be dropped, and the problem can be compactly formulated as

$$
\begin{aligned}
minimize_{\{x,r\}} \quad & F = r \\
subject\ to \quad & x_{11} + x_{12} = 60 \\
& x_{11} \le 100\,r \\
& x_{12} \le 100\,r \\
& x_{11}, x_{12} \ge 0.
\end{aligned}
\tag{7.4.5}
$$

Intuitively, it is optimal to split the demand evenly on both paths, i.e., $x_{11}^* = x_{12}^* = 30$. In fact, this is the optimal solution we would get from solving Problem Eq. (7.4.4) if we use an LP solver. Here, the optimal link utilization is $r^* = 30/100 = 0.3$.   ▲

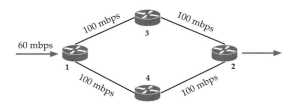

FIGURE 7.5   A four-node network example with the same link capacity.

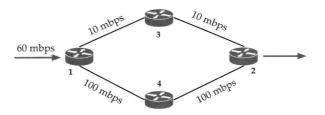

FIGURE 7.6   A four-node network example with different link capacity.

Example 7.2 *Links with different capacity (Figure 7.6).*
Suppose that the demand volume remains at $h_1 = 60$, but capacities of links on path 1-3-2 are 10 Mbps each while those of path 1-4-2 are still at 100 Mbps. Then, it makes sense to send more traffic on path 1-4-2 than on path 1-3-2. How much to send depends on the optimal balance. This can be obtained by noting that at the optimal solution, we must have

$$\frac{x_{11}^*}{10} = \frac{60 - x_{11}^*}{100}. \tag{7.4.6}$$

This implies that path 1-3-2 will be assigned flow $x_{11}^* = 60/11$ and path 1-4-2 will be assigned flow $x_{12}^* = 600/11$. In this case, $r^* = 6/11 \approx 0.5454$. Again, the same optimal solution can be obtained by using an LP solver. Note that while both paths have the same utilization, path 1-4-2 is allocated approximately 90% of the traffic volume.
▲

From the above examples, we can say that network flow optimization of a single-commodity network flow problem results in proportional flow allocations at optimality. What if we use shortest path routing? In the next section, we will discuss the connection between shortest path routing and network flow optimization.

7.4.2 Shortest Path Routing and Network Flow

In an IP network based on the OSPF or IS-IS protocol, the shortest path is computed based on the link weight (cost or metric) that is exchanged through flooding. It is important to note that this computation does not consider the traffic volume or capacity of the network. Thus, the general question is: how is shortest path routing related to network flow modeling? To understand this problem, we consider the four-node problem again, first starting with special cases of capacity illustrated above. We will denote the link weights by the notation w; thus, w_1 is the link weight for link-id 1 (i.e., link 1-3) , w_2 for link-id 2 (i.e., link 3-2), and so on.

Example 7.3 *Optimal flow decision with shortest path routing.*
Consider again the case in which all links have the same capacity, i.e., 100 Mbps. In this case, the optimal decision from network flow optimization was to split the traffic volume

equally among the two paths. Recall that OSPF allows equal-cost multipath (ECMP) (see Section 6.2.7); thus, if we can pick the link weight in such a way that an equal splitting of traffic can be achieved, we achieve the same optimal flow as network flow optimization. This can be realized if we pick the link weight to be 1 on each link (i.e., $w_1 = w_2 = w_3 = w_4 = 1$); we then achieve this splitting since each path cost is 2 and with ECMP, the traffic volume will be equally split. In other words, a hop-based metric will work in this case.

What if the links are of different size, i.e., 10 Mbps for links 1-3 and 3-2 and 100 Mbps for links 1-4 and 4-2? If we still keep the link weight at 1 each, then of the total traffic volume of 60 Mbps, half will try to use path 1-3-2 due to ECMP. However, the capacity limit on this path is 10 Mbps; that is, a 30 Mbps data traffic on this path will cause massive overflow! This means that we need to use some other link weights so that this does not happen. Essentially, we want traffic flow to veer *away* from path 1-3-2 since the capacity of this path is much smaller than the other path. A way to accomplish this would be to set the link weight as the inverse of the link capacity, i.e.,

$$w_1 = w_2 = 1/10,$$
$$w_3 = w_4 = 1/100.$$

This would imply that the path cost for path 1-3-2 is 2/10 while the path cost for path 1-4-2 is 2/100; thus, all of traffic volume, 60 Mbps, will be allocated to path 1-4-2 since 1-4-2 is the shortest path with this set of link weights. In fact, under shortest path routing this is the best we can do, and the maximum link utilization is $60/100 = 0.6$. In other words, it is not possible to achieve the optimality that was achieved with pure network flow optimization where the optimal flows were proportional flows.   ▲

From the above illustration, we can see that link weights is really driving the flow. While it seems that x_{11} is dependent only on link weights w_1 and w_2, the actual values of the other link weights w_3, w_4 do also matter in the allocation of flow to x_{11}; this is similar for x_{12}. That is, flows x_{11} and x_{12} are really dependent on all link weights w_1, w_2, w_3, w_4. If we use $w = (w_1, w_2, w_3, w_4)$ to denote the array of link weights for all links, then we can write the dependency as $x_{11}(w)$ and $x_{12}(w)$. Since the total flow needs to be equal to the traffic demand volume, we can write

$$x_{11}(w) + x_{12}(w) = h_1. \tag{7.4.7}$$

Now compare Eq. (7.4.7) to Eq. (7.4.1); they are almost the same except for the dependency on w. Similarly, the link-flow requirements can be written for dependent flow variables as

$$x_{11}(w) \le c_1, \qquad x_{11}(w) \le c_2, \qquad x_{12}(w) \le c_3, \qquad x_{12}(w) \le c_4. \tag{7.4.8}$$

Again compare Eq. (7.4.8) to Eq. (7.4.2), and note the difference due to dependency on w. In regard to the link weight system w, there are some restrictions on what values a link metric can take; for example, in OSPF, the range is from 1 to $2^{16} - 1$ while in IS-IS the range is from 0 to 63. While in the above illustration we chose link metric also as the inverse of link speed, we can use a normalization factor to change the link metrics to an acceptable range; for example, if we multiply by 100, then the metric for a link with speed 10 Mbps would be 10 ($=100/10$) while the metric for a link with speed 100 Mbps would be 1 ($=100/100$). For simplicity, we denote the set of allowable values for link metrics as $\mathcal{W}$. Thus, similar to Eq. (7.4.4), we can write the following optimization problem:

$$
\begin{aligned}
&minimize_{\{w,\,r\}} &&F = r \\
&subject\ to &&x_{11}(w) + x_{12}(w) = h_1 \\
& &&x_{11}(w) \le c_1\,r, \quad x_{11}(w) \le c_2\,r, \\
& &&x_{12}(w) \le c_3\,r, \quad x_{12}(w) \le c_4\,r \\
& &&x_{11}(w) \ge 0, \quad x_{12}(w) \ge 0 \\
& &&w_1, w_2, w_3, w_4 \in \mathcal{W}.
\end{aligned}
\tag{7.4.9}
$$

We will refer to the above formulation as the *single-commodity shortest path routing–based flow (SCSPRF)* problem, to distinguish it from the *single-commodity network flow* problem presented in Eq. (7.4.4). As noted earlier for Eq. (7.4.4), capacity constraints $x_{11}(w) \le c_1\,r$ and $x_{11}(w) \le c_2\,r$ can be combined into a single constraint by considering the smaller of the link capacities c_1 and c_2 indicating the tighter constraint; this is similar for the other two capacity constraints. It is important to note the following observations in regard to Eq. (7.4.4) and Eq. (7.4.9):

- In Eq. (7.4.9), the main variables are link weights w and maximum link utilization, r; flow variables $x(w)$ are *dependent* variables while in Eq. (7.4.4), the main variables are x and r.

- If we denote the optimal objective cost for the network flow problem Eq. (7.4.4) by $F^*_{netflow}$, and the optimal objective cost due to shortest path–based flow problem Eq. (7.4.9) by F^*_{SPR}, then we can write

$$
F^*_{netflow} \le F^*_{SPR}.
\tag{7.4.10}
$$

Intuitively, we can see this relation since the restriction on flow variables due to the link weight can be thought of as additional restrictions/constraints on the original network flow problem; thus, any additional restrictions can/may increase the optimal cost of the original network flow problem.

In addition, there is an important difference to note. While Eq. (7.4.4) is a linear programming problem, Eq. (7.4.9) is *not*. In addition, Eq. (7.4.9) is not a standard nonlinear programming problem due to implicit functional dependency of flows, x, on the link weight system w. More importantly, Eq. (7.4.9) cannot be directly solved as we have already noticed. Note that so far, we have used two simple rules for choosing link weights in illustrating Example 7.3. We start by summarizing these two rules:

Rule-1: Choose the link weights to be based on hop count, to be referred to as a *hop-based* metric

Rule-2: Choose the link weights to be based on the inverse of the link speed, to be referred to as an *inverse-of-the-link speed* metric.

We digress a bit further with these two rules in relation to earlier examples by considering certain variations.

Example 7.4 *Changing the traffic volume in Example 7.3.*

Consider reducing the traffic volume for pair 1:2 from 60 Mbps to 5 Mbps. In this case, we note that even if we use Rule 1 on the link capacity as given in Figure 7.6, we do not face the overflow problem discussed earlier in Example 7.3. This would cause the maximum link utilization to be 0.25 since 2.5 Mbps of traffic volume would be allocated to path 1-3-2 with the remaining 2.5 Mbps allocated to path 1-4-2 due to ECMP. If Rule 2 is used, then all traffic would be allocated to path 1-4-2, and thus, the maximum link utilization in this case is 0.05. With either rule, we can see that when traffic volume is low compared to network capacity, the utilization remains low.

Now consider the topology with all links being 100 Mbps (Figure 7.5). In this case with 5 Mbps of network traffic, we will arrive at the same maximum link utilization (= 0.025) with either rule for link weight.   ▲

From the above illustration, it is easy to see that if all links in a network are of the same speed, then the link weight based on the inverse-of-the-link speed is the same as the hop-based metric. Thus, Rule 1 can be thought of as a special case of Rule 2. We next consider an example where due to an anticipated increase in network traffic, a new link is added; this actually falls under the network dimensioning problem. It should be noted that there are systematic ways to address the network dimensioning problem in terms of where to add capacity and if new links should be added; for a detailed discussion, see [564]. Here, we present a simple illustration to address the impact on the link weight selection rules.

Example 7.5 *Topology change in anticipation of increase in traffic volume.*

Consider the network shown in Figure 7.7. Note that there is one key change from the topology shown in Figure 7.5: a new direct link between node 1 and node 2 is added; we will identify this link as link number 5, and the link weight as w_5, while keeping the link numbering for other links as before. Suppose that this link was added in anticipation of an increase in traffic volume. Now we have one more path possible from node 1 to node 2; we label this path as path number 3 with the flow variable labeled as x_{13}. Formulation (7.4.9) will now change to the following:

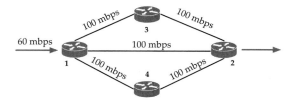

FIGURE 7.7   A four-node network example with five links.

$$\begin{aligned}
\textit{minimize}_{\{w, r\}} \quad & F = r \\
\textbf{subject to} \quad & x_{11}(w) + x_{12}(w) + x_{13}(w) = 60 \\
& x_{11}(w) \leq 100\,r \\
& x_{12}(w) \leq 100\,r \\
& x_{13}(w) \leq 100\,r \\
& x_{11}(w), x_{12}(w), x_{13}(w) \geq 0 \\
& w_1, w_2, w_3, w_4, w_5 \in \mathcal{W}.
\end{aligned}$$
(7.4.11)

Note the difference between Eq. (7.4.9) and Eq. (7.4.11). In the latter, specific values of capacity and traffic volumes are shown along with the new path; furthermore, since the link capacities are the same, only a single-capacity constraint is shown for path number 1 and path number 2.

We can easily see that regardless of whether we use Rule 1 or Rule 2, all traffic will be allocated to the direct link path on link number 5 with shortest path routing since the direct link is the shortest path under either rule for the link weight. In this case, the maximum link utilization is $r = 60/100 = 0.6$.   ▲

From the above illustration, we note that Rule 2 does not always work well since in this instance, by adding a new link, we have syphoned all traffic to take the new link, instead of equally splitting flow allocation among the three paths. This means that we have *increased* maximum link utilization, thereby *increasing* average network delay for a traffic volume of 60 Mbps by *adding* new link/capacity compared to when the direct link did not exist. This is certainly counterintuitive. In road transportation networks, an analogous situation has long been known; it states that under certain load conditions the travel cost can *increase* with the addition of a new link (road), which is known as *Braess' Paradox* ([91], [280], [447]). A phenomenon similar to Braess' paradox can be *induced* in IP networks if link weights are *not* chosen properly.

Going back to the network shown in Figure 7.7, an important question is: can we choose a better set of link weights that will reduce maximum link utilization compared to when Rule 2 is used, i.e., a better solution to Eq. (7.4.11)? This is possible if we choose the link weights as follows (see Figure 7.8):

$$w_1 = w_2 = w_3 = w_4 = 1, \quad w_5 = 2.$$

This way, the path cost for all paths are 2; thus, due to ECMP, traffic volume will be equally split among the three paths, thereby reducing maximum link utilization, r, to 0.2

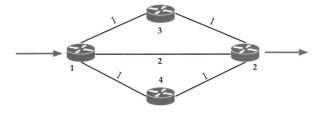

FIGURE 7.8   A four-node network example with link weights.

from 0.6. In fact, this set of link weights is optimal for the SCSPRF problem, Eq. (7.4.11). It should be noted that the optimal link weights are not unique; for example, if we choose

$$w_1 = w_2 = w_3 = w_4 = 2, \quad w_5 = 4,$$

it will result in the same optimal maximum link utilization value, $r^* = 0.2$.

While for Eq. (7.4.11), we have found an optimal set of link weights by inspecting the data for the problem, this is not always easy, especially for a large network problem. In the next sections, we discuss the general problem and possible solutions.

7.5 Link Weight Determination Problem: Preliminary Discussion

In the previous section, we emphasized the need to determine a good set of link weights by considering a single-commodity problem. In an IP network, there can be traffic volume between any pair of routers that serves IP subnets; thus, the general problem is *multicommodity* in nature. Our goal here is to determine link weights for given traffic volume demand and capacity limits where a certain objective is optimized. A good candidate for this objective/goal is to minimize the maximum link utilization in the network; thus, we will start with this objective; later, we will consider other objectives as well.

From the discussion in the previous section for the single-commodity example, you can see the similarity between the *multicommodity network flow (MCNF)* problem discussed earlier in Section 4.4 and the *multicommodity shortest path–based routing flow (MCSPRF)* problem we face for IP network traffic engineering; the key difference is that the MCSPRF problem is required to have the link weight as the main variables. Thus, analogous to Formulation (4.4.10) for the MCNF problem with minimizing maximum link utilization, we can state the MCSPRF formulation with the same objective as follows:

$$
\begin{aligned}
&\text{minimize}_{\{w,r\}} && F = r \\
&\text{subject to} && \sum_{p=1}^{P_k} x_{kp}(w) = h_k, && k = 1, 2, ..., K \\
& && \sum_{k=1}^{K} \sum_{p=1}^{P_k} \delta_{kp\ell} x_{kp}(w) = y_\ell, && \ell = 1, 2, ..., L \\
& && y_\ell \le c_\ell\, r, && \ell = 1, 2, ..., L \\
& && w_1, w_2, ..., w_L \in \mathcal{W} \\
& && x_{kp}(w) \ge 0, && p = 1, 2, ..., P_k, \quad k = 1, 2, ..., K \\
& && y_\ell \ge 0, && \ell = 1, 2, ..., L \\
& && r \ge 0.
\end{aligned}
\tag{7.5.1}
$$

Notations for this problem and other related problems are summarized in Table 7.1. If we *relax* the requirement on path flow being subject to link weights, then the corresponding MCNF problem can be written as

TABLE 7.1 Summary of notation used in MCNF and MCSPRF formulations.

Notation	Explanation
K	Number of demand pairs with positive demand volume
L	Number of links
h_k	Demand volume of demand index $k = 1, 2, ..., K$
c_ℓ	Capacity of link $\ell = 1, 2, ..., L$
P_k	Number of candidate paths for demand k, $k = 1, 2, ..., K$
$\delta_{kp\ell}$	Link-path indicator, set to 1 if path p for demand pair k uses the link ℓ; 0, otherwise
ξ_{kp}	Unit cost of flow on path p for demand k
$\hat{\xi}_\ell$	Unit cost of flow on link ℓ
w_ℓ	Link weight for link $\ell = 1, 2, ..., L$
$x_{kp}(\boldsymbol{w})$	Flow amount on path p for demand k for given link weight system $\boldsymbol{w}$
x_{kp}	Flow amount on path p for demand k
y_ℓ	Link flow variable for link ℓ
r	maximum link utilization variable
*	Use as a superscript with a variable to indicate optimal solution, e.g., x_{kp}^*

$$\textit{minimize}_{\{x,y,r\}} \quad F = r$$

$$\textit{subject to} \quad \sum_{p=1}^{P_k} x_{kp} = h_k, \qquad k = 1, 2, ..., K$$

$$\sum_{k=1}^{K} \sum_{p=1}^{P_k} \delta_{kp\ell} x_{kp} = y_\ell, \quad \ell = 1, 2, ..., L \qquad (7.5.2)$$

$$y_\ell \le c_\ell r, \qquad \ell = 1, 2, ..., L$$

$$x_{kp} \ge 0, \qquad p = 1, 2, ..., P_k, \quad k = 1, 2, ..., K$$

$$y_\ell \ge 0, \qquad \ell = 1, 2, ..., L$$

$$r \ge 0.$$

If we denote the optimal objective cost for MCNF Formulation (7.5.2) by F^*_{MCNF}, and the optimal objective cost for MCSPRF Formulation (7.4.9) by F^*_{MCSPRF}, we can write

$$F^*_{\text{MCNF}} \le F^*_{\text{MCSPRF}}. \qquad (7.5.3)$$

That is, this relation holds much like the single-commodity illustration given earlier.

Why is it important to consider the relaxed problem shown by Eq. (7.5.2)? It so happens that the relaxed problem, which is an LP problem, has an equivalent LP problem called the *dual* that allows us to obtain a set of link weights; not only that, commercial LP solvers can be used on the network flow problem, without needing to develop a specialized algorithm, to obtain link weights; at least, this is doable for networks of reasonable size. In other words, we cannot completely rule out development of specialized algorithms for determining link weights. In any case, link weights so obtained may not be from the allowable range; thus, some transformation/scaling might be necessary. Once we have made this adjustment on link weights, we can determine flows and compute the objective cost of Eq. (7.5.1) to see how far this is from Eq. (7.5.2). It may be noted that flows so obtained based on dual-based link weights may not necessarily match the flow directly obtained from solving the original MCNF problem since the optimal solution to the original problem Eq. (7.5.2) can result in proportional flows. Furthermore, flow allocation based on a dual-based weight can be different depending on whether the network has the ECMP feature activated. Finally, the dual-based approach holds for any MCNF problem; that is, this result is *not* dependent on the specific objective function discussed above as long as the objective function is *linear*. This will be clear from the discussion in the next section.

7.6 Duality of the MCNF Problem

We will now consider the MCNF problem with different objective functions and the corresponding dual problems for different objectives. We first start with minimum cost routing for a three-node network to illustrate how dual problems are formulated.

   While the minimum cost routing objective is not an appropriate objective for the IP traffic engineering problem, it is a good one to help understand the dual problem; later, we will consider objective functions that are appropriate for IP traffic engineering, discuss how the dual changes, and the related impact on link weights.

7.6.1 Illustration of Duality Through a Three-Node Network

Consider minimum cost routing for the three-node MCNF problem discussed earlier in Section 4.3.1 in its index-based formulation presented in Section 4.4. We reproduce Problem (4.4.3) below:

$$
\begin{aligned}
&\text{minimize}_{\{x\}} \quad F = \xi_{11}x_{11} + \xi_{12}x_{12} + \xi_{21}x_{21} + \xi_{22}x_{22} + \xi_{31}x_{31} + \xi_{32}x_{32} \\
&\text{subject to} \\
&\quad x_{11} + x_{12} \qquad\qquad\qquad\qquad\qquad\;\; = \; h_1 \\
&\qquad\qquad\;\; x_{21} + x_{22} \qquad\qquad\qquad\;\; = \; h_2 \\
&\qquad\qquad\qquad\qquad\quad x_{31} + x_{32} \;\; = \; h_3 \\
&\quad x_{11} \qquad\qquad\quad + x_{22} \qquad + x_{32} \;\; \le \; c_1 \\
&\qquad x_{12} + x_{21} \qquad\qquad\quad + x_{32} \;\; \le \; c_2 \\
&\qquad x_{12} \qquad\quad + x_{22} + x_{31} \qquad\quad\;\; \le \; c_3 \\
&\quad x_{11},\, x_{12},\, x_{21},\, x_{22},\, x_{31},\, x_{32} \ge 0.
\end{aligned}
\tag{7.6.1}
$$

We assume that a unit cost of a path is the sum of the unit cost of the links of which this path is made (i.e., $\xi_{11} = \hat{\xi}_1$, $\xi_{12} = \hat{\xi}_2 + \hat{\xi}_3$, $\xi_{21} = \hat{\xi}_2$, $\xi_{22} = \hat{\xi}_1 + \hat{\xi}_3$, $\xi_{31} = \hat{\xi}_3$, $\xi_{32} = \hat{\xi}_1 + \hat{\xi}_2$). We can then rewrite Problem (7.6.1) by first changing less-than-equal-to constraints to greater-than-equal-to constraints, and associating a dual variable with each constraint (indicated on the right side in parentheses), as

$$\text{minimize}_{\{x\}} \quad F = \hat{\xi}_1 x_{11} + (\hat{\xi}_2 + \hat{\xi}_3) x_{12} + \hat{\xi}_2 x_{21} + (\hat{\xi}_1 + \hat{\xi}_3) x_{22} + \hat{\xi}_3 x_{31} + (\hat{\xi}_1 + \hat{\xi}_3) x_{32}$$

subject to

(dual variables)

$$
\begin{array}{rcll}
x_{11} + x_{12} & = & h_1 & (\nu_1) \\
x_{21} + x_{22} & = & h_2 & (\nu_2) \\
x_{31} + x_{32} & = & h_3 & (\nu_3) \\
-x_{11} \quad - x_{22} \quad - x_{32} & \geq & -c_1 & (\pi_1) \\
-x_{12} - x_{21} \quad - x_{32} & \geq & -c_2 & (\pi_2) \\
-x_{12} \quad - x_{22} - x_{31} & \geq & -c_3 & (\pi_3)
\end{array}
$$

$$x_{11},\ x_{12},\ x_{21},\ x_{22},\ x_{31},\ x_{32} \geq 0.$$

(7.6.2)

We now assign a dual variable with each constraint—an unrestricted variable if it is an equality constraint and a non-negative variable if it is a greater-than-or-equal-to constraint. Then, the dual LP problem is:

$$\text{maximize}_{\{v,\pi\}} \quad F_D = h_1 \nu_1 + h_2 \nu_2 + h_3 \nu_3 - c_1 \pi_1 - c_2 \pi_2 - c_3 \pi_3$$

subject to

$$
\begin{array}{rcl}
\nu_1 \quad - \pi_1 & \leq & \hat{\xi}_1 \\
\nu_1 \quad - \pi_2 - \pi_3 & \leq & \hat{\xi}_2 + \hat{\xi}_3 \\
\nu_2 \quad - \pi_2 & \leq & \hat{\xi}_2 \\
\nu_2 \quad - \pi_1 \quad - \pi_3 & \leq & \hat{\xi}_1 + \hat{\xi}_3 \\
\nu_3 \quad - \pi_3 & \leq & \hat{\xi}_3 \\
\nu_3 - \pi_1 - \pi_2 & \leq & \hat{\xi}_1 + \hat{\xi}_2
\end{array}
$$

$$\nu_1,\ \nu_2,\ \nu_3 \text{ unrestricted}$$
$$\pi_1,\ \pi_2,\ \pi_3 \geq 0.$$

(7.6.3)

While it may not be obvious, there is a pattern to writing the dual. First, the cost coefficients from the original problem go on the right-hand side of the constraints; thus, coefficients $\hat{\xi}$ from Eq. (7.6.2) are on the right-hand side of constraints in Eq. (7.6.3). The right-hand side constraints of the original problem, Eq. (7.6.2), become coefficients in the objective for the dual, Eq. (7.6.3); that is, h and c from Eq. (7.6.2) are coefficients in the objective in Eq. (7.6.3). Finally, coefficients 1, 0, or -1, associated with *rows* on the left-hand side of a constraint for Eq. (7.6.2), show up in *columns* on the left-hand side of constraints for the dual given by Eq. (7.6.3); that is, this is a *transposed* view. Rewriting, we have

$$\text{maximize}_{\{v,\pi\}} \quad F_D = h_1 \nu_1 + h_2 \nu_2 + h_3 \nu_3 - c_1 \pi_1 - c_2 \pi_2 - c_3 \pi_3$$

subject to

$$\nu_1 \leq \hat{\xi}_1 + \pi_1$$
$$\nu_1 \leq (\hat{\xi}_2 + \pi_2) + (\hat{\xi}_3 + \pi_3)$$
$$\nu_2 \leq \hat{\xi}_2 + \pi_2$$
$$\nu_2 \leq (\hat{\xi}_1 + \pi_1) + (\hat{\xi}_3 + \pi_3)$$
$$\nu_3 \leq \hat{\xi}_3 + \pi_3$$
$$\nu_3 \leq (\hat{\xi}_1 + \pi_1) + (\hat{\xi}_2 + \pi_2)$$

$$\nu_1,\ \nu_2,\ \nu_3 \text{ unrestricted}$$
$$\pi_1,\ \pi_2,\ \pi_3 \geq 0.$$

(7.6.4)

7.6.2 General Case: Minimum Cost Routing

Recall that minimum cost routing for the MCNF problem was discussed earlier in Section 4.3.1 in its index-based formulation presented in Section 4.4. Notations for this and other related problems are summarized in Table 7.1. Thus, we start with the general formulation corresponding to Eq. (7.6.1):

$$minimize_{\{x\}} \quad F = \sum_{k=1}^{K} \sum_{p=1}^{P_k} \xi_{kp} x_{kp}$$

$$subject \ to \quad \sum_{p=1}^{P_k} x_{kp} = h_k, \qquad\qquad k = 1, 2, ..., K$$

$$\sum_{k=1}^{K} \sum_{p=1}^{P_k} \delta_{kp\ell} x_{kp} \leq c_\ell, \quad \ell = 1, 2, ..., L$$

$$x_{kp} \geq 0, \qquad\qquad p = 1, 2, ..., P_k, \quad k = 1, 2, ..., K. \qquad (7.6.5)$$

The above LP problem is then the network flow relaxation of the following MCSPRF problem:

$$minimize_{\{w\}} \quad F = \sum_{k=1}^{K} \sum_{p=1}^{P_k} \xi_{kp} x_{kp}(\boldsymbol{w})$$

$$subject \ to \quad \sum_{p=1}^{P_k} x_{kp}(\boldsymbol{w}) = h_k, \qquad\qquad k = 1, 2, ..., K$$

$$\sum_{k=1}^{K} \sum_{p=1}^{P_k} \delta_{kp\ell} x_{kp}(\boldsymbol{w}) \leq c_\ell, \quad \ell = 1, 2, ..., L$$

$$w_1, w_2, ..., w_L \in \mathcal{W}$$

$$x_{kp}(\boldsymbol{w}) \geq 0, \qquad\qquad p = 1, 2, ..., P_k, \quad k = 1, 2, ..., K. \qquad (7.6.6)$$

We consider the unit path flow cost ξ_{kp} to be the summation of the unit flow cost on links that make up the path. Suppose we denote the unit link-flow cost to be $\hat{\xi}_\ell$ on link $\ell (\ell = 1, 2, ..., L)$. Then, ξ_{kp} for path p for demand k can be given by

$$\xi_{kp} = \sum_{\ell=1}^{L} \delta_{kp\ell} \hat{\xi}_\ell. \qquad\qquad\qquad (7.6.7)$$

To apply LP duality, we associate dual variables v_k with demand k ($k = 1, 2, ..., K$), and dual variables π_ℓ with each link ℓ. First, we rearrange the capacity constraints as greater-than-or-equal-to constraints; due to LP duality theory, this then makes the associated dual variables π_ℓ non-negative. In a similar manner for equality constraints, LP duality theory says that dual variables become unrestricted. Also note that all terms associated with a variable are written on the left-hand side of the constraint while constants are written on the right-hand side of the constraints; this helps in properly identifying and writing the dual. Thus, we rewrite Eq. (7.6.5) as

$$\textit{minimize}_{\{x\}} \quad F = \sum_{k=1}^{K} \sum_{p=1}^{P_k} (\sum_{\ell=1}^{L} \delta_{kp\ell} \hat{\xi}_\ell) x_{kp}$$

$$\textit{subject to} \quad \sum_{p=1}^{P_k} x_{kp} = h_k, \qquad\qquad k = 1, 2, ..., K \qquad (v_k)$$

$$-\sum_{k=1}^{K} \sum_{p=1}^{P_k} \delta_{kp\ell} x_{kp} \geq -c_\ell, \qquad \ell = 1, 2, ..., L \qquad (\pi_\ell)$$

$$x_{kp} \geq 0, \qquad\qquad\qquad\qquad p = 1, 2, ..., P_k,$$
$$k = 1, 2, ..., K.$$

(7.6.8)

You may compare this problem with the counterpart for the three-node network given in Eq. (7.6.2). The original problem when discussed with its dual problem is referred to as the *primal* problem; thus, we will refer to Eq. (7.6.8) as the primal problem. Then the dual LP problem of primal problem Eq. (7.6.8) can be written as the following maximization problem:

$$\textit{maximize}_{\{v, \pi\}} \quad F_D = \sum_{k=1}^{K} h_k v_k - \sum_{\ell=1}^{L} c_\ell \pi_\ell$$

$$\textit{subject to} \quad v_k - \sum_{\ell=1}^{L} \delta_{kp\ell} \pi_\ell \leq \sum_{\ell=1}^{L} \delta_{kp\ell} \hat{\xi}_\ell, \quad p = 1, 2, ..., P_k, \quad k = 1, 2, ..., K$$

$$v_k \text{ unrestricted}, \qquad\qquad k = 1, 2, ..., K$$
$$\pi_\ell \geq 0, \qquad\qquad\qquad\quad \ell = 1, 2, ..., L.$$

(7.6.9)

This general formulation then corresponds to the dual formulation for the three-node network example given in Eq. (7.6.3). Note that with duality, coefficients $\hat{\xi}$ from the original problem show up on the right-hand side of the constraint in the dual problem and vice versa. The information about coefficient in the constraints appears in transposed form in the dual. By rearranging, we can write the dual as

$$\textit{maximize}_{\{v, \pi\}} \quad F_D = \sum_{k=1}^{K} h_k v_k - \sum_{\ell=1}^{L} c_\ell \pi_\ell$$

$$\textit{subject to} \quad v_k \leq \sum_{\ell=1}^{L} \delta_{kp\ell} (\hat{\xi}_\ell + \pi_\ell), \qquad p = 1, 2, ..., P_k, \quad k = 1, 2, ..., K$$

$$v_k \text{ unrestricted}, \qquad\qquad k = 1, 2, ..., K$$
$$\pi_\ell \geq 0, \qquad\qquad\qquad\quad \ell = 1, 2, ..., L.$$

(7.6.10)

Note that this formulation is corresponding the model shown for the three-node network example in Eq. (7.6.4).

There is an important relation between the objective function value of the primal and the dual. Note that

$$
\begin{aligned}
F &= \sum_{k=1}^{K} \sum_{p=1}^{P_k} \left(\sum_{\ell=1}^{L} \delta_{kp\ell} \hat{\xi}_\ell \right) x_{kp} \\
&\geq \sum_{k=1}^{K} \sum_{p=1}^{P_k} \left(v_k - \sum_{\ell=1}^{L} \delta_{kp\ell} \pi_\ell \right) x_{kp} \\
&= \sum_{k=1}^{K} v_k \sum_{p=1}^{P_k} x_{kp} - \sum_{\ell=1}^{L} \left(\sum_{k=1}^{K} \sum_{p=1}^{P_k} \delta_{kp\ell} x_{kp} \right) \pi_\ell \\
&\geq \sum_{k=1}^{K} h_k v_k - \sum_{\ell=1}^{L} c_\ell \pi_\ell \\
&= F_{\mathrm{D}}.
\end{aligned}
\tag{7.6.11}
$$

That is, the primal objective is greater than or equal to the dual objective; in fact, this property holds for any LP problem and is known as the *weak duality theorem*. In light of the MCSPRF problem and its LP relaxation MCNF problem, and now to the above duality result, and by denoting the objective function values as F_{MCSPRF}, F_{MCNF}, and $F_{\mathrm{Dual\text{-}of\text{-}MCNF}}$, respectively, we can write $F_{\mathrm{DUAL\text{-}of\text{-}MCNF}} \leq F_{\mathrm{MCNF}} \leq F_{\mathrm{MCSPRF}}$. Furthermore, at optimality, assuming that the primal problem is feasible, the following holds:

$$
F^*_{\mathrm{DUAL\text{-}of\text{-}MCNF}} = F^*_{\mathrm{MCNF}} \leq F^*_{\mathrm{MCSPRF}}.
\tag{7.6.12}
$$

Since the dual is a maximization problem, this means that for any dual variable values that satisfy the constraints in the dual problem, we can compute the objective function, which can serve as a lower bound to the MCSPRF problem, and we can determine the gap by determining the difference.

We now go back to the general formulations: Eq. (7.6.8) and its dual Eq. (7.6.10). Why is the dual important to consider? The *optimality conditions* for LP problems state that if $\boldsymbol{x}^*$ is optimal for primal problem Eq. (7.6.8), and $\boldsymbol{v}^*$ and $\boldsymbol{\pi}^*$ are optimal for dual problem Eq. (7.6.10), then the following must be satisfied:

1. Primal solutions $\boldsymbol{x}^*$ must satisfy constraints in Eq. (7.6.8).

2. Dual solutions $\boldsymbol{v}^*$, $\boldsymbol{\pi}^*$ must satisfy constraints in Eq. (7.6.10).

3. The following *complementary slackness condition* must be satisfied:

$$
x_{kp} \left(v_k - \sum_{\ell=1}^{L} \delta_{kp\ell} (\hat{\xi}_\ell + \pi_\ell) \right) = 0, \quad p = 1, 2, ..., P_k, \quad k = 1, 2, ..., K.
\tag{7.6.13a}
$$

$$
\pi_\ell \left(c_\ell - \sum_{k=1}^{K} \sum_{p=1}^{P_k} \delta_{kp\ell} x_{kp} \right) = 0, \quad \ell = 1, 2, ..., L.
\tag{7.6.13b}
$$

That is, the product of a primal (dual) constraint and its associated dual (primal) variable is zero. Here, the first one is shown for primal variable x_{kp} and its associated dual constraint and the second one for dual variable π_ℓ and its associated primal constraint. Note that there is none listed for dual variables $\mathbf{v}$ since its associated primal constraints are equality constraints ($\sum_{p=1}^{P_k} x_{kp} = h_k$); thus, the product is zero and does not need to be listed.

First note that due to the second condition, that is, satisfying dual constraints in Eq. (7.6.10), we can say that the modified path cost, $\sum_{\ell=1}^{L} \delta_{kp\ell}(\hat{\xi}_\ell + \pi_\ell^*)$, for each path for demand k must be at least as large as the *commodity cost* reflected by dual variable v_k for demand k. Furthermore, condition (7.6.13a) indicates that if x_{kp}^* for any path p for demand k is positive, i.e., if a path for a demand has a positive flow, then the path cost, $\sum_{\ell=1}^{L} \delta_{kp\ell}(\hat{\xi}_\ell + \pi_\ell^*)$, for this path must be equal to the *commodity cost* v_k^*. Note that $\delta_{kp\ell}$ defines which links are using this optimal path; thus, $\hat{\xi}_\ell + \pi_\ell$ is the *modified link cost* for link ℓ. This modified link cost then takes us back to the link weight, w_ℓ, for the original shortest path routing problem. To summarize, we have the following important result [6]:

Result 7.1. *For the MCNF problem given by Eq. (7.6.8) and its corresponding dual, Eq. (7.6.10), the commodity cost, v_k^*, is the shortest distance for demand k with respect to the link weight $w_\ell = \hat{\xi}_\ell + \pi_\ell^*$, and at optimality, every path for demand k that carries a positive flow must be a shortest path with respect to the link cost system given by*

$$w_\ell = \hat{\xi}_\ell + \pi_\ell^* \tag{7.6.14}$$

for $\ell = 1, 2, ..., L$.

Based on the above result, we make the following important remark:

Remark 7.3. *Implications of Result 7.1.*
    We have the following observations:

1. If we can find the dual optimal solution π_ℓ^*, we then have a link weight system available, given by $w_\ell = \hat{\xi}_\ell + \pi_\ell^*$, for $\ell = 1, 2, ..., L$ for the MCSPRF problem.

2. In most LP solvers, it is in fact *not* necessary to transform Problem (7.6.8) to its dual. Problem Eq. (7.6.8) can be directly solved and the dual solution, π_ℓ^*, is readily available, which can be used in turn to obtain w_ℓ^*.

3. If two paths for the same demand identifier k have positive flows, it does not mean that they will be equal in the primal MCNF problem where optimal flows can be proportional. On the other hand, if flow is *allocated* based on the solution of the dual problem, this flow allocation will follow the MCSPRF problem, with ECMP being an added feature.

4. Multiple flows being positive for a demand k does not mean that the flows would be equal since optimal MCNF can result in proportional flows; this is an important difference com-

pared to the MCSPRF problem since w_ℓ would tell the MCSPRF problem to allocate flow based on shortest path routing (along with ECMP).   ◆

7.6.3 Minimization of Maximum Link Utilization

We next consider the objective to be minimization of maximum link utilization. How does the link weight selection change for a different objective function? The MCNF problem presented earlier Eq. (7.5.2) can be written in the following format where the dual variables are also identified:

$$
\begin{aligned}
&minimize_{\{x,r\}} && F = r \\
&subject\ to && \sum_{p=1}^{P_k} x_{kp} = h_k, && k = 1, 2, ..., K && (v_k) \\
& && -\sum_{k=1}^{K}\sum_{p=1}^{P_k} \delta_{kp\ell}\, x_{kp} + c_\ell\, r \geq 0, && \ell = 1, 2, ..., L && (\pi_\ell) \\
& && x_{kp} \geq 0, && p = 1, 2, ..., P_k, \\
& && && k = 1, 2, ..., K. \\
& && r \geq 0.
\end{aligned}
\tag{7.6.15}
$$

Note that in this case there are no coefficients associated with flow variables x in the objective, which means that they are zero; thus, this will show up as zeros on the right-hand side in the dual. There is, however, a coefficient with r in the objective, which is 1; this must be accounted for in the dual. Note that the right-hand side with capacity constraints is zero; thus, this part would not contribute to the objective function in the dual. Thus, the dual can be written as

$$
\begin{aligned}
&maximize_{\{v,\pi\}} && F = \sum_{k=1}^{K} h_k v_k \\
&subject\ to && v_k - \sum_{\ell=1}^{L} \delta_{kp\ell} \pi_\ell \leq 0, && p = 1, 2, ..., P_k, \quad k = 1, 2, ..., K \\
& && \sum_{\ell=1}^{L} c_\ell \pi_\ell \leq 1 \\
& && v_k\ \text{unrestricted}, && k = 1, 2, ..., K \\
& && \pi_\ell \geq 0, && \ell = 1, 2, ..., L.
\end{aligned}
\tag{7.6.16}
$$

If we move the term associated with π_ℓ in the first set of constraints to the right-hand side, we can re-write the dual as

$$
\begin{aligned}
&maximize_{\{v,\pi\}} && F = \sum_{k=1}^{K} h_k v_k \\
&subject\ to && v_k \leq \sum_{\ell=1}^{L} \delta_{kp\ell} \pi_\ell, && p = 1, 2, ..., P_k, \quad k = 1, 2, ..., K \\
& && \sum_{\ell=1}^{L} c_\ell \pi_\ell \leq 1 \\
& && v_k\ \text{unrestricted}, && k = 1, 2, ..., K \\
& && \pi_\ell \geq 0, && \ell = 1, 2, ..., L.
\end{aligned}
\tag{7.6.17}
$$

If $\mathbf{v}^*$ and $\boldsymbol{\pi}^*$ are the optimal solutions to this problem, then by comparing this to our discussion about minimum cost routing, we can see that in this case the link weights would be $w_\ell = \pi_\ell^*$, $\ell = 1, 2, ..., L$ with the requirement that each π_ℓ satisfies $\sum_{\ell=1}^{L} c_\ell \pi_\ell \le 1$. In fact, at optimality $\sum_{\ell=1}^{L} c_\ell \pi_\ell^* = 1$, i.e., this constraint is said to be a *binding* constraint at optimality. This is easy to see from the complementary slackness condition at optimality. For Eq. (7.6.15), the condition related to this constraint would take the following form:

$$r^* \left(\sum_{\ell=1}^{L} c_\ell \pi_\ell^* - 1 \right) = 0. \tag{7.6.18}$$

If $\sum_{\ell=1}^{L} c_\ell \pi_\ell^* < 1$ at optimality, then this would mean that $r^* = 0$ at optimality; this is not possible since this would mean that the maximum link utilization is zero (rather, this is theoretically possible only if traffic flow for every demand pair is on an infinite capacity link).

Thus, if the objective is to minimize the maximum link utilization, we can summarize the following result:

Result 7.2. *For MCNF Formulation (7.6.15) and its corresponding dual given by Eq. (7.6.17), the commodity cost, v_k^*, is the shortest distance for demand k with respect to link weight $w_\ell = \pi_\ell^*$, and at optimality, every path for demand k that carries a positive flow must be a shortest path with respect to the link cost system given by*

$$w_\ell = \pi_\ell^* \tag{7.6.19}$$

for $\ell = 1, 2, ..., L$ where $\sum_{\ell=1}^{L} c_\ell \pi_\ell^* = 1$.

Based on the above result, we make the following remark:

Remark 7.4. *Comparison of dual-based link weights based on Eq. (7.6.8) and Eq. (7.6.15).*

Comparing Result 7.2 to Result 7.1 from the previous section, we can see how the link weight selection can change depending on the objective function used and the form of the constraints. In either case, it is important to note that dual solution π_ℓ^* takes non-negative values. However, routing protocols such as OSPF and IS-IS allow non-negative *integer* values. Thus, some adjustments from the solution from the dual are required to obtain integer weights. Furthermore, OSPF does not allow any link metric to be zero since its range starts from 1, unlike IS-IS, which starts from 0. Yet zero is a possible link weight for a link if the objective chosen is the minimization of maximum link utilization; thus, for this objective, an additional adjustment would be needed to avoid a link being assigned metric zero by the link weight determination procedure if the weight so determined were to be used in an OSPF environment.   ◆

Finally, in Problem (7.6.15), scaling can be directly addressed by changing the objective function from just r to βr where β is a large positive number. This then changes the constraint

$\sum_{\ell=1}^{L} c_\ell \pi_\ell^* = 1$ to $\sum_{\ell=1}^{L} c_\ell \pi_\ell^* = \beta$ for the dual problem; thus, πs need not be restricted to less than 1 if scaling is addressed.

7.6.4 A Composite Objective Function

A composite objective function that combines minimum cost routing with minimization of maximum link utilization can also be considered by allocating positive weights α and β, respectively; such a composite objective is referred to as a *utility function*. The MCNF problem with this composite objective can be written as

$$
\begin{aligned}
&\textit{minimize}_{\{x,r\}} \quad F = \alpha \sum_{k=1}^{K} \sum_{p=1}^{P_k} \left(\sum_{\ell=1}^{L} \hat{\xi}_\ell \delta_{kp\ell} \right) x_{kp} + \beta r \\
&\textit{subject to} \quad \sum_{p=1}^{P_k} x_{kp} = h_k, \qquad\qquad\qquad k = 1, 2, ..., K \qquad (v_k) \\
&\qquad\qquad\quad -\sum_{k=1}^{K} \sum_{p=1}^{P_k} \delta_{kp\ell} x_{kp} + c_\ell r \ge 0, \qquad \ell = 1, 2, ..., L \qquad (\pi_\ell) \\
&\qquad\qquad\quad x_{kp} \ge 0, \qquad\qquad\qquad\qquad p = 1, 2, ..., P_k, \quad k = 1, 2, ..., K. \\
&\qquad\qquad\quad r \ge 0.
\end{aligned}
\tag{7.6.20}
$$

Then, the dual problem becomes

$$
\begin{aligned}
&\textit{maximize}_{\{v,\pi\}} \quad F = \sum_{k=1}^{K} h_k v_k \\
&\textit{subject to} \quad v_k - \sum_{\ell=1}^{L} \delta_{kp\ell} \pi_\ell \le \alpha \sum_{\ell=1}^{L} \delta_{kp\ell} \hat{\xi}_\ell, \quad p = 1, 2, ..., P_k, \quad k = 1, 2, ..., K \\
&\qquad\qquad\quad \sum_{\ell=1}^{L} c_\ell \pi_\ell \le \beta \\
&\qquad\qquad\quad v_k \text{ unrestricted}, \qquad\qquad k = 1, 2, ..., K \\
&\qquad\qquad\quad \pi_\ell \ge 0, \qquad\qquad\qquad \ell = 1, 2, ..., L.
\end{aligned}
\tag{7.6.21}
$$

On simplification, we can rewrite as

$$
\begin{aligned}
&\textit{maximize}_{\{v,\pi\}} \quad F = \sum_{k=1}^{K} h_k v_k \\
&\textit{subject to} \quad v_k \le \sum_{\ell=1}^{L} \delta_{kp\ell} (\alpha \hat{\xi}_\ell + \pi_\ell), \quad p = 1, 2, ..., P_k, \quad k = 1, 2, ..., K \\
&\qquad\qquad\quad \sum_{\ell=1}^{L} c_\ell \pi_\ell \le \beta \\
&\qquad\qquad\quad v_k \text{ unrestricted}, \qquad\qquad k = 1, 2, ..., K \\
&\qquad\qquad\quad \pi_\ell \ge 0, \qquad\qquad\qquad \ell = 1, 2, ..., L.
\end{aligned}
\tag{7.6.22}
$$

This time, by inspecting and comparing previous results, we can easily see that the link weight would be $w_\ell = \alpha \hat{\xi}_\ell + \pi_\ell^*$, $\ell = 1, 2..., L$. Thus, we can summarize the following result:

Result 7.3. *For MCNF Formulation (7.6.20) and its corresponding dual, Eq. (7.6.22), the commodity cost, v_k^*, is the shortest distance for demand k with respect to link weight $w_\ell = \alpha \hat{\xi}_\ell + \pi_\ell^*$, and at optimality, every path for demand k that carries a positive flow must be a shortest path with respect to the link cost system given by*

$$w_\ell = \alpha \hat{\xi}_\ell + \pi_\ell^* \tag{7.6.23}$$

for $\ell = 1, 2, ..., L$ where $\sum_{\ell=1}^{L} c_\ell \pi_\ell^* = \beta$.

7.6.5 Minimization of Average Delay

The average delay in a network is another commonly considered objective for IP traffic engineering. In Section 4.3.3, we presented the minimization of the average delay problem through a three-node example; refer to Eq. (4.3.14). On generalizing, we can write the average delay minimization problem as

$$
\begin{aligned}
&\text{minimize}_{\{x,y\}} \quad F = \sum_{\ell=1}^{L} \frac{y_\ell}{c_\ell - y_\ell} \\
&\text{subject to} \quad \sum_{p=1}^{P_k} x_{kp} = h_k, && k = 1, 2, ..., K \\
&\qquad\qquad \sum_{k=1}^{K} \sum_{p=1}^{P_k} \delta_{kp\ell} x_{kp} = y_\ell, && \ell = 1, 2, ..., L \\
&\qquad\qquad y_\ell \leq c_\ell, && \ell = 1, 2, ..., L \\
&\qquad\qquad x_{kp} \geq 0, && p = 1, 2, ..., P_k, \quad k = 1, 2, ..., K
\end{aligned}
\tag{7.6.24}
$$

A known difficulty with this formulation, as discussed earlier in Section 4.3.3, is that the objective function is nonlinear and is discontinuous at $y_\ell = c_\ell$. However, using a piecewise linear approximation of the objective function, such a problem can be transformed to an LP problem; see Section 4.3.3. We again take the same approach. Here, we will illustrate using a piecewise linear convex function due to Fortz and Thorup [233], useful in the IGP link weight determination problem. For a link load y and capacity c, the Fortz–Thorup (FT) function is given by

$$
\phi(y; c) = \begin{cases}
y & \text{for } 0 \leq \frac{y}{c} < \frac{1}{3} \\
3y - \frac{2}{3}c & \text{for } \frac{1}{3} \leq \frac{y}{c} < \frac{2}{3} \\
10y - \frac{16}{3}c & \text{for } \frac{2}{3} \leq \frac{y}{c} < \frac{9}{10} \\
70y - \frac{178}{3}c & \text{for } \frac{9}{10} \leq \frac{y}{c} < 1 \\
500y - \frac{1468}{3}c & \text{for } 1 \leq \frac{y}{c} < \frac{11}{10} \\
5000y - \frac{16318}{3}c & \text{for } \frac{11}{10} \leq \frac{y}{c} < \infty.
\end{cases}
\tag{7.6.25}
$$

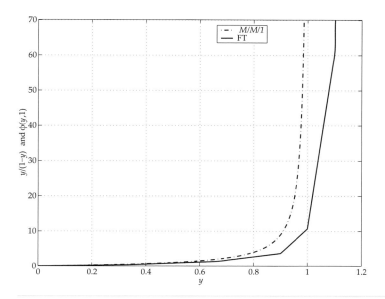

FIGURE 7.9   The Fortz–Thorup function and the load latency function (when $c = 1$).

The load latency function for the $M/M/1$ queueing model is given by $y/(c - y)$; the FT function is a piecewise linear envelope of the load latency function, divided by c (see Figure 7.9 when $c = 1$). For a network, the FT function is considered separately for each link since c_ℓ would be different. Thus, by incorporating Eq. (7.6.25) and accounting for different capacity c_ℓ for each link ℓ, we can consider the following formulation in place of Eq. (7.6.24):

$$
\begin{aligned}
\text{\textit{minimize}}_{\{x,y\}} \quad & F = \sum_{\ell=1}^{L} \frac{\phi(y_\ell : c_\ell)}{c_\ell} \\
\text{\textit{subject to}} \quad & \sum_{p=1}^{P_k} x_{kp} = h_k, && k = 1, 2, ..., K \\
& \sum_{k=1}^{K} \sum_{p=1}^{P_k} \delta_{kp\ell}\, x_{kp} = y_\ell, && \ell = 1, 2, ..., L \\
& x_{kp} \geq 0, && p = 1, 2, ..., P_k, \quad k = 1, 2, ..., K \\
& y_\ell \geq 0, && \ell = 1, 2, ..., L.
\end{aligned}
\tag{7.6.26}
$$

It is important to note that Eq. (7.6.24) and Eq. (7.6.26) differ in the following way: since Eq. (7.6.25) is defined beyond the capacity of the link, the capacity constraint, $y_\ell \leq c_\ell$, is not required to be included in Eq. (7.6.26).

To convert Eq. (7.6.26), we introduce a variable z_ℓ for each link. Then, we can write Eq. (7.6.26) as

$$
\begin{aligned}
\text{minimize}_{\{x,y,z\}} \quad & F = \sum_{\ell=1}^{L} \frac{z_\ell}{c_\ell} \\
\text{subject to} \quad & \sum_{p=1}^{P_k} x_{kp} = h_k, & k = 1, 2, ..., K \\
& \sum_{k=1}^{K}\sum_{p=1}^{P_k} \delta_{kp\ell} x_{kp} = y_\ell, & \ell = 1, 2, ..., L \\
& z_\ell \geq y_\ell, & \ell = 1, 2, ..., L \\
& z_\ell \geq 3y_\ell - \tfrac{2}{3}c_\ell, & \ell = 1, 2, ..., L \\
& z_\ell \geq 10y_\ell - \tfrac{16}{3}c_\ell, & \ell = 1, 2, ..., L \\
& z_\ell \geq 70y_\ell - \tfrac{178}{3}c_\ell, & \ell = 1, 2, ..., L \\
& z_\ell \geq 500y_\ell - \tfrac{1468}{3}c_\ell, & \ell = 1, 2, ..., L \\
& z_\ell \geq 5000y_\ell - \tfrac{16318}{3}c_\ell, & \ell = 1, 2, ..., L \\
& x_{kp} \geq 0, & p = 1, 2, ..., P_k, \quad k = 1, 2, ..., K \\
& y_\ell \geq 0, z_\ell \geq 0, & \ell = 1, 2, ..., L.
\end{aligned}
\tag{7.6.27}
$$

To avoid cluttering, we will use a compact representation for the slopes and the intercept of each segment of the FT function, i.e., $a_1 = 1, b_1 = 0$ for the first segment, $a_2 = 3, b_2 = \frac{2}{3}$ for the second segment, and so on for a total of $I = 6$ segments. Then, we can write

$$
\begin{aligned}
\text{minimize}_{\{x,y,z\}} \quad & F = \sum_{\ell=1}^{L} \frac{z_\ell}{c_\ell} \\
\text{subject to} \quad & \sum_{p=1}^{P_k} x_{kp} = h_k, & k = 1, 2, ..., K \\
& \sum_{k=1}^{K}\sum_{p=1}^{P_k} \delta_{kp\ell} x_{kp} = y_\ell, & \ell = 1, 2, ..., L \\
& z_\ell \geq a_i y_\ell - b_i c_\ell, & i = 1, 2, ..., I, \ell = 1, 2, ..., L \\
& x_{kp} \geq 0, & p = 1, 2, ..., P_k, \quad k = 1, 2, ..., K \\
& y_\ell \geq 0, z_\ell \geq 0, & \ell = 1, 2, ..., L.
\end{aligned}
\tag{7.6.28}
$$

Due to minimization, the above problem remains the same if the second constraint is changed to less-than-equal-to constraints. Again, transforming to a standard representation, we can rewrite it as

$$\text{minimize}_{\{x,y,z\}} \quad F = \sum_{\ell=1}^{L} \frac{z_\ell}{c_\ell}$$

$$\text{subject to} \quad \sum_{p=1}^{P_k} x_{kp} = h_k, \qquad k = 1, 2, ..., K$$

$$-\sum_{k=1}^{K}\sum_{p=1}^{P_k} \delta_{kp\ell}\, x_{kp} + y_\ell \geq 0, \quad \ell = 1, 2, ..., L \qquad (7.6.29)$$

$$-a_i y_\ell + z_\ell \geq -b_i c_\ell, \qquad i = 1, 2, ..., I, \ell = 1, 2, ..., L$$

$$x_{kp} \geq 0, \qquad p = 1, 2, ..., P_k, \quad k = 1, 2, ..., K$$

$$y_\ell \geq 0, z_\ell \geq 0, \qquad \ell = 1, 2, ..., L$$

For each new constraint, $z_\ell \geq a_i y_\ell - b_i c_\ell$, we will associate the non-negative dual variable $\gamma_{\ell i}$. The dual LP problem becomes

$$\text{maximize}_{\{v,\pi,\gamma\}} \quad \sum_{k=1}^{K} h_k v_k - \sum_{\ell=1}^{L}\sum_{i=1}^{I} b_i c_\ell \gamma_{\ell i}$$

$$\text{subject to} \quad v_k \leq \sum_{\ell=1}^{L} \delta_{kp}^{\ell} \pi_\ell, \qquad p = 1, 2, ..., P_k, \quad k = 1, 2, ..., K$$

$$\sum_{i=1}^{I} a_i \gamma_{\ell i} \geq \pi_\ell, \qquad \ell = 1, 2, ..., L$$

$$\sum_{i=1}^{I} \gamma_{\ell i} \leq \frac{1}{c_\ell}, \qquad \ell = 1, 2, ..., L \qquad (7.6.30)$$

$$v_k \text{ unrestricted}$$

$$\pi_\ell, \gamma_{\ell i} \geq 0.$$

While the relation between the primal and dual seems more complicated for the FT function than the previous illustrations, there is in fact a nice observation between the link weights and the *slopes* of the FT function, which is summarized below:

Result 7.4. *For each link* $\ell = 1, 2, .., L$, *assume that constraint* $z_\ell \geq a_i y_\ell - b_i c_\ell$ *for Problem Eq. (7.6.28) is binding for a unique i [denote by $i'(\ell)$] at optimality. Then an optimal link weight system is given by*

$$w_\ell^* = \pi_\ell^* = a_{i'(\ell)}, \quad \ell \in \mathcal{L}. \qquad (7.6.31)$$

Uniqueness is, however, not always possible for every link; the general result then is as follows:

Result 7.5. *For each link* $\ell = 1, 2, .., L$, *constraint* $z_\ell \geq a_i y_\ell - b_i c_\ell$ *for Problem Eq. (7.6.28) can be binding for at most two consecutive i's [denote by $i'(\ell)$ and $i'(\ell) + 1$]. Furthermore, an optimal link weight system is given by*

$$w_\ell^* = \pi_\ell^* = a_{i'(\ell)}\gamma_{\ell,i'(\ell)}^* + a_{i'(\ell)+1}\gamma_{\ell,i'(\ell)+1}^*, \quad \ell \in \mathcal{L}, \qquad (7.6.32)$$

where $\gamma_{\ell,i'(\ell)}^* + \gamma_{\ell,i'(\ell)+1}^* = 1/c_\ell, \ \gamma_{\ell,i'(\ell)}^*, \gamma_{\ell,i'(\ell)+1}^* \geq 0.$

The above results and proofs are described in [663], and are derived from complementary slackness conditions at optimality. Briefly, complementary slackness conditions lead to the realization that $\sum_{i=1}^{I} \gamma_{\ell i}^{*} = 1/c_{\ell}$, i.e., the third constraint of dual problem Eq. (7.6.30) is binding. Then, if just one $\gamma_{\ell i}$ is positive for a link, then it must be equal to 1, which in turn leads to the first result based on the second constraint of the dual. Furthermore, if more than one $\gamma_{\ell i}$ is positive for a link ℓ, then it must be for two consecutive segments since slopes are strictly increasing in nature from one segment to the next; this then leads to the second result. An illustration of these results is presented later in Section 7.7.1.

From the above results, we make an interesting observation. If a piecewise linear function is used as the objective, the slopes of this function appear as the link weight directly or through convex combination of consecutive slopes; the actual values depend on the load of traffic on the link; this will become clear through our illustration in the next section. Furthermore, the above results hold for any piecewise linear increasing function that is similar to the FT function. Thus, other similar functions including a slight modification of the FT function can be incorporated to obtain better link weights from the slopes [663].

7.7 Illustration of Link Weight Determination Through Duality

In this section, we will present two case studies based on topologies discussed earlier—one for a four-node, five-link network with all links of the same capacity and the other for a four-node, four-link network (with the direct link removed).

7.7.1 Case Study: I

First, we consider determination of the link weight for the four-node network with five links shown earlier in Figure 7.7. We will use the commercial LP solver, CPLEX, and show how to obtain the dual solution. Note our earlier remark that a problem need not be written in the dual form since by solving the original (primal) problem, dual solutions are readily available from such commercial solvers. However, it is important to write the original problem carefully so that dual variables are easy to identify and signs of variables are easy to follow.

OBJECTIVE: MAXIMUM LINK UTILIZATION

If we drop the dependency on link weight w, then we can write the LP relaxation of Eq. (7.4.11) by following the representation given in Eq. (7.6.15):

$$
\begin{aligned}
\text{minimize}_{\{x,\, r\}} \quad & F = r \\
\text{subject to} \quad & x_{11} + x_{12} + x_{13} = 60 && (v_1) \\
& -x_{11} + 100\, r \geq 0 && \text{(for link ID 1)} && (\pi_1) \\
& -x_{11} + 100\, r \geq 0 && \text{(for link ID 2)} && (\pi_1) \\
& -x_{12} + 100\, r \geq 0 && \text{(for link ID 3)} && (\pi_1) \\
& -x_{12} + 100\, r \geq 0 && \text{(for link ID 4)} && (\pi_1) \\
& -x_{13} + 100\, r \geq 0 && \text{(for link ID 5)} && (\pi_1) \\
& x_{11}, x_{12}, x_{13} \geq 0 \\
& r \geq 0.
\end{aligned}
\tag{7.7.1}
$$

There are two important points to note: (1) capacity constraints are represented in the greater-than-or-equal-to format, and (2) although redundant, all capacity constraints are listed. This is done so that the result from CPLEX is easily identifiable. CPLEX requires a name for each constraint to be listed on the left side when a problem is specified; in fact, these constraint identifiers are none other than the dual variable identifiers. Thus, we can represent the above problem in CPLEX as

```
Minimize r
subject to
    nu_1:  x_11 + x_12 + x_13 = 60
    pi_1:  - x_11 + 100 r >= 0
    pi_2:  - x_11 + 100 r >= 0
    pi_3:  - x_12 + 100 r >= 0
    pi_4:  - x_12 + 100 r >= 0
    pi_5:  - x_13 + 100 r >= 0
Bounds
    0 <= x_11
    0 <= x_12
    0 <= x_13
End
```

On solving the above problem using the CPLEX command **optimize**, we obtain optimal r^* to be 0.2. Note that although we listed non-negativity of the variables through **Bounds**, this is not necessary since by default CPLEX assumes the variables to be non-negative; thus, in subsequent listing, this part will be omitted. From CPLEX, we can obtain dual solutions (which CPLEX lists as dual price) as follows:

```
CPLEX> display solution dual -
Constraint Name         Dual Price
nu_1                    0.003333
pi_2                    0.003333
pi_4                    0.003333
pi_5                    0.003333
All other dual prices in the range 1–6 are zero.
```

Recall from Result 7.2 that here $w_\ell^* = \pi_\ell^*$. This means that the link weights are as follows: $w_1^* = 0$, $w_2^* = 0.003333$, $w_3^* = 0$, $w_4^* = 0.00333$, and $w_5^* = 0.00333$. Furthermore, v_1 gives the value of the total path cost.

There are two important observations to note here: (1) link weights for some links are zero, and (2) due to constraint $\sum_{\ell=1}^{L} c_\ell \pi_\ell^* = 1$ in Eq. (7.6.17), weights are all smaller than one. Nevertheless, the cost of each path based on these link weights is the same. If we now scale the objective function from $F = r$ to $F = \beta r$ using $\beta = 1000$, then the dual solution scales to the following: $\pi_2^* = 3.33$, $\pi_4^* = 3.33$, $\pi_5^* = 3.33$, while the rest of the πs are zero. The cost of each path remains equal while we obtained weights that can be rounded off to obtain integer weight value 3, with two of the πs still zero. Thus, link weights so obtained are

$$w_1^* = 0, \ w_2^* = 3, \ w_3^* = 0, \ w_4^* = 3, \ w_5^* = 3.$$

Thus, through scaling and rounding, integer weights can be obtained; however, this does not rule out that some link weights are zero. This means that the overall link weight systems can be used for IS-IS, but not for OSPF (unless other adjustments are done).

OBJECTIVE: COMPOSITE FUNCTION

We next consider the composite function as the objective. Assume here that $\hat{\xi}_\ell = 1$, $\ell = 1, 2, 3, 4, 5$, and that $\alpha = 1$ and $\beta = 1000$. Thus, the problem in CPLEX would take the following form:

```
Minimize 2 x_11 + 2 x_12 + x_13 + 1000 r
subject to
      nu_1:  x_11 + x_12 + x_13 = 60
      pi_1:  - x_11 + 100 r >= 0
      pi_2:  - x_11 + 100 r >= 0
      pi_3:  - x_12 + 100 r >= 0
      pi_4:  - x_12 + 100 r >= 0
      pi_5:  - x_13 + 100 r >= 0
End
```

On solving the above, we obtain dual solutions as follows:

```
CPLEX> display solution dual -
Constraint Name        Dual Price
nu_1                   5.000000
pi_2                   3.000000
pi_4                   3.000000
pi_5                   4.000000
All other dual prices in the range 1–6 are zero.
```

From Result 7.3 for the composite objective, we note that $w_\ell = \alpha \hat{\xi}_\ell + \pi_\ell$. Since $\alpha = 1$ and $\xi_\ell = 1, \ell = 1, 2, ..., 5$, and π_ℓs are already integers, we thus have

$$w_1^* = 1, w_2^* = 4, w_3^* = 1, w_4^* = 4, w_5^* = 5.$$

That is, in this case, no additional adjustment is necessary to obtain integer weights. Furthermore, we do get all paths to be of equal cost and thus the flow would be optimal in the sense of the MCSPRF problem. This is *not* to say that this will always be the case when the composite function is used. Rather, a good choice of α and β can make it easier to obtain integer link weights. Second, as long as $\hat{\xi}_\ell$ is greater than or equal to 1, the link metric so obtained would at least have the minimum value 1; thus, this satisfies the requirement of OSPF that the link weights must have a minimum value of 1. An important comment about the use of the composite function is that it does not directly address an objective that is of interest in IP traffic engineering. Thus, in this case, after solving the problem, we calculate the maximum link utilization and/or the average delay function to see whether these values are in an acceptable range.

OBJECTIVE: PIECEWISE LINEAR APPROXIMATION OF AVERAGE DELAY

Recall from Result 7.4 and Result 7.5 that the dual solutions take values from the slopes of the FT function. We illustrate this through the same example. Note that since all links have the same capacity, we can ignore c_ℓ from the objective function given in Eq. (7.6.29), but not from the constraints; as a consequence, we have $\sum_{i=1}^{I} \gamma_{\ell i} \leq 1, \ell = 1, 2, ..., L$ instead in the dual given by Eq. (7.6.30)—based on the discussion earlier, these constraints would be binding at optimality for each link ℓ. The original problem in CPLEX would then take the following form:

```
Minimize z_1 + z_2 + z_3 + z_4 + z_5
subject to
    nu_1:   x_11 + x_12 + x_13 = 60
    pi_1:   - x_11 + y_1 >= 0
    pi_2:   - x_11 + y_2 >= 0
    pi_3:   - x_12 + y_3 >= 0
    pi_4:   - x_12 + y_4 >= 0
    pi_5:   - x_13 + y_5 >= 0
    gamma_1_1: z_1 - 1 y_1 >= - 0
    gamma_1_2: z_1 - 3 y_1 >= - 66.6667
    gamma_1_3: z_1 - 10 y_1 >= - 533.333
    gamma_1_4: z_1 - 70 y_1 >= - 5933.33
    gamma_1_5: z_1 - 500 y_1 >= - 48933.3
    gamma_1_6: z_1 - 5000 y_1 >= - 543933
    gamma_2_1: z_2 - 1 y_2 >= - 0
    gamma_2_2: z_2 - 3 y_2 >= - 66.6667
    gamma_2_3: z_2 - 10 y_2 >= - 533.333
    gamma_2_4: z_2 - 70 y_2 >= - 5933.33
    gamma_2_5: z_2 - 500 y_2 >= - 48933.3
    gamma_2_6: z_2 - 5000 y_2 >= - 543933
    gamma_3_1: z_3 - 1 y_3 >= - 0
    gamma_3_2: z_3 - 3 y_3 >= - 66.6667
    gamma_3_3: z_3 - 10 y_3 >= - 533.333
    gamma_3_4: z_3 - 70 y_3 >= - 5933.33
    gamma_3_5: z_3 - 500 y_3 >= - 48933.3
    gamma_3_6: z_3 - 5000 y_3 >= - 543933
    gamma_4_1: z_4 - 1 y_4 >= - 0
    gamma_4_2: z_4 - 3 y_4 >= - 66.6667
    gamma_4_3: z_4 - 10 y_4 >= - 533.333
    gamma_4_4: z_4 - 70 y_4 >= - 5933.33
    gamma_4_5: z_4 - 500 y_4 >= - 48933.3
    gamma_4_6: z_4 - 5000 y_4 >= - 543933
    gamma_5_1: z_5 - 1 y_5 >= - 0
    gamma_5_2: z_5 - 3 y_5 >= - 66.6667
    gamma_5_3: z_5 - 10 y_5 >= - 533.333
    gamma_5_4: z_5 - 70 y_5 >= - 5933.33
    gamma_5_5: z_5 - 500 y_5 >= - 48933.3
    gamma_5_6: z_5 - 5000 y_5 >= - 543933
End
```

On solving, we obtain dual solutions as follows:

```
CPLEX> display solution dual -
Constraint Name          Dual Price
nu_1                      2.000000
pi_1                      1.000000
pi_2                      1.000000
pi_3                      1.000000
pi_4                      1.000000
pi_5                      2.000000
gamma_1_1                 1.000000
gamma_2_1                 1.000000
gamma_3_1                 1.000000
gamma_4_1                 1.000000
gamma_5_1                 0.500000
gamma_5_2                 0.500000
All other dual prices in the range 1–36 are zero.
```

Here, for links 1, 2, 3, and 4, we can see that dual solutions take the unique slope value of 1 from the first segment of the FT function, as discussed in Result 7.4. Note that $\sum_{i=1}^{I} \gamma_{\ell i} \leq 1$ is binding, and is, in fact, unique for these links. For link 5, the dual solution is a convex combination of the slopes of the first and the second segment; the dual variables γ associated with the links give the weights to be given to the slope values (see Result 7.5). That is, we can recalculate and check that

$$\pi_5^* = a_1 \gamma_{51} + a_2 \gamma_{52} = 1 \times 0.5 + 3 \times 0.5 = 2.$$

As a further illustration, consider the same problem, but this time with increased traffic volume at 150 Mbps. This changes the first constraint to

nu_1:   x_11 + x_12 + x_13 = 150

On solving, we obtain the dual solutions as

```
CPLEX> display solution dual -
Constraint Name          Dual Price
nu_1                      6.000000
pi_1                      3.000000
pi_2                      3.000000
pi_3                      3.000000
pi_4                      3.000000
pi_5                      6.000000
gamma_1_2                 1.000000
gamma_2_2                 1.000000
gamma_3_2                 1.000000
gamma_4_2                 1.000000
gamma_5_2                 0.571429
gamma_5_3                 0.428571
All other dual prices in the range 1–36 are zero.
```

Thus, we can see that the three paths are still of equal cost (6 this time) with link weights as $w_1^* = w_2^* = w_3^* = w_4^* = 3$, and $w_5^* = 6$. Here, we can see that the link weight for link 5 is the convex combination of the slopes of the second and the third segment of the FT function:

$$\pi_5^* = a_2\gamma_{52} + a_3\gamma_{53} = 3 \times 0.571429 + 10 \times 0.428571 = 6.$$

7.7.2 Case Study: II

In this case, we consider the four-node network with the direct link removed; furthermore, the capacity of links on path 1-3-2 is reduced to 10 Mbps. This is also another topology we have discussed earlier in this chapter (see Figure 7.6).

We fist consider minimization of maximum link utilization. The associated network flow problem (including scaling the objective function) can be stated in CPLEX as follows:

```
Minimize 1000 r
subject to
      nu_1:  x_11 + x_12  = 60
      pi_1:  - x_11 + 10 r >= 0
      pi_2:  - x_11 + 10 r >= 0
      pi_3:  - x_12 + 100 r >= 0
      pi_4:  - x_12 + 100 r >= 0
End
```

For this problem, the optimal maximum link utilization is 0.5454, and MCNF produces proportional flow at optimality, 10% of the traffic volume on path 1, and the rest 90% of the traffic volume on path 2. At optimality, we obtain the following dual solutions:

```
CPLEX> display solution dual -
Constraint Name       Dual Price
nu_1                  9.090909
pi_1                  9.090909
pi_3                  9.090909
```

This implies that link weights would be $w_1^* = 9.09$, $w_2^* = 0$, $w_3^* = 9.09$, and $w_4^* = 0$. However, if we were to allocate flow based on shortest path routing with ECMP, both paths being of equal cost, we will see that path 1 would overflow due to lack of capacity! This indicates that link weights determined from duality may not always produce a good result. In fact, if we were to use the piecewise linear objective approximation of the average delay as the objective, we face the same problem (this is left as an exercise).

Now consider the composite objection function. An advantage of this function is that it allows a user to play with providing weights in two ways: costly paths based on *a priori* knowledge of link speed and weights between the minimum cost part and the maximum link utilization part. Suppose we assign $\hat{\xi}_1 = \hat{\xi}_2 = 10, \hat{\xi}_3 = \hat{\xi}_4 = 1$, and weights $\alpha = 1, \beta = 100$. Thus, we have the following problem

```
Minimize  20 x_11 +  2 x_12 + 100 r
subject to
      nu_1:  x_11 + x_12  = 60
      pi_1:  - x_11 + 10 r >= 0
      pi_2:  - x_11 + 10 r >= 0
      pi_3:  - x_12 + 100 r >= 0
      pi_4:  - x_12 + 100 r >= 0
End
```

On solving the above, we obtain the following dual solution:

```
CPLEX> display solution dual -
Constraint Name        Dual Price
nu_1                   3.000000
pi_4                   1.000000
All other dual prices in the range 1–5 are zero.
```

Since, here, $w_\ell = \alpha \hat{\xi}_\ell + \pi_\ell$, link weights would be $w_1^* = 10$, $w_2^* = 10$, $w_3^* = 1$, and $w_4^* = 2$. Thus, path 2 is the shortest path where all flow can be allocated. Thus, we do not face the overflow problem as we did with other objectives for this network.

7.8 Link Weight Determination: Large Networks

It is important to note that the dual-based approach is not the only approach to determine link weights. This is an active area of research; many methods have been proposed. Nevertheless, irrespective of deciding on an objective function, several performance measures are often of interest to service providers. Consider the following measures:

1. Maximum Link Utilization (ML) captures the utilization of the link that is maximum loaded in the entire network.

2. Fraction of Used Capacity (FU) captures the total used capacity in the final solution as a fraction of the total capacity in the network.

3. Number of Overloaded Links (NOL) refers to the number of links that requires extra capacity to make the solution feasible. This metric is important only when the obtained solution is infeasible.

4. Fraction of Required Extra Capacity (FE) captures the additional capacity required to make the solution feasible as a fraction of the total capacity of the network. This is relevant only when the solution is infeasible.

In addition, the FT Function in its normalized form that captures the total congestion cost is also a good measure. The scaled (normalized) FT function cost ($\sum_{\ell=1}^{L} \phi_\ell / \varphi$) is the ratio of total cost of current allocation ($\sum_{\ell=1}^{L} \phi_\ell$) for the given capacitated network as compared to the cost in case the network was uncapacitated (φ). Observe that for an uncapacitated network with convex link cost function, cost is minimal when flows are allocated to hop count–based shortest paths.

In large problems, say a network consisting of at least 50 routers, using a commercial general-purpose LP solver can be challenging since the time taken to compute results can be quite high. Thus, an efficient method to obtain link weights through the dual-based approach is required. Such an efficient approach that also incorporates the ECMP functionality uses an iterative approach on the dual variables through a decomposition method [564], [664]; we reproduce here results from [664] for randomly generated 100-router networks with a different number of links in Table 7.2 and Table 7.3 for the minimize cost routing objective and the composite objective function, respectively, along with the computing time. Studies show that the composite objective function is good in capturing different performance measures. In case of duality framework, an additional measure is important to consider is the solution gap or the duality gap as highlighted earlier in Eq. (7.6.12); that is, the gap between the dual solution and the objective value based on the weights determined by the specialized algorithm is a useful indicator of the quality of the solution. As we can see from the tables, these two gap measures were less than a fraction of 1%. There are a few instances where the gaps were about 6% when the maximum iteration count for the dual iteration is reached. Thus, in general, the convergence property is found to be good and the method is efficient in determining link weights that work with the ECMP principle within an acceptable tolerance.

In large IP networks, some aggregation of information is also possible and might be necessary. For example, in a geographically distributed network, there are often many points of presence (PoPs) where a provider locates several routers in each PoP (refer to Section 9.6). Thus, an abstraction is possible where such PoP locations can be thought of as a supernode [308]; this then reduces the size of the problem for the purpose of determining the link weight since it is then sufficient to consider the PoP-to-PoP traffic matrix instead of the router-to-router traffic matrix; once such weights are determined, then proper mapping back to the actual network is required.

Suppose we consider the hypothetical situation in which traffic matrices can be frequently determined due to, say, some new, efficient methodology. If we now determine link weights with each such traffic matrix estimated in each time window, we might possibly have frequent weight changes. In operational networks, it is desirable to minimize the number and frequency of change in link weights [595] since each link weight change (unless due to a failure) can lead to flooding, and consequently, the packet loss or long delays during this transition cannot be completely ruled out. There has been a recent effort to reduce transition time. Another issue is that for a small change in the traffic matrix, the link weight should

TABLE 7.2   Results using the minimum cost routing objective for 100-node networks.

Nodal Degree (Number of Links)	ML	FU	FT	Solution Gap	F/I(NOL, FE)	Computing Time
2 (197)	0.79	0.25	1.22	0.6%	F(-)	44 sec
3 (294)	0.68	0.24	1.15	0.5%	F(-)	13 sec
4 (390)	0.48	0.18	1.04	0.6%	F(-)	60 sec
5 (485)	0.50	0.19	1.04	0.2%	F(-)	20 sec
6 (579)	0.49	0.17	1.02	0.4%	F(-)	23 sec

TABLE 7.3   Results using the composite objective function for 100-node networks.

Nodal Degree (Number of Links)	(α, β)	ML	FU	FT	Solution Gap	F/I(NOL, FE)	Computing Time
2 (197)	(0.9, 32)	0.67	0.25	1.15	0.1%	F(-)	5 sec
3 (294)	(0.9, 11585)	0.66	0.24	1.15	4.4%	F(-)	21 sec
4 (390)	(0.9, 2896)	0.37	0.18	1.00	0.7%	F(-)	19 sec
5 (485)	(0.9, 2896)	0.36	0.19	1.00	0.5%	F(-)	16 sec
6 (579)	(0.9, 16)	0.39	0.17	1.00	0.1%	F(-)	5 sec

not be sensitive. To circumvent this issue, a stable traffic matrix can be used as input to the weight determination process. For example, for a 24-hour cycle, the maximum traffic load can be used so that it accommodates traffic variation during the day. Note that such a decision can vary depending on the size and traffic pattern of a particular service provider's network and should be arrived at by analyzing factors such as traffic fluctuations, and impact of weight change on utilization, and so on. Later in Section 9.7, we will discuss traffic engineering implications for large tier 1 ISPs.

Note that most link weight determination schemes (other than simple hop-based, or inverse-of-the-link-speed weights) require that the traffic matrix is given. In IP networks, estimating the traffic matrix based on measurements is itself a difficult, time-consuming, and costly process. While large ISPs can use methodologies such as the ones described in [219] for estimating traffic demand volume, many medium-scale and small-scale network providers (for example, an enterprise network) may not have the resources to devise a full monitoring system to determine the traffic demand volume/traffic matrix. In such an environment, if link utilization can still be assessed periodically through tools such as MRTG [533], then such information can be used to identify highly loaded links and such links can be given a high link weight value so that traffic can be moved away from such links. This process is, however, ad hoc and still requires a certain amount of fine-tuning, and no general link weight determination method is known that works in the absence of the availability of a traffic demand volume.

An alternative option, again in the absence of a complete measurement system, is to use enhanced OSPF or IS-IS, which allows functionalities to facilitate traffic engineering, especially useful in an integrated IP/MPLS environment. In this environment, the traffic engineering enhanced capability may be used to query link bandwidth; this can be followed by setting MPLS tunnels on the command line of a router to set up traffic engineering tunnels for controlled traffic engineering, at least to a certain extent. This aspect and other related aspects on IP/MPLS traffic engineering will be discussed later (see Section 18.3.4 and Section 19.1).

7.9 Summary

Traffic engineering of IP networks is an important problem in operational IP networks. While protocols such as OSPF and IS-IS define how routers communicate among themselves to update information such as link weights, they are silent on how to pick good link weights. Thus, mechanisms are needed to determine good link weights. To do so, a critical component is to recognize that this leads to first identifying how to estimate traffic in the network, as well as what performance measures might be of interest in IP networks. Through our initial discussion, we show that there is a direct relation between average delay and average utilization; thus, the traffic engineering goal is to keep either one at a minimum by obtaining optimal link weights. Certainly, there is a connection to the network dimensioning problem.

The framework for determining link weight when the routing is based on shortest paths has an important relation to the MCNF problem. In this chapter, we show the connection between shortest path routing, link weights, and the MCNF problem; furthermore, we have indicated that the IP traffic engineering problem can be considered as the MCSPRF problem. In addition, we show here how LP duality can be used to determine link weights. A nice advantage of this approach is that commercial LP solvers can be used to find dual solutions;

this is especially attractive for network providers who do not want to develop any meta-heuristic-based link weight determination algorithms.

Further Lookup

In general, determining good link weights through various methods including meta-heuristics have been addressed by many researchers; this started in 2000 with independent works by different researchers [69], [233], [242]. Other early works are [67], [235], [532], [574]. As a matter of fact, there have been numerous works in IP traffic engineering in the last several years that consider different approaches to the link weight determination problem. That is, we do not want to give the impression that the dual-based approach is the only approach for the link weight determination problem.

Failure in a network, such as a link or a line card failure, is another important factor to consider in determining link weights. The general question is: can we determine a robust link weight system that works both under normal operating condition *and* also under a failure. The benefit of such link weights is that the transient behavior after a failure can be minimized. Recently, integrated models have been developed to consider such situations; for example, see [470], [532], [564] for additional discussions.

Some network providers prefer to obtain link weights that result in *unique* shortest paths for all demand pairs, the primary reason being the ability to easily troubleshoot a network [698]. In light of Result 4.1 and Example 4.5 discussed earlier, it is important to note that most demand pairs in large networks are likely to have flows taking single paths at optimality. While several methods have been proposed [67], [69], [564], [663], [698], obtaining link weights that lead to unique shortest paths (without significantly increasing the total cost, average delay, or maximum link utilization) remains a difficult problem.

Note that while this chapter is primarily about IP traffic engineering, there is a connection between flow control and traffic engineering; for example, see [289], [581] and Chapters 22 and 23. Later in Chapter 11, we will discuss the connection between control and traffic engineering for voice engineering.

Exercises

7.1. Review questions:

(a) What is traffic engineering?

(b) What does the bandwidth-delay product signify?

(c) What is the difference between the multi-commodity network flow problem and the multi-commodity shortest path routing problem.

(d) How does the buffer size of a router impact traffic engineering?

7.2. Refer to Exercise 4.3 in Chapter 4. Now, determine optimal link weights for the two objectives described there.

7.3. Consider the nine-node network [261] shown in Figure 7.10 where the number next to a link represents the capacity of the link, and the table shows the traffic volume for three

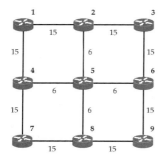

FIGURE 7.10   A nine-node network example.

different demand pairs. Determine best link weights using duality (i.e., using CPLEX or any other LP solver) when the following objective options are considered: (1) minimization of maximum link utilization, (2) minimum cost routing, (3) piece-wise linear envelope, (4) a composite cost function. For the composite cost function, test for different values of α and β to see how the link weights determined might change.

7.4. Consider the nine-node Manhattan street network shown in Figure 7.10.

   (a) Assume that all links have 100 units of capacity, and the goal is to do load balancing the network. Find the best link weights (i) if a single demand between node 1 and 9 with volume of 60 units is considered, (ii) if two demands, one between 1 and 9 and another between 3 and 7, each of volume 60, are considered.

   (b) Assume that all links have 100 units of capacity except for links 2-5, 4-5, 5-6, 5-8, which have 60 units of capacity. The goal is to do load balancing the network. Find the best link weights (i) if a single demand between node 1 and 9 with volume of 60 units is considered, (ii) if two demands, one each between 1 and 9 and another between 3 and 7, each of volume 60, are considered.

7.5. Consider the same demand/capacity scenarios described in Exercise 7.4; find the best link weights if minimum cost routing is the objective used instead (assume unit cost of flow on each link).

7.6. Consider the same demand/capacity scenarios described in Exercise 7.4; find the best link weights if a composite objective function that combines minimum cost routing with load balancing is used. Determine how the solution changes as the parameters associated with the cost components of the composite object are varied.

7.7. In Chapter 4, you will find exercises that are similar to the above three exercises. Compare your results and draw your conclusions.

8

BGP

All truths are easy to understand once they are discovered; the point is to discover them.

Galileo Galilei

Reading Guideline

The chapter starts with the basic conceptual idea behind BGP. Several details are then introduced one at a time. This chapter is helpful in reading Chapter 9. BGP uses the path vector protocol approach. You may note that the concept of a path vector protocol and its behavior has been discussed in depth in Section 3.5; thus, it is helpful to refer to this discussion in parallel with this chapter.

The Border Gateway Protocol (BGP) plays a critical role in making communication on the Internet work. It facilitates exchange of information about networks, defined by IP address blocks, between entities, known as *autonomous systems* (ASes), so that one part of the Internet knows how to reach another part. BGP is thus an inter-AS routing protocol. It does, however, allow intra-AS exchanges in certain situations as will be described later.

In this chapter, we describe BGP and its operational characteristics. The current BGP standard is known as version 4, with its most recent specification described in RFC 4271 [591]; we will simply use BGP to refer to BGP4 since our entire discussion here about BGP is about BGP4. In Chapter 9, we will cover the Internet routing architecture where we will show how BGP is used. The evolutionary path to the development of BGP will also be described in that chapter (refer to Section 9.1).

It is important to note that BGP is an excellent example of a *work-in-progress* protocol. In the early 1990s, BGP went through multiple versions to arrive at version 4; yet many issues were addressed as add-on features to version 4 by taking the operational experience of the Internet into account. As we move through this chapter, we will point out a few of these issues.

8.1 BGP: A Brief Overview

The BGP protocol is used to *communicate* information about *networks* currently residing (or homed) in an autonomous system to other autonomous systems. The term *network* has a specific meaning in regard to BGP, which we will describe in the next section; in this section, we will italicize it to avoid confusion with the general use of the term network. The exchange of *network* information is done by setting up a communication session between bordering autonomous systems. For reliable delivery of information, a TCP-based communication session is set up between bordering autonomous systems using TCP listening port number 179. This communication session is required to stay connected, which is used by both sides to periodically exchange and update information. When this TCP connection breaks for some reason, each side is required to stop using information it has learned from the other side. In other words, the TCP session serves as a *virtual link* between two neighboring autonomous systems, and the lack of communication means that this virtual link is down. Certainly, this virtual link will be over a physical link that connects the border routers between two autonomous systems; it is important to note that if a virtual link is broken, it does not necessarily mean that the physical link is broken. Now imagine that each autonomous system is a *virtual supernode*; then the entire Internet can be thought of as a graph connecting virtual supernodes by virtual links.

Example 8.1 *BGP topology illustration.*
    In Figure 8.1, we have shown six virtual supernodes (autonomous systems), AS1 to AS6, connected by virtual links, i.e., TCP-based BGP sessions for communication between two adjacent virtual supernodes. Each virtual supernode then contains one or more *networks* identified as N1, N2, N3 in AS1, and so on. From the figure, we can see that there is more than one possible path between certain ASes. It is also possible to have a supernode at the edge of the entire network such as AS6. Furthermore, multiple virtual links between two neighbor-

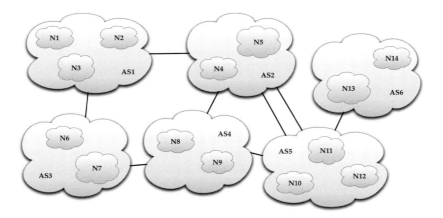

FIGURE 8.1 Internet: a conceptional graph view through clouds of autonomous systems (virtual super-nodes) connected by BGP sessions (virtual links).

ing ASes are allowed/possible; for example, in this figure, we have shown that there are two virtual links between AS2 and AS5.   ▲

For the supernetwork of ASes that connects virtual supernodes through virtual links, we need a mechanism for routing information updates about *networks* to be exchanged. Recall from Chapter 3 that routing information in a network can be essentially disseminated in two different ways: using a distance vector approach or using a link state approach. A difficulty with a link state concept is that it is not scalable in its normal form when the number of nodes grows, although a link state protocol can be made scalable through extended mechanisms— ATM Private Network-to-Network Interface (PNNI) is such an example. However, a nice property of a distance vector protocol is that a node does not need to maintain the entire topology; the supernetwork (Internet) consisting of supernodes made of ASes is very large and, thus, a distance vector approach is appealing. However, the difficulty with a distance vector protocol is that looping can occur and unreliable delivery of routing information is not desirable. BGP follows a *path vector* routing protocol approach that is roughly based on a distance vector–type approach where looping is avoided through path tagging and where reliable sessions for information exchange are used. The basic concept behind a path vector protocol, without cluttering its description with BGP, was described earlier in Section 3.5 in Chapter 3. While it is tempting to refer to BGP as *the* path vector protocol, it is more appropriate to refer to BGP as an *instance* of a path vector protocol family. As an example, in BGP, the cost between two adjoining ASes is implicitly assumed to be just a hop; any local virtual link selection decision can be specified if there are parallel links.

Since BGP uses a hop count metric as the distance between two adjoining ASes, the shortest path from an AS to a distant AS is essentially counted in terms of the shortest number of AS hops; note that it is *not* in terms of number of routers along this path. BGP also allows parallel virtual links between adjoining ASes; thus, a mechanism is also provided for local exchange of information to decide on a preferred link. Visits through ASes as information about *networks* is propagated is prepended using the BGP protocol. We provide a simple illustration here.

Example 8.2 *Prepending of AS Paths in BGP.*

Consider again Figure 8.1. Here, we can clearly see that the shortest AS hop path from AS1 to AS6 is AS1 to AS2 to AS5 to AS6. There are two parallel links between AS2 and AS5; the choice of either link is a local decision and is immaterial to ASes outside this part such as AS1 and AS6.

As part of the BGP protocol, AS6 will let its neighboring AS, AS5, know that it is home for N13 and N14 by broadcasting the AS identifier with the *network* identifier, i.e., as (AS6) $\mapsto$ N13 and (AS6) $\mapsto$ N14. It is easy to see that instead of generating a separate message for N13 and N14, a combined message would conserve the number of announcements since both *networks* are served by AS6 with one outlet. Thus, a common way is to announce a *set* of *networks* such as {N13, N14}; thus, we can write the prepended announcement as (AS6) $\mapsto$ {N13, N14}. Through a series of exchanges, AS1 will receive the prepended path information (AS2, AS5, AS6) $\mapsto$ {N13, N14} from AS2 and (AS3, AS4, AS5, AS6) $\mapsto$ {N13, N14} from AS3. Thus, AS1 can decide that route (AS2, AS5, AS6) is the shortest AS hop-based path to reach destinations N13 and N14.   ▲

Remark 8.1. *Explicit announcement of home AS with a network.*

There is an important basic issue: if we know that N1 belongs to AS1, could we communicate the path information to AS5 simply as (AS2) $\mapsto$ N1 instead of as (AS2, AS1) $\mapsto$ N1? That is, could we drop AS1 entirely from the path information as being the home of N1? The answer is no. If we were to do that, it would *implicitly* suggest that N1 *always* belongs to AS1. Instead, we want the flexibility of a *network* that can be homed off of another AS if the owner of the address space chooses to do so in the future. Second, the ability to detect looping will be lost. Thus, the explicit announcement of AS1 as the *current* home of N1 as in (AS2, AS1) $\mapsto$ N1 automatically provides these flexibilities.   ♦

Essentially, BGP chooses a path between two ASes in terms of the shortest number of AS hops. However, BGP allows an AS number to be repeated during the announcement for the benefit of inter-AS traffic engineering. This is illustrated below.

Example 8.3 *Repetition of an AS number in AS paths for inter-AS traffic engineering.*

Consider Figure 8.1 again. Suppose that AS1 would prefer that traffic be routed via AS3 instead of AS2 to its own *networks*. Thus, AS1 can send the announcement about N1, N2, N3 to AS2 with AS1 repeated three times as (AS1, AS1, AS1) $\mapsto$ {N1, N2, N3}, but to AS3 once as (AS1) $\mapsto$ {N1, N2, N3}; after prepending, each of these announcements will then arrive at AS4 as (AS2, AS1, AS1, AS1) $\mapsto$ {N1, N2, N3} and as (AS3, AS1) $\mapsto$ {N1, N2, N3}, respectively. Thus, the route chosen by AS4 for destinations N1, N2, and N3 would be via AS3 since the announcement indicates that this is the shortest number of AS-hops. Assuming no repeats are done by other ASes, traffic originating in AS5 and AS6 would also go via AS3 to destinations N1, N2, and N3.

The case for AS2 is interesting to note. AS4 would forward the announcement about *networks* in AS1 to AS2. Thus, AS2 will receive it as (AS4, AS3, AS1) $\mapsto$ {N1, N2, N3}, while AS2 has already received it from AS1 as (AS1, AS1, AS1) $\mapsto$ {N1, N2, N3}. Thus, for AS2, reaching *networks* N1, N2, and N3 is the same length in terms of AS-hops. Here, additional tie-breaking rules discussed later would be applied to determine the preferred route.   ▲

The actual best path decision in BGP has far more details than the simple shortest-hop idea described so far; this will be covered in detail later along will more information concerning about BGP operations.

While we have so far given a fairly simplistic view of BGP, several important points have been covered such as: (1) what types of information to communicate between neighboring autonomous systems, and the format and types of messages, (2) how to ensure that the virtual link connectivity is maintained, and (3) how to react if the virtual link breaks down; these are important in maintaining relations among autonomous systems so that a packet can move from an end host to a distant end host through the Internet.

8.2 BGP: Basic Terminology

We have so far described the basic notion of BGP using virtual supernodes and virtual links, where each supernode contains one or more *networks*. It should be clear by now that the virtual link functionality is provided by a *BGP session* and supernodes are referred to as *autonomous systems*. That is, the Internet is composed of *autonomous systems* that connect to one or more autonomous systems while an autonomous system contains one or more *networks*. The term *network* has a specific meaning; more accurately, it refers to an *IP prefix–defined network*; for brevity, this notion of a network is referred to as an IP prefix, which we described briefly in Chapter 1. Recall that an IP prefix is identified through the CIDR notation, A.B.C.D/n, where/n refers to the network mask. For example, an IP prefix is listed as 134.193.0.0/16 where the network mask is/16; this means that this address block identifies a certain *network*. In BGP folklore, an IP prefix is often referred to simply as a network, or a route, or an IP prefix, and sometimes all three are used interchangeably. For clarity, we will use the term IP prefix henceforth. The term *route* has a specific meaning as defined in the BGP specification; a *route* is a unit of information that associates an IP prefix destination or a set of IP prefix destinations with the attributes of an AS-path that has been visited so far, as seen by a receiving AS through an UPDATE message. For example, in Figure 8.1 from the viewpoint of AS6, a route to the IP prefix destination N1 is (AS5, AS4, AS3, AS1) $\mapsto$ N1.

In each AS, certain entities are designated as BGP agents for communication with neighboring ASes. These agents are specially designated routers, commonly referred to as *BGP speakers*. This means that the TCP-based BGP session is in fact set up between two adjoining BGP speakers, and thus, each speaker is considered the peer of the other. Since the BGP protocol is meant for use in the global Internet, an identification number for each AS has been defined and is tracked for determining a path between two ASes. This identifier is a unique 16-bit *autonomous system number*, assigned to each AS. Thus, each AS is assigned a globally unique number in the range 1 to 64511; the rest, 64512 to 65535, are reserved as private AS numbers. Originally, the private AS numbers were not defined; they first became necessary when the *AS confederation* approach, discussed later in Section 8.8.2, was introduced.

Finally, it is important to note that an AS is not the same as an ISP. An ISP can have multiple ASes contained in it; conversely, an AS can be made up of multiple providers where different IP prefixes are provided by each provider. We will consistently use the term AS and discuss specifics in regard to an ISP as and when required.

8.3 BGP Operations

To facilitate learning about routes to reachable IP prefixes, a BGP speaker is engaged in exchange of network reachability information with its neighboring BGP speakers. During an exchange, a BGP session may go down; thus, the basic BGP operation needs to also address how to handle such situations. To enable various BGP activities, the BGP protocol defines four key message types: OPEN, UPDATE, KEEPALIVE, and NOTIFICATION, and an optional message type ROUTE–REFRESH. Details on the message formats for these BGP messages will be described later in Section 8.12. Here, we will concentrate on the BGP operational functions for which these messages are used, and the operations require certain timers.

8.3.1 Message Operations

The OPEN message is the first message sent to establish a BGP session after the TCP connection has been established. This is started by the BGP speakers that act as designated agents of autonomous systems to talk to other neighboring BGP speakers. Often in practice, each BGP speaker is configured in advance with the IP address of the other BGP speaker so that either end can initiate this TCP connection. It is quite possible that different BGP speakers use different BGP version numbers; thus, the OPEN message contains the version number as well as the AS number.

The UPDATE message, the key message in BGP operations, is sent between two BGP speakers to exchange information regarding IP prefixes; this message type works in a push mode, i.e., whenever a BGP speaker has new information regarding an IP prefix to communicate to its peering BGP speaker, an UPDATE message is sent. In steady state, the BGP speakers generate UPDATE messages whenever either end has determined a new best AS route for any specific IP prefix. More importantly, if one end of the BGP session was the announcer of a route to a particular IP prefix to its other end, it must generate a withdrawal if this speaker can no longer reach this particular IP prefix. The reason for this withdrawal announcement is that the path through the AS where the sending speaker is homed could be the best path for the receiving speaker, yet the sending speaker has no way of knowing if it is otherwise.

Once a BGP session is up and running, the KEEPALIVE message is sent periodically between two BGP speakers as a confirmation that the session is still alive. Each end learns and agrees on a maximum acceptable time, known as the *hold time*, during the initial exchange of OPEN messages. The KEEPALIVE messages are then generated approximately once every third of the hold time, but no more than once every second. The KEEPALIVE messages should not be generated if the hold time is agreed to be zero; this case assumes that somehow the session is completely reliable.

The NOTIFICATION message is sent to close a BGP session; this is done when some error occurs requiring closing down of the session. Thus, a virtual link between two BGP speakers is considered to be unavailable (1) when the NOTIFICATION message sent by one end leads to a graceful close of a BGP session, or (2) when there is an absence of KEEPALIVE or UPDATE messages within a hold time.

Besides the four mandatory message types, an additional optional message type, ROUTE–REFRESH, has also been added [129]. For example, at any instant during a session, one end can send ROUTE–REFRESH to its neighboring BGP speaker requesting to readvertise all its IP prefix entries in its routing table; thus, the ROUTE–REFRESH, can be thought of as a pull request that is responded using an UPDATE message. Its usefulness will be discussed later in Section 8.7.

Since ROUTE–REFRESH is an optional type, how does a BGP speaker know whether its neighboring BGP speaker supports this feature? To make such optional functionalities work, BGP defines a parameter called *Capabilities* [113], [114] which is carried in the initial OPEN message for capabilities negotiation. Thus, ROUTE–REFRESH is sent as an optional capability to be negotiated in the initial OPEN message. If the receiving speaker does not support the Capabilities option or ROUTE–REFRESH option, it sends a NOTIFICATION message back to the sending BGP speaker to close the session. In this situation, the sending speaker would need to send a new OPEN message without the Capabilities option so that normal establishment can be accomplished; the session would continue without the ROUTE–REFRESH option.

8.3.2 BGP Timers

We have so far described the use of different message types. For proper functioning of BGP, several timers are also defined. It is important to understand the need and the role of the timers. For example, how long should a BGP speaker try to set up a connection with a neighbor before giving up? How often should two neighbors exchange KEEPALIVE messages? How often should routes to a particular IP prefix be announced or withdrawn? and so on. Implicit in defining such timers is the need to limit link bandwidth consumption between two BGP neighbors as well as to limit the processing of resources at the BGP speaker related to BGP traffic. After all, for a link that connects two neighboring ASes through the border BGP speakers, the main role is to push actual user traffic, not be consumed/dominated by the BGP protocol-related traffic. To address these points, the BGP protocol has five required timers and two optional timers; with each timer, a time parameter is assigned. We describe them and their roles:

- *ConnectRetryTimer:* This timer defines the timeout interval before retrying a connection request. While the recommended ConnectionRetryTime value is 120 sec, it can set to zero for certain event conditions.

- *HoldTimer:* This timer indicates the maximum time (in seconds) that is allowed to elapse without receiving an UPDATE or KEEPALIVE message from a peering BGP speaker before declaring that the peer is not reachable. That is, the expiration of HoldTimer indicates that the virtual link between these two BGP speaker is down. The recommended value for HoldTime is set to 90 sec while the minimum positive value must be 3 sec. The time is allowed To be set to zero, which is used as the indicator that the session is never to expire.

- *KeepAliveTimer:* This timer relates to the frequency of generating KEEPALIVE messages; the timer value is set to one-third of the value of HoldTime. For example, if HoldTime is agreed to be 90 sec through the exchange of OPEN messages at the beginning of a BGP connection, then KeepAliveTime is set to 30 sec.

- *MinRouteAdvertisementIntervalTimer:* This timer refers to the minimum time that must expire before a BGP speaker can advertise and/or withdraw routes to a peering BGP speaker in regard to a particular IP prefix destination. While the timer can be defined on a per-IP prefix destination basis, the value is maintained on a per–BGP peer speaker basis. If it is an intra-AS peer, then the recommended value is 5 sec; for an external peer, the value is set at 30 sec.

- *MinASOriginationIntervalTimer:* this timer indicates the minimum time that must expire before a BGP speaker can report changes in its *own* autonomous system through another UPDATE message. The recommended value for MinASOriginationIntervalTime is 15 sec.

Note that a BGP speaker may be involved in setting up peering sessions with multiple BGP peers. Thus, it is possible that certain timers expire at about the same time, causing a spike in activity at a BGP speaker; second, even if they are originally set to expire at a different time, it is possible that some eventually synchronize; that is, the *pendulum effect* much like RIP (refer to the discussion on timer adjustment Section 3.3.3) cannot be completely ruled out. Thus, jitter is required to be implemented on the following four timers: ConnectRetryTimer, KeepAliveTimer, MinASOriginationIntervalTimer, and MinRouteAdvertisementIntervalTimer. The recommended value is obtained by determining a random quantity that is uniformly distributed from the range 0.75 to 1.0 of the base value, and a new random quantity is determined each time.

There are two additional optional timers: DelayOpenTimer and IdleHoldTimer. The DelayOpenTimer may be used once a TCP connection is set up for a BGP session to indicate wait time before the OPEN message is to be sent. The IdleHoldTimer is used to determine how long to wait in the idle state of the BGP protocol by a BGP speaker before triggering restart of a BGP session to a particular peer; this factor is used to dampen any oscillatory behavior.

8.4 BGP Configuration Initialization

In this section, we discuss BGP initial configuration in some detail. To do this, we first consider two commonly used approaches: one in which two ASes are connected directly through their respective BGP speakers and one in which a border BGP speaker is connected to multiple neighboring BGP speakers.

Consider two ASes with AS numbers 64516 and 64521 wanting to set up a BGP neighboring relation (see Figure 8.2). In this case, a common approach is to set a direct physical interface between two bordering BGP speakers, thus forming a point-to-point link; then, a subnet address block (IP prefix) is defined where both interfaces have addresses from this address block. For example, if we use the IP prefix 10.6.17.0/30 block to describe this subnet, then 10.6.17.1 can be assigned as the interface address (serial: s0) to the BGP speaker in AS64516 and 10.6.17.2 as the interface address (serial: s1) to the BGP speaker in AS64521. Once this is configured, the neighboring relation can be established, for example, by the BGP speaker in AS64516 indicating that 10.5.21.2 is the IP address for the neighboring BGP speaker in AS64617; similarly, the other end can issue the neighboring relation. This method of configuration then avoids the *chicken and egg* problem of how each BGP speaker determines how to reach its neighboring BGP speaker so that they can exchange routing information.

Now consider the case in which an AS has more than one neighboring ASes (see Figure 8.3); that is, AS number 64516 has two neighboring ASes, AS64521 and AS64822. One possibility is to take the same approach as the first case, i.e., define separate subnet address blocks for each neighboring relation such as address block 10.6.17.0/30 between AS64516 and AS4521 and address block 10.6.17.4/30 between AS64516 and AS64822. In this case, the BGP speakers in AS64521 and AS64822 will see the border BGP speaker in AS64516 with different addresses. While this is workable, this is not preferable, since it can be hard to manage the different address assignments and can be cumbersome when a BGP speaker has many neighboring ASes.

The configuration just described raises the following question: can we configure in such a way that each neighbor can use the *same* IP address for a particular border BGP speaker in AS64516? In fact, this is possible. To do it, a *loopback address*, that is, a *loopback interface-based approach* is taken. Consider again just two neighboring ASes, AS64516 and AS64521.

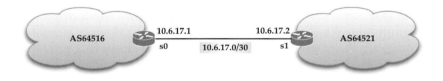

FIGURE 8.2   BGP session setup: direct interface between two ASes.

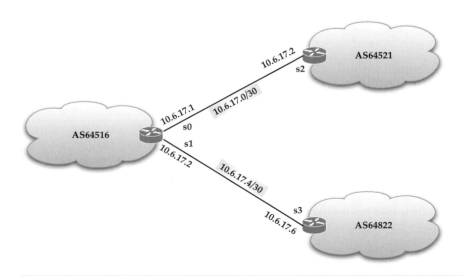

FIGURE 8.3   BGP session setup: direct interface between an autonomous system (AS) and its two neighbors.

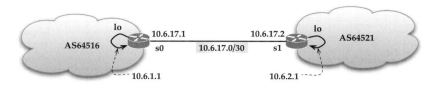

FIGURE 8.4   BGP session setup: use loopback interface between two ASes.

As shown in Figure 8.4, at the BGP speaker 10.6.17.1 in AS64516, a loopback interface (lo) is created with IP address 10.6.1.1 while at the BGP speaker 10.6.17.2 in AS64521, a loopback is created with IP address 10.6.2.1. Now, BGP speaker 10.6.17.1 indicates 10.6.2.1 as the remote end for AS64521 and then indicates that the route to 10.6.2.1 is to use the serial interface s0. Thus, when a BGP packet is generated at the BGP speaker 10.6.17.1 for AS64521, the packet will take interface s0 to reach BGP speaker 10.6.17.2 where it will now loop back to feed to the BGP session. Similarly, the other end is configured; that is, BGP speaker 10.6.17.2 indicates 10.6.1.1 as the remote end for AS64516, and then indicates that the route to 10.6.1.1 is to use the serial interface s1. Similarly, a third neighbor of AS64516 can be connected without the necessity of changing the IP address of the border BGP speaker 10.6.17.1.

There is also a third possible approach for the initial configuration; this approach is used when two BGP speakers are not directly connected; this is often encountered in regard to internal BGP, to be discussed in the next section. In this situation, an interior dynamic routing protocol can be used where one end learns about the other end dynamically.

It should be noted that whether the direct interface-based or the loopback interface–based approach is used, the time-to-live (TTL) field in the IP packet that contains BGP information is set to 1 when loopback addressing is not used; when loopback addressing is used, the TTL is set to 2, and for a multihop environment to 255. Limiting TTL then helps prevent BGP packets from spreading beyond where they need to be contained, and serves as a basic security mechanism [252].

8.5 Two Faces of BGP: External BGP and Internal BGP

Our discussion so far has exclusively concerned routing information exchanges *between* ASes through BGP speakers. In fact, BGP is also used to set up peer (neighbor) connections between two BGP speakers *within* an AS, known as internal BGP (IBGP) speakers. The question is why such an arrangement is needed and in what scenarios. Before we delve into this, we clarify a curiosity that remains: how does a BGP speaker find out whether it is communicating with an external peer BGP speaker or an internal peer BGP speaker? This can be determined by comparing the AS number communicated in the OPEN message by its peer BGP with that of its own internal value; if it matches, then this neighbor is an IBGP speaker, and if it does not, then it is an EBGP speaker.

To consider why IBGP is needed and for what types of scenarios, we begin with an illustration consisting of four ASes as shown in Figure 8.5. Here, AS64777, AS64999, and AS65222 are referred to as *stub* ASes since they each of them has one BGP speaker as an outlet. We start with AS64777. There are three IP prefixes in this AS: N1, N2, and N3. To route user traffic from

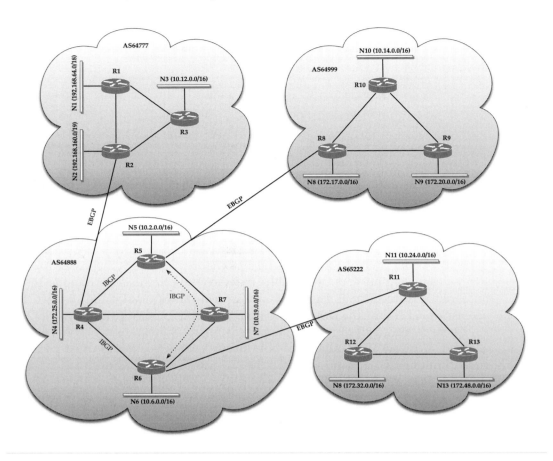

FIGURE 8.5   External BGP and internal BGP example.

one of them to another one *within* AS64777, an interior gateway protocol is sufficient to determine routing. Similarly, the internal routing scenario is handled within the other ASes as well.
We assume that AS64777 uses OSPF within its domain. What about inter-AS? For example,
how does R7 in AS64888 learn about network N1, and conversely, R1 in AS64777 learn about
network N7? Note that neither R1 nor R7 is a BGP speaker; they are interior routers within
their respective ASes. One possible way is that BGP speaker R2 learns about N5 from AS64888
and then communicates this information through OSPF protocol to routers R1 and R3. Thus,
R2 can learn about all external IP prefix networks from its neighboring AS and communicate to R1 and R3 through OSPF. The difficulty is that this immediately creates a scalability
problem at routers R1 and R3 since they will need to maintain link state database entries for
such external IP prefixes *and* compute the shortest paths to all such IP prefixes. Furthermore,
if router R3 were to populate the external IP prefixes to each internal router, it defeats the
purpose of BGP. In any case, while theoretically possible, BGP speaker R2 in AS64777 does
not inform other interior routers within AS64777 about external IP prefixes about which it
has learned. Instead, BGP speaker R2 becomes the *default* gateway for all external IP prefixes;

a common way to configure in a stub AS; routers R1 and R3 do not need to maintain routing table entries for external IP prefixes, thus reducing the routing table size at R1 and R3 as well.

The above discussion gives the impression that the border BGP speaker such as R2 is the stopping point for external IP prefixes it has learned from its neighboring AS, i.e., through incoming BGP messages. This is, however, true only if the AS is a stub AS with only a single BGP speaker to its neighboring AS; in fact, this is the case with R2. This situation, however, no longer holds when we consider AS64888. If BGP speaker R4 were to stop distributing IP prefixes N1 and N2 it has learned from BGP speaker R2, then the third AS, AS64999, would have no way of knowing that N1 and N2 actually exist. Thus, a mechanism is needed so that an AS that has connectivity to multiple ASs through multiple BGP speakers such as AS64999 can communicate information about network reachability. This is where an *internal* BGP session is required between two BGP speakers *within* an AS so that network reachability information can be exchanged. In our example, such an internal BGP session is required between BGP speakers R4 and R5 so that R4 can learn about IP prefixes N8, N9, and N10 while R5 can learn about IP prefixes N1, N2, and N3. That is, an important BGP rule is as follows:

> Rule 1: *A BGP speaker can advertise IP prefixes it has learned from an EBGP speaker to a neighboring IBGP speaker; similarly, a BGP speaker can advertise IP prefixes it has learned from an IBGP speaker to an EBGP speaker.*

Note that due to the second part of Rule 1, it is acceptable for R4 to advertise to R2 in AS64777 about IP prefix N7, which is part of AS64888.

While internal BGP works very much the same way as external BGP, there is an important difference. Consider network N6 in AS64888; within this AS, BGP speaker R4 learns about it internally through OSPF from internal router R6; similarly, BGP speaker R5 also learns about N6 from R6. Should R4 and R5, both IBGP speakers, advertise N6 to each other through the IBGP session? The answer is no. That is, a second important rule is as follows:

> Rule 2: *An IBGP speaker cannot advertise IP prefixes it has learned from an IBGP speaker to another neighboring IBGP speaker.*

The primary reason for this rule requires some explanation. The AS number is prepended only when an advertised IP prefix crosses an AS boundary (Example 8.2). Recall that a BGP speaker prevents looping by checking if its own AS number is on the path for any network reachability received from another BGP speaker. When the communication is *within* an AS, and since the AS number is *not* prepended in this scenario, looping is possible! This mandates the need for Rule 2. Furthermore, the routers within an AS are supposed to handle internal routing through the interior gateway protocol for all its internal IP prefixes; this is not the role of an IBGP.

There is, however, an important implication of Rule 2 in regard to external IP prefixes when there are more than two IBGP speakers in an AS. To understand this issue, consider again AS64888. Due to Rule 1, BGP speaker R4 will learn about N1, N2, and N3 from EBGP speaker R2 and let IBGP speaker R5 know so that R5 can communicate this information to AS64999. Also, by Rule 1, BGP speaker R4 will also learn about N11, N12, N13 located in AS65222 from IBGP speaker R6 and let AS64777 know about existence of N11, N12, and N13.

However, due to Rule 2, R4 cannot inform IBGP speaker R5 about N11, N12, N13. How then would AS64999 know about N11, N12, N13? As a consequence of Rule 2, AS64999 would not know unless that is *also* an IBGP session between IBGP speakers R5 and R6. That is, if there are more than two IBGP speakers in an AS, there must be an IBGP session between *each* pair of IBGP speakers; thus, this leads to the case of *full-mesh* IBGP connectivity. This certainly raises the scalability issue, which is discussed later in Section 8.8.

To summarize, IBGP is required whenever an AS has multiple EBGP speakers. Certainly, a stub AS that has only a single BGP speaker does not need to consider IBGP. The basic mechanism for IBGP and EBGP is the same as long as the two rules discussed above are addressed properly. Note that IBGP is a situation in which the third approach for initial configuration, mentioned earlier in Section 8.4, is often used for connecting two IBGP speakers. IBGP speakers are also configured using loopback addressing for ease of configuration manageability. In this case, the interior gateway protocol is used for routing BGP-related data from one IBGP speaker to reach the other IBGP speaker.

8.6 Path Attributes

A critical part of BGP operation is route advertisement; as a part of route advertisement, specific information about a route to an IP prefix destination or a set of IP prefix destinations is also distributed; this set of information, known as *path attributes*, is then used in the BGP routing decision process. BGP path attributes are classified into the following four categories:

- *Well-known mandatory:* All BGP implementations must recognize such an attribute and it must appear in an UPDATE message.

- *Well-known discretionary:* All BGP implementations must recognize such an attribute; however, it may not be included in an UPDATE message.

- *Optional transitive:* A BGP implementation might not support such an attribute, but it must forward it to its BGP peers.

- *Optional nontransitive:* A BGP implementation might not support such an attribute; it should not forward it to its BGP peers.

We now describe several key path attributes while identifying the category to which they belong.

ORIGIN

This well-known mandatory attribute identifies the mechanism by which an IP prefix is first announced into BGP, commonly referred to as *injected into BGP*. It can be specified as IGP, EGP, or Incomplete. IGP means that the IP prefix was learned from an interior gateway protocol such as OSPF; EGP means that it is learned from an exterior gateway protocol such as BGP; Incomplete refers to the case when the IP prefix is unknown, often the case for static routes. The value assigned by the originating BGP speaker is not allowed to be changed by any subsequent speaker, although, in practice, it can be. This, this attribute is not always meaningful in practice.

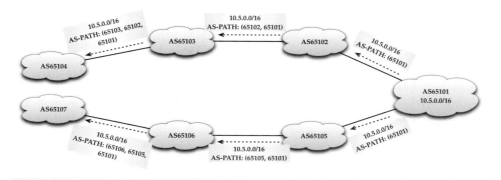

FIGURE 8.6   BGP path Attribute: AS–PATH example.

AS–Path

This well-known mandatory attribute stores a sequence of AS numbers that identifies the ASes a route has visited so far. This is accomplished using the UPDATE message; whenever an UPDATE message is communicated from one BGP speaker to another BGP speaker as it crosses an AS boundary, its AS number must be prepended to the AS–PATH.

Consider Figure 8.6. Here, the IP prefix 10.5.0.0/16 originates in AS65101. Thus, the BGP speaker on the border in AS65101 includes its AS number in the UPDATE message sent to its BGP speakers located at neighboring ASes, AS65102 and AS65105. When the route for 10.5.0.0/16 is advertised to other ASes, the AS–PATH in the UPDATE message is prepended with the AS number of the leaving AS. Thus, on receiving the UPDATE message in AS65107, the BGP speaker at AS65107 will learn that 10.5.0.0/16 has originated in AS65101 and that it has since passed through AS65105 and AS65106 by inspecting the AS–PATH attribute, while recognizing that AS65106 is the most recent AS visited. An interesting feature is that the bordering BGP speaker may prepend its own AS number more than once in the AS–PATH attribute; this was illustrated earlier in Example 8.3.

Next–Hop

This well-known mandatory attribute identifies the IP address of the next hop router to the IP prefix destination in the UPDATE message. Note that an IP prefix is advertised outside of an AS so that others are aware of it; thus, for the rest, this IP prefix is the destination they are now aware of and to which they want to send user traffic. Thus, when user traffic is forwarded, the next hop router is the *final* router in the BGP domain, which knows how to forward it to the IP prefix destination. Since the view is from the incoming direction, a name such as last hop or final hop might have sounded more appropriate since next hop is also commonly used to mean the next router for an *outgoing* direction.

The next hop router identified is dependent on from where it is advertised and whether it is internal or external to the originating AS. We illustrate NEXT–HOP through three scenarios (see Figure 8.7). In scenario 1, IP prefix 10.12.0.0/16 homed in AS64777 is advertised to AS64888 by EBGP speaker 10.6.17.1 to EBGP speaker 10.6.17.2; in this case, the NEXT–HOP to IP prefix destination 10.12.0.0/16 is 10.6.17.1. This attribute value is advertised outside its home AS. In scenario 2, the announcement is entirely within an AS between IBGP speakers through a TCP-based BGP session that is set up between IBGP speakers 10.5.16.1 and

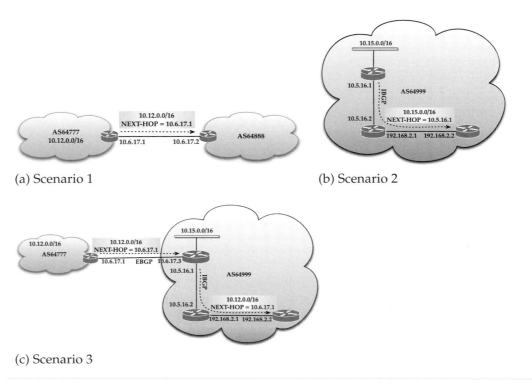

FIGURE 8.7 BGP path attribute: NEXT–HOP examples.

192.168.2.2; note that this TCP is routed internally via the interior router 192.168.2.1. Since IP prefix 10.15.0.0/16 is off the IBGP speaker 10.5.16.1 that is advertising it, the NEXT–HOP to IP prefix destination 10.15.0.0/16 is 10.5.16.1, *not* 192.168.2.1. Finally in scenario 3, IP prefix 10.12.0.0/16 is advertised by BGP speaker 10.6.17.1, which is passed on from AS64777 to AS64999; then, within, AS64999, the UPDATE message that contains this information is forwarded from one IBGP speaker to another IBGP speaker. Since network 10.12.0.0/16 originated in AS64777 at BGP speaker 10.6.17.1, the NEXT–HOP value will remain at 10.6.17.1.

As you can see, NEXT–HOP in BGP can be somewhat confusing. NEXT–HOP in BGP really needs to be defined because of the basic concept of IP routing that a destination network address must have a next hop entry in a routing table. In an IGP environment, this is not a problem since a router can compute the next hop based on the shortest path first algorithm. Thus, NEXT–HOP in BGP can be thought of as a recursive idea; that is, it is listed for the purpose of following the next hop notion with a pointer; when the actual traffic arrives, this pointer would know how to route through IGP.

MULTI–EXIT–DISCRIMINATOR (MED)

This optional nontransitive attribute is a metric meant for use when there are multiple external links to a neighboring AS; in this case, the exit point with the lowest metric value is preferred by the neighboring AS. The MED attribute is allowed to be sent to other IBGP speakers in the same AS; it is, however, never propagated to other ASes beyond that. Thus,

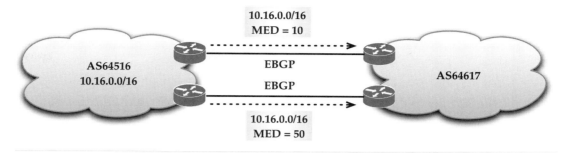

FIGURE 8.8 BGP path attribute: MED example.

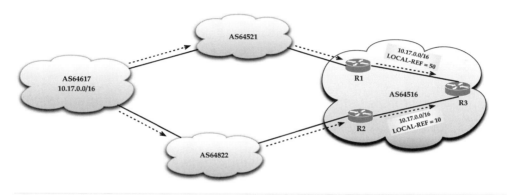

FIGURE 8.9 BGP path attribute: LOCAL–PREF example.

the border BGP speaker must have the ability to delete the MED attribute from a route before passing the UPDATE message to its neighboring AS. MED is typically prevalent when the multiple exit links are of a different bandwidth so that the link with a higher bandwidth can be preferred by setting a lower metric value to this link, or to control the entry point into the next AS, sometimes referred to as *cold potato routing*.

Consider Figure 8.8. Here, IP prefix 10.16.0.0/16 is advertised by AS64516 with different MED values for each EBGP session to AS64617. Thus, when AS64617 sends user traffic to AS64516, it will use the link where the MED value is smaller.

LOCAL–PREF

This well-known discretionary attribute is a metric used internally within an AS between BGP speakers; this is also helpful in selecting an outgoing BGP speaker when an AS has connectivity to multiple ASes or multiple BGP routes even with the same next hop AS.

Consider Figure 8.9. Here, IP prefix 10.17.0.0/16, originated from AS64617, is advertised to AS64521 and AS64822. The intermediate ASes, AS64521 and AS64822, in turn, advertise to AS64516, which arrives at BGP speakers R1 and R2, respectively. Now, AS64516 internally wants to introduce a local preference for this route due to the preference to use AS64521 for outgoing traffic. Thus, BGP speakers R1 and R2 are configured with local preference values that are internally communicated to IBGP speaker R3. Thus, when user traffic arrives at R3 destined for IP prefix 10.17.0.0/16, it will prefer to use the outgoing BGP speaker R1 since the local preference metric value is higher for this router than the other one.

8.7 BGP Decision Process

The BGP decision process can be divided into two parts: (1) path selection and (2) aggregation and dissemination. It may be noted that the BGP decision process is used interchangeably with the path selection process in the common literature. We make a subtle distinction here to separate out the role of aggregation and dissemination.

8.7.1 BGP Path Selection Process

The BGP path selection process has the responsibility of selecting routes to various IP prefix destinations for use locally by a BGP speaker. The BGP path selection process is part of the overall BGP decision process, which also handles route dissemination to its neighboring BGP peer speakers; the route dissemination process will be discussed along with route aggregation later in Section 8.7.2. To accomplish route selection, each BGP speaker maintains two *routing information bases* (RIBs):

- *Adjacent RIBs-In (Adj-RIBs-In)* is the information base that stores AS-level routing information for each IP prefix it has learned about from its neighbors through inbound UPDATE messages. From its different neighbors, a BGP speaker may learn about more than one AS path for a particular IP prefix; in most implementations of BGP, information learned from different neighbors for a particular IP prefix destination is cached. While the BGP specification does not require a BGP speaker to cache more than one path learned for a particular IP prefix, most BGP implementations do cache paths so that it can use one of the cached paths as the preferred path if the one currently used is no longer available. Caching and impact on route selection in case of a failure are illustrated in detail in Section 3.5.

- *Loc-RIB* is the information base that stores the routes that have been determined locally by its decision process, which is to be used for determining the forwarding table.

   The purpose of ROUTE–REFRESH becomes clear when we consider Adj-RIBs-In. Suppose that a BGP speaker, while keeping a cached path, might have changed certain attribute values in its memory compared to what it received from its neighbor, for example, due to a configuration change by the network administrator; thus, to check/verify what values the neighbor originally communicated, the network administrator can take advantage of the ROUTE–REFRESH message to request the neighboring BGP to communicate its data again using the UPDATE message.

   It may be noted that each BGP speaker also maintains the following RIB:

- *Adjacent RIBs-Out (Adj-RIBs-Out)* is the information base that stores the routes for advertisement to its neighboring BGP speakers through outbound UPDATE messages.

This RIB is used in route dissemination and will be discussed later in Section 8.7.2. The route selection process at a BGP speaker can be categorized into two phases:

1. *Import policy and filtering phase:* When a BGP speaker receives an UPDATE message from a peering BGP speaker, this phase is activated. Note that such an announcement can be about a new route, a replacement route, or withdrawn routes. An import policy is maintained by the BGP speaker to filter out IP prefixes it does not want to support [260]; for example, it may choose to filter out an IP prefix from nonallowable address space such as a private IP address block or a route that contains a private AS number. Furthermore, for each feasible route learned, the BGP speaker locally assesses a degree of preference. This assessment can be based either on LOCAL–PREF if the announcement is received from an IBGP speaker or any locally preconfigured decision rule. BGP specification leaves any such preconfigured decision rule as a local matter.

2. *Best route determination phase:* This phase determines the best path for each distinct IP prefix of which it is aware based on certain tie-breaking rules described later; the result is then recorded in Loc-RIB. In this process, if the NEXT–HOP attribute is not found to be resolvable for a particular IP prefix, such a route must be dropped during this decision phase.

Note that the best route phase is started after the completion of the import policy and the filtering phase. The routing decision criteria, which involve tie-breaking rules, are applied to each IP prefix destination or a set of IP prefix destinations as received through the UPDATE message. For clarity, we will present the description below in terms of determining the AS–PATH to a specific IP prefix destination:

1. If the IP prefix destination is unwanted due to import policy and filtering, discard the route.

2. Apply the degree of preference with the highest LOCAL–PREF or preconfigured local policy, if applicable.

3. If there is more than one route to the IP prefix destination, select the route that originated locally at the BGP speaker.

4. If there is still more than one route to the destination IP prefix, select the one with the smallest number of AS numbers listed in the AS–PATH attribute.

5. If there is still more than one route to the destination IP prefix, select the one with the lowest ORIGIN attribute. Thus, this selection will follow the order: IGP, then EGP, then Incomplete.

6. If there is still more than one route to the destination IP prefix, select the route with the lowest MULTI–EXIT–DISCRIMINATOR.

7. If there is still more than one route to the destination IP prefix, select the route received from EBGP over IBGP.

8. If there is still more than one route to the destination IP prefix, select the route with minimum interior cost to the NEXT–HOP that is determined based on the metric value.

9. If there is still more than one route to the destination IP prefix, select the route learned from the EBGP neighbor with the lowest BGP identifier.

10. If there is still more than one route to the destination IP prefix, select the route from the IBGP neighbor with the lowest BGP identifier.

Best AS paths to IP prefix destinations that result from the above process are then stored in Loc-RIB, locally by each BGP speaker.

8.7.2 Route Aggregation and Dissemination

An important component of the BGP decision process is route dissemination. This phase, which comes after completion of the route selection process, entails route aggregation along with application of export policy.

We first start by discussing route aggregation at a BGP speaker; in fact, a critical ability of BGP 4 is the handling of route aggregation that is made possible due to CIDR. The basic idea is to combine IP address blocks for networks from two or more ASes through supernetting at a downstream AS. In a sense, the newly announced supernetted address block is less specific and it announces the AS number only of the AS where supernetting is done. To do that, two path attributes, ATOMIC–AGGREGATE and AGGREGATOR, have been defined. ATOMIC–AGGREGATE is a well-known discretionary attribute that is attached to a route out of the AS where supernetting is done, and the BGP identifier of speaker where this aggregation is done is indicated through the attribute AGGREGATOR, an optional transitive attribute.

Example 8.4 *Route aggregation.*

Consider Figure 8.10. Here AS64822 announces IP prefix 10.5.160.0/19 to AS64617; AS64617, in turn, announces this one and IP prefix 10.5.224.0/19 that it hosts to AS64701. Note that AS64701 also receives the announcement about 10.5.192.0/19 from AS64816. Furthermore, AS64701 houses 10.5.128.0/19. By inspecting these four IP prefixes, the BGP speaker in AS64701 determines that these can be combined to go from a /19 address block to a /17 address block 10.5.128.0/17.

Thus, AS64701 announces downstream about 10.5.128.0/17 with itself as the AS host, but indicating that ATOMIC–AGGREGATE is set and that the AGGREGATOR, the BGP speaker where this aggregation is done, is identified through the BGP identifier of the router as 192.168.4.18. Clearly, this information is less specific since 10.5.128.0/17 is advertised at the /17 netmask; furthermore, this speaker lists its own AS number due to aggregation, to serve the role of a proxy, instead of listing the AS number of one of the actual originating ASes.   ▲

An important advantage of route aggregation is that it reduces the number of routing table entries that need to be maintained in downstream ASes. For the example just considered, the immediate downstream AS needs to maintain a single entry about 10.5.128.0/17 with AS64701 as the home, instead of maintaining four routes identified at the /19 address block along with the appropriate AS number entries. In essence, route aggregation addresses scalability. There are certain exceptions when route aggregation should not be performed: for example, when two routes have a different MED, when routes have different attribute values for one of the attributes, and so on.

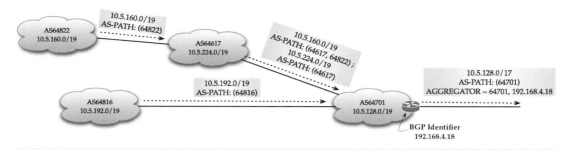

FIGURE 8.10   Route aggregation example using ATOMIC–AGGREGATE and AGGREGATOR.

Along with route aggregation, a BGP speaker also applies *export policy* before propagating routes to other BGP speakers. It may be noted that the export policy can contain separate requirements for each neighboring BGP speaker. Thus, the output of this process is not the same for every BGP speaker for which this BGP speaker is connected; thus, a separate Adj-RIB-out is created for each such speaker in order to maintain different rule with different speakers.

8.7.3 Recap

We now summarize the entire BGP decision process. It involves best route selection at a BGP speaker by applying import policy to Adj-RIB-In and by applying routing decision criteria to determine Loc-RIB; in turn, export policies and optionally route aggregation are applied, different for different peers, to determine Adj-RIB-Out separately for each peering BGP speaker. This aspect is depicted in Figure 8.11.

In Table 8.1, we list samples of import and export policies at BGP speaker, AS64701 (Figure 8.10). Thus, when an update is received from BGP peer AS64617, the speaker at AS64701 will store it in Adj-RIB-In; this will be separate from the update received from another BGP peer in AS64816. Now, AS64701 will compute best routes taking into account import policies and the criteria described earlier in Section 8.7.1. The output will be stored in Loc-RIB, which then will be subject to the export policy for AS64999 (not shown in figure) to arrive at different Adj-RIB-Out.

By this time, it should be apparent why routing in BGP is often referred to as *policy-based routing*. In fact, import and export policies are critical components in the BGP routing decision process, which are not seen in other routing protocols. Note that import and export policies are placed at a BGP speaker by a network administrator due to business relations or peering arrangement, i.e., external factors. Router vendors provide user interfaces to be able to enter policy rules; also, Routing Policy Specification Language (RPSL) (RFC 2622 [8]) is a generic platform to describe policies. Later in Section 9.5, we will delve more into policy-based routing.

8.8 Internal BGP Scalability

Earlier in Section 8.5, we introduced the notion of IBGP. We mentioned that IBGP requires full mess connectivity among IBGP speakers as a consequence of Rule 2. It is easy to see that this

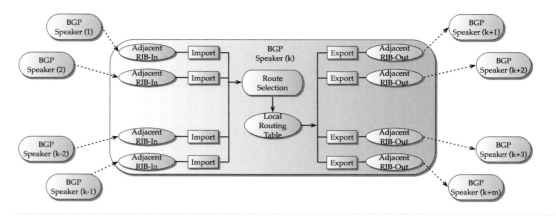

FIGURE 8.11   BGP decision process.

TABLE 8.1   Examples of import and export policies at a BGP speaker.

Import Policy	Export Policy
— Do not accept default 0.0.0.0/0 from AS64617.	— Do not propagate default route 0.0.0.0/0 except to internal peers.
— Assign 192.168.1.0/24 coming from AS64617 preference to receiving it from AS64816.	— Do not advertise 192.168.1.0/24 to AS64999.
— Accept all other IP prefixes.	— Assign 172.22.8.0/24 a MED metric of 10 when sent to AS64999.

raises a scalability problem. If there are n IBGP speakers, then $n(n-1)/2$ total IBGP sessions would be required with each speaker handling $n-1$ sessions. If n is small, this is not an issue. However, when n is large, an IBGP speaker is required to maintain a large number of IBGP sessions. There are two well-known approaches to handle this scalability issue among IBGP speakers: route reflection approach and AS confederation approach. They are described below.

8.8.1   Route Reflection Approach

The concept of route reflector [59], [60], [62] has been developed to address the scalability problem of full-mesh IBGP sessions. The idea is fairly simple: have one or more IBGP speakers act as *concentration* routers, commonly known as *route reflectors*. Introduction of route reflectors then creates a hierarchy among the IBGP speakers by *clustering* a subset of IBGP speakers with each route reflector. IBGP speakers associated with a route reflector in a cluster are referred to as *route reflector clients*; IBGP speakers that are not clients are referred to as *nonclients*. Note that a client is not aware that it is talking to a route reflector and assumes that it is as if like a full-mesh configuration. In Figure 8.12, we show IBGP session connectivity under full mesh and when there are one, two, or three route reflectors. For example, in Figure 8.12(c),

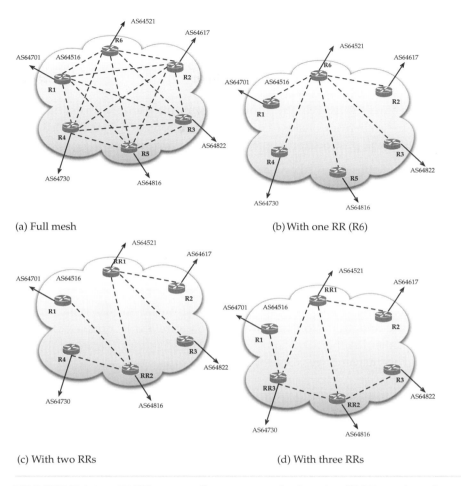

(a) Full mesh

(b) With one RR (R6)

(c) With two RRs

(d) With three RRs

FIGURE 8.12 IBGP route reflector example showing IBGP sessions (route reflectors are identified by RR).

there are two clusters: one cluster is RR1 with R2 and R3, and another cluster is RR2 with R1 and R4. While we show here just one route reflector in each cluster, for redundancy, a cluster may have multiple route reflectors. Each cluster is identified by a CLUSTER–ID. If there is only one route reflector in a cluster, then CLUSTER–ID is the BGP identifier of the route reflector; otherwise, a common CLUSTER–ID can be defined for use by multiple route reflectors within a cluster. Note that if there is only one route reflector (see Figure 8.12(b)), then it creates the hub-and-spoke connectivity where the route reflector connects to each route reflector client (see Figure 8.12(b)). In this case, the lone route reflector is still required to maintain $n-1$ sessions with the other IBGP speakers; that is, for this route reflector the processing overhead is no different than if it were under the full-mesh scenario. Thus, often, it is better to deploy two or more route reflectors to form clusters so that each route reflector has a reasonable number of IBGP sessions to handle.

It is important to note that route reflectors must form full mesh among themselves and each client peer with only its route reflector. Full mesh among route reflector is not apparent until there are at least three route reflectors (see Figure 8.12(d)). There are certain rules to follow in regard to announcements:

- If an announcement is received by a route reflector from another route reflector, then reflect/pass it to its clients. Consider Figure 8.12(c); if route reflector RR2 learns from route reflector RR1, it will pass on to route reflector clients R1 and R4.

- If an announcement is received by a route reflector from a route reflector client, then reflect it to another route reflector. Consider Figure 8.12(d); if route reflector RR3 learns from client R1, it will pass on to route reflectors RR1 and RR2.

- If an announcement is received by a route reflector from an EBGP speaker, reflect it to all other route reflectors and its clients. Consider Figure 8.12(d); if route reflector RR1 learns about external IP prefixes from AS64521, it will pass on to route reflectors RR2 and RR3, and route reflector client R2.

From the above discussion, you might realize that Rule 2 discussed earlier in Section 8.5 is relaxed since the route reflectors are now allowed to reflect IP prefixes they have learned from an IBGP speaker to other IBGP speakers. The question then is: can we detect and avoid routing loops? The answer is yes, but the solution requires two additional attributes as described below:

- ORIGINATOR–ID: This attribute identifies a route reflector through its 4-byte router ID; it is given type code 9 and is optional and nontransitive. This attribute is added only by the originating route reflector. That is, when a route reflector learns about an IP prefix from one of its clients, it adds the ORIGINATOR–ID attribute before reflecting to other speakers. Note that a BGP speaker should not create an ORIGINATOR–ID if one is already created by another speaker. If a route reflector receives an announcement about an IP prefix with the ORIGINATOR–ID that matches its router ID, it should ignore this announcement. Consider Figure 8.12(d); if route reflector RR3 learns about an IP prefix from R1 that is advertised by AS64701, it will add the ORIGINATOR–ID attribute with its router ID and announce to route reflectors RR1 and RR2. If somehow RR3 learns about the same IP prefix, it will check the ORIGINATOR–ID attribute and recognize that it is the same value as its router ID; thus, it will not forward to client R1.

- CLUSTER–LIST: This list stores a sequence of 4-byte CLUSTER–ID values to indicate the path of clusters that an advertised IP prefix has visited. The role of CLUSTER–ID is similar to AS number; this is used to identify each cluster uniquely within an AS. Thus, when a route reflector reflects an IP prefix, it is required to prepend the local CLUSTER–ID to the CLUSTER–LIST; thus, CLUSTER–LIST is similar to the function of AS–PATH attribute and is used for detecting and avoiding looping.

With the introduction of ORIGINATOR–ID and CLUSTER–LIST, the BGP route selection process described in Section 8.7.1 requires the following modification:

9′. Use ORIGINATOR–ID as the BGP IDENTIFIER.

9.1   Prefer a route with the shortest CLUSTER–LIST length.

That is, this new step 9′ replaces the previously described step 9 in Section 8.7.1 and step 9.1 is a new step inserted before step 10. Note that CLUSTER–LIST length is assumed to be zero when a route does not include the CLUSTER–LIST attribute. A final comment is that the ORIGINATOR–ID and CLUSTER–LIST are not advertised outside its AS.

8.8.2 Confederation Approach

In lieu of the route reflection method, another method known as the *AS confederation* approach [701], [702] can be used to address IBGP scalability. The basic idea is fairly simple: use a divide-and-conquer approach to break the entire AS into multiple sub-ASes where IBGP full mesh is maintained only *within* each sub-AS and sub-ASes connected by *exterior* IBGP sessions. The entire AS is then known as a confederated AS. While the entire confederation has a unique AS number, sub-ASes may have AS numbers obtained and assigned from the public AS number space, or use AS numbers from private AS number space. Consider Figure 8.13 where AS64516 is divided into two sub-ASes, AS65161 and AS65162. Here, IBGP speakers R1, R4, and R5 are fully meshed in sub-AS, AS65161, and IBGP speakers R2, R3, and R6 are fully meshed in sub-AS, AS65162, and the two sub-ASes maintain a BGP session, referred to as exterior IBGP session, between R5 and R6.

For the confederation concept to work within an AS without looping, two segments types, AS–CONFED–SET and AS–CONFED–SEQUENCE, which parallel AS–SET and AS–SEQUENCE, respectively, have been defined as part of the AS–PATH attribute. Then, each sub-AS talks to another sub-AS within the same AS, somewhat similar to the way that two EBGP speakers talk to each other across ASes; however, there are the following additional requirements:

- LOCAL–PREF attribute for a route is allowed to be carried from one sub-AS to another sub-AS. This is required since the LOCAL–PREF value for this route is meant for the entire AS. Consider Figure 8.13; for a route learned from AS64701, BGP speaker R1 might set LOCAL–PREF; this information is to be carried when going from sub-AS AS65161 to sub-AS AS65162. Recall that in regular EBGP, LOCAL–PREF is ignored.

- NEXT–HOP attribute for a route set by the first BGP speaker in the entire AS is allowed to be carried from one sub-AS to another sub-AS. Consider Figure 8.13 again; for a route learned from AS64701, BGP speaker R1 sets the NEXT–HOP attribute; this information is to be carried when going from sub-AS AS65161 to sub-AS AS65162.

- When advertising a route from one sub-AS to the next sub-AS, insert AS–CONFED–SEQUENCE to the AS–PATH attribute along with the AS number of the sub-AS. This then helps detect and prevent looping.

Note that before leaving a confederated AS, any AS–CONFED–SEQUENCE or AS–CONFED–SET information is removed from the AS–PATH attribute and AS numbers of sub-ASes are not advertised outside of the confederation.

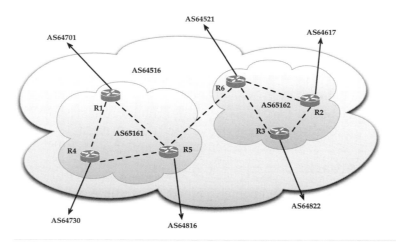

FIGURE 8.13   IBGP confederation example.

Similar to the route reflection approach, the confederation approach requires a slight modification to the BGP route selection process described earlier in Section 8.7.1 as follows:

7′. If there is still more than one route to the IP prefix destination, select the route received from EBGP over confederation EBGP, which in turn over IBGP.

That is, step 7′ replaces step 7 in the standard BGP route selection process.

It is easy to see that the confederations approach requires having an AS number due to the creation of sub-ASes. This had the potential of using many AS numbers if multiple providers choose to use this option. To alleviate the AS number exhaustion problem, a private AS number was introduced so that private AS numbers can be used by a provider to number their sub-ASes for the purpose of confederation.

Deployment of confederation requires a methodical hierarchical approach. If confederations are created in an unorganized manner, it can also create unnecessary message exchanges affecting the performance of the network. For instance, imagine the scenario in which each BGP speaker in an AS is designated as a separate sub-AS and one of them goes downs; it can create long convergence problems.

8.9 Route Flap Dampening

In BGP, an UPDATE message is sent from a BGP speaker to its neighboring speaker whenever any change to a route of an IP prefix destination occurs. Furthermore, a speaker that announces such a route to a neighboring BGP speaker is also responsible for reporting any changes, including withdrawal, to the same neighboring BGP speaker subsequently, irrespective of where it has learned from. As you can imagine, it is quite possible for a BGP speaker to announce a new route regarding an IP prefix destination, then almost immediately withdraw it a few seconds later, and then announce it again, and so on. For example, such a situation can occur if there is some problem maintaining the BGP session between two speakers; this is compounded by the fact that when a BGP session is initialized, full route exchange is typically done between two BGP speakers.

The unpleasant situation regarding announcement and withdrawal of a route to an IP prefix destination from a BGP speaker to its neighboring speaker is that there is a ripple effect since the receiving BGP speaker, unless it is in a stub AS, announces this route to its downstream BGP speaker in another AS, and this one in turn to its downstream BGP speaker, and so on. Frequent changes of routes, commonly known as *route flapping*, can result in creating a cascading storm of updates through ASes. A consequence is that it causes BGP instability along with computational overhead incurred by the downstream BGP speakers as well as additional bandwidth consumption to report changes. Suppose that a BGP speaker in the core of the Internet handles 196,000 IP prefix destinations; if 1% of them are flapping every couple of minutes, this speaker would have difficulty handling the CPU load.

It is important to recognize that such instability must be addressed at the granularity of an IP prefix destination, not at the level of a BGP neighbor. Note the problem could be at the origin when an IP prefix destination is first injected into the world of ASes, for example, the change in NEXT–HOP attribute; thus, this change will ripple through ASes, while other IP prefixes are not affected at all. Similarly, on a route to an IP prefix destination, there could be a change in the AS path. Thus, for each IP prefix destination, any change due to either AS change or NEXT–HOP change is minimally considered as a change.

To minimize the impact of possible instability caused by UPDATES messages in regard to an IP prefix destination, a route flap dampening principle has been introduced in BGP [720] to determine *when* next to advertise the announcement or withdrawal about this destination to a BGP peer. A side effect of route flapping is an increase in convergence time. Thus, the main objective of a route dampening approach is to reduce the route update load in such a way that well-behaved and poorly behaved routes are treated with some fairness.

The basic principle is that at each BGP speaker, as it learns about an IP prefix destination that has been announced, a default *figure-of-merit* penalty metric per IP prefix destination is assigned. Whenever the announcing BGP peer sends a change or withdrawal, i.e., flaps in regard to this IP prefix, the penalty is increased by a fixed amount from its current value. The penalty then decays exponentially to half its value at the end of each *half-life*. There are two additional conditions: (1) If the penalty crosses a specified upper threshold known as the *suppress* or *cutoff* limit, the speaker suppresses the view of this IP prefix and its associated AS path, i.e., it is not announced downstream; (2) the speaker frees this route from the suppressed state when the penalty goes below the *reuse* limit, or when the time since the last time of announcement exceeds a certain length of time such as four times that of the half-life.

Consider that the route flap occurs at $t_0, t_1, t_2, \ldots$. Then, given half-life quantity H, and the penalty amount P_{inc} per flap, the dynamic penalty assessment over time, t, can be expressed as follows:

$$P(t) = \begin{cases} 0, & t < t_0 \\ P_{inc}, & t = t_0 \\ P(t_i) \times e^{\frac{-(t-t_i)\log 2}{H}}, & t > t_0, \end{cases}$$

$$P(t_i) = P(t_i) + P_{inc}, \quad i = 1, 2, \ldots. \tag{8.9.1}$$

This says that until the first time announcement of an IP prefix destination received at a BGP speaker, there is no penalty. At t_0 when the first announcement arrives, the assigned penalty is the penalty per flap, P_{inc}. Then the exponential decay starts with this penalty as the starting value; it continues until the next flap at t_1, when the penalty is increased by the increment P_{inc}, and so on. You might wonder why there is a "log 2" in the exponential decay expression; this factor balances out the requirement that after passage of half-life since last flap, the penalty should be half its value. To see it, note that half-life since last flap at t_i means $t - t_i = H$. Then the main expression reduces to $P(t_i + H) = P(t_i) e^{-\log 2}$; since $e^{-\log 2} = \frac{1}{2}$, we thus have $P(t_i + H) = P(t_i)/2$.

In Figure 8.14, we show a simulated behavior of dynamic penalty given by Eq. (8.9.1) with half-life (H) set to 7.5 min and the penalty per flap (P_{inc}) set to 1000; suppress limit of 2000 and reuse limit of 750 are also marked. Two plots are shown: one where flapping occurs every 2 min for five times and another where flapping occurs every 4 min for three times. Consider the case when it flaps every 2 min. When the first announcement is received at a speaker at 2 min, a penalty of 1000 is assigned and it is accepted and is announced downstream; the next flap at 4 min is also accepted and downstream since the value after adjustment is still below 2000. At the next flap at 6 min, the penalty value now crosses 2000; thus, this update is suppressed and so are the next two flaps. After no further update is received, the penalty decays until it crosses the reuse limit around 27 min, when it is ready to advertise again. For the second plot with flaps every 4 min, the third flap at 12 min is suppressed and is not available for consideration until about 24 min.

From this illustration, we can see that there are two possible side effects of route dampening: (1) a legitimate update on a route to an IP prefix destination that has been received is not considered since this IP prefix is in the suppressed state; and (2) communication in regard to a legitimate change is delayed to a neighboring speaker. Note that a receipt is still possible since the receiving speaker cannot control when it wants to receive from a sending BGP speaker as long as the sending BGP speaker satisfies the MinRouteAdvertisementInterval-Time requirement. A positive benefit of route flap dampening occurs when route aggregation

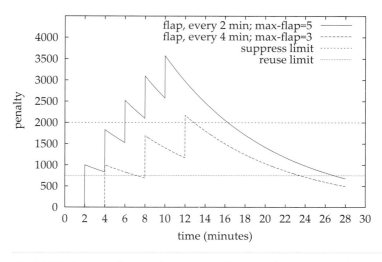

FIGURE 8.14   Route dampening figure-of-merit penalty.

is done at a downstream speaker. The aggregated route is then like a new announcement; any prior knowledge about flap is not carried over and is lost.

We briefly discuss why route dampening was introduced. When the Internet connected different ASes between Internet service providers around the mid-1990s, it was often on low-speed links such as 56 Kbps. Thus, when route flapping occurred, it consumed a significant part of such low-bandwidth links. However, at that time, BGP had no knobs to control excessive BGP-related traffic. Thus, route dampening was introduced as an add-on feature to provide a relief mechanism. In today's environment, most ASes are connected by high-speed links and, thus, it is not always necessary to invoke route flapping, since the proportion of bandwidth required for BGP update is much less in a high-speed environment. Second, the exact parameter values might need to be different than the ones set by vendors with their BGP release; this requires some amount of fine-tuning to avoid any undesirable behavior.

8.10 BGP Additional Features

There are a variety of extensions proposed to BGP. Here we discuss two well-known features.

8.10.1 Communities

This feature [115] was developed to incorporate a desirable property observed from deployment of BGP. From a deployment perspective, BGP is heavily policy dependent; for example, peering neighboring ASes might have a certain agreement that goes into the BGP decision process. In the Internet, BGP is used by transit ISPs who provide transit services to other ASes. In case some control is needed over distribution of routing information, it is based on either any IP prefix destination allowed or not allowed, or the AS–PATH attribute. The communities attribute was developed to simplify such control. For example, for a certain set of IP prefix destinations, irrespective of their path attribute, a community can be defined so that it is easier to handle/configure this group.

Three well-known communities values have been defined: (1) IP prefixes marked such are not allowed to be advertised outside an AS ("no-export"), (2) IP prefixes marked such are not to be advertised beyond the receiving BGP speaker to other BGP speakers ("no-advertise"), and (3) IP prefixes marked such may be advertised by a receiving BGP speaker, but not outside the current AS confederation ("no-export-subconfd").

8.10.2 Multiprotocol Extension

Originally, BGP was defined assuming that it will be used only for the IPv4 address family. An important question is how to extend BGP if it were to be used for other address families such as IPv6, especially without changing the version number. For this purpose, the optional parameter field in the OPEN message turned out to be helpful [113], [114]. Using this feature, BGP multiprotocol extension has been developed [63], so that network layer reachability information (NLRI) field can be used for other address families such as IPv6, or IPv4-based virtual private networks (VPNs) [602], and so on. Later, extensions presented in [602] for virtual private networking will be discussed in Section 18.5.1.

8.11 Finite State Machine of a BGP Connection

When a BGP speaker sets up a BGP session with a peer BGP speaker, several different types of messages are communicated during the session starting with the OPEN messages. How to handle different types of messages depends on triggering a number of events and what state to move to depends on the action—which is a stateful mechanism. Thus, a finite state machine is used to describe the relation of a BGP speaker's state to each of its BGP peers; that is, a speaker needs to maintain a separate finite state machine with each of its BGP peering speakers. The basic states are classified into the following: Idle state, Connect state, Active state, OpenSent state, OpenConfirm state, and Established state. BGP specification [591] documents 28 different types of possible events: 16 of them are mandatory and the rest are optional; they are listed in Table 8.2. In the following, we discuss each state and provide a general overview about state transition. The finite state machine is shown in Figure 8.15; for details, refer to RFC 4271 [591].

IDLE STATE

This is the initial state of a BGP speaker. In this state, the BGP speaker is not yet ready to accept a BGP connection. At the occurrence of either a manual start (ME01) or automatic start (OE03) event, the BGP speaker initializes BGP resources, starts the ConnectRetryTimer, starts a TCP connection to its BGP peer speaker, and also listens for any incoming BGP connection. It then moves to the Connect state. However, if this BGP speaker were to take the passive role in the sense of event OE04 or OE05, the initialization is similar to when either ME01 or ME02 occurs, except that it moves to the Active state instead of Connect state. If dampening of peer oscillation is activated, then three additional events, OE06, OE07, and OE13, may occur; in this situation, the local BGP speaker tries to prevent peer oscillations using the dampening principle.

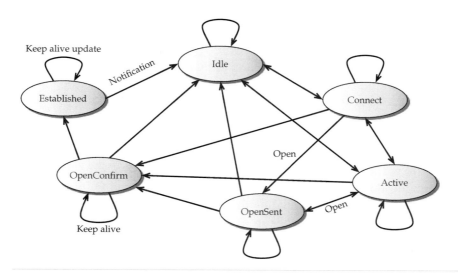

FIGURE 8.15   Finite state machine of a BGP speaker for connection to a peer.

TABLE 8.2   BGP events.

Event	Status	Remark
ME01	Mandatory	Local BGP administrator manually starts the BGP connection to a peer BGP speaker
ME02	Mandatory	Local BGP administrator stops the BGP connection to a peer BGP speaker
OE03	Optional	Local system automatically starts the BGP connection
OE04	Optional	A local BGP administrator is to manually start the BGP connection to a peer BGP speaker, but it first listens to an incoming BGP connection before starting the BGP connection
OE05	Optional	Local system automatically starts the BGP connection, but it first listens to an incoming BGP connection before starting the BGP connection
OE06	Optional	Local system automatically starts the BGP connection with the damping of oscillation activated
OE07	Optional	Local system automatically starts the BGP connection with the damping of oscillation activated and also first listens for an incoming BGP connection
OE08	Optional	Local system automatically stops the BGP connection
ME09	Mandatory	Indication that the ConnectRetryTimer has just expired
ME10	Mandatory	Indication that the HoldTimer has just expired
ME11	Mandatory	Indication that the KeepAliveTimer has just expired
OE12	Optional	Indication that the DelayOpenTimer has just expired
OE13	Optional	Indication that the IdleHoldTimer has just expired
OE14	Optional	Indication that the local system has received a valid TCP connection
OE15	Optional	Indication that the local system has received a TCP connection for the BGP session with either an invalid source IP address or port number, or an invalid destination IP address or port number
ME16	Mandatory	Indication that the local system has successfully set up a TCP connection to its remote BGP speaker that it initiated
ME17	Mandatory	Indication that the local system has confirmed the TCP connection initiated by a remote BGP speaker
ME18	Mandatory	Indication that the local system has received a notice about failure of a TCP connection to a peer BGP speaker
ME19	Mandatory	Indication that a valid OPEN message is received from the remote BGP speaker
OE20	Optional	Indication that a valid OPEN message has been received, but is delaying the sending of an OPEN message due to DelayOpenTimer running
ME21	Mandatory	Indication that the header of a received BGP message is not valid
ME22	Mandatory	Indication that there is some error with the OPEN message received
OE23	Optional	Indication of the detection of a connection collision during processing of an incoming OPEN message and this connection is to be disconnected

TABLE 8.2   (Continued.)

Event	Status	Remark
ME24	Mandatory	Indication that a version error code has been received with a NOTIFI-CATION message
ME25	Mandatory	Indication that an error code other than version error code has been received with a NOTIFICATION message
ME26	Mandatory	Indication that a KEEPALIVE message has been received
ME27	Mandatory	Indication that a valid UPDATE message has been received
ME28	Mandatory	Indication that an invalid UPDATE message has been received

CONNECT STATE

In this state, the BGP speaker is now waiting for the TCP connection to be established. Several different actions are possible depending on the triggering of events. If manual stoppage is invoked (event ME02), the connection is dropped, resources are released, and the BGP speaker then moves to the Idle state. If ConnectRetryTimer expires (event ME09), the BGP speaker drops the connection, restarts the ConnectRetryTimer, stops the DelayOpenTimer if this was activated earlier, starts a new TCP connection while listening to a connection initiated by the other side, and remains in the Connect state.

From a receiving point of view, if the TCP connection is valid (event OE14), it is processed and it stays at the Connect state. If the TCP connection is successful (ME16 or ME17), then the DelayOpen value is checked. If it is true, then timers are reset and it remains at the Connect set; however, if DelayOpen is not set, then the local BGP speaker stops the ConnectRetry-Timer, completes BGP initialization, sends an OPEN message to the remote BGP speaker, sets the HoldTimer to a large value, and moves to the OpenSent state.

If the TCP connection request is invalid (OE15), it is rejected and stays at the Connect state. However, if it is valid but the connection fails (ME18), the system checks if DelayOpen-Timer is running. If so, the timers are reset, and it continues to listen to a connection from its peer, and then moves to the Active state. If DelayOpenTimer is not running, the connection is dropped, BGP resources are released, and it moves to the Idle state.

It is possible to receive an OPEN message while the DelayOpenTimer is still running (OE20). If so, an OPEN message is sent in response and a KEEPALIVE message is also generated, and the state is changed to the OpenConfirm state. This also occurs when the autonomous system number on the received OPEN message is checked to determine if the peer is external or internal.

If there is any error while checking BGP header (ME21) or OPEN message (ME22), all BGP resources are released and the connection is dropped.

In response to an OPEN message, it is possible to receive a NOTIFICATION message with a version error (ME24); if so, BGP resources are released, the TCP connection is dropped, and it changes to the Idle state. For the rest of the events, if any of them occurs, the same handling procedure is used.

ACTIVE STATE

The role of this state is to acquire a BGP peer. If manual stoppage occurs (ME02), the connection is dropped and the state is changed to the Idle state. If the ConnectRetryTimer expires, a new TCP connection is initiated and, at the same time, listens to one, then moves its state to the Connect state. If the DelayOpenTimer expires or delayed open is not set, an OPEN message is sent to its BGP peer, and it moves to the OpenSent state. If, however, an OPEN message is received and the delay open timer is running (OE20), an OPEN message is sent and a KEEPALIVE message is sent.

If there is an error detected when checking the BGP message header or OPEN message, the connection is dropped and the state changes to the Idle state. In case of a NOTIFICATION message with version error (ME24), the handling is the same.

OPENSENT

Normally, the speaker arrives at the OpenSent state from the Active state; in this state, an OPEN message is sent immediately or later depending on the DelayOpenTimer value and at the same time it waits for an OPEN message from its BGP peer. If an OPEN message is received and there is no error in the message (ME19), a KEEPALIVE message is sent, the KeepAliveTimer is activated, and it moves to the OpenConfirm state. In all other events, the connection is dropped while sending a NOTIFICATION message as applicable and the state is changed to Idle, except if TCP connection fails (ME18); in this case, the state is changed to the Active state.

Once the optional local DelayOpenTimer expires, the speaker sends the OPEN message and waits to hear an OPEN message from its neighboring BGP speaker. If an OPEN message is received, fields are checked; if errors occurs such as a bad version number or an unacceptable AS number, it sends a NOTIFICATION message and moves to the Idle state.

Note that if the BGP speaker supports the Capabilities option, it can advertise this information when it sends the OPEN message and inquire, for example, if the ROUTE–REFRESH capability is supported by the receiving speaker. If the other end responds using a NOTIFICATION message stating that it does not support the ROUTE–REFRESH capability, a new OPEN message is generated in which the optional capability is turned off.

OPENCONFIRM STATE

In this state, the speaker waits for either a KEEPALIVE or a NOTIFICATION message, or generates a KEEPALIVE message. An important action here is when the HoldTimer expires; in this case, a NOTIFICATION message is sent and the TCP connection is dropped. If the KeepAliveTimer expires, a KEEPALIVE message is sent and a new KeepAliveTime value is generated, and the state is changed to the Established state.

If a NOTIFICATION or a NOTIFICATION with a version error (ME24) is received, the connection is dropped and the state changes to the Idle state. If an OPEN message is received (ME19), a NOTIFICATION is sent and the connection is closed and moves to the Idle state. Note that at this state, a NOTIFICATION message from its peer BGP speaker is also possible if the peer does not support the ROUTE–REFRESH capability; this situation arises only if it has included ROUTE–REFRESH capability in the optional capability parameter when it sent out its OPEN message.

ESTABLISHED STATE

In the established state, a BGP speaker normally exchanges UPDATE, NOTIFICATION, and KEEPALIVE messages with a peering BGP speaker. If a stoppage event occurs (ME02, OE08), a NOTIFICATION is sent, the connection is closed, and it moves to the Idle state.

If HoldTimer expires (ME10), a NOTIFICATION is sent, the TCP connection is dropped, peer oscillation damping is performed, and the system changes to the Idle state. If the KeepAliveTimer expires, a KEEPALIVE message is sent and a new KeepAliveTime value is generated, and the state is changed to the Established state. Note that each time a KEEPALIVE or UPDATE is received, the KeepAliveTimer is re-initialized with a new value.

8.12 Protocol Message Format

In this section, we provide detailed information about BGP message formats.

8.12.1 Common Header

BGP4 has a common header of 19 bytes that consists of 16 bytes of marker, followed by 2 bytes of length field, and 1 byte of type field (see Figure 8.16). The intent of the marker field is to do synchronization, although it can also be used for the security option. For example, this field contains all 1s when used for synchronization, especially when no security options are used. The length of the entire BGP message, including header, is indicated through the length field in byte count; although this field is 2 bytes long, any BGP message cannot be longer than 4096 bytes.

8.12.2 Message Type: OPEN

The message type, OPEN, follows the common header where the type field in the header indicates that this is a message type with value 1 (Figure 8.17). The OPEN message is sent at the start of a BGP session between two peering BGP speakers. The OPEN message has a required field of 10 bytes followed by an optional parameter field; thus, including the common header, the OPEN message is at least 29 bytes long. The main fields are as follows:

- *Version* (1 byte): This field indicates the BGP protocol version, currently set to version 4.

- *Autonomous System Number* (2 bytes): This field is used to declare the AS number to which the sending BGP speaker belongs.

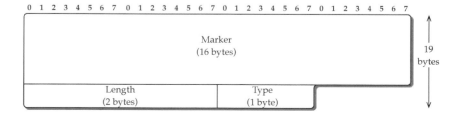

FIGURE 8.16   BGP4 common header.

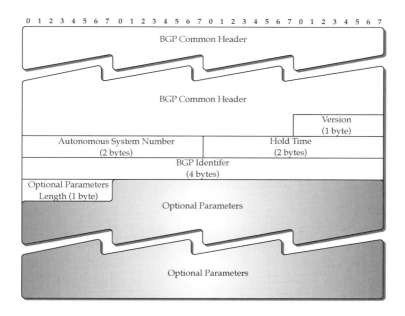

Optional Parameter Encoding:

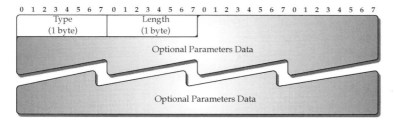

FIGURE 8.17 BGP4 OPEN message.

- *Hold Time* (2 bytes): The maximum time (in seconds) before a BGP speaker assumes that its peering BGP speaker is down, the time before the virtual link is assumed down. Since this value is advertised as part of the OPEN message by each side, the smaller of the two values is agreed upon as the HoldTime for the rest of this BGP session. Hold Time *must* be either zero or at least 3 sec, while the recommended value is 90 sec. The value zero is meant to indicate that the BGP session is never to go down.

- *BGP Identifier* (4 bytes): This is an identifier for the sending BGP speaker, serving as this router's ID. Usually, it is set to the highest value of all the sending BGP speaker's interfaces.

- *Optional Parameter Length* (1 byte): The parameter length is used to indicate optional parameters. When this field is set to zero, no optional parameter follows.

- *Optional Parameters:* The optional parameter is expressed using the type-length-value (TLV) format. Optional Parameter Type 1 (Authentication Information), originally defined in RFC 1771, has now been deprecated in RFC 4271. If the parameter type is 2, then capabilities such as the following have been defined so far: 1 for Multiprotocol Extension, which allows for addresses to be other than IP addresses, 2 for indicating route refresh capability, 4 for Multiple route to a destination capability, 64 for Graceful restart, 65 for Support for 4-byte AS number capability, and so on; the updated list is maintained at [314].

8.12.3 Message Type: UPDATE

Once a BGP session is established, the UPDATE message is sent to withdraw or announce IP prefixes with their route information, as and when needed (Figure 8.18). The message type

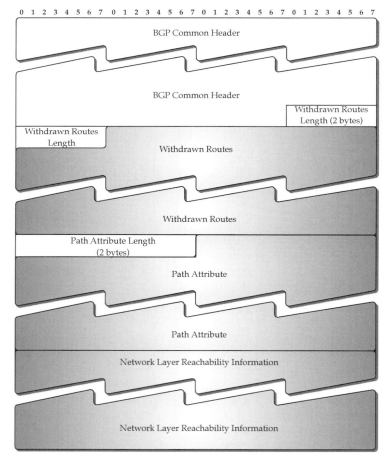

Withdrawn Routes (repeat pattern):

F I G U R E 8.18   BGP4 UPDATE message.

value in the common header is set to 2 to indicate that this is an UPDATE message. This message type has three subparts: withdrawal route part, path attribute part, and network layer reachability information part. An UPDATE message may have content for any or all of these subparts. The minimum size of an UPDATE message is 23 bytes consisting of 19 bytes of the common header, 2 bytes for Withdrawn Routes Length, and 2 bytes for Total Path Attribute Length.

An UPDATE message can contain just one IP prefix destination as advertisement along with its path attributes, the exception being that it can include a set of IP prefix destinations if they share the same path attributes. An update message can list multiple IP prefix destinations that are to be withdrawn; however, both withdrawal and advertisement may be combined in a single UPDATE message.

- *Withdrawn Routes Length* (2 bytes): This field indicates the total length of the Withdrawn Routes field in bytes. When this field is set to zero, it means that there is no announcement about withdrawal of routes in this particular UPDATE message.

- *Withdrawn Routes* (variable): This field is not present when Withdrawn Routes Length is zero. Otherwise, this field has the format ⟨prefix length, IP prefix⟩. Here, the prefix length is always 1 byte long and indicates the length in bits of the IP prefix address of a routable network address while the IP prefix field is of variable length, which is required to carry the IP prefix with the additional requirement that it must fit into a byte boundary; to do so, trailing bits are added. For example, if the IP prefix to be withdrawn is 134.193.0.0/16, the length field will have the value 16, followed by the IP prefix taking 16 bits, which represents 134.193, i.e., 1000 0110 1100 0001 (in bits). However, if a BGP speaker *were* to withdraw prefix 134.193.128.0/17, the length field would have the value 17, followed by the IP prefix 134.193.128, taking 24-bit space instead of 17-bit space, i.e., 1000 0110 1100 0001 1000 0000 (in bits), to align with the byte boundary; note that only the first 17 bits are important and the trailing bits beyond 17 bits would be ignored when reading this field at the receiving BGP speaker. The ⟨prefix length, IP prefix⟩ pattern is repeated for all routes withdrawn in an OPEN message while the total length in bytes (*not* in number of routes) is indicated through the Withdraw Routes Length.

- *Path Attribute Length* (2 bytes): This field indicates the total length of the Path Attributes field in bytes. If this value is zero, this UPDATE message does not include the Path Attribute and the NLRI field.

- *Path Attributes* (variable): The format for this field is described later in Section 8.12.7. Earlier in Section 8.6, we described the role played by different attributes.

- *Network Layer Reachability Information* (variable): This field contains one or more IP prefix destinations. Each IP prefix is encoded in the format of ⟨prefix length, prefix⟩ where prefix length is 1 byte which indicates the length of the IP prefix in number of bits, and the prefix is of variable length, which is derived from the number of bits in the length rounded up to the byte boundary. The encoding is similar to the illustration shown above in Withdrawn routes. There is no length field for NLRI; this is determined from the total length of the UPDATE message by subtracting the length for Withdrawn Routes and the length for Total Path Attributes.

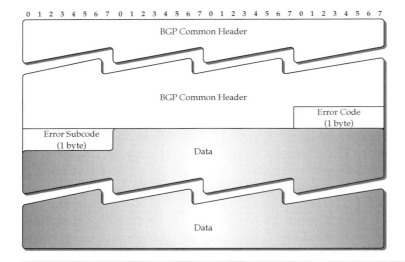

FIGURE 8.19   BGP4 NOTIFICATION message.

8.12.4 Message Type: NOTIFICATION

The role of the NOTIFICATION message is to indicate if an error has occurred. The message type value in the common header is set to 3 to indicate that this is a NOTIFICATION message. The message format for NOTIFICATION has three parts: Error Code (1 byte), Error Subcode (1 byte), and Data (variable), as shown in Figure 8.19. Error codes have been categorized into six parts, where the first three categories have their own subcodes. The error codes/subcodes are summarized in Table 8.3. The Data field is typically used to provide additional information about an error; for example, the Data field includes the erroneous Length field value if the error code is 1 and the error subcode is 2. It may be noted that if there is an error in the NOTIFICATION message itself sent by a BGP speaker, the receiver speaker is not allowed to respond with another NOTIFICATION message.

8.12.5 Message Type: KEEPALIVE

The role of the KEEPALIVE message is to indicate that the BGP session is active. To ensure that the HoldTimer does not expire, a KEEPALIVE message is sent that is approximately one-third the HoldTime value as long as it is not sent more than once per second. If, however, the HoldTime is agreed to be zero, which is to mean that the BGP session is to remain alive, then KEEPALIVE messages are not generated. A KEEPALIVE message does not have any data on its own; the common message header is sent with the message type value set to 4.

8.12.6 Message Type: ROUTE–REFRESH

The ROUTE–REFRESH message is an optional BGP message that is generated only if this capability is negotiated through the exchange of initial OPEN messages. Operationally, on receiving a ROUTE–REFRESH message from a peer, a BGP speaker would need to send the content of Adj-RIB-Out for this peer using an UPDATE message. The message type value

TABLE 8.3 Error Codes/Subcodes with BGP NOTIFICATION message.

Code	Subcode	Remark
1		Error detected in the BGP message header
	1	Marker field is not synchronized with all 1s
	2	Message Length is not valid. This can happen if (1) the message length is smaller than 19 bytes or larger than 4096 bytes, (2) the Length field of the OPEN (UPDATE/KEEPALIVE/NOTIFICATION) message does not meet the minimum length of an OPEN (UP-DATE/KEEPALIVE/NOTIFICATION) message
	3	The Message Type is not recognized
2		Error in the OPEN message content as specified through subcodes
	1	Version number is not supported
	2	AS number of the peer is not acceptable
	3	The BGP Identifier is not a valid unicast IP address
	4	The Optional Parameter is not supported. However, if the Optional Parameter is recognized, but is malformed, then the Subcode is set to 0
	5	Deprecated
	6	Unacceptable Hold Time; this occurs if the value is announced to be either 1 or 2 sec
	7	Capability not supported (in response to Capabilities advertisements discussed in [113])
3		Error in the UPDATE Message
	1	Attribute List is nonconforming
	2	Well-known Attribute is not recognized
	3	A mandatory Well-known Attribute is missing
	4	Error with Attribute Flags
	5	Error with Attribute Length
	6	The value in ORIGIN Attribute is not valid
	7	Deprecated
	8	NEXT–HOP Attribute value is not valid
	9	Error in Optional Attribute
	10	The content in the NLRI field is not correct
	11	AS–PATH is malformed
4		To indicate expiration of Hold Timer
5		Error in Finite State Machine for the BGP connection
6		This allows a BGP connection to close a session normally

in the common header is set to 5 to indicate that this is a ROUTE–REFRESH message. The message format for ROUTE–REFRESH has three parts: Address Family Identifier (2 bytes), Reserved (1 byte), and Subsequent Address Family Identifier, as shown in Figure 8.20. The information about address family identifier (AFI) is included since BGP is now extended with multiprotocol capability [63]; thus, an address family can be properly identified. Note that the AFI for the IPv4 address is 1; the list of AFIs is regularly updated and is maintained at [313].

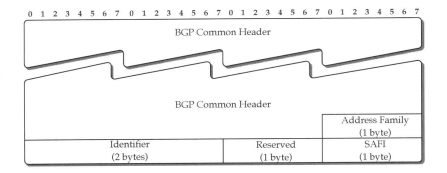

FIGURE 8.20   BGP4 ROUTE–REFRESH message.

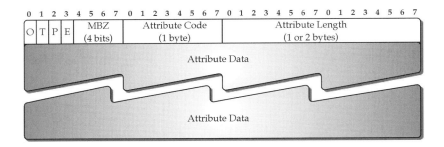

FIGURE 8.21   Path Attribute in UPDATE message.

8.12.7 Path Attribute in UPDATE message

Path Attribute appears in a BGP UPDATE message if an announcement or a change in regard to an IP prefix destination or a set of IP prefix destinations is advertised. If the Path Attribute Length field contains a nonzero value, a variable length Path Attributes field follows that has the TLV format: ⟨attribute type, attribute length, attribute value⟩. The attribute type is 2 bytes long and has two subparts: attribute flags (1 byte), and attribute type code (1 byte). Attribute flags consists of four higher-order bits: O, Optional (0 for well-known or 1 for optional); T, Transitive (0 for nontransitive or 1 for transitive); P, Partial (0 for complete or 1 for partial); and E, Extended (0 for one byte 1 for two bytes); the lower 4 bits must be zero, denoted by MBZ in Figure 8.21. W-bit indicates whether the attribute is well-known and is supported by the receiving BGP speaker, or the attribute is optional and may not be supported by the receiving BGP speaker. If E is set to 0, then the attribute length is 1 byte long; if E is set to 1, then the attribute length is 2 bytes long.

The path attributes that are significant have already been discussed in Section 8.12.7 and the ones related to the route reflector option have been discussed in Section 8.8.1. Note that for a confederation approach in IBGP, only the AS–PATH attribute is extended with new types. The communities attribute, discussed in Section 8.10.1, is optional transitive and is assigned a type value of 8. Each community identifier is 4 bytes consisting of two parts: 2 bytes for AS number, and 2 bytes for indicating communities such as the three described in Section 8.10.1.

TABLE 8.4 BGP Path Attributes.

Type Code	Type Name	OT-bits	Remark
1	ORIGIN	01	Well-known mandatory; indicates IGP (0), EGP (1), or INCOMPLETE (3); see RFC 4271 [591]
2	AS–PATH	01	Well-known mandatory; indicates AS–SET (1), AS–SEQUENCE (2), AS–CONFED–SET (3), AS–CONFED–SEQUENCE (4); see RFC 4271 [591]
3	NEXT–HOP	01	Well-known mandatory; includes 4-byte IP address; see RFC 4271 [591]
4	MED	10	Optional nontransitive; 4-byte MED identifier; see RFC 4271 [591]
5	LOCAL–PREF	01	Well-known discretionary; 4-byte LOCAL–PREF identifier; see RFC 4271 [591]
6	ATOMIC–AGGREGATE	01	Well-known discretionary; indicated when aggregation is done at BGP speakers; see RFC 4271 [591]
7	AGGREGATOR	11	Optional transitive; AS number and the IP address of the BGP speaker aggregator; see RFC 4271 [591]
8	COMMUNITIES	11	Optional transitive; 4-byte community identifier; see RFC 1997 [115]
9	ORIGINATOR–ID	10	Optional nontransitive used for route reflector; 4-byte ID of originator; see RFC 4456 [62]
10	CLUSTER–LIST	10	Optional nontransitive used for route reflector; variable length; see RFC 4456 [62]
16	Extended Communities	11	Optional transitive; see RFC 4360 [621]

The Extended Communities Attribute [621] extends Communities Attribute by allowing an extended range for covering a large number of different usages. The updated list of BGP parameters such as Path Attribute types is maintained at [315].

8.13 Summary

BGP is like a glue that helps connect the Internet together. It is an interdomain routing protocol that is used between two autonomous systems; it is also used in internal BGP mode when an autonomous system has multiple BGP speakers talking to the outside of this autonomous system. Between two BGP speakers, a TCP-based BGP session is set up. Using UPDATE message type, at first complete BGP routes are exchanged; after that, only incremental changes such as a new announcement, withdrawal, or change in path attributes, are exchanged.

BGP is a path vector protocol where granularity of information is at an IP prefix level, served by autonomous systems. Each IP prefix is attached with its *home* AS number, which is disseminated from one autonomous system to another by prepending path attributes; an exception is when route aggregation is done through supernetting when a set of IP prefixes can be combined and the aggregated information is forwarded downstream where the point of aggregation serves as the "care of" home for the supernetted address block.

BGP is used in two ways: external BGP and internal BGP. While the basic protocol messaging is the same, there are certain restrictions/rules imposed on IBGP. For a large-scale IBGP scenario, approaches such as route reflectors or confederations may be used.

Announcement and withdrawal of an IP prefix can lead to route flapping; to minimize/avoid this flap, a route flap dampening approach can be used.

Finally, the finite state machine of a BGP speaker to a peer speaker is quite elaborate. Any state transition is triggered through a well-defined set of events.

Further Lookup

The initial version of BGP was described in RFC 1105 [428]. BGP4, the current version of BGP, was first described in RFC 1771 [590] and has been recently updated in RFC 4271 [591]; this RFC includes a summary on changes compared to earlier versions of BGP. The fifth message type, ROUTE–REFRESH, is described in RFC 2918 [129], which uses Capabilities advertisements described in [114], which makes [113] obsolete.

The concept of route reflection for internal BGP was first described in RFC 1966 [59], which was subsequently updated in RFC 2796 [60], and further updated in RFC 4456 [60]. A formal confederation approach was first proposed in RFC 975 [490] for circumventing certain restrictions of the Exterior Gateway Protocol; we will describe EGP briefly in Section 9.1. The confederation concept for internal BGP was presented in RFC 1965 [701], which has been updated in RFC 3065 [702].

Currently, the AS number field is 2 bytes long. In anticipation of running out of AS numbers, a 4-byte AS number is being currently proposed. As of this writing, this proposal remains as an Internet draft.

There are several books that treat BGP extensively [188], [282], [301], [546], [571], [669], [709], [738], [764]. BGP routing table analysis reports are available at [303]. There are many resources on the Internet about BGP; see excellent central resource sites such as [170] and [704].

Like any protocol, BGP has vulnerabilities; see [513] for a discussion.

Exercises

8.1 Review questions:

  (a) What are the different BGP timers?

  (b) What are the different states in the BGP finite state machine?

  (c) What are the different BGP message types?

8.2 How is looping avoided in BGP?

8.3 What would happen if an IBGP speaker does advertise IP prefixes it has learned from an IBGP speaker to another IBGP speaker?

8.4 Suppose an autonomous system is set up with a single route reflector. What would be the consequence if the route reflector fails?

8.5 Analyze the route flap dampening concept by trying out different penalty values and flap time intervals.

8.6 How is the route reflector approach different from the confederation approach? Explain.

9

Internet Routing Architectures

Architecture is the will of an epoch translated into space.

Ludwig Mies van der Rohe

Reading Guideline

This chapter may be read without much dependence on other chapters. However, knowledge of routing protocols such as BGP (refer to Chapter 8) and OSPF/IS-IS (refer to Chapter 6) helps facilitate better understanding of the content presented in this chapter.

Internet routing depends heavily on the Border Gateway Protocol (BGP) for inter-AS relations. At the same time, because of business relations among Internet service providers, Internet routing architectures have evolved to include public and private peering among providers and transit issues. In addition, the growth of IP address space allocation and AS number allocation constitutes additional factors to be understood in the context of Internet routing architectures. In this chapter, we discuss these aspects in details.

9.1 Internet Routing Evolution

We first briefly discuss the evolution of Internet architecture from a historical perspective. Note that we focus on Internet *routing* rather than the Internet as a whole; for an excellent summary on Internet history, refer to [403].

Until the early 1980s, the ARPANET served the role of interconnecting various sites with a rigid two-level hierarchy where the ARPANET nodes were at the top level. In 1983, ARPANET was split, resulting in two networks: ARPANET and MILNET (see Figure 9.1); this was the birth of the two separate networks talking to each other in case one host in one network wants to communicate with another host in the other network, and vice versa. This also resulted in the *need* to have a mechanism by which separate networks could talk to each other. Here "separate networks" means that they are run by different entities.

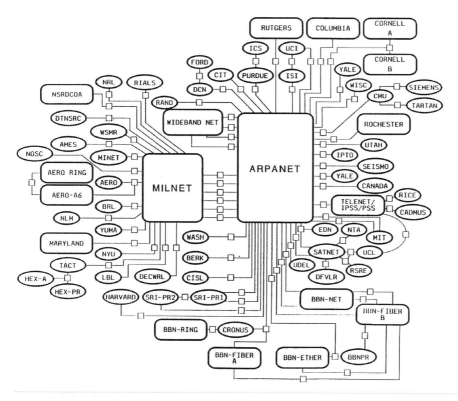

FIGURE 9.1   ARPANET and MILNET, circa 1983 (courtesy [488]).

Exterior Gateway Protocol (EGP), developed in 1982–1984 (refer to RFC 827 [600], RFC 888 [623], RFC 904 [489]) describes how separate networks that are autonomous can talk to each other. Along with EGP, the term *autonomous system* (AS) and the notion of a 16-bit *autonomous system number* (ASN) were introduced in [600]. Briefly, EGP defined a two-level strict hierarchy view with the top level labeled as the core backbone and the bottom level being the level at which the different networks, defined through ASes, were connected. NSFNET deployed first in 1984 relied on EGP. The architecture of and experience with NSFNET and EGP have been documented in [94], [587].

It is important to note that in EGP, nonbackbone ASes were not allowed to be directly connected; this is a direct consequence of the strict two-level hierarchy imposed by EGP. Another consequence was that the structure allowed only a single provider at the top level, i.e., the NSFNET. Furthermore, unlike BGP, EGP messages were sent directly over IP without invoking any reliable transport protocol. Thus, if the exchange of information required a large message to be generated, this needed to be handled by fragmentation and reassembly at the application layer of the TCP/IP protocol stack.

In essence, while EGP provided a much needed transitional platform to go from the ARPANET to the NSFNET, it had several restrictions not desirable for longer term growth. For example, EGP did not allow ASes to be directly connected. Thus, a network that is located in an AS would need to go through the top level, i.e., the NSFNET, to reach another network in another AS. However, NSFNET faced the situation that certain networks that belonged to different ASes had backdoor connectivity. Thus, EGP's strict requirement could not be directly applied or enforced in the NSFNET. It may be noted that to circumvent the limitation of EGP, a formal confederation approach was suggested in RFC 975 [490]. An important lesson learned from NSFNET in regard to the routing architecture is that no single entity would be managing the global Internet. Each system that is a component of the global Internet will have its own routing paradigm that can be driven by economics and other factors; each such system would have its own interest to connect to other systems directly, instead of using a global core such as the one suggested by EGP. As a corollary, global consensus from the deployment point of view is hard to arrive at while mutual bilateral agreement is possible. Since ASes use a common prefix address space (i.e., IPv4 address space), and an AS cannot control what an upstream AS announces, it became necessary to take a policy-driven approach; for example, how routing is done to handle packets from certain networks based on import policy of an AS. It is to be noted that some rudimentary policy-based routing was done so that certain rule checking can be invoked in the NSFNET as noted in RFC 1092 [587] in order to handle import and export policies.

EGP and, particularly, NSFNET experiences led to the recognition that any future routing architecture must be able to handle policy-based routing (see RFC 1102 [144], RFC 1104 [93], RFC 1124 [402]), and any newly developed exterior gateway protocol must have the ability to handle policy decisions. That is, experience and realization served as the impetus to the development of BGP, which was first introduced in 1989 through RFC 1105 [428]. To summarize, BGP tried to address the following issues: (1) avoiding a strict two-level hierarchy like EGP, (2) allowing multiple levels such that any AS has the option to connect to another AS, (3) using TCP for reliable delivery of BGP data, and (4) making policy-based routing possible.

By 1991, BGP was expanded to BGP, version 3 (see RFC 1267 [429]). At about the same time, it was recognized the implicit address block classification of an IP address under Class A, Class B, and especially Class C, i.e., classful addressing, would cause a significant growth in the routing table entries at core backbone routers; thus, some mechanisms to avoid/minimize assigning address block straight at Class C were needed. This has led to consider address aggregation through supernetting [240], which subsequently led to the development of *classless interdomain routing (CIDR)*.

While BGP, version 4 (BGP4) has resulted in several improvements over BGP, version 3, it is clear that use of CIDR was one of the most significant changes that required communicating netmask information to be advertised along with an IP address block during a BGP announcement; that is, the addressing structure played a critical role in routing. Before we further discuss Internet routing architecture, we present a brief background discussion on IP addressing and routing.

Finally, it is worth noting that the notion of dividing a network into hierarchical structure of intradomain and interdomain and allowing each intradomain to define its own routing can be traced backed to the OSI routing model developed in the 1980s; see [567] for further details.

9.2 Addressing and Routing: Illustrations

Routing in the Internet is fundamentally impacted by IP addressing. A unique feature of the Internet is that the end hosts and routers alike share from the same addressing family, and this has a profound impact on routing. The address family is known as the IPv4 address family, and its recent version is known as IPv6. We will focus our discussion here specifically on IPv4 addressing.

The IPv4 address family is a 32-bit address that is typically written in dotted decimal format, A.B.C.D, where each part represents the decimal value for 8 bits. Routing benefits form an important requirement in regard to address space allocation; that is, the address space is compacted through subnet masking, and addresses are assigned in *contiguous* blocks for a specific network. For example, contiguous addresses 192.168.1.0 to 192.168.1.255 would be assigned to a network (or subnet); similarly, contiguous addresses 192.168.2.0 to 192.168.2.255 would be assigned to another network, and so on. To reiterate, address block contiguity to define a network is a fundamental requirement in IP that impacts routing. For example, because of this contiguity, a routing table at a router needs only one entry for an address block such as 192.168.1.0 to 192.168.1.255, instead of 256 separate address entries for each of these IP addresses from this range. If each router were required to keep an entry for all 2^{32} IP addresses, this would simply not scale! There is, however, an important trade-off, due to contiguous address blocks—not all addresses can be assigned to end hosts. For example, if we consider the address block from 192.168.1.0 to 192.168.1.255 to identify a subnet, then two addresses at the extreme ends are reserved to identify the network and for the broadcast purpose, respectively; specifically, the "0" address, i.e., 192.168.1.0, will be reserved to identify the subnet and and the "255" address, i.e., 192.168.1.255, will be reserved as the broadcast address.

We, however, need a simple mechanism to define contiguous address blocks that may fall at a different bit boundary level. Originally, IPv4 unicast addressing was allocated through *implicit* bit boundaries for network block addresses at an 8-bit, 16-bit, and 24-bit boundary, known as Class A, Class B, Class C addresses, respectively. The difficulty with implicit boundary, at least for routing purpose, is that at the 24-bit level boundary, the number of address blocks is too huge to handle if all were advertised! This is mainly because of another important imposition on IPv4 addressing; that is, a network address block follows a simple flat addressing principle. This means that if we want to route a packet from network 134.193.0.0 to network 134.194.0.0, we cannot count on the most significant 8 bits, i.e., 134, as some hierarchical indicator to make a local/hierarchical routing decision; instead, we need to keep both entries 134.193.0.0 and 134.194.0.0—this is known as *flat addressing*. Similarly, if all 24-bit network address blocks are to be considered, then we need 2^{24} entries for a routing decision due to flat addressing. Instead of the implicit network boundaries at an 8-bit, 16-bit, and 24-bit level, the *explicit* network boundaries through network masking, referred to as *classless inter-domain routing* (CIDR), were found to be more flexible in reducing the need to assign IP address blocks for networks at a 24-bit boundary or the other implicit boundaries at 8-bit and 16-bit level.

The basic idea behind CIDR is that along with the address of a specific host, an explicit net masking is also applied that defines the network where this host resides. For example, if the host that we want to reach is 192.168.40.49, and if the address block is netmasked at the 21-bit boundary level, all a router needs to know is that 192.168.40.49 is identified as being on a network defined on the 21-bit boundary. Typically, this is indicated through the CIDR notation 192.168.40.0/21 where /21 indicates the network address block netmask. How do we arrive at 192.168.40.0/21 from 192.168.40.49? It is easy to understand when we look at the bit level information. Note that /21 means that the first 21 most significant bits in a 32-bit address are 1s and the rest are 0s, i.e., 11111111 11111111 11111000 00000000; this 32-bit netmask can also be written in the dotted decimal IP address format as 255.255.248.0. That is, a netmask written in CIDR notation /21 and its IP address notation, 255.255.248.0, are interchangeable. As a convention, the CIDR netmask is used in identifying IP prefix-level networks, while the format such as 255.255.248.0 is used in a subnet mask on a computer when comparisons are required for packet forwarding in a subnet.

When we consider 192.168.40.49 with /21 in the CIDR notation, we can perform a bitwise "AND" operation as shown below:

```
          11000000 10101000 00101000 00110001   → 192.168.40.49
    AND   11111111 11111111 11111000 00000000   → netmask (/21)
          11000000 10101000 00101000 00000000   → 192.168.40.0
```

That is, the bitwise AND operations result in obtaining the net address 192.168.40.0, which is tied with /21 so that the network boundary is understood.

Thus, when a host in another network that has the IP address, say, 10.6.17.14 wants to send a packet to, say, 192.168.40.76, it needs to send to network 192.168.40.0/21, hoping that once it reaches this destination network, i.e., 192.168.40.0/21, it knows how to handle the delivery to the final host. This is analogous to the postal system; it is similar to sending a letter that needs to reach a postal code, and hoping that once it reaches that postal code, it can be delivered to the actual house address.

A question is how does the originating host know how to get the packet out of its own network so that it can then traverse the global Internet toward its destination. Second, is it different and/or where is it different if the packet had a destination that happens to be in the same network, for example, if 192.168.40.49 were to send a packet to 192.168.40.76, or beyond that. We will consider routing a packet under three different scenarios.

9.2.1 Routing Packet: Scenario A

The first scenario we consider is a subnet defined by an IP address block through a standard subnet masking. That is, consider sending a packet from a host with IP address 192.168.40.49 ("49ers") to another host with IP address 192.168.40.76 ("76ers"). *The first requirement is that each host along with its IP address must have a subnet mask associated with it.* In this example, we assume the subnet mask for this subnet is 255.255.255.0 which is indicated in the configuration profile of these two hosts. The sending host first determines its own subnet by comparing the subnet mask and the IP address of the destination (192.168.40.76) with its stored subnet mask 255.255.255.0 through the bitwise AND operation as shown below:

	11000000 10101000 00101000 00110001	→ 192.168.40.49 ("49ers")
AND	11111111 11111111 11111111 00000000	→ subnet mask (255.255.255.0)
	11000000 10101000 00101000 00000000	→ 192.168.40.0 (/24)

	11000000 10101000 00101000 01001100	→ 192.168.40.76 ("76ers")
AND	11111111 11111111 11111111 00000000	→ subnet mask (255.255.255.0)
	11000000 10101000 00101000 00000000	→ 192.168.40.0 (/24)

Thus, Host "49ers" realizes that the destination host, Host "76ers," belongs to the same subnet. Assume that subnet 192.168.40.0 is served by an Ethernet LAN (see Figure 9.2). Thus, to send a packet, the host with address 192.168.40.49 is required to rely on the Ethernet interface for packet delivery; for that, a protocol called the *Address Resolution Protocol* (ARP), which does the function of mapping the IP address to the Ethernet address, is first invoked. Through this process, Host "49ers" finds the Ethernet address of the destination IP address 192.168.40.76. Once the Ethernet address of the destination is determined, the packet is sent as an Ethernet frame with the destination address set to this Ethernet address. In a sense, we can say that in the same Ethernet LAN, "routing" a packet does not really involve routing.

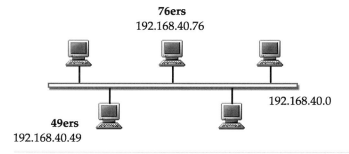

FIGURE 9.2   Host "49ers" (192.168.40.49) and Host "76ers" (192.168.40.76) on an Ethernet subnet with mask 255.255.255.0.

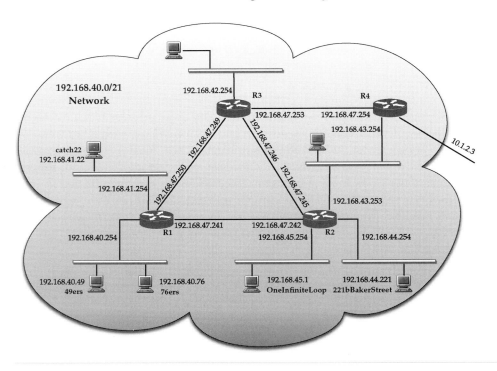

FIGURE 9.3   Network 192.168.40.0/21 with subnets and routers.

9.2.2   Routing Packet: Scenario B

The second scenario is where communication is *not* limited to the same subnet, but it is in the same network in the sense that it is provided by the same provider, such as a campus or an enterprise network. We identify this network as 192.168.40.0/21, which covers the address range 192.168.40.0–192.168.47.255. We assume that it consists of Ethernet segments where subnets are allocated and that all Ethernet-based subnets use subnet mask 255.255.255.0. Serial links are used between three of the four routers where subnet mask 255.255.255.252 is used. The topology of network 192.168.40.0/21 with all its subnets is shown in Figure 9.3. We assume that this intradomain network uses the OSPF protocol among its routers. For illustration, consider sending a packet again from Host "49ers," with IP address 192.168.40.49 and subnet mask 255.255.255.0, to a third host with IP address 192.168.41.22 ("catch22"). Note that Host "49ers" does not know about the subnet mask of Host "catch22." Based on its own subnet mask, Host "49ers" compares and determines that Host "catch22" is on a different subnet, 192.168.41.0, as shown below:

	11000000 10101000 00101000 00110001	→ 192.168.40.49 ("49ers")
AND	11111111 11111111 11111111 00000000	→ subnet mask (255.255.255.0)
	11000000 10101000 00101000 00000000	→ 192.168.40.0 (/24)

	11000000 10101000 00101001 00010110	→ 192.168.41.22 ("catch22")
AND	11111111 11111111 11111111 00000000	→ subnet mask (255.255.255.0)
	11000000 10101000 00101001 00000000	→ 192.168.41.0 (/24)

R1:

Net	Mask	NextHop	Interface
192.168.40.0	255.255.255.0	direct	en0
192.168.41.0	255.255.255.0	direct	en1
192.168.42.0	255.255.255.0	192.168.47.249	sl2
192.168.43.0	255.255.255.0	192.168.47.242	sl1
192.168.44.0	255.255.255.0	192.168.47.242	sl1
192.168.45.0	255.255.255.0	192.168.47.242	sl1
192.168.47.240	255.255.255.252	direct	sl1
192.168.47.248	255.255.255.252	direct	sl2
0.0.0.0	0.0.0.0	192.168.47.242	sl1

R2:

Net	Mask	NextHop	Interface
192.168.40.0	255.255.255.0	192.168.47.241	sl1
192.168.41.0	255.255.255.0	192.168.47.241	sl1
192.168.42.0	255.255.255.0	192.168.47.246	sl0
192.168.43.0	255.255.255.0	direct	en2
192.168.44.0	255.255.255.0	direct	en0
192.168.45.0	255.255.255.0	direct	en1
192.168.47.240	255.255.255.252	direct	sl1
192.168.47.244	255.255.255.252	direct	sl1
0.0.0.0	0.0.0.0	192.168.43.254	en2

R3:

Net	Mask	NextHop	Interface
192.168.40.0	255.255.255.0	192.168.47.250	sl0
192.168.41.0	255.255.255.0	192.168.47.250	sl0
192.168.42.0	255.255.255.0	direct	en0
192.168.43.0	255.255.255.0	192.168.47.254	sl2
192.168.44.0	255.255.255.0	192.168.47.245	sl1
192.168.45.0	255.255.255.0	192.168.47.245	sl1
192.168.47.244	255.255.255.252	direct	sl0
192.168.47.248	255.255.255.252	direct	sl1
192.168.47.252	255.255.255.252	direct	sl2
0.0.0.0	0.0.0.0	192.168.47.254	sl2

R4:

Net	Mask	NextHop	Interface
192.168.40.0	255.255.255.0	192.168.43.253	en0
192.168.41.0	255.255.255.0	192.168.43.253	en0
192.168.42.0	255.255.255.0	192.168.47.253	sl0
192.168.43.0	255.255.255.0	direct	en0
192.168.44.0	255.255.255.0	192.168.43.253	en0
192.168.45.0	255.255.255.0	192.168.43.253	en0
192.168.47.252	255.255.255.252	direct	sl0
0.0.0.0	0.0.0.0	10.1.2.3	sl1

FIGURE 9.4 Routing tables (with interfaces) at routers in Network 192.168.40.0/21 (see Figure 9.3).

Now Host "49ers" has a decision to make since it realizes that Host "catch22" is not on the same subnet. To make this decision, Host "49ers" must be equipped with a mechanism for handling such packet arrival; this mechanism is provided through a *default gateway* address. This means that if a packet's destination is not on the same subnet, the default gateway will be the agent that will be the recipient of this packet, which, in turn, hopefully knows how to handle it. The knowledge of this default gateway is known to the sending host either by static configuration or through the Dynamic Host Configuration Protocol (DHCP). In general, the following must hold for a host to communicate on the Internet:

> *Either through static configuration or through DHCP, a host must have three key pieces of information: its host IP address, the subnet mask, and the default gateway. Note that the default gateway is not needed for scenario A; however, since a host must reside in an interconnected environment where it will invariably want to send a packet to a destination such as an email server or a web server that is outside its subnet, the default gateway information becomes a necessity. For a host to use Internet services, it is also required to have information about the IP address of at least a DNS server so that this server can be queried to find the actual IP address of a specific domain name. Thus, a host typically requires four pieces of information: its host IP address, the subnet mask, the default gateway, and a DNS server's IP address.*

Now going back to our example, the IP address of the default gateway must fall on the same subnet as the host. Here, the default gateway for Host "49ers" is assigned the address 192.168.40.254, and this happens to be an interface to a router that has an interface to subnet

192.168.41.0, and also to other subnets; see Figure 9.3, where this is depicted with the router marked as R1.

With the availability of the default gateway information at Host "49ers," and on recognizing that this packet is to be sent to the gateway, it would do an ARP request to find the Ethernet interface address for 192.168.40.254, so that the packet can be sent as an Ethernet frame to router R1.

Once the packet arrives at router R1, the router is now required to make a decision on which interface to send it out since it has multiple interfaces. Based on the destination networks it has learned about, a router would maintain a routing table so it knows how to handle an arriving packet. Furthermore, based on the routing table, the forwarding table information is derived to determine which interface is to be used for packet forwarding. In Figure 9.4, we show a routing table view with interface information for router R1. From the table for R1, we can see that the packet that originated at Host "49ers" will be sent on the interface with IP address 192.168.41.254 for delivery to Host "catch22" on the Ethernet segment.

We next consider the case in which Host "49ers" has a packet to send to Host "221bBakerStreet" (with IP address 192.168.44.221). This packet will first arrive at router R1 since by inspecting the address of Host "221bBakerStreet", Host "49ers" would realize that the host does not belong to this subnet. At router R1, by inspecting the routing table, the packet would be forwarded to router R2, whereupon it will be sent on the Ethernet interface with IP address 192.168.44.254.

We can thus see that for any packet that is meant for a host within network 192.168.40.0/21, routers would be required to have next-hop information for different subnet segments. The process of learning about different subnets within this domain can be accomplished using OSPF flooding. For instance, each router can learn about different subnets from a link state advertisement (LSA) that would contain subnet information with mask by using the network link-type LSA (refer to Chapter 6). Once announcements about various subnets are received, each router can use shortest path routing to determine the appropriate next hop to reach different subnets; the tables at each router are shown in Figure 9.4. You may note that the table at router R1 does not show an entry for the serial link subnet 192.168.42.0/30; the assumption here is that serial link subnets are not to be advertised in the LSA; this is similar for other tables. Furthermore, note that each table contains a 0.0.0.0/0 entry, which is referred to as the *default route*. The default route is similar to the default gateway maintained by each host; this entry points to a next-hop for forwarding a packet that lists a destination not listed in the routing table.

In essence, through scenario B, we have illustrated how to route a packet from a host to another host in a different subnet but within an administrative domain. Typically, such an administrative domain is defined by an AS, or a provider.

9.2.3 Routing Packet: Scenario C

In this scenario, we consider routing a packet that is generated at Host "49ers" meant for a host outside of network 192.168.40.0/21 to, say, host 10.5.16.60 where each network is served by a different AS. By using the next hop for the default route at each router, the packet generated at Host "49ers" would be forwarded from R1 to R2 to R4. Note that R4 is the border router in this domain that can speak OSPF to interior routers, but can also speak BGP to its

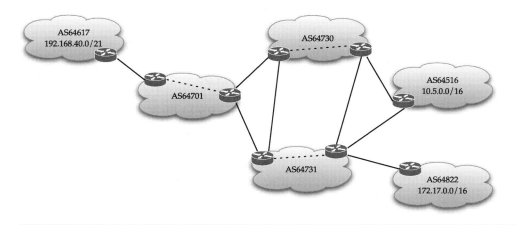

FIGURE 9.5 AS routing path of a packet from 192.168.40.0/21 to 10.5.0.0/16 through intermediate ASes.

peering EBG speaker. If R4 were to maintain a default route entry in its routing table, the packet that originated at Host "49ers" will be forwarded outside network 192.168.40.0/21 on the external link.

In Figure 9.5, we show connectivity from network 192.168.40.0/21 to network 10.5.0.0/16 that requires traversing through intermediate ASes. Suppose that each AS maintains an entry for default route to the next AS. Then the packet from 192.168.40.49 will be routed from AS64617, its home AS, to its neighboring AS, AS64701. The border router at AS64701, on receiving this packet, will check if it is meant for an address block that is internal to it, and will realize that it does not; thus, it will forward it to its other border router through internal routers. Assume that this border router in AS64701 has set up the default route to be to AS64730. Thus, the packet will eventually arrive at the border router in AS64516, the home to network 10.5.0.0/16. The border router will recognize that it supports this network and would then use interior routing protocol to deliver to the final destination host.

Note that we have assumed that everyone uses the default route concept. There are, however, certain problems with every AS using the default route to send a packet out if its AS. For example, if the destination host is from an IP address block that has not yet been allocated by the Internet registry, the packet would keep hopping from one AS to another until the age field (also known as the time-to-live field) in the IP packet header expires. This indicates that there are two possibilities: (1) at least one of the intermediate ASes maintains all *default-free* entries, that is, every valid IP prefix is explicitly listed, meaning there is *no* default route entry, 0.0.0.0/0, or (2) the originating AS at its border router maintains the list of every valid IP prefix assigned thus far so that it can filter this packet (refer to Chapter 16) and drop it, preventing from going into the next AS.

We first discuss the first possibility. An intermediate AS such as AS64730, shown in Figure 9.5, is known as a *transit AS*. Note that there is no such thing as *the* core transit AS; in fact, that would be restrictive, like EGP, which we discussed earlier. BGP provides the flexibility that there can be different transit ASes that serve as peers to each other. Since different stub ASes can be connected to different transit ASes, any transit ASes at the core need to exchange the entire routing table information about IP prefixes with other peering ASes using the BGP protocol. Thus, typically the core backbone Internet service providers (ISPs) maintain default-free routing tables for all valid IP prefixes so learned.

Consider Figure 9.5 again. Here network 172.17.0.0/16 homed in AS64822 would become known to transit AS, AS64731, which in turn would share this information with transit AS, AS64730. This way, BGP routers at transit ASes can build a complete default-free routing table. Now if Host "49ers" in network 192.168.40.0/21 in AS64617 generates a packet to an IP destination that is from an IP address that is not valid, then the border router in AS64701 would note this and drop this packet. In other words, a default-free routing table allows an IP packet with a destination in nonallocated address blocks from being forwarded further by dropping it. Transit ASes commonly employ a default-free routing table for such reasons.

We next discuss the second possibility. This option is possible since the BGP protocol announces IP prefixes through UPDATE messages that would traverse through and reach every *edge* or *stub* AS, such as AS64617 and AS64516. Typically, most stub ASes use the default route entry, i.e., they do not store all IP prefix entries, partly because they usually have single outlets to the rest of the Internet, and because it puts more than the required work on its border router, which may not be able to handle the load if it is not a powerful router with required memory. That is, it is not necessary to maintain a full default-free table at the border router of a stub AS. However, more and more stub ASes now maintain a full IP prefix table at their border routers. This can be driven by local needs in a stub AS, for example, to perform unicast reverse-path-forwarding (uRPF) checks as a measure for IP address spoofing. Thus, while from a BGP perspective, a stub AS is not required to maintain a default-free routing table, it has essentially become a necessity because of issues such as spoofing attacks.

To summarize, routing a packet that has originated in a network (IP prefix) in a stub AS with destination in another network (IP prefix) in another stub AS, would hop through at least a transit AS. Any path selection decision at a BGP speaker when there is connectivity from one AS to the next at multiple border BGP speakers, or from one AS to multiple ASes, would be based on the BGP path selection algorithm described in Section 8.7.1 in Chapter 8. Certainly, before route selection can be invoked about an IP prefix , the BGP UPDATE message would be used to announce this IP prefix ; the AS number is prepended as necessary when the information about this IP prefix crosses from one AS to another.

Before concluding this section, we discuss another term in regard to ASes. So far, we have mentioned two types of ASes: stub AS and transit AS. An AS can also be multihomed. Briefly, a *multihomed AS* connects to two or more transit ASes. As an example, consider Figure 9.5 again. Here, AS64617 is a stub AS; AS64701 is a transit AS that is also multihomed while AS64516 is a stub AS that is multihomed. Thus, ASes can be classified into three categories: stub singlehomed, stub multihomed, and transit multihomed.

9.3 Current Architectural View of the Internet

In the previous sections, we discussed packet routing in an IP addressing structure for a set of scenarios and also allocation of IP addresses. In doing that, we also discussed the role of ASes and BGP in routing. We next consider how ASes are related to ISPs and the role of IP address space in the context of Internet routing architecture.

9.3.1 Customers and Providers, Peering and Tiering, and Exchange Points

In the world of Internet routing connectivity, the term *customer* typically refers to an organization that has an IP address block; it relies on a *provider* for Internet connectivity; note that owning an AS number is not necessary since you can have an address block and be a part of existing AS number. For ease of discussion here, we will restrict to those customers that have their own AS numbers. The customer/provider relation is hierarchical and is sometimes described also as *downstream ISP/upstream ISP* relation. At the top of the hierarchy is *tier 1 Internet service providers* (tier 1 ISPs). Each tier 1 ISP has its own AS number. It is certainly possible to have more than one AS number belong to an ISP, for example, due to the merger of companies. For simplicity, we will assume that each ISP has its own unique AS number. A tier 1 ISP provides a large network spanning a geographic region such as the entire country, and sometimes across countries; such networks are often referred to as *Internet backbone* networks where link speeds can be in the order of 10 Gbps with the most advanced routers deployed. All tier 1 ISPs are at the same peering level. Typically, tier 1 ISPs peer privately with each other at one or more points. It used to be the case that tier 1 ISPs meet at *network access points* (NAPs) to exchange traffic from one network to another. In Figure 9.6, we show a generic example with four tier 1 ISPs meeting at an NAP; note that this is not common any more, it is shown here for illustration only. It may be noted that NAPs are also known as *Internet exchange points* (XP, or, EP in short), or Metropolitan Area Exchanges (MAEs). Furthermore, such arrangements are known as *public peering* since they are neutral meeting points. First, it should be noted that exchange points are operated by neutral entities that play the role of providers for traffic exchange services to tier 1 ISP customers. During transition from the NSFNET, the notion of NAPs was conceived when it became clear that one core network would not be the carrier for all Internet traffic. Initially, there were four NAPs that were connected to the NSFNET during 1994–1995. Currently, there are more than 175 exchange points around the world.

There is also *private peering* between two tier *n* ISPs where they connect directly to each other and agree to exchange traffic with each other; this then can serve as a bypass from congested exchange points, which some ISPs prefer. In Figure 9.6, we show that two tier *n*

FIGURE 9.6 ISP connectivity through public peering at an exchange point and through private peering (left: used to be more common among tier 1 ISPs; right: now seen more commonly at other tiering ("tier-N") levels).

ISPs are directly connected to each other through private peering while they are also part of the common exchange points with two other ISPs. For example, this would be a scenario where two tier n ISPs that have private peering as well as public peering would use the AS-path count to choose the private peering as the better path since they can use the exchange point as another AS in the path length between them. It may be noted that private network exchange points are also possible.

Exchange points provide physical connectivity to customers using technologies such as Gigabit Ethernet, ATM, and SONET, where customers' routers for connectivity are collocated in the same physical facility. Mostly, exchange point provides a meeting place for layer 2 connectivity. Layer 2 connectivity can give the impression that a simple Ethernet environment with every ISP's router attached to this Ethernet facility is probably sufficient. The difficulty is that the sheer volume of traffic each ISP generates is so high that such a simple environment is not possible in practice. Thus, you see a combination of sophisticated technologies with functionalities for peer management at most of the exchange points. In any case, at an exchange point, each ISP's BGP speaker can set up a BGP session to all other ISPs that have a BGP speaker collocated. In recent years, some exchange points have become popular for content delivery network providers since they can be directly connected to various major ISPs.

It is important to note that exchanges points have fairly well-defined policies while such policies can vary from one exchange point to another and certainly can evolve over time. Some examples of policies are: (1) an ISP must have its own AS number and use BGP to become a member of a exchange point, (2) the exchange point cannot be used for transit, (3) the exchange point policy requires full peering among all parties, or, each ISP can choose a different policy from a set of acceptable policies. Depending on policy, some large ISPs might or might not want to joint an exchange point; for example, if some large ISPs do not want to peer with smaller ISPs, they might not join an exchange point that stipulates that they must peer with all parties. In some instances, ISPs of different tiers, including tier 1 ISPs, do meet at large exchange points that may not require that all parties must peer with everyone. In such cases, each ISP has the option of not peering with everyone that is a member of this exchange point. Currently, Amsterdam Internet Exchange (AMS-IX) [14], considered the largest exchange point, has a flexible policy; it lets providers of different size to connect to its exchange point allowing each provider to set their own peering restrictions, including allowing private interconnects between two members.

In essence, an exchange is a giant traffic switching point. Some of the large exchange points push traffic in the order of 135 Gbps. It is not hard to imagine that such a high data push requirement can be taxing even with the top of the line inter-connecting hardware; in fact, this is no longer possible to do on a single hardware device. Thus, such exchange points must set up their own internal topology in such a way that multiple hardware devices are used for efficient traffic flow.

Now we move to consider multiple tiers. Tier 1 ISPs, in turn, provide connectivity to tier 2 ISPs; thus, in this case tier 2 ISPs are the customers and tier 1 ISPs are the providers. Tier 2 ISPs use tier 1 ISPs for transit service, but may peer with other tier 2 ISPs as well, for example, either through regional exchange points or private peering. Typically, tier 2 ISPs do not have international coverage—they are either at regional or national levels. It may be noted that a tier 1 ISP provides transit service to many tier 2 providers at certain meeting points; these meeting points are commonly referred to as *Points of Presences (PoPs)*. We will discuss PoPs more later in Section 9.6.

Tier 3 ISPs are the ones that seek transit service only from either tier 2 or tier 1 providers; they are typically not involved in public peering, although they may do some private peering. At the same time, tier 3 providers usually do not provide direct internet connectivity to users.

Beyond tier 3 ISPs, it becomes a bit murky in regard to the role of lower tiers or how many more tiers there are. To limit our discussion, we will stop at tier 4 ISPs by assuming that they provide local access to the Internet for institutions, enterprises, and residential users. Note that tier 4 ISPs require transit connectivity from tier 3 providers.

Although we have discussed various tier levels, there is no clear rule that indicates who is or is not a certain tier provider. Certainly, this is more clear in the case of a tier 1 ISP. However, consider content delivery providers who want to be located close to tier 1 ISPs' peering points. They usually have two options: (1) have their web servers hosted directly on one of the tier 1 ISPs; in this case, no AS number is necessary, or (2) have their series of web servers connected through routers to form a network with their own AS number, and then have peering with every major provider at major peering points or through private peering. If they choose option 2, they do not exactly fall into one of the tiering providers—we label them as *content delivery service* (CDS) ISPs. Examples of CDS ISPs are Google, Yahoo!, and Akamai.

There is also some difference in peering arrangements which varies from one country to another. For example, private peering at tier 1 level is now common in US, while public peering in other countries often includes some tier 1 providers. The largest public peering point now is considered to be Amsterdam Internet Exchange, AMS-IX [14]. As of this writing, AMS-IX has about 250 members which includes some large tier 1 ISPs as well; the peak rate is 150 Gbps. London Internet exchange, LINX [419] has over 200 members with peak traffic of 130 Gbps and Japan Internet exchange, JPIX [338] has 100 members with the peak rate at approximately 65 Gbps.

Since the Internet is made up of many providers with different relations and tiers, the obvious question is: what possible traffic exchange and payment relation do ISPs agree on? Here are a representative set of possible options [692]:

- Multilateral agreement: Several ISPs build/use shared facilities and share cost; for example, this agreement can be possible with public exchange points or private exchange points.

- Bilateral agreement: Two providers agree to exchange traffic if traffic is almost symmetric, or agree on a price, taking into account the imbalance in traffic swapped; for example, in a private peering setting.

- Unilateral agreement for transit: A customer pays its provider an "access" charge for carrying traffic; for example, a tier 4 ISP would pay a charge to tier 3 ISP.

- Sender Keeps All (SKA): ISPs do not track or charge for traffic exchange; this is possible in private peering and in some public peering.

Along with agreements, especially the ones that involve payment, it is common to also write up *service level agreements (SLAs)*. SLAs refer to an agreement on performance that is to be met on a course time scale; for example, the average delay between entry and exit points not to exceed 20 millisec, when averaged over a certain time period such as a week or a month. Typically, SLAs do not include performance requirement on a short time window such as in seconds. Thus, SLAs can be thought of more as a coarse grain quality-of-service requirement than a fine grain quality-of-service requirement. Furthermore, SLAs may also cover issues such as *demarcation points*; this refers to the line that indicates who manages what on a day-to-day basis. When a customer connects to a provider, there are three points involved: the routers at each end (one for the customer and the other for the provider), and the physical connectivity that connects them, such as a physical wire or a layer-2 connectivity. In some cases, the demarcation point is where the customer connects to a layer-2 switch in the physical connectivity part; in other cases, the customer's router is completely located at the provider's site; and yet in other cases, the provider's access router may be physically at the customer's site. Sometimes suitability of a demarcation point can be a factor for a customer in deciding which one to choose as a provider.

9.3.2 A Representative Architecture

In Figure 9.7, we show a representative view of connecting ISPs of different tiers, including CDS ISPs; routers shown are all BGP speakers in each ISP and only one exchange point is

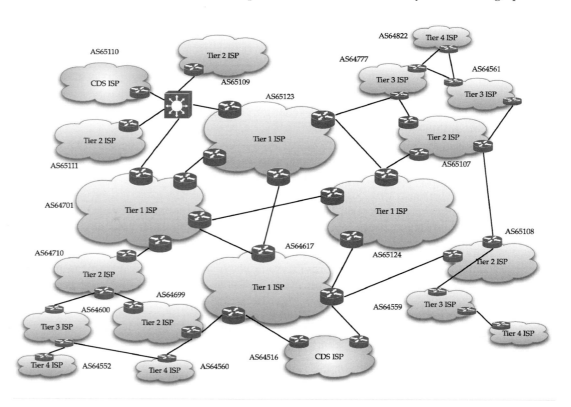

FIGURE 9.7   Interconnection of ISPs of different tiers: a representative view.

shown; peering at different tiers are also indicated. Note that private AS numbers are used to illustrate this example. In some cases, ISPs of different tiers are specifically identified using an AS number for ease of discussion.

We will illustrate three representative sessions using this architecture; these three sessions are shown in Figure 9.8: (1) from user U1 to server S1, (2) from user U2 to server S2, and (3) from user U3 to server S2 where the actual flow paths are shown. As you can see from the the original topology that there are multiple paths for each of these sessions. In regard to path selection, we note the following:

- For the session from U1 to S1, there are two AS-paths. In this case, the AS-path with the shorter AS length is chosen since there is private peering between respective pier-2 ISPs.

- For the session from U2 to S2 located at CDS ISP AS64516, there also appears to be two AS-paths. However, the one through Tier 1 ISPs would be taken that is based on the policy in place at the Tier-2 ISP upstream from U2.

- For the session from U3 to S2, there appears to be a second path from AS64552 to AS64600 to AS64560 to AS64699 to AS64617 to AS64516. However, this path would not be advertised at all. Note that AS64560 is a tier 4 ISP. While it will learn about AS64516 from AS64699, this would not be advertised to tier 3 provider AS64600, due to the stub rule described earlier. Thus, AS54552 (and AS64600) would not know about this connectivity.

In addition, the following observations are worth noting:

- In the case of server S2 located in the CDP ISP (AS64516), we have noted that it is multi-connected to tier 1 ISP AS64617. From the perspective of the tier 1 ISP, it has multiple egress points to the CDP ISP, AS64516. How does AS64617 choose one egress point over the other for the sessions U3 → S2 and U2 → S2. Typically, this depends on a rule called the *early-exit routing* rule within AS64617; this will be discussed further in Section 9.3.3. An important consequence is that intra-domain routing optimization in AS64617 would need to address this issue; this will be discussed later in Section 9.7.

- U2 is a user in a tier 4 ISP (AS64822), which is a stub AS. AS64822 is dual-homed from a single BGP speaker to two different tier 3 ISPs, AS64777 and AS64561. The tier 4 ISP, AS64822, has a couple of different options to prefer accessing one tier 3 ISP over the other: (1) set the local pre-configured priority values to access, say, AS64777, as opposed to the other, since this factor is given higher priority in the BGP route selection process, or (2) insert its AS number, AS64822, more than once when advertising IP prefixes it houses to the ISP with the less preferred route, i.e., to AS64561.

- From the topology, it gives the appearance that the tier 4 ISP that houses U2, AS64822, might be able to provide transit service to tier 3 ISPs to which it is connected. This is where we need to make an important point about a stub AS such as AS64822. While a stub AS learns about outside IP prefixes from BGP UPDATE messages it receives from both its tier 3 providers, AS64777 and AS64561, it should not advertise what it learns from one to the other. Note that usually tier 4 ISPs would connect to their providers on a low data rate link such as T1. Thus, advertising what it has learned from one to the

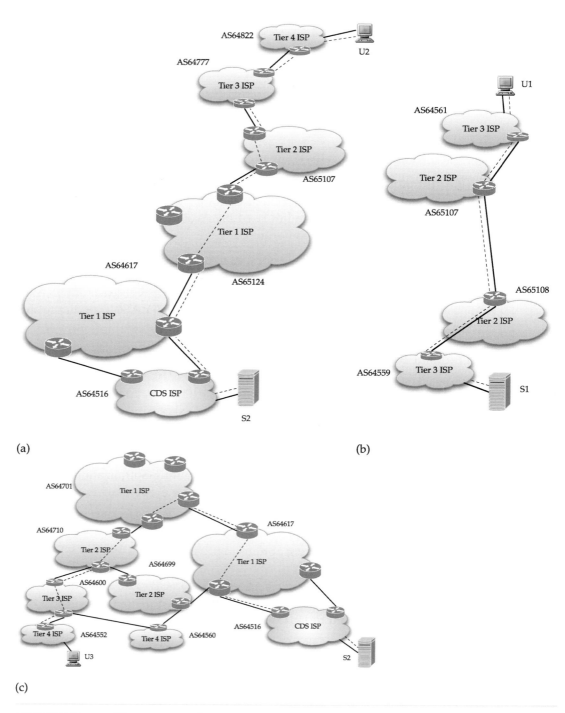

(a)

(b)

(c)

FIGURE 9.8   Three session flows in the architecture presented in Figure 9.7.

other would open up the possibility of being the best path to some IP prefix destinations; as a result, its access links can clog up with unnecessary traffic routing! this behavior is known as blackholing and should be avoided by carefully programming a BGP speaker with appropriate policies when it is multihomed to multiple providers; see Section 9.5 for policy examples.

- An ISP that has multiple BGP speakers would set up IBGP sessions among themselves so that routes can be exchanged internally.

9.3.3 Customer Traffic Routing: A Geographic Perspective

Customer traffic routing leads to interesting scenarios when observed from the geographic perspective as it depends on how and where customers are connected to tier 1 ISPs. In this section, we will illustrate two cases using Figure 9.9 in which we list three customers off of two different tier 1 ISPs in three different locations: San Francisco, New York, and Amsterdam. Note that these customers can be transit providers to other customers; for our illustration, this relation suffices. In addition, there is a fourth customer of CDS type that has connectivity to a tier 1 ISP at all three locations.

Clearly, traffic from Customer 1 would transfer at San Francisco through the tier 1 ISPs if it is meant for Customer 3. This also illustrates why tier 1 ISPs peer at multiple geographic

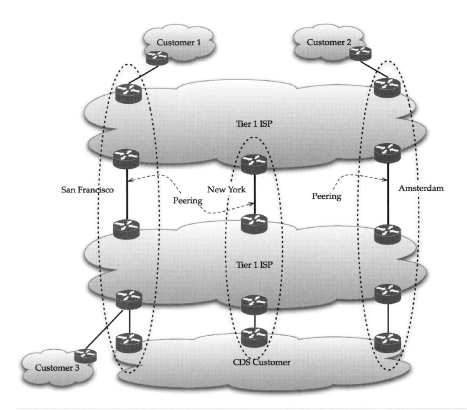

FIGURE 9.9   Customer traffic routing from a geographic perspective.

locations. In this example, if the tier 1 ISPs were not have peering at San Francisco, the traffic between Customer 1 and Customer 3 would transfer in New York. This means that each tier 1 ISPs would need to carry the traffic cross-continent unnecessarily to bring it back to customers located off of San Francisco.

Now consider traffic between Customer 2 and Customer 3. It is easy to see that the tier 1 provider for Customer 2 would let the traffic transfer at Amsterdam to the other tier 1 provider so that the second tier 1 provider would need to carry this traffic through its own network all the way to San Francisco; this is known an *early-exit routing*. Similarly, for the other direction, for traffic received from Customer 3, its tier 1 provider would transfer the traffic to the other tier 1 provider at San Francisco so that the second tier 1 provider would need to carry the traffic in its network all the way to Amsterdam to deliver to Customer 2. Thus, it is easy to see that because of early-exit routing, traffic flows can be on asymmetric paths. Note that early-exit routing is not necessarily a problem if both the tier 1 providers have an agreement in place (such as Sender Keeps All) because their overall transfer from one network to the other and vice versa is somewhat comparable.

Finally, we consider the case of traffic being routed to the CDS customer who is connected to a tier 1 ISP in all three locations where they have mirror sites so that a user's request can be handled by the nearest site. That is, requests from users in Customer 1 and Customer 2's network would be directly handled off to the CDS customer in San Francisco, while requests from users in Customer 2's network would be handed off to the CDS customer in Amsterdam. It may be noted that the network of CDS is shown to be connected among all three locations. While a CDS customer may not own a facility, it can use an IP virtual private network (VPN) to transfer high-volume data between its data centers in different cities, rather than using the public Internet; more discussion about IP VPNs can be found in Section 18.5.

9.3.4 Size and Growth

In this section, we portray the current size and growth of the Internet in terms of AS numbers and IP prefixes. There are currently more than 23,000 active ASes with about 195,000 IPv4 prefixes advertised externally. A sample summary is shown in Table 9.1, drawn from a web site that reports BGP routing table analysis [303]; this site reports an external BGP view of certain ASes obtained using *Route Views* [707]. Briefly, Route Views is a special-purpose AS (AS6447) that uses multihop BGP sessions to peer with several BGP speakers at well-known ASes. While it learns about IP prefixes from each of these ASes, it does not forward them to others—that is, it serves an important role as a sink for BGP information, thus helping to understand BGP growth. Table 9.1 also lists information for Telstra-i (AS1221) and Route Views itself (AS6447); there is an explanation for why the IP prefix counts for these two ASes are significantly different from others [302]. Telstra-i includes an internal view, including a significant number of more specific prefixes that are yet to be aggregated before being announced for the external view. AS6447, being the sink, receives different information from its peer ASes, some feeding only external views and others feeding local specific information—the number for AS6447 reflects the sum of unique ones learned from all sources. Also, there are currently over 850 IPv6 routes [303].

From Table 9.1, we can see that of the total number of active ASes, about 70% are originating-only (stub) ASes; a significant portion of them are originating ASes with only

TABLE 9.1 IPv4 Route and AS Data, as of September 30, 2006 (courtesy [303]).

Name	AS Number	IP Prefixes	AS Count	Originating AS		Originating AS with Single Prefix		Transit Only		Mixed ASes	
Telstra-i (Australia)	AS1221	266,837	23,123	16,251	70.28%	9,707	41.98%	78	0.34%	6,794	29.38%
Telstra-e (Australia)	AS1221	195,322	23,081	16,228	70.31%	9,684	41.96%	79	0.34%	6,774	29.35%
Sprint (USA)	AS1239	192,925	22,975	16,329	71.07%	9,675	42.11%	73	0.32%	6,573	28.61%
RIPE (Europe)	AS3333	197,323	23,172	15,954	68.85%	9,684	41.79%	74	0.32%	7,144	30.83%
Reach Network	AS4637	196,319	23,125	16,240	70.23%	9,709	41.98%	79	0.34%	6,806	29.43%
Oregon Route Views	AS6447	212,368	23,423	15,747	67.23%	9,687	41.36%	66	0.28%	7,610	32.49%
AT&T Worldnet	AS7018	192,708	23,055	16,229	70.39%	9,678	41.98%	79	0.34%	6,747	29.26%

TABLE 9.2 Prefix Length Distribution of the Top Five Prefix Lengths at AS4637 (September 30, 2006).

Prefix boundary	Number	Percentage
Prefix /24	105,987	53.99
Prefix /23	16,817	8.57
Prefix /22	15,407	7.85
Prefix /20	13,870	7.07
Prefix /19	12,275	6.25
Total	196,319	100

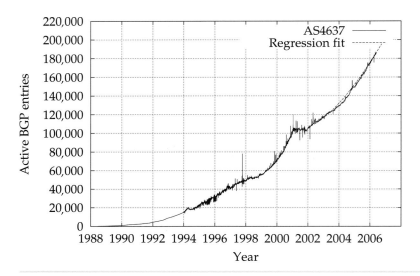

FIGURE 9.10   BGP routing table growth at AS4637 and regression fit.

a single prefix advertised. Most of the remaining 30% are mixed ASes; i.e., they originate and also provide transit services. The remainder (less than 0.5%) are "pure" transit ASes. Reportedly, as of 2004, about 60% of stub ASes are multihomed [381]; however, this is a difficult number to compute since there is no field in the BGP protocol to indicate if an AS is multihomed; thus, extensive study/assessment is required to identify the proportion of ASes that are currently multihomed. You may note that the numbers on routes and ASes polled at different ASes do not match; this is certainly possible since at the time of polling each such AS may have a slightly different view to announcements and withdrawals, while including counts for some internal IP prefixes as well.

In Figure 9.10, we present the growth in the number of advertised IP prefixes since 1989 as obtained for AS4637. The current value of more than 195,000 IP prefixes serves as a gauge of lookup size that a BGP speaker faces when a packet arrives and requires forwarding. By considering growth data since the beginning of 2002, we have performed a nonlinear regression analysis. If we denote Y for year (where $Y \geq 2002$) and N for the number of IP prefixes, their relation can be roughly estimated by the following regression fit (see Appendix B.7):

$$N = 105625 + 920.608\,(4\,Y - 8007)^{1.52894}, \quad Y \geq 2002. \tag{9.3.1}$$

Using this relation and if no other aggregation is assumed, it can be estimated that N will grow to about 232,000 by the beginning of 2008. It has been found that many ASes do not aggregate routes before announcing outside; such an aggregation could significantly reduce the BGP routing table entry. The recent assessment is that it can reduce the size from 195,000 to 127,000 entries, a 35% potential reduction [304]. In Table 9.2, we show the prefix length distribution of the top five prefix lengths in terms of counts for AS4637. Note that almost 54% of the total of more than 196,000 are at /24 prefixes. In regard to AS hop distances, about 77% of addresses can be reached in within two AS hop distances, and 99% of all addresses can be reached within four AS hop distances. We include a picture of AS-level connectivity, created by CAIDA, to provide some idea about the connectivity view (Figure 9.11).

An important reason to understand growth in IP prefixes and ASes is the impact on memory requirement at a BGP speaker. The following are the main factors that impact memory requirement at a BGP speaker:

N = Number of IP prefixes

M = Average AS distance (in terms of number of AS hops)

A = Total number of unique ASes

P = Average number of BGP peers per BGP speaker.

Then, the memory growth complexity can be given by ([485], [593], [703]):

$$\text{Memory Growth Complexity} = O((N + MA)P). \tag{9.3.2}$$

Note that M has a slow growth and N is more dominant than A. Thus, the complexity growth can be approximated as $O(NP)$, that is, the memory growth can be estimated as the product of the number of IP prefixes and the average number of BGP peers per BGP speaker.

9.4 Allocation of IP Prefixes and AS Number

So far, we have discussed AS numbering, IP addressing, customer and provider relationships, and so on. An important question remains: how does an organization obtain an IP address block? In this section, we answer this question.

Internet Corporation for Assigned Names and Numbers (ICANN) is the organization that handles global coordination of unique identifiers used in the Internet. Through agreements, IP address block assignments have been distributed to five different Regional Internet Registries (RIRs). The five RIRs are geographically organized as follows:

- American Registry for Internet Numbers (ARIN) (http://www.arin.net/) to serve the North American region

- RIPE (Réseaux IP Européens) Network Coordination Centre (http://www.ripe.net/) to serve the European and the West Asian region

- Asia Pacific Network Information Centre (APNIC) (http://www.apnic.net/) to serve the South/East Asian and the Pacific region

- Latin American and Caribbean Internet Address Registry (LACNIC) (http://www.lacnic.net/) to serve the Latin and South American region

- African Network Information Center (AfriNIC) (http://www.afrinic.net/) to serve the African region.

Each registry has its own rules and pricing in regard to IP address block allocation; this allocation depends on allocation size as well, as indicated through netmask boundary such as /19. For example, ARIN's current policy is that the minimum allocation size is a /20, while for multihomed organizations, the minimum allocation size is a /22. This means that if an organization needs only a /24 allocation, it cannot obtain it directly from ARIN; instead, it

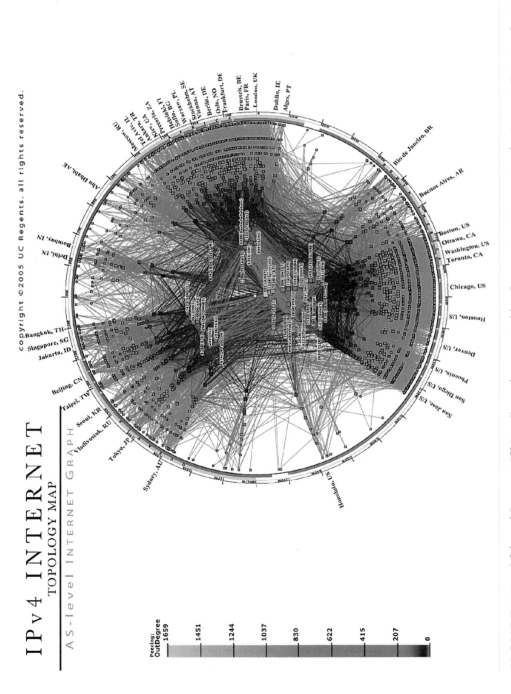

FIGURE 9.11   AS-based Internet "Skitter" graph generated by Cooperative Association for Internet Data Analysis (CAIDA), 2005. (Copyright © 2005. The Regents of the University of California. All Rights Reserved. Used by permission.)

must obtain it from an upstream ISP (provider) who has been already allocated at least a /20 address block by ARIN. Similarly, registries put restrictions on allocation of an AS number. For example, currently ARIN would allocate an AS number to an organization only if it plans to do multihomed connectivity to two ISPs or can justify that it has a unique routing policy requirement. Note that allocation polices, both for IP prefix and AS number, do change from time to time. For recent polices, you may check the web site of the respective registries.

Suppose that an organization obtains an IP address block along with an AS number from ARIN. It would then need to establish multihomed connectivity to two upstream ISPs who would have their respective AS numbers. Once the physical connectivity is set up, the BGP speaker at the organization establishes a BGP session with the BGP speakers of its upstream ISPs to announce its address blocks. This information is then propagated throughout the Internet so that the rest of the Internet would know how to reach a host in this address block. Note that the organization may have separate BGP speakers, one each for connecting to its upstream ISPs; in this case, the organization would need to run IBGP between its two BGP speakers in order to establish rules on how to handle routing of outgoing traffic.

Now suppose that an organization obtains an IP prefix from one of the regional Internet registries but does not obtain an AS number. In this case, at first it would then need to set up an agreement with an ISP that has an AS number; this ISP would then serve as the "home" AS for this address block. Once connectivity and agreements are put in place, this ISP would then announce this IP prefix along with other IP prefixes that are in its AS to its upstream provider(s). Once this announcement is propagated throughout the Internet, the newly announced IP prefix becomes known to the rest of the Internet. We discuss below two possibilities of how the connectivity between an organization (customer) and its provider can be set up when a customer does not own a public AS number:

- The ISP may set up private AS numbering to divide its customers into different ASes. Thus, each customer (organization) has the flexibility to choose a routing protocol of its choice internally and then use a BGP session to talk to the provider's BGP and announce its IP prefixes.

- If the provider uses OSPF protocol, then it can use Not-so-stubby Areas (NSSA) LSA (refer to Section 6.2.8) to allow external routes from its customer to be advertised into the OSPF autonomous system, while the customer may run its own routing protocol.

Choosing one over another or using any other mechanisms depends on the size of an ISP, as well as its internal network management philosophy and policy, and its business strategy. Furthermore, note that route redistribution (refer to Section 5.7) is a common mechanism to exchange prefixes among different administrative organizations that fall within an AS number.

It may be noted that a customer who obtains an IP address block from a provider may choose to switch to a different provider after some time. Suppose that a provider has the address block 192.168.40.0/24, and it has allocated 192.168.45.0/24 to a customer. Initially, through route aggregation the provider will announce 192.168.40.0/21 with its AS number. Now the customer wants to move to a different provider keeping the address block. Thus, the address block, 192.168.45.0/24, would now need to be announced with the AS number of the new provider. This then creates a situation, known as a *hole* since the *more-specific prefix*

(192.168.45.0/24) creates a hole in the *aggregated prefix* (192.168.40.0/21). However, both the aggregated prefix and the more-specific prefix would need to reside in the global routing table at a BGP speaker; this is so that packets can be forwarded properly to the right destination. This means that the IP address lookup process at a router needs to work very efficiently for handling holes as well. Details on IP address lookup algorithms will be covered later in Chapter 15.

9.5 Policy-Based Routing

Earlier in Section 8.7, we presented the BGP routing decision process; therein, we indicated why policy-based routing is needed in an interdomain environment. By now, you probably have realized that policy-based routing is an extremely critical component used at the BGP speakers for handling inbound and outbound traffic. For example, in Section 9.3.2, we highlighted examples to show why import and export policies must be maintained at a BGP speaker. We also noted earlier that NSFNET necessitated the need for policy-based routing. In this section, we briefly explore why policy-based routing is needed, and how it may impact customer provisioning.

Policy-based routing emerged because in an interdomain environment, announcements received from a neighboring AS through an exterior routing protocol may contain IP prefixes that the receiving AS may not want to handle or forward. Note that the receiving AS has no control over what IP prefixes it receives from its neighbor, but it *can* control which ones it does not want to handle/forward. Furthermore, due to the business agreement with a certain neighboring AS, the receiving AS might want to give preference to a particular IP prefix received from this neighbor compared to other neighbors.

It is important to realize that policy-based routing has three phases: (1) determine the list of policies and load them to a BGP speaker; (2) when BGP messages arrive, apply policies to update Routing Information Base (RIB) and Forwarding Information Base (FIB); and (3) when an actual user packet arrives that affects a certain policy, take action as per policy in real-time through FIB.

To determine and specify policies, it is imperative to have a generic routing policy language that can work in a vendor-independent environment; then, from this format, a vendor-specific format can be generated. Routing Policy Specification Language (RPSL), described in RFC 2622 [8], is a language for declaring routing policy of an AS in public registry in order to provide a common interface mechanism that others can easily verify. RPSL serves the purpose of a vendor-independent language to describe policies. Usually, most providers use vendor-dependent policy tools provided by a router vendor in its software platform. We use RPSL for the purpose of illustration.

Here, we will illustrate a few examples adapted from [486]. Consider four ASes, AS65001, AS65200, AS65201, and AS65202, as shown in Figure 9.12. Here, AS65201 and AS65202 are customers of AS65200, while AS65001 is a transit provider for AS65200. Here, AS65200 will accept any announcement from AS65201 if it has originated at AS65201, since AS65201 is a stub AS; this rule also protects from any misannouncement by AS65201 or AS65202 such as the blackholing scenario discussed earlier. Thus, AS65200 can set up a policy as shown below:

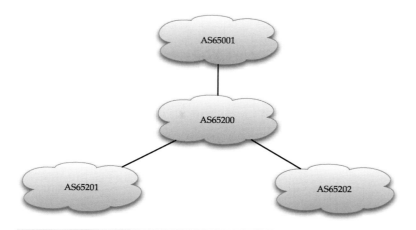

FIGURE 9.12   AS topology example for policy specification.

```
import:   from AS65001 accept ANY
import:   from AS65201 accept <^AS65201+$>
import:   from AS65202 accept <^AS65202+$>
export:   to AS65201 announce ANY
export:   to AS65202 announce ANY
export:   to AS65001 announce AS65200 AS65201 AS65202
```

This means that AS65200 accepts any announcements from AS65001, the transit provider, and, in turn, will export these announcements to AS65201 and AS65202. However, it will import from AS65201 only paths made with AS65201 as the first member of the path; "+" means that AS65201 may appear more than once, and "$" means that any AS listed after AS65201 will not be considered. The same principle is applied to AS65202. In turn, AS65200 will export to transit provider AS65001 by announcing routes that originated in itself and the ones from its customers' ASes, AS65201 and AS65202.

If AS65200 has many customers similar to AS65201 and AS65202, a compact representation can be done instead of creating an entry for each one; this helps in minimizing errors as well as in letting AS65001 know its consistent export policy. To do so, "**as-set**" and "**members**" can be used as shown below:

```
as-set:    AS65200:AS-CUSTOMERS
members:   AS65201 AS65202

import:   from AS65001 accept ANY
import:   from AS65200:AS-CUSTOMERS accept <^AS65200:AS-CUSTOMERS+$>
export:   to AS65200:AS-CUSTOMERS announce ANY
export:   to AS65001 announce AS65200 AS65200:AS-CUSTOMERS
```

This then requires just updating the member list, as needed.

We next consider the case where AS65200 wants to specifically allow the IP address space assigned to, say, AS65201. If AS65001 is assigned the space 10.10.0.0/16, then AS65200 can include a more specific rule as follows:

```
import:     from AS65201 accept { 10.10.0.0/16^16-19 }
```

This means that AS65200 will accept announcements from AS65201 if the netmask starts at /16 but not longer than /19. However, if it would receive any address block except for, say, 10.20.0.0/16, then this can be stated as:

```
import:     from AS65201 accept ANY AND NOT {10.20.0.0/16}
```

We have shown some simple rules to illustrate how import and export policies can be enabled in policy-based routing using RPSL. RPSL provides a rich set of commands to create fairly complex rules. Later in Chapter 16, we will discuss packet classification handling by a router when an actual packet arrives; we will then present algorithms implemented in a router for efficient packet processing due to such rules.

It is important to note that policy-based routing is quite complex, not as simple as the example rules listed above. Since different ASes have different policies, it is also possible to have oscillatory behavior, unintended behavior, loss of reachability, and so on. As an example, we discuss BGP wedgies below to show how unintended behavior is possible.

9.5.1 BGP Wedgies

An undesirable consequence of BGP policy-based routing is that it can lead to stable but unintended routing, known as *BGP Wedgies* [264] . We will illustrate through an example (see Figure 9.13(a)). Consider an ISP with AS number 65101; its primary provider is AS65301, but it also has a backup provider, AS65201. In turn, AS65201 uses AS65302 as a provider who has a peering relation with AS65301.

The primary/backup relation can be implemented in two ways: one way is to prepend AS65101 twice (instead of the normal one time) in the announcement to the backup provider, and just once to the primary provider. In this way, the path through the backup provider is intended to be longer than the path through the primary provider in terms of AS hops. However, this may not be fully guaranteed if its primary provider, AS65301, also does extra AS hop prepending to any upstream backup provider. Thus, it is still possible that traffic originating beyond such an upstream provider may still traverse via AS65201, rather than taking the path via AS65301.

In order to be not affected by the decision of providers further upstream, a second alternative is possible. This option uses BGP communities (refer to Section 8.10.1). In this approach, a provider announces community values to its neighbors; thus, customer AS65101 can select the provider's local preference setting. There is another reason for preferring this option over the AS-path prepending option. This is since, in BGP path calculation, local preference is given higher priority over AS-path (refer to Section 8.7.1). Note that the community values marked by the customer must be understood and supported by both the providers.

Through this process, intended routing is that any traffic takes the primary path through AS65301 to destination AS65101, including from AS65201 via AS65302, AS65301 to AS65101 (see Figure 9.13(b)). In order to achieve the intended routing, AS65101 needs to announce its routes on the primary path AS65301 *before* announcing its backup routes to AS65201. However, the intended outcome may not work if after path priorities are established, the BGP session between AS65101 and AS65301 fails. This would result in AS65301 generating a

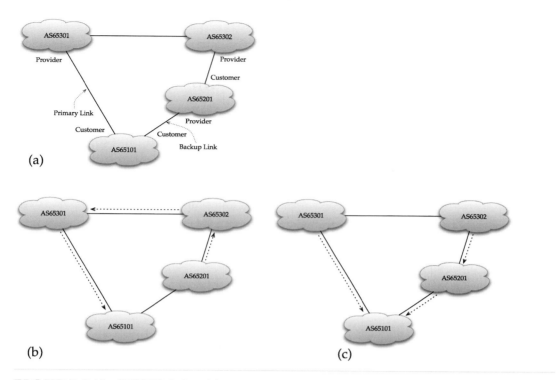

FIGURE 9.13   BGP Wedgies: (a) connectivity setup, (b) intended routing, (c) unintended routing.

withdrawal message indicating nonavailability of routes in AS65101; this message will reach AS65302, which, in turn, will announce to AS65201. Now the BGP speaker at AS65201 will look up its cached AS-paths in Adj-RIBs-in and will find that there is a path available from AS65201 to AS65101. Thus, AS65201 will advertise the availability of this backup path to AS65302, which in turn will inform AS65301. Now all traffic to AS65101 will take the backup path via AS65201.

The problem is that once the BGP session between AS65101 and AS65301 is again back in service, the original path is not restored. This is because AS65302 would enforce the policy that it prefers customer-advertised routes (i.e., from AS65201) over routes learned from peer ASes (in this case, from AS65301). Thus, the untended routing, shown in Figure 9.13(c), occurs. The only way to revert to the primary path is to intentionally bring down the BGP session between AS65101 and AS65201.

The above BGP Wedgie example is known as "3/4" wedgie. For additional examples, see [264].

9.6 Point of Presence

Earlier in Section 9.3.2, we briefly discussed demarcation points. In this section, we discuss the general architecture of access points to large ISPs such as tier 1 ISPs where the demarca-

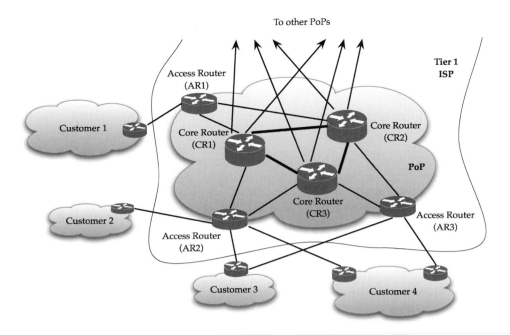

FIGURE 9.14   Tier 1 ISP's Point of Presence (PoP) connectivity architecture.

tion point between customers and providers lies. The meeting point at which many customers connect to a large ISP is often known as a point of presence (PoP).

In Figure 9.14, we show the topological architecture of a PoP, found to be common with tier 1 ISPs [308]. Typically, customers have their own AS numbers and routers. For redundancy, it is common for a customer to connect to multiple routers in the provider's network; this, however, depends on how much a customer is willing to pay for redundancy. The PoP architecture of a provider's network has two sets of routers: core routers that connect to other PoPs and access routers that serve as the ingress points for connectivity to customers' routers. Again, access routers are dual-homed to core routers for redundancy. It is important to note that the access routers in the tier 1 ISP's network serve as BGP speakers; they talk to the border routers at customers' networks, which serve as BGP speakers for the respective customers' networks.

The allocation of customers to an access router is a critical network design problem for a tier 1 ISP. Customers can have varied requirements: single connectivity, connectivity to two access routers, multiple routers to multiple access routers, as well as different access data rate requirements. From the perspective of the tier 1 ISP, they need to consider the number of physical ports of different access routers, not overloading any access router with too many connections to all their customers' routers since each such connection also results in a BGP session and the agreed upon maximum data rate to control traffic flow is to be taken into account. Thus, given various constraints, the general goal of the provider is to load balance access routers, keeping in mind future growth. Furthermore, the situation becomes even more complicated since there are also layer 2 technologies to manage for physical connectivity. To solve these types of access network design problems, optimization models such as the ones presented in [564, Chapter 6] would need to be customized/adapted.

An important issue to consider is whether the architecture can be simplified. The PoP architecture originally emerged because traditionally routers were not always reliable; thus, it provides reliability through redundancy. However, the loss of an access router, say AR2, rather than just the route controller hardware/software failure alone, would result in loss of all BGP sessions to associated customers' routers. Thus, customers' routers will delete routes to IP prefixes learned from failed routers; this will result in other peers withdrawing routes, which can lead to route flap with a network-wide effect (refer to Section 8.9). Traditionally a router can take 3–10 minutes to restore service for a router controller failure [347]. Due to recent router technologies such as nonstop routing, a peer would not know of a failure since the TCP connection for the BGP session would not be lost, thus avoiding BGP flapping. This has another benefit in that customers' routers need not connect to two access routers; this can reduce configuration complexity and cost, the impact of a protocol, and the number of access routers required.

There is another issue to consider from the perspective of traffic engineering and OSPF/IS-IS routing. This will be discussed in the next section.

9.7 Traffic Engineering Implications

In Chapter 7, we presented intradomain IP traffic engineering. While the methodologies discussed there work in general, there are certain issues to specifically consider in the presence of BGP interactions with neighbors and the size of the overall topology, especially for large tier 1 ISPs or transit ISPs.

Typically, large ISPs deploy IS-IS or OSPF as routing protocols within their networks. This requires computation of the shortest path by a router. From Figure 9.14, we can see that a tier 1 ISP has many routers in a PoP (we have shown six in the figure). If a tier 1 ISP has k routers at a PoP and there are m PoPs in its network, then there are $k \times m$ routers in the entire network. At least in theory, each router would then need to do shortest path computation on a $k \times m$-router network. In practice, this can be avoided for certain routers; for example, an access router such as AR1 has a designated primary and a designated secondary connectivity to core routers; in addition, they never connect outside the PoP. Thus, for these routers, link weights can be set up by setting a lower value to one link, say AR1 to CR1, which is meant as the primary link than the other link; use of the static route option is also advisable here. For core routers, it is clear that the primary intent is to route traffic to another router at another PoP, *not* to another intra-PoP core router; such intra-PoP connectivity is for redundancy in the event of a failure. Thus, typically, the link weights between intracore links at a PoP should have very high link weights so that such links can be avoided for intra-PoP traffic, unless there is a link failure. In essence, what it means is that the traffic demand can be abstracted on an inter-PoP pair basis, not at a per-router pair basis. A desirable consequence is that the link weight determination traffic engineering problem need *not* be considered for the $k \times m$-node problem; instead, each PoP can be abstracted as a single node leading to an m-node traffic engineering problem, a much smaller optimization problem to solve than the $k \times m$-node problem.

In addition to using traffic engineering models presented in Chapter 7, certain variations are necessary to address practical issues. From the discussion above, we see that many core providers have multiple egress points for *early-exit* routing to its customers; we illustrated this aspect in Section 9.3.3. This requires some tweaking to the link-weight setting problem from the point of view of modeling. Refer to Table 7.1 in Chapter 7, in which we summarized the notation for the IP traffic engineering problem. Specifically, we have defined K as the number of demand pairs with positive demand volume for ingress-egress node pairs, and h_k as the demand volume of demand index $k = 1, 2, \ldots, K$. In light of early-exit routing, several extensions are required [218], [595], [688]:

- *Offered traffic from ingress to IP prefix destinations:* This means that traffic demand be represented more accurately at the IP prefix destination level rather than for egress router.

- *Egress links connecting to neighboring domains:* External links to routers in neighboring ASes should be considered explicitly in the modeling framework, and objective function should consider the load on this link as well.

- *Set of egress links for each IP prefix destination:* This set can be modeled as a logical node in the formulation, but without considering them in the objective function for traffic load utilization.

- *Selection of the closest egress point:* Based on link weights, the closest egress point can be identified for each ingress point.

These extensions can be modeled in the framework presented in Chapter 7 since K is an abstract notion for the number of demand pairs; instead here, K needs to consider the above variations along with the creation of new link entities in the overall problem formulation. That is, the rest of the modeling framework described in Chapter 7 remains the same; in this sense, the framework presented there is powerful in addressing such variations.

There are, however, additional important considerations. Typically, any traffic matrix–based approach considers only a single traffic demand matrix. Since the actual traffic varies from one time instant to another within a 24-hour cyclical pattern, this variation is important to consider. There are typically two ways to consider this time-dependent variation factor: (1) consider the peak demand during the day, factor in any day-to-day fluctuations, and use this traffic demand matrix in the link weight determination problem, (2) take multiple snapshots of traffic demands during the day, and solve the link-weight determination problem independently for each such demand matrix, and compare any differences in link weights obtained for each snapshot—if they are comparable with an acceptable tolerance, then a robust set of link weights can serve any traffic variation during the day. Whether one works better than the other requires computing and analyzing link weights, customizing/tailoring for a particular service provider's network. Another important consideration is equipment/line-card failures. The question then is how does traffic rerouting impact the overall performance. To avoid link weight re-optimization in the event of a failure, an integrated link weight can be determined with failure restoration also as a goal; for example, see [308], [470], [532], [564], for additional details.

It may be noted that many large ISPs use IP with Multiprotocol Label Switching (MPLS). MPLS allows flexible options for traffic engineering controls in order to route customers' traffic. We will present MPLS in Section 18.3 and discuss MPLS traffic engineering in Section 19.1.2.

We conclude this section by briefly discussing *interdomain* traffic engineering. While each ISP is motivated by its own interest in optimizing its intradomain network, the need for interdomain traffic engineering, especially among neighboring ASes, is motivated by issues such as managing inbound traffic, managing outbound traffic, and selecting peering points optimally rather than relying on early-exit routing for a sub-optimal solution [217], [439]. In general, interdomain traffic engineering may involve a contractual peering agreement. In general, interdomain traffic engineering is an emerging area that requires further research.

9.8 Internet Routing Instability

Routing instability in the global Internet is caused by a variety of factors. A series of studies based on actual measurements in the mid-1990s first pointed out routing instability on the global Internet ([389], [391]). We start with an example from this series of work. Suppose that the CPU at a BGP speaker is overloaded. Then, it is possible that KEEPALIVE messages are not communicated on a timely basis; this can make the BGP speaker on one end think that the other end is not available any more, much as if the link between the routers has gone down. Each router can then generate a series of messages to indicate nonreachability, which can cascade from one AS to another one, thus causing a *network storm* effect. A second related problem is when the CPU overload subsides and both the routers determine that they can talk to each other again (that means the link is up again); this can cause another storm of activities in regard to re-updating reachability information, thus causing unstable behavior. Since that time, router vendors have made significant progress on how to handle KEEPALIVE messages during CPU overload, for example, by giving priority to such messages over regular traffic. Furthermore, route flap dampening (refer to Section 8.9) has been added to the BGP protocol to minimize any storm effect that can be created.

For a more recent example of CPU overload, we consider the impact of a virus on the routing infrastructure. Routers usually cache forwarding path for commonly routed IP addresses to provide fast lookup and fast packet forwarding. During the Code Red-II and Nimda virus attack [156], [157], [728], instances of the virus started random IP address scanning; this resulted in cache misses as well as generation of router error messages through ICMP error messages, leading to extensive CPU overload in routers, thus causing further storm like instabilities.

The CPU overload problem provides an example of what types of routing instability can/might occur. There are multiple sources of problems that can lead to instability; for example, (1) layer 2 data link failure or layer 2 timer device failure, (2) virus attacks, (3) software bugs, and (4) router configuration error. Note that *not* all of these can be labeled as failures. Some of the incidents can cause the entire BGP session to be disconnected for a long time, while others may result in intermittent problems where the session goes down and comes back up again. Because of the BGP protocol's reachability concept, withdrawals and announcements are generated. If the entire architecture was built on the notion of single homing with just transits for connectivity to other ASes, this would not be a major concern. However, in light of multihoming AS, and a transit AS being connected to multiple ASes due to either public or private peering, the effect can be magnified. Due to the path vector protocol nature of the BGP protocol and to avoid a route looping problem, finding another path through other ASes can take a long time, and in fact, can require many exchange of messages;

we have illustrated such a behavior earlier in Section 3.5 when we illustrated an example in which a multihomed AS loses connectivity to its multiple providers (ASes).

Another effect of a BGP session drop/restart is that it can lead to duplicate announcements due to the operation of the timers. This can be addressed by BGP speakers maintaining partial state information about announcements to its peers.

A recent study has looked into the BGP storm issue and also tried to understand if it affects the data plane, i.e., does a control plane problem cause a data plane problem? This study observed that it may or may not. While during Code Red and Nimda viruses, the BGP update storm was prevalent, it did not necessarily affect the data plane, while during another virus, known as Slammer virus, did affect the data plane performance. For additional detail, see [607].

Finally, a general concern is that some small unknown problems in one part of the Internet routing architecture could cause significant instability to the overall system; such an effect is often referred to as *the butterfly effect* [424]. We have already seen some examples that can be labeled the butterfly effect. Certainly, there have already been many checks and balances introduced to the routing system to avoid/minimize such behavior; however, butterfly effects in the routing system in the future due to yet unknown factors cannot be completely ruled out. In general, understanding Internet routing instability and finding good solutions continue to be an active research area.

9.9 Summary

In this chapter, we present Internet routing architecture, starting with its initial evolution. Clearly, Internet routing architecture has experienced tremendous growth and changes in the past decade since the introduction of NAPs. Business agreements have played critical roles in public and private peering in order to provide efficient traffic movement through the Internet. In parallel, we have seen the emergence of the role of policy-based routing.

Traffic engineering objectives are also somewhat different for certain ISPs due to the service they provide to their customers. Thus, appropriate adjustments are needed. Furthermore, global Internet routing instability remains a concern in light of conflicting goals and unknown factors.

A general question remains about viability/scalability of BGP as Internet continues to grow; thus, the exploration of new routing paradigms for large, loosely connected networks is an important research direction.

Further Lookup

There are many foundational problems associated with interdomain routing as the Internet grows [216]. They can be classified broadly into two categories: (1) policy-induced problems, and (2) scalability-induced problems. Examples of policy-induced problems are policy disputes between ASes, policy enforcement, and secure route advertisement. Examples of scalability-induced problems are nonvisibility of paths when route reflection does not distribute all routes, IBGP/EBGP interactions causing loops and oscillations, difficulty to determine the cause of update triggers, and a BGP speaker's ability to handle routes as the number of routes grows. This is an interesting class of problem domains that requires further inquiry.

Another problem of importance is whether the current Internet architecture provides structural incentive for competitive providers at the access network level for interdomain routing so that multiple choices are available instead of a single way out. This area also has received attention recently; for example, see [753], [754].

For traffic engineering in the presence of inter-domain issues and to address restoration of the tier 1 backbone, see [217], [308], [439], [595]. For provisioning of BGP customers, see [260]. Understanding the AS-level topology is an active area of research; for example, see [438], [506].

Policy-based routing is an important topic that has emerged as a result of experience with the NSFNET. Early discussions on policy-based routing can be found in [93], [144]. The BGP convergence issue in the presence of policy-based routing has raised significant interest in recent years, for example, in regard to the stable paths problem and inter-domain routing [243], [265], [388], [728].

Many providers maintain web sites that announce their policies on peering; for examples of current peering policies by large ISPs, see [20], [40], [656], [718]. It is, however, not always easy to determine who peers with whom, and whether through public or private peering; nonetheless, some information on peering relations can be found at [552]. Settlements on payment between customers and providers were described in [692]; for a recent discussion, see [305].

For other general issues related to the Internet, see [636] for routing stability, [168], [182] for interdomain routing history and requirements, [445] for routing design, [550] for end-to-end routing behavior, and [256], [749] for a detailed discussion on network neutrality.

Exercises

9.1. Review questions:

  (a) What is the relation between an AS and an ISP?

  (b) Is policy routing checked on the inbound or the outbound interface of a BGP speaker?

9.2. Suppose that you manage an ISP that has its own AS number and your domain serves as a stub AS. Occasionally, your AS receives traffic that does not belong to your AS. Identify possible cause(s) for this behavior.

9.3. For a given IP address, how would you find out its home AS number?

9.4. Why would a stub AS use uRPF at its border router? Explain.

9.5. While inspecting your BGP speaker, you found that an AS number shows up more than once for certain destination IP prefixes. Why is this possible?

9.6. Consider policy-based routing. Investigate possible scenarios in which oscillator behavior and/or loss of reachability might occur.

Part III: Routing in the PSTN

In this part, we present routing in the PSTN.

In Chapter 10, we start with a discussion of hierarchical call routing. We then delve into a number of dynamic call routing schemes, and present a qualitative discussion on their similarities and differences.

The notion of traffic in the telephone network is presented in Chapter 11. A set of control schemes that works in tandem with routing is discussed. In this framework, we then describe voice traffic engineering and present performance behavior of dynamic routing schemes by considering traffic load and capacity.

To facility call routing in PSTN, the SS7 network and its services play important roles. Thus, in Chapter 12, we present SS7 networking and discuss how the service functionality it provides through ISUP is used in call set up and control. We also discuss the SS7 protocol stack architecture, as well as message routing in the SS7 network.

Finally, in Chapter 13, we present PSTN routing taking into account E.164 addressing. Our treatment gradually changes from a nation-wide single provider environment to multi-provider scenarios, while introducing how SS7 messaging facilitate call routing decision in a multi-provider environment. We also discuss number portability and its impact on changes in call routing decision.

10

Hierarchical and Dynamic Call Routing in the Telephone Network

A good hockey player plays where the puck is. A great hockey player plays where the puck is going to be.

Wayne Gretzky

Reading Guideline

Understanding hierarchical routing gives you an idea about the issues involved in doing loop-free routing in an information-less setting. The section on dynamic routing can be read independently, although reading about hierarchical routing first provides a better perspective. Related traffic engineering problems will be discussed in Chapter 11. The discussion on dynamic call routing is also helpful in understanding quality of service routing presented later in Chapter 17.

Routing is a critical function in the global switched telephone network. The routing archi-tecture in the switched telephone network is based on the notion of hierarchical routing that was originally designed a half a century ago, and the hierarchical concept as it was thought of is still in place in the overall global switched telephone network architecture. In addition, dynamic call routing schemes have been introduced in the past 25 years that can function in this hierarchical architecture.

In this chapter, we will present both hierarchical routing and dynamic routing. The reader might want to note that the term *dynamic routing* used in this chapter refers to dynamic *call* routing in the telephone network; it should not be confused with dynamic routing in IP net-works.

We start with a few definitions. *Circuit switching* is used for call routing in the telephone network. Circuit switching refers to the mechanism of communication in which a dedicated path with allocated bandwidth is set up in an on-demand basis before the actual communi-cation can take place. On-demand means that the path is set up quickly when the request is made. The dedicated path is released immediately when the communication is over. The most well-known application of circuit switching is telephone network calls. The call band-width for a wire-line telephone circuit is 4 kilohertz in the analog mode or 64 Kbps in the digital mode. That is, a voice connection in the wired telephone network takes up a voice circuit established through circuit switching, requiring 64 Kbps of bandwidth. When a circuit is considered on a link, it is also referred to as a *trunk*. Thus, the terms circuit and trunk will be used interchangeably. The term *trunkgroup* refers to a group of circuits or trunks on a link between two directly connected switches; a trunkgroup is also referred as an *inter-machine trunk (IMT)* while considered in the context of connecting two switches. A switch in digital telephony is a time-division–multiplexed (TDM) switch. In this chapter, we will use node, switch, and TDM switch interchangeably.

10.1 Hierarchical Routing

We first start by describing hierarchical routing in a telephone network.

10.1.1 Basic Idea

Telephone networks have been around for over a century. However, the need for any form of routing did not arise until the 1930s. Until then, essentially point-to-point direct links (trunk-groups) were set up to connect calls between different places; there was no routing involved. The need for routing arose for two primary reasons: (1) point-to-point links lead to the N^2 problem, i.e., if there are N nodes in a network, we need $N(N-1)/2$ directly connected links; thus, as more and more cities (with multiple switches) offer telephone services, this problem grows significantly, and (2) it was recognized that some trunkgroups were less utilized com-pared to others; thus, if there were any way to take advantage of this by routing calls through less utilized trunkgroups, capacity expansion could be avoided. Capacity expansion used to be very costly and still is in many cases. There is another impetus to arriving at some form of routing: as the switching technology started to move from old mechanical switches to electro-mechanical switches, the possibility of switching being capitalized to perform some form of routing became more than a thought.

This is where we need to understand something important. Unlike routers for the Internet (as discussed elsewhere in this book) that have the ability to compute and store routing tables, telephone switches did not have this ability in the early years. Thus, routing was to be performed in an age when neither information storage nor information exchange was possible. When you think about it, this is a complicated problem. This problem becomes more pronounced when you add an important requirement of routing: *looping must be avoided*. There is another important point to note here. With the technology available at that time, the call setup was accomplished through *progressive call control (PCC)* by forwarding setup signaling from one switch to the next; this is to be done in an information-*less* setting—that is, nodes did not have any ability to exchange any status information. Thus, a call control cannot get back to a switch from where it started; there was no way to look backward. The question is: how can looping be avoided and yet provide some form of routing by forwarding a call from one trunkgroup to another as the call goes from one switch to another in such an information-less setting? The Bell System came up with an innovate idea for routing without looping. The basic idea was to introduce *hierarchy* among network nodes and still use PCC. To describe it, we start with a simple illustration.

10.1.2 A Simple Illustration

We start with a four-node illustration (Figure 10.1). In this example, switches are divided into two levels: switches 1 and 4 are at the lower level and switches 2 and 3 are at a higher level; furthermore, switches 1 and 2 are in the same ladder of the hierarchy, while 4 and 3 are on another ladder.

First consider Figure 10.1(a). A call from switch 1 to switch 4 can take the direct link (if it exists). However, if the direct link 1-4 is busy, the call overflows and is attempted on link 1-3. The important thing is that the control of the call is forwarded on this link with the call attempt. Once the call reaches switch 3, the call can only go toward its destination, which means taking link 3-4. If capacity is available on link 3-4, the call is carried through on link 3-4. However, if there is no capacity available on link 3-4 when the call arrives at node 3, the call is considered *lost*, and the network cannot retry through another path; because of PCC. A lost call means that users hear a fast busy signal, and the user has to hang up and redial the number. It is, however, important to recognize that the scheme still provides alternate routing. If the call cannot find an available circuit on the outgoing link 1-3, the call can be attempted on link 1-2 as the *final* trunkgroup, where switch 2 is the switch above switch 1

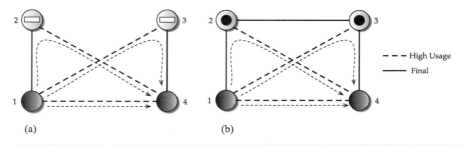

FIGURE 10.1 Hierarchical routing example.

in its direct hierarchy. Thus, the trunkgroups in hierarchical routing networks are classified into two groups: *high-usage (HU)* trunkgroups and *final* trunkgroups. So, which ones are HU groups? In this example, trunkgroups 1-4, 1-3, and 2-4 are HU groups since they are not necessary but are installed because of the high volume of traffic between those end nodes connecting such trunkgroups. A final trunkgroup means that there is no other trunkgroup to overflow to if a call does not find a circuit on a final trunkgroup. Thus, for a call from switch 1 to switch 4 (for Figure 10.1(a)), route attempt options are in the following fixed order: $1 \to 4$, $1 \to 3 \to 4$, and $1 \to 2 \to 4$. Usually, such usage of switches 2 and 3 is called *tandem* switches, which create the opportunity to provide alternate routing paths as transiting nodes.

Now consider Figure 10.1(b) where trunkgroup 2-3 is now added, compared to Figure 10.1(a). Since this is a two-level hierarchy example where there are no switches above switches 2 and 3, trunkgroup 2-3 is also a final trunkgroup. In this network, a call from switch 1 to switch 4 has the following fixed order for attempting to route a call: $1 \to 4$, $1 \to 3 \to 4$, $1 \to 2 \to 4$, and $1 \to 2 \to 3 \to 4$. In this network, a call can originate at either switch 2 or 3 as long as it has a lower layer interface for call origination; thus, in the figure a dark circle (indicating a switch at the lowest level) is embedded into the second-level switch. Now, for a call from switch 2 to switch 4, there are two routes in the following order: $2 \to 4$ and $2 \to 3 \to 4$; the first route is allowed since the HU group 2-4 is the destination switch. For a call from switch 2 to switch 3, there is only one route: $2 \to 3$.

How does the hierarchy of nodes help? Consider a call originating at switch 2 for either 3 or 4. It is not allowed to go down toward its immediate lower-level switch for routing; for example, $2 \to 1 \to 3$ and $2 \to 4 \to 3$ are *not* valid routes for calls from switch 2 to 3. Otherwise, in a PCC environment, a looping can take place since a call originating at switch 1 will go to switch 2, which will send it back to switch 1!

In summary, the main rules for routing (while avoiding looping) in a hierarchical routing environment can be summarized as follows:

- A switch in a higher level must have the switching function of the lower level in a nested manner. This is known as the *multiple switching function rule*. In Figure 10.1(b), switches 2 and 3 internally have switching functionalities of the lower-level switches.

- Calls must be routed through the direct switch hierarchy, both at the originating switch and the destination switch. This is known as the *two-ladder limit rule*. In Figure 10.1(b), the direct switches hierarchically above switch 1 and switch 4 at the next level are switch 2 and switch 3, respectively. Now, imagine a fifth switch (switch 5) at the same level as switches 2 and 3 in Figure 10.1(b), and a HU trunkgroup between 1 and 5, and also another HU trunkgroup between 4 and 5. A call from switch 1 to 4 is not allowed to take the route $1 \to 5 \to 3 \to 4$ or the route $1 \to 5 \to 4$, since switch 5 is not in an originating or destination switching hierarchy of switch 4.

- For a call from one area to another, a HU trunkgroup from a switch in the originating area to a switch at the next higher level in the destination area is a preferred selection over the final trunkgroup to the switch at a level directly above it. This is known as the *ordered routing rule*. In other words, the route order of attempts is predefined and consistent when multiple routes exist and is based on the level and location of switches in different areas.

Thus, in Figure 10.1(b), for a call from switch 1 to 4, the route $1 \to 3 \to 4$ is preferred over route $1 \to 2 \to 4$. Using the same rule, for a call from switch 4 to 1, route $4 \to 2 \to 1$ is preferred over $4 \to 3 \to 1$.

10.1.3 Overall Hierarchical Routing Architecture

In the previous section, we discussed hierarchical routing using switches at two levels. In a national network, there are actually five levels defined in the hierarchy. At the bottom are the end office switches; as we move up, we go from toll switching centers to primary switching centers to secondary switching centers to regional switching centers. The five levels of switching hierarchy are shown in Figure 10.2.

From a geographic perspective, there is another way to view the network that takes a planar view. This is shown in Figure 10.3. We can see that the part of the network that is under a regional switching center is essentially a tree-based network except for any HU trunkgroups (marked by a dashed line) that connect a switch under one regional switch to another switch in the same region or a different regional switch. The network at the regional switching center level (or the highest level if all five levels are not used) is fully connected.

To summarize, through the introduction of a hierarchy of switching nodes, several issues were addressed simultaneously:

- The scalability issue of full connectivity or N^2 growth in number of links in a network at the end-office level is addressed. Full connectively is needed only for a handful of switches at the highest level of the hierarchy. To obtain some perspective [697, §4.1.5], by 1981 (before the breakup of the Bell System), there were 20,000 end offices in the United states;

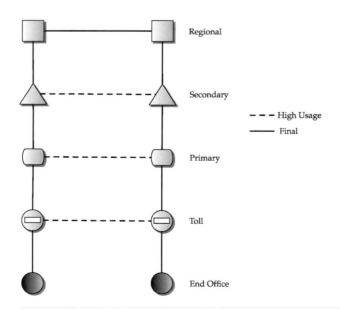

FIGURE 10.2   Switching hierarchy in hierarchical routing.

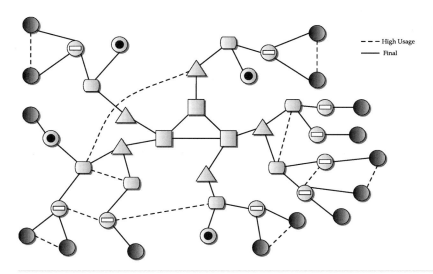

FIGURE 10.3   Geographical perspective.

imagine trying to link all of them directly in a fully connected network! However, there were only 10 regional switching centers needed.

- Multiple alternate paths were available in many cases between end offices where the call was attempted in a predefined order. An important point is that routing was accomplished without requiring any information exchange between switches, that is, in an information-*less* environment.

- Looping was avoided by defining carefully rules for switching hierarchy and forwarding of calls.

A final note is how to address routing of international calls from one country to another. In the hierarchical routing structure, another switching level is defined above the regional switching center to connect trunkgroups from one country to another country.

Thus, hierarchical routing can be briefly summarized in the following way: switches in the network are placed at different levels; a call can move up a trunkgroup from a lower-level switch to a higher-level switch unless the call is going from a higher-level switch directly to the final destination switch; a call can go from one switch to another in the same level if the second switch is in the "destination region."

10.1.4 Telephone Service Providers and Telephone Network Architecture

Until the divestiture of the Bell System in 1983, the entire hierarchy of the telephone network was provided by the same telephone service provider (TSP) in the United States. In fact, in most nations across the world, the telephone network is still provided by the same provider. With the breakup of the Bell System in the United States, different TSPs play different roles in carrying a call. A call originates in an access TSP, a "local exchange carrier (LEC)," where the call starts from the end office. If the call is destined for another user in the same LEC, the

FIGURE 10.4   LEC/IXC architecture.

call is routed within its network. When a call starts from one LEC and is destined to terminate in another LEC, the call typically goes through an inter-exchange carrier (IXC) before entering the destination LEC. The relation between LEC and IXC is shown in Figure 10.4. From a routing hierarchy point of view, IXC enters at the level of the primary switching centers.

In most cases, LECs use a two-level fixed hierarchical routing architecture with call overflow from the lower level to the upper level (see Figure 10.1(a)). An IXC can deploy either a fixed hierarchical routing or dynamic routing. Unless a call terminates in a different country, there is usually at most one IXC involved between the access LECs. For a call going from one country to another country, the call may go through the second country's interexchange provider or equivalent before reaching the destination address in another access carrier. In many countries, both the access service and the interexchange/long-distance service are provided by the same provider; regardless, a fixed hierarchical routing is commonly deployed in access LECs.

10.2   The Road to Dynamic Routing

Now that we have learned about hierarchical routing, we are almost ready to discuss dynamic routing. Before we go from hierarchical routing to dynamic routing, we need to note and understand a few critical issues.

10.2.1   Limitation of Hierarchical Routing

The need for dynamic routing is better understood if we understand the limitations of hierarchical routing. Recall that while hierarchical routing avoided the looping problem by clever use of nodes at different levels along with a set of rules, it also led to situations in which some trunkgroups could not be used for routing even though capacity was available.

Consider Figure 10.5, where switches 1 and 4 are at a lower level and switches 2 and 3 are at a higher level. We can see from the figure that a call originating in switch 1 destined for switch 4 can use the HU trunkgroup 1-4 or overflow the call to routes 1-3-4, or 1-2-4 or finally to 1-2-3-4. However, a call from switch 2 to switch 3 can only use the final trunkgroup 2-3; it cannot use a path such as 2-1-3 or 2-4-3, although at the time of the arrival of the call there might be plenty of trunks available on links 2-1, 1-3, 2-4, and 4-3. Thus, you can see the inefficiency in how hierarchical routing works.

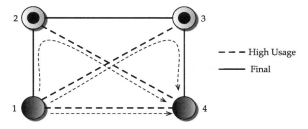

FIGURE 10.5 Limitation due to hierarchical routing.

10.2.2 Historical Perspective

In the 1970s, the idea of being able to have some flexibility in routing that can use unused capacity, rather than the limitation imposed by hierarchical routing, was explored. One important issue that needed to be addressed was the looping problem—that is, it could not be done within the framework of the hierarchical routing since a major reason for using hierarchy was to avoid the looping problem. In the 1970s, some important developments took place in parallel: the first was the ability to use stored program control (SPC) in a switch, and the second was the development of common channel interoffice signaling (CCIS). SPC provided the software functionality for switching control. CCIS provided the ability to exchange control information such as call setup and tear-down through out-of-band signaling instead of using in-trunk signaling for such functions, which was becoming noticeably slower when a call needed to go over multiple trunkgroups in serial to its destination; this out-of-band communication was a data communication service, meaning that information was exchanged as data packet services. Through evolution, CCIS, which used to be referred to as CCIS6, became CCS and eventually SS7; see Chapter 12 for further details.

In addition, there was another important observation, especially in the continental United States. Due to different time zones in the country, there were times when certain trunkgroups were idle or had very little utilization, but again, due to hierarchical routing, these trunkgroups could not be used. An example involves calls between New York and Atlanta, which are located in the Eastern time zone, at 8:00 AM. If the trunkgroup between New York and Atlanta is fully occupied at that time, a newly arrived call between these two cities could be alternately routed via Los Angeles, located in the Pacific time zone, which is 3 hours behind the Eastern time zone. It is then 5:00 AM in Los Angeles and it is less likely that there will be many calls between New York and Los Angeles or between Atlanta and Los Angeles at that time. This means the new call can conceivably use New York–Los Angeles and Los Angeles–Atlanta trunkgroups to route this call. While this is an extreme example, it suggests that at least the set of routes a call between two switches can attempt can possibly be different depending on the time of the day, that is, the routing can be time-dependent rather than having the same or a fixed order at all times of the day. In essence, we can start to see that some form of dynamic routing that is at least time-dependent can be of benefit to the network in terms of carrying more calls. Actually, the potential benefit of dynamic routing is often credited to Nakagome and Mori [519], who first discussed the benefits of flexible routing.

As you can see, in the above illustration, all the routes are of a maximum of two links (trunkgroups). We want to clarify that these two links are only in the core of the network. The actual call dialed by a user arrives at an end office from which the call is forwarded to the ingress switch in the core network. Similarly, from the egress switch, the call goes to the end office at the other end before reaching the actual receiving user. Thus, the maximum two-links part is addressed only between the ingress and the egress switch in the core network. Obviously, more than two links in this part can be possible. However, there are three important drivers that led to all dynamic routing methods to limit calls to a maximum of two links:

- An issue was how to handle the looping problem. It is easy to see that the looping problem can be easily handled with a maximum of two links for a call: a call can be going directly from the ingress switch to the egress switch on a direct link; if this link is busy, the call can try another route going from the ingress switch to an intermediate switch. The intermediate switch on receiving the call knows that the call needs to be sent directly to the egress switch, not another intermediate switch, due to the limit on the number of links. This then automatically avoids any looping problem.

- A second issue was the complexity of software implementation of the dynamic routing function. Note that the concept of dynamic routing arose toward the end of the 1970s and early 1980s when software for telephone switches was still in its nascent stage, not to mention the high cost of implementing a complex function. The goal was to keep the complexity down, for example, if the looping problem could be addressed easily without introducing software complexity.

- There is minimal incremental gain from allowing more than two links. Common sense indicates that if more than two links are allowed, a network will certainly have more paths to the destination, and thus would have the ability to complete more calls. However, a telephone network is required to maintain an acceptable grade-of-service (GoS); in the United States, this was mandated by the Federal Communication Commission (FCC). An acceptable level of GoS was to maintain average call blocking at 1% or lower; an additional discussion of call blocking is presented later in Chapter 11. What we need to understand is that if a network is provisioned with a bandwidth to meet 1% call blocking GoS in the presence of dynamic routing where a call is limited to a maximum of two links, how much incremental gain can we gain if we were to have dynamic routing with more than two links? It was reported in [30] that this gain was not significant, i.e., the blocking would go down from 1% to about 0.96%.

Now, from the first two items, we can see that the software complexity can be minimized if a route is limited to a maximum of two-link paths. From the third item, when considered along with the software complexity issue, we can see that the gain in reduction in blocking can come at a very heavy price in terms of increased software complexity. Rather, if keeping GoS low is an important goal, it can be achieved by other means, for example, adding more capacity to the network. At the same time, it is easy to recognize that if we can provide many alternate paths between two switches, we have the opportunity to reduce call blocking. In a network with a maximum of two links for a path, the simplest way to achieve this is to make the topology of the network fully connected or nearly fully connected. For example, in a fully connected network with N switches, there are $N - 2$ two-link paths in addition to the direct link path.

10.2.3 Call Control and Crankback

Hierarchical routing uses a progressive call control (PCC) mechanism. This means that the call control is forwarded from one switch to another until it reaches its destination unless the call cannot find any outgoing trunk at an intermediate trunk; in this case, the call is lost. In other words, the control of the call is not returned to the originating switch to try another possible path.

Suppose we could return the control of a call from an intermediate switch to the originating switch. This would mean that the network is providing originating call control (OCC); the functionality of returning a call to the originating switch and trying another route is called *crankback*. With the advent of the dynamic call routing, the question of whether the network should provide PCC or OCC and whether it should provide crankback also arises.

Figure 10.6 illustrates how crankback works and its relation to OCC and PCC. Consider a call arriving at switch 1 destined for switch 2. It can try the direct link path 1-2. Suppose there is no bandwidth available on link 1-2 when the call arrives. The call will then attempt to use the next route in the routing table 1-3-2. If link 1-3 has no available capacity, the call will attempt the next route in the routing table 1-4-2; this overflow attempt is, however, not a crankback. So what is a crankback? Consider a slightly different situation. Suppose when the call attempted the second route 1-3-2, it found bandwidth on the first link 1-3 and thus the control of the call is forwarded to node 3; however, on arriving at node 3 it was discovered that there is no bandwidth available on link 3-2 for this call. There are two possibilities: either send the control of the call back to the originating switch 1 and let the originating switch decide what to do next (for example, try another route such as 1-4-2), or drop the call. The control of the call can be sent back to the originating switch 1 if the network has OCC, the process of reverting back to switch 1 and trying another route is called *crankback*. If the network does not have OCC, it must act as PCC. Thus, drop the call means that the call on arriving at node 3 is lost due to nonavailability of capacity on link 3-2; this occurs due to PCC, the call control cannot be returned to switch 1. As you will see later, some dynamic routing schemes provide OCC while others do not.

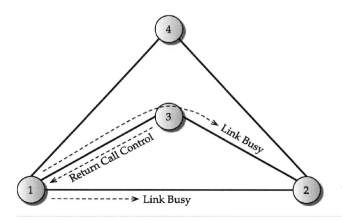

FIGURE 10.6   Illustration of crankback.

10.2.4 Trunk Reservation

Trunk reservation, also known as state protection, refers to logical reservation of a part of a capacity on a link for its own direct traffic. Note that trunk reservation does not mean physical reservation of a trunk. In this sense, this is a misnomer, and state protection is a better name. We have decided to retain the term *trunk reservation* because of its historical use and the prevalence of the use of this term in a large body of literature over the years.

Trunk reservation refers to a threshold on a trunkgroup; if a trunkgroup is not filled with calls before this threshold is reached, a call between other origin-destination pairs can still use a trunk from this trunkgroup. Consider trunkgroup ℓ connecting switching nodes i and j with capacity c_{ij}, which is given in a number of circuits. Suppose the trunk reservation parameter is given by r_{ij}, also given in number of circuits. If $r_{ij} = 0$, then no trunk is reserved. However, if $r_{ij} = c_{ij}$, trunkgroup i-j does not allow any alternate routing; certainly, in real networks this condition is never used. Typically, r_{ij} is close to zero; it should not be too low or too high. A rule of thumb is that $r_{ij} \approx \frac{\sqrt{c_{ij}}}{2}$; later, in Section 11.8, we will illustrate the impact on performance for different values of trunk reservation.

Another interpretation of trunk reservation is that a call that connects the ends of trunkgroup i-j is given access to all capacity c_{ij}, while a call for another origin-destination pair that is using link i-j can have access to effective capacity, $c_{ij} - r_{ij}$. You may wonder why we need to do this. While dynamic routing provides flexibility to use multilink paths for routing, under certain loads it may not be desirable to route calls on multilink paths for the benefit of the network. Consider a path made up of two links; it can route a call for the end nodes of this path, or, from the point of view of the network, each link can be used to carry a call for the end nodes of each link. Thus, a network carrying two calls, one call each on direct link paths instead of one call carrying on the two-link path made up of the same two direct links, has more call-carrying capacity. To illustrate this, consider a three-node network where links between each node have one unit of capacity each. We number the nodes as 1, 2, and 3. Here, the maximum call carrying capacity is three—one each for each pair of nodes on the corresponding direct link, that is, 1-2, 1-3, and 2-3. However, if we allow a pair of nodes to use the alternate path, say for pair 1:2 we allow a call to take the route 1-3-2, then the network would have only two call-carrying capacity—one on the direct path for pair 1:2, another on the alternate path 1-3-2, and none for other pairs.

It so happens that in the absence of trunk reservation, dynamic routing can exhibit bistability in certain load conditions (refer to Section 11.8). That is, a network can have different blocking levels for the *same* offered load, sometimes staying in one for a certain amount of time and then moving to another due to fluctuation in load. This is also referred to as *metastability* (or bistability). By using trunk reservation, this metastable behavior can be minimized and often avoided. A formal definition of offered load will be presented later in Chapter 11; furthermore, in Section 11.8.4, we will discuss the implication of no trunk reservation and metastable behavior.

10.2.5 Where Does Dynamic Routing Fit with Hierarchical Routing?

When a hierarchical routing architecture already exists, the question of where to fit in dynamic routing arises. When dynamic routing is introduced as the routing scheme within an

IXC's network, the switching level of switches in the dynamic routing network can be thought of as if it is at the primary switching center level. In other words, toll switches and end-office switches are considered to be in a level below the switch level of dynamic routing switches. This is illustrated in Figure 10.7.

It may be noted that both an end-office switch or a toll switch may be connected by a trunkgroup to a switch in the dynamic routing network. Consider end-office switch 9, which is connected to the dynamic routing core via toll switch 8. However, end-office switches 6 and 7 are directly connected to the dynamic routing core. Furthermore, we can now see that a call from one end office to another end office can traverse at least three trunkgroups (for example, 6-1, then 1-3, and finally 3-7), or it can possibly be five trunkgroups where at most two trunkgroups are in the dynamic routing core (for example, 9-8-5-4-3-7).

10.2.6 Mixing of OCC and PCC

It is possible to mix OCC and PCC from the perspective of an end office to another end office. The edge networks, where a call starts and ends, have PCC while the dynamic routing core has OCC. Consider again Figure 10.7. A call originating in end-office switch 9 and destined for switch 7 uses PCC to forward the call to switch 8, which forwards it to switch 5. Then switch 5, being the originating node in the dynamic routing core, may hold the control of the call and try alternate routes within the dynamic routing network until the path is established to switch 3, the destination node within the dynamic routing core for this call. Once the call is established within the dynamic routing core to switch 3, the call control is forwarded from switch 5 to switch 3 so that progressive call control can be used for completing the call to end-office switch 7.

10.2.7 Recap

We now summarize a few key points about dynamic routing. All dynamic routing schemes for the telephone networks allow at most two links for a call. Often, the network is fully-

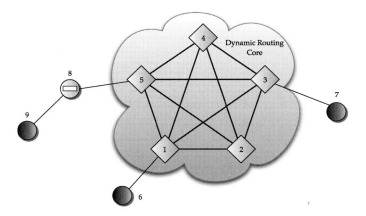

FIGURE 10.7 Dynamic call routing in conjunction with hierarchical routing.

interconnected, or nearly fully-interconnected. Also, all schemes include trunk reservation. They all differ in the following areas:

- Progressive or originating call control, and crankback.

- Time-dependent, or adaptive.

- Off-line computation, or near on-line computation.

- Routing calculation.

- Link information used and how it is used.

10.3 Dynamic Nonhierarchical Routing

Dynamic Nonhierarchical Routing (DNHR) is the first implemented dynamic routing scheme. It was first deployed in AT&T's long-distance telephone network in 1984 and was retired in 1991. We discuss it here primarily for its historical context and its evolution to RTNR, which is discussed later in Section 10.6.

DNHR is a time-dependent routing. This means that the set of routes available (and their order) at different times of the day is different. In the case of DNHR, the 24-hour time period spanning a 7-day week was divided into 15 load set periods: 10 for the weekdays and 5 for weekends. The different number of load set periods was determined based on understanding traffic patterns. For example, the same routing pattern can be used from midnight to 8 AM due to low traffic volume.

For each load set period, based on traffic projection, a set of routes is computed ahead of time. Typically, traffic projection and routing computation were computed off-line 1 week in advance and the routing table is then uploaded to each switch in the network. The routes so computed ahead of time are referred to as *engineered paths*. When an actual call arrives at a switch, the switch first determines the correct routing table based on the time of arrival and tries the various paths in the order shown in the routing table. Certainly, the actual traffic would be different than the projected traffic demand volume. Thus, the routes computed and the order of routing provided from off-line computation may not be optimal at the actual time. One way for the network to obtain some flexibility in such a situation is to allow the crankback option. For example, if there are three engineered paths between a source and a destination, a call can first attempt the first path (which is often the direct link path). If no bandwidth is available on the first path, the call tries the second path, and so on, as described earlier in Section 10.2.3.

While engineered paths can provide an acceptable GoS under normal operating conditions, they may not always be well suited if a traffic overload occurs; this is partly because the engineered paths are computed based on traffic projection. To circumvent this situation, DNHR allows additional paths to be considered almost on a real-time basis that are appended to the list of engineered paths; these additional paths are referred to as *real-time paths*.

If the blocking between a pair of switching nodes goes beyond an acceptable threshold, a new estimation of traffic over every 5-min window is invoked; based on this 5-min short-

Time Period	Routing Sequence						
Morning	7	3	6	4	*2*	*1*	
Afternoon	7	6		3	4	*2*	*1*
Evening	7	4	6		3	*2*	*1*
Weekend	6	4		*7*	*3*	*2*	*1*

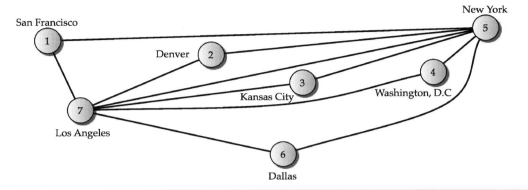

FIGURE 10.8   Engineered and real-time paths in DNHR between New York (switch 5) and Los Angeles (switch 7). Engineered and real-time paths are partitioned using a vertical line, whereas real-time paths are listed in italics in the routing table; note: switch 7 in the routing table indicates a direct link path. (Adapted from [27].)

time traffic snapshot, a centralized computation at the network management center seeks trunkgroups that have available capacity to determine real-time paths. These newly computed paths are then loaded into the network switches. The use of engineering and real-time paths is illustrated in Figure 10.8.

We now need to understand how many alternate paths we can store (cache) for such engineered and real-time paths. Note that for an N-node network that is fully connected, a pair of switches has $(N - 1)$ possible paths that are made of a maximum of two links. For example, for a 100-node network, there are 99 possible paths. Now, if we consider that an ordered routing list for each of the 15 load set periods is needed and that a switch needs to keep such routes for every destination switch, the list of paths that need to be loaded in a switch can be quite large. This is a major issue that the designers of DNHR faced, especially given that this development was done in the early 1980s when processors were not as fast as today's processors and memory was very expensive as well. The solution that was arrived at was to allow each source destination switch pair to have a maximum of 15 routes in the routing list per load set period per switch pair; of these, a maximum of 10 could be for engineered paths and the other 5 for real-time paths. In practice, it was often found that three to four engineered paths were enough.

To summarize, DNHR allows at most two links for a call within its network, as do the other dynamic routing schemes. DNHR is based on OCC, and it allows crankback. DNHR employs trunk reservation. It is a time-dependent routing scheme in which some routes ("engineered paths") are computed off-line ahead of time, while other routes ("real-time paths") can be computed and appended on a near real-time basis when congestion occurs.

10.4 Dynamically Controlled Routing

Dynamically Controlled Routing (DCR) was originally developed by Bell-Northern Research (which became Nortel Networks) [66], [680]. DCR is an adaptive routing scheme that can be updated frequently (usually every 10 sec) based on the status of the network links. The computation of routes to be cached has been done through a centralized route processor (Figure 10.9). Routes take at most two links to complete a call, and crankback is not implemented in this scheme. Thus, with PCC, if a call is blocked on the second leg of a two-link call, the call is lost; this means that the user has to try again. DCR has two fall-back mechanisms: (1) in a situation in which the route processor is down or cannot compute routes in a timely manner, or does not communicate back to the switched nodes in a timely manner, DCR continues to operate using the last known routing table, and (2) if a switch loses dynamic routing functionality for some unknown reason, the network can still operate as a two-level hierarchical routing system in which certain nodes are labeled ahead of time as nodes in the second level of the hierarchy.

To understand DCR, consider Figure 10.10 where we have indicated that the centralized route processor is where the link state information is updated regularly. We want to update the routing list for traffic between nodes i and j; in addition, Consider two possible intermediate nodes, k_1 and k_2, are to be considered for alternate routes. Periodically, the switching node reports its available capacity to the centralized route processor; we will denote dependency on time using the parameter t. To determine this we need to consider capacity, currently used capacity, and trunk reservation at time t shown below:

Link ID	Capacity	Currently Used	Trunk Reservation
i-j	$c_{ij}(t)$	$u_{ij}(t)$	$r_{ij}(t)$
i-k_1	$c_{ik_1}(t)$	$u_{ik_1}(t)$	$r_{ik_1}(t)$
i-k_2	$c_{ik_2}(t)$	$u_{ik_2}(t)$	$r_{ik_2}(t)$
k_1-j	$c_{k_1j}(t)$	$u_{k_1j}(t)$	$r_{k_1j}(t)$
k_2-j	$c_{k_2j}(t)$	$u_{k_2j}(t)$	$r_{k_2j}(t)$

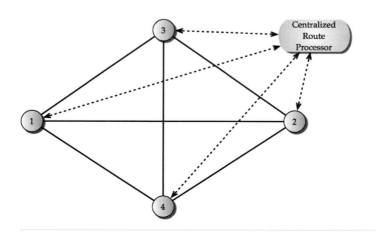

FIGURE 10.9 DCR architecture.

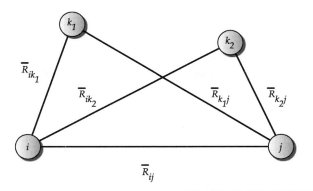

FIGURE 10.10   DCR: available capacity on links.

If we denote the effective residual capacity on any link *l-m* at time *t* by $R_{lm}(t)$, it is clear that

$$R_{lm}(t) = \max\{c_{lm}(t) - u_{lm}(t) - r_{lm}(t), 0\}. \tag{10.4.1}$$

Since an alternate path is made up of two links, the availability of capacity on a path would be the minimum of the effective residual capacity on each link of the path. Thus, the available capacity on path *i-k_1-j* and *i-k_2-j* can be written as follows:

Path	Available Capacity
i-k_1-j	$\overline{R}_{ij}^{k_1}(t) = \min\{R_{ik_1}(t), R_{k_1 j}(t)\}$
i-k_2-j	$\overline{R}_{ij}^{k_2}(t) = \min\{R_{ik_2}(t), R_{k_2 j}(t)\}$

DCR uses the availability information to compute the probability of choosing an alternate path. If a path has zero availability, there is no reason to consider this path as a possible alternate path (at this time instant). Thus, choice of path is considered only for a path's positive availability, i.e., in this case, if $\overline{R}_{ij}^{k_1}(t) > 0$ and $\overline{R}_{ij}^{k_2}(t) > 0$. Then, the probability of selecting each path is computed as follows:

$$p_{ij}^{k_1} = \frac{\overline{R}_{ij}^{k_1}(t)}{\overline{R}_{ij}^{k_1}(t) + \overline{R}_{ij}^{k_2}(t)}, \quad p_{ij}^{k_2} = \frac{\overline{R}_{ij}^{k_2}(t)}{\overline{R}_{ij}^{k_1}(t) + \overline{R}_{ij}^{k_2}(t)}. \tag{10.4.2}$$

Note that this is shown for two paths for node pair *i* and *j*. In a fully connected network, a node pair has up to $N-2$ two-link alternate paths. Thus, the general expression for choosing a path via node *k* is given by

$$p_{ij}^{k} = \frac{\overline{R}_{ij}^{k}(t)}{\sum_{\{m \neq i, j \text{ and } \overline{R}_{ij}^{m} > 0\}} \overline{R}_{ij}^{m}(t)}. \tag{10.4.3}$$

There are two important points to note: (1) the above expression should be considered only for paths with positive availability capacity, and (2) the same probability calculation is performed for all demand pairs in the network, which means that no capacity is specifically curved out for a particular demand pair in this probability calculation. The second point implies that the residual capacity of a link $i\text{-}k$, $\overline{R}_{ik}(t)$, can be used by a demand pair connecting node i and m where k is also a possible intermediate node. Finally, note that the computation of route probabilities is performed based on the information available at t. At every Δt unit of time, the switching nodes update the centralized route processor with the new status of link information for use in computing updated routes. In practice, Δt is set to be 10 sec. Thus, the routing in DCR is very adaptive to short-term link status fluctuations.

The actual call routing in DCR uses the routing probability computation for selecting alternate paths. For each probability value computed for a path with intermediate node k, it imagines the probability range to be divided as follows for a set of alternate paths identified by the intermediate node identifier k:

$$(0, p_{ij}^1], \ (p_{ij}^1, p_{ij}^1 + p_{ij}^2], \ \ldots, \ \left(\sum_{m=1}^{k-1} p_{ij}^m, \sum_{m=1}^{k} p_{ij}^m \right], \ \ldots, \ \left(\sum_{m=1}^{K-1} p_{ij}^m, 1 \right].$$

When a call arrives at node i destined for node j, the call first tries the direct link $i\text{-}j$. If there is no capacity available in the direct link path at that instant, the call then generates a uniform random number between 0 and 1. Depending on where this number falls, the appropriate alternate path is chosen to try the call. For example, if the probability value is between 0 and p_{ij}^1, an alternate path with intermediate node identifier 1 is attempted. Similarly, if the random number falls between p_{ij}^1 and $p_{ij}^1 + p_{ij}^2$, an alternate path with intermediate node identifier 2 is attempted, and so on.

Example 10.1 *Illustration of DCR path computation.*

We will now illustrate DCR by using a four-node example. For simplicity, we show the residual capacity of each link in Figure 10.11. For demand pair 1:2, there are two alternate routes: 1-3-2, and 1-4-2. Here, the availability for each path is as follows:

Path	Available Capacity
1-3-2	$\overline{R}_{12}^3(t) = \min\{10, 7\} = 7$
1-4-2	$\overline{R}_{12}^4(t) = \min\{3, 4\} = 3$

Then the probability of choosing each path is

$$p_{12}^3 = \frac{7}{7+3} = 0.7, \quad p_{12}^4 = \frac{3}{7+3} = 0.3.$$

The probability range for call selection is then set up as $(0, 0.7]$ and $(0.7, 1.0]$ for paths 1-3-2 and 1-4-2, respectively. For example, if an arriving call (after trying the direct link path) randomly picks the value 0.3 from a uniform random number distribution, the call will be attempted on path 1-3-2; if the random pick is 0.8, the call will attempt 1-4-2. ▲

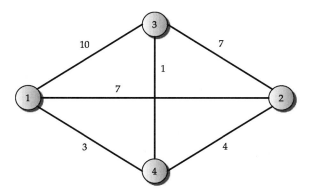

FIGURE 10.11 DCR example.

It is important to note that an arriving call may not find an alternate path despite the probability of choosing a positive path as computed above. There are a few reasons for this. First, the probability computation is based on a snapshot taken every Δt sec; when a new call arrives, the network state may have changed since the last computation was done. Second, a link is also a candidate for alternate paths for other demand pairs in addition to being the direct path for a call that is between the two end nodes of the link.

Consider again the four-node network of Figure 10.11, this time for calls between nodes 2 and 3. Its two alternate paths 2-1-3 and 2-4-3 have probability 0.875 ($= \frac{7}{8}$) and 0.125 ($= \frac{1}{8}$), respectively. Thus, if there is a sudden surge of calls (before the next update) between nodes 2 and 3 requiring alternate routing, they are more likely to take path 2-1-3, thus using up the available capacity on each leg of this path, that is, on links 2-1 and 1-3, quickly before calls for nodes 1 and 2 have a chance to use the residual capacity on link 1-3 for its alternate route 1-3-2.

To summarize, DCR allows at most two links for a call within its network. DCR is based on PCC and does not allow crankback. DCR employs trunk reservation. It is an adaptive routing scheme in which the probability of selecting alternate routes is computed at a centralized route processor every 10 sec based on the state of the network links obtained by the centralized system. This scheme has been deployed in the following networks: Stentor, MCI, and Sprint.

10.5 Dynamic Alternate Routing

Dynamic Alternate Routing (DAR) [251] was a project initiated by British Telecom. This is a distributed, adaptive routing scheme. Like other dynamic routing schemes, DAR is limited to maximum two-link routing and employs trunk reservation. DAR has no crankback and PCC is used.

For each destination switch j, the originating switch i maintains an alternate path k in its cache. A newly arrived call first attempts the direct link i-j. If there is no capacity available on the direct link, it tries the alternate path in its cache. If the call succeeds on this alternate path, this alternate path remains in the cache. However, if the call cannot set up a call through the alternate path due to nonavailability of capacity, the call is lost, meaning the user has to retry it. In addition, the originating node i picks an intermediate node randomly and sticks this in the cache for the next call to use. This means an alternate path remains in the cache as long as any calls using this alternate path are successfully connected; a new alternate path is picked randomly the instant the current alternate route cannot connect a call using this path.

The elegance of this routing scheme lies in its simplicity. How could such an algorithm work well in practice? To see why this algorithm works, think of calls arriving over a window of time: an alternate path that is successful remains in the cache; if an alternate path is not successful, it is probably because the links in this path are already congested. Thus, over time, a least-loaded alternate route is likely to stay in the cache for a higher percentage of time. Since network traffic can change over time, the routing automatically adapts to another least-loaded alternate path.

There is another important observation about this routing scheme: it does not require any network link status to be updated for computing routes. Such approaches are sometimes referred to as *learning automata* [520], [521], [657]. Sometimes, a routing scheme such as DAR is referred to as event driven routing.

10.6 Real-Time Network Routing

Real-Time Network Routing (RTNR) [34] is the successor to DNHR and was deployed in the AT&T long-distance network in 1991. It is still in use today. Unlike DNHR, RTNR is an adaptive routing scheme. The routing table for alternate routes can be updated almost on a per call basis. Consider an RTNR network with N nodes. When a call arrives at node i and is destined for node j at time t, node i queries node j seeking information about the status of outgoing links from node j to a switching node k other than node i, i.e., the status of link j-k. Note that node i knows the status of all its outgoing links i-k. Because a path is limited to a maximum of two links and circuit-switched links being bidirectional, node i can then determine the status of all two-link paths i-k-j to destination j by combining information for link i-k and link j-k. Now, knowing this information, i can decide on choosing an alternate route for this call.

Now we discuss the type of information sought from node j. In case of RTNR, node i requests (in its simplest form) the status of availability of all outgoing links as binary status: 1 if the link is available, 0 if it is not available. In essence, RTNR does not quite care about how much is available on outgoing links from node j as long as there is available capacity. That is, if the effective residual capacity R_{jk} as given in Eq. (10.4.1), which takes into account any trunk reservation, is positive, link j-k is available. Thus, availability of outgoing links j-k is

$$I_{jk} = \begin{cases} 1, & \text{if } R_{jk} \geq \widehat{R} \\ 0, & \text{otherwise} \end{cases}$$

for $k = 1, 2, \ldots, N$, $k \neq i$, $k \neq j$, where $\widehat{R}$ is a predefined positive integer. While $\widehat{R} = 1$ would be the lowest value for which this relation works, for practical purposes, it would be safer to use a higher value. In other words, $\widehat{R} = 1$ would mean the link is almost congested; thus, it may not be preferable. Thus, $\widehat{R}$ in RTNR defines the threshold for indicating where a link is heavily loaded (utilized) or lightly loaded. Similarly, node i can determine the status of all its outgoing links, i.e.,

$$I_{ik} = \begin{cases} 1, & \text{if } R_{ik} \geq \widehat{R} \\ 0, & \text{otherwise} \end{cases}$$

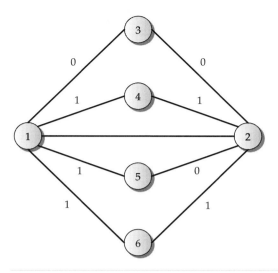

FIGURE 10.12 RTNR example.

for $k = 1, 2, \ldots, N, k \neq i, k \neq j$. Once node i receives the availability information from node j, it performs the boolean "AND" ($\wedge$) operation for all intermediate nodes k, i.e.,

$$I_{ik} \wedge I_{jk}, \quad k = 1, 2, \ldots, N, \ k \neq i, \ k \neq j$$

to determine the availability of alternate paths. For the subset of paths for which the result is 1, i.e, the path is available, node i selects one of the paths at random as the alternate route for this call. In reality, RTNR also does an additional "AND" operation to check preference on availability of switches, referred to as *node availability*, since a switch's CPU can be overloaded ("switch congestion") or a switch might not be preferred due to maintenance, before identifying usable paths.

We now illustrate RTNR through a simple example, where node availability is also incorporated (Figure 10.12). Consider a call arriving at node 1 destined to node 2, and suppose that the network consists of 6 nodes. Then, the status of the availability of node 1 for its outgoing links and of node 2 for its outgoing links is shown in Table 10.1.

In this example, we see that paths 1-4-2 and 1-6-2 are both usable; node 1 can then randomly choose one of them. It is important to note that if the direct link exists, the call attempts

TABLE 10.1 RTNR route availability computation.

	Node k	1	2	3	4	5	6
Outgoing from node 1	I_{1k}	—	—	0	1	1	1
Outgoing from node 2	I_{2k}	—	—	0	1	0	1
Path availability 1-k-2		—	—	0	1	0	1
Node availability	I_{2k}	—	—	1	1	0	1
Usable paths 1-k-2		—	—	0	1	0	1

the direct link route first before attempting the alternate route decided by the above mechanism.

From the basic operation of the above method, we can see that crankback is not essential since the information about availability is fresh since switch i is making this decision based on querying after the *arrival* of the call. However, an important issue in circuit-switched voice networks is to keep the call setup time to a minimum. We can see that if a query is generated *after* the arrival of the call, the call has to wait for a response from the destination node *and* computation of path availability before the call can attempt an alternate route; this time adds up to the call setup time, i.e., post-dial delay. In general, it is desirable to avoid this delay. In practice, RTNR uses a slight variation. To choose an alternate path, an arriving call uses the query and path availability result *already* obtained for the last call for the same source–destination pair; it still generates a query to its destination node for this call, which is meant for use by a future call arriving for the same source–destination pair. This process reduces the impact on the call setup delay. However, this "one-call-old" notion introduces the possibility that a path randomly chosen from the set of available paths so obtained can possibly be stale for a new arriving call. Thus, in this case, having the crankback function allows a call to try another path from it availability list (for last call) if the first one is not available any more. In the worst case, if none of the paths from the one-call-old list works, then a call can wait for the result for its own query to see if a better path is available. However, a word of caution: such a crankback from one to the next can also induce an additional delay on call setup time; thus, this should be used only for a certain number of attempts on alternate paths for a specific call.

RTNR also allows for a network status map query at different congestion levels. Path selection for different levels is done similar to what we discussed above, except that different availability maps are generated based on a different congestion threshold values. In all cases, the query and query exchange for such information are done using the out-of-band SS7 network (see Section 12.10 in Chapter 12). Finally, RTNR can handle calls with different bandwidth requirements ("multi-rate") and thus provide different classes of services; the multi-rate, multi-service case is discussed later in Chapter 17.

10.7 Classification of Dynamic Call Routing Schemes

While there are other dynamic routing schemes (see Section 10.8 and Section 11.7), the above four schemes give us a good sense about different ways to classify dynamic routing. That is, in general, dynamic routing schemes can be classified in the following ways:

- *Centralized vs. Distributed:* In the centralized system, routing computation is performed at a centralized processor. For example, DNHR and DCR fall into this category. While DNHR is done on a weekly basis based on traffic projection, DCR is done on a 10-sec interval. In the case of distributed routing, the switch itself does the routing computation, presumably relying on status obtained or observed about the network. Both RTNR and DAR are distributed routing schemes.

- *Time-Dependent vs. Adaptive:* Time-dependent routing refers to updating of the routing table at a certain time (once a day or once a week) for a preplanned set of routes a call can attempt while the set of routes may vary from one time period to another. Adaptive routing refers to frequent update (in min or sec) of the routing table based on traffic measurements or on events. DNHR falls under time-dependent routing, while DCR, RTNR, and DAR can all be considered adaptive routing.

- *Periodic vs. On-demand:* Within adaptive routing, the routing table update can be performed on a periodic basic, or on-demand. DCR is an example of routing that is updated every 10 sec. Both RTNR and DAR fall under on-demand since in the case of RTNR, the routing table update can be on a call-by-call basis (whether current or one-call-old), while in the case of DAR, a routing table update is performed whenever the current alternate route blocks a call.

- *State-Dependent vs. Event-Dependent:* State-dependent refers to considering the state of a network in determining a routing decision; typically, *state* here refers to the state of a link such as available capacity or just availability, or some notion of a cost. DCR and RTNR fall into this category. Event-dependent refers to selection of a new alternate route if a certain event occurs; DAR falls into this category.

- *OCC vs. PCC:* DNHR and RTNR are based on OCC, while DAR and DCR are based on PCC. An intertwined feature with OCC and PCC is crankback.

Despite the above classification, all dynamic routing schemes have two things in common: (1) all routing schemes allow at most two links for routing a call, and (2) they all provide trunk reservation. If the network is fully connected, a call tries the direct link route first, although DNHR or RTNR does not require it; if it is not fully connected, a preferred two-link route is usually determined as the first attempt path for the demand pair that does not have a direct link. However, the actual computation of alternate routes and how many alternate routes to keep in the routing table differs from one scheme to another.

10.8 Maximum Allowable Residual Capacity Routing

From the discussion of the above routing schemes, it is apparent that other routing schemes can be designed; the most obvious one is maximum allowable residual capacity (MARC) [680]. Recall that MARC is a form of widest-path routing discussed earlier in Chapter 2. To describe MARC routing, we will use the notation described earlier in Section 10.4 and repeat some common descriptions. To be consistent with the previous routing schemes, we consider a network where at most two-link paths are allowed, and that is almost fully connected. Consider a pair of nodes i and j. Suppose that at time t, we have the following link information regarding two possible alternate routing nodes k_1 and k_2 for the pair of nodes i and j:

Link ID	Capacity	Currently Used	Trunk Reservation
i-j	$c_{ij}(t)$	$u_{ij}(t)$	$r_{ij}(t)$
i-k_1	$c_{ik_1}(t)$	$u_{ik_1}(t)$	$r_{ik_1}(t)$
i-k_2	$c_{ik_2}(t)$	$u_{ik_2}(t)$	$r_{ik_2}(t)$
k_1-j	$c_{k_1j}(t)$	$u_{k_1j}(t)$	$r_{k_1j}(t)$
k_2-j	$c_{k_2j}(t)$	$u_{k_2j}(t)$	$r_{k_2j}(t)$

Then allowable residual capacity of any generic link *l-m* is given by

$$R_{lm}(t) = \max\{c_{lm}(t) - u_{lm}(t) - r_{lm}, 0\}. \tag{10.8.1}$$

Knowing the above quantity for different links, the available capacity via node k_1 and k_2 can be given as follows:

Path	Path Available Capacity
$i\text{-}k_1\text{-}j$	$\overline{R}_{ij}^{k_1}(t) = \min\{R_{ik_1}(t), R_{k_1j}(t)\}$
$i\text{-}k_2\text{-}j$	$\overline{R}_{ij}^{k_2}(t) = \min\{R_{ik_2}(t), R_{k_2j}(t)\}$

Now, instead of choosing the probability of selecting an alternate path as done in DCR, another rule can be applied in which the path with MARC is chosen. As for the example with the two alternate paths considered above, via node k_1 would be chosen if $\overline{R}_{ij}^{k_1}(t)$ is larger than $\overline{R}_{ij}^{k_2}(t)$; otherwise, k_2 will be chosen.

The same argument can be applied if more than two alternate paths are considered. When you look carefully, you notice that this scheme uses the shortest-widest path computation described earlier in Section 2.6. The major difference is that the direct link between the source and the destination node is always tried first, before trying the most preferred alternate path based on MARC. It is important to note that the direct path is the attempted path even if it has less available capacity than the alternate path with MARC at that time instant. You might wonder why we need to give priority to routing a call first on the direct link; this again goes back to the bandwidth used by a call. If a call is alternate routed, this will use up two units of capacity, one from each leg, as opposed to one unit of capacity for a direct link call. Because of this, the network can also show bistable behavior if naively the path with the maximum available capacity is chosen regardless of whether this is a direct or an alternate path. In other words, attempting a call on the direct link path first is important.

To compute the most preferred alternate path, a network may choose to do periodic update of allowable residual capacity either in a centralized manner similar to DCR or a distributed manner similar to RTNR. If a periodic update/computation window, say Δt sec, is used, there is a change in available capacity between the updates due to newly arrived calls and also due to calls that have been completed. Thus, a preferred alternate route at a particular instant of time may not be preferred after the next computation. Thus, MARC routing has the potential of oscillating the preferred alternate route from one time window to another. Thus, to avoid the impact of sudden short-term fluctuations, the *exponential weighted moving average method* (see Appendix B.6) can be employed to obtain a smoothed allowable residual capacity that can be used instead of Eq. (10.8.1).

Finally, it is important to note that MARC seems to indicate that an alternate route is preferable if it was the least busy. This, however, brings up whether load matters; in the case of MARC, it implies allowable residual capacity. However, a least-loaded path, instead of a least-busy path, can be considered where the traffic load is directly taken into account; a state-dependent dynamic routing scheme based on this least-loaded concept is described later in Section 11.7.

10.9 Dynamic Routing and Its Relation to Other Routing

We will now discuss how dynamic routing for telephone networks is related to other routing covered in this book.

10.9.1 Dynamic Routing and Link State Protocol

When we consider a dynamic routing scheme that is based on the state of the network, be it centralized or distributed, the routing decision can be based on the state of the link such as availability capacity, or load of the link. Such routing schemes are in essence then based on some link state routing protocol that allows exchange of information about "cost" of links. An obvious question that comes to mind is how dynamic routing is similar or different from the link state protocol-based routing discussed earlier in general (see Section 3.4 in Chapter 3).

For link state protocol-based routing for IP networks, the cost of a link is the unit of information that is used in routing computation using shortest-path routing. Dynamic state-dependent routing also uses the "cost" of the link in its routing decision while this cost may be dependent on available capacity (as in DCR), availability (as in RTNR), or load on link. In the case of link state protocol-based routing for IP networks, the link state advertisement is accomplished through a flooding mechanism on an in-band hop-by-hop basis (see Section 3.4.1). For dynamic state-dependent routing (whether centralized or distributed), the link information is exchanged through an out-of-band network; typically, this is done either through dedicated data circuits or through an SS7 network. Finally, note that RTNR also uses node availability; thus, RTNR employs an extended link state concept that includes node-state information. Current IP routing protocols do not have the ability to communicate the state of a router, e.g., if it is highly congested due to, say CPU overload; this would need to be manifested through link cost by factoring in queueing delay in some manner.

10.9.2 Path Selection in Dynamic Routing in Telephone Networks and IP Routing

If we observe the state-dependent routing schemes, we can see that they all have a preference for alternate routes that indicate least or low load in some sense. Consider, for example, DCR. In this routing scheme, path availability is determined based on the probability of available bandwidth (subject to trunk reservation) of each link. In the case of RTNR, path availability is defined based on doing an "AND" operation of lightly loaded links.

In the case of link state routing for IP networks, the link cost information is used to compute shortest-path routing using Dijkstra's algorithm (see Section 2.3), where path cost is computed using an additive property. However, for dynamic state-dependent routing, the link state is used to compute one or more alternate paths while a pre-assigned direct path is usually assumed; also, *none* of the schemes uses Dijkstra's algorithm. More importantly, path determination is often done in a nonadditive manner, e.g., as in DCR. This then takes us back to the notion of nonadditive routing, such as maximum free capacity or shortest-widest path routing discussed earlier in Section 2.6 and Section 2.7. However, we must point out that in schemes such as DCR and RTNR, the path "cost" computation is different from "plain"

TABLE 10.2   Comparison between intradomain link state–based routing in IP networks and dynamic routing in telephone networks.

	Link State Protocol for IP Networks	Dynamic State-Dependent Routing for Telephone Networks
Link state advertisement	Through flooding	Via dedicated data circuits or SS7 network
Link "cost"	Link cost used for shortest path routing	Link state (e.g., available capacity) used for computing alternate routes
Route computation algorithm	Dijkstra's algorithm	Variety of algorithms
Additive/Nonadditive	Additive property of link cost used	Nonadditive property of link "cost" used

shortest-widest path computation, while accounting for trunk reservation is also a critical issue to consider for network stability. In Table 10.2, we summarize the differences between link state protocol-based routing in IP networks (used in intradomain) and for telephone networks (used in an IXC environment).

In essence, there are important lessons to learn when we start comparing and contrasting dynamic state-dependent routing and shortest-path routing in IP networks. For example, use of a link state protocol concept does not necessarily mean that the path selection is to be solely based on additive link cost property; furthermore, that flooding is not the only way to communicate link state information. We have discussed this earlier in Section 3.4.2.

10.9.3 Relation to Constraint-Based Routing

Later, we will discuss constraint-based routing for quality-of-service routing (see Chapter 17) and for MPLS networks (see Chapter 19). In fact, dynamic routing in telephone networks is a form of constraint-based routing.

In the case of dynamic routing in the telephone network, the critical constraint is that the path must have at least one unit of bandwidth available to be able to route a call. This is really the idea of a constraint. Furthermore, we can also argue that trunk reservation is a form of constraint since it is preferable to have it to maintain the stability of the network. As discussed in RTNR, sometimes it is not preferable to have certain switches on the routing path. Thus, in essence, alternate route selection in dynamic routing requires taking into consideration constraints such as availability of bandwidth on a path, any restriction due to trunk reservation, switch preference, and so on.

10.10 Summary

In this chapter, we have presented routing in the telephone network while taking an evolutionary view starting with hierarchical routing to various forms of dynamic routing.

There are several important aspects to understand from this chapter:

- It is possible to have routing functionalities in a network without introducing loops even in the *absence* of any information exchange as long as nodes are labeled differently; this is the case with hierarchical routing; in a sense, this is a remarkable achievement.

- Most dynamic call routing schemes for the telephone network require exchange of link state information—such information exchange does not use flooding; instead separate dedicated channels or a signaling network are used.

- The route computation is primarily based on bandwidth availability while the exact computation is different for different routing. There is, however, at least one dynamic routing scheme (DAR) that does not require any information exchange to do route computation.

- While routing can address certain congestion issues in a network, it cannot by itself take care of all types of congestion; rather, a good network requires proper control schemes in addition to routing schemes for efficient workings.

We have not discussed how traffic load is taken into account in the routing decision or how a network is traffic engineered; these aspects are covered in Chapter 11.

Further Lookup

The classical book [596] gives a broad overview of a telephone network under the old Bell System, including hierarchical routing. There are many rules for hierarchical routing but only the main ones are discussed here; the interested reader is referred to [27] for all the different rules.

The set of studies on routing goes back to the work of J. H. Weber in the 1960s [735], [736]; he is often credited with identifying the concept and need for trunk reservation. The notion of flexible dynamic routing and its benefit was first articulated by Nakagomi and Mori in the early 1970s [519]. Since then, with the advent of switching technology, dynamic routing research gained momentum, leading to different approaches to dynamic routing; for example, DNHR ([30], [36]), DCR ([680]), DAR ([251]), and RTNR ([34]). Another dynamic routing scheme, called state-dependent routing [539], [763], is discussed later in Section 11.7 in Chapter 11, since this scheme depends on understanding load and blocking in a telephone network, which are also covered in Chapter 11. Dynamic call routing based on learning automata [520] has also been proposed; for example, see [521], [657]. Note that dynamic alternate routing (DAR) can also be classified as a learning automata-based routing scheme. There have been numerous studies about dynamic routing, including early work, such as [7], [128], [197], [385].

To learn more about dynamic routing in the telephone network, the reader is directed to the books by Ash [27] and by Girard [253]. A 20-year overview of dynamic routing has been recently presented in [33].

For all about telephone switching systems and its many details, see the book by Thompson [697].

Exercises

10.1 Review questions:

  (a) Explain how hierarchical routing works.

  (b) Explain how real-time network routing (RTNR) works.

  (c) What is crankback?

  (d) What is trunk reservation?

10.2 What are the primary differences between telephone routing and Internet routing?

TABLE 10.3   Network Data for a 5-node network.

Link ID	Capacity	Currently Used	Trunk Reservation
1-2	34	30	1
1-3	116	93	5
1-4	25	20	1
1-5	41	27	2
2-3	61	61	3
2-4	76	43	4
2-5	33	30	3
3-4	97	81	5
3-5	141	118	7
4-5	110	102	5

10.3 Compare DCR and DAR, and determine their similarities and differences.

10.4 When is crankback helpful? When would it be not beneficial?

10.5 Dynamic routing schemes discussed in this chapter allow at most two links for a path. Discuss whether the schemes presented here can be extended to the multilink path case.

10.6 Consider a 5-node fully-connected network, with current network data as given in Table 10.3.

  (a) Assume DCR is used. For each demand pair, compute the probability of selection of all possible paths subject to trunk reservation.

  (b) Assume RTNR is used in which it was decided that if a link has five or less units of capacity are available, these will be marked as no-available links (i.e., as 0), and the rest are available links (i.e., as 1). For each demand pair, determine the valid paths chosen by RTNR.

  (c) Identify the demand pairs that would not allow any alternate routing through its direct link.

11

Traffic Engineering in the Voice Telephone Network

There are no traffic jams when you go the extra mile.

Anonymous

Reading Guideline

Understanding voice traffic engineering gives you an idea on how traffic load, call blocking, and routing are related. The first part of this chapter can be read independently of Chapter 10; this material is also helpful for quality-of-service routing (Chapter 17) and VoIP routing (Chapter 20). Detailed analytical modeling for dynamic call routing is presented in the second half of the chapter; this assumes some knowledge of the fixed-point equation and the queueing theory. Regardless, illustrations such as in Section 11.8.4 can be read independently and are helpful in understanding routing dynamics and performance.

In Chapter 10, we presented hierarchical and dynamic routing in the telephone network. In this chapter, we introduce the notion of traffic load in a telephone and discuss how it plays an integral part in routing from a traffic engineering perspective.

11.1 Why Traffic Engineering?

The goal of traffic engineering for the telephone network is to attain optimal performance of the network in an operational environment. We decouple traffic engineering from medium to long-term network planning when capacity can be added to the network. Note that if there is not enough capacity in the network to provide an acceptable performance guarantee, there is not much that traffic engineering mechanisms can do. Thus, with regard to traffic engineering, there is an implicit assumption that the network is engineered with enough capacity to provide an acceptable performance under normal operating conditions; to understand and learn more about capacity planning and design, you may refer to [564].

This brings up an important question: why do we need to consider any traffic engineering issues if the network has enough capacity to begin with? There are multiple reasons for this. For network planning, projected traffic is considered as input and the design methods used for estimating capacity usually use coarse-grain approximations; such approximations may not necessarily be a good way to do fine-grain evaluation of network performance on a near-term basis. In addition, we need to consider the fact that from the time network planning is done (that is, the capacity was last adjusted), traffic as seen in the network might be quite different than what it was projected to be. Furthermore, there is short-term traffic fluctuation/overload (either network wide or in a focused area) that cannot be avoided; in such cases, we need to determine whether the network is performing properly and/or if there are some short-term measures that can help the network perform at its best. This is where routing plays an important role. Thus, we need to understand the performance of a network due to various routing schemes. Furthermore, there are various control mechanisms that are needed, often to address overload situations.

In a telephone network, there are two high-level parameters that impact the performance of a network: traffic load and capacity. Since capacity is assumed to be given for the purpose of traffic engineering, the key parameter is traffic load. The main performance metric for the telephone network is call blocking. It is important to note that traffic load is not the only parameter; routing as well as various controls play roles in performance. The functional relation between these various components can be summarized as follows:

$$\text{Call blocking } = \mathcal{F}(\text{traffic load, capacity, routing, controls}). \tag{11.1.1}$$

To understand the above relation, it is important to define traffic load. Obviously, traffic load is dependent on traffic measurements as observed from a network. We will, however, not go into the details of traffic measurement issues. However, traffic load (especially projected traffic load) can be influenced by business decisions as well. Thus, we need to have a general notion of traffic load for the telephone network that can be usable from traffic measurements as well as for projected load. In the next section, we discuss traffic load and then tie it to call blocking.

11.2 Traffic Load and Blocking

We will first start with the simple case of a single network link (ignoring routing and control), i.e., to understand the following functional relation:

$$\text{Call blocking } = \mathcal{F}(\text{traffic load, capacity}). \tag{11.2.1}$$

Voice telephone networks operate on the following basic principle: there is a finite amount of capacity (bandwidth) and each arriving call must be allocated dedicated bandwidth for the duration of the call; if bandwidth for this call is not available, the call must be blocked. Thus, the user is required to retry when blocking occurs. Such systems are also referred to as *loss systems*.

The relationship between call arrival and blocking and capacity is an important traffic engineering issue in the voice telephone network. A key result in this regard is attributed to A. K. Erlang for his seminal work on how to compute blocking, almost a century ago. We need to explain a few things before we are ready to present his results.

For the purpose of this discussion, we will consider a network link in which calls are arriving on either end of the link destined for the other end of the link; this also reflects the fact that call bandwidth in the voice telephone network is bidirectional.

Call arrivals in the telephone network are random. However, to make it simple, we will assume temporarily that calls arrive in a deterministic fashion and that we are considering only a single voice circuit. First, suppose that calls arrive in a deterministic fashion at the start of an hour and the user talks for exactly an hour. Thus, one user occupies the circuit for an hour and no one else can use it. Now suppose the user talks for only 10 min and then hangs up. The circuit is free for others to use for the rest of the hour. In fact, if another user arrives at that instant and occupies the circuit for, say an additional 10 min, then a third user can start using the circuits 20 min into the hour. Thus, if we slice the length of the calls to fixed 10-min windows, the system can accommodate six calls; that is, this looks like the system can handle six arrivals per hour (each of 10 minutes' duration) as opposed to one arrival per user (using the entire hour for talking), while in either case just one circuit was considered! Simply put, this intuitively says that an increase in the call arrival rate does not necessarily mean that we need more circuits (or in general bandwidth) since the call duration is also a critical factor.

From the above illustration, it is clear that we need to consider call arrival as well as call duration to understand the notion of traffic. However, note that both call arrival as well as call duration are actually *random*, not deterministic as we have used in the simple illustration above. In Table 11.1, we have listed the call arrival time, duration, and end time of seven randomly arriving calls; in Figure 11.1, we have plotted this information in terms of number of busy circuits; note that the number of busy circuits also accordingly has a nonuniform behavior. In analysis, however, we use the *average* call arrival rate and the *average* duration. However, to account for any random event, we need to see what type of statistical distribution is appropriate. It has been found that interarrival time between calls is exponentially distributed, which is equivalent to saying that the call arrival follows the *Poisson process* (see Appendix B.10); thus, we sometimes loosely refer to call traffic as Poisson traffic. Furthermore, the call duration time is found to follow the exponential distribution. Given the average call arrival rate and the average duration of a call, a good way to capture the traffic demand volume for telephone traffic is to consider the product of these two terms. If λ is the

TABLE 11.1 Call information.

Call Number	Start Time (in sec)	Duration (in sec)	End Time (in sec)
1	2.3	145.3	147.6
2	6.7	128.8	135.5
3	45.2	18.4	63.6
4	62.2	512.5	574.7
5	73.2	96.2	169.4
6	94.1	1045.7	1139.8
7	196.6	15.2	211.8

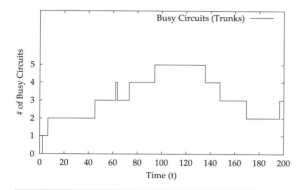

FIGURE 11.1 Number of busy circuits (trunks) as calls arrive and leave over time (shown up to 200 sec) for calls listed in Table 11.1.

average call arrival rate, and τ is the average duration of a call, the traffic demand, referred to as *offered load* or *offered traffic*, is given by

$$a = \lambda\tau. \tag{11.2.2}$$

This dimensionless quantity is given the unit name *Erlang* (or *Erl*), in honor of A. K. Erlang. This quantity is hard to visualize since it is a product of two terms. However, there is a nice physical interpretation: this quantity refers to the average number of ongoing (busy) calls offered to a network link if we were to assume the link to have *infinite* capacity. In practice, we do not have infinite capacity; thus, there is always a chance that (some) calls will be blocked. This also means that we need to consider another entity called *carried load* or *carried traffic* to refer to traffic that is carried (not blocked) due to finiteness of capacity. It is easy to see that in the case of infinite capacity, carried load is the same as the offered load.

A. K. Erlang's work is profound in that he determined how to compute call blocking probability when average offered load and capacity are given. If c is the capacity of a link in terms of number of voice circuits, then call blocking for offered load a is given by the following *Erlang-B loss formula*:

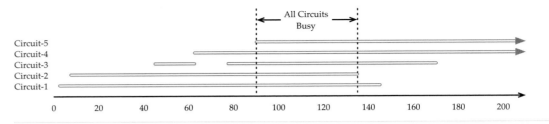

FIGURE 11.2   Call sequence and circuit occupancy for calls listed in Table 11.1.

$$B(a, c) = \frac{\frac{a^c}{c!}}{\sum_{k=0}^{c} \frac{a^k}{k!}}. \tag{11.2.3}$$

This then is the relationship we noted in a generic way in Eq. (11.2.1). Several different aspects can be learned from the above formula. We present two illustrations.

Example 11.1   *Erlang blocking illustration.*

Consider a T1-link (1.54 Mbps) where calls are offered. Since each digitized voice circuit requires 64 Kilobits per second (Kbps) $(= K)$, the bandwidth in terms of call units is $c = C/K$ where C is the raw data rate of the link; in the case of a T1-link, we have $c = 24 (\approx 1.54/.064)$ voice circuits. Suppose that a load of 20 *Erl* is offered to this link. Then, the call blocking probability obtained using Eq. (11.2.3) is 6.6%.

We note the following important aspects: (1) it is possible to offer a load higher than 24 *Erl* to a link of capacity 24 since this is a *loss* system, and there is still a chance that some calls will go through; in fact, the call blocking probability for an offered load of 24 *Erl* to a capacity of 24 units, is 14.6%; and (2) blocking also benefits from scaling due to nonlinearity of the blocking formula; for example, if we increase the capacity 10 times to 240 circuits and the load 10 times to 240 *Erl*, the blocking drops to 4.9% (from 14.6%).   ▲

It is also important to understand what happens to load and blocking when the average call duration is changed. The following example illustrates that.

Example 11.2   *Change in average holding time and impact on blocking.*

Recall that offered load is a combination of two parameters: average call arrival rate and average call duration. Suppose that the offered load is 15 *Erl*; this load offered to a link with 24 units of bandwidth results in 0.8% call blocking. If the average call duration is 3 min, the average call arrival rate is 5/min. Suppose in a network the average call arrival rate stays the same (at 5/min), and the average call duration increases from 3 to 6 min; we can then see that the newly determined offered load is 30 *Erl* $(= 5 \times 6)$; then, call blocking for 30 *Erl* offered to a link of 24 voice circuits is 27.1%! That is, without any increase in the call arrival rate, the offered load increases if the average call duration increases, thus impacting the call blocking rate.   ▲

So far, we have talked about traffic, but we have not talked about directional traffic while telephone trunks are most often bidirectional. In the following, we illustrate what it means to be directional traffic in terms of blocking.

Example 11.3   *Bidirectional trunks, directional traffic, and blocking.*

Consider the simple case of a network link where traffic arrives on either end destined for the other end; thus, whenever a circuit is found free, either side can grab it. The question then is: what is the blocking observed by each side?

Let us consider this problem with some numbers. Let the number of circuits on a link be 100 between two nodes 1 and 2 connected directly by a bidirectional trunk group. Let the offered load from node 1 to node 2 be 85 *Erl* and let the offered load from node 2 to node 1 be 5 *Erl*. You might be misled to think that node 1 to node 2 should face more blocking since it has more traffic than the other direction (in fact, 17 times more). Actually, both sides will face the *same* blocking. Due to Poisson arrival of traffic and since either side can grab a circuit as long as there is one available without any preferential treatment, the sum traffic remains Poisson (see Appendix B.10) and we can sum up the loads. Thus, this situation is equivalent to the case of a total of 90 *Erl* being offered to a link with 100 circuits; the call blocking from Erlang-B loss formula is then found to be 2.69%.

In fact, if 45 *Erl* is offered from one side and another 45 *Erl* from the other side, the blocking value will remain the same since the total offered load is still 90 *Erl*.   ▲

Remark 11.1.   *Blocking with multiple traffic classes.*

The scenario for two traffic classes follows the same principle described in the above example. That is, if there are two different traffic classes using the same link, both classes will observe the same blocking value regardless of each one's specific load *as long as* the per-call bandwidth for both classes is the same. If the per-call bandwidth is different for different traffic classes, the situation changes; in fact, the Erlang-B formula is not applicable any more. This situation will be discussed later in Section 17.6.   ◆

In the examples discussed so far, we have computed results using the Erlang-B loss formula (11.2.3); a closer look reveals that it includes both factorial ($c!$) and exponential (a^c) terms, which can cause numerical difficulty for large numbers. Below, we describe a simple way to compute this formula.

11.2.1 Computing Erlang-B Loss Formula

It may be noted that the Erlang-B loss formula given in Eq. (11.2.3) can also be expressed through the following recurrence relation:

$$\mathcal{B}(a, c) = \frac{a\,\mathcal{B}(a, c-1)}{c + a\,\mathcal{B}(a, c-1)} \tag{11.2.4}$$

with the initial condition $\mathcal{B}(a, 0) = 1$.

If we write $\mathcal{B}(a, c) = 1/d(a, c)$, then we can rewrite the above recurrence relation as follows:

$$d(a, c) = c\, d(a, c - 1)/a + 1.$$

Using this result, we can develop an iterative algorithm, as given in Algorithm 11.1, for computing call blocking when the offered load and the number of circuits are given. While this is the basic idea of the algorithm, in an actual implementation, some numerical round-off issues should also be addressed for extreme cases of load and capacity.

ALGORITHM 11.1   **Computing blocking using Erlang-B loss formula**

procedure erlangb (a,c)
if $(a \leq 0$ or $c < 0)$ return 'input not valid'
$d = 1$
for $i = 1, \ldots, c$ do
    $d = i * d/a + 1$
endfor
$b = 1/d$
return(b)
end *procedure*

11.3 Grade-of-Service and Trunk Occupancy

An important measure in a telephone network is the *grade-of-service* (GoS) provided by the network. GoS is a form of quality-of-service (QoS) where the quality perceived by users is in terms of average call blocking probability below a threshold. Usually, such acceptable blocking is 1% or less, while sometimes 0.1% or less is preferable.

For a network link, the question is how to determine the number of circuits needed if the offered load, a, and GoS, b, are given. By considering Erlang-B loss formula (11.2.3), this would mean that we need to determine the smallest integer c such that

$$\mathcal{B}(a, c) \leq b. \tag{11.3.1}$$

This is equivalent to saying that we want to find an inversion function $\mathcal{B}^{-1}(a, b)$ of the Erlang-B loss formula to obtain c. However, Erlang-B loss formula is not invertible analytically. Thus, we can take an algorithmic approach to determine the minimum integer c for which Eq. (11.3.1) is satisfied; such an approach is presented in Algorithm 11.2. Note that you can choose the starting c_{guess} to be 1, and the algorithm still works; however, the convergence would be long. Thus, a good starting point c_{guess} is preferable. For example, if a is less than 1500 *Erls*, and GoS is smaller than 0.02, then we can use starting c_{guess} to be $\lfloor a \rfloor$, i.e., the largest integer smaller than or equal to a. It may be noted that Algorithm 11.2 determines the number of circuits as an integer quantity; if the number of circuits needs to be a modular value such as a multiple of 24 circuits (for a T1-link), then proper rounding off to a multiple of 24 must be addressed.

Using Algorithm 11.2, we have plotted the minimum number of circuits required for two GoS values, 0.01 and 0.001, as the offered load changes from 1 to 100 in Figure 11.3(a). This figure should help you obtain some perspective on the difference in the number of circuits required for different GoS values.

ALGORITHM 11.2 **Determining the minimum number of circuits, given a and GoS**

procedure DetermineCircuits (a, GoS)
if ($a \leq 0$ or GoS < 0.0 or GoS > 1.0) return 'input not valid'
Pick a starting c_{guess}
$b_{\text{temp}} = \text{erlangb}(a, c_{\text{guess}})$
if ($b_{\text{temp}} \geq \text{GoS}$) then
    while ($b_{\text{temp}} \geq \text{GoS}$) do
        $c_{\text{guess}} = c_{\text{guess}} + 1$
        $b_{\text{temp}} = \text{erlangb}(a, c_{\text{guess}})$
    end do
else
    while ($b_{\text{temp}} \leq \text{GoS}$) do
        $c_{\text{guess}} = c_{\text{guess}} - 1$
        $b_{\text{temp}} = \text{erlangb}(a, c_{\text{guess}})$
    end do
    $c_{\text{guess}} = c_{\text{guess}} + 1$
endif
return(c_{guess})
end *procedure*

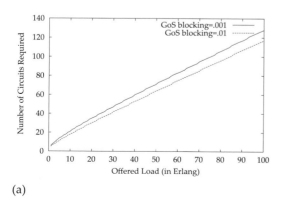

(a)

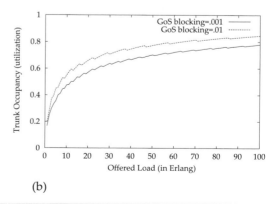

(b)

FIGURE 11.3 (a) Number of circuits needed, (b) trunk occupancy.

There is also another measure besides GoS that is of importance in telephone networking; it is called *trunk occupancy*. In essence, this represents trunk utilization for traffic that is carried by the trunkgroup. Carried traffic is, however, not the same as offered traffic. The offered load, a, and the carried load, $\hat{a}$, are related as follows:

$$\hat{a} = a(1 - \mathcal{B}(a, c)). \tag{11.3.2}$$

That is, an offered load that is not blocked is the carried load. We have stated earlier that a physical interpretation of offered load is the average number of ongoing calls if we were to have infinite capacity. This can be seen as follows: at infinite capacity, there is no blocking, i.e., $\mathcal{B}(a, \infty) = 0$; thus, we have $\hat{a} = a$. This also means that another way to understand carried load is the average number of ongoing calls if there is finite capacity, i.e., the average number of busy trunks. Now trunk occupancy is the ratio, the average number of busy trunks divided by the number of trunks in the trunkgroup, i.e.,

$$\eta = \frac{a(1 - \mathcal{B}(a, c))}{c}. \tag{11.3.3}$$

It is helpful to understand trunk occupancy as the traffic load changes while trying to maintain a particular GoS. In Figure 11.3(b), we have plotted trunk occupancy as the offered load changes from 1 to 100 for two different GoS values, 0.01 and 0.001, corresponding to the respective capacity determined, shown in Figure 11.3(a). We note that for small trunkgroup (i.e., a trunkgroup with a small number of trunks), the trunk occupancy is quite low; for example, for a T1 trunkgroup that supports 24 voice circuits, the trunk occupancy is about 68% for 0.01 GoS while for a trunk group with 96 circuits together, the trunk occupancy increases to 87% at the same GoS. This shows that there is a multiplexing gain with a larger trunkgroup.

11.4 Centi-Call Seconds and Determining Offered Load

Traditionally, it was not possible to measure offered load directly, only whether a circuit is busy can be checked. Thus, the question is how to determine the offered load from such measurements.

If we can measure the number of busy trunks over a period of time, we can then estimate the average number of busy trunks. We have already indicated above that the average number of busy trunks is none other than the carried load. Traditionally, voice circuits busy/idle status was checked every 100 sec resulting in the measure called *centi-call second* (CCS). Thus, in an hour, measurements are done 36 times. Suppose that we obtain a measure of 540 CCSes by checking 24 voice circuits over an hour. Then, the average number of busy trunks (circuits) is $540/36 = 15$. Since we noted earlier that the average number of busy trunks is the same as the carried load, we can equate this quantity, 15, to 15 *Erl* of carried load. Given a carried load of 15 *Erl* for 24 voice circuits, it is not obvious how we can compute offered load using Eq. (11.3.2); in this case, $\hat{a}$ and c are given while offered load, a, is the unknown. Since the Erlang-B loss formula cannot be analytically inverted, an algorithmic approach can be used, much like what we discussed in the last section. To do that, we note that Eq. (11.3.2) can be rewritten as follows:

$$a = \hat{a} + a\mathcal{B}(a, c). \tag{11.4.1}$$

Since $\hat{a}$ and c are known, it then takes the fixed-point equation form (see Appendix B.2):

$$a = \mathcal{F}(a).$$

Thus, a fixed-point–based algorithm, as described in Algorithm 11.3, can be used for computing offered load. An important point to note is that since the average number of busy trunks can never be larger than the number of trunks, the following must hold: $\hat{a} < c$. This means that for bad inputs that do not follow this requirement, it must be specially taken care of. For our illustration of 15 *Erls* of carried load on a trunkgroup with 24 trunks, we can find that the offered load is 15.138 *Erls*.

ALGORITHM 11.3   Computing offered load from carried load

---

procedure OfferedLoadFromCarriedLoad $(\hat{a}, c)$
if $(\hat{a} \geq c)$ return 'input not valid'
Initialize $a_{\text{new}} = \hat{a}$
Set $a_{\text{old}} \neq a_{\text{new}}$
while $(|a_{\text{new}} - a_{\text{old}}| > \varepsilon)$ do
    $a_{\text{old}} = a_{\text{new}}$
    $a_{\text{new}} = \hat{a} + a_{\text{old}}\mathcal{B}(a_{\text{old}}, c)$
endwhile
return(a_{new})
end *procedure*

There is also a direct connection between the units in *Erls* and CCS. Recall that the offered load in Erl reflects the average number of busy trunks if there were infinite capacity. Thus, if a trunk is always busy every time the trunk is checked for busy/idle status, we arrive at 36 CCSes in an hour. Thus, we can see that an *Erl* and CCS are related as follows:

$$1 \; Erl = 36 \; \text{CCS}. \qquad (11.4.2)$$

With modern switching equipment, the actual number of accepted calls as well as the time length of all calls can be measured; this is in fact used for billing later. A simple measure from the network perspective that is often collected is called the *Minutes of Usage* (MoU), which is computed over a period of an hour; it refers to the total amount of time all trunks in a group are occupied for the calls accepted by the system. Thus, it is easy to see that the average number of busy trunks during the duration, i.e., $\hat{a}$, can be obtained from MoU as follows:

$$\hat{a} = \frac{\text{MoU}}{60}. \qquad (11.4.3)$$

From our earlier discussion, we know that the average number of busy trunks is the carried load; thus, this gives us the carried load in *Erls*. With such a measurement system, during an hour's worth of measurement, the number of calls accepted by the system, sometimes also referred to as *call (or trunk) seizure*, denoted by s, is also available. Then, the average call duration, τ, in seconds can be computed as follows:

$$\tau = \frac{\text{MoU} \times 60}{s} \text{ sec.} \tag{11.4.4}$$

MoU is also sometimes obtained in different time windows such as on a daily, weekly, or monthly basis to obtain a sense of network usage and revenue generated. In such cases, the value would need to be divided by the appropriate time window to obtain the number of busy trunks/carried load for this time window.

In summary, we have shown here the relationship among carried load, offered load, CCS, and minutes of usage.

11.5   Economic CCS Method

So far, we have discussed a single-link network case to understand traffic offered load, capacity and blocking. Now, we will introduce the concept of an alternate path and its implication through a classical method called the Economic CCS (ECCS) method.

This method allows you to determine the number of circuits on the direct link when there is an alternate shared path. In this method, it is not important to exactly know how many circuits are in the shared alternate path; instead, the notion of trunk occupancy, defined in Section 11.3, is used.

Recall that trunk occupancy reflects both utilization and acceptable grade of service; certainly, the average value can be dependent on the total number of trunks due to non-linearity of the Erlang-B loss function, which was illustrated earlier in Figure 11.3(b). When the ECCS method uses the notion of an alternate shared path, it typically refers to a large group of circuits that serves as an alternate path for many different pairs. Thus, a trunk occupancy value of at least around 0.85 is more useful for such a calculation. In our discussion of the method below, we will use the generic notation, η, to refer to trunk occupancy.

We assume that the unit cost of circuits on the direct and the alternate paths is given as ξ_D and ξ_A, respectively. Also, we are given that a Erls of load is offered to the direct link. Our unknown here is the number of circuits needed on the direct link path, denoted by x_D.

With a Erls of offered load to the direct link with x_D circuits, the blocking will be $\mathcal{B}(a, x_D)$; consequently, $a\,\mathcal{B}(a, x_D)$ of load will overflow from the direct link to the alternate shared path. If a trunk occupancy of u is maintained for the alternate path, this would mean $a\,\mathcal{B}(a, x_D)/u$ units of circuits would be occupied on the alternate path on average. Now, given the unit cost of the direct and alternate path for circuits, the total cost can be written as

$$\mathcal{C}(x_D) = \xi_D x_D + \xi_A \frac{a\,\mathcal{B}(a, x_D)}{\eta}. \tag{11.5.1}$$

We now want to determine x_D for which this total cost is minimum. Since x_D takes an integer value, this means at optimality we want to determine the *smallest* x_D^* that satisfies

$$\mathcal{C}(x_D^*) < \mathcal{C}(x_D^* + 1). \tag{11.5.2}$$

ALGORITHM 11.4   Algorithm for ECCS method

---

procedure ECCS (a, η, R)
set $x_D = 0$
$H = \eta/R$
$d_l = a * \text{erlangb}(a, x_D);\ d_r = a * \text{erlangb}(a, x_D + 1)$
while $(d_l - d_r \geq H)$ do
    $x_D = x_D + 1$
    $d_l = d_r$
    $d_r = a * \text{erlangb}(a, x_D + 1)$
end while
return(x_D)
end *procedure*

Using Eq. (11.5.1), we can write this condition as

$$\xi_D x_D^* + \xi_A \frac{a\,\mathcal{B}(a, x_D^*)}{\eta} < \xi_D(x_D^* + 1) + \xi_A \frac{a\,\mathcal{B}(a, x_D^* + 1)}{\eta}. \tag{11.5.3}$$

On simplification, this implies that

$$a(\mathcal{B}(a, x_D^*) - \mathcal{B}(a, x_D^* + 1)) < \eta \frac{\xi_D}{\xi_A}. \tag{11.5.4}$$

Observe from the right-hand side that instead of requiring a separate unit cost for direct and alternate paths, the ratio, R, of alternate path to direct path cost (i.e., $R = \frac{\xi_A}{\xi_D}$), along with the trunk occupancy threshold, η, is sufficient to check for optimality. Thus, in order to satisfy Eq. (11.5.4), we can start with $x_D = 0$ as the starting point until the smallest value of x_D that satisfies Eq. (11.5.4) is reached; this is described in Algorithm 11.4. It may be noted that for most real networks, $R > 1$ and η is 0.75 or above.

Example 11.4 *Illustration of ECCS.*
    We consider three values of offered load a (= 20, 30, 40) and three values of η (= 0.65, 0.8, 0.85). We then vary the cost ratio, R, from 1 to 5 to determine x_D^*; this is shown in Figure 11.4.
    For a given offered load, we note that the change in trunk occupancy, η, is not necessarily a dominant factor; the dominant factor happens to be the change in the cost ratio, R.   ▲

    The ECCS method is commonly used in determining circuits needed on a direct link when a shared alternate path is available. We have presented the method using integer values of circuits. However, network link capacity comes in modular units such as T1s (= 24 circuits), and T3s (= $28 \times 24 = 672$ circuits). This is where the illustration above is helpful. For example, we can see that for an offered load of 20 *Erls*, a T1 link capacity can be installed to usually meet the requirement; similarly, for an offered load of 40 *Erls*, two T1 link capacity units must be installed. However, for an offered load of 30 *Erls*, a T1 capacity module would not be enough

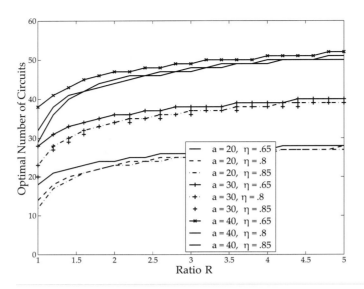

FIGURE 11.4   Determination of the optimal number of circuits using the ECCS method.

while two T1 capacity is more than enough; thus, in this case, installing two T1s would be necessary.

While the ECCS method is fairly simple to understand, it reveals only how the availability of alternate routes as well as a high usage link can impact a system in terms of capacity and blocking.

11.6 Network Controls for Traffic Engineering

Network controls play an important role in traffic engineering of the telephone network. While this book is about network routing, we do not want you to think that routing can take care of *all* problems. Traffic engineering of a telephone network requires more than just routing; there are many other measures or controls that need to be in place in the network for effective traffic engineering, especially to handle congestion. Weber, in his seminal work done in the early 1960s [735], [736], observed that (1) when a network with alternate routing capabilities becomes overloaded, the throughput can decrease if alternate routing is *not* reduced, and (2) trunk reservation is an effective control in an overload situation to maintain throughput. Note that trunk reservation has been described earlier in Section 10.2.4. Somewhat related to these observations, the bistability issue in a dynamic routing network was analyzed in the early 1980s [7], [385]. Bistability refers to the condition in a network when for a particular load offered to a capacitated network, the blocking observed takes two different values, i.e., it is not unique. That is, at a certain load, it is possible that the dynamic routing can show poor performance leading to bistability behavior unless appropriate trunk reservation control is introduced to alleviate this problem.

Any measures for network control to handle congestion have two distinct parts: detection and action. First, there must be ways to detect that the network is congested or affected, and second, appropriate action (controls) is then invoked. The telephone switching system has extensive pattern methodologies to detect congestion and what controls or a combination of controls to invoke; for example, see [283]. In the following, we will present a few highlights to provide an idea about detection and controls based on patterns.

11.6.1 Guidelines on Detection of Congestion

To detect congestion, it is important to recognize that there are two places where congestion in a telephone network can occur [27], [279], [596], [700]: trunkgroup (trunk congestion) and switch (switch congestion).

Detection of trunk congestion can be based on data collected on the following: (1) *attempts per circuit per hour* (ACH), and (2) *connections per circuit per hour* (CCH); here *connection* refers to the ability to find an outgoing trunk free, not whether it made it all the way to the destination, i.e., this is local information. Such data are typically collected every 5 min and a proper statistical measure on the reliability of the measured data is important. For example, if CCH is significantly low compared to ACH, this means that the call completion rate is low. If ACH is higher than normally expected, but CCH is at a normal level, this would mean that the load is heavy while call completion is still at an acceptable level. However, another pattern emerges if CCH is combined with trunk holding time; for example, if CCH on a trunk is higher than the normal threshold value, and if the trunk holding time is short, then call connections are not being effective. This can happen, for example, if the switch downstream is congested, or trunks outgoing from the downstream switch are congested; this means as a precautionary measure, some controls may need to be invoked.

To discuss switch congestion, it is helpful to first differentiate it based on the location of a switch in the network, i.e., whether a switch is a central office switch or a tandem/toll switch. If the switch is a central office switch where a call originates or terminates, this is the only switch in the network that needs to provide dial-tone service. For such switches, dial-tone delay is a critical indicator about switch congestion on the originating side; if users cannot get a dial-tone immediately, it signifies that the switch is congested. Dial-tone delay usually signifies full switch-level congestion. Besides dial-tone delay, it is also important to monitor the call completion rate to a code, either at the central office level (i.e., at NPA-NXX level in the North American numbering plan; see Section 13.1.1) or at the destination number level (i.e., at the NPA-NXX-XXXX level in the North American Numbering Plan); monitoring of such completion rates leads to recognizing hard-to-reach (HTR) codes.

11.6.2 Examples of Controls

There are several possible controls that can be invoked when a congestion is detected. In general, controls can be classified as (1) trunk reservation, (2) dynamic overload control, (3) code control and HTR capability, (4) reroute control, (5) directionalization of a link (trunkgroup), and (6) selective dynamic overload control. Since we have already discussed trunk reservation, we will not discuss it any further here. We do need to point out that in general controls are of two types: *restrictive* and *expansive*. Restrictive limits traffic from reaching congested location(s); expansive means allowing new paths to explore to avoid congestion. Of the above classifications, only reroute control falls under expansive while the rest are restrictive controls.

DYNAMIC OVERLOAD CONTROL

Dynamic overload control (DOC) is used for *sensing* switch congestion (Figure 11.5). When a switch is congested, it makes sense to reduce traffic being directed to this switch by other switches. So that the other switches can reduce traffic, the congested switch would need to inform other switches by sending a "machine congestion" message. Once this message is received by other switches, these switches reduce the amount of traffic directed to the congested switch; usually, alternate-routed traffic is controlled first. When the congestion in a switch goes below a congestion threshold, it sends another message to other switches to inform them about the new states so that the other switches can send alternate-routed traffic again at a normal level.

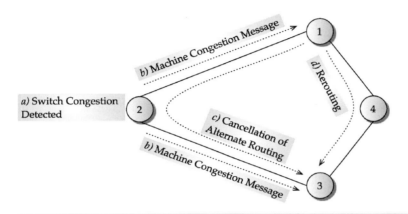

FIGURE 11.5   Illustration of dynamic overload control, from detection to action.

CODE CONTROL AND HARD-TO-REACH CAPABILITY

This is another control related to congestion that takes a different perspective. In this case, instead of looking at the view of congestion from the perspective of a switch, the view is taken at the destination code level. Destination codes can be at the central-office level and/or at the actual destination number level; for example, in the North American numbering plan, this means either at the NPA-NXX or at the NPA-NXX-XXXX level, respectively (see Section 13.1.1 for a discussion on numbering plans). For a variety of reasons, a destination code may suddenly become popular for a short duration of time. The need for code control can be better understood from the following real-world example from the mid-1980s [700]. The concert tickets for well-known rock singer Bruce Springsteen went on sale on July 19, 1985, and interested buyers could call a specific telephone number to order tickets. The central office switch that served the ticket office received 10 million attempts over a 24-hour period while normal call volume for this switch was around 500,000. Certainly, there were not enough tickets or ticket agents to handle even a small fraction of these calls. On top of that, such an overload can overwhelm the destination switch, thus affecting accessibility to *other* telephone numbers served by that switch. Furthermore, incoming trunkgroups to this switch also became clogged; then, through a domino effect, this event can affect other switches downstream, thus partially affecting/paralyzing the overall network.

It is clear from this example that any code control information to such a destination code needs to be pushed to where calls have originated; this way, a call to such a destination code does not even enter the network since the hope of completion of such a call is very small, that is, pushing such information on code control to originating switches throughout the network helps in this regard. The receiving switch must be able to track call completion rates to different codes in order to provide the HTR capability. Once the switch realizes that the completion rate to a particular code has dropped below a threshold value, it can generate a HTR signaling message and inform other switches. Code control and HTR capability are illustrated in Figure 11.6.

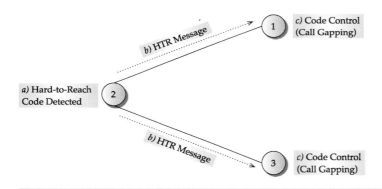

FIGURE 11.6   Code control and hard-to-reach (HTR) capability, from detection to action.

When a switch in a network receives a HTR code from another switch, it employs code control. There are two forms of code control: *call gapping* and *code blocking*. Call gapping means that calls made to a particular destination are *gapped* every Δt sec; that is, only one call is allowed to pass through every Δt sec. Code blocking allows calls to go through on a percentage basis. While conceptually they sound similar, the difference is that with code blocking, two or more back-to-back calls may still be allowed to go through since probability is used; in a congested situation, this may not be desirable. However, with call gapping, allowed calls are paced evenly.

REROUTE CONTROL

This is an expansive control that is used for overflowing traffic to a new route/trunkgroup. It may be noted that traffic in the network is not uniformly loaded; thus, it is not unusual to find underloaded trunkgroups in the network. Through the reroute control, traffic is routed through such trunkgroups as long as the hierarchical routing rule allows it. Note that reroute controls can be activated in a code-specific basis as well. In dynamic routing networks, call-routing logic handles all reroute controls; thus, the traditional reroute controls are not used separately in dynamic routing networks.

DIRECTIONALIZATION OF A LINK

Recall that links or trunkgroups in a circuit-switched telephone network are bidirectional. As we have illustrated earlier in Example 11.3, the call blocking depends on the total offered load, regardless of the amount from each direction. Now imagine a focused overload situation where traffic from one end is excessive compared to the other. To take the previous example a bit further, consider that the offered load from node 1 to node 2 increases excessively from 85 *Erls* to 195 *Erls* while the other direction remains at 5 *Erls*. Now, we have a total of 200 *Erl* offered to a link with 100 circuits; the Erlang-B loss formula tells us that the blocking increases to 50.48%.

We can say that it is unfair that the other side with only 5 *Erl* of offered traffic has to face over 50% blocking. Furthermore, we can argue that it is important to keep blocking not high on the low-traffic side. Let us consider a real-world situation. Suppose the low-traffic side is in a natural disaster-struck area. In this case, the high-traffic side is where almost everyone wants to call their friends and families who live or are visiting the disaster-struck area; many may not get through partly due to high load and also partly due to any call gapping being invoked at originating nodes based on destination codes being affected in the disaster-struck area. Thus, it is desirable to let some calls out from the disaster-struck area so that affected people can let their friends and families know on the other side that they are doing fine.

Thus, the question is: how do we let the lower-traffic end have lower blocking? First, it is important to understand that call gapping can only help in terms of reducing traffic from entering in one direction, but it cannot solve the trunk congestion problem. Thus, we need a measure to reduce trunk congestion. This measure, called *directional* trunk reservation, is invoked from the high-traffic side to the low-traffic side, but no trunk reservation from the low-traffic side to the high-traffic side; the process is illustrated in Figure 11.7. Another way to say this is that from the high-traffic side a newly attempted call is not allowed to enter the link if the link has r units of free trunks left, where r is a parameter value that can be appropriately set; calls are allowed from the other side as long as a free circuit is available. This preferential treatment is helpful in lowering blocking for the low-traffic end; this will certainly increase blocking for the high-traffic end since it has access to less capacity.

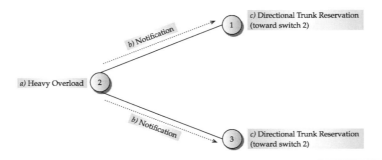

FIGURE 11.7   Directionalization of a trunkgroup, from detection to action.

SELECTIVE DYNAMIC OVERLOAD CONTROL

Dynamic overload control, described earlier in this section, acts at a generic level; that is, it cannot distinguish between codes with high-completion probability and codes with low-completion probability. Selective DOC (SDOC) is a merger of DOC and code blocking; thus, traffic cancellation can be activated for HTR codes.

11.6.3 Communication of Congestion Control Information

In the previous section, we discussed several control schemes; some are for switch congestion, while others are for trunk congestion. Clearly, information that a congestion is detected must be communicated to other switching nodes so that they can take proper action. However, note that telephone circuits are primarily for voice communication; such information communication, which is really data communication, is not possible through trunk circuits as is.

Such communication is accomplished using a *separate* data network that connects the switching node. This specialized network is called the signaling system 7 (SS7) network; we will discuss SS7 networking in Chapter 12 where different message types will be described; in particular, the SS7 message type, REL, that is used to release a call, is used with certain embedded code values to indicate congestion. In the absence of an SS7 network, dedicated circuits can be set up between switches to communicate such information; however, this is less common these days with extensive deployment of SS7 networks.

Finally, when we combine routing update information, especially in dynamic routing networks, and network controls, it is clear that an SS7 network carries two types of information for the purpose of traffic engineering: routing update information and network control information. As discussed earlier in Section 10.9.1, for routing, typically link state information is required. For network control, however, it needs to contain not only link-level information (trunk congestion), but also switch load–level information, thus requiring the use of the *extended* link state concept. Since switch congestion information is communicated, a switch can make a temporarily *taboo* list of congested switches in a routing decision; at least this is possible in dynamic routing networks.

11.6.4 Congestion Manifestation

While most congestion in the network is legitimate congestion, congestion can be manifested in a telephone network, partly due to the involvement of the SS7 network, the signaling network that supports the telephone network (SS7 networking is described in Chapter 12). First, the signaling network is also a network; it has its own issues about its links being congested when too many routing and control messages are generated. Thus, such an SS7 network needs to be properly engineered as well so that it can induce congestion on the telephone network.

Despite traffic engineering of the SS7 network, manifestations of congestion in the telephone network are still possible through failure propagation. In the following, we will discuss a real-world incident that occurred in AT&T's long-distance telephone network in 1990 [298]; this serves as an example of lessons we can learn about control principles from an operational environment.

The problem started with a switch (say, "Switch-A") in the dynamic routing network taking itself out of service due to a trunk interface issue. Thus, it informed the other switches through network management messages using the SS7 network. After Switch-A recovered, it started processing calls again; this resulted in generating call setup messages to another switch, say Switch-B. Unfortunately, there was a software flaw that occurred only if two call setup messages were generated within 0.01 sec to the same switch, such as Switch-B; the processor in Switch-B tried to execute an illegal instruction that resulted in telling the switch that the processor was faulty. Thus, Switch-B took itself out of service and informed the other switches through the SS7 network. When Switch-B recovered, it did what Switch-A did earlier, i.e., try to do call setup messages within 0.01 sec to another switch. Thus, this started a chain reaction of out-of-service for all switches.

Also, when a switch was isolated from "failure," its home signal transfer point (STP) would return a network management message for every message to this switch. Thus, STPs in the SS7 network become heavily loaded with network management *and* routing retry/crankback messages.

This entire cascade of events resulted in 98% of the switches being out of service in the first 30 min; the network continued to have over 50% call blocking for 9 hours! Note that during this entire period the trunk capacity was there to carry the call volume; instead the switches going out of service along with the SS7 network being congested caused this high blocking.

It is important to note that telephone switching systems operate on two basic principles [1], [283]: (1) minimize human intervention, and (2) people do know best. The first one says that the system should be automated, including controls for overload and routing, while the second one says that the system should be built with the provision that human oversight and altering normal actions are possible. This second principle means that the capability to deactivate certain network management/control logics is provided; furthermore, such deactivation should be possible to do in a layered structure, i.e., not all controls needed to be turned off at the same time.

After learning from the incident described above, an additional principle was added [298]: every network element must have adequate overload control so that it can still function if network management controls are deactivated.

11.7 State-Dependent Call Routing

Recall that we have discussed several dynamic routing schemes in Chapter 10; there is yet another dynamic routing scheme that uses the knowledge from the Erlang-B loss formula (11.2.3). This is commonly known as the state-dependent routing scheme [539]. To understand this scheme, we start by considering a generic link i-j with call capacity c_{ij} where the offered load is A_{ij}. Suppose at a particular instant in time, we know the actual number of busy trunks, which is denoted by n_{ij}. Now, consider a new definition of a link cost given by the ratio

$$d_{ij} = \frac{\mathcal{B}(A_{ij}, c_{ij})}{\mathcal{B}(A_{ij}, n_{ij})}. \tag{11.7.1}$$

First, note that $n_{ij} \leq c_{ij}$. If $n_{ij} = c_{ij}$, then clearly $d_{ij} = 1$. Consider when $n_{ij} < c_{ij}$; in this case, Erlang-B loss formula tells us that $\mathcal{B}(A_{ij}, c_{ij}) < \mathcal{B}(A_{ij}, n_{ij})$ since fewer circuits means more blocking—thus, $d_{ij} < 1$. Also, note that when $n_{ij} = 0$, i.e., no circuits are busy, the quantity d_{ij} is smallest for a given A_{ij} and c_{ij}. Thus, the notion of cost given by this ratio indicates that if all circuits are busy, then the cost is 1, while if no circuits are currently being used, the cost is closest to 0.

Now consider routing a call between switches i and j, and a two-link path through switch k consisting of links i-k and k-j. The capacity on each link i-k and k-j, in terms of number of voice circuits, is denoted by c_{ik} and c_{kj}, respectively. We assume that we know the offered load to each link is A_{ik} and A_{kj}, respectively. Note that such an offered load can be computed from the carried load as described earlier in Algorithm 11.3. Furthermore, suppose at a particular instant of time, the actual number of busy circuits is obtained from measurements and is denoted by n_{ik} and n_{kj} for links i-k and k-j, respectively. Based on the above discussion, we can then compute the link cost of each leg i-k and k-j as

$$d_{ik} = \frac{\mathcal{B}(A_{ik}, c_{ik})}{\mathcal{B}(A_{ik}, n_{ik})}, \quad d_{kj} = \frac{\mathcal{B}(A_{kj}, c_{kj})}{\mathcal{B}(A_{kj}, n_{kj})}. \tag{11.7.2}$$

In state-dependent routing, the cost of path i-k-j via node k for pair i:j is given by $d_{ij} + d_{kj}$. Since this quantity is dependent on the number of calls on each link, the actual call routing decision would be based on using the count on the number of calls at the instant the new call arrived followed by computing the link cost as described above. Thus, in the state-dependent routing (with at most two-link routing), a call is routed on the path with the least cost among all the alternate paths and the direct link path.

There are two important points to note: (1) unlike the routing schemes described in Chapter 10, this routing is based on the *additive* cost property; and (2) the link cost determination requires computing the Erlang-B loss formula; this is an expensive function to do on a per-call basis. In practice, to avoid Erlang-B loss computation on a per call basis, the link cost d_{ij} can be computed on a periodic basis, say every 5 min. Furthermore, the exponential moving average method (see Appendix B.6) can be employed for smoothing, instead of using the average over the previous 5 min. Certainly, the cost value would not be as accurate at the time of making a decision on a call; such smoothing, however, avoids route fluctuation, somewhat similar to ARPAnet routing metric discussed earlier in Section 3.6.1. Finally, based on the cost estimate for each link, multiple paths may be computed and stored, and crankback may be employed to try another alternate path if the first alternate tried is no longer available.

A final important point to note is that trunk reservation need not be explicitly engaged in this routing scheme. If trunkgroup i-j has only a few circuits left, $n_{ij} \approx c_{ij}$; from Eq. (11.7.1), this means that link cost $d_{ij} \approx 1$; due to additive cost of a path, such links will be less preferred in path selection over links that have link cost closer to zero (meaning more capacity available on such links).

11.8 Analysis of Dynamic Routing

In Chapter 10, we discussed dynamic routing. In this section, we will discuss how to analyze maximum two-link dynamic routing in general.

11.8.1 Three-Node Network

We will start with a three-node fully-connected network where offered load between two nodes i and j is given by a_{ij}. Since traffic is bidirectional, we need to consider three offered loads, a_{12}, a_{13}, and a_{23}. Assume that the capacity of a link i-j is given by c_{ij} circuits. It is important to distinguish the difference in traffic between a pair of nodes and the capacity on a link. For example, consider nodes 1 and 2; the traffic between nodes 1 and 2 can use the direct link 1-2 or overflow to the alternate path 1-3-2 if there is no capacity on the direct link. At the same time, the direct link 1-2 can carry alternate routed traffic (however small) between node pairs 1 and 3, and also between node pairs 2 and 3, since link 1-2 is on the alternate path for both these pairs.

Each link has a link blocking b_{12}, b_{13}, and b_{23}; this is, however, not pairwise blocking. We will soon define what pairwise blocking would be. But first we need to go through a few steps. The first one we need to determine is path blocking. Consider the direct link path 1-2; obviously, in this case, path blocking is the same as link blocking, i.e., b_{12}. Now consider the alternate two-link path 1-3-2. If we assume that link blocking is independent of one another, then the path blocking (see Appendix B.8) for path, $\mathcal{P}_{132}$, is given by

$$\mathcal{P}_{132} = 1 - (1 - b_{13})(1 - b_{23}). \tag{11.8.1}$$

Intuitively, this can be seen as not blocked on link 1-3 and not blocked on link 2-3 would not be blocked on path 1-3-2, which is $(1 - b_{13})(1 - b_{23})$.

Now consider the pairwise offered load a_{12} between node 1 and node 2. On the direct link path 1-2, the amount of blocked traffic would be $a_{12}b_{12}$; this blocked traffic will then overflow and try the two-link path 1-3-2. As a result, the blocked traffic after attempting either path would be $a_{12}b_{12}\mathcal{P}_{132}$. Thus, the pairwise blocking for node pair 1:2 is defined as

$$z_{12} = b_{12}\mathcal{P}_{132} = b_{12}\left(1 - (1 - b_{13})(1 - b_{23})\right). \tag{11.8.2}$$

Similarly, for other pairs, we can write path blocking and pairwise blocking as:

$$\mathcal{P}_{123} = 1 - (1 - b_{12})(1 - b_{23}). \tag{11.8.3}$$

$$\mathcal{P}_{213} = 1 - (1 - b_{12})(1 - b_{13}). \tag{11.8.4}$$

$$z_{13} = b_{13}\mathcal{P}_{123} = b_{13}\left(1 - (1 - b_{12})(1 - b_{23})\right). \tag{11.8.5}$$

$$z_{23} = b_{23}\mathcal{P}_{213} = b_{23}\left(1 - (1 - b_{12})(1 - b_{13})\right). \tag{11.8.6}$$

To differentiate from pairwise offered load a_{12}, we denote the total offered load on link 1-2 by A_{12}. It is important to note that $A_{12} \neq a_{12}$, which will be clear soon. There are two ways to determine carried load on link 1-2 to be denoted by $\hat{A}_{12}$. One way is to consider offered load, A_{12}, and link blocking, b_{12}; then, we have

$$\hat{A}_{12} = A_{12}(1 - b_{12}). \tag{11.8.7}$$

Another way to consider this is to see what is being carried due to traffic from direct and alternate routing. Certainly, due to blocking being b_{12}, link 1-2 will carry the load $a_{12}(1 - b_{12})$

for offered load a_{12} between nodes 1 and 2 due to direct link routing. In addition, link 1-2 will also *attempt* to carry any traffic that is alternate routed from node pair 1 and 3 and node pair 2 and 3. Offered traffic alternate routed for pairs 1:3 and 2:3 would be $a_{13}b_{13}$ and $a_{23}b_{23}$, respectively; however, these paths (1-2-3 and 2-1-3) have path blocking $\mathcal{P}_{123}$ and $\mathcal{P}_{213}$, respectively. This means that any alternate routed traffic would be carried that is not blocked. Thus, we can write

$$\hat{A}_{12} = \underbrace{a_{12}(1-b_{12})}_{\text{carried direct traffic}} + \underbrace{a_{13}b_{13}(1-\mathcal{P}_{123})}_{\text{carried load for overflow traffic for 1:3}} + \underbrace{a_{23}b_{23}(1-\mathcal{P}_{213})}_{\text{carried load for overflow traffic for 2:3}}$$

(11.8.8)

Since Eq. (11.8.7) and Eq. (11.8.8) are the same, we can write

$$A_{12}(1-b_{12}) = a_{12}(1-b_{12}) + a_{13}b_{13}(1-\mathcal{P}_{123}) + a_{23}b_{23}(1-\mathcal{P}_{213}). \qquad (11.8.9)$$

Now using path probabilities given by Eq. (11.8.3) and Eq. (11.8.4), we can rewrite the above as

$$A_{12}(1-b_{12}) = a_{12}(1-b_{12}) + a_{13}b_{13}(1-b_{12})(1-b_{23}) + a_{23}b_{23}(1-b_{12})(1-b_{13}). \quad (11.8.10)$$

Now, dividing each side by the common term $(1-b_{12})$, we get

$$A_{12} = a_{12} + a_{13}b_{13}(1-b_{23}) + a_{23}b_{23}(1-b_{13}). \qquad (11.8.11)$$

This shows then that the link offered load, A_{12}, on link 1-2 is more than the pairwise load offered to node pair 1:2; this is correct since there is some alternate routed traffic offered that needs to be accounted for. Observing the pattern, we can write the link offered loads for links 1-3 and 2-3 as follows:

$$A_{13} = a_{13} + a_{12}b_{12}(1-b_{23}) + a_{23}b_{23}(1-b_{12}), \qquad (11.8.12)$$

$$A_{23} = a_{23} + a_{12}b_{12}(1-b_{13}) + a_{13}b_{13}(1-b_{23}). \qquad (11.8.13)$$

Now, from the Erlang-B loss formula, we know that for an offered load on a link to a given capacity is the link blocking, i.e.,

$$\begin{aligned} b_{12} &= \mathcal{B}(A_{12}, c_{12}) \\ b_{13} &= \mathcal{B}(A_{13}, c_{13}) \\ b_{23} &= \mathcal{B}(A_{23}, c_{23}). \end{aligned} \qquad (11.8.14)$$

Since we have derived the quantity for link-offered load, we can then write

$$\begin{aligned} b_{12} &= \mathcal{B}(a_{12} + a_{13}b_{13}(1-b_{23}) + a_{23}b_{23}(1-b_{13}), c_{12}) \\ b_{13} &= \mathcal{B}(a_{13} + a_{12}b_{12}(1-b_{23}) + a_{23}b_{23}(1-b_{12}), c_{13}) \\ b_{23} &= \mathcal{B}(a_{23} + a_{12}b_{12}(1-b_{13}) + a_{13}b_{13}(1-b_{23}), c_{23}). \end{aligned} \qquad (11.8.15)$$

Thus, we can see that for a given offered load and capacity, the link blocking is iteratively related through a set of three nonlinear equations in three unknowns b_{12}, b_{13}, b_{23}; this is a fixed-point equation system; see Appendix B.2 for a short tour of the fixed-point equation problem and how to solve it. In the above system, there are three unknowns with three equations. Due to the need to use the Erlang loss formula, the system is often referred to as the *Erlang fixed-point equation*.

Example 11.5 *Symmetric three-node network.*

To make it easier to see, consider that the three-node network is symmetric in traffic as well as capacity, i.e., $a_{12} = a_{13} = a_{23}(= a)$, and $c_{12} = c_{13} = c_{23}(= c)$. It is not hard to recognize that link blocking will be symmetric as well, i.e., $b_{12} = b_{13} = b_{23}(= b)$. Thus, link offered load as given in Eq. (11.8.11), Eq. (11.8.12), and Eq. (11.8.13) reduces to just the following single expression:

$$A = a + 2ab(1 - b). \tag{11.8.16}$$

Similarly, the set of equations given in Eq. (11.8.15) will reduce to just the following one:

$$b = \mathcal{B}\left(a + 2ab(1 - b), c\right). \tag{11.8.17}$$

Now, this is an Erlang fixed-point equation with just one unknown, b. This can be solved using the fixed-point iteration described in Algorithm B.1 in Appendix B.2. The basic procedure is to begin with a starting b, for which A is computed; this, in turn, is used in Eq. (11.8.17) to obtain a new b; the process is continued until the difference between two successive b's is within a specified tolerance.

Note that, for the three-node symmetric case, pairwise blocking, (11.8.2), reduces to

$$z = b\left(1 - (1 - b)^2\right). \tag{11.8.18}$$

Thus, once we solve the fixed-point equation (11.8.17), we can compute the pairwise blocking using Eq. (11.8.18).   ▲

11.8.2 *N*-Node Symmetric Network

In this section, we will generalize the three-node symmetric network case to N-node symmetric networks. We assume that the network is fully connected and that alternate paths are limited to at most two links.

Let a denote the pairwise offered load, while A denotes the link offered load. As we know from the three-node symmetric example, they are different. We will show below the connection between them. To do that, consider the number of circuits on a link to be c.

We define b as the link-blocking probability. For brevity, we will use $q = 1 - b$ as the probability of not being blocked on a link. Then, the probability of not being blocking on a two-link path is q^2. This, in turn, means that the probability of being blocked on a two-link path is $1 - q^2$.

Assume that we consider M independent two-link paths for a node pair. Then the probability of being blocked on M independent paths is $(1 - q^2)^M$. Since any calls blocked on the direct link path are alternate routed to the M two-link paths, the pairwise blocking, z, for a node pair is

$$z = b(1 - q^2)^M. \tag{11.8.19}$$

The total carried load on a link for two-link paths is $2a(b - z)$ while for direct traffic it is $a(1 - b)$. Thus, we can equate

$$A(1 - b) = a(1 - b) + 2a(b - z).$$

This implies that

$$A = a + \frac{2a(b - z)}{1 - b} = a + \frac{2ab[1 - (1 - q^2)^M]}{q}. \tag{11.8.20}$$

Note that when $M = 1$ this is then equivalent to a three-node network, and Eq. (11.8.20) reduces to Eq. (11.8.16). From the Erlang-B loss formula, we know that $b = \mathcal{B}(A, c)$. Replacing A, we have the following fixed-point equation

$$b = \mathcal{B}\left(a + \frac{2ab[1 - (1 - q^2)^M]}{q}, c\right). \tag{11.8.21}$$

Pairwise blocking z is derivable once we know b due to Eq. (11.8.19). Note that Eq. (11.8.21) reduces to Eq. (11.8.17) for the three-node symmetric network case since the number of alternate routes $M = 1$. This approach is summarized in Algorithm 11.5.

ALGORITHM 11.5   **Computing Network Blocking for Symmetric Networks**

---

procedure ComputeNetworkBlocking (a,c,M)
Initialize b_{new}
Set $b_{old} \neq b_{new}$
while $(|b_{new} - b_{old}| > \varepsilon)$ do
    $b_{old} = b_{new}$
    $q = 1 - b_{old}$
    $z = b_{old}(1 - q^2)^M$
    $A = a + 2a(b_{old} - z)/q$
    $b_{new} = \mathcal{B}(A, c)$
endwhile
return(z)
end *procedure*

11.8.3 *N*-Node Symmetric Network with Trunk Reservation

In Section 10.2.4, we discussed the role of trunk reservation in dynamic routing. Here, we will consider how to incorporate trunk reservation in analysis. To do this analysis, we need to consider the *general* form of blocking on a link for two different states (compared to "pure" link blocking through the Erlang-B loss formula used in Section 11.8.2). The derivation is shown in Appendix B.12.3.

When we have a link with c circuits, and r ($r < c$) as the circuits reserved for trunk reservation, then how the call arrival is handled can be divided into two categories: (1) if the call is a direct link call, it can be attempted as long as there is a circuit left out of the total capacity c, and (2) if the call is an alternate call, then it is allowed to be routed if fewer than $c - r$ circuits are busy; in other words, if r or fewer circuits are free, an alternate routed call is not allowed on this link.

To be consistent with the previous section, let a be the pairwise offered load. We use $\underline{A}$ to denote the load offered to the link *subject to* trunk reservation r. The probability, p_j, that j circuits are busy (subject to trunk reservation parameter r) is given by

$$
p_j = \begin{cases} \dfrac{\underline{A}^j}{j!} p_0, & \text{for } j = 0, \dots, c - r - 1 \\[2ex] \dfrac{\underline{A}^j}{j!} \left(\dfrac{a}{\underline{A}} \right)^{j-(c-r)} p_0, & \text{for } j = c - r, \dots, c, \end{cases}
\tag{11.8.22a}
$$

where

$$
p_0 = \left[\sum_{k=0}^{c-r-1} \frac{\underline{A}^k}{k!} + \underline{A}^{c-r} \sum_{k=c-r}^{c} \frac{a^{k-(c-r)}}{k!} \right]^{-1}.
\tag{11.8.22b}
$$

If we denoted the link blocking by b, then $b = p_c$, and we can write

$$
b = \frac{\underline{A}^{c-r}}{c!} a^r p_0.
\tag{11.8.23}
$$

If $\hat{q}$ denotes the probability that no more than $(c - r - 1)$ circuits are busy, then $\hat{q}$ can be obtained from Eq. (11.8.22a) by summing over states 0 to $c - r - 1$; i.e., it is given by

$$
\hat{q} = \sum_{j=0}^{c-r-1} \frac{\underline{A}^j}{j!} p_0.
\tag{11.8.24}
$$

In this case, the total carried load on a link is $a(1 - b) + (\underline{A} - a)\hat{q}$. Similar to the case without trunk reservation, this carried load can also be expressed as $a(1 - b) + 2a(b - z)$, where z is the pairwise blocking. Equating them and simplifying, we get

$$
\underline{A} = a \left[1 + \frac{2(b - z)}{\hat{q}} \right].
\tag{11.8.25}
$$

This time we have two unknowns, b and $\hat{q}$; thus, we have the set of Erlang fixed-point equations over two variables connected by the relations Eq. (11.8.23) and Eq. (11.8.24), where $\underline{A}$ is given by Eq. (11.8.25). Solving the Erlang fixed-point equation, we can obtain pairwise blocking as described in Algorithm 11.6.

ALGORITHM 11.6   Computing Pairwise Blocking for Symmetric Networks with Trunk Reservation.

---

procedure ComputeNetworkBlockingW-TR (a,c,M,r)

Initialize $b_{new}, \hat{q}_{new}$

Set $b_{old} \neq b_{new}$; $\hat{q}_{old} \neq \hat{q}_{new}$; $d = 1.0$

while $(d > \varepsilon)$ do

    $b_{old} = b_{new}$

    $\hat{q}_{old} = \hat{q}_{new}$

    $z = b_{old}(1 - \hat{q}_{old}^2)^M$

    $A = a + 2a(b_{old} - z)/\hat{q}_{old}$

    $b_{new} = \frac{A^{c-r}}{c!}a^r p_0$

    $\hat{q}_{new} = \sum_{j=0}^{c-r-1} \frac{A^j}{j!}p_0$

    $d = \max\{|b_{old} - b_{new}|, |\hat{q}_{old} - \hat{q}_{new}|\}$

endwhile

return(z)

end *procedure*

Finally, it is worth comparing this analysis to the case when there is no trunk reservation. Note that when there is no trunk reservation, $r = 0$, then $\hat{q}$ becomes $q = 1 - b$ and Eq. (11.8.23) becomes Eq. (11.8.21).

11.8.4 Illustration Without and with Trunk Reservation

In the previous two sections, we explained how to determine call blocking in a symmetric network without and with trunk reservation [385]. For this illustration, we consider a symmetric network with 100 units of capacity in each link. The offered load is varied from 70 *Erls* to 100 *Erls* for each pair of nodes.

First, consider the case of no trunk reservation. We have computed carried load for different load values using Algorithm 11.5; the results for a different number of alternate routes considered are shown in Figure 11.8. Note that $M = 0$ means no alternate routing allowed, which can be obtained directly from the Erlang-loss formula; $M = 1$ means the three-node network discussed earlier. We can easily see that as the load increases, there are multiple carried load values for the same offered load showing bistable behavior; the difference is more pronounced when the number of alternate paths increases. This indicates that in an overloaded situation, it is not a good idea to allow too many alternate routes since they compete with other traffic pair paths, thus almost nullifying any benefit of alternate routing.

We next consider trunk reservation, turned on with different values. The results are shown separately for different numbers of alternate paths considered, $M = 1, 2, 4, 8$. These results are obtained using Algorithm 11.6 and are shown in Figure 11.9.

We can see that with trunk reservation, the drop in carried load for a high offered load is not as severe as without trunk reservation; in fact in certain cases, the carried load continues to increase, even for a large number of alternate paths. It is, however, important to note that trunk reservation cannot always avoid bistability; the actual trunk reservation parameter value matters. With an appropriate value, bistability can be avoided and carried load drop

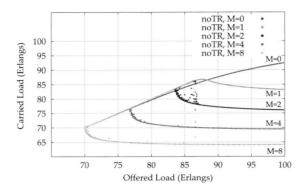

FIGURE 11.8   Symmetric network, no trunk reservation.

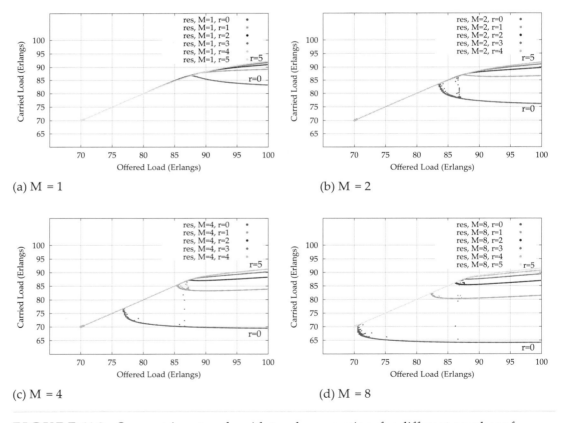

(a) M = 1

(b) M = 2

(c) M = 4

(d) M = 8

FIGURE 11.9   Symmetric network, with trunk reservation, for different number of alternate paths (*M*).

is avoided as well; the rule of thumb mentioned earlier as $\sqrt{c}/2$ appears to be a good one. Second, by comparing performance for different values of *M*, we can see that under heavy load, even with trunk reservation, it is beneficial to limit to a small number of alternate paths than many alternate paths.

There is an important point to note: different values of capacity and offered load with the ratio being the same as discussed above would provide similar behavior, but not exact behavior due to the nonlinear property of blocking.

The above result is for a generic call routing scheme. Obtaining analytical formulas for different routing schemes is not easy. In most cases, different routing schemes are evaluated using call-by-call routing simulation. A general observation from simulation of different routing schemes is that performance can vary depending on the load, and capacity, especially for the asymmetric case, with some looking better under one condition but so under another condition. Second, manageability of a routing scheme and impact on signaling are also important factors to consider. Later in Section 17.7.4, we will present results for different call routing schemes in which we also consider several extensions including multiservice environments.

11.9 Summary

In this chapter, we have introduced voice traffic engineering basics. We explained several key concepts and presented analytical models for analyzing dynamic call routing along with numerical results to understand the bistability issue.

We also discussed various network controls for traffic engineering. In fact, various controls are an integral part of an effective routing system. Thus, for good traffic engineering of a network, it is important to understand performance impact of routing along with different controls. Often, such an analysis requires simulation modeling to capture details of each routing scheme and control mechanisms.

Further Lookup

A. K. Erlang's seminal work almost a century ago established the basic foundation for voice traffic engineering. In [204], Erlang showed that voice traffic follows the Poisson process, while in [205] he derived the Erlang-B loss formula along with several other key results.

In his seminal work done in the early 1960s, Weber [735], [736] considered understanding alternate routing and observed that (1) when a network with alternate routing capabilities becomes overloaded, the throughput can decrease if alternate routing is not reduced, and (2) trunk reservation is an effective control in an overload situation to maintain throughput. In essence, the notion of trunk reservation can be attributed to Weber.

Engineering and Operations in the Bell System [596] is one of the most comprehensive books on how the entire telecommunication infrastructure works. After almost a quarter century of publication, it still remains a valuable reference guide, even though many things have changed since then. Network management controls for the telephone network were comprehensively presented in the seminal paper [279], while an updated summary can be found in [700]. Manifestation of congestion from the signal network to the telephone is well documented in [298], along with additional measures to consider.

With interest in dynamic call routing starting in the late 1970s, various analyses have been done over the years in regard to dynamic routing and control. Bistability analysis for symmetric dynamic routing networks and the issue of stabilization can be found in [385]; analytical modeling for the general case of hierarchical and nonhierarchical routing can be found in [7]. Other analyses, including simulation, of different routing schemes can be found in works such as [30], [36], [250], [251], [357], [358], [360], [382], [496], [539], [763]. For extensive models and results on dynamic call routing, the reader is referred to the books by Girard [253], Ash [27], and Conte [153].

Exercises

11.1 Review questions:

   (a) What is 1 Erlang of offered load?

   (b) What is ECCS?

   (c) What is the relation between offered load in Erlangs and CCS?

   (d) What is the relation between offered load in Erlangs and minutes of usage (MoU)?

   (e) List factors that impact call blocking.

   (f) What is grade-of-service?

   (g) What is trunk reservation?

   (h) In what situation(s), is a hard-to-reach code invoked?

11.2. Why is the arrival rate for telephone calls not sufficient to capture the traffic load?

11.3. Consider the symmetric case with and without trunk reservation (see Section 11.8.4). Determine how the bistability behavior changes for different number of alternate paths and trunk reservation values, if the capacity is changed to 50 and the offered load is proportionately varied. Do the same exercise also when the capacity is changed to 200. Compare the results and tabulate your observations.

11.4. Explain the role of different controls with dynamic call routing in improving network performance.

11.5. Calculate call blocking when 40 *Erls* is offered to a link with 50 circuits. Now change the number circuits to 100 circuits and recalculate blocking. How does the result change?

11.6. Consider a three-node network numbered 1, 2, and 3. Suppose that that the voice circuit capacity of the links and the pair-wise offered load are given as follows:

Link ID	Capacity		Pair ID	Offered load
1-2	50		1:2	40
1-3	40		1:3	20
2-3	60		2:3	60

   (a) Determine link call blocking probability and pair-wise call blocking probability. Investigate if this system has the bi-stability problem.

   (b) For this load and capacity, does the network need to invoke any of the control schemes?

11.7. What are the pros and cons of state-dependent call routing? Is this scheme similar to any of the dynamic call routing schemes discussed in Chapter 10.

11.8. Implement a call-by-call routing simulator that takes Erlang offered load for different node-pairs and capacity for different links as input, and also allows to select a trunk reservation parameter value and any of the dynamic call routing schemes. Refer to Appendix B.10 for a discussion on how to generate Poisson call arrival and exponential call duration time.

11.9. Identify a network situation in which more than one of the controls described Section 11.6 might be invoked.

12

SS7: Signaling Network for Telephony

A doctor can bury his mistakes, but an architect can only advise his client to plant vines.

Frank Lloyd Wright

Reading Guideline

Understanding SS7 provides you with an idea about how call control function is accomplished in the telephone network. This chapter can be read independently. The material is organized so that the basic idea about how an SS7 network works is presented first, and then the chapter delves into issues such as SS7 point code addressing and the protocol stack. Several topics presented in this chapter are helpful in understanding Chapter 13 and Chapter 20.

SS7 can be thought of as the nerve behind the workings of the telephone network. It can also be thought of as a shadow network since it shadows the telephone network for functions such as call setup and call tear-down. More specifically, SS7 is an independent data network that is strongly tied to the telephone network. For example, intelligent network services such as 1-800-number translation uses SS7. While SS7 networks have been deployed worldwide, the standards vary in different parts of the world, including the addressing scheme, known as the point codes.

Keeping with the theme of this book, we present SS7 networking here primarily from a routing perspective. We also discuss key concepts behind SS7 networking that are helpful in understanding call routing in the presence of SS7. Its usage in call routing by the telephone network is discussed later in Chapter 13.

12.1 Why SS7?

Until the early 1970s, all signaling related to setting up a telephone call was done within the voice circuit of a telephone network in which the call is placed—this is known as *in-band* signaling. This in-band signaling, however, produces noticeable delay in setting up a call after the numbers are dialed; this delay is known as the *postdial delay* or the *call setup time*. From hierarchical routing discussed earlier in Chapter 10, we know that due to switching hierarchy a call can take up to nine consecutive circuits in the worst case (refer to Figure 10.2 in Chapter 10). In such a situation, the time it takes to set up a call can be significantly high if in-band signaling is used. It was recognized that the postdial delay was noticeable that a user, after dialing a number, would think that the call is not connected, although the call set up is still in progress due to transfer of signals through a series of switches; that is, such a delay should be bounded to a threshold value acceptable to user perception.

With the advances in data communication, it was realized that a separate data signaling mechanism can be used outside the voice network by sending short data messages for call set up and call tear-down. This process was found to be much faster than sending such signals in analog mode directly on a voice circuit. Such observations have led to the development of early versions of separate signaling mechanisms, which eventually led to *common-channel interoffice signaling* (CCIS), with the well-known version named CCIS No. 6. This then evolved to common-channel signaling 7, or CCS7, and finally to signaling system 7, or SS7.

Besides call setup and call tear-down, SS7-based services have been found to be useful for the toll-free 1-800-number lookup in the United States leading to intelligent network services, and more recently, local number portability. Features such as Caller ID, Call Waiting, and 3-Way Calling are all available as a result of SS7. Furthermore, the SS7 network has been used to exchange information for dynamic call routing techniques such as real-time network routing (RTNR), described earlier in Section 10.6 of Chapter 10.

12.2 SS7 Network Topology

We start by describing the elements of an SS7 network topology. Like any network, it has nodes and links; the nodes and links have specific roles in an SS7 network. This is described next.

12.2.1 Node Types

There are three types of nodes in the SS7 architecture:

- *Service Switching Point (SSP):* This node type is associated with a switch in a telephone network. Recall from Section 10.1 in Chapter 10 that a telephone network is hierarchical with several different levels of switches. Thus, an SSP can be associated with a switch at any level as long as the switch supports SS7 functionality. Thus, a switch that supports SS7 functionality can be thought of as having two interfaces; on one side is the connectivity for voice functionality and on the other side is the connectivity for SS7 data functionality. For simplicity, switches are thus synonymous with SSPs. That is, an SSP converts signaling for a voice call into SS7 signaling messages. The primary responsibilities of an SSP are call processing, call management, and helping route calls to their proper destination.

- *Service Control Point (SCP):* This type of node provides database access services to the telephone network. Using appropriate upper-layer SS7 protocols, to be discussed later, database services are accessed. Example of services provided by an SCP are 1-800 (toll-free) number translation services in the United States and local number portability (LNP); such services are often also clubbed under intelligent network (IN) and advanced intelligent network (AIN) services. It should be noted that an SCP acts as the interface to computers that store databases.

- *Signal Transfer Point (STP):* These nodes are the routers in an SS7 network. Their function is to route messages between two SSPs, or between an SSP and an SCP. It is important to note that it is not always necessary to have an STP for two SSPs to talk to each other; however, a message from an SSP is required to go through an STP to reach an SCP, especially in North America; outside North America, the STP function is often integrated with an SSP. For redundancy, STPs are deployed in mated pairs—thus, mated STPs are sometimes referred to as 2STPs.

In general, an SS7 network can be thought of as providing either a client-server service when an SSP talks to an SCP, or a client-to-client communication service when two SSPs talk to each other; STPs are used to route messages for such communication. Usually, there are two different types of SSPs: national and international. It is possible that an SSP can serve both national and international functions, thus providing hybrid services.

12.2.2 SS7 Links

Six different types of links have been defined for SS7 networking. They are as described below:

- A-link (Access link): A-links connect SSPs to STPs, or SCPs to STPs, or sometimes SSPs directly to SCPs.

- B-link (Bridge link): B-links connect nonmated STPs; these links are usually needed if an SS7 network is large enough to route messages to go through more than a pair of mated STPs.

- C-link (Cross link): C-links connect mated STPs. These links are useful, for example, in rerouting call setup-related signaling messages.

- D-link (Diagonal link): D-links are used to connect local mated STPs to regional mated STPs.

- E-link (Extended link): E-links can be used to connect an SSP directly to an STP that is not in its home STP.

- F-link (Fully associated link): F-links are used to connect two SSPs directly, thus not necessitating going through an STP to send signaling messages.

There can be multiple links between any two nodes; because of that, such a set of links is referred to as a *linkset*. For example, there can be up to 16 parallel links in a linkset between an SSP and an STP. Thus, from an SSP to its mated STP pairs, there can be up to a maximum of 32 links. These links are then identified through a signaling link code (SLC).

It should be noted that a link's naming such as A-link, B-link, and so on, is mainly for the purpose of understanding connectivity between different types of nodes; it does not have any other significance. At the same time, this naming is helpful in describing SS7 routing. Note that it is not necessary to deploy all different link types in a specific network. For example, a network can be deployed with just F-links that connect each pair of switching nodes directly.

A sample SS7 network topology with all types of representative links and nodes is shown in Figure 12.1. Recall that STPs are deployed in mated pairs; this does not imply that an SS7 network needs twice the number of telephone switching nodes. Instead, multiple telephone switches can be served by just a pair of mated STPs as we can see from Figure 12.1.

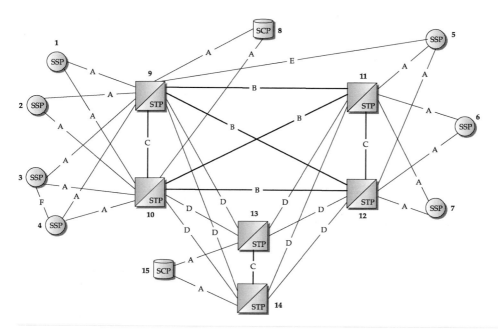

FIGURE 12.1   A sample SS7 network topology where all link and node types are shown.

12.3 Routing in the SS7 Network

Routing of signaling messages in an SS7 network is hop-by-hop and is based on a set of rules. In fact, link naming described in the previous section is helpful in describing these rules. Furthermore, for the purpose of routing, of the two mated STP pairs, one STP is assigned as the *primary home STP* for a set of SSPs. The routing rules are listed below:

- A message generated by an SSP to a directly connected SSP must take the F-link first, if it exists between them, and if the directly connected SSP is the destination for this message.

- If an F-link does not exist, a message generated from an SSP to another SSP takes the A-link for routing to the originating SSP's primary home STP; the A-link to the secondary STP is taken if the A-link to the primary STP has failed or is congested, or the primary STP has failed or is congested.

- A message from an SSP to another SSP that is served by a different pair of mated STPs is routed on the A-link by the originating SSP to its primary home STP for forwarding further.

- A message that has already arrived at an STP for a destination SSP served by this STP takes the A-link; alternately, the message takes the C-link to its mated STP if the direct A-link to destination SSP has failed or is congested.

- A message that has already arrived at an STP destined for an SSP that is served by different mated STP pairs takes the E-link directly to the destination SSP if it exists; alternately, the message is routed to the primary home STP of the destination SSP using the B-link. The third alternate option is to use the B-link to the secondary home STP of the destination SSP. The fourth alternate option is to take the C-link to its mated STP for further forwarding of the message.

- A message from an SSP destined for an SCP takes the F-link if it exists. Otherwise, the message is routed to its primary home STP on the A-link for forwarding to the SCP.

Based on the above rules and given the topological redundancy shown in Figure 12.1, we can see that an SS7 network has built-in redundancy to allow multiple paths between two SS7 nodes. This is in fact important due to the critical role SS7 plays in the workings of the telephone network.

A route in an SS7 network is a sequence of linksets that defines a path from a source SSP to a destination SSP. Similar to a linkset, a *routeset* is defined as a collection of routes. A routeset must consist of at least two routes: a primary route and a secondary route; this can be ensured by providing appropriate alternate next-hop options at each node.

TABLE 12.1   Routesets for two source-destination pairs.

Source:Destination	Routeset
3:4	3-4; 3-9-4; 3-10-4; 3-9-10-4
2:9	2-9-5; 2-9-11-5; 2-10-11-5

Example 12.1   *Illustration of SS7 routing.*

To illustrate SS7 routing, we consider again Figure 12.1. Suppose that a signaling message is to be sent from SSP 3 to SSP 4. Since these two nodes are directly connected by an F-link, it will take the direct link 3-4 as its first route. For this pair of SSPs, the alternate path would be 3-9-4 where node 9 is the primary home STP for SSP 3; this path, thus, consists of just two consecutive A-links. A third diverse path for the same pair of nodes is 3-9-10-4 consisting of an A-link followed by a C-link followed by an A-link.

Now consider sending a signaling message from SSP 2 to SSP 5; the first path would be 2-9-5 consisting of an A-link followed by an E-link. The second path is 2-9-11-5, and the third path is 2-10-11-5. Note that a link can appear in multiple paths. Routesets for these two source-destination pairs are summarized in Table 12.1.   ▲

Example 12.2   *Illustration of a routing table.*

In the above example, we have given the path view through routesets. To accomplish such routesets, the routing table at a node must keep entries for the next hop for each destination. Rather, to provide redundancy and diverse paths, two different next-hops for each destination are usually maintained in the routing table.

To have a routeset from SSP 3 to SSP 4, the routing table at node 3 (SSP) and the routing table at node 9 (STP) are shown in Table 12.2, under the assumption that node 9 is the primary home STP for SSP 2 and SSP 3.   ▲

TABLE 12.2   Routing Table entries for different destinations at node 3 and node 9.

View at Node 3:			*View at Node 9:*		
Destination	Next Hop (Primary)	Next Hop (Secondary)	Destination	Next Hop (Primary)	Next Hop (Secondary)
1	9	10	1	1	10
2	9	10	2	2	10
3	—	—	3	3	10
4	4	9	4	4	10
5	9	10	5	5	11
6	9	10	6	11	12
7	9	10	7	11	12
8	9	10	8	8	10

12.4 Point Codes: Addressing in SS7

Since an SS7 network is similar to other data networks, its various nodes must have addresses so that messages can be forwarded properly. So far in our discussion, we have used generic node numbering to describe an SS7 network. Node addresses in SS7 are called *point codes*; they are logical addresses and they do not identify the physical location of an SS7 node. In the SS7 protocol stack, this addressing is considered in layer 3.

When a user dials a number to reach a destination, the originating SSP generates a call setup message. The point code associated with the originating SSP is often referred to as the *Originating Point Code* (OPC), and the point code associated withe the destination SSP is referred to as the *Destination Point Code* (DPC). Note that the DPC need not be the ultimate destination of the dialed call; we will illustrate this later in Example 12.4. A message going from an originating SSP to a destination SSP might need to be routed via an STP; the point code for this STP is referred to as the *Adjacent Point Code* (APC).

There are mainly two point code formats currently used around the world: one is known as the North American format defined by the *American National Standards Institute* (ANSI) and the other is known as the ITU format defined by the *International Telecommunication Union* (ITU).

12.4.1 North American Point Code

ANSI has standardized the North American point code format to be 24 bits in length partitioned into three 8-bit bytes. Thus, each byte range is from 0 to 255. In dotted-decimal notation, a point code can be written in a manner similar to an IP address; the difference is that since this is a 24-bit address, it has just three parts; for example, a point code would look like 5.7.15 in the dotted-decimal notation. There is no consistent notation for writing a point code. For instance, point codes are sometimes written in dashed-decimal notation, i.e., as in 5-7-15; at other times, point codes are written in a nine-digit format where the first three digits identify the first byte, the second three digits identify the second byte, and the third three digits identify the third byte without any marker in between, i.e., as in 005007015. In this book, we will use the dotted-decimal convention for point codes to be consistent with the convention used for writing IP addresses.

We now discuss the structure of the SS7 point code addressing scheme. The first byte identifies an SS7 network provider (or network number), the second byte identifies a cluster or a region, and the third byte identifies a member within a cluster. Member identifier 0 is usually reserved for STPs. Thus, for the address 5.7.15, the SS7 provider identifier is 5, the cluster identifier is 7, and the member identifier is 15; such an address very likely identifies an SSP. However, 5.7.0 is usually assigned to an STP that serves the SSP node 5.7.15 as its primary home STP. Based on this information, we can summarize the point code format as follows:

Point Code (North American): | network number (8 bits) | cluster (8 bits) | member (8 bits) |

It is important to note that the first byte identifies an SS7 network provider, not a carrier that provides telephone services. In other words, it is possible that a local telephone service provider might use an SS7 network provider for SS7 functionality. In this case, the SS7 provider is likely to either assign a cluster group to this telephone service provider and then the member code for each SSP that the provider has, or directly assign a member identifier for all its SSPs.

12.4.2 ITU Point Code

Point codes in the ITU format are only 14 bits long. An ITU point code also has three parts: the first part consisting of 3 bits identifies a world zone, the second part consisting of 8 bits identifies a network, while the third part consisting of 3 bits is the signaling point identifier (SPID). The entire address then identifies a node in an SS7 network.

The world zone identifiers are listed in Table 12.3; as of now, 0 and 1 are not assigned as zone identifiers. The network identifier is typically assigned at the country level, while a particular country can be assigned multiple network identifiers. Note that network ID is 8 bits long; thus, it is restricted to the range 0 to 255. The general format can be written as:

Point Code (ITU): | zone identifier (3 bits) | network identifier (8 bits) | SSP ID (3 bits) |

The network identifier together with the zone identifier is referred to as a *Signaling Area/Network Code* (SANC) designation. We will again use the dotted-decimal notation to write SANC designations and the complete point codes for the ITU format as well. However, keep in mind that the first part can take a value up to 7. For example, 4.164 is the SANC designation for Uzbekistan where 4 identifies the world zone, and 164 is the network ID for Uzbekistan. Indonesia, for example, has been assigned three SANC designations: 5.020, 5.021, and 5.022. Thus, an SSP in 5.020 can be identified as 5.020.1 where the third part is the SPID. In recent years, SANC codes have been assigned to countries that do not follow the original world zone designations; this is a result of the growth in SS7 deployment and unavailability of SANC codes in certain zones. For example, the United Kingdom falls under world zone 2. Thus, 2.068, a SANC code conforming to this rule, has been assigned to the United Kingdom. Since 2.x address space is already fully assigned, 4.253 has also been assigned to the United Kingdom, although 4.x is in world zone 4 (parts of Asia). A list of assigned SANC designations can be found at [322].

Note that not all countries follow the ITU format. As already discussed, North America has its own point code format that is based on ANSI. China also uses a 24-bit format, somewhat similar to the ANSI point code, while Japan uses its own point code format that is based on 16 bits.

TABLE 12.3   World Zone IDs for ITU Point Code Format.

Zone Identifier	Geographic Region
2	Europe
3	North American, Mexico, the Caribbean, Greenland
4	Middle East, North and East Asia
5	South Asia, Australia, New Zealand
6	Africa
7	South America

12.5 Point Code Usage

In this section, we discuss how point codes are used for routing table aggregation, its relation to telephone switches, and on interworking between different SS7 networks. Certainly, a telephone service provider can be its own SS7 provider, which is often true for large to medium-sized providers in North America. In such cases, the telephone service provider is assigned its own unique SS7 provider identifier.

12.5.1 Address Assignment

Actual assignment of point codes to nodes is usually done in a systematic manner so that the routing table size can be minimized; this can be done by assigning the member level entry for the primary home STP to zero. For example, consider Figure 12.2; the primary home STP for the SSPs on the left side on this figure is assigned the point code 1.4.0 while the SSPs are numbered 1.4.2, 1.4.3, 1.4.4, and 1.4.5; this is similar for the right-hand side. At the SSP with point code 1.4.2, the routing table entries for destination SSPs with point codes 1.5.2, 1.5.3, and 1.5.4 can be minimized by creating a single entry as 1.5.0 for the primary home STP for SSPs 1.5.2, 1.5.3, and 1.5.4; if a message from 1.4.2 reaches STP 1.5.0, it is this STP's responsibility to deliver it to the appropriate destination SSP.

12.5.2 Relationship Between a Telephone Switch and an SSP

If a telephone network provider uses SS7 for signaling, all its switches must have an SSP interface with a point code associated with it; this point code is then at the level of member (third byte) in the SS7 addressing scheme in the North American point code format and at the level of SPID in the ITU format. In this way, a telephone switch is synonymous with an SSP.

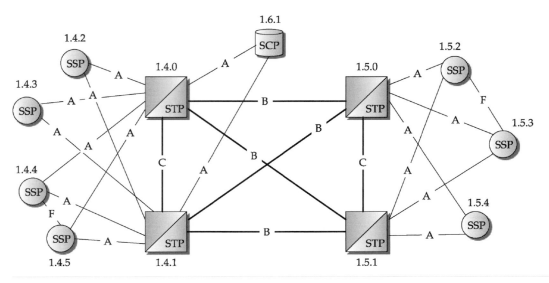

FIGURE 12.2   An SS7 network example with point codes identified.

To see their relation, consider Figure 12.3 where two switches, Switch A and Switch B, are directly connected by a voice trunkgroup. Here, Switch A's SS7 interface, the service switching point, is shown with a half-circle marker. As a general convention, we will use this mixed node picture to denote a switch with an SS7 interface whenever two switches are required to be shown along with their SS7 connectivity. Here, Switch A's interface SSP is identified by point code 1.4.4 while Switch B's SS7 interface SSP is identified by point code 1.4.5; they are connected by SS7 signaling links. This linkset is logically completely separate from the bearer circuits used for the voice service; it forms the F-link in this case.

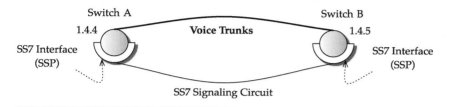

FIGURE 12.3   Switch–SS7 interface.

12.5.3 Interworking of SS7 Networks with Different Addressing Schemes

Recall that there are primarily two point code formats: North American and ITU. How does it work when a message is to be routed from a network with one type of point code format to another network with another type of point code format? First, each country must have at least one SSP ("a hybrid SSP") that acts as an international gateway for routing messages from its national network to other nations' networks. In countries where multiple providers are allowed to have international call handling capability, each such provider would have a separate international gateway. Thus, if an SS7 network uses the North American address format and it needs to communicate a signaling message to another SS7 network that uses the ITU format, the hybrid SSP changes the address format from one to the other format before forwarding the message. There is also the concept of an international SSP that serves as an exchange point between several national networks; this then avoids the need for a bilateral arrangement between every two countries. Such an international SSP essentially requires functionality similar to an STP since its responsibility is to forward messages to different gateway SSPs, albeit by also taking care of address conversion.

Consider Figure 12.4 where we have shown the SS7 network connectivity spanning four countries. Assume that Country A uses North American point codes for its SS7; for transfer of signaling messages to another country, the gateway SSP (hybrid SSP) in Country A is required to have an ITU-based address that can be understood by SS7 nodes in other countries; this node has 1.6.1 as its point code in the North American format for the internal network and 3.021.1 as its point code in the ITU format where the address block, i.e., the SANC code, 3.021, is assigned to the United States under the ITU-based point code addressing. All SS7 SSPs in the rest of the countries use ITU-based point codes; thus, communication is now possible. Certainly, this gateway node needs to regenerate every message going from one network to the other by replacing North American–based point codes by ITU-based point

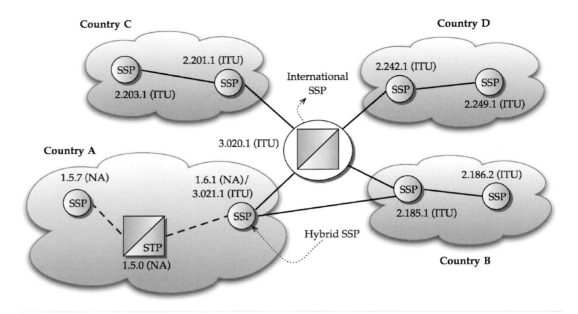

FIGURE 12.4   Interworking of SS7 networks with different point code systems.

codes; this is similar for a message going in the other direction. Note that we have also shown an international SSP with point code 3.020.1; this node serves as the exchange point between different countries.

12.6 SS7 Protocol Stack

So far, we have described SS7 network topology and addressing. Being a data network, SS7 also has a protocol stack; this protocol stack, however, predates TCP/IP. A schematic diagram of the protocol stack with certain key protocols identified is presented in Figure 12.5, along with a comparison to the OSI reference model.

12.6.1 Lower-Layer Protocols: MTP1, MTP2, and MTP3

The lower three layers of the SS7 protocol stack are labeled *Message Transfer Part-Level 1* (MTP1), *Message Transfer Part-Level 2* (MTP2), and *Message Transfer Part-Level 3* (MTP3), corresponding to physical, data link, and network layers of the OSI reference model, respectively.

MTP1 defines the physical and electrical interfaces of the SS7 protocol stack. The following interfaces are defined: DS0A (56 Kbps), DS0 (64 Kbps), DS1 (1.544 Mbps), and E1 (2.048 Mbps), with DS0A as the mostly commonly deployed one due to historical reasons.

MTP2 takes care of typical data link layer issues such as error detection, bit stuffing, flow control, and so on. There are two key signaling units primarily for MTP2: (1) the *fill-in signaling unit* (FISU)—this message is sent continuously whenever there are no other messages to send, and (2) the *link status signaling unit* (LSSU)—this message is sent to provide the status of links connected to STPs. The FISU is 6 bytes long while an LSSU can be either 7 or 8 bytes

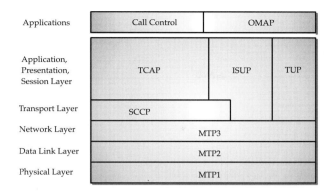

Applications

Application,
Presentation,
Session Layer

Transport Layer

Network Layer

Data Link Layer

Physical Layer

MTP : Message Transfer Part
TUP: Telephone User Part
ISUP: ISDN User Part
SCCP: Signaling Connection User Part
TCAP: Transaction Capabilities Applications Part
OMAP: Operations, Maintenance, and Administration Part

FIGURE 12.5   SS7 protocol stack.

long. FISU is sent as filler for idle time. The most important field in FISU is the frame check sequence (FCS) field. By monitoring FISU and observing any errors based on FCS, the MTP can assess the health of a link and take it out of service, if needed. LSSU is not normally sent; it is sent to indicate status such as out of service, link alignment problem, or busy.

MTP3, which is the equivalent of the network layer in OSI, provides addressing and routing. As described earlier, addressing is based on 24-bit point codes in the North American format and 14-bit point codes in the ITU format; routing rules are defined by labeling different link types in the network so that communication between one and another SSP, and between an SSP and an SCP can take place using STPs for routing. The basic messaging unit for MTP3 is called the *message signaling unit* (MSU); it is of variable length not to exceed 278 bytes.

We will discuss the MSU here in detail due to its role in call processing and routing, and because the FISU and the LSSU can be thought of as subsets of the MSU. The basic format of an MSU message is shown in Figure 12.6. The format for the FISU in MTP2 is similar to an MSU except that service information octet (SIO) and service information field (SIF) are not included and the length indicator (LI) is set to zero. The format for an LSSU is also similar to an MSU except that instead of the SIO and SIF, the link status indicator (LSI) field is included in their place; this field indicates the status of the link. The LI field is set to either 1 or 2 to indicate 1- or 2-byte lengths of the link status indicator field. For an MSU, the link indicator field is set to 3 or more to indicate the length of SIF in bytes. Since the LI is 6 bits long, this value is limited to 63; this means that even if SIF is more than 63 bytes, the value is still set to 63. It may be noted that all signaling units have the following fields in common: Flag, Backward Information Bit (BIB), Backward Sequence Number (BSN), Forward Information Bit (FIB), Forward Sequence Number (FSN), Spare bits, Length Indicator (LI), and Frame Check Sequence (FCS) (see Figure 12.6). The Flag field at the beginning of a signaling message indicates the start of the packet; its binary value is 0111 1110; thus, any time the remaining data in a signaling unit have more than five 1s, bit stuffing is used to avoid confusion with the Flag field. The frame check sequence is appended as the trailer for error detection. It should now be apparent that the left byte (Flag) shown in the MSU in Figure 12.6 is the first byte sent while FCS is the last part of the message. Within each byte, we use the convention that the leftmost bit is marked as the most significant bit.

 Byte transmission order

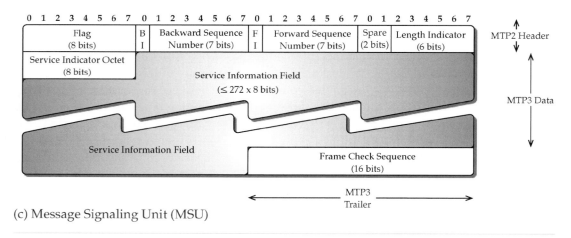

(a) Fill-In Signaling Unit (FISU)

(b) Link Status Signaling Unit (LSSU)

(c) Message Signaling Unit (MSU)

F I G U R E 12.6   FISU, LSSU, and MSU in SS7 (BI, short for BIB; FI, short for FIB).

In the MTP2 header part, FSN is used as the sequence number of a signaling unit; this field is incremented by the sender each time a new MSU is generated. BSN is used as the acknowledgment by the receiving side by indicating the FSN received for a new MSU. In case the FCS is incorrect, the receiving end sets the BIB field to 1 so that the sending end knows which FSN to retransmit. When the sender sides retransmit a particular FSN, it sets the FIB field so that the receiving end knows that it is a retransmitted packet. In other words, for normal transmission and acknowledgment, FIB and BIB bits are always set to zero. Note that for the FISU and LSSUs, FSN is not incremented; the FSN for the last MSU is kept the same on the FISU or LSSUs that are generated

T A B L E 12.4   Sample Values in Service Information Octet (SIO) in an MSU.

Bit Position 87	Value	Bit Position 4321	Value
00	International network	0000	Network management message
01	Reserved (for international use)	0011	SCCP
10	National network	0100	TUP
11	Reserved (for national use)	0101	ISUP

before another MSU is generated; note that neither the FISU nor LSSUs are retransmitted.

The SIO field in an MSU is further divided into two subfields: 4 bits for indicating the network type, although the two most significant bits are often used in practice, and the other 4 bits for indicating the service type. Sample values are shown in Table 12.4.

The SIF in MTP3 is the critical field where detailed information needed for a service such as call processing is encapsulated, where the SIO subfield described above provides the indicator for the transport protocol. However, the beginning part of the SIF field in MTP3 includes two pieces of information that are needed for the network layer: destination point code (DPC) and originating point code (OPC). The DPC and OPC fields are followed by the signaling link selection (SLS) field. In ANSI-based networks, the SLS field is used in selecting an outgoing link from a linkset when coupled with the DPC field; in ITU-based systems, the SLS field is used to identify the signaling link code that is influenced by the upper layer protocol. The three fields, DPC, OPC, and SLS, are collectively known as the *routing label* in SS7 networks. We have already described the point code structure for the North American and the ITU format. Taking the point code structure into account, the routing label in SIF has the following form:

Routing Label (North American):   DPC (24 bits) | OPC (24 bits) | SLS (8 bits)

Routing Label (ITU):   DPC (14 bits) | OPC (14 bits) | SLS (4 bits)

Thus, the routing label in the North American format requires 7 bytes while in the ITU format it requires just 4 bytes. The SIO field along with the routing label can be thought of as the core of the network layer header; the routing label identifies source and destination SSPs, which are similar to the source and destination IP addresses in an IP datagram header, and part of the SIO field identifies the upper layer protocol, similar to the protocol type field in an IP datagram header.

From Figure 12.6, we can see that the SIF field can be up to 272 bytes. The routing label is the first entity in this field and occupies only a small part of this length. The space provided by the remaining bytes is left for use by upper-layer protocols as and where needed.

12.6.2 Upper-Layer Protocols

Immediately above MTP3 in the protocol stack there are three well-known protocols, especially as related to call routing: *Telephone User Part* (TUP), *Integrated Services Digital Network (ISDN) User Part* (ISUP), and *Signaling Connection Control Part* (SCCP). From Figure 12.5, you can see that part of ISUP is also shown to be above SCCP—this is because ISUP can be implemented above SCCP while it is most commonly done directly over MTP3. In this section, we briefly describe TUP; ISUP and SCCP are discussed in subsequent sections.

TUP: Telephone User Part

TUP is a legacy protocol used for voice call control. Although it is no longer widely used, many countries around the world still use it. TUP is used to control only analog voice circuits. TUP provides the basic functions of call control such as call setup and tear-down; it also allows identification of the voice circuit to be used for a call using a field called *Circuit Identification Code*. To avoid confusion with *Carrier Identification Code*, to be discussed later in Chapter 13, we will refer to Circuit Identification Code as *Trunk Circuit Identification Code*, or TCIC. That is, the TCIC of a call is the code that identifies the logical circuit number used for voice transmission on a trunkgroup. It may be noted that TUP uses the TCIC in the routing label in MTP3 in place of the SLS code.

The initial setup message, known as the Initial Address Message (IAM), is generated by the original switch through its SS7 interface SSP identifying the destination SSP based on the leading digits dialed. For simplicity, consider again Figure 12.3, where we have shown a direct trunkgroup between an originating switch and a destination switch along with its SS7 interfaces. When the originating switch identifies which idle voice circuit is to be used from the trunkgroup that connects to the destination switch, it is translated to the TCIC code. Thus, in the IAM message, this TCIC code is included to inform the destination switch of the voice circuit chosen along with the OPC and DPC of the originating and destination switches.

If the circuit is successfully set up, the destination switch then generates an Address Complete Message (ACM) that is sent to the originating switch, again using the SS7 interface. Basic usage of IAM and ACM is similar to ISUP call processing described later in Section 12.8.

TUP is compatible with ISUP, while ISUP provides an extended set of parameters. Due to compatibility, an IAM message generated in an SS7 network that supports only TUP can be forwarded to an SS7 network that supports ISUP; certainly, the gateway SS7 node is required to reformat the message before forwarding it to the other network.

12.7 SS7 Network Management

Management of an SS7 network is also a critical function the SS7 protocol stack is required to provide. In fact, the SIO subfield can be used to indicate network management messages for the SS7 network (see Table 12.4). Once it is indicated that the message is a network management message through this field, the SIF field is used to provide further guidance on the details of the network management message types.

Network management messages are used primarily for routing management in the SS7 network. For example, if a certain linkset is congested in an SS7 network, a network management message can be generated to inform the associated SS7 nodes that it is temporarily unavailable so that rerouting of signaling messages can be performed. To illustrate this usage, consider again Figure 12.2. Suppose that the A-link between the STP with PC 1.5.0 and the SSP with PC 1.5.3 is unavailable due to congestion; the STP with PC 1.5.0 will generate a transfer-prohibited (TFP) message identifying the unavailability of 1.5.3 and send on B-links to inform STPs with PCs 1.4.0 and 1.4.3. This way, a new call setup message generated the next time at PC 1.4.3 destined for PC 1.5.3 can take the alternate route, 1.4.3 to 1.4.0 to 1.5.1 to 1.5.3. When the A-link between the STP with PC 1.5.0 and the SSP with PC 1.5.3 becomes available again, the STP with PC 1.5.0 can generate a transfer-allowed (TFA) message along the B-links, so that the downstream nodes can now perform normal routing of signaling messages.

Due to the cluster concept discussed earlier, a cluster level congestion message, known as the transfer cluster-prohibited (TCP) message, can also be generated. Consider again Figure 12.2. Suppose that the primary home STP 1.5.0 is not able to access SSPs 1.5.2, 1.5.3, and 1.5.4. Then STP 1.5.0 can generate a TCP message to inform distant STPs 1.4.0 and 1.4.1 that SSPs in cluster 1.5.0 are not accessible through 1.5.0. Similar to TFA, the network management message type, transfer cluster-allowed (TCA), is generated when this cluster becomes available again.

There are many other messages defined for the purpose of SS7 network management and testing. We have discussed only a few critical ones to show their need and how their usage can be helpful in routing management of the SS7 network. Finally, it is important to note that these messages are meant to manage the SS7 network, *not* the telephone network trunks. Management of telephone network trunkgroups is often done using an ISUP message; this is discussed later in Section 12.9.

12.8 ISUP and Call Processing

While TUP has certainly provided key functionalities for setting up a call using the SS7 network, it was still a basic protocol. ISUP was designed to overcome limitations of TUP; an important aspect of ISUP is that it is an extensible protocol—thus, it can be customized for local need within a country. ISUP is now extensively used around the world for call processing in the PSTN, although it was originally defined for ISDN services.

ISUP defines about 100 different message types. We have summarized several key messages types such as Initial Address Message (IAM), Address Complete Message (ACM), and so on, in Table 12.5. Similar to TUP, the routing label and the TCIC code are included in all ISUP messages; the major difference here is that in the case of an ISUP message, the routing label consists of the DPC, the OPC, and the SLS fields while the TCIC code is included in a separate field following the routing label. Furthermore, in ISUP, IAM is generated only after all the dialed digits are entered by the user, while in TUP, IAM can be generated based on leading digits dialed.

The IAM message is the central message and carries much information; in fact, it has 39 different fields, of which 6 fields are mandatory fields. Table 12.6 lists key fields in the IAM message, including the mandatory fields. The IAM message carries all the critical information in regard to setting up a call; it includes information such as the TCIC field, and the Called Party Number and Calling Party Number. It is important to note that Calling Party Number is not a mandatory field. This means that the starting OPC is not required to include this field

TABLE 12.5   Sample ISUP messages.

Message Type	Full Name	Usage
IAM	Initial Address Message	Used for establishing a call
ACM	Address Complete Message	Indicates that the other end is processing the call
CPG	Call Progress	Call in progress message used for alerting, etc.
ANM	Answer Message	Indicates that the called party has answered the call
EXM	Exit Message	Indicates that the IAM message has been passed to another network when internetworking is required to establish a call
REL	Release	Indicates that the call is being terminated
RLC	Release Complete	Indicates that the REL message has been received and the voice circuit can be released
INR	Information Request	Used for requesting additional information about a call in progress
INF	Information	Used as a response to the INR message to provide information requested
BLO	Blocking	Blocking on a circuit by a switch so that other end does not use this circuit
BLA	Blocking Acknowledgment	Acknowledgment of a BLO message
UBL	Unblocking	Unblocking a circuit previously blocked using BLO
UBA	Unblocking Acknowledgment	Acknowledgment of a UBL message
CGB	Circuit Group Blocking	For blocking a range of circuits
CGBA	Circuit Group Blocking Acknowledgment	Acknowledgment of a CGB message
CGU	Circuit Group Unlocking	For unblocking of a range of circuits blocked earlier using CGB
CGBA	Circuit Group Unlocking Acknowledgment	Acknowledgment of a CGU message

when an IAM message is generated. An advantage of not including the Calling Party Number is avoiding automatic number identification by the receiving party; however, a difficulty is that if an error code is to be generated along a call path by an intermediate SSP, it would not know what calling party it is for—certainly, it is debatable whether the intermediate SSP needs to know this information. Thus, leaving it as an optional field was probably the best solution.

In an IAM message, the Calling Party's Category field is used to indicate the type of subscriber originating the call. This field is used to indicate whether the originator is an ordinary calling subscriber, a call from a pay phone, or a call from a special-language operator.

TABLE 12.6 ISUP Initial Address Message (IAM) and Release Message (REL): Key fields (Mandatory fields are marked with *).

INITIAL ADDRESS MESSAGE (IAM):	RELEASE MESSAGE (REL):
*Routing Label**	*Routing Label**
*Trunk Circuit Identification Code (TCIC)**	*Trunk Circuit Identification Code (TCIC)**
*Message Type (IAM)**	*Message Type (REL)**
*Nature of Connection Indicator**	*Cause Indicator**
*Forward Call Indicator**	*...*
*Calling Party's Category**	*Call Reference*
*User Service Information**	*...*
*Called Party Number**	
...	
Call Reference	
...	
Calling Party Number	
...	
Carrier Identification Code	
...	
Generic Address	
...	

The User Service Information field is usually not needed for a regular telephone call. It is used when the subscriber is requesting data transmission such as ISDN that does not use a modem. The Called Party Number contains the number that the originating user dials, excluding any leading digits used to indicate whether it is a long distance, or an international, or an operator-assisted call. Since the number can be for a variety of services, we will discuss it further after a discussion of numbering plans in Chapter 13. Call Reference is an optional field in an IAM message. This is assigned to a call for the purpose of identification of the call; this is not a global value and is used primarily for tracking. The Carrier Identification Code field identifies the network carrier; its use will be discussed in Chapter 13. Furthermore, the role of the fields, Forward Call Indicator and Generic Address, will be discussed later in Section 13.11.4, when we introduce number portability.

An ISUP REL message is an important message that is often associated with an IAM message. Key fields for an ISUP REL message, including 4 mandatory fields, are listed in Table 12.6. In the normal mode, this message is used to indicate the release of a call. However, REL is used for much more than that. It includes a cause indicator field; for a normal call release, cause identifier 16 is used. If a call is not accepted during setup due to lack of an available voice circuit, a REL message is generated by specifying this cause in the cause indicator field. Automatic congestion level is also indicated using the cause indicator field in the REL message. Two levels of congestion are used, which depends on the threshold value specified; on receiving the level of congestion information, a source switch may choose to active hard-to-reach code (refer to Section 11.6.2). In fact, the REL message type is used in numerous ways [331]; for example, for indicating if the receiving user's line is busy (cause identifier 17), or if the number at the receiving end is an unallocated/unassigned number (cause identifier 1), or if no circuits are available (cause identifier 34), or if switching equipment is congested (cause identifier 42).

There are some differences between the ISUP messages in the ANSI format and the ITU format; for example, the TCIC field in the ANSI-based IAM message is 14 bits long while it is 12 bits long in the ITU-based IAM message. Similarly, values for certain types or identifiers can be different.

Example 12.3 *Call setup and tear-down process using ISUP messages.*
   We first present a simple illustration of the call setup process using ISUP messages. Consider a call from a switch with point code 1.4.4 to a destination switch with point code 1.4.5 (see Figure 12.7). Assume that by checking idle voice circuits, it has identified the TCIC code of the idle voice circuit as 51.
   A call process starts with the IAM message from PC 1.4.4 to PC 1.4.5. At this point, the voice circuit number 51 is tagged for use by this call, but not fully reserved yet. If the user on the receiving side is available, PC 1.4.5 generates an ACM message to the PC 1.4.4 of the originating switch, and the ring is started while voice circuit number 51 is fully reserved for the call. On receiving the ACM message at the originating switch, user B starts hearing the ring. The destination switch generates the CPG message periodically until the user on the receiving side picks up the phone. Once the user picks up the phone, the ANM message is generated.

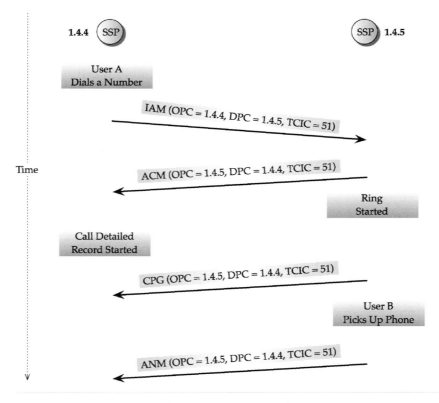

FIGURE 12.7   ISUP call setup message exchanges.

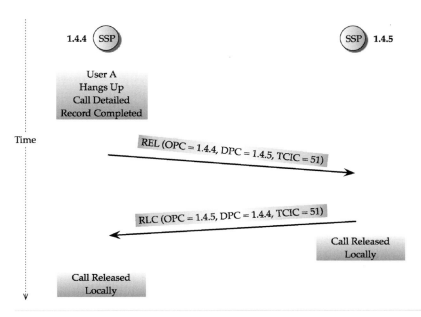

FIGURE 12.8   ISUP call tear-down message exchanges.

Assume that the call originator is the one who hangs up the phone at the end of the conversation (see Figure 12.8). A REL message is then generated at this end for the other end with cause identifier 16 for normal call clearing [331]; this REL message is then responded to using an RLC message to indicate complete call tear-down.   ▲

We make the following important remarks:

- Before the trunkgroup is initialized between two adjacent telephone switches, they must agree on the logical voice circuit numbering that is to be viewed consistently from both sides so that there is no ambiguity in the TCIC code included in the initial IAM message. That is, if TCIC is set to 51, the receiving switch knows which voice circuit it is, so that it does a tagging on this circuit and does not assign this same TCIC code to a newly arrived call at this switch going in the other direction. This is important since voice circuits are often bidirectional. There is a general rule about how to assign the TCIC code. Usually, two telephone switches are connected by T1 (with 24 voice circuits) or E1 (with 30 voice circuits) circuit groups; multiple circuit groups can make up the entire trunkgroup. Each T1/E1 circuit group is first numbered from 0 to a maximum of 127; instead of using 24 or 30, the circuit bundle for each link unit is often set to 32 following an ITU convention. Thus,

$$\text{TCIC code} = \text{circuit\_group\_number} \times 32 + \text{circuit\_id}.$$

Suppose that the circuit group to be used is numbered 20 and the circuit number assigned within this circuit group is 15. The TCIC code for this circuit will then be 655. With this rule, the maximum value possible is 4096, which is also the maximum value allowed in an ITU IAM packet due to 12 bits assigned to the TCIC field. Although the TCIC field in ANSI IAM is 14 bits long, i.e., the TCIC field can take a value up to 16,384, such a high value is rarely required if we consider the traffic demand factor. Using the Erlang-B loss formula given by Eq. (11.2.3), we can see that 3000 Erlangs of offered load at 1% call blocking requires 3023 trunks, which can be easily accommodated in the TCIC code range. Not only that, it is extremely rare that two adjacent switches have more than 3000 Erlangs of offered load.

- There are many functions that can occur as soon as the ACM message is received by the originating switch. As an example, we show the initiation of the *call detail record* (CDR) for this specific call at the originating switch; this record is closed when the REL message is generated at the completion of the call. CDRs are used for the purpose of billing as well as for other usage. Note that not every CDR entry results in a billable entry; for example, a call may not be answered by the receiving side; a CDR is generated for this call but it is not a billable call.

ISUP currently uses the *pass-along method* for sending signaling messages from an originating telephone switch to the ultimate destination telephone switch; that is, signaling messages are handed on a hop-by-hop basis from one telephone switch to the next telephone switch until the ultimate destination is reached; this is consistent with progressive call control (PCC) discussed earlier (refer to Section 10.1 in Chapter 10). This is not to be confused with actual routing of such messages within the SS7 network.

Example 12.4 *Call routing and SS7 ISUP IAM message routing.*
To understand the above discussion about the pass-along method/progress call control, consider Figure 12.9. We assume here that the telephone switch associated with point code 1.4.2 for which the call originates ("O-switch") and the telephone switch associated with point code 1.4.3 for which the call is destined ("D-switch") are not directly connected by a direct trunkgroup. A call between them is required to use the tandem switch identified by point

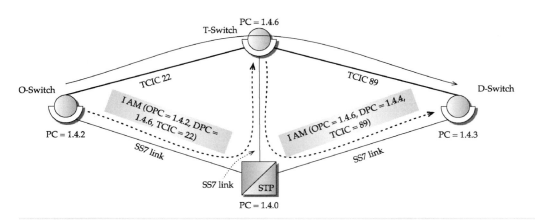

FIGURE 12.9   Call routing and SS7 message routing illustration.

code 1.4.6 ("T-Switch"). That is, there is a trunkgroup between the O-Switch and T-Switch, and also a trunkgroup between the T-Switch and D-Switch—these are marked by bold lines. We assume that when a call arrives at an O-switch, there are circuits available between the O-Switch and T-Switch, and between the T-Switch and D-Switch.

From an SS7 network perspective, assume that there is no direct F-link between SSPs 1.4.2 and 1.4.6, and between SSPs 1.4.6 and 1.4.3. All of them use STP with PC 1.4.0 to route signaling messages; for simplicity, we show only one STP instead of a mated pair of STPs.

Now consider a user associated with an O-Switch who dials a number to another user associated with a D-Switch. The IAM message (not shown in the figure) will first be generated at PC 1.4.2 with OPC as 1.4.2 and DPC as 1.4.6 while including called and calling party number information; the TCIC code of the voice circuit assigned on the trunkgroup between the O-Switch and T-Switch is, say, 22. This IAM message will be routed from SSP with PC 1.4.2 to STP with PC 1.4.0 to SSP with PC 1.4.6. On receiving this IAM message by PC 1.4.6, it will realize that the call is not terminating at this switch by inspecting the called party number. Thus, PC 1.4.6 will regenerate an IAM message for this call keeping the called and calling party information but changing OPC to 1.4.6 and DPC to 1.4.3; this IAM message will be routed from the SSP with PC 1.4.6 to the STP with PC 1.4.0 to the SSP with PC 1.4.3. The TCIC code of the voice circuit assigned on the trunkgroup between the T-Switch and D-Switch is, say, 89 and is not related to the TCIC code on the first leg.   ▲

12.8.1 Called/Calling Party Number Format

So far, we have not discussed actually referencing to any specific Called or Calling Party Number format. This requires familiarity with E.164 numbering, which will be discussed later in Chapter 13, where we will again bring back SS7 call processing for a complete understanding of call routing in the public switched telephone network. Here, we present a brief description of the format and a simple illustration.

The Called Party Number parameter contains two main subfields, one to indicate the nature of the address, and then the called number. The first subfield, Nature of Address Indicator, is encoded to indicate if it is a national number or an international number, whether an operator is requested, and which numbering plan is used. Certainly, the most common numbering plan is E.164. Once the numbering plan is indicated, then the called party number is encoded using four bits for each digit while all 1s in the four bits indicates the end of the number. The nature of address indicator can be used to indicate if the called number is a ported number; this will be discussed later in Section 13.11.4. We illustrate below Called and Calling Party Number as used, for example, in an IAM message.

```
Routing Label: OPC=1.4.4 DPC=1.4.5
TCIC: 55
Message Type: IAM
CalledPartyNumber:
  NatureofAddressIndicator: National
  NumberingPlan: E.164
  Digit: 816-344-2525
CallingPartyNumber:
  NatureofAddressIndicator: National
  NumberingPlan: E.164
  Digits: 816-328-2208
```

It may be noted that ISUP messages use bit/byte–level encoding for different information. The above is listed in a textual mode for ease of understanding.

12.9   ISUP Messages and Trunk Management

ISUP messages are also used for trunk management. If you look at Table 12.5, you will notice that it contains several message types that have little to do with either call setup or call teardown. We have listed several of them for illustration such as BLO, BLA, CGB, and CGBA.

The message type BLO can be used to block a particular bidirectional voice circuit on the direct trunkgroup connecting two telephone switches when the telephone switch on one end invokes it so that the telephone switch on the other end cannot use the same circuit for a call. In this case, the TCIC field indicates the specific circuit to be blocked. When this message is sent, the other end responds by sending the BLA message to acknowledge the BLO message. The originating switch can unblock this circuit later by sending message type UBL; certainly, the TCIC field on this message would need to match what was announced earlier. The receiving end then acknowledges by sending a UBA message.

The functions of message types CGB, CGBA, CGU, and CGUA are similar to BLO, BLA, UBL, and UBA, respectively. The main difference is that these are used in regard to a range of circuits on a trunkgroup instead of just a single circuit. That is, to block a range of circuits by one end, the message type CGB can be used; in this case, the payload part of the message will include the range of circuits to be blocked from this trunkgroup. The other end acknowledges it by responding with the message CGBA. The message type CGU is sent later by the originating switch to deactivate one directional blocking of the range of circuits; the receiving end then acknowledges by responding with the message CGUA.

These functions are useful for maintenance of a voice circuit or a range of circuits; for example, a circuit failure can be indicated by using the BLO message.

12.10   ISUP Messages and Dynamic Call Routing

Earlier in Chapter 10 we presented several different dynamic call routing schemes that have been deployed in operational networks. Some of them use SS7 messaging for status related to call routing while others use dedicated circuits. Specifically, dynamic nonhierarchical routing (DNHR) and real-time network routing (RTNR) use SS7 ISUP messages for dynamic routing controls. Note that such messages are not standardized messages. We discuss them here to show how ISUP can be useful for the exchange of control messages for dynamic call routing.

Recall from Section 10.2.3 that telephone call control is usually based on progress call control (PCC). However, DNHR implemented originating call control (OCC) within its network. Thus, it requires a certain functionality to do OCC. Furthermore, crankback also plays a role in this regard. Also recall that the dynamic call routing network discussed in Chapter 10 is limited to using at most two links for call routing *within* its routing domain, where it is often the middle network connecting a local exchange carrier (LEC) to another local exchange carrier (see Figure 12.10). For example, a call that originates at switch 6 in a LEC's network that is destined for switch 7 in another LEC's network uses the dynamic call routing network as the middle network. Thus, within the dynamic call routing domain, we can

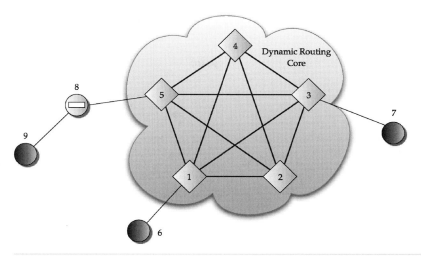

FIGURE 12.10   Dynamic call routing along with hierarchical routing.

say that Switch 1 is the originating switch and Switch 3 is the destination switch. For alternate routing, we need to keep this in mind while considering the functionalities discussed below.

12.10.1 Functionalities

To set up a call in a dynamic call routing network, four types of messages are helpful [27]:

- Traveling class mark—to control the selection of the route at a switch within the dynamic call routing network.
- Crankback—to return control of the call to the originating switch.
- Status Query—to request link status (trunkgroup status) from a switch.
- Status Response—to respond to a link status query that contains link status information.

The first two message types can be used during the call setup phase within the domain of the dynamic call routing network. The last two message types are not required to be tied up directly to the call setup phase.

TRAVELING CLASS MARK

This message type can be used to control routing choices at different switches and is typically implemented as part of the ISUP IAM message by including this information in the optional part of the payload. Thus, when the originating switch within the dynamic call routing network receives its call control from the access network that is based on hierarchical routing, it needs to decide by consulting the routing table whether to route the call on the direct link path or on a two-link path to the destination switch within its network. Within its network, there are three traveling class mark (TCM) tags, EXIT1, VIA1, and EXIT2, required for call handling in an IAM message. In essence, an IAM message is extended with additional information for the purpose of controlling routing choices.

CRANKBACK

Recall that we discussed crankback and call control in Section 10.2.3 in Chapter 10. To accomplish crankback, a new ISUP message type for crankback is used. Its role comes within the dynamic call routing network at an intermediate switch. When the intermediate switch cannot find an available trunk to the destination switch, it has to return the control of the call to the originating switch within the dynamic routing network, for example, in a DNHR-based routing environment where originating call control is used. The returning control of the call is accomplished by sending an ISUP crankback message to the originating switch.

STATUS QUERY

A Status Query message can be generated depending on the actual implementation of the dynamic call routing schemes. In case of RTNR (refer to Section 10.6), the originating switch within the dynamic call routing network can initiate a status query to the destination switch to inquire about the link status of all outgoing trunksgroups from the destination switch that are within its routing domain. For RTNR, the query is about the availability of trunkgroups based on a certain load threshold value.

A Status Query message can be used in other dynamic call routing frameworks as well whenever the status of a trunkgroup is needed. For example, the query can be about available bandwidth, in a periodic or aperiodic manner, to a neighboring switch.

STATUS RESPONSE

A Status Response message is generated as a response to a Status Query message. For RTNR, the terminating switch responds with the link state status of outgoing trunkgroups. While for RTNR this information concerns availability at a certain threshold, the status can be about available bandwidth in the case of an appropriate dynamic call routing framework.

In general, a status response would be generated as a response to a status query. However, a status response can also be generated by a switch to inform its neighbor about a change in trunk availability status based on an increase or decrease in call traffic. Either way, the status response message provides the functionality of the link state routing concept in a dynamic call routing environment.

12.10.2 Illustration

We now illustrate usage of the above messages in a dynamic call routing environment. Consider Figure 12.10 again; suppose the call originated in the access network at Switch 6 and is destined for Switch 7 in another access network. Switch 6, on recognizing that this call needs to go through another network, consults its translation table to determine that the call must be forwarded to Switch 1. Although the figure shows only the telephone switches, imagine that there is a parallel SS7 network that shadows the switched network. We consider the following possible scenarios:

1. If Switch 1 can route on the direct trunkgroup to Switch 3 due to availability of a circuit, then its IAM message includes the TCM tag EXIT1 so that Switch 3 knows that it is the destination switch *within* its network.

2. Suppose that Switch 1 needs to use the alternate route 1-2-3 to Switch 3 due to non-availability circuits on the direct trunkgroup 1-3. In this case, the IAM message is generated by Switch 1 where the TCM tag is set to VIA1. On receiving this switch, Switch 2 recognizes that it is the intermediate switch within the dynamic routing network. By inspecting the called party number in this IAM message, Switch 2 determines that the destination switch is Switch 3 and finds an available circuit on the 2-3 link; Switch 2 then generates a new IAM message for this call with the TCM tag set to EXIT2, which is sent to Switch 3. On receiving this IAM message by Switch 3, it recognizes that it is the destination of the call by inspecting the tag field of the IAM message.

3. When Switch 2 receives an IAM message with the TCM tag set to VIA1, it is possible that there are no trunks available on the trunkgroup from Switch 2 to Switch 3 (subject to trunk reservation) for this call. Two situations are possible:

   (a) If the dynamic routing network has crankback capability, Switch 2 will generate a crankback message to Switch 1 identifying the handling of the call and also specifying that the TCIC code of the circuit on link 1-2 identified through the IAM message is no longer needed. On receiving this crankback message at Switch 1, Switch 1 may try another alternate path such as 1-4-2 if this path is listed as the next route in its call routing table.

   (b) If the dynamic routing network does *not* have crankback capability, Switch 2 will generate an ISUP network trunk congestion (NTC) message to Switch 1 so that Switch 1 in turn can indicate to the originating switch (Switch 6) that the call cannot be completed.

4. On arrival of a call within the dynamic routing core, Switch 1 (the ingress switch) may request Switch 3 (the egress switch) provide a status report for all outgoing trunkgroups by sending a Status Query message; on receiving such a message, Switch 3 may respond by sending a Status Response message that reports the availability status of all outgoing links.

It is easy to see that all dynamic call routing schemes that allow at most two-link call routing within its network can invoke step 1 and step 3. Any scheme that has crankback functionality can invoke step 3(a); otherwise, step 3(b) is invoked. A scheme such as the RTNR scheme would invoke step 4, which essentially enables the link state routing concept for dynamic call routing networks. Finally, a Status Response message can be generated by any switch to update information about its outgoing links to others, either periodically or based on load-based triggers; this functionality also applies to link state routing where each switch upon receiving such information may decide to update its routing.

12.11 Transaction Services

Transaction services that use SS7 networks require access to SCPs. The SCP nodes in an SS7 network provide an interface to database services. Message routing for such services does not follow the pass-along method (progressive call control) that is used for ISUP messages over MTP3. Instead, message routing for transaction services uses a different transport protocol, the *Signaling Connection Control Part* (SCCP). Transaction-based applications then use SCCP as the transport layer protocol. TCAP and SCCP information is embedded in the SIF field along with the routing label (see Figure 12.11).

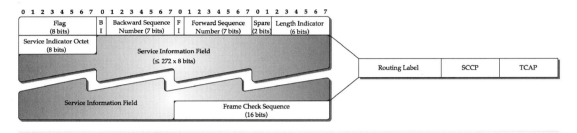

FIGURE 12.11 SCCP and TCAP in Service Information Field of MSU.

12.11.1 SCCP: Signaling Connection Control Part

SCCP provides both connectionless and connection-oriented network services. Connectionless services are divided into two categories: basic connectionless service and sequenced connectionless service. The basic connectionless service serves as the transport for TCAP messages that fit within a single message. If a particular transaction requires more than one message, the sequenced connectionless service is used. Connection-oriented services are rarely supported in the current deployment of SS7. In fact, the connectionless service is the only one used/deployed in the North American networks.

To indicate that a packet is an SCCP packet at an MTP3 level, the SIO subfield is used (refer to Table 12.4). SCCP information is carried in the SIF field as shown in Figure 12.11. There are different SCCP message types due to different services provided by SCCP. The most commonly used message type among SCCP messages is an SCCP unitdata message; its schematic format is shown in Figure 12.12.

Here, we will describe several critical fields of this message. Two important fields in SCCP unitdata messages are *Called Party Address* and *Calling Party Address*; in fact, they both include further detailed information. For example, the first byte of called/calling party address contains the following information: routing indicator (2 bits), global title indicator (4 bits), subsystem indicator (1 bit), and point code indicator (1 bit). Routing indicator indicates whether to use national or international networks and whether to route using global title or point code/subsystem number. If the global title bit in the routing indicator is turned on, the global title indicator field is used to provide further information about the transaction type, numbering plan, and so on. The global title is used, for example, in 1-800-number translation. The subsystem indicator is used to indicate if the called/calling party address contains the subsystem number; if so, the next byte indicates the subsystem number. Finally, the point code

Routing Label (DPC, OPC, SLS)
Message Type (Value = 9)
Protocol Class (Value = 5)
Called Party Address
Calling Party Address

FIGURE 12.12   SCCP Unitdata message.

indicator is used to indicate if the called/calling Party Address contains the point code; if so, the point code is included after the subsystem number field.

It is important to keep in mind that understanding the SCCP fields described above requires some knowledge of a telephone numbering plan, the role of certain numbers in a numbering plan, and their usage. This will be discussed further in Chapter 13.

12.11.2 TCAP: Transaction Capabilities Application Part

TCAP uses SCCP as the transport layer for a variety of applications such as 1-800-number lookup (for global title translation), calling card validation, roaming functionality for wireless services, and so on. For example, for global title translation, the routing indicator field in SCCP is set to zero. We will discuss global title translation in further details in Chapter 13 after we introduce the E.164 numbering plan.

A TCAP message has two key parts, the transaction portion and the component portion, and may include an optional part, the dialog portion. The transaction portion contains information about a specific transaction, such as a transaction identifier and the packet type, that can be used by an STP to route to the proper destination. The component portion contains the data received from the application. The optional dialog portion is used, for example, for encryption information.

There are several TCAP message types that can be identified in the transaction portion: (1) Unidirectional—sent in one direction without expecting a reply; (2) Query with Permission/Query without Permission—sent when accessing information stored in a database; also to separately indicate whether it is to be done with permission or without permission; (3) Response—used in responding to a query; (4) Conversation with Permission/Conversation without Permission—generated as a follow-up to a query; again, with or without permission can be indicated; and (5) Abort—sent when an originating entity wants to end a transaction. As an example, an 1-800-number query with permission will result in a response that contains the routing number(s) associated with the 800-number. We will discuss this further and illustrate its usage in Chapter 13 after discussing numbering plans.

12.12 SS7 Link Traffic Engineering

It is extremely important that a message generated at an SSP destined for another SSP reaches there quickly. In SS7 networks, this is governed by a simple rule of thumb: the utilization on any link of a date rate of 56 Kbps should not exceed 40%. Recall that between two adjacent nodes in an SS7 network, we have a linkset consisting of up to a maximum of 16 links. Here the rule of thumb refers to a link, not the linkset.

What does 40% utilization threshold on a 56-Kbps link mean? To understand that, we need to understand the signaling message arrival process. Note that call arrivals in the telephone network trigger the generation of associated ISUP IAM messages. Since call arrival is assumed to follow the Poisson process (Appendix B.10), we can assume that ISUP IAM message arrivals also follow the Poisson process. Indeed, there are other SS7 messages that can traverse an SS7 link in addition to ISUP IAM messages. For simplicity, we assume that all SS7 message arrivals to a link follow the Poisson process. We consider the 40% utilization as the rule of thumb for all such arrivals. To see the impact on delay, we will use the $M/M/1$ queueing result discussed earlier in the context of IP traffic engineering (see Section 7.1.4 in Chapter 7). Let the Poisson arrival of SS7 messages be denoted by λ, and the message service rate of a 56-Kbps link by μ. Then the utilization for a 56-Kbps link, ρ, is given by

$$\rho = \frac{\lambda}{\mu}.$$

Since $\rho = 0.4$, thus, we have $\lambda = 0.4\,\mu$. The service rate of messages is related to the link speed, c, and is given by the following relation:

$$c = \kappa\mu$$

where κ is the average message size (in bits). In SS7 networks, the average message size is around 40 bytes for most ISUP messages. For a link speed of 56 Kbps, we can determine the average message service rate of a 56-Kbps link as

$$\mu = c/\kappa = 56{,}000/(40 \times 8) = 175 \text{ messages/sec.}$$

Thus, the average message arrival rate at 40% utilization is then

$$\lambda = 0.4\,\mu = 70 \text{ messages/sec.}$$

Using the $M/M/1$ queueing delay formula, given by Eq. (7.1.4), we can find that the average queueing delay, τ, on a 56-Kbps link at 70 messages/sec average arrival rate is

$$\tau = \frac{1}{\mu - \lambda} = \frac{1}{175 - 70} = \frac{1}{105} \approx 9.5 \text{ millisec.}$$

Now let us examine how this impacts the end-to-end delay in an SS7 network. Note that the delay on a link also depends on the propagation, the transmission delay, and the node processing delay (refer to Appendix B.9). We assume here that the node processing delay for an SS7 packet is comparatively negligible compared to the other three factors. For a 56-Kbps link traversing 300 km in length, the delay due to the propagation and the transmission delay is about 7.2 millisec for a 40-byte packet. Thus, we can consider the delay on a link to be not more than 17 millisec when we also include the queueing delay factor.

Consider now Figure 12.2. Sending a message from an SSP to another SSP in this network can take up to three links under the normal routing rule: an A-link followed by a B-link followed again by an A-link if the SSPs are served by different mated home STPs. For simplicity, we will assume that all links are 300 km long. Thus, to traverse these three links, the total delay, based on 40% utilization on a 56-Kbps link speed for each leg, will be 51 millisec.

When a call goes over multiple network providers, say three network providers, the call setup message will be transferred from one provider's SS7 network to the next and so on. Assume this time that the message must go over two SS7 links in each provider's network; then, the end-to-end delay for a message would be about 102 millisec ($= 17 \times 2 \times 3$) under the 40% utilization rule on 56-Kbps link speed. This illustrates two important aspects: (1) why it is good to keep link utilization low, and (2) why the number of links in SS7 networks should be kept to a minimum to contain the end-to-end delay at an acceptable level since this counts toward postdial delay.

We now briefly comment on why 40% utilization is a good rule on an SS7 link of low speed such as 56 Kbps. Note that an SS7 is subject to two additional functionalities at the MTP2 level: (1) frame check sequencing catches any transmission error through checksum check, and (2) if a packet is garbled when recognized through checksum, retransmission needs to be indicated. Thus, the sending node is required to keep an MSU in its retransmission buffer until an acknowledgment is received from the other end. If there is nothing new to send, the FISU is transmitted. In addition, SS7 message size in general has minimal variability. Due to these factors, the message processing behavior is much more complicated than the exponential service distribution assumed above. Even with Poisson packet arrival behavior, such systems, known as the M/G/1 system in queueing theory due to such a general service process, have a more complicated queueing delay behavior. Rather the system can potentially become unstable at utilization over 50% if there are any transmission errors; as a result of this instability, the delay curve rises steeply near 50% utilization; for a detailed analysis, the reader is referred to [646]. Thus, the 40% utilization rule is a good rule of thumb for low-speed links such as a 56-Kbps link.

An SS7 network can now allow higher link speeds such as 1.544 Mbps. It is easy to see that due to multiplexing gain on such a higher speed link, with 40% utilization, the queueing delay will be much less. Note that the service rate of a 1.544-Mbps link with an average message size of 40 bytes is 4825 messages/sec ($= 1,544,000/(40 \times 8)$); at 40% utilization, the allowable arrival rate will be up to 1930 messages/sec; again using the $M/M/1$ queuing delay formula given by Eq. (7.1.4), we can find that the average delay will now be less than 1 millisec. Thus, through higher speed links, the end-to-end delay can be reduced significantly if 40% utilization threshold is still used as the rule of thumb. Alternately, a somewhat higher link utilization threshold can be permissible on a 1.544-Mbps SS7 link while the delay is still predictably bounded.

12.12.1 SS7 Network Performance Requirements

For operational environments, performance parameters for specifying signaling delays are specified in terms of a hypothetical signaling reference connection (HSRC). The aim is to contain the postdial delay incurred during SS7 message transmission, i.e., signaling delays in

T A B L E 12.7   Hypothetical signaling reference connection for 95% of signaling connections: delay requirement and nodes in a path.

Network Type	Time (in millisec) Simple (complex) messages	Number of Nodes
National Component (large)	520 (800)	8
National Component (average)	390 (600)	6
International Component (large to large)	410 (620)	7
International Component (large to average)	540 (820)	9
International Component (average to average)	690 (1040)	12

regard to the time taken for the transfer of SS7 messages from the originating SSP to the destination SSP in the HSRC. The reference value is given for 95% of the signaling connections in terms of large or small networks, and whether it is for national or international networks, including recommendation for the maximum number of nodes on an SS7 path [85] (Table 12.7).

12.13 Summary

SS7 networking is critical to the workings of the public switched telephone network. It has two key dimensions: its network topological architecture and its protocol stack. We have indicated how routing works in an SS7 network given its topological architecture. The topological architecture is built in such a way that there is enough redundancy so that a message has multiple paths to reach its destination. SS7 protocol stack provides several application functions such as call processing and service lookup that are used by the telephone networks. We discussed in detail ISUP messaging that can be used for call processing, trunk management, and in a dynamic call routing environment. We presented the difference between call routing and SS7 message routing. Finally, we covered SS7 link traffic engineering.

An important point to note about SS7 is that many message types have been defined for call setup, maintenance, and call release using ISDN user part (ISUP). Note that while ISUP was originally defined for ISDN, it is now commonly used for regular telephone call management. Furthermore, some message types are used for trunk management, while others can be used for congestion management.

Further Lookup

The best source of material on SS7 is actual specifications such as [327], [329], [331], [689]; for example, for a complete list of cause identifiers with an ISUP REL message, refer to [331]. There are several books and web-sites that provide significant details about SS7 networking; for example, see [102], [183], [190], [194, Chapter 2], [311], [556], [557], [613].

To further understand the use of signaling messages in dynamic call routing, see [27, §16.4]. For SS7 network engineering and performance analysis, the reader may refer to [297], [516], [612], [646]. For performance requirements of SS7 networks, see [85], [325], [326], [634]. Like any other network, SS7 also is found to have security issues; for example, see [425], [431], [500], [522] for a discussion. In recent years, there have been efforts in the IETF community to standardize SS7 over IP [247], [501], [642].

Exercises

12.1. Review questions:

   (a) What are the three key components of an SS7 network?

   (b) Describe all the different types of links possible in an SS7 network.

   (c) Identify the common fields in FISU, LSSU, and MSU.

   (d) Describe the various components of the North-American point code addressing.

12.2. Why are 2STPs associated with each SSP in an SS7 network?

12.3. Compare the SS7 protocol stack to the TCP/IP protocol stack, and identify similarities and differences.

12.4. Ignoring regional STPs, determine the minimum and maximum number of hops from an SSP to another SSP.

12.5. Investigate point code formats used in China and Japan, and compare them to the North American and the ITU formats.

12.6. Why is it necessary to have an F-link?

12.7. Explain all the valid routing rules in the SS7 network.

13

Public Switched Telephone Network: Architecture and Routing

Watson come here, I want you!

Alexander Graham Bell

Reading Guideline

The first part of this chapter can be read independently. Knowledge of SS7 (Chapter 12) and call routing (Chapter 10) is necessary for understanding various scenarios discussed as we progress through this chapter. In turn, this chapter helps in understanding IP-PSTN routing presented later in Chapter 20.

In this chapter, we introduce E.164, the telephone number addressing scheme used worldwide. Within this plan, each country can determine its own numbering scheme. The Public Switched Telephone Network (PSTN) architecture relies on this numbering scheme along with routing hierarchy to properly deliver calls from one part of the world to another. With the advent of SS7 messaging, the SS7 network takes an important role in call routing, especially due to the numbering plan, while maintaining the hierarchical structure. While there are still parts of the world that do not use SS7, we still assume prevalence of SS7 as the basis for discussing call routing. The absence of SS7 in call routing reverts back to the hierarchy rule as described earlier in Chapter 10 and is not discussed here.

After introducing the number plans and the dialing plans, we then progressively consider call routing in the presence of the numbering plan, as we move from a single nationwide network, to the multiple long-distance service provider environment, to the multiple local exchange provider environment, and, finally, to the scenario of number portability. In most cases, we use the North American numbering plan for illustration; the basic ideas are applicable in other numbering plans as well. As appropriate, we present variations due to differences in the numbering scheme.

Finally, as you read through this chapter, an important point to keep in mind is that the overall architecture should be efficient so that postdial delay/call setup time is kept to a minimum without customers noticing this factor.

13.1 Global Telephone Addressing

While telephone services have been around for more than 100 years, a global telephone addressing architecture is only about 40 years old. The telephone numbering format currently used is known as E.164 [323]; it has evolved from E.29 to E.161/Q.11 to E.163 to its current form E.164. The E.164 addressing scheme E.164 has the following general format:

$$\boxed{\text{Country Code | National Destination Code | Subscriber Number}} \qquad (13.1.1)$$

A country code can be one to three digits and a national destination code is generally two to four digits; the total number of digits is of variable length with a maximum limit of 15 digits. In practice at present, the entire number, including the country code, is not longer than 13 digits, while lengths of 10 and 11 digits are the most common. To avoid confusion between country and national destination codes, and to give global significance, the complete telephone address is now listed starting with "+" sign followed by the country codes to help quickly identify the country codes and then the rest. The national destination code is also referred to as the area code in countries such as the United States and Canada, and city codes in other countries. Usually, for the purpose of human processing, country/city/subscriber numbers are either separated by space and/or a dash or a dot or parentheses. For example, for country code 1 used in North America, area code 816, and subscriber number 328-9999, it can be listed as: +1 816 328-9999 or +1 (816) 328-9999, or +1.816.328.9999, all meaning the same. Note a convention for human understanding. In fact, only the actual number sequence matters; dashes or periods are not entered. The call processing module in the originating time-division multiplexed (TDM) switch would understand and handle only the dialed numbers. For simplicity and consistency, we will use the dash convention in this book, as in +1-816-555-1212, which is then an 11-digit number in E.164 format.

TABLE 13.1   World zone for the first digit of country codes, based on E.164 addressing.

World zone	Regions/Countries
1	North America, Caribbean Countries, and US territories
2	Africa
3 and 4	Europe
5	South/Latin America
6	Oceania, South East Asia, South Pacific
7	Russia and Kazakhstan (originally assigned to former Soviet Union)
8	East Asia and Special Services
9	Middle East, West Asia, South Asia

We will start with a brief history of the world telephone numbering scheme. In the mid-1960s, the CCITT Blue Book (also known as Recommendation E.29) first presented the initial partitioning of the world into several code zones and described the initial set of country codes. Zones are the single number to identify large geographic regions. While much has remained the same as was assigned in the beginning, some adjustments have been made since then. The current breakdown of the zones can be best described as shown in Table 13.1.

Essentially, a zone signifies the *first* digit of a country code; a country code often starts in the world zone in which the country is geographically located and has a maximum of three digits. While in most cases the country code of a country is located in the geographic zone, this is not always the case; for example, Mexico is allocated a country code +52 from World Zone 5, not from World Zone 1; Greenland is allocated country code +299, not from World Zone 3 or 4. There are other exceptions as well. Some countries do have the world zone as the country code; furthermore, a country code can be shared by different countries. For example, the United States and Canada share country code "1," while Russia and Kazakhstan share country code "7." It may be noted that the former Soviet Union was allocated zone 7; all of the rest of the countries that were part of the former Soviet Union have moved since then to either zone 3 or zone 9. Another exception, particularly in World Zone 1, is that the Caribbean countries do not really have country codes; these countries are assigned area codes under the North American Zone; for example, Jamaica has the country identifier +1-876, Barbados has the identifier +1-246, and so on. Note that a zone can be defined to start with "0;" thus, a country code may start with "0." However, as of now, no country codes starting with "0" have been assigned. In Table 13.2, the country codes for a representative set is listed.

There is an important point to note about the assignment of country codes; the entire addressing scheme must have *no* ambiguity from the point of view of dialing and, thus, for processing of dialed numbers at a TDM switch. From Table 13.2, we can see that there are country codes that are two digits long, and there are others that are three digits long, yet there is a special pattern to it. Consider, for example, the country code for India; it is +91. Can we then have a country country code +913? The answer is no. Since India is already assigned +91, the third number after 9 and 1 would indicate the *first* number of a city code in India that starts with 3. This also means that, besides two regions with one-digit country codes, not many countries should be assigned a two-digit country code so that all countries in the world can be covered with a maximum of three digits. Sometimes, there are interesting extensions to

TABLE 13.2 Examples of country codes from different zones.

Country	Country Code	Country	Country Code	Country	Country Code
Zone 1		Slovak Republic	+421	*Zone 7*	
USA & Canada	+1	United Kingdom	+44	Russia and Kazakhstan	+7
Jamaica	+1-876	Poland	+48	*Zone 8*	
Trinidad & Tobago	+1-868	Germany	+49	International Freephone	+800
Zone 2		*Zone 5*		Japan	+81
Egypt	+20	Nicaragua	+505	South Korea	+82
Nigeria	+234	Costa Rica	+506	Hong Kong	+852
Kenya	+254	Peru	+51	China	+86
South Africa	+27	Mexico	+52	Bangladesh	+880
Zone 3		Brazil	+55	Taiwan	+886
Greece	+30	Ecuador	+593	*Zone 9*	
France	+33	*Zone 6*		Turkey	+90
Iceland	+354	Malaysia	+60	India	+91
Lithuania	+370	Australia	+61	Pakistan	+92
Italy	+39	Indonesia	+62	Lebanon	+961
Zone 4		New Zealand	+64	Palestine	+970
Romania	+40	Thailand	+66	Israel	+972
Switzerland	+41	Kirbati	+686	Iran	+98
Czech Republic	+420	Marshall Islands	+692	Uzbekistan	+998

three-digit country codes from a two-digit country code. For example, Czechoslovakia was originally assigned country code +42. After the breakup of Czechoslovakia into two countries, country code +42 is no longer valid; instead, the Czech Republic has been assigned country code +420 and the Slovak Republic has been assigned country code +421.

Finally, +388 is a *virtual* country code space that can be used by a *group* of countries. Within this address space, +388-3 has been assigned to European Telephone Numbering Space (ETNS), intended for services Europe-wide [207], [526].

13.1.1 National Numbering Plan

Beyond the country code, each country must have a numbering plan so that the number dialed by users can be of variable length, yet there is no ambiguity. Below, we will discuss briefly three different national numbering plans. It is important to note that the dialing plan is adapted from the numbering plan; the dialing plan need *not* be exactly as specified in the numbering plan and should not be confused with the numbering plan; this will become clear when we discuss dialing examples later.

NORTH AMERICAN NUMBERING PLAN

The North American Numbering Plan (NANP) that spans the United States, Canada, and Caribbean countries follows a fixed 10-digit format for a normal phone number, while allowing certain exceptions for special purposes such as 911 for emergency services and 411 for directory services. Recall that they are in World zone 1 with country code designation +1. The NANP is often referred to as the NPA–NXX–XXXX format, where NPA is a three-digit numbering plan area, also commonly referred to as the area code; NXX is a three-digit numbering for local exchange; and XXXX is a four-digit station code that identifies a subscriber. For example, for telephone number 816-235-2006, 816 is the NPA, 235 is the NXX, and 2006 is the station code.

Specifics of which combination of three digits is allowed to be in an NPA or an NXX have evolved over time. For example, until a decade ago, the middle digit in NPA could only be 0 or 1; this is no longer the case (as of 1995). In the current form, the first digit N, however, cannot be 0 or 1, i.e., it can take 2 to 9, regardless of whether the first digit is the numbering plan area or the exchange code, while all the other positions can be any digits from 0 to 9, which is often indicated by X. Thus, the current format can be best described as NXX–NXX–XXXX, although we will refer to it as NPA–NXX–XXXX so that it is easy for us to distinguish between the numbering plan area code and the local exchange code. Within the numbering format, N11 is saved for special services such as 911, and, thus, cannot be used as area codes or an exchange code. The reason why no NPA or NXX starts with 1 or 0 is that 1 as the first digit dialed is reserved to recognize that it is a toll call, i.e., the subscriber pays, usually on a per-call basis. Thus, the first dialed digit being 1 is referred to as the national direct dialing (NDD) prefix and 0 as the first digit followed by any digits other than 1 is reserved for operator-assisted calls, while 0 followed by 11 is used for international direct dialing (IDD) prefix. There are also special purpose area codes 800, 888, 877, and 866 that are allocated for toll-free (freephone) services; for simplicity, the numbers in this group will be referred to as *1-800-numbers* or *toll-free numbers*. The area code or exchange code 555 is reserved for informational purpose. The area codes N9X are known as expansion/reserved codes; they are slated for use when the current 10-digit NANP format would need to be expanded. Furthermore, area codes 37X and 96X are set aside in case it is desirable to have 10 contiguous address blocks for any unanticipated need.

In the NANP, there is no special numbering plan for mobile/cellular services. Mobile numbers are also of NPA–NXX–XXXX format. While specific NXX's within a particular NPA has been assigned to mobile providers, such assignments are no longer necessary with local number portability (refer to Section 13.9).

The NANP is currently administered by the North American Numbering Plan Administration (NANPA) [531]. NANPA follows the NPA Allocation Plan and Assignment Guidelines as prepared by the Industry Numbering Committee (INC) of Alliance for Telecommunications Industry Solution (ATIS) [309], [310]. The Canadian Steering Committee on Numbering (CSCN) is in charge of allocation within Canada [109].

NUMBERING PLAN IN INDIA

In April 2003, India (country code +91) did its most recent and extensive revision of the national numbering plan, with the expectation that this plan will serve the country for another 30 years [491].

The numbering plan in India now conforms uniformly to 10 digits where the national destination code and subscriber numbers can be of different lengths as long as they add up to 10 digits; thus, in E.164 format, the total length is 12 digits including two digits for the country code. The national destination code in India is referred to as the trunk code or area code that identifies a short-distance charging area (SDCA); it can be of two, three, or four digits in length, which means that the subscriber numbers in corresponding cases are of 8, 7, or 6 digits, respectively. For example, the SDCA for Delhi/New Delhi is 11 (just two digits); thus, all subscriber numbers are of eight digits. The first two digits of trunk codes range from 11 to 89, while the trunk codes can be of two to four digits in length. In other words, no trunk codes can start with 0 or 9. Note that the NDD prefix in India is 0, and the IDD prefix is 00. If a trunk code were to start with 0 as the first digit, then there would have been a conflict while trying to dial such a number with NDD prefix 0 since IDD is 00. A trunk code that starts with 9 is primarily reserved for mobile services with certain exceptions (see below for further details).

Codes 100, 101, and 102 are assigned for police, fire, and ambulance services, respectively. Thus, no subscriber number can start with these as the first three digits. Subscriber numbers can start in the range of 2 to 6; currently, only 2 is used in practice as the first digit for a subscriber number. Digits 0, 1, 7, 8, and 9 cannot be used as the first digit for telephone exchange codes in basic services. In fact, the new guidelines say that a number that starts with 1 is for special services; there are many special services defined beyond police, fire, and ambulance services. An amusing pair of numbers is 116 for wakeup call registration and 117 for wakeup call cancellation; see [491] for a detailed list. A number that starts with 1 such as 116 might seem contradictory since the SDCA code for Delhi is 11 and a subscriber number can possibly start with 6. Note that while the SDCA code for Delhi is 11, a telephone number in Delhi is dialed *without* 11 when dialed within the Delhi SDCA area, i.e., just the subscriber number is dialed, which starts in the range 2 to 6. From outside Delhi but *within* the rest of India, a telephone number located in Delhi is dialed with 0 as the prefix and then the SDCA code 11, i.e., as 011, which is then followed by the actual subscriber number. From outside India, a number in Delhi will have the dialing format +91-11. Note that the special services numbers that are dialed starting with 1, such as 116 and 117, cannot be dialed or accessed from outside India; thus, there is no ambiguity.

It may be noted that the notion of the national destination code/subscriber number is for the basic PSTN service, that is, for the landline number. Mobile services in India use a three-part format keeping the total number of digits also at 10. A mobile number has the following structure: a two-digit *Public Land Mobile Network* (PLMN) number, a three-digit *Mobile Switching Center* (MSC) code, followed by a five-digit subscriber number. The range of numbers for two-digit PLMN starts with 9, except that 90, 95 and 96 are reserved for other purposes, mainly due to historical reasons. Thus, valid PLMN numbers are 91, 92, 93, 94, 97, 98, and 99. As an example, a valid 10-digit mobile number is of the format 98-965-827XX where PLMN is 98, MSC is 965, and the subscriber number is 827XX. In "+" notation, this can be written as +91-98-965-827XX.

NUMBERING PLAN IN CHINA

The numbering plan in China (+86) has certain similarity to the numbering plan in India. Landline numbers, excluding country code, can be of 10 or 11 digits in total while mobile numbers are always of 11 digits. The national destination code in China is referred to as the area code; it can be of two, three, or four digits in length while subscriber numbers are either eight digits or at minimum seven digits. Area code allocation is divided by geographic areas:

Area 1 is Beijing with area code 10; Area 2 is for large Chinese cities that are given two-digit area codes such as Shanghai – 21, Tianjin – 22, Chongqing – 23, and Nanjing – 25. Area 3 is for the Hebei, Shanxi, and Henan area codes, and so on. Thus, an example of a number in Beijing in E.164 notation is +86-10-8230-XXXX. It may be noted that mobile services nationwide start with code 13. An example of a mobile phone number with "+" notation would be: +86-13-81782-XXXX.

Codes 110, 119, and 120 are assigned to police, fire, and ambulance services, respectively. The national direct dialing prefix in China is 0, while 00 is the international direct dialing prefix, similar to India and most countries.

13.1.2 Dialing Plan

The E.164 plan shown in Format (13.1.1) describes the full format of international telephone addressing. For historic reasons, the full format is not required for dialing unless a call is made to an international location. Most countries allow essentially four forms of dialing: a call to a local number within a city or geographic proximity, a local call for special services such as for police, a call to another city/region outside the local dialing region, and a call to an international location; except for the first and the second cases, prefixes are required to be prepended to a number such as the NDD prefix that is used for within a country and the IDD prefix that is used for international calls. Mobile phones have somewhat different dialing plans than residential landline phones. Even with this variety, there must be no ambiguity in dialing. We will illustrate two examples for dialing from residential landline phones.

Example 13.1 *Dialing from a residential landline phone in the United States.*

Consider a user making a call from the landline[1] number +1-816-328-9999, shown in E.164 format; note that it is located in area code 816 (Kansas City area). For emergency service calls to 911, the user simply dials 911. A local call to a number in the same area code, say, +1-816-235-2006, is dialed simply as 235-2006. Consider now calling two different numbers, +1-913-235-2006 and +1-351-235-2006, in two different area codes. The first number, +1-913-235-2006, is also considered a local number in the Kansas City area, although it has a different area code; thus, the user can simply dial this number as 913-235-2006 without prepending 1 for the NDD prefix; note that there is no ambiguity in dialing 913-235-2006 compared to dialing just 235-2006 since the second one means that the number is already in area code 816 where the call origination is also from a number with 816 as the area code. The other number, +1-351-235-2006, is a number in area code 351, which is assigned to the state of Massachusetts, i.e., in a different geographic region; thus, the user is required to dial 1-351-235-2006; here, the first digit "1" indicates that it is a long-distance call, i.e., "1" is the NDD prefix. There are also area codes that are used for a variety of services; the best known is the toll-free 1-800-number calls; for example, the user would dial such a number as 1-800-213-XXXX; this number is then routed to a *routable number*, i.e., a valid line number for actual call routing.

---

[1] At present, central office codes such as 816-328 and 816-367 have not been assigned, according to NANPA [531]; we are using them here for illustration assuming that they are associated with landline TDM switches.

TABLE 13.3   Numbers as dialed from +1-816-328-9999: the originating TDM switch's view.

```
911
2352006
9132352006
13512352006
1800213XXXX
011911162345678
011919896582779
011861381782XXXX
```

Now consider that the user wants to make a call to the following international number: +91-11-623-45678; this number would be located in India (country code: +91), in Delhi (cite code: 11), and the local subscriber number is 623-45678. In this case, the user in Kansas City dials 011-91-11-623-45678, where "011" indicates the international access code, i.e., the IDD prefix. Similarly, to dial a mobile number in India, for example, +91-98-965-82779, the user would need to dial 011-91-98-965-82779. To dial a mobile number in China such as of the form +86-13-81782-XXXX, the user would need to dial 011-86-13-81782-XXXX.

We can see that although dialing a telephone number is of variable length, there is no ambiguity in dialing so that the originating TDM switch can properly process the dialed numbers. From the originating TDM switch's perspective in regard to the example numbers discussed so far, it receives the digits as shown in Table 13.3. We can see from this table that the originating TDM switch, based on the first few digits and/or the number of digits dialed, would need to know how to handle/route a call without ambiguity.

To summarize, the first number is handled for emergency services based on the numbers dialed. The next two are based on the number of digits dialed, with the first three digits indicating that it is not a call for a special service. The fourth one is handled for long-distance, based on the first digit being 1; the last three are handled for an international connection based on the first three digits being 011.   ▲

Example 13.2   *Dialing from a residential landline phone in India.*

Consider a user making calls from the residential landline[2] number +91-11-623-45678; this number is located in the city code 11 (Delhi). Thus, a local call to another number within the same city code, say, to +91-11-676-54321, can be dialed simply as 676-54321. Now consider calling a number within India to a different city code, say, dialing +91-361-673-0710, which would be located in city code 361 (Guwahati), or dialing to +91-452-667-1203, which would be located in city code 452 (Madurai); these are long-distance (or trunk) calls and, thus, the user is required to dial 0-361-673-0710 and 0-452-667-1203, respectively. Here, the first "0" followed by a digit other than 0 indicates that it is a long-distance (trunk) call since the NDD prefix in India is "0." Now consider that the user wants to dial a call to the following international

---

[2]This is valid, but is not an assigned number yet; all subscriber numbers in Delhi currently start with 2 while the new numbering plan allows that the first digit of a subscriber number can be in the range of 2 to 6 [491].

TABLE 13.4   Numbers as dialed from +91-11-623-45678: the originating TDM switch's view.

```
67654321
03616730710
04526671203
09896582779
0018163289999
009221585XXXX
00861381782XXXX
```

number: +1-816-328-9999, located in the United States, in city code 816. In this case, the user in Delhi would dial 00-1-816-328-9999, where "00" indicates the international access code, i.e., the IDD prefix in India, followed by "1" for the country code for the United States or Canada, and then the rest of the digits. Similarly, to reach the international number, +92-21-585-XXXX, which is located in Pakistan, the user would dial 00-92-21-585-XXXX, or to reach the mobile number +86-13-81782-XXXX in China, the user would dial 00-86-13-81782-XXXX.

A mobile number in India can be a bit confusing, partly because the number does not appear to indicate locality based on the initial few digits. Consider the number +91-98-965-82779. If it were a local number in Delhi, the user in Delhi would dial 98-965-82779, while if it is a number from another geographic area, then the user in Delhi would need to dial 0-98-965-82779 to indicate that it is a long-distance call. A common user may not know that the mobile switching center code part 965 is not located in Delhi. In this specific example, it is indeed a long-distance call from Delhi and the landline user is required to dial 0-98-965-82779.

We can again see that there is no ambiguity in dialing. The originating TDM switch in Delhi based on the first few digits and/or number of digits dialed knows how to handle a call. From the perspective of the TDM switch, the TDM switch receives the digits as shown in Table 13.4. Here, the first one is based on the number of digits dialed; the next three are handled for long-distance based on the first digit being 0; the last two are handled for an international connection based on the first two digits dialed being 00.   ▲

Dialing from a phone other than a residential landline phone varies. For example, from a corporate environment, it is common to include the prefix 9 to dial outside and then dial the digits as shown in earlier illustration with residential landline numbers. In the case of mobile phones, it is often not necessary to dial the NDD prefix; for example, in North American, most wireless network providers allow the user to dial a number without the NDD prefix.

Later, we will discuss how routing is accomplished based on a dialed number, especially dialing of *leading digits*, i.e., the first few digits dialed, at the originating TDM switch. The need for determining routing at the originating switch based on leading digits partly arises from the the fact that the numbers dialed can be of variable length, thus, not knowing when it ends, and also since there is not really a concept of an "enter" button on telephone dialing, although the # key on telephones sometimes serves as the enter button. That is, preprocessing of dialed digits is required before deciding on routing.

13.2 Setting Up a Basic Telephone Call and Its Steps

We will illustrate a basic telephone call; this process helps in understanding how digits dialed leads to a routing decision. While there have been many different types of switches over the years, we will assume in this illustration that the switches have the ability to store and process dialed digits without any out-of-band signaling functionality; that is, call routing that requires SS7 signaling will be discussed later.

Consider a telephone call from user A to user B who are located in the same local area, requiring local dialing. In this illustration, we will start with the assumption that both numbers are served by the same service providers. For this illustration, we will use the telephone numbering system in North America. Consider user A's number to be (816) 328-9999 and user B's number to be (816) 235-2006. In this case, the called party number is (816) 235-2006 and the calling party number is (816) 328-9999. Note that both numbers are located in the same city served by the area code 816, but served by different central office codes; thus, user A is required to dial only 235-2006.

User A picks up the receiver:

1. User A first lifts the receiver; this off-hook status lets the central office/exchange 328 know that user A requires the phone service, based on the presence of direct current in subscriber line interface (known as DC signaling).

2. Central office/exchange 328 determines whether it has the *originating register* to store digits dialed.

3. Once central office/exchange 328 verifies the availability of the originating register, it then provides dial tone functionality to user A.

User A dials a number:

1. Once user A hears the dial tone, she starts dialing user B's number 235-2006.

2. As soon as user A dials the first digit, the dial tone is disconnected, and the exchange starts storing the digits dialed.

3. By examining the leading digits dialed, 235 in this case, the exchange realizes that it is a local *interoffice* call.

4. The originating exchange does a lookup process, known as *translation*, to determine the trunkgroup identifier that connects exchange 328 to exchange 235, and starts scanning for an idle trunk on this trunkgroup.

5. If an idle trunk is found, then this trunk is *seized* and is marked as unavailable to other calls originating at the same exchange; furthermore, this seizure also causes the incoming register to be seized at exchange 235.

6. Once the incoming register at exchange 235 indicates readiness to receive dialed information to exchange 328, outpulsing of digits dialed by user A begins.

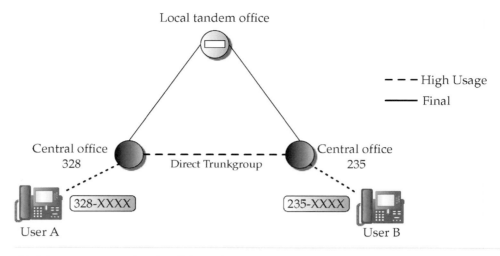

FIGURE 13.1   A local call from (816) 328-9999 to (816) 235-2006.

7. Before sending the last digit, the originating exchange checks if user A is still off-hook since otherwise there is no reason to connect the call. Assuming user A is still off-hook, incoming digits are then stored at the terminating exchange.

8. The terminating exchange then determines if user B's phone is on or off-hook. If it is idle, then the ringing register at the terminating exchange is seized, releasing the incoming register; the audible ring is then returned to the originating exchange, while it also starts ringing user B's telephone.

This, thus, completes the call establishment.

There is an important point to note here. Suppose after translation, and scanning the trunkgroup to the terminating exchange, it was found that all trunks in this trunkgroup are busy. In this case, the originating exchange will identify the local tandem office and look for an idle trunk to this switch. If this trunkgroup is busy, then a fast busy tone would be generated and the call will be cleared from the system. If an idle trunk is found on this trunkgroup, then this trunk is seized, and the incoming register is seized at the local tandem office. The dialed digits are pulsed to the local tandem office, and the control of the call is forwarded to the local tandem office; the local tandem office then seeks to find an idle trunk on the trunkgroup that connects the local tandem office to the destination exchange (see Figure 13.1); the rest of the process is the same as the attempt on the direct trunkgroup described above.

To summarize, as soon as the user starts dialing, the originating exchange's immediate responsibility is to identify the appropriate outgoing trunkgroup. In general, an originating exchange have many outgoing trunkgroups to cater for connecting to different providers and services. Starting with Section 13.5, we will take this aspect into account as well as out-of-band signaling through SS7 in determining call routing. To concentrate on these aspects, functions such as having an originating register for holding dialed digits will be implicitly assumed from now on.

13.3 Digit Analysis versus Translation

In a digital switched network, digits dialed by a user are analyzed for the purpose of routing. There are two terminologies that require clarification in this regard.

Digit Analysis usually refers to inspection of leading digits dialed at the originating TDM switch to identify the type of dialed call, e.g., national, international, emergency, mobile call. Typically, digit analysis is done first on the three to six digits. This process is first executed in order to determine the next step to perform.

Translation usually refers to the set of steps necessary to perform to route a call. Steps may include, for example, lookup tables inside a switch as well as lookup in a service control point (SCP) database from a dialed number such as a 1-800-number. However, the term translation, as in *global title translation*, is also used when referring just to the lookup from the SCP database for a 1-800 number to determine a routable number.

13.4 Routing Decision for a Dialed Call

From Example 13.1 and Example 13.2 discussed earlier, we can see that there are several types of calls possible based on the numbers dialed. Regardless of where in the world a call originates with a residential phone, it can be essentially classified into the following basic categories:

- *Local calls*
- *Long-distance (trunk) calls*
- *International calls.*

We also note that in certain parts of the world the following two additional categories of calls are also prevalent:

- *Special services calls*
- *Toll-free calls.*

While the leading digits dialed provide some indication on how to handle a dialed number, this is not enough information for a routing decision at the originating TDM switch in a multiprovider setting, or with number portability (discussed later). Thus, before we go further, we first consider the simplest case in which the phone service within the national network is provided by a single provider, and for international calls, a handoff is done. Deviations from this scenario are discussed in subsequent sections.

13.5 Call Routing: Single National Provider Environment

The single national provider environment is still prevalent in many parts of the world. In fact, in many national networks, for both local and nationwide long-distance calls, the phone service is provided by the post and telecommunication department of the country. Note that for international calls, the national network forwards the call to a gateway TDM switch for international connectivity. We assume here that an SS7 network to support call routing is in place.

13.5.1 Handling Dialed Numbers

Based on the leading digits dialed, the originating TDM switch is required to identify the appropriate outing trunkgroup, which is often the case in the absence of SS7 signaling, or pre-SS7 days. When we take SS7 into account, especially ISUP messaging, the steps work somewhat differently. We assume that the reader is already familiar with SS7 terminology, especially point code (PC) addressing, SSPs, and ISUP messaging (refer to Chapter 12).

DIALED NUMBER IS A SUBSCRIBER NUMBER

We first consider the case in which the dialed number is a subscriber number. Under ISUP call processing, the originating TDM switch waits until all the digits are dialed (within an allowable time limit). Based on the dialed number, it parses and identifies the destination point code of the SS7 node to which the call is to be handed off. It is important to note that the destination point code (DPC) is not necessarily the final destination TDM switch of the call; it only signifies the SS7 node to which the call processing message is to be forwarded.

1. User dials a number.

2. The originating TDM switch performs digit analysis to determine how to handle the call. If it is an intraswitch call, it rings the subscriber's number; otherwise, it identifies the directly connected DPC of the TDM switch to which this call should be forwarded. Based on the identified DPC, it performs lookup of a destination point code to trunkgroup ID. This part can be, however, vendor implementation specific; in some switch implementation, from dialed digits, trunkgroup ID is first extracted and then the DPC is found. For uniformity in the rest of the discussion, we will assume that DPC is identified first and then the trunkgroup ID is identified.

3. For the trunkgroup ID, it seeks to identify an idle trunk (circuit) if it is available. It tags the identified available circuit for this call without fully reserving it, determines the trunk circuit identification code (TCIC) for this circuit, and generates an ISUP initial address message (IAM) that contains the TCIC code and the called number; this IAM is then sent to the DPC. It starts a timer and waits to hear either a response back such as an address completion message (ACM) or a release (REL) message.

   The TDM switch that receives this IAM message performs the following functions:

- To process the IAM message to identify the called number and to determine if it is itself the final destination TDM switch for this call.

  - If this TDM switch is the final destination, then it rings the called number and generates an ACM message to send back to the SSP from which it has received the IAM message.

  - If this TDM switch is NOT the destination, then it performs a lookup service based on the called number to consult the routing table to determine the next TDM switch's DPC, and identifies an idle outgoing circuit to this switch and its TCIC code (refer to Section 12.8). It starts a call state for this call (along with a timer), and *regenerates* an IAM message with itself as the originating point code (OPC) and the newly determined TDM switch as the destination code along with the new TCIC value just determined; certainly the original called number remains unchanged. This newly generated IAM message is forwarded to the DPC identified.

If a TDM switch, on receiving an IAM message, cannot find an available circuit to any of the TDM switches that are listed in the routing table, then it returns a REL message to the OPC from which it received this IAM message; in this REL response message, the cause is also included. According to ITU-T Recommendation Q.850 [331], the cause indicator for no circuit/channel availability is 34. In fact, there are numerous cause indicators specified for the REL message [331]. For example, a normal call is released when either party of a call decides to end the conversation; in this case, the REL message is generated with cause indicator field set to 16.

There are many functions that can occur as soon as the ACM message is received by the originating TDM switch. As an example, the *call detail record* (CDR) for this specific call at the originating switch is opened; this record is closed when the REL message is generated at the completion of the call. The CDR entry for a call stores information such as the called number, the calling number, the time the call started, and the time the call ended. In the rest of the discussion, we assume that a CDR is created for a call without explicitly mentioning it for every case.

DIALED NUMBER REQUIRING TRANSLATION TO A ROUTABLE NUMBER

This situation arises when the user dials, for example, a 1-800 toll-free number that is available; similar services are now available in many countries around the world. Such a number does not indicate where the destination number is. Thus, a lookup service using the 1-800-number database is required.

In a single-provider setting, we assume that the point code of the SCP node in the SS7 network that holds the 1-800-number translation database is *known* to the originating TDM switch. For the lookup service, the originating TDM switches recognizes that the dialed number requires title translation; thus, it creates a TCAP message and routes this message using the SS7 network to the SCP to request a routable number that should be used in place of the 1-800-number. Once the SCP responds with a routable number, which then becomes the called number, the originating TDM switch performs the same steps as the ones described above when the dialed number is a subscriber number. That is, dialed numbers such as 1-800 numbers require an extra phase before the normal call processing is done.

13.5.2 Illustration of Call Routing

We will illustrate the above concept through the use of the NANP, dialing from a residential phone. While this scenario is not applicable in practice in the United States, we use it purely for illustration and also for further discussion in the multiprovider environment later.

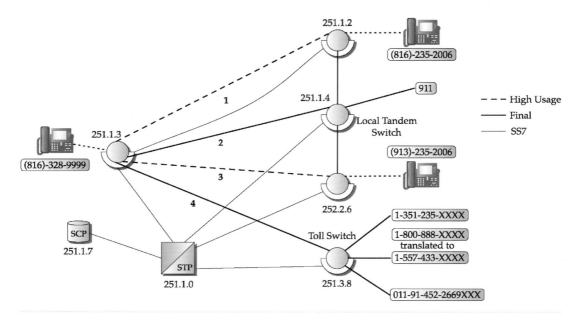

FIGURE 13.2   Call and message routing for calls dialed from 816-328-9999.

First, it is important to note that this is a *conceptual* logical view of the steps taking place *inside* a TDM switch while noting that different switch vendors may implement the entire call processing steps/logics somewhat differently. This illustration builds upon Example 13.1 for dialing from a residential phone in the United States. Recall that the call has originated from the number +1-816-328-9999. The user dials basic categories of calls such as local calls 235-2006, 913-235-2006, long-distance calls 1-351-235-2006, and international calls 011-91-11-623-45678, in addition, to special service calls 911, and toll-free calls—1-800-212-XXXX.

The network architecture as viewed from the originating TDM switch, taking into account SS7 networking and point code addressing, is presented in Figure 13.2. For simplicity, only one signal transfer point (STP) is shown to be connected to each SSP; that is, the mated STP is not shown. Assume that the point code for the originating SSP is 251.1.3. When the user (i.e., +1-816-328-9999) dials a telephone number for processing at the originating TDM switch (PC 251.1.3), it must follow the routing rule for call routing that is based on hierarchical routing; that is, first do direct trunkgroup routing if there is a high usage trunkgroup and the overflow to an alternate routing. For a local call, the final trunkgroup is the trunkgroup to the local tandem switch for alternate routing; for calls outside its region, the final trunkgroup is the trunkgroup to an appropriate toll switch. For convenience, this view of hierarchical routing is shown in Figure 13.4 using the routing rule and node notations discussed earlier in Section 10.1 in Chapter 10. To mimic this behavior in an SS7-enabled environment, the originating TDM switch maintains a lookup table for leading digits dialed to immediate next-hop SSP, along with an alternate SSP for overflow routing. In our example, SSP 251.1.4 is a local tandem switch for overflow routing for local traffic. In Figure 13.3, we present the example tables maintained at the originating TDM switch for call processing; their usage will be discussed below through a set of scenarios.

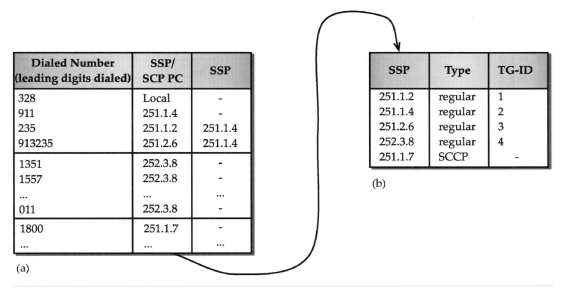

Dialed Number (leading digits dialed)	SSP/ SCP PC	SSP
328	Local	-
911	251.1.4	-
235	251.1.2	251.1.4
913235	251.2.6	251.1.4
1351	252.3.8	-
1557	252.3.8	-
...	...	...
011	252.3.8	-
1800	251.1.7	-
...	...	...

(a)

SSP	Type	TG-ID
251.1.2	regular	1
251.1.4	regular	2
251.2.6	regular	3
252.3.8	regular	4
251.1.7	SCCP	-

(b)

FIGURE 13.3   Call routing decision: conceptual view through table lookup.

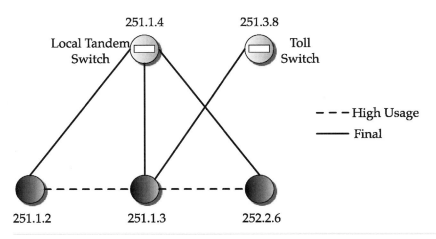

FIGURE 13.4   Hierarchical routing view at the TDM switch level.

DIALING 235-2006

Consider first that the user dials 235-2006. Then, the originating TDM switch identifies the next-hop SSP to be 251.1.2 using Figure 13.3(a); it then uses Figure 13.3(b) to identify the trunkgroup ID as 1 and checks if an idle circuit is available. We assume that there is a circuit available so that direct routing can be performed. The TCIC code for this circuit is then generated; the IAM message originating at SSP 251.1.3 includes this TCIC; in addition, this IAM message contains considerable information, most critical of which is that it includes the called number 816-235-2006. The prepared IAM message is sent to SSP 251.1.2. From Figure 13.2, note that these two SSPs are shown to be connected by an F-link in the SS7 network; thus, this IAM message is routed on this link. On receiving this message at SSP 251.1.2, it checks the content of the message and determines that this is the termination TDM switch for this call.

DIALING 913-235-2006

Now consider dialing 913-235-2006, a local number within geographic proximity of the originating number. First, assume that the call would be directly connected like the call to 816-235-2006. In this case, the destination SSP is identified as 251.2.6, and thus, the originating TDM switch will identify the trunk group identifier as 3 and check if there are circuits available on trunkgroup identifier 3. One difference, however, is that there is no F-link in the SS7 network between SSP 251.1.3 and SSP 252.2.6. Thus, the IAM message on the SS7 network would be forwarded to the STP 251.1.0, for further forwarding to SSP 252.2.6.

We now consider the next possible scenario for the call to 913-235-2006, i.e., if there are no circuits available on the high-usage, direct trunkgroup (identifier 3) from the originating TDM switch to the destination TDM switch. Based on the routing rule and lookup (see Figure 13.3), the originating TDM switch would try the trunkgroup to the tandem switch with PC 251.1.4 to see if there are any available circuits. We assume that a circuit is available on the trunkgroup to the tandem switch. Then, due to the progressive call control (PCC) functionality, the control of the call would need to be forwarded to SSP 251.1.4. This can be accomplished using the IAM pass along method, i.e., the IAM message is sent to SSP 251.1.4 indicating in its content that the called number is 913-235-2006 with OPC as 251.1.3 and DPC as 251.1.4. On receiving this IAM message, SSP 251.1.4 would check if it is meant for itself; based on its own routing table, it would know that the call setup message must be forwarded to the final destination, SSP 252.2.6. Thus, SSP 251.1.4 will create a *new* IAM message indicating itself as the OPC and setting the DPC as 252.2.6; this message will contain the newly determined TCIC code for the available circuit on the trunkgroup between SSP 252.1.4 and SSP 252.2.6 and will also contain the same called number 913-235-2006. Furthermore, SSP 251.1.4 will maintain in its memory a connectivity/mapping state table between the IAM message received from SSP 251.1.3 and the one sent to SSP 252.2.6.

When SSP 252.2.6 receives the IAM message from SSP 251.1.4, it determines that it is indeed terminating at this node, and it then sends an indication to telephone 913-235-2006 about ringing, and returns an ACM message to 251.1.4 with OPC 252.2.6 and DPC 251.1.4. When SSP 251.1.4 receives this ACM message, it checks its connectivty/mapping state table; thus, SSP 251.1.4 will in turn generate an ACM message with OPC 251.1.4 and DPC as 251.1.3. When the originating TDM switch 251.1.3 receives this ACM message, it knows that the call is connected.

From the above discussion, we see that IAM/ACM message content in regard to OPC and DPC changes at the intermediate TDM switch; this is essential due to hierarchical call routing with PCC and to avoid any confusion at the destination TDM switch. For instance, suppose that SSP 251.1.4 did *not* change the OPC field but changed only the DPC field to 252.2.6 in the IAM message sent to 251.1.6. Then SSP 252.2.6, on receiving this message, would see the OPC as 251.1.3 and would think that the TCIC value is on the trunkgroup between the originating SSP 251.1.3 and itself, which is certainly not the case.

DIALING EMERGENCY SERVICE NUMBER 911

Next, we consider dialing 911. In this case, we assume that the 911 operator trunks are connected off the local tandem switch with PC 251.1.4. Thus, in this case, the IAM message would need to be routed through STP 251.1.0 to SSP 251.1.4 since there is no direct F-link between SSP 251.1.3 and SSP 251.1.4. Once the IAM message arrives at SSP 251.1.4, it determines that the call is to be put on the trunkgroup to the 911 operator center, known as the Public Service Access Point (PSAP).

DIALING 1-351-235-2006 AND 011-...

Next we consider dialing a long-distance or an international number. By inspecting Figure 13.3, we see that for both cases the toll switch with SSP 251.3.8 is the next TDM switch, and the call will be attempted on the final trunkgroup to this switch (there is no high-usage trunkgroup in this case). The originating switch (SSP) will create an IAM message with OPC as 251.1.3 and DPC as 251.3.8. Since there is no F-link between these to SSPs, this IAM will be forwarded to STP 251.1.0, for further forwarding to 251.3.8.

DIALING 1-800 NUMBER

Finally, consider the case where the user dials an 1-800 number. From Figure 13.3(a), the originating switch would know that the next-hop SSP is 251.1.7. By consulting Figure 13.3(b), the switch would know that it is required to generate an SCCP message to SCP node 251.1.7 to obtain title translation for the 1-800 number. This SCCP message will be forwarded to STP 251.1.0, which will in turn be forwarded to PC 251.1.7 (see Figure 13.2). In the response message to SSP 251.1.3, SCP 251.1.7 will indicate that the mapping for this 1-800 number is 1-557-433-XXXX. On receiving this message, the originating TDM switch again consults Figure 13.3(a) and finds out that the SSP is 252.3.8 (same as the number that starts with +1-351). It will then handle the call in the same way as it would handle to call to 1-351-235-2005 described earlier.

Today's 800 number handling is much more sophisticated than what is described so far. This will be discussed later in Section 13.10.

13.5.3 Some Observations

From the above discussions and the illustration, we make several observations.

First, there is no direct relation between the telephone numbering plan, that is, how the end device (phone) is numbered, and the TDM switch number; the TDM switch number is based on the SS7 point code addressing. This is unlike IP networks where the end device numbering and the router numbering come from the same scheme.

Second, because telephone call routing is based on the TDM switching hierarchy with PCC, message routing in the SS7 network must follow this requirement by sending a message from an SSP to another SSP, sort of on a hop-by-hop basis in the TDM switched network. Consequently, it is important to properly take care of changes in OPC and DPC values at each SSP as well as TCIC value computation and change; from a performance point of view, this requires that an SS7 SSP does fast packet header change processing before forwarding. Routing between two SSPs in the SS7 network follows SS7 network routing rules.

Third, the originating TDM switch must maintain appropriate lookup tables to map from a dialed number to an SSP node, and then from an SSP node to a trunkgroup. Beyond that, a trunk-hunting algorithm is required to identify an available circuit, which in turn is used for generating the TCIC code; this TCIC code is then included in the ISUP IAM message. Beyond determining the TCIC code of a circuit, the actual physical circuit must be identified for actual routing of a call.

Finally, there is a subtle difference between a dialed number and the called number encoded in an IAM message. For example, for a local call from +1-816-328-9999 to +1-816-235-2006, the user simply dials 235-2006. When the IAM message is generated, the called number can be filled up based on several addressing rules, one of which is based on E.164; thus, if E.164 is used, then the called number field in the IAM message will contain all of 18162352006 instead of just the dial digits 2352006.

13.6 Call Routing: Multiple Long-Distance Provider Case

From the single network provider (SNP) environment discussed above, the first restriction we relax is to allow multiple providers to coexist and operate long-distance services (including international calls) while each region has only one specific local/access provider. If the call starts from one region to another region that falls under the long-distance dialing, the destination region can be served by another local/access provider. Thus, in the simplest form, there are two types of providers: local exchange carriers (LEC), which provide the local/access telephone service, and interexchange carriers (IXC), which carry the long-distance part of a call. There are multiple IXCs that operate in the long-distance part. A subscriber can choose any of the IXCs as their main carrier for long-distance calls, but can change to another IXC later by requesting a change; the general assumption is that such change is not very frequent, perhaps only a few times every year by each subscriber. Regardless, the subscriber must be given the opportunity to use another IXC on a per-call basis by entering an access code. In the rest of the discussion, we will use the terms *interexchange carrier* and *long-distance provider* interchangeably; similarly, we will use the terms *local exchange carrier* and *access provider* interchangeably.

Example 13.3 *Call Traversal through LECs and IXCs.*

First we illustrate an important effect on how a call traverses from one LEC to another LEC through an IXC. The conceptual picture is shown in Figure 13.5 with two LECs, LEC-A and LEC-B, and two IXCs, IXC-1 and IXC-2. Suppose that the number +1-816-328-9999 is served by LEC-A and the number +1-351-235-2006 is served by LEC-B. A subscriber can choose any one of the IXCs for the long-distance service. Suppose that, at a particular instant of time, the subscriber with the number +1-816-328-9999 is registered with IXC-1 as its long-distance carrier, while the subscriber with the number +1-351-235-2006 is registered with IXC-2 as its long-distance carrier.

If the user with the number +1-816-328-9999 calls the user with the number +1-351-235-2006, then the call would start at the originating TDM switch in LEC-A, and is handed off to the toll switch at the border of LEC-A to IXC-1 since user +1-816-328-9999 uses IXC-1 as the long-distance carrier. The hand-off point between a LEC and an IXC is often referred to as

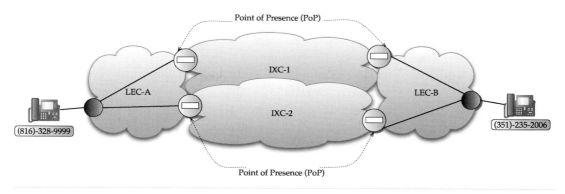

FIGURE 13.5   Calling between +1-816-328-9999 and +1-351-235-2006 through LEC/IXC networks.

the Point of Presence (PoP). From the PoP, IXC-1 routes the call through its network using its own method; for example, dynamic call routing methods discussed earlier in Chapter 10 can be used in the IXC's network. The call is then handed off at the TDM switch on the border of IXC-1 and LEC-B. From there, LEC-B routes the call to user +1-351-235-2006. Overall, we can see that this call goes from LEC-A to IXC-1 to LEC-B.

Now consider the call originating at the other end: the user with number +1-351-235-2006 calls the other user with number +1-816-328-9999. Note that user +1-351-235-2006 uses IXC-2 as it long-distance carrier. Thus, LEC-B will route the call to the toll switch at the border between LEC-B and IXC-2. In turn, IXC-2 would decide its own routing method for routing this call all the way to the border of IXC-2 and LEC-A, where it will hand off to LEC-A. From there, LEC-A routes the call to user +1-816-328-9999. This time, the call traverses from LEC-B to IXC-2 to LEC-A.

Thus, we can see that depending on who initiates the call, the call path can traverse through different carriers or providers.   ▲

An important question is: for a particular subscriber, how would LEC-A or LEC-B know which IXC to use for long-distance call forwarding, especially since each subscriber can use a different IXC? A follow-up question is: how would the subscriber still access an alternate IXC on a per-call basis if he or she wants to?

To address both these issues, a basic requirement is to assign a network number to each IXC; this number is known as the *Carrier Identification Code*, or CIC in short. The CIC number is a four-digit code that any provider who wants to provide the interexchange long-distance service must have. The four-digit CIC code can consist of any of the digits 0 to 9; but, X411 and 411X, where X represents 0 to 9, are not assigned as CIC codes. Furthermore, the range 9000–9199 is set aside for intranetwork CICs. As an analogy, a CIC code can be thought of as similar to an autonomous system (AS) number in the Internet, except that CIC codes are restricted within a particular numbering administration while AS numbers are worldwide. Similar to private AS numbers (between 64512 and 65535), a certain set of CICs is set aside for intranetwork purposes. An additional point of clarification is that CIC numbers for usage by IXCs are also referred to as *Feature Group-D* (FG-D) CIC numbers since a protocol known as FG-D protocol is used to describe interconnection rules between a LEC and an IXC.

Now, to address the question about how each subscriber is associated with a CIC number, a database lookup table is required to be maintained that maps each valid subscriber number to a CIC number. In North America, this table is maintained as part of the *Line Information Database (LIDB)*. This database also includes other information about the subscriber and the subscriber number such as whether the subscriber has signed up for features such as call forwarding or call waiting and whether the number is a residential or a business subscriber. LIDB is a network database that is not directly attached to the originating TDM switch. It is usually connected via the SS7 network off an SCP node that is in close proximity to the originating switch. Thus, whenever a user makes a long-distance call, the originating TDM switch sends a TCAP message on the SS7 network to the LIDB database to look up the CIC number. A subscriber can bypass the current IXC it subscribes to by first dialing an access code; this access code has the format 101XXXX where XXXX is the CIC number of the long-distance provider the subscriber wants to use for this specific call; this addresses the second question posed above. From a digit analysis point of view at the originating TDM switch, this access code does not cause any ambiguity since 1 followed by any digits other than 0 or 1 would signify a long-distance call. In other words, 1 followed by 0, i.e., the first two dialed digits being 10, is distinct enough to avoid any ambiguity. Originally, such alternate access functionality to IXCs did indeed require 10 followed by a three-digit CIC number. This was later extended to the start code being 101 followed by the four-digit CIC number. As of now, the first two dialed digits being 11 is not used for any service and is left for future extensions.

Thus, for a dialed long-distance call, the originating switch first needs to determine the CIC number by consulting the LIDB database; it then needs to determine the appropriate trunkgroup associated with the long-distance provider that owns this CIC number. Since a call setup message is to be routed on the SS7 network, the next step is to identify the SSP associated with this CIC number; the SSP in turn identifies the appropriate trunkgroup ID. These are then the steps needed for eventually identifying an available circuit to route a long-distance call.

An important point to note is that along with the demarcation of a LEC and an IXC for the trunk network, demarcation/handoff for SS7 is also required. That is, each provider has its own SS7 network and connects to another provider in accordance with the rule of local/interexchange connectivity. This has an important implication in regard to the 1-800 number that is homed off a particular IXC. Recall that the 1-800 number translation request is first generated as a TCAP message by the originating TDM switch (SSP) to send to the SCP that handles 1-800 number translation. Suppose that the IXC does not want to divulge the point code of the SCP for the 1-800 number database, or might want the flexibility of moving the 1-800 number from one point code to another point code without having to notify the access provider. To address this issue, the IXC is allowed to provide only the point code of the PoP (border) STP of its SS7 network to the LEC for 1-800 number translation, *not* necessarily the point code of the actual SCP. In fact, this works within the SS7 protocol stack through global title translation (refer to Section 12.11). Certainly, the IXC would need to maintain a routing table entry at the PoP (border) STP for forwarding such TCAP messages properly to the appropriate SCP.

13.6.1 Illustration of Call Routing

We will now illustrate the multiple IXC environment by extending the illustration presented earlier in Section 13.5.2. First, we note that local calls still stay within a LEC's network; thus, the call routing remains the same as discussed earlier in Section 13.5.2 and is not discussed further. We now make an important clarification about calling long-distance numbers. An LEC is, in fact, allowed to provide long-distance services *within* a certain geographic proximity referred to as *local access and transport area* (LATA). LATA is the area in which a LEC is allowed to operate fully; such intra-LATA long-distance calls will be dialed starting with 1 and are handled for routing as described previously in the single-provider case. That is, the originating TDM switch must still function according to the old rules for local calls and intra-LATA long-distance calls; this can be handled through digit analysis by inspecting the leading digits and is based on the same logic as described for the single-provider case earlier. Thus, in the rest of this section, we discuss primarily the impact of a long-distance call that requires using an IXC; furthermore, when we say a long-distance call, we mean a call that requires using an IXC.

Here, we list two subscribers with numbers +1-816-328-9999 and +1-816-328-0001 that are connected to the TDM switch point code 251.1.3 that use different IXCs as their default long-distance providers. The view from switching hierarchy is shown in Figure 13.6, which extends the case for the single provider shown earlier in Figure 13.4. The network architecture from the view of the originating TDM switch for handling both local and long-distance calls through IXCs for these two subscribers is shown in Figure 13.7; this extends the single-provider architectures shown earlier in Figure 13.2. First note that each IXC would have different PoPs to serve as IXC access tandem switches; thus a trunkgroup is required from the originating TDM to each IXC PoP. Here, trunkgroup ID 6 is the IXC with CIC number 3773 and trunkgroup ID 7 is the IXC with CIC number 7337. In Figure 13.7, two IXCs are shown. Each PoP has its point code for its SSP interface. A regional STP is shown for routing SS7 messages to each IXC SSP.

The changes required at the originating TDM switch are fairly significant. The initial logic for how to handle a call would start with determining the type of operation based on whether it is a local call or a long-distance call. Thus, instead of the simple table shown in

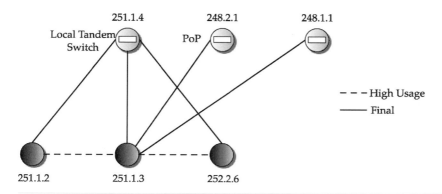

FIGURE 13.6   Hierarchical routing view at the TDM switch level in a LEC/IXC environment.

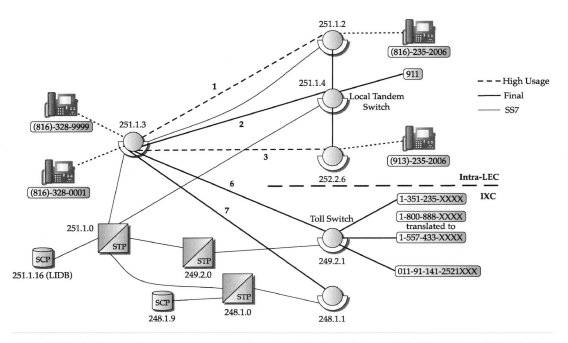

FIGURE 13.7   Call and message routing for calls dialed from SSP 251.1.3 in a LEC/IXC environment.

Figure 13.3(a), an additional column is needed for the corresponding table as shown in Figure 13.8(a). Note that local calls still require accessing two table lookups as before. However, by inspecting Figure 13.8, for the long-distance call that uses an IXC, we can see that it requires up to four table lookup operations. It may be noted that in Figure 13.8(c), the lookup information obtained from the LIDB database located at the SCP node 251.1.16 (see Figure 13.7) is shown; this is not done internally at the originating switch.

Now consider that the subscriber with number +1-816-328-9999 is signed up with an IXC with CIC number 3773, and the subscriber with number +1-816-328-0001 has signed up with another IXC with CIC number 7337. If the subscriber with number +1-816-328-0001 wants to access the other IXC to make a long-distance call, then it would first need to dial 1013773 before dialing +1-351-235-2006. From Figure 13.8, we can see how the originating switch can handle this request going from (a) to (d) to (b). However, if it has direct dialed +1-351-235-2006, then the originating switch would need to request a LIDB lookup to obtain the CIC number for this subscriber; this one uses (a) to (c) to (d) to (b).

How about the SS7 network? It typically requires a regional STP for routing messages from the LEC's SS7 network to the IXC's SS7 network—see STP 249.2.0 in regard to the IXC with CIC number 3773 and STP 248.1.0 in regard to the IXC with CIC number 7337 (see Figure 13.7 and Figure 13.8(d)).

CALLING A 1-800 NUMBER

Calls to a 1-800 number require some explanation. This is somewhat different than a long-distance call to an IXC. Like a regular number subscriber, there is also a subscriber with

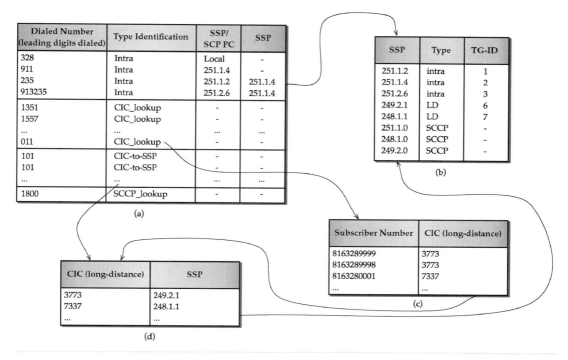

FIGURE 13.8   Table lookup phases in a LEC/IXC environment.

every valid/activated 1-800 number; these subscribers are usually companies, organizations, or even a person. So as not to confuse with regular telephone subscribers, we will refer to this group of subscribers as entities. An entity can choose who is going to be the provider for delivery of calls to its premise: it can be any of the IXCs or the LEC itself. This means that any routable number for a 1-800 number can be homed off either the LEC or any of the IXCs. Thus, when a user dials a particular 1-800 number, the originating TDM switch needs to consider yet another lookup function to find out who the provider for this 1-800 number is. To identify the provider, an ISUP TCAP message is sent to the 800-number SCP in the SS7 network that stores the lookup information.

On receiving this message at the 800-number SCP, it is mapped to a routable number and the CIC number of the provider who is to handle delivery of the call to the routable number. On receiving this response from the SCP, the originating TDM switch identifies the appropriate trunkgroup to PoP switch of the service provider based on the CIC number (and the associated TCIC code for the circuit) if it is other than the LEC itself; if the LEC itself is supposed to be the provider in this case, then the call is routed to the routable number as it would route any intra-LEC call. Now the originating TDM switch generates an ISUP IAM message with the called number being the routable number obtained from the 800-number SCP.

In Figure 13.9 we show the message flow and translation along with call routing. Here, once a regular subscriber dials the number 1-800-213-XX01, the originating TDM switch with PC 251.1.3 would send a TCAP message to the 800-number SCP with PC 242.3.4. The SCP will

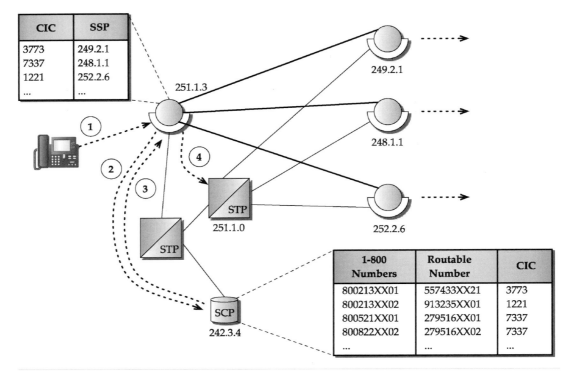

FIGURE 13.9   Lookup up for 1-800 number routing.

return the information that the routable number is 557-433-XXXX and that the CIC number is 3773. The originating switch will then determine the appropriate SSP by doing a CIC to SSP lookup (PC 249.2.1 is assumed in this case); then the SSP to trunkgroup ID mapping will be done (not shown in the figure) to further calculate TCIC for the circuit ID. The ISUP IAM message will be routed to this SSP for further handling by the provider to complete the call setup process.

Finally, note that in Figure 13.9 we show only one routable number per 800-number. With intelligent network capability, the mapping and lookup is much more sophisticated than this simplistic view; this will be discussed later along with number portability in Section 13.10.

13.6.2 Impact on Routing

With multiple IXC environments, end customers/users are given the opportunity to select any IXC as their long-distance service provider. From an economic point of view, this definitely opens up the market and competition. From a technical point of view, this adds additional complexity to the system as we can see going from Figure 13.3 and Figure 13.8. However, this complexity should not be considered in isolation. That is, added complexity to a system does not necessary imply that the efficiency is always impacted. Since the opening of the IXC market, there has been significant research and progress in understanding software process control and implementation, hardware architecture, and network engineering so that even services such as 1-800 lookup are possible without users perceiving any delay.

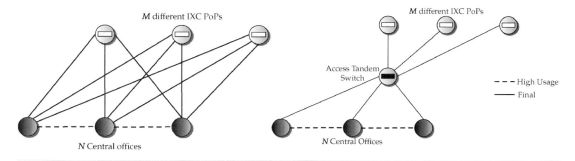

F I G U R E 13.10   LEC/IXC access topology scenarios: (a) full connectivity requires $N \times M$ trunkgroups, (b) access-tandem based connectivity requiring $N + M$ trunkgroups ("hub-and-spoke topology").

From a network topology point of view, a LEC would be required to connect as many IXCs that have requested connectivity. If there were only a few IXCs, then the number of trunkgroups to such IXCs would remain limited. If, however, the number of IXCs increases significantly, then the LEC faces a critical network engineering problem that impacts routing as well. Consider that a LEC has N central offices, and there are M IXCs to which the LEC needs to connect. An option is that each central office has connectivity to each IXC's PoP for long-distance call forwarding; this connectivity would result in $N \times M$ trunkgroups for long-distance services to all carriers. Clearly, this does not scale well from a network configuration management point of view. Furthermore, there may not be enough offered load to justify a direct trunkgroup from each central office to each IXC; this is where the ECCS method[3] can be applicable in considering the trade-off on a per-trunkgroup basis between the LEC and an IXC, leading us to consider an access tandem through which all central offices can connect from the point at which traffic is forwarded to IXCs. If a single tandem office is used for all IXC traffic from the central offices, then this network architecture requires just N trunkgroups from the central offices to the tandem office, and M from this tandem office to the IXC PoPs, resulting in a hub-and-spoke topology (also known as star topology). This then leads to only $N + M$ trunkgroups, which is a much better option than the scalability issue faced with the fully connected case, which requires $N \times M$ trunkgroups; note that this is from the point of view of the LEC. From each IXC's point of view, both scenarios require the same N trunkgroups to the LEC (Figure 13.10). Certainly, the $N + M$ solution considers that the access tandem is assumed to be at the middle of the switching hierarchy between the central office level and the toll-switch/PoP level; this access tandem switch must maintain proper routing tables for call forwarding in either direction. Note that the access switch would require a total of $N + M$ trunkgroups.

Eventually, there are two factors that can impact the actual layout decision: (1) the optimal economic access network design that considers the ECCS method or other similar methods, and (2) any bi-lateral business agreement between a LEC and each IXC.

---

[3]Refer to Section 11.5 for the ECCS method.

13.7   Multiple-Provider Environment: Multiple Local Exchange Carriers

Multiple LECs are currently allowed to operate for landline local phone services in a geographic region. Traditionally, there has been only a single LEC in each geographic region; in the United States, this LEC is often referred to as the *incumbent* LEC (ILEC), and any new entrants into the same geographic region as *competitive* LECs (CLECs). For simplicity, we will refer to all local providers simply as LECs, such as LEC A, LEC B. We first assume that each LEC in a particular geographic region is allocated separate local exchange codes (this assumption will be relaxed in a subsequent section). This then requires handling of routing and handoff between LECs properly. Thus, we start with the inter-LEC routing scenario by using the NANP.

Example 13.4   *Inter-LEC routing illustration.*

To understand the impact of multiple LECs, we start with a simple illustration. Consider exchange codes 816-328 and 913-235 that belong to a LEC ("LEC-A"), while in the same geographic area exchange codes 816-334 and 913-237 are allocated to another LEC ("LEC-B"). Then, the central office TDM switch that serves 816-328 would need to recognize that 816-334 is not an intranetwork call and it must know the SSP point code of this local exchange to which the call setup message should be forwarded. Instead of directly connecting local exchange switches between two LECS, these LECs might decide to share an inter-LEC PoP tandem switch. This interconnection concept is shown in Figure 13.11.

The routing table entry at the switch that is home to 816-328-9999 is shown in Figure 13.12 for local calls. When a user with the number 816-328-9999 dials 334-1990, the originating TDM switch will recognize that it is an inter-LEC call by inspecting the central office code 334 and would identify that the inter-LEC SSP (PoP switch) has the point code 247.3.1. By inspecting the SSP to trunkgroup ID table, the trunkgroup ID will be identified and TCIC will be determined; this information will be inserted in the IAM message to SSP

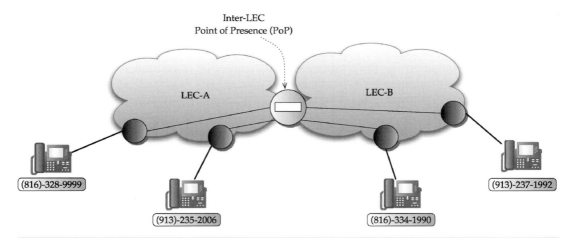

FIGURE 13.11   Inter-LEC connectivity through a PoP tandem switch.

Dialed Number (leading digits dialed)	Type Identification	SSP/ SCP PC	SSP
328	Intra	Local	-
911	Intra	251.1.4	-
913235	Intra	251.2.6	251.1.4
334	Inter-LEC	247.3.1	251.1.4
913327	Inter-LEC	247.3.1	-
...	...	...	...

SSP	Type	TG-ID
251.1.2	intra	1
251.1.4	intra	2
251.2.6	intra	3
247.3.1	Inter-LEC	9
...	...	...

FIGURE 13.12   Tables at a central office switch in LEC A serving 816-328 (entries shown for local calls only).

247.3.1. When the inter-LEC PoP switch receives this message, it will route to the destination SSP.   ▲

Note that the introduction of multiple LECs in a geographic area does not affect or change the depiction of connectivity and routing from these LECs to multiple IXCs, nor the need for change in LIDB to look up the CIC number for IXC, from the description in the previous section. That is, outgoing call routing architecture calls entering an IXC remain the same as in the previous section. However, an incoming call from an IXC to any of the LECs requires proper handling. The last switch on the call path in an IXC's network needs to determine to which LEC to send the call. To do that, this TDM switch must maintain a lookup table based on the local exchange code to determine the appropriate destination LEC and, consequently, the egress trunkgroup to this LEC's access tandem switch.

13.8 Routing Decision at an Intermediate TDM Switch

So far our discussion on call routing has been from the point of view of the originating switch. A routing decision at an intermediate switch along the call path is somewhat similar. A switch through its SSP interface first receives the ISUP IAM message and proceeds to check the Called Party Number field (refer to Section 12.8.1). First it checks to identify the nature of the address field in the Called Party Number field to determine if it is a national number or an international number; this information then instructs the switch to process the digits included in the rest of the Called Party Number field. For a national number, the high-order digits in the Called Party Number field indicate the destination switch for the call, for example, the NPA–NXX part in the NANP. Thus, it can now do a routing table lookup to determine which next switch to use and then identify the associated trunkgroup. Certainly, anytime a switch receives an IAM message from an upstream switch, by inspecting the Called Party Number field it first checks to see if this is the destination for this call. If so, it rings the user and generates an ACM message and sends it upstream.

Number portability is presented in the next section. This results in changes to the IAM message that will be discussed later in Section 13.11.4.

13.9 Number Portability

Number portability is perhaps the most significant factor that is dramatically changing network routing in the public switched telephone network. In this section, we start with an introductory discussion followed by portability classification. In subsequent sections, we discuss the impact of number portability on architecture and routing in detail.

13.9.1 Introduction

Number portability often refers to a subscriber's ability to change a network service provider while retaining the same number. Before proceeding, we introduce two term: *nonported number* and *ported number*. A nonported number indicates that a subscriber number has remained with the original network provider; a ported number indicates a number that is no longer with the original network provider.

Keeping with the main theme of this book, we will consider the implications for network routing due to number portability; yet, it is important to understand the key driver for number portability. Historically, telephone services, especially fixed/landline services, have been provided in a monopolistic environment; so that consumers will not suffer bad service, appropriate regulatory agencies have mandated maintaining a certain grade of service, such as designing networks for 1% call blocking probability, and also ensuring that the pricing is reasonable. The key driver for mandating number portability is to allow competition in providing telephone services in order to provide flexible choice and benefit to consumers. For example, each time a consumer changes the service provider, no number change is required.

From another perspective, number portability means that network service providers do not have ownership of large blocks of contiguous numbers allocated to them; at most, each service provider has the privilege of *hosting* a number until the subscriber of the number decides to move on and sign up with another provider; that is, the ownership of the number moves from the network service provider to the consumer or user. A critical implication for number portability is that we are moving from an *ownership-based* environment to a *hosting-based* environment. A simple analogy from the Internet is helpful here. Consider the domain name system in the Internet; you can own a domain name, but might choose to change your web-hosting providers over time.

To move from an ownership-based environment to a hosting-based environment, there is a significant cost that would need to be incurred up front by a network service provider, along with the impact on the overall architecture. In this regard, it is sensible to introduce number portability in a phased manner so that smaller providers have a longer time window in which to introduce number portability. That is, any number portability environment must coexist with the possibility that some exchange codes may not be ported and/or that some countries may not have introduced number portability such that the telephone call flow remains seamless to the ordinary user across the world.

We now start with an illustration to show the implication of number portability in regard to a landline number in a LEC environment, which is often referred to as *local number portability* (LNP).

Example 13.5 *Illustration of local number portability using the North American numbering plan.*
Consider two subscriber numbers 816-328-9999 and 816-328-0001, both served originally by LEC A from the the central office/exchange code 816-328. With LNP, the user with the number 816-328-0001 now decides to change the LEC to LEC B. Thus, 816-328-9999 remains a nonported number, while 816-328-0001 is now a ported number. If the user with the number 816-328-9999 calls the user with the number 816-328-0001, it is no longer an intraswitch call!   ▲

Thus, a significant impact of LNP is that the implicit assumption that the central office code part of a subscriber number is the home central office code is *not* always true anymore. Another impact is how to handle call routing when a distant user tries to reach a ported number, say, 816-328-0001. That is, consider the user with the number +1-351-235-2006 from another geographic region or an international caller wanting to reach +1-816-328-0001; how should this call be handled for routing to the correct subscriber? Thus, LNP has two significant impacts: (1) the need to handle calls to ported numbers at a switch that used to be intra-switch calls for these ported numbers, and (2) the need to route a call from another region to the correct local exchange carrier that serves a ported number. Furthermore, another requirement must be satisfied: a user with a ported number might choose to move to a third LEC, or move back to the original carrier sometime after. How should such a change be structured and portability be communicated in a timely manner? These are some of the key questions that arise with number portability. These will be considered in the next section.

13.9.2 Portability Classification

There are three types of numbers that are of interest as far as number portability is concerned: landline telephone numbers, mobile telephone numbers, and nongeographic toll-free numbers. Implicit in number portability is a subscriber's ability to change a network service provider while retaining the same number; thus, all number portability can be broadly described as *service provider number portability* (SPNP), or *operator portability*. Similarly, implicit in a fixed/landline or mobile telephone number is that fixed number service is provided by a fixed network service provider (operator) such as a local exchange carrier, while a mobile number service is provided by a mobile or wireless carrier; however, there is no such implicit understanding when referring to nongeographic toll-free numbers such as 1-800 numbers.

We start with fixed/mobile number portability. Within the broad definition of service provider number portability, several situations are possible:

1. Local fixed-to-fixed porting—a subscriber retains the number moving from a fixed-number operator to another fixed-number operator *within* a local/geographic area.

2. Fixed-to-fixed porting across geographic regions—a subscriber retains the number moving from one geographic area to another geographic area, often changing the operator as well.

3. Mobile-to-mobile porting—a subscriber retains the number when moving from a wireless carrier to another wireless carrier.

4. Fixed-to-mobile porting—a subscriber retains the number moving from a fixed LEC to a wireless carrier.

5. Mobile-to-fixed porting—a subscriber retains the number moving from a wireless carrier to a fixed LEC.

Collectively, the above five situations will be referred to as *regular subscriber number portability*, to distinguish them from nongeographic toll-free number portability. There are also three additional terms associated with number portability: (1) *local number portability*, (2) *location portability*, and (3) *service portability*. Typically, local number portability (LNP) refers to local fixed-to-fixed porting, and location portability refers to fixed-to-fixed porting across geographic regions; however, local number portability can also be fixed-to-mobile or mobile-to-fixed; this will be discussed later. In addition, *mobile number portability* refers to mobile-to-mobile porting. Service portability refers to changing a service, for example, from plain old telephone service (POTS) to ISDN services; this, however, may or may not result in changing an operator. Service portability is still not well understood and will not be considered in the rest of our discussion.

It is helpful to quickly revisit E.164 numbering. Note that E.164 numbering has a well-defined format; however, once the country code is allocated, it is left up to each country to decide on the numbering plan within this framework. Earlier in Section 13.1.1, we have given examples of numbering plans that differ from one country to another. Of particular interest is how fixed and mobile number space is allocated in different countries. For instance, in the NANP, fixed and mobile numbers have the same NPA–NXX–XXXX format; however, in countries such as India and China, mobile numbers are allocated in an address space that has specific starting digit strings such as the ones in India that start with 9 (with certain exceptions) while all mobile numbers in China start with 13. In either type of numbering plan, number portability can potentially lead to consumer confusion of varying degrees, at least initially, depending on the type of portability. Certainly, local fixed-to-fixed and mobile-to-mobile portability situations are the least confusing ones, regardless of the numbering scheme. However, fixed-to-mobile and mobile-to-fixed portability are perhaps the most confusing one to consumers in countries where different numbering formats for land and mobile numbers are used, besides possible implications on charging. These issues as well as regulatory issues are important under the umbrella of number portability; they can potentially play a role in the eventual routing architecture; we believe it is important for the reader to be aware of these issues since the primary focus in the rest of the discussion is from a technical perspective. That is, because of the numbering plan and to avoid confusion, some countries may or may not allow fixed-to-mobile porting or mobile-to-fixed porting.

13.10 Nongeographic or Toll-Free Number Portability

To understand nongeographic or toll-free number portability, we first need to understand what a subscriber means in regard to a nongeographic/toll-free number. Like a regular fixed or mobile telephone number subscriber, there is also a subscriber associated with every valid/activated toll-free number; these subscribers are usually companies, organizations, or even a person. Not to confuse them with regular telephone subscribers, we will refer to these types of subscribers as *entities*. An entity can choose who is going to be the provider for *delivery* of calls to its premise. In a number portability environment, it is further desirable that while maintaining the same toll-free number, an entity might want to have multiple providers handle its call delivery based on percentage traffic allocation and/or time of the day.

Our illustration here for nongeographic number portability is based on the North American environment. A similar concept can be deployed in other countries. First recall from the earlier discussion about 1-800 number routing that the 800-number SCP is involved in looking up a routable number. Thus, first we need to understand the process of an entity requesting change in number portability, which is eventually communicated to the 800-number SCP. That is, there are two distinct phases in regard to number portability: the first is the process in regard to requesting a change and how this is communicated to the 800-number SCP, and the second is the actual routing when a regular subscriber dials an 1-800 number.

13.10.1 800-Number Management Architecture

An independent administration for 800-number maintenance, to be referred to as an *800-Service Management System (SMS/800) Administration Center* is responsible for maintaining the most up-to-date record in regard to 800-number translation. A responsible organization, or RespOrg in short, that has been certified is allowed access to the SMS/800 database located at the SMS/800 administration center. A RespOrg can request a change for an entity for changing service provider or allocation. Once this information is recorded at the SMS/800 database

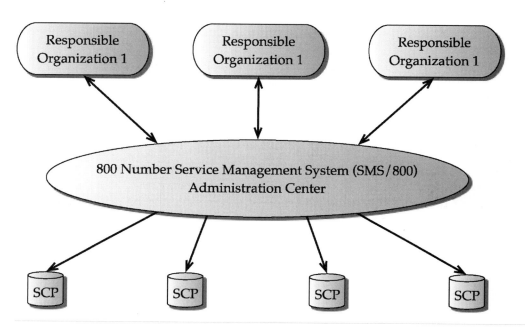

FIGURE 13.13   Service Management System/800-number architecture: process and communication to SCPs.

and tested, a copy of the current information in the SMS/800 is downloaded to Service Control Points (SCPs). There are currently approximately 20 mated pairs of 800-number SCPs across North America. The basic architecture is shown in Figure 13.13.

13.10.2 Message and Call Routing

A LEC accesses the closest 800-number SCP pair using the SS7 network. Thus, when a regular subscriber dials a 1-800 number, the originating TDM switch (SSP) generates a TCAP query that is routed to the 800-number SCP using the SS7 network. The 800-number SCP looks up the entry applicable at that instant for the dialed 800-number and returns the CIC number along with a routable number or the dialed toll-free number. If the CIC number happens to be for the LEC itself, then the originating switch identifies the SSP for the switch (either tandem or terminating switch) to which the ISUP IAM message is to be sent that contains the TCIC code. If the CIC number happens to belong to an IXC, then the originating switch identifies IXC's PoP SSP and determines the trunkgroup ID and TCIC to be used for this call; an ISUP IAM message is then generated that includes the TCIC information; the called number field is the routable number obtained from the 800-number SCP and this message is sent to the PoP SSP. Note that the ISUP IAM message can include the dialed 800-number as the called number if the 800-number SCP returns this number; in this case, the carrier that receives this message does further translation internally for ISUP IAM message routing and call routing; this phase is completely transparent to the originating TDM switch.

For illustration, we first note that the basis architecture for call routing remains the same as shown earlier in Figure 13.9. Thus, we concentrate here on the message routing part of the call setup phase from the point of view of the originating TDM switch as shown in Figure 13.14. At the 800-number SCP database, we show the conceptual view of information ac-

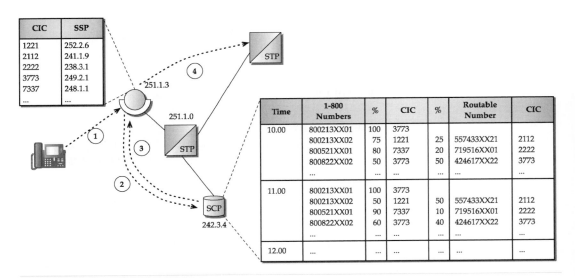

FIGURE 13.14   800-number portability, lookup, and ISUP message routing: 1—call initiated by user; 2, 3—TCAP messages; 4—IAM message.

cessed in regard to the dialed 800-number, which can be mapped to *multiple* routable numbers with percentage allocation; the allocation may be based either on *different* service providers or multiple numbers for the same service provider. It should be noted that in actuality more than two routable numbers can be listed and the time granularity for a routable entry is not fixed; here, for brevity, we show a maximum of two entries for each 800-number and change in entries on an hourly basis. This is *time-of-the-day* routing. There is another important point to note; it is possible that at a particular time instant, calls arriving at central offices in different geographic locations that are destined for a particular 800-number are routed to different routable numbers or different providers. In other words, the entry need not be unique nationwide at a particular point of time.

13.11 Fixed/Mobile Number Portability

In Section 13.9.2, we identified five different situations in regard to fixed/mobile number portability. While some countries are limiting number portability to only mobile number portability, other countries are exploring all five situations. In this section, we present the number portability architecture and discuss various possible routing schemes.

13.11.1 Portability Architecture

A key aspect of number portability is the process architecture that allows porting of a number from one network service provider to another network service provider.

Prior to regular subscriber number portability, there are several possible operator scenarios depending on how the telephone service is provided in a country: (1) local landline telephone service is provided by a single network service provider in the entire country, (2) local landline telephone service is provided by different network service providers in different parts of a country but a single provider in each region, (3) local landline telephone service *within* a local calling region is provided by multiple different network service providers, or (4) mobile telephone service is provided by different network service providers regardless of any geographic region in a country. Regardless of these scenarios, each provider, whether fixed or mobile, is assigned telephone numbers in blocks, typically at the exchange code level, by the numbering administration that has jurisdiction over it. This is then the ownership-based model for telephone numbering.

With number portability, a subscriber can decide to take his or her number and request service from another provider. The network provider that originally owned this number as part of a numbering block assignment is referred to as the *donor network*. The network provider that the subscriber approaches to host its number is called the *recipient network*. Note that the subscriber might want to switch service provider again after some time. Thus, the recipient network can change over time, as often as a subscriber changes the provider; however, the donor network is always considered to be the original network that owned the address block from which this subscriber number came. For example, if a provider was originally assigned the exchange code +1-816-328, then this provider is called the donor network in the number portability architecture. If the subscriber with the number +1-816-328-5001 chooses to move to another provider, then this provider is referred to as the recipient network. If the subscriber moves again to a third provider, the donor network remains the same but the third provider is now the recipient network.

In the number portability architecture, a third party organization called the *network portability administration center* (NPAC) is in charge of coordinating the number portability. The NPAC is given jurisdiction depending on the scope of the portability. That is, this jurisdiction is typically dependent on whether the portability is local number portability or location portability. If it is local number portability (fixed-to-fixed, or fixed-to-mobile, or mobile-to-fixed, or mobile-to-mobile), then there is a certain geographic region within which the portability is allowed, for example, within a city or local number calling area; in this case, the NPAC would need to cover all network providers that serve this area. A much clearer way to understand NPAC jurisdiction is to consider the portability at the telephone number address code level. For example, if subscriber +1-816-328-5001 is allowed to port the number to a provider only in the local calling region, then an NPAC's jurisdiction is more regional based. If however, the subscriber with the number +1-816-328-5001 is allowed to take the number to another geographic region ("location portability"), then the NPAC must have the ability to coordinate over this geographic area, too. Regardless of the reachability of number portability from a geographic perspective, the three basic components—the donor network, the recipient network, and the NPAC—remain valid.

An important requirement as far as impact on routing is concerned is that with number portability a call cannot be routed based solely on the called number. Note that prior to number portability, the called number indicates the destination of the call; specifically the destination TDM switch can be identified from the first few digits of the called number, such as the NPA–NXX part of the called number in the NANP. Such identification based on the called number is no longer possible with number portability if a number is ported. Thus, for each number ported, we need a way to identify where this call is going to be routed to, that is the *address* of the destination TDM switch that currently hosts the ported number. Thus, for number portability to be possible, each destination TDM switch (or point of interconnection) requires an address, which address is referred to as the *location routing number* (LRN) or the *network routing number* (NRN).

In the NANP, an LRN is a 10-digit number taking the form NPA–NXX–XXXX, where the last four digits are typically set to zeros. It is a unique number that identifies either a local-exchange TDM switch or a point of interconnection (POI) in each LATA. Usually, the first six digits of an LRN of a TDM switch is the original LEC of the switch. Since some carriers may not have implemented number portability yet, the local exchange switches in that carrier's network are still required to have an LRN, but they are marked as nonported LRNs. The benefit of switches so marked will be discussed later Section 13.11.2.

To port a number then requires several steps that involve the recipient network, the donor network, and the NPAC. From the point of view of each network provider, any portability request must be handled appropriately, which also includes, for example, a billing operations support system. The schematic diagram of the process flow architecture for the *first-time* move from a donor network to a recipient network is shown in Figure 13.15; the steps are described below:

1. The Point of Sale sends the order request to the Operations Support System (OSS) for billing of the recipient network.

2. The recipient network's billing OSS sends a message to the donor network's billing OSS to indicate the order request.

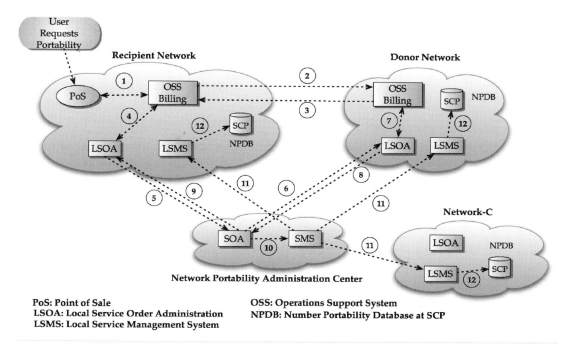

FIGURE 13.15   Number portability architecture: initial porting from a donor network to a recipient network.

3. The donor network's billing OSS responds with approval for this request if everything is in order. For example, if there are any billing issues, the donor network's billing OSS can request a delay in transfer to the recipient network. Until this request is processed with approval, the recipient network cannot move forward with porting the number.

4. The recipient network's OSS sends the porting request information to the Local Service Order Administration (LSOA) in its network. This request includes information such as the subscriber number, location routing number (LRN) of the donor network, and the date/time the number is to be ported to the recipient network.

5. The recipient network's LSOA sends the porting request to the SOA at the NPAC.

6. The SOA at the NPAC records this request and forwards it to the LSOA of the donor network.

7. On receipt of this request, the LSOA of the donor network informs its billing OSS when to stop billing and receives confirmation.

8. The LSOA of the donor network now confirms the SOA at the NPAC that it has agreed to the porting request along with confirmation on the date and time of the actual change.

9. The SOA at the NPAC sends a confirmation to the LSOA at the recipient network in regard to the porting.

10. The SOA at the NPAC now communicates the porting information to its service management system (SMS) in regard to the number, the LRN, and the effective time/date of change.

11. The SMS at the NPAC is now responsible to inform local service management systems (LSMSs) at the recipient network, donor network, and any other networks under its jurisdiction about the new routing information for the ported number.

12. The LSMS in each network communicates the ported number along with routing information to its number portability database (NPDB) SCP. This then completes the portability.

Note that it is left to each network provider to decide whether it wants to maintain a separate NPDB SCP for each originating TDM switch, or to have just one combined NPDB SCP for all its originating TDM switches.

If the subscriber moves from its current network provider ("current recipient network") to a new network provider, then the new provider becomes the recipient network. In this case, the process flow will involve the new recipient network, the old recipient network (especially for billing purposes), and the original donor network.

We now comment on the portability architecture if the portability is for location portability. The basic concept as described above remains the same. However, the NPAC would need to be a larger area, such as the entire nation. Certainly, in this case, for scalability, a recipient network can send the portability request to a regional NPAC, which in turn coordinates with a centralized NPAC—this option also has the advantage when the proportion of location portability request is small compared to local number portability request.

13.11.2 Routing Schemes

Once a number is ported as described above, we consider how to route a call to the ported number. This is where the LRN for a ported number is needed for proper delivery of a call. There are essentially four different schemes that can be invoked in regard to routing decision of a call to a ported number; all of these address how and when to determine the LRN, each in its own way. To consider these schemes, we assume that once a number is ported, a central NPDB stores the mapping information that maps a ported number to a location routing number. In addition, we assume that a donor network has an internal database that stores the mapping from a ported number to an LRN; this can be only for the number for which the donor network used to be owner before number portability was introduced.

To discuss the four schemes, we consider an environment in which a caller calls a subscriber ("callee"). The network that receives the call from the caller will be referred to as the originating network. For this discussion, we assume that there is *no* intermediate or transit network between the call originating network and the recipient network, or between the call originating network and the donor network.

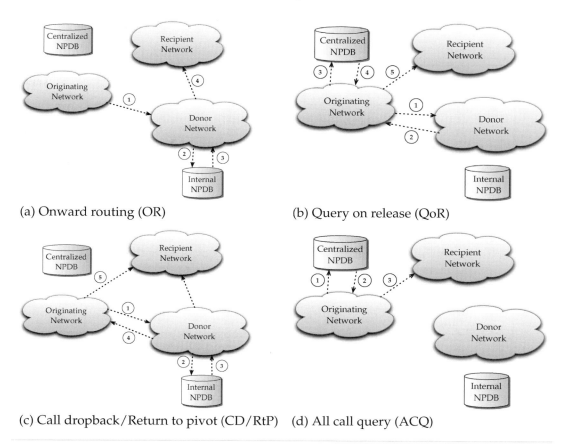

(a) Onward routing (OR)   (b) Query on release (QoR)

(c) Call dropback/Return to pivot (CD/RtP)   (d) All call query (ACQ)

FIGURE 13.16   Call routing schemes, with number portability.

ONWARD ROUTING (OR)

In this scheme, a call from the originating network is routed based on dialed digits. This then implicitly assumes that the call is sent to the donor network, i.e., as if the porting has not occurred yet. This scheme has the following steps (see Figure 13.16(a)):

1. On receiving a call from the caller, the originating network routes the call to the donor network like a regular call, i.e., circuits are allocated.

2. The donor network's terminating switch recognizes that the number is ported; it holds the call in memory and queries the internal NPDB to learn about the recipient network.

3. The internal NPDB does a called-number-to-location routing number lookup and returns the location routing number to the donor network.

4. Using the location routing number, the donor network routes the call to the recipient network.

It is easy to see that the onward routing scheme can be thought of as call forwarding.

QUERY ON RELEASE (QOR)

This procedure initially starts with the assumption that the porting has not occurred, and thus, call routing involves the donor network. This scheme has the following steps (see Figure 13.16(b)):

1. On receiving a call from the caller, the originating network routes the call to the donor network like a regular call, i.e., circuits are allocated.

2. The donor network's terminating switch recognizes that the number is ported and informs the originating network that the number is ported. Note that the donor network does not track where the number is ported.

3. On learning from the donor network that the number is ported, the originating network queries the centralized NPDB.

4. The centralized NBDB does a called-number-to-location routing number lookup and returns the location routing number to the originating network.

5. The originating network identifies the recipient network based on the location routing number and routes the call accordingly.

Note that QoR does not need to use the internal NPDB at the donor network.

CALL DROPBACK OR RETURN TO PIVOT (CD/RTP)

Call dropback, also known as Return to Pivot, is somewhat of a hybrid between the onward routing scheme and the QoR scheme. Unlike onward routing, the call dropback scheme allows the call control to be cranked back to the originating network that is now in charge of handling routing to the recipient network. Similar to the QoR scheme, the originating network routes the call to the recipient while each finds out about porting differently: QoR relies on the centralized NPDB, while CD/RtP depends on the donor network to query its internal NPDB. This scheme has the following steps (see Figure 13.16(c)):

1. On receiving a call from the caller, the originating network routes the call to the donor network like a regular call, i.e., circuits are allocated.

2. The donor network's terminating switch recognizes that the number is ported; it holds the call in memory temporarily and queries the internal NPDB to learn about the recipient network.

3. The internal NBDB does a called-number-to-location routing number lookup and returns the location routing number to the donor network.

4. The donor network releases the call to the originating network and includes the location routing number.

5. On receiving the release message from the donor network with the location routing number, it releases the circuits already assigned to the donor network, and now routes the recipient network.

Note that the dropback feature is similar to the crankback feature discussed earlier with dynamic routing (refer to Section 10.2.3). Fundamentally, this means that dropback feature is like an originating call control feature, instead of a progressive call control feature.

ALL CALL QUERY (ACQ)

In this scheme, a query is generated for every call, regardless of whether the called number is ported or not. This scheme has the following steps (see Figure 13.16(d)):

1. On receiving a call from the subscriber, the originating networks sends a query to the centralized NBDB.

2. The centralized NBDB returns the location routing number to the originating network.

3. The originating network identifies the recipient network based on the location routing number and routes the call accordingly.

Note that the donor network (and its internal NPDB) is not involved at all in the routing decision.

DISCUSSION

We have presented the above four schemes using the notion of networks and databases in a somewhat simplistic manner without providing several details. We now provide some more details to show where the circuit is actually set up and the ISUP message types involved in different segments.

Consider onward routing. In this case, the call is routed assuming no number portability. Assuming SS7 is used for call setup messages, the ISUP IAM message will be routed from the originating TDM switch (SSP) in the originating network, through a series of SSPs, including the PoP switch between the originating and the donor network, to the destination TDM switch that is supposed to serve the called number; the trunk circuit is assigned/reserved along the path so far. The destination switch will be the one that will now do the query to the internal database to determine the location routing number and will further create an ISUP IAM message to send to the recipient network to set up a series of circuits to the actual terminating switch. Once the eventual destination accepts the call, the ISUP ACM will be generated, which will then hop through the switches, including the donor network all the way back to the originating TDM switch. Thus, referring back to Figure 13.16(a), segments 1 and 4 involve both SS7 messaging and call circuits, while segments 3 and 4 involve ISUP TCAP message query and responses to internal NPDB.

Now consider QoR. In this case, the originating network routes the ISAM IAM message and circuits are allocated along the path to the donor network's destination switch, much like onward routing. The destination TDM switch in the donor's network will check its internal table to recognize that the number has been ported; thus, it will return an ISAM REL message. This message will hop through back to the originating switch, releasing the circuit that was reserved along the way. Now, the originating TDM switch will query the centralized NPDB to determine the location routing number for the called number. It will then create a new ISUP IAM message that will be sent toward the actual termination switch that has this LRN. Once accepted, the ACM message will be generated by the termination switch that has this LRN. In this case, referring to Figure 13.16(b), for segment 1, both SS7 IAM messaging and circuit setup are completed; in segment 2, anSS7 REL message is communicated and the circuit is released; segments 3 and 4 involve TCAP message exchanges; finally, segment 5 involves both ISAM IAM messaging (ACM in return) as well as circuit setup for the actual call.

Call dropback is somewhat similar to QoR. The main difference is that it is the donor network that initiates the called-number-to-location routing number query. In this case, referring to Figure 13.16(c), for segment 1, both SS7 IAM messaging and circuit setup are completed; segments 2 and 3 involve TCAP message exchanges; in segment 4, an SS7 REL message is communicated that includes the LRN information and the circuit is released; finally, segment 5 involves both ISAM IAM messaging (ACM in return), as well as circuit setup for the actual call.

All call query starts with the TCAP message sent to the centralized database and a response obtained with the LRN information. Then, the call IAM message and circuit setup are completed to the recipient network. In this case, referring to Figure 13.16(d), segments 1 and 2 involve TCAP message exchanges, and segment 3 involves both ISAM IAM messaging (ACM in return) as well as circuit setup for the actual call. There is another slight variation in the use of ACQ. In the second case, the originating switch first checks an internal table to see if the called number falls in the range of the ported number; this then saves sending a query to the NPDB if the number happens to be from a range that is not ported.

13.11.3 Comparison of Routing Schemes

From the discussion above, we can compare benefits and drawback of each routing scheme. This is summarized in Table 13.5. The main comment we would like to add is that number portability comes at the price of the setup cost for the portability architecture and the database needed; this setup cost can vary depending on the scheme. In addition, circuits tied up for a call (and thus cost) is also impacted, for example with the onward routing scheme.

It is important to keep in mind that even with number portability, it is unlikely that all or most subscribers will move out of the donor network. That is, in practice, the percentage of portability can vary considerably from one country to another. Keeping all these factors in mind, it is safe to say that there is no a single "best" solution. In any case, it is important to keep in mind that the postdial delay incurred due to any number portability scheme must be as small as possible; otherwise, customer experience is likely to suffer. The postdial delay can be minimized by effective traffic engineering of the signaling network, including minimizing the number of SS7 links and nodes traversed, as well as ensuring that the transaction rate of the NPDB is high.

13.11.4 Impact on IAM Message

Recall that in a standard IAM message, the caller party number contains the actual destination telephone number ("directory number") of the call. The call party number is used by a

TABLE 13.5 Comparison of routing schemes for number portability.

Method	Benefits	Drawbacks
Onward Routing (OR)	1. No centralized database needed 2. Internal NPDB can be stand-alone and contains only the ported number from the donor network 3. Good solution for short term, or if a small percentage of subscribers chooses to do number portability	1. Completely relies on the donor network during call setup 2. Requires setting up two physical call segments
Query on Release (QoR)	1. Centralized number portability database used for call routing decision	1. Involves the donor network during call set-up 2. Circuits are reserved temporarily to the donor network
Call dropback/ Return to Pivot (CD/RtP)	1. Centralized number portability database not needed 2. Internal NPDB can be stand-alone and contains only ported numbers from the donor network	1. Involves the donor network during call setup 2. Circuits are reserved temporarily to the donor network
All Call Query (ACQ)	1. Centralized number portability database used for call routing decision 2. Does not involve the donor network 3. Efficient in usage of switch ports and circuits 4. Good long-term solution, especially when most subscribers choose number portability	1. Relatively high portability setup cost 2. High ISUP TCAP traffic to NPDB from originating switches

switch to determine routing toward the destination switch; for example, in the North American format, this means that the NPA–NXX part of the called number is used for routing a call. With number portability, the NPA–NXX part of the called party parameter no longer indicates the destination switch. To accommodate routing properly, several fields (and subfields) in the IAM message have been modified to indicate that it is a ported number. The changes are as follows:

- In the Forward Call Indicator parameter, a subfield called the Ported Number Translation bit (PNTI) is set.

- The Nature of Address subfield in the Called Party Number field is set to be a ported number.

- The Called Party Number parameter now contains the LRN, and the actual destination telephone number is moved to the Generic Address field.

Note that while an exchange code is classified as ported, not all numbers in its address block would be ported; this is because some, in fact most, customers remain with the original LEC. Thus, if a user dials a directory number that is not ported, then this must be handled properly as well. In this case, the NPDB query will return the directory number itself, not the LRN. The IAM message generated will still set the PNTI bit; however, the Nature of Address subfield in the Called Party Number field is set to the normal mode, and then the called party parameter contains the directory number and the Generic Address field is not used.

The change in IAM is somewhat different in some European countries. For example, some countries do not use the Generic Address field in the IAM message if the dialed directory number is a ported number. Instead, a routing prefix is added with the directory number in the Called Party Number parameter itself, while the Nature of Address subfield indicates that the routing prefix is related to porting. For example, the United Kingdom uses a six-digit routing prefix that has the format 5XXXXX. The routing number (prefix) can be either at the level of identifying the recipient network, or the TDM switch within the recipient network that hosts the ported number. The benefit of the first option is that it allows a network provider to block the internal view of its network to the outside world; however, it is necessary to do an additional within-its-network lookup once a call arrives at its point of presence.

The main message to learn from this section is that there are multiple ways to extend the IAM message to address number portability. This is partly possible because IAM has been designed to be extensible, and also because the Called Party Number field is a variable-length field (refer to Section 12.8.1) that is not limited to the allowable length of E.164 addressing; see [236], [330] for additional details.

13.11.5 Number Portability Implementation

Decisions on number portability are country specific where regulatory bodies as well as the telecommunication industries are involved in agreeing on a workable scheme. In Table 13.6, we show a representative set of countries where number portability has been deployed. Keep in mind that the list is current at present; many countries are still in the investigative stage.

13.11.6 Routing in the Presence of Transit Network

Finally, we comment on the change in role for the network from which the call originates if there is at least a transit network between where the call originates and the recipient/donor network. This raises the following question: which network will trigger an NPDB query? Recall that in the case of OR routing, the donor network is responsible for the NPDB query; thus, there is no change in this case. Thus, the question is really for the CD/RtP, QoR, and ACQ schemes. In the United States, the ACQ scheme is used and there is an industry agreed-upon $(N-1)$ querying policy protocol to handle the NPDB query issue if there are transit networks along the call path. This means that the $(N-1)$ network, or the network before the destination network, on the path of a call from the actual origination to the destination, is required do to the NPDB query. Note that network here means call network, not the SS7 network. Theoretically, it is possible that the same SS7 network provider handles call signaling messaging routing for both the $(N-1)$ network and the donor network.

The $(N-1)$ policy is considered to be a good policy as it does not place the responsibility on the donor network when the ACQ scheme is used since the $(N-1)$ network does all

TABLE 13.6 Routing methods for number portability currently used in various countries (adapted from [236], [679], [691]).

Country	Scheme	Remark
Austria	OR	Routing prefix: 86XX where XX specifies recipient network
Belgium	ACQ	Routing prefix: CXXXX where XXXX specifies the destination TDM switch in the recipient network; use C00XX to specify the recipient network
Denmark	ACQ	Routing number is not communicated between providers; Nature of Address indicator is set; QoR possible through bilateral agreements
Finland	ACQ	Routing prefix: 1DXXY where XX specifies the recipient network, Y specifies service type
France	OR	Routing prefix: Z0XXX where XXX specifies the destination TDM switch in the recipient network
Germany	ACQ	Deutsche Telekom uses ACQ while other providers can select another scheme
Italy	OR	Routing prefix: C600XXXXXX where XXXXXX specifies the destination TDM switch in the recipient network
Japan	OR	The donor network performs SS7/IN lookup to obtain routing number
Netherlands	QoR/ACQ	Operators decide
Norway	OR, ACQ	OR is short-term, ACQ is long-term; QoR is optional; Nature of Address field used as indicator along with routing prefix
Spain	ACQ	QoR used internally by Telefonica; Routing prefix: XXYYZZ that specifies the recipient network along with Nature of Address field indicator
UK	OR	Routing prefix: 5XXXXX where XXXXX specifies the destination TDM switch in the recipient network; in parts of the network, British Telecom also uses dropback scheme
US	ACQ	PNTI bit set in Forward Call indicator; location routing number takes the place of Called Party Number field; the directory number is placed in Generic Address Parameter

the NPDB queries. Furthermore, the entire world numbering plan and routing architecture can still work globally without requiring fundamental changes to the current architecture. We describe the possible scenarios below assuming the donor network is marked as a local exchange carrier, LEC-Z.

- A geographic local call: Consider a call that is local where LEC-Z operates that originates from a local exchange carrier, LEC-A. In this, the $(N-1)$ network is LEC-A where the call originates and there is no transit network. Thus, LEC-A itself is required to do the NPDB query.

- A long-distance call that involves an IXC: Consider a call from a geographic area originating from a local exchange carrier, LEC-B, to a ported number where the donor network

("LEC-Z") is in another geographic area where the call must traverse through an IXC. That is, the call path (before number portability) is LEC-B to IXC to LEC-Z. In this case, IXC is considered the $(N - 1)$ network and is required to do the NPDB query.

- An international call from outside the United States to the donor network, LEC-Z: The provider outside the United States would not be able to know which LEC is ported and which is not. In this case, the call is handed over from the other nation's network provider to the IXC that provides service within the United States. If this carrier is the same carrier for long-distance service before reaching LEC-Z, then this carrier is required to do the NPDB query.

In all cases, since multiple switches are involved along the call path, a mechanism is still needed to ensure that a downstream switch does not perform an NPDB lookup if it has already been performed by an upstream switch. Thus, if a TDM switch performs an NPDB lookup to obtain an LRN, the ensuing IAM message that is sent out of this switch for this call is modified. This modification includes setting the PNTI bit in the Forward Call Indicator parameter in the IAM message; furthermore, the LRN code is placed in the Called Party Number parameter, and the called number is included in the Generic Address Parameter field. By inspecting the PNTI bit, a downstream TDM switch knows that the NPDB is already performed. This feature is also helpful in case $(N - 1)$ network fails to perform the NPDB query, for instance, due to a technical failure. Thus, the call setup IAM message without this modification will be routed all the way to the donor network's supposed terminating switch; that is, this will look like the onward routing case where it is left to the switch that would have been the terminating switch prior to number portability deployment to perform the NPDB query. Such default-routed calls are then the burden of the donor network to perform the query. Incidentally, in the United States, the Federal Communication Commission's policy allows the donor network to charge the $(N - 1)$ network for this extra work and even allows it to "block default-routed calls, but only in special circumstances when failure to do so is likely to impair network reliability" [417].

In the case of the long-distance and international call scenarios above, there is another decision the IXC, the $(N - 1)$ network, is required to address. For instance, the call in the IXC's part of the network usually traverses through multiple TDM switches. For instance, if dynamic call routing is used in the IXC's network, there can be at most three TDM switches involved, the ingress switch, an intermediate (via) switch, and the egress switch. The decision partly depends on whether portability is local number portability or location portability. If it is local number portability, then the LRN will be adjacent to the donor network for the ported number that has been dialed; thus, in this, the egress switch in the IXC's network is the appropriate switch to do the NPDB query and then identify the PoP for the network that serves this LRN. However, if it is for *location* portability, then we need to consider the possibility that the ported number might have moved anywhere in the entire country; thus, it would be more appropriate to do the NPDB query at the ingress switch, especially to avoid any crankback. Furthermore, with location portability, for an intracountry call, it might be more appropriate to do all call query (ACQ) at the originating LEC. We note that the idea of location portability is still in its infancy; further development is required to understand all its possible implications.

13.12 Multiple-Provider Environment with Local Number Portability

We will illustrate the entire call routing scenario in a local number portability environment. Our illustration primarily focuses on a geographically contained area where local number dialing is sufficient. We assume that the ACQ scheme is used with the variation that the originating switch first checks an internal table to see if it is still host to the dialed number; such numbers will referred to as native numbers. Thus, when a user calls a number that used to be homed off this switch, then the originating TDM would first check internally to determine if it is a native number. If it is native, then it will do intraswitch routing as it used to do prior to LNP implementation. If the dialed number is not a native number, then the switch sends a TCAP message to the NPDB that is connected to the SS7 network in order to determine the LRN. Once LRN is determined, the trunkgroup ID and the TCIC on the appropriate outgoing trunkgroup for this call must be identified.

We will now illustrate the above scenario using Figure 13.17. Here, we consider three LECs—LEC A, LEC B, and LEC C; each can be the donor network for the number range originally assigned to them. We assume that originally the exchange code 816-328 belonged to LEC A. The user with the number 816-328-0001 decided to migrate to LEC B for local telephone services and is homed to the TDM switch with LRN 816-342-0000, and the user with the number 816-328-0002 to LEC C homed to the TDM switch with LRN 816-367-0000. Thus, LEC A is the donor network in both cases, and LEC B is the recipient network in the first case, while LEC C is the recipient network in the second case. Thus, when the user with the number 816-328-9999 who has remained with LEC A calls the number 328-0001, the originating TDM switch with LRN 816-328-0000 checks if the number dialed is in its dialed number inventory to see if it is a native or a ported number. It will recognize that it is a ported number, thus as shown Figure 13.17, it will identify the LRN for the ported number using the centralized NPDB, and in turn, SSP and the trunkgroup ID for ISUP IAM message routing and call routing, respectively.

For the above scenario, the network topology architecture encompassing all three LECs with both TDM switching and SS7 nodes is shown in Figure 13.18. For simplicity, in each LEC, only a single TDM switch is shown with connectivity between switches to different LECs; SS7 STP nodes are similarly kept simple. Consider now a call from the user with the number 816-367-2525 homed off LRN 816-367-0000 in LEC-B to the user with the number 816-328-0002. From Figure 13.18, we can see that this is really an intraswitch call. However,

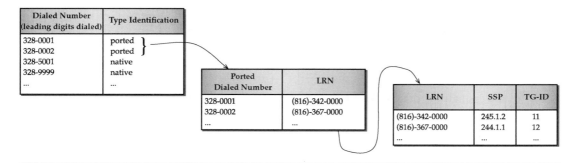

F I G U R E 13.17   Lookup for ported local numbers for ISUP message and call routing.

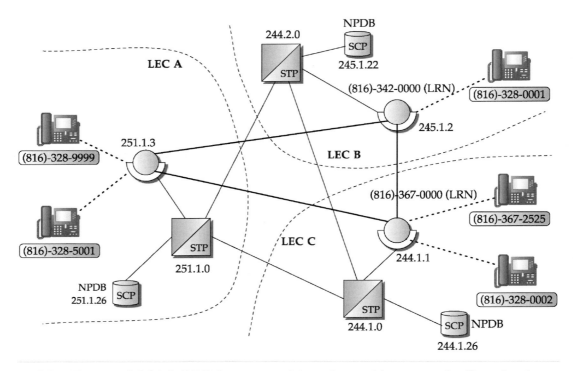

FIGURE 13.18   Multiple LECs in a geographic region: architecture and call routing for ported numbers.

it is important to avoid this call being first routed to LRN 816-328-0000 and then routed back to LRN 816-367-0000 (which can lead to looping). With ACQ, the query to the centralized database will be performed to find out that it is itself indeed the host of this number. It might be advisable for each switch to maintain entries for ported numbers for which it is now the home switch to avoid this lookup; this should be done cautiously and in proper coordination with the operations support system that talks to the local service order administration and local service management system since this user might choose to move out of this switch (provider) to another provider over time. If local marking inside the switch is done for the numbers it hosts now, then when 816-367-2525 calls 328-0002, the originating TDM switch (with LRN 816-367-0000) immediately recognizes that this is an intraswitch call.

We now make an important comment about the topology architecture shown in Figure 13.18. It is easy to see that as the number of LEC grows, inter-LEC connectivity grows quadratically; certainly, this is not a scalable solution. Thus, topology aggregation can possibly be achieved by introducing inter-LEC PoP tandem switches for call transfer; that is, this is analogous to the multiple LECs with the common interconnect point described earlier in Figure 13.11. Such topology aggregation would certainly require appropriate call routing table changes at the originating TDM switches in each LEC. Similarly, the number of STPs can be reduced if an inter-LEC exchange point STP is introduced to serve as a point of presence for SS7 message forwarding.

```
if ( dialed number is not N11 call and dial number is NOT 1-8XX number ) then
    if ( dialed number is a local number ) then
        if ( an intra-switch call ) then
            Complete the call
        else // not interswitch call
                if ( NPA-NXX is not ported) then
                    // handle as before pre-LNP days
                    Determine hand-off SSP/PoP for this CIC based on routing/forwarding
                    table
                    Determine TrunkGroup ID to this SSP and available circuit TCIC
                    Generate standard ISUP IAM message with this SSP as DPC and TCIC
                    value
                else // NPA-NXX is ported
                    Query NPDB/SCP with NPA-NXX to obtain LRN
                    Determine SSP for this LRN
                    Determine trunkgroup ID for this SSP
                    Generate modified ISUP IAM message with this SSP as DPC and TCIC
                        value, with PNTI set, Nature of Address field indicated,
                        Called Party Number field containing
                        LRN, and the dialed number in generic address field
                endif
        endif
    else
        // dialed number is a long-distance/international number
        // no NPDB lookup required
        Query LIDB database and determine CIC for this subscriber
        Determine hand-off SSP/PoP for this CIC based on routing/forwarding rule
        Determine TrunkGroup ID to this SSP and available circuit TCIC
        Generate standard ISUP IAM message with this SSP as DPC and TCIC value
    endif
endif
```

FIGURE 13.19   Call processing logic at the originating switch with number portability: a conceptual view.

Now, suppose that a call originates from another geographic area to 816-328-0001 that requires traversing through an IXC's network. In this case, $(N-1)$ policy is used; this means that the IXC is required to do NPDB lookup; thus, in this case, IXC's process will be similar to LEC-A's process.

We conclude this section by providing the call processing logic at the originating switch that incorporates number portability. This is shown in Figure 13.19.

13.13 Summary

In this chapter, we presented PSTN architecture and routing. We have started with numbering plan E.164 and discussed how a dialing plan is different from a numbering plan; because of this, the originating TDM switch has the responsibility for digit analysis so that a call can be routed properly.

Originally, there was a direct connection between call routing and addressing with the one provider per nation scenario, where outgoing trunkgroups from a TDM switch were identified based on the numbers dialed. In the past two decades, there has been significant impact on call routing decisions as new requirements have been placed such as allowing multiple long-distance providers. We have illustrated how such additional requirements can be addressed by introducing new functionalities such as the CIC code in order to properly route a call. After multiprovider environments, we presented call routing and its impact on the underlying system due to number portability.

It is important to note that while originally addressing and routing were directly related, in a multiprovider environment with number portability, the role of E.164 addressing has changed to being a universal identifier, rather than remaining as a physical locator for routing. That is, number portability thus has necessitated a number to be mapped to a routable entity through a lookup process; this is an important consequence of number portability. Later in Chapter 20, we will present routing for voice over IP, and additional issues will be discussed.

Another important point to note is the regulatory factor and how it can impact changes in protocol, architectural functions, and routing. For example, if a nation has a single body running the network nationwide, then clearly a field such as the CIC code is not needed in the ISUP IAM message. In a multiprovider environment, the CIC code is needed and it plays a critical role in routing decision. Number portability has also necessitated additional changes in the ISUP IAM message. Furthermore, the need for a neutral party to do certain centralized coordination in a multiprovider setting is important to note.

Further Lookup

ITU-T recommendation covers call routing, ISUP messaging, and so on; see [328], [324] for details. WTNG [746] maintains an up-to-date website for numbering plans for different countries around the world. For an international dialing sequence between any two countries, see [312]. North American numbering plan adminstration maintains an informative website that contains NPA–NXX availability, CIC number, and so on; see [531]. Note that 555 numbers such as 555-1212 require special handling; you may consult [10] for a detailed discussion about 555 handling. For the new numbering plan in India (as of April 2003), see [491].

For additional material on number portability, refer to [236], [679], [690], [691], [716], [717]. The local number portability working group [416] maintains a website that discusses $(N-1)$ carrier architecture [417]. Furthermore, best practices for number portability have been tabulated [418]. The 1-800 toll-free service was first introduced in 1967; for a comprehensive historical background on 800-service, see [652]. The emergency 911 service is currently addressing enhanced services [523]. An analysis of call routing using SS7 data can be found in [100].

Exercises

13.1. Review questions:

   (a) What is number translation?

   (b) What is Carrier Identification Code?

   (c) What is local number portability?

13.2. Why does the subscriber's telephone number not need an SS7 point code address?

13.3. Explore the numbering plan deployed in various countries (for example, select five different countries, other than the three already mentioned in this chapter). Compare their similarities and differences.

13.4. In Section 13.12, we illustrated local number portability. Introduce multiple long-distance providers and discuss how call routing will be handled.

13.5. Suppose that we want to architect worldwide number portability. Investigate what types of functions and architectural components would be needed, and the impact these factors would place on call routing.

13.6. Explain how dialed number, point code, CIC code, LRN, and trunkgroup ID are related for each of the following scenarios: (i) single national provider, (ii) multiple local exchange providers, (iii) local number portability.

Part IV: Router Architectures

An important component of routing is routers. A router's job is to do efficient packet processing of any incoming packet and to track information due to exchange of routing protocol messages; it also must support network management functions. However, how a router is architected can strongly impact overall packet processing. In this part, we present three chapters that encompasses routers.

In Chapter 14, we present a general overview of different needs and requirements of a router. This is then followed by a classification of different routing architectures. In general, this chapter serves as a road map.

An important function of a router is to do IP address lookup. Due to classless interdomain routing (CIDR), the lookup function must take netmask into consideration. Not only that, if a subset of address blocks from a contiguous address block plans to move to be provided by a different provider, the router must have the ability to handle processing of such exceptions efficiently. In Chapter 15, we present a variety of IP lookup algorithms along with a discussion of their strengths and limitations.

Another important function of a router is packet filtering and classification. This means that beyond address lookup, a router is often required to handle packets differently depending on customer requirements. These additional constraints must be handled efficiently by a router. In Chapter 16, we present a variety of algorithms for packet filtering and classification. As with lookup algorithms, their strengths and limitations are also highlighted.

Finally, a router does much more than lookup and classification. For this the switching backplane inside the router must be efficient, the packet queueing and scheduling must address a variety of requirements, and often some traffic conditioning is required due to service level agreements. These topics are covered later in Part VI.

14

Router Architectures

Architecture starts when you carefully put two bricks together. There it begins.

Ludwig Mies van der Rohe

Reading Guideline

This chapter serves as the platform for understanding the basics of routers and types of routers, and as the background material to understand more details about a router's critical functions, such as address lookup and packet class classification, which are discussed in subsequent chapters.

In Part II of this book, we presented IP network routing, focusing primarily on routing protocols and their usage, with only cursory remarks about routers. In this chapter, we present an overview of routers and how they are architected for the purpose of packet forwarding and handling routing protocols.

Traditionally, routers have been implemented purely with software running on a general purpose personal computer (PC) with a number of interfaces. Such a device can receive packets on one of its interfaces, perform routing functions, and send packets out on another of its interfaces. As the Internet grew over the years, the type and size of routers changed, since routers based on general-purpose PC architectures are limited by the performance of the central processor and memory. Fortunately, advances in silicon technology have made it possible to build hardware-based routers capable of handling high data rates.

In this chapter, we describe various IP router architectures and highlight their advantages and disadvantages. In addition, we examine the performance trade-offs imposed by the architectural constraints of these routers. We start our discussion with a high-level overview of the routing process and describe the functions a router should implement.

14.1 Functions of a Router

Broadly speaking, a router must perform two fundamental tasks: *routing* and *packet forwarding*, as shown in Figure 14.1. Based on the information exchanged between neighboring routers using routing protocols, the routing process constructs a view of the network topology and computes the best paths. The network topology reflects network destinations that can be reached as identified through IP prefix-based network address blocks. The best paths are stored in a data structure called a *forwarding table*. The packet forwarding process moves a packet from an input interface ("ingress") of a router to the appropriate output interface ("egress") based on the information contained in the forwarding table. Since each packet arriving at the router needs to be forwarded, the performance of the forwarding process determines the overall performance of the router.

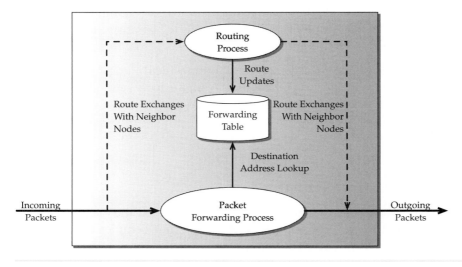

FIGURE 14.1 Routing and packet forwarding process.

The functions of the packet forwarding process can be categorized into two subgroups: basic forwarding and complex forwarding. Basic forwarding defines the minimal set of functions a router should implement in order to transfer packets between interfaces. Complex forwarding functions represent the additional processing required by the routers, depending on their deployment environments and their usage.

14.1.1 Basic Forwarding Functions

For forwarding an IP packet from an incoming interface to an outgoing interface, a router needs to implement the following basic forwarding functions [46], [50], [98]:

- *IP Header Validation:* Every IP packet arriving at a router needs to be validated. Such a test ensures that only well-formed packets are processed further while the rest are discarded. This test also ensures that the version number of the protocol is correct, the header length of the packet is valid, and the computed header checksum of the packet is same as the value of the checksum field in the packet header.

- *Packet Lifetime Control:* Routers must decrement the time-to-live (TTL) field in the IP packet header to prevent packets from getting caught in the routing loops forever. If the TTL value is zero or negative, the packet is discarded; an ICMP message is generated and sent to the original sender.

- *Checksum Recalculation:* Since the value of the TTL is modified, the header checksum needs to be updated. Instead of computing the entire header checksum again, it is more efficient to compute it incrementally [444]; after all, the TTL value is always decremented by 1.

- *Route Lookup:* The destination address of the packet is used to search the forwarding table for determining the output port. The result of this search will indicate whether the packet is destined for the router or to an output port (unicast) or to a set of multiple output ports (multicast).

- *Fragmentation:* It is possible that the maximum transmission unit (MTU) of the outgoing link is smaller than the size of the packet that needs to be transmitted. This means that the packet would need to be split into multiple fragments before transmission.

- *Handling IP Options:* The presence of the IP options field indicates that there are special processing needs for the packet at the router. While such packets might arrive infrequently, a router nonetheless needs to support those processing needs.

When there are routing or packet errors, routers use ICMP messages to communicate the information.

14.1.2 Complex Forwarding Functions

Besides the basic functions, the marketplace has necessitated the need for additional, complex functions. That is, with the popularity of the Internet, complex issues such as security, different user requirements, and service guarantees based on different service level agreements have become paramount and need to be addressed. These issues translate to additional processing when forwarding a packet, without essentially increasing overall packet processing *time* at a router. To cite an example of service differentiation, consider a scenario where customers are interested in watching a high-definition movie streaming directly over the Internet. Such a streaming requires not only high bandwidth but timely delivery of the data. The router needs to distinguish such packets so that it can forward them earlier. This results in the notion of differentiated services, and consequently requires that routers support a variety of mechanisms such as the following:

- *Packet Classification:* For distinguishing packets, a router might need to examine not only the destination IP address but also other fields such as source address, destination port, and source port. The process of differentiating the packets and applying the necessary actions according to certain rules is known as packet classification.

- *Packet Translation:* As the public IPv4 address space is being exhausted, there is a need to map several hosts to a single public address. Thus, a router that acts as a gateway to a network needs to support network address translation (NAT). NAT maps a public IP address into a set of private IP addresses and vice versa. This requires a router to maintain a list of connected hosts and their local addresses and to translate the incoming and outgoing packets.

- *Traffic Prioritization:* A router might need to guarantee a certain quality of service to meet service level agreements. This involves applying different priorities to different customers or data flows and providing a level of performance in accordance with the predetermined service agreements. For example, the agreement might specify that a fixed number of packets must be delivered at a constant rate, which is necessary for real-time streaming multimedia applications such as IPTV, or real-time interactive applications such as VoIP.

14.1.3 Routing Process Functions

Besides packet forwarding, i.e., the data plane function, a router needs to ensure that the contents of the forwarding table reflect the current network topology. For this, a router also needs to provide control plane and management plane functions. In particular, a router needs to handle:

- *Routing Protocols:* Routers need to implement different routing protocols, such as OSPF, BGP, and RIP for maintaining peer relationships by sending and receiving route updates from adjacent routers. These route updates are sent and received as normal IP packets. But the key difference between these packets and the packets that transit through the router is the destination address, which is the router itself for route update packets. Once the updates are received, the forwarding table is modified so that subsequent packets are forwarded to the correct outgoing links.

- *System Configuration:* Network operators need to configure various administrative tasks such as configuring of interfaces, routing protocol keep alives, rules for classifying packets. Hence, a router needs to implement various functions for adding, modifying, and deleting these configuration data, as well as persistently storing them for retrieval later.

- *Router Management:* In addition to the configuration tasks, the router needs to be monitored for continuous operation. These functions include supporting various management functions that are implemented using protocols such as simple network management protocol (SNMP).

14.1.4 Routing Table versus Forwarding Table

As described earlier, the packet forwarding function directs an incoming packet to the appropriate output interface based on the results of looking up a forwarding table. The routing function builds a routing table that is used in the construction of forwarding tables. Often, in the literature, the terms *routing table* and *forwarding table* are used interchangeably to refer to the data structures in a router for forwarding packets. In this section, we highlight the differences between those tables.

The routing table is constructed by the routing algorithms based on the information exchanged between neighboring routers by the routing protocols. Each entry in the routing table maps an IP prefix to a next hop. The forwarding table, on the other hand, is consulted by the router to determine the output interface an incoming packet needs to be forwarded. Thus, each entry in the forwarding table maps an IP prefix to an outgoing interface. Depending on the implementation, the entries might contain additional information such as the MAC address for the next hop and statistics about the number of packets forwarded through using the interface.

While a single table for routing and forwarding is possible, most implementations tend to keep these two separate for the following reasons. First, the forwarding table is optimized for searching a destination IP address against a set of IP prefixes, while the routing table is optimized for calculating changes in the topology. Second, as every packet needs to examine the forwarding table, it is implemented in a specialized hardware for high-speed routers. However, the routing tables are usually implemented in software. An instance of a routing table and forwarding table is shown in Table 14.1. The routing table in the figure indicates the next-hop IP address for a destination IP prefix. The forwarding table tells us a packet bound to the network identified by the IP prefix should be forwarded to interface eth0 with the appropriate MAC address.

TABLE 14.1   Routing table and forwarding table.

(a) Routing table		(b) Forwarding table		
IP prefix	Next hop	IP prefix	Interface	MAC address
10.5.0.0/16	192.168.5.254	10.5.0.0/16	eth0	00:0F:1F:CC:F3:06

14.1.5 Performance of Routers

The performance of a router is stated in terms of throughput expressed in bits per second. The throughput characterizes how much data the router can transfer per second from input network interfaces to an output network interfaces. A router throughput T is calculated as

$$T = P \times R, \tag{14.1.1}$$

where P represents the number of ports or interfaces feeding the router and R represents the line rate of each port. For instance, a router containing 16 ports with each port running at a line rate of 40 Gbps has a throughput of 640 Gbps. However, the throughput is not a measure of the real capability of the router. As routers forward packets, it is more important to know how many packets they are capable of forwarding in a second, which is referred to as packets per second (pps). For instance, a router throughput of 640 Gbps could mean packets of size 40 bytes forwarded at 2 billion pps or packets of size 80 bytes forwarded at 1 billion pps. Obviously, a router capable of handling more packets per second for the same packet size is considered better from performance perspective.

Now let us try to express the router throughput in terms of packet size and packets per second. If S is the packet size and P_s represents packets per second, then the line rate R can be expressed as

$$R = S \times P_s. \tag{14.1.2}$$

Substituting for R in Eq. (14.1.1), the throughput can be reformulated as

$$T = P \times S \times P_s. \tag{14.1.3}$$

The next logical question is: what should be the packet size used for this assessment? In a decade-old study, the average packet size was found to be 300 bytes [696]. In recent observations, commonly seen packet sizes are 40 bytes (due to TCP acknowledgments), 576 bytes (due to RFC 879, which is now outdated), 1500 bytes (due to Ethernet MTU size), and 1300 bytes (due to VPN software). If a router is designed with any of these sizes other than the smallest size, it might not be able to sustain a long sequence of shorter packets. Thus, most router designers use the minimum of 40 bytes as the standard packet size for such assessment.

14.2 Types of Routers

Routers can be of different complexity based on where in the network they are deployed and how much traffic they need to sustain. Naturally, this means that routers can be of different types. In this section, we describe three types of routers: core routers, edge routers, and enterprise routers and outline their requirements [367].

CORE ROUTERS

Core routers are used by service providers for interconnecting a few thousand small networks so that the cost of moving traffic is shared among a large customer base. Since the traffic arriving at the core router is highly aggregated, it should be capable of handling large amounts of traffic. Hence, the primary requirements for a core router are *high speed* and *reliability*. While it is important to keep the cost of a core router to be reasonable, the cost is a secondary issue.

The speed at which a core router can forward packets is mostly limited by the time spent to look up a route in the forwarding table. On receiving a packet from an ingress interface, the forwarding table entries need to be searched to locate the longest prefix match. The prefix represents the target IP network the packet is destined for. The matching prefix determines the egress interface. With the increase in the number of systems connected to the Internet and the associated surge in traffic growth, demand is placed on core routers to forward more packets per second. Hence, specialized algorithms implemented in hardware are required for fast and efficient lookups. These algorithms are the focus of Chapter 15.

Since core routers form the critical nodes in the network, it is essential that these routers do not fail under any conditions. The reliability of a router depends on the reliability of physical elements such as the line cards, switch fabric, and route control processor cards. The reliability of these physical elements is achieved by full redundancy—dual power supplies, standby switch fabric, and duplicate line cards and route control processor cards. Moreover, the software is enhanced so that when one of the elements fails, the packet forwarding and the routing protocols continue to function.

EDGE ROUTERS

Edge routers, also known as access routers, are deployed at the edge of the service provider networks for providing connectivity to customers from home and small businesses. The first generation of edge routers were really remote access servers attached to terminal concentrators that aggregated large number of slow-speed dial-up customers. However, this is not the case anymore. First, the need for more bandwidth has led to the introduction of a variety of access technologies such as high-speed modems, DSL, and cable modems. Hence the edge routers need to support an aggregation of customers using different access technologies. Second, in addition to legacy remote access protocols, these routers need to implement newer protocols such as point-to-point tunneling protocol (PPTP), point-to-point protocol over Ethernet (PPPoE), and IPsec that support VPNs. These protocol implementations should also scale as they need to be run on every port. Finally, these routers should be capable of handling a large amount of traffic. This is necessary as many customers are migrating from dial-up access to high-speed modems. These trends suggest that the edge routers support a large number of ports capable of different access technologies and many protocols operating at each port.

ENTERPRISE ROUTERS

Enterprise networks interconnect end systems located in companies, universities, and so on. The primary requirement of routers in these networks is to provide connectivity at a very low cost to a large number of end systems. In addition, a desirable requirement is to allow service differentiation to provide quality of service (QoS) guarantees for different departments of an enterprise.

A typical enterprise network is built using many Ethernet segments interconnected by hubs, bridges, and switches. These devices are inexpensive and can be easily installed with limited configuration effort. A network built using such inexpensive devices tend to degrade in performance as the size of the network increases. Hence, using routers in these networks to divide the end systems into hierarchical IP subnetworks is desirable. Moreover, it scales the network better.

In addition to providing the basic connectivity, there are several additional design requirements for the enterprise routers. First, these routers require efficient support for multicast and broadcast traffic as applications such as video broadcasting are more predominantly used in the enterprise. Second, these routers need to implement many legacy technologies that are still in use in the enterprises. The third requirement is the extensive support for security firewalls, filters, and VLANs. Finally, as these routers must connect many LANs, they are required to support large number of ports.

For enterprises, the network is considered as an operational expense and the goal is to minimize this expense. Hence, the routers targeted for enterprise deployment are required to have low cost per port, a large number of ports, and the ease of maintenance. Hence, it is challenging to design an enterprise router that satisfies these requirements for every port and still keep the cost low per port.

14.3 Elements of a Router

So far, we have discussed the functions and types of a router; we next discuss the elements needed in a router to provide these functions. For this purpose, a router can be viewed from two different perspectives. From a *functional* perspective, it can be logically viewed as a collection of modules where each module implements a set of related functions to achieve the overall goal of forwarding packets. From an *architectural perspective*, a router can be considered as an interconnection of different types of cards running specialized software. We discuss the functional perspective before examining the architectural perspective.

A router can be divided into several modules from a functional point of view. These components implement the various requirements of a router described in the previous sections. A generic router consists of six major functional modules: network interfaces, a forwarding engine, a queue manager, a traffic manager, a backplane, and a route control processor. These functional modules are shown in Figure 14.2.

- **Network Interfaces:** A network interface contains many ports that provide the connectivity to physical network links. A port terminates a physical link at the router and serves as the entry and exit point for incoming and outgoing packets, respectively. A port is specific to a particular type of network physical medium. For instance, a port can be an Ethernet port or a SONET interface. In addition, a network interface provides several functions. First, it understands various data link protocols so that when the packet arrives it can decapsulate the incoming packets by stripping the Layer 2 (L2) headers. Second, it extracts the IP headers, i.e., the Layer 3 (L3) headers, and sends them to the forwarding engine for route lookup while the entire packet is stored in memory. Collectively, this processing is referred to as L2/L3 processing. Further, it provides the functionality of encapsulating L2 headers before the packet is send out on the link.

- **Forwarding Engines:** These are responsible for deciding to which network interface the incoming packet should be forwarded. When a port receives a new packet, it deencapsulates L2 headers and sends the entire IP packet, or just the packet header, to the forwarding engine. The forwarding engine consults a table, i.e., engages in a route lookup function, and determines to which network interface the packet should be forwarded. This

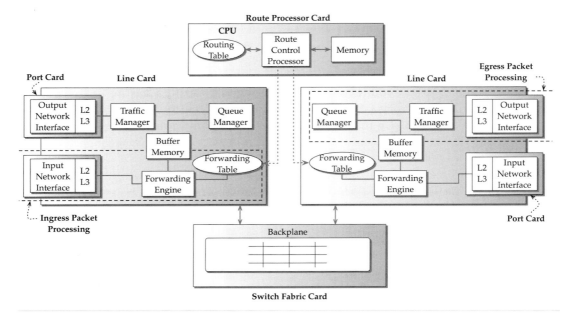

FIGURE 14.2   Components of a router.

table is called the *forwarding information base* or simply the *forwarding table*. Algorithms for route lookup can be implemented in custom hardware or software running on a commodity hardware. Depending on the architecture, the lookups can occur in the custom hardware or in a local route cache in the line card. Futhermore, to provide QoS guarantees, forwarding engines may need to classify packets into predefined service classes.

- **Queue Manager:** This component provides buffers for temporary storage of packets when an outgoing link from a router is overbooked. When these buffer queues overflow due to congestion in the network, the queue manager selectively drops packets. Thus, the responsibility of this component is to manage the occupancy of the queue and implement policies about which packets to drop when the queues are about to be fully occupied.

- **Traffic Manager:** This component is responsible for prioritizing and regulating the outgoing traffic, depending on the desired level of service. This is necessary as routers carry traffic from different subscribers and it is important to ensure that they get the level of service for which they pay. The traffic manager shapes the outgoing traffic to the subscriber according to the service level agreement. Similarly, when a router receives traffic from a subscriber, the traffic manager ensures that it does not accept more than what is specified in the contract. Sometimes the functionality of the queue manager and the traffic manager are merged into a single component.

- **Backplane:** This component provides connectivity for the network interfaces so that packets from an incoming network interface can be transferred to the outgoing network interface card. The backplane can be either *shared*, where only two interfaces can communicate at any instant, or *switched*, where multiple interfaces can communicate simultaneously. The aggregate bandwidth of all the attached network interfaces defines the bandwidth required for the backplane.

- **Route Control Processor:** The control processor is responsible for implementing and executing routing protocols. It maintains a routing table that is updated whenever a route change occurs. Based on the contents of the routing table, the forwarding table is computed and updated. In addition, it also runs the software to configure and manage the router. A route control processor also performs complex packet-by-packet operations like errors during packet processing. For example, it handles any packet whose destination address cannot be found in the forwarding table in the line card by sending an ICMP packet to its source of origin indicating the error. These functionalities are typically implemented in software running on a general-purpose microprocessor.

We next consider the architectural perspective and how the functional modules are implemented in practice. Figure 14.2 shows the various architectural components of a router and the functional modules each implements. They are:

- **Port Cards:** A port card implements the network interfaces. Each port card is capable of handling only a specific medium. For instance, a port card will support only Ethernet while the other can handle only SONET. The port cards contain L2 processing logic that understands the L2 packet format specific for that medium. In addition, the port cards perform accounting about the incoming and outgoing packets. Such cards are given different names by different vendors; for example, Juniper networks refers to them as Physical Interface Cards (PICs), whereas Cisco refers to them as Physical Layer Interface modules (PLIMs) in CRS-1 routers.

- **Line Cards:** A line card implements a majority of the functional components, forwarding engine, queue manager, and traffic manager. It parses the IP payload and uses the contents of the header to make decisions about forwarding, queueing, and discarding during periods of link congestion. It also contains memory buffers for storing the packet during processing and queueing. The line card houses port cards and connects to the backplane and ultimately to another line card. Sometimes, the line cards include the ports specific to certain media rather than using port cards.

- **Switch Fabric Cards:** While a line card implements the packet processing functions, a switch fabric card serves as the backplane for transferring packets from the ingress line card to the egress line card. In high-end routers, multiple switch fabric cards are used for increased throughput and redundancy.

- **Route Processor Cards:** These cards implement the functionality of the route control processor. The routing protocols and the management software run on these cards. In high-end routers, these cards use general-purpose processors with a large amount of memory running a commodity operating system.

From the discussion of the above two perspectives, we can see that there is a relationship between them. Many functional modules are directly mapped to physical components (such as cards, other hardware, and software). Note that there can be numerous ways to map functional modules to physical components, which leads to different router architectures. Before delving into different router architectures, let us understand how a packet is processed in a generic router.

14.4 Packet Flow

The packet flow in a generic router is shown in Figure 14.3. The processing steps can be broadly grouped into ingress packet processing and egress packet processing.

14.4.1 Ingress Packet Processing

When an IP packet arrives from the network, it first enters the network interface. For the sake of discussion, let us assume that the packet is received on an Ethernet port. The network interface interprets the Ethernet header, detects frame boundaries, and identifies the starting point of the payload and the IP packet in the frame. The L2 processing logic in the card removes the L2 header and constructs a *packet context*. A packet context is a data structure that essentially serves as a scratch pad for carrying information between different stages of packet processing inside the router. The L2 processing logic appends to the packet context information about L2 headers, for instance, in the case of Ethernet, the source and destination MAC address. In addition to L2 information, the packet context can carry additional information as shown in Figure 14.4. Use of other fields in the packet context will be revealed later in the discussion.

Now the L2 processing logic peels off the payload, which is an IP packet, and along with the packet context sends it to the L3 processing logic. The L3 processing logic locates the IP header and checks its validity. It extracts the relevant IP header information and stores it in the packet context. The header information includes the destination address, source address, protocol type, DSCP bits (for differentiated services), and if the IP packet is carrying TCP or UDP payload, the destination and the source ports as well.

At this point, the packet context contains enough information for route lookup and classification of the packet. Next, the entire packet context is sent to the forwarding engine in the line card. The forwarding engine searches a table (the forwarding table) to determine the next hop. The next-hop information contains the egress line card and the outgoing port the packet needs to be transferred. This information is populated in the packet context.

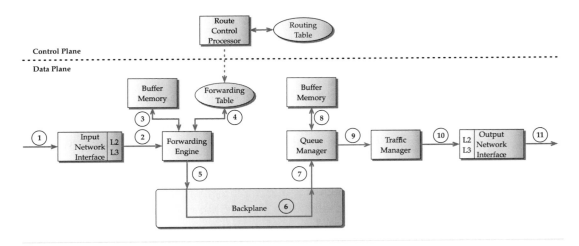

FIGURE 14.3   Packet flow in a router.

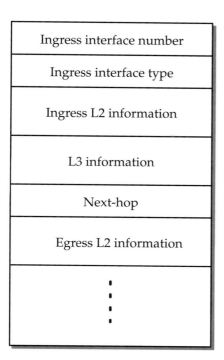

Ingress interface number
Ingress interface type
Ingress L2 information
L3 information
Next-hop
Egress L2 information
⋮

FIGURE 14.4   Typical fields of a packet context.

While the forwarding engine is determining the next hop using the packet context, the L3 processing logic sends the IP packet to be stored in the buffer memory temporarily. When the forwarding engine completes, the packet context is appended with the address of the packet in memory and is sent to the backplane interface.

From the packet context, the backplane interface knows to which line card the packet needs to be transferred. It then schedules the packet for transmission along with the packet context over the backplane. Note that the priority of the packet is taken into account while transmitting on the backplane: higher-priority packets need to be scheduled ahead of lower-priority packets.

14.4.2 Egress Packet Processing

When the packet reaches the egress line card, the backplane interface on the egress line card receives the packet and stores it in the line card memory. Meanwhile, the received packet context is updated with the new address of the memory location and sent to the queue manager. The queue manager examines the packet context to determine the packet priority. Recall that the priority was determined by the forwarding engine in the ingress line card during packet classification. Next the queue manager inserts the context of the packet in the appropriate queue.

As different queues, depending on the priority, consume different amounts of bandwidth on the same output link, the queue manager implements a scheduling algorithm.

The scheduling algorithm chooses the next packet to be transmitted according to the bandwidth configured for each queue. In some instances, the queues could be full because of congestion in the network. In order to handle such cases, the queue manager implements packet dropping behavior to proactively drop packets when the router experiences congestion.

Once the packet is scheduled to be transmitted, the traffic manager examines its context to identify the customer and if there are any transmit rate limitations that need to be enforced according to the service contract. Such a mechanism is referred to as *traffic shaping*. If the traffic exceeds any rate limitations, the traffic manager delays or drops the packet in order to comply with the agreed rate.

Finally, the packet arrives at the network interface where L3 processing logic updates its TTL and updates the checksum. The L2 processing logic adds the appropriate L2 headers and the packet is transmitted.

14.5 Packet Processing: Fast Path versus Slow Path

The tasks performed by a router can be categorized into *time-critical* and *non–time-critical* operations depending on their frequency; they are referred to as *fast path* and *slow path*, respectively. The time-critical operations are those that affect the majority of the packets and need to be highly optimized in order to achieve gigabit forwarding rates. The time-critical tasks can be broadly grouped into *header processing* and *forwarding*. The header processing functions include packet validation, packet lifetime control, and checksum calculation, while forwarding functions include destination address lookup, packet classification for service differentiation, packet buffering, and scheduling. Since these tasks need to be executed for every packet in real time, a high performance router implements these fast path functions in hardware.

Non–time-critical tasks are typically performed on packets destined to a router for maintenance, management, and error handling. Such tasks include, but are not limited to:

- Processing of data packets that lead to errors in the fast path and and generation of ICMP packets to inform the originating source of the packets

- Processing of routing protocol keep-alive messages from adjacent neighbors and sending of these messages to the neighboring routers

- Processing of incoming packets that carry route table updates and sending messages to neighboring routers when network topology changes

- Processing of packets pertaining to management protocols, such as SNMP, and the associated replies

These slow-path tasks are integrated so that they do not interfere with the fast-path mechanism. In other words, time-critical operations must have the highest priority under any circumstances. The fast path and slow path are identified in Figure 14.5. As shown in the figure, a packet using the fast path is processed only by the modules in the line cards as it traverses the router. On the other hand, a packet on the slow path is forwarded to the CPU, as many of the slow path tasks are implemented by the software running on it. Such an implementation is advantageous, as there is a clear separation between fast path and slow path. Consequently, there is no interference with the performance of packets on the fast path.

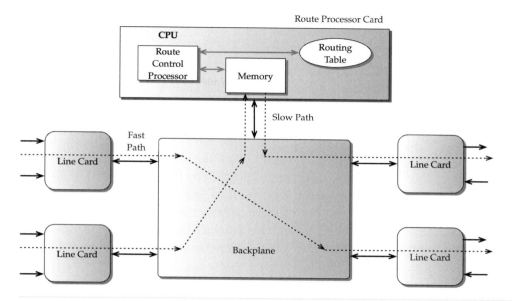

FIGURE 14.5 Router components.

In the figure, the route processor card is directly attached to the backplane, as are the line cards. The packets on the slow path from line cards are forwarded through this link to the CPU in the router processor card. The CPU dispatches them to the appropriate protocol handlers for processing. Similarly, the protocols running on the CPU can generate IP packets to be transmitted to the network. These are forwarded to the appropriate line card, as if they were from another line card. For these packets, the CPU needs to perform the route lookup; otherwise, it cannot deduce to which line card the packet needs to be forwarded. Therefore, the CPU needs to maintain its own routing table. Alternatively, instead of using a separate routing table, it can consult the master routing table of the router that it maintains.

Having delineated the distinction between fast path and slow path, the next logical question is: which router functions need to be implemented in the slow path, and which need to be implemented on the fast path? Many of the forwarding functions are typically implemented in the fast path, while the routing protocol and management functions are implemented in the slow path. However, for certain functions, it is not obvious how they should be implemented. In the following two sections, we will study in detail some of the fast path and slow path functions.

14.5.1 Fast Path Functions

In the fast path, the packets are processed and transferred from the ingress line card to the egress line card through the backplane. To achieve high speeds, the fast path functions are implemented in custom hardware, such as ASICs. While such custom implementations are less flexible, the increasing need for more packet processing at the router, and the relatively small changes in IP packet format, makes the custom hardware implementation attractive. Now let us examine some of the fast path operations in detail.

IP HEADER PROCESSING

As soon as an IP packet enters a router, it is subjected to a set of validity checks to ensure that the packet is properly formed and the header is meaningful. Only well-formed packets can be further processed; otherwise the packet is discarded.

The processing begins with a verification of the protocol version, as routers can support either IPv4 or both IPv4 and IPv6. If the version number does not match, then the packet could be malformed. The second step is for the router to check whether the length of packet reported by the MAC or the link layer is at least the minimum legal length of an IP packet. This test ensures that the IP header is not truncated by the MAC layer and filters packets less than the minimum intended length. Next, for IPv4, the value of the IP header checksum must equal the calculated header checksum computed by the router.

The routers must decrement the TTL field in the IP header to prevent packets from getting caught in routing loops forever and consuming network resources. A packet destined for the local address of the router will be accepted by the router if it has zero or a positive value of TTL. On the other hand, the packets that are being forwarded by the router should have their TTL value decremented and checked whether the TTL value is positive, zero or negative. A positive value of TTL indicates that the packets have more life left and such packets are actually forwarded. The remaining packets that have a TTL value equal to or less than zero are discarded and an ICMP error message is sent to the original sender.

Since the TTL field has been modified, the IP header checksum must be recalculated. A naive approach is to compute the checksum over the entire IP packet again, which could be computationally expensive. An efficient method to compute the Internet checksum on the entire packet is described in RFC 1071 [89]. However, as the checksum algorithms exhibit the nice properties of being commutative and associative, it is possible to compute the checksum in an incremental fashion. Such an approach is attractive and computationally less intensive, which is vital because routers have to change the TTL field of every packet that they forward. A fast approach to incrementally update the checksum is described in RFC 1141 [444] (assuming the only change to the IP header is TTL).

PACKET FORWARDING

The function of packet forwarding is to determine on which network interface a packet needs to be transmitted out of the router. The forwarding engine module controls this function using a forwarding table. The forwarding table is a summary of the routing table created by the route control processor. The router extracts the destination IP address from an incoming packet and performs a lookup in the forwarding table to determine the next-hop IP address for the packet. This procedure also decides which output port and network interface should be used to send the packet. The result of the lookup could lead to three possibilities:

- *Local:* If the IP packet is destined for the router's local IP addresses, it is delivered to the route control processor. For example, the destination for the packets carrying routing protocol keep alives and route updates is the router itself.

- *Unicast:* The packet needs to be delivered to a single output port on a network interface, either to a next-hop router or to the ultimate destination. In the latter case, the router is directly connected to the destination network.

- *Multicast:* The IP packet is delivered to a set of output ports on the same or different network interfaces, based on multicast group membership, which is maintained by the router.

As the volume of data traffic grows, routers are expected to forward more packets per second. Hence, the budget of time allowed per lookup gets reduced, and fast, efficient algorithms are required. Such algorithms are discussed in detail in Chapter 15.

PACKET CLASSIFICATION

In addition to forwarding packets, the routers need to isolate different classes, or types, of IP traffic, based on information carried in the packet. Subsequently, depending on the the type of IP traffic, an appropriate action is applied. This process of selectively identifying packets and applying the necessary actions according to certain rules is known as *packet classification*. A set of such rules is referred to as a *classifier*. A router should be capable of discriminating packets not only with the destination address, but also with the source address, source port, destination port, and protocol flags, commonly referred to as a *5-tuple*. The source and destination addresses identify the participating endpoints, the protocol flags identify the type of payload, and the source and destination ports identify the application (assuming the payload is TCP or UDP).

The packet classification function should be fast enough to keep up with the line rate. Hence, the algorithms for classification need to be fast and efficient. A detailed discussion of these algorithms and their complexities can be found in Chapter 16.

PACKET QUEUEING AND SCHEDULING

As routers keep forwarding packets, there can be an instance where multiple packets arriving on different ingress network interfaces need to be forwarded to the same egress network interface simultaneously. Such burstiness in the Internet traffic requires buffers which serve as a temporary waiting area for packets to queue up before transmission. The order in which they are transmitted is determined by various factors such as the service class the packet, the service guarantees associated with the class, etc.

Therefore, routers not only provide buffers but also require sophisticated scheduling function. The scheduling function prioritizes the traffic based on the bandwidth requirements and tolerable amount of delay by choosing the appropriate packet from these buffers. Without such options, packets simply line up and are transmitted in the order in which they are received (FIFO). Many data applications like file transfers and web browsing can tolerate some delay. However, for delay-sensitive applications such as VoIP, FIFO behavior is not clearly desirable. Chapter 22 discusses in detail various scheduling algorithms and their advantages and disadvantages.

When a network is congested, traffic arriving at the router could fill up its buffers, thereby dropping subsequent packets. If congestion could be detected before it actually occurs, proactive measures can be taken for prevention. Some of these measures include packet dropping when the occupancy of buffers reaches a predefined threshold. As the packet dropping function needs to determine whether a packet needs to be dropped, it is considered as a fast path function. Such congestion control mechanism are described in detail in Chapter 22.

14.5.2   Slow Path Operations

The packets following the slow path are partially processed by the ingress line card before forwarded to the CPU for further processing. Once the CPU completes processing, it directly sends those packet to the egress line card. Some of the slow path functions are highlighted below.

ADDRESS RESOLUTION PROTOCOL PROCESSING

When a packet needs to be sent on an egress interface, the router needs to determine the data link or the MAC address for the destination IP address or the next-hop IP address. This is because the network interface hardware on the router to which the packet needs to be forwarded understands only the addressing scheme of that physical network. Hence, a mechanism is needed to translate the IP address to a link-level address (for Ethernet, it is 48-bit MAC address). Once the link-level address is determined, the IP packet can be encapsulated in a frame that contains the link-level address and transmitted either to the ultimate destination or to the next-hop router.

A router that forwards IP packets to the destination address must either maintain these link-level addresses or dynamically discover them. The mechanism for discovering dynamically requires the use of address resolution protocol (ARP). ARP assumes that the underlying network supports link-level broadcasts and sends a query ARP request containing the IP address. When the ARP reply comes in from the host with the link-level address, it is maintained as a part of the forwarding table in the router. These entries are timed out periodically and removed and rediscovered again since the mappings change over time (possibly, the media card could have been changed).

When a packet needs to be forwarded, these link-level addresses are obtained as a result of the address lookup operation on the forwarding table along with the outgoing interface. Hence a router designer might choose to implement ARP processing in the fast path for two reasons: performance and the need for direct access to the physical network. Other designers might choose to implement ARP in the slow path, since it does not occur very frequently. When implemented in the slow path, an IP packet arriving in the router whose link-level address is not known is forwarded to the central CPU. The CPU initiates an ARP request and once the ARP reply arrives, the IP packet is forwarded. The CPU updates the forwarding tables in the line cards with the link-address for future packets.

Another variation of a slow path implementation is to initiate a link-level address request notification to the CPU from the line card. The CPU issues an ARP request and upon the arrival of the ARP reply, the CPU updates the forwarding table in the line cards with the link-level address for future packets. Meanwhile, the IP packet that triggered the notification is discarded.

FRAGMENTATION AND REASSEMBLY

Since a router connects disparate physical networks, there can be scenarios in which the message transfer unit (MTU) of one physical network is different from the other. When this happens, an incoming IP packet can be fragmented into small packets by the router if the output port is incapable of carrying the packet with its original length, that is, the MTU of the output port is less than that of the input port. Thus fragmentation enables transparent connectivity even across physical networks with different MTU sizes. However, the downside of fragmentation is that it adds more complexity in the design of the router and reduces the overall data throughput since the entire IP packet needs to retransmitted, even if a fragment is lost.

As the fast path is implemented in hardware in high-speed routers, adding support for fragmentation in hardware could be complex and expensive. The need to fragment packets is often an exceptional condition. When path MTU discovery is used, meaning that the smallest MTU size in a path is discovered before packet transmission, the need for fragmentation is very rare. Therefore, fragmentation is usually implemented in the slow path.

For further efficiency, fragmented packets transiting through a router are not reassembled, even if the output port is capable of supporting a higher MTU. The rationale is that it makes the design of the router complex, especially in the fast path, and the end system will be capable of reassembling it anyway. Implementing reassembly in the fast path requires handling of packets arriving out of order, detecting lost fragments and discarding the remaining fragments in the buffers. Such tasks are complex to implement in hardware. However, packets destined for the router should be reassembled and usually it is implemented in software. Fragment reassembly can consume substantial amounts of both CPU and memory resources. The percentage of packets sent to the router is normally quite low relative to the packets transiting through the router, which is another argument for fragmentation to be implemented in the slow path.

ADVANCED IP PROCESSING

Some of the advanced IP options include source routing, route recording, time stamping, and ICMP error generation. Source routing allows the sender of a packet to specify the route it should take to reach the destination. The main argument for implementing these functions in the slow path is that the packets requiring these functions are rare and can be handled as exceptional conditions. Hence, these packets can be processed in the control processor in the slow path.

For reporting errors about IP packets with invalid headers, the control processor can instruct the ingress network interface to discard the packet. Another alternative is to discard the packet in the fast path and send a notification to the control processor that generates an ICMP message. Some designers consider that it is more efficient to store templates of various errors in the forwarding engine, and then combine them with the IP header of the invalid packet to generate a valid ICMP message immediately.

14.6 Router Architectures

Many discussions in the literature about router architectures provide a historical perspective [365], [459], and [712]. Based on this, router architectures were classified as first generation, second generation, and so on. However, such a classification does not capture any information about commonalities, differences, and functionalities of different routers.

We present a new classification of router architectures that differs from the traditional classification. Our classification scheme is based on how the packet forwarding function is implemented from the view point of a line card. This classification was inspired by a similar scheme described in [176] and [672] in the context of parallel database systems. In the new scheme, the router architectures are broadly classified into the following:

- Shared CPU architectures

- Shared forwarding engine architectures

- Shared nothing architectures

- Clustered architectures.

Furthermore, each of these architecture as can be considered as an instance of mapping various routing functional modules to architectural components. In the next few sections, we examine each of these architectures in detail.

14.6.1 Shared CPU Architectures

This architecture is built around a conventional computer architecture; a CPU with memory and multiple line cards are connected by a shared backplane. Each line card implements a network interface to provide connectivity to the external links. The CPU runs a commodity real-time operating system and implements the functional modules, including the forwarding engine, the queue manager, the traffic manager, and some parts of the network interface, especially L2/L3 processing logic in software. In addition, the same CPU also incorporates the functionality of the route control processor that implements the routing protocols, route table maintenance, and router management functions. All the line cards share the CPU for their forwarding function; hence, the name shared CPU architecture.

An instance of this architecture is illustrated in Figure 14.6. Note that the figure also captures the flow of a packet and each step is indicated by a number enclosed in a circle. When a packet arrives at the line card, it raises an interrupt to the CPU. The interrupt service routine schedules a transfer of the packet to the buffer memory through the shared backplane. Once the transfer is complete, the CPU extracts the headers of the packet and uses the forwarding table to determine the egress line card and the outgoing port. The packet is subsequently prioritized by the queue manager and shaped by traffic manager. Finally, the packet is transferred from the memory to the appropriate output port in the egress line card. As one can see, each packet is transferred twice over the shared backplane—once from the ingress line card to the shared CPU and once from the shared CPU to the egress line card.

While most cycles of the CPU are used for packet forwarding, it spares some of its cycles running the routing protocols. It periodically exchanges protocol keep alive messages with the neighbor routers; whenever a route change occurs it incrementally updates the routing table and the forwarding table. In addition, the CPU also executes management functions for configuring and administering the router. A significant design issue, in this architecture, is how the CPU divides its execution cycles between control path and data path software.

The main advantages of this architecture are the simplicity and the flexibility of implementation. However, the following bottlenecks present in the system limit the performance of this architecture.

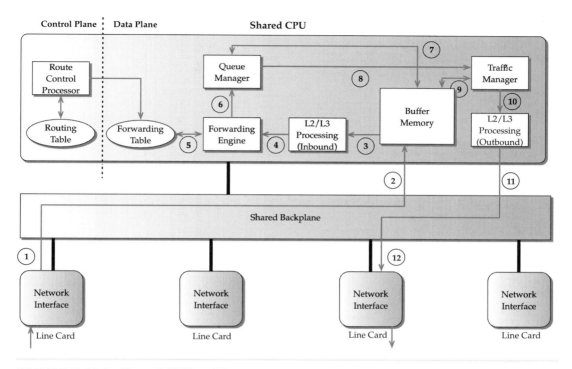

FIGURE 14.6   Shared CPU architecture.

- Each packet entering the system has to traverse the CPU; thus, the limited number of CPU cycles results in a processing bottleneck.

- The packet forwarding functions, such as forwarding table lookup, buffering and retrieval of the packet involve accessing memory. Due to mismatch in speed between memory and CPU, access to memory contributes to a larger amount of overhead. The memory access speeds have increased little over the last few years.

- The shared backplane becomes a severe limiting factor as each packet has to traverse the backplane twice. This effectively reduces the throughput by a factor of two.

To summarize, the performance of this architecture depends heavily on the throughput of the shared backplane, the speed of the shared CPU, and the cost of memory access. Hence, this architecture does not scale well to meet increasing throughput requirements. However, for lowend access and enterprise routers, where the throughput requirements are less than 1 Gbps, this architecture is still used.

Assuming the CPU speed and the cost of memory access remain the same, the throughput of shared CPU architecture can be increased if the packet traverses the shared backplane once instead of twice. If the functionality of the forwarding engine can be offloaded to the line cards, the packets need to be transferred through the backplane only once (just to the egress line card). Such an architecture is shown in Figure 14.7. The basic idea is that caching the results of the route lookup in the line card allows many of the incoming packets to be transferred directly to the egress line card; thus increasing the throughput.

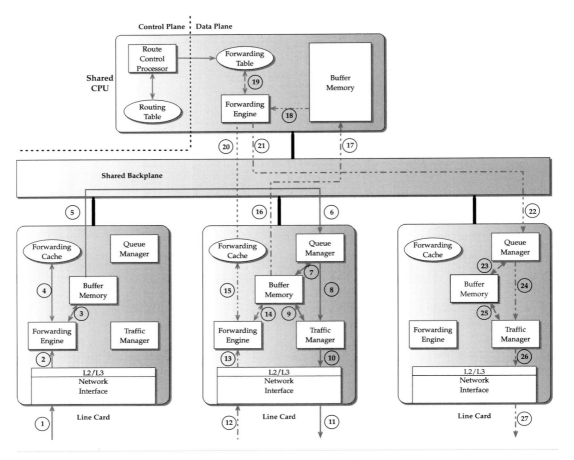

FIGURE 14.7   Shared CPU architecture with route caches.

As shown in the figure, this architecture also consists of a CPU with buffer memory and line cards connected to a shared backplane. Unlike the previous architecture, more intelligence is added to the line cards, with processor, memory and forwarding caches. The CPU maintains the central forwarding table and the line cards cache a subset of the master forwarding table based on recently used routes. When a first packet from a new destination arrives in the line card, it is sent to the shared CPU, which looks up its route using the central forwarding table. The result of the lookup is then added to the forwarding cache in the ingress line card. This allows subsequent packet flows to the same destination match the cached route entry, and the packet is directly transferred to the egress line card.

Figure 14.7 identifies the flow of two different packets. The first, indicated by steps 1 through 11, shows the case when the destination address of the incoming packet is found in the forwarding cache. The second identifies the case when the search for the destination address fails in the forwarding cache and the central forwarding table in the CPU needs to be consulted, which is indicated by steps 12 through 27. Since cache memory is limited, the entries are discarded based on LRU (least recently used) or FIFO (first-in first-out) to make space for new entries. The cache entries are periodically aged out to keep the forwarding cache current and, in the case of a route change, immediately invalidated.

The advantage of this architecture is the increased throughput because the forwarding cache of frequently seen addresses in the line card allows to process packets locally most of the time. However, the throughput is, in fact, highly dependent on the incoming traffic. The traffic arriving at a core router is an aggregation of traffic from various users and hence it exhibits little cache locality. As a result, most of the packets have to be sent to the shared CPU for route lookup. Hence, there might be little increase in throughput compared to the shared CPU architecture without forwarding caches. The performance of this architecture can be improved by increasing the memory for storing forwarding cache to include the entire forwarding table. This is feasible as the cost of memory has substantially reduced. However, the shared backplane still presents a bottleneck. Due to these drawbacks, this architecture can neither scale to high-capacity links nor provide traffic pattern-independent throughput.

14.6.2 Shared Forwarding Engine Architectures

In the shared CPU architecture, we identified that the shared CPU is one of the major bottlenecks, as it is in the path of every packet flow. The shared forwarding engine architecture is an attempt to mitigate the bottleneck by offloading the functionality of the forwarding engine to a dedicated card called *forwarding engine cards*. Each forwarding engine card contains a dedicated processor executing the software for route lookup and memory for storing the forwarding table. With multiple such cards, many packets can be processed in parallel, which considerably scales the packet forwarding speed. The shared forwarding engine architectures were used in [38], [547] to build routers capable of forwarding gigabits per second.

In this architecture, multiple line cards are connected through a shared backplane through which the packets are transferred from one line card to another. Line cards and forwarding engine cards are connected through a separate shared backplane called *forwarding backplane*. The rationale behind using two different backplanes is to separate the data traffic from the traffic generated for the forwarding engine cards, thereby improving throughput. This architecture is shown in Figure 14.8, which also illustrates the packet flow. In the figure, the numbers enclosed in circles indicate the steps of packet processing in order.

When the ingress line card receives packets, the IP header is stripped and augmented with the packet context containing a identifying tag. The packet context and the IP header are sent to a forwarding engine through the forwarding backplane for IP header validation and route lookup. Since the forwarding engine is responsible for route lookup, sending only IP headers eliminates the unnecessary overhead of transferring the packet payload over the forwarding backplane.

While the forwarding engine is performing the lookup, the packet payload is buffered in the memory of the ingress line card. The result of route lookup determines the egress line card and the interface where the packet needs to be transmitted. This information is stored in the packet context which is followed by decrementing the TTL and updating the checksum in the IP header. The updated header along with the packet context containing the tag is sent to the ingress line card. Upon examining the packet context, the ingress line card transfers the packet from its buffer memory through the shared backplane to the egress line

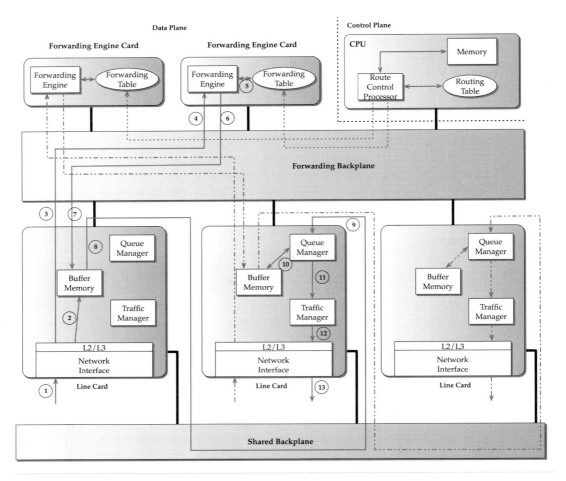

FIGURE 14.8   Shared forwarding engine architecture using two shared backplanes.

card. Subsequently, it is queued in the buffer memory of the egress line card until the queue manager and traffic manager decide to transmit on the outgoing link.

The route control processor maintains the routing table by exchanging route update messages and computes the forwarding table. The forwarding table is propagated to all the forwarding engines when a new route is added or an existing route is updated or deleted. Since the forwarding table at the forwarding engines has the same contents, their consistency needs to be maintained.

Since there are multiple forwarding engines, multiple IP headers can be processed in parallel. This could lead to the situation where packets that arrived later might finish their route lookup earlier than the packets that entered the router earlier. Subsequently, these packets will depart from the router earlier, causing packet reordering. The routers need to maintain packet ordering as sequencing of packets in a TCP connection needs to be maintained, which otherwise can trigger retransmits and degrade the performance of the overall network.

The L2/L3 packet processing logic in the ingress line card removes the IP headers and assigns these headers to forwarding engines in a round robin fashion. To ensure packet ordering, the packet processing logic in egress interface also goes round robin, guaranteeing that packets are sent out in the order in which it is received.

The time required to process each packet depends on the actual load of the forwarding engine. Hence, instead of round robin, a better load-balancing algorithm that assigns each header to the lightly loaded forwarding engine can be used. To maintain packet ordering in a TCP connection, the load-balancing algorithm should assign individual TCP connections to forwarding engines rather than packets. All the packets belonging to the connection need to use the same forwarding engine. But there are scenarios in which, once a connection is assigned to a forwarding engine, the load could increase as it is hard to predict the packet arrivals for other connections after the load-balancing decision. This could lead to more packets being queued at the forwarding engine and cause delay. However, such scenarios could be minimized by increasing the number of forwarding engine cards, which increases the probability there will be a free forwarding engine when a new connection arrives. But this might not be cost effective. Furthermore, from design perspective, the line card should have the capability to recognize the packets that signal the start and end of a TCP connection and also needs to maintain state about which forwarding engine the connection has been assigned.

The main advantage of this architecture is the ability to scale to higher forwarding speeds. Another advantage of this architecture is that it provides flexibility; the forwarding engine cards can be added whenever needed so that the necessary forwarding speed can be achieved for high-speed core routers.

A key drawback is the use of a shared backplane that does not provide sufficient bandwidth for transmitting packets between line cards and limits the router throughput. Hence, in order to remove this bandwidth limitation, the shared backplane is replaced by a switched backplane. As a switched backplane has higher bandwidth, a separate forward backplane is not required. Instead, both the line cards and forwarding engine cards are directly connected to the switched backplane, thus providing a communication path in which each line card can reach any forwarding engine. The control processor is also attached to a switched backplane, which provides a path for updating the forwarding tables in the forwarding engine cards. It is shown in Figure 14.9. Such an architecture is used in the building of a multigigabit router described in [547].

14.6.3 Shared Nothing Architectures

With increasing link speeds, the architectures described so far are stretched to their limits. First, in the shared forwarding engine architecture, forwarding a packet requires traversing the backplane twice, irrespective of using two shared backplanes or a single switched backplane as shown in Figures 14.8 and 14.9. This reduces the available backplane bandwidth for forwarding packets. Second, the use of general-purpose processors in the forwarding engine cards further limits the number of packets that can be processed.

A closer look at the shared forwarding engine architecture indicates that the extra hop through the backplane can be eliminated if the forwarding engine is incorporated into the line card. As routers are dedicated systems not running any specific application tasks, off-loading processing to line cards can increase the overall router performance. Further, more

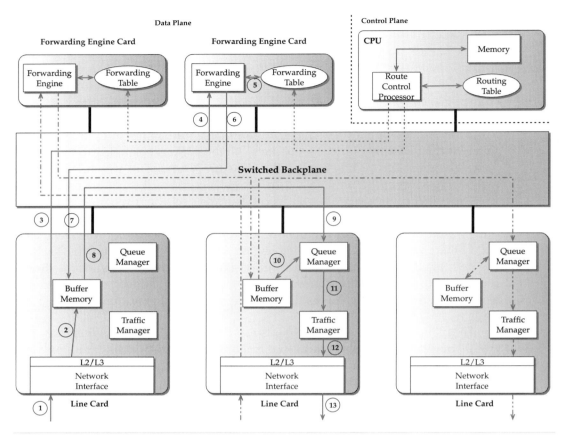

FIGURE 14.9   Shared forwarding engine architecture using a switched backplane.

processing power can be added by implementing each functional module in hardware such as high speed FPGA (field-programmable gate arrays) or ASICs (application-specific integrated circuits). To achieve high performance, these hardware components are interconnected by high-speed links embedded in the line card.

A shared nothing router architecture offloads all the packet forwarding functions to the line cards. The line cards implement these functions using custom hardware for high performance and do not share any of these components with other line cards. Hence, this architecture is named as *shared nothing*. Now since the line cards are capable of handling large number of packets, the backplane should be fast enough to handle aggregate input from all the line cards. Hence, this architecture employs switched backplanes, which makes this setup capable of multiple transfers simultaneously. An instance of this architecture is illustrated in Figure 14.10, which also depicts the packet flow.

As you can see from Figure 14.10, all the line cards are connected to a high-speed switched backplane. A packet enters the router through the network interface in the line card. It is subjected to L2/L3 processing logic, which peels off L2 header and creates a packet context. The L2/L3 processing logic appends L2 information such as source MAC and destination MAC to the packet context. In addition, the packet context is appended with the IP header of

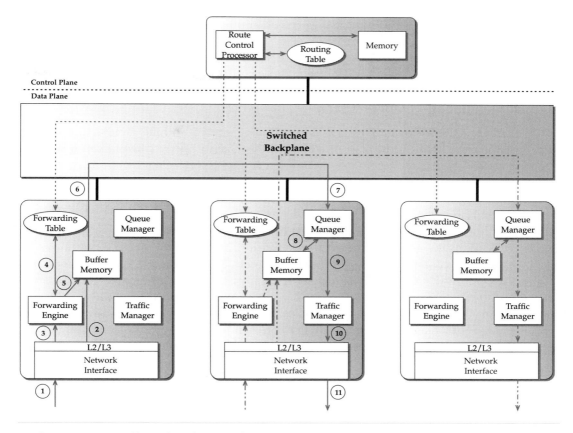

FIGURE 14.10   Shared nothing architecture.

the packet. The L2/L3 processing module deposits the packet payload to buffer memory and in parallel sends the packet context along with the header to the forwarding engine.

The forwarding engine consults the forwarding table for route lookup and determines the outgoing port and egress line card. In addition to route lookup, the forwarding engine classifies the packet into a service class based on the contents of the packet header. This service class information is stored in the packet context. Depending on the result of the route lookup, the packet is extracted from the buffer memory and transmitted to the egress line card through the switched backplane. The packet is received by the queue manager in the egress line card, which stores it in the buffer memory. Depending on the priority of the packet, it is scheduled for transmission to traffic manager. The traffic manager, based on the agreed rate of transmission, might delay the packet further or transmit it immediately on the appropriate egress interface.

The route control processer is implemented on a separate card using a general-purpose processor. In some of the architectures, this card is attached to the switched backplane for communicating to the line cards. In other architectures, a separate path for communicating to the line cards is implemented. The processor runs routing protocols and when a route update occurs, the forwarding table is computed and propagated to the forwarding tables maintained in the line cards.

Many high-end routers from router vendors use this architecture. These routers have been demonstrated to achieve throughputs greater than 640 Gbps.

14.6.4 Clustered Architectures

One of the major limitation of routers using shared nothing architecture is the number of line cards that can be supported in a single chassis. Two factors affect this limitation. First, such routers are used in the core and at higher layers of aggregation where the number of links required is small but the bandwidth per link increases. Second, the packaging density possible within the racks used in central offices is limited to 19 inches (NEBS standards). In addition, a spacing of 1 inch is needed between line cards for air flow that limits the number of line cards to 16, assuming the line cards are being arranged vertically.

With the advent of dense wave-division multiplexing (DWDM) technology, each fiber can now contain many independent channels. The data rate on each channel can be as high as OC-48 (2.4 Gbps). These channels are separated and terminated by the router with one port per channel. Hence, support for a large number of ports is required. With each line card carrying only a fixed number of ports, a router needs to support large number of line cards.

For increasing the number of line cards and the aggregate system throughput, major vendors use a clustering approach. This approach uses multiple chassis–containing line cards connected to central switch core, as shown in Figure 14.11. A variation of this approach is the use of multiple independent routers connected to a central switch core but function as a single router. In these architectures, the chassis-containing line cards are connected to the switch

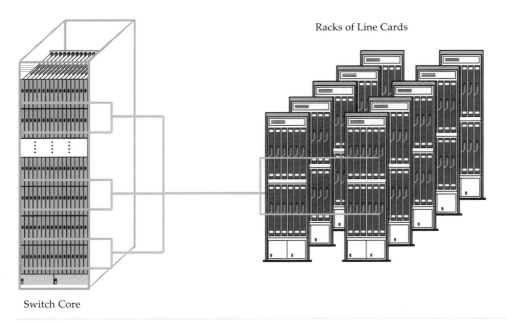

Switch Core

FIGURE 14.11   Clustered router architecture with a central switch core (adapted from [569]).

core using very-high-speed optical links. A packet entering a network interface in a line card, depending on the result of route lookup, can be destined to a line card in the same chassis or a line card in a different chassis. In the latter case, the packet has to be forwarded through the switch core that sends it to the correct chassis. Once the packet reaches the chassis, it is forwarded through the appropriate egress line card.

The advantage of this architecture is the ability to incrementally add the line card chassis depending on the need. A disadvantage of this architecture is that the switch core is a single point of failure. Hence, for high-availability routers, a second switch core is needed that increases the cost.

14.7 Summary

In this chapter, we presented an overview of router architectures. We have pointed out the critical functions of a router. Certainly, the main function is to forward packets as efficiently as possible. In addition, a router must be able to support different routing protocols and must provide interface for network administrators to configure a number of tasks including which routing protocol(s) to invoke.

For different network sizes deployed and depending on whether deployed by an access provider, transit provider, or core Internet provider, routers are designed to be of different sizes and handling capability in terms of address lookup. As you can see, a single, general-purpose architecture does not fit all requirements. Broadly, we can classify routers into four categories, which we discussed in this chapter.

In subsequent chapters, we will be delving into more details about various components and functions of a routers, and the types of operations they are required to handle.

It is important to point out that discussions presented in this chapter have been for "pure" IP routers. Routers designed for multiprotocol label switching (MPLS), often referred to as MPLS routers, do require additional functionalities for tracking and management of labels, in addition to the typical functions already discussed; we will present MPLS router architecture later, when we cover MPLS (refer to Chapter 18).

Further Lookup

In the past two decades, router architectures have seen tremendous changes, dictated by the exponential growth of the Internet. A good introduction to routing can be found in the white-paper from Juniper Networks [629]. The survey papers [46], [98] provide a detailed introduction to various router architectures. The requirements for different types of routers are described in [367]. A shorter introduction to the historical perspective of router and switch architectures can be found in [121], [459], and [712]. The tutorial on router architectures [294] provides an elegant introduction and discusses the advantages and disadvantages. Experiences about building a multibit gigabit router using shared forwarding engine architectures can be found in [547]. The best way to understand different types of routers (and their architectures) is to read products developed by various vendors. In this regard, we recommend that reader explore the web-sites of router vendors such as Cisco Systems (http://www.cisco.com/), Juniper Networks (http://www.juniper.net/), and Avici Systems (http://www.avici.com/).

We refer you to the chapters that follow (and the Further Lookup section in these chapters) for details about various elements and functions of routers.

Exercises

14.1. Review questions:

(a) What are the basic functions of a router?

(b) What fields in an IP packet header are typically changed for every incoming packet before it is being forwarded?

(c) Why is it important to distinguish between basic forwarding functions and complex forwarding functions

(d) What are the key elements of a router?

(e) What is packet context? Why is it necessary?

(f) What is the difference between a shared and switched backplane?

14.2. Discuss the strengths and weaknesses of various router architectures presented in this chapter.

14.3. Give an argument why ARP processing should be in the fast path and not in the slow path.

14.4. How is shared nothing architecture different from shared forwarding engine architecture?

14.5. Investigate various router products from different vendors, and determine which of them fall into the four router architecture classifications presented in this chapter.

14.6. Refer to Exercise 14.5. Do you find any routers that do not fall into the classification presented in this chapter? Why do you think they do not fall into any of these classifications?

15

IP Address Lookup Algorithms

I'm so fast that last night I turned off the light switch in my hotel room and was in bed before the room was dark.

Muhammad Ali

Reading Guideline

Address lookup is an important function of a router. At first, the need for address lookup is presented followed by why this function needs to be efficient. We then present a series of different approaches for address lookup; they are organized to capture different needs that must be addressed as the routing table size grows. This material assumes that the reader has some familiarity with data structures, notion of computational complexity, and hardware architecture.

The primary function of routers is to forward packets toward their final destination. To accomplish this, a router must decide for each incoming packet where to send it next. To be precise, the forwarding decision consists of two components: (1) finding the address of the next-hop router to forward the packet to, and (2) determining the egress interface through which the packet should be sent. This forwarding information, referred to as *next-hop* information, is stored in a forwarding table populated by information gathered from the routing protocols. This forwarding table is consulted using the packet's destination address as the key. Such an operation is called an *address lookup*. Once the next-hop information is retrieved, the router can transfer the packet from the ingress interface to the egress interface.

There are four key factors that affect a routing system: link speeds, router data throughput, packet forwarding rates, and quick adaptation to routing changes. The link speeds and router data throughput have kept pace with the increase in traffic demands because of advances in fiber-optic cables and fast switching technology. However, the major challenge has been to increase the packet forwarding rates to keep up with the increased data rates. Even though the packet forwarding operation consists of other chores, such as updating time-to-live (TTL) fields, these are computationally inexpensive compared to the address lookup operation. The challenge, therefore, is to develop algorithms that can perform millions of address lookup operations per second. In this chapter, we look at such algorithms in detail.

15.1 Impact of Addressing on Lookup

The addressing architecture is of fundamental importance to the routing architecture and tracing its evolution will make it clear how it impacts the complexity of the lookup mechanism. As discussed earlier in Section 1.1, in the early days the Internet used a *classful* addressing scheme, known as Class A, Class B, and Class C addresses.

With the classful addressing scheme, the forwarding of packets is straightforward. Routers need to examine only the network part of the destination address to forward it to the destination. Thus, the forwarding table needs to store a single entry (the network part) for routing the packets destined to all the hosts attached to a given network. Such a technique is called *address aggregation* and uses prefixes to represent a group of addresses. As described earlier in Section 14.1.4, a forwarding table entry consists of a prefix, the next-hop address, and the outgoing interface. Finding the forwarding information requires searching the prefixes in the forwarding table for the one that matches the same set of bits in the destination address.

The lookup operation in a classful addressing scheme proceeds as shown in Figure 15.1. The forwarding table is organized into three separate tables: one for each of the three allowed lengths: 7 bits, 14 bits, and 21 bits for classes *A*, *B* and *C*, respectively. As shown in Figure 15.1, first the address class is determined from the first few bits of the destination address. Based on this information, one of the three tables is chosen to search. Meanwhile, the network part of the destination is extracted based on the class. Then the chosen table is searched for an exact match between the network part and the prefixes present in the table. The search for an exact match can be performed using well-known algorithms such as binary search or hashing.

The class-based addressing scheme worked well in the early years of the Internet. However, as the Internet started growing, this scheme presented two problems:

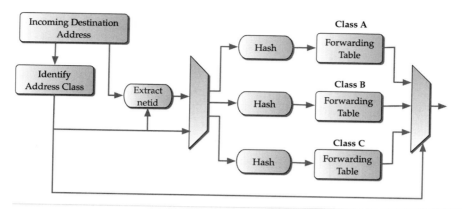

FIGURE 15.1　Lookup operation in a classful IP addressing scheme (adapted from [273]).

- *Depletion of IP Address Space:* With only three different network sizes to choose from, the IP address space was not used efficiently and it was being exhausted very rapidly, since only a fraction of the addresses allocated was actually in use (approximately 1%). For example, a class *B* netid (good for 2^{16} hosts) had to be allocated to any organization with more than 254 hosts.

- *Exponential Growth of Routing Tables:* The route information stored in the forwarding tables of core IP routers grew in proportion to the number of networks. As a result, routing tables were growing exponentially. This led to higher lookup times on the processor and higher memory requirements in the routers.

In an attempt to allow more efficient use of IP address space and to slow down the exponential growth of forwarding tables in routers, a new scheme called *classless interdomain routing* (CIDR) was introduced [239]; see Section 1.3.3 for an introduction about CIDR.

15.1.1 Address Aggregation

Because of CIDR, address aggregation is possible so that a router can maintain one entry instead of its constituents before aggregation; however, sometimes it is not possible if an address block is missing. To understand aggregation and exception in aggregation, consider the following example.

Example 15.1　*Address aggregation and exception in address aggregation.*
First, we consider address aggregation. Assume that ISP1, a service provider, connects three customers—C1, C2, and C3—with the rest of the Internet; see Figure 15.2(a). ISP1 is, in turn, connected to some backbone provider through router R1. The backbone can also connect other service providers like ISP2. Assume that ISP1 owns IP prefix block 10.2.0.0/22 and partitions it among its customers. Let us say that prefix 10.2.1.0/24 has been allocated to C1, 10.2.2.0/24 to C2, and 10.2.3.0/24 to C3. Now the router in the backbone R1 needs to keep only a single forwarding table entry for IP prefix 10.2.0.0/22 that directs the traffic bound

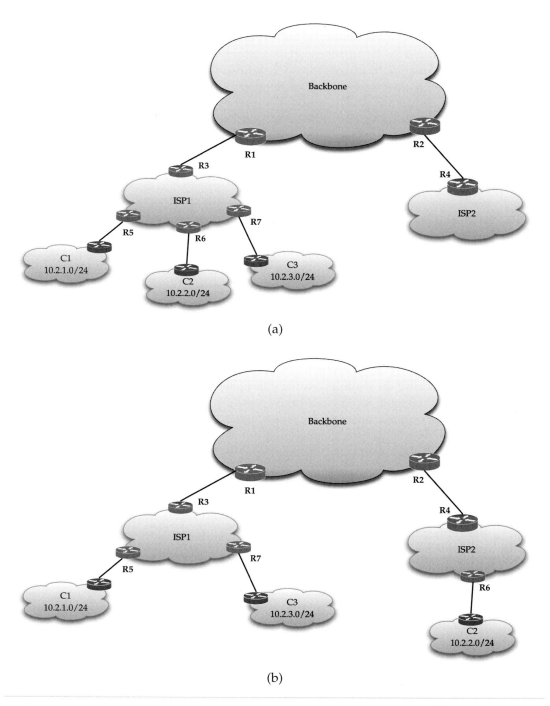

(a)

(b)

FIGURE 15.2   Examples of (a) address aggregation, (b) exception in address aggregation.

to C1, C2, and C3 through router R3. As you can see, the hierarchical allocation of prefixes obviates the need for separate routing table entries for C1, C2, and C3 at router R1. In other words, the backbone routes the traffic bound for ISP1 to R3 and it is the responsibility of the routers within ISP1 to distinguish the traffic between C1, C2, and C3.

Next, assume that customer C2 would like to change its service provider from ISP1 to ISP2, but does not want to renumber its network. This is depicted in Figure 15.2(b). Now all the traffic in the backbone bound to prefix 10.2.0.0/22 would need to be directed to ISP1 except for prefix 10.2.2.0/24. We cannot perform aggregation as before at router R1 since there is a "hole" in the address space. This is called *exception to address aggregation*.

▲

Example 15.1 demonstrates how route aggregation leads to a reduction in the size of backbone routing tables. A great deal of aggregation can be achieved if addresses are carefully assigned. However, in some situations, a few networks can interfere with the process of aggregation as we illustrated above. The question is: how can we accommodate such a hole? A straightforward solution is to deaggregate and increase the number of entries in the backbone router R1 to three, one for each of the customers. This is not desirable since the number of entries in the backbone will increase dramatically.

Can we come up with a solution that preserves the benefits of aggregation and still accommodates such exceptions? Yes, using the following. Keep the existing prefix 10.2.0.0/22 and add the exception prefix 10.2.2.0/24 to the forwarding table in router R1. The next-hop information for exception prefix will direct the traffic to router R2. This will result in only two entries in the forwarding table. Note, however, that some addresses will match both the entries because the prefixes overlap. To make the correct forwarding decision, routers must do more than look for an exact prefix match. Since exceptions in the aggregations may exist, it needs to find the most specific match that is the longest matching prefix. Now we are ready to define the longest prefix matching problem and describe why it is difficult.

15.2 Longest Prefix Matching

The problem of identifying the forwarding entry containing the longest prefix among all the prefixes matching the destination address of an incoming packet is defined as the longest matching prefix problem. This longest prefix is called the *longest matching prefix*. It is also referred to as the *best matching prefix*.

Example 15.2 *Identifying the longest matching prefix.*

Consider the forwarding table at a router, as shown in Table 15.1. Each entry contains the prefix and the name of the outgoing interface. Entry 1 indicates that the packet matching

TABLE 15.1   A forwarding table.

Entry Number	Prefix	Next-Hop
1	98.1.1.1/24	eth3
2	171.1.0.0/16	so6
3	171.1.1.0/24	fe5

TABLE 15.2   Prefix table example.

Prefix label	Prefix
P_1	0*
P_2	00001*
P_3	001*
P_4	1*
P_5	1000*
P_6	1001*
P_7	1010*
P_8	1011*
P_9	111*

prefix 98.1.1.1/24 will go out on interface *eth3*. If the destination address of the incoming packet is 171.1.1.2, then it will match prefix 171.1.0.0/16 in entry 2 as well as 171.1.1.0/24 in entry 3. Since prefix 171.1.1.0/24 is the longest matching prefix, the packet will be forwarded on the outgoing interface *fe5*.   ▲

The difficulty of the longest prefix matching arises because of the following reasons. First, a destination IP address does not explicitly carry the netmask information when a packet traverses through. Second, the prefixes in the forwarding table against which the destination address needs to be matched can be of arbitrary lengths; this could be as a result of an arbitrary number of network aggregations. Thus, the search for the longest prefix not only requires determining the length of the longest matching prefix, but also involves finding the forwarding table entry containing the prefix of this length. To conclude, the adaptation of CIDR has made route lookups more complex than they were when the classful addressing scheme was used.

Throughout the rest of the chapter, we will be using the following two parameters:

N = Number of prefixes in a forwarding table

W = Maximum length of the prefixes in bits

Unless otherwise specified, we use the prefixes shown in Table 15.2 as a running example for various algorithms discussed in later sections. For ease of reading, this table is subsequently shown next to each figure that is associated with a specific lookup algorithm.

15.2.1 Trends, Observations, and Requirements

It is imperative for any algorithm designer to understand the requirements of the problem and how these requirements are expected to evolve. The basic requirements for the longest prefix matching include the following:

- *Lookup Speed:* Internet traffic measurements [696] show that roughly 50% of the packets that arrive at a router are TCP-acknowledgment packets, which are typically 40-byte packets. As a result, a router can be expected to receive a steady stream of such minimum size packets. Thus, the prefix lookup has to happen in the time it takes to forward a minimum-size packet (40 bytes), known as wire speed forwarding. At wire speed forwarding, the amount of time that it takes for a lookup should not exceed 320 nanosec at 1 Gbps ($= 10^9$ bps), which is computed as follows:

$$\frac{40\,\text{bytes} \times 8\,\text{bits/byte}}{1 \times 10^9\,\text{bps}} = 320\,\text{nanosec}.$$

Similarly, the lookup cannot exceed the budget time of 32 nanosec at 10 Gbps and 8 nanosec at 40 Gbps. The main bottleneck in achieving such high lookup speed is the cost of memory access. Thus, the lookup speed is measured in terms of the number of memory accesses.

- *Memory Usage:* The amount of memory consumed by the data structures of the algorithm is also important. Ideally, it should occupy as little memory as possible. A memory-efficient algorithm can effectively use the fast but small cache memory if implemented in software. On the other hand, hardware implementations allow the use of fast but expensive on-chip SRAM needed to achieve high speeds.

- *Scalability:* The algorithms are expected to scale both in speed and memory as the size of the forwarding table increases. While core routers presently contain as many as 200,000 prefixes, it is expected to increase to 500,000 to 1 million prefixes with the possible use of host routes and multicast routes. When routers are deployed in the real network, the service providers expect them to provide consistent and predictable performance despite the increase in routing table size. This is expected since a router needs to have a useful lifetime of at least five years to recuperate the return on investment.

- *Updatability:* It has been observed in practice that the route changes occur fairly frequently. Studies [387] show that core routers may receive bursts of these changes at rates varying from a few prefixes per second to a few hundred prefixes per second. Thus, the route changes require updating the forwarding table data structure, in the order of milliseconds or less. These requirements are still several orders of magnitude less than the lookup speed requirements. Nonetheless, it is important for an algorithm to support incremental updates.

To summarize, the important requirements of a lookup algorithm are speed, storage, update time, and scalability. We ideally require algorithms to perform well in the worst case. However, exploiting some of the following practical observations to improve the expected case is desirable.

- Most of the prefixes are 24 bits or less in core routers, while more than half are 24 bits (see Table 9.2).

- There are not very many prefixes that are prefixes of other prefixes. Practical observations show that the number of prefixes of a given prefix is at most seven.

These practical observations can be further exploited to come up with efficient schemes.

15.3 Naïve Algorithms

The simplest algorithm for finding the best matching prefix is a linear search of prefixes. It uses an array in which the prefixes are stored in an unordered fashion. The search iterates through each prefix and compares it with the destination address. If a match occurs, it is remembered as the best match and the search continues. The best match is updated as the search walks through each prefix in the array. When the search terminates the last prefix remembered is the best matching prefix. The time complexity for such a search is $O(N)$. Linear search might be useful if there are very few prefixes to search; however, the search time degrades as N becomes large.

Some researchers proposed the idea of route caching in conjunction with linear search to speed up the lookup time. A cache is a fast buffer for storing recently accessed data. The main use of the cache is to speed up subsequent access to the same data if there is a sufficient amount of locality in data access requests. The average time to access data is significantly reduced since access to cache takes significantly less time than access to SRAM or other storage media [293].

For lookup the cache stores the recently seen 32-bit destination addresses and the associated next-hop information. When a packet arrives at the router, the destination address is extracted and the route cache is consulted. If an entry exists in the cache, then the lookup operation is completed. Otherwise, the linear search discussed above is invoked and the result is cached, possibly replacing an existing entry. Such a caching scheme is effective only when there is *locality* in a stream of packets, i.e., a packet arrival implies another packet with the same destination address will arrive with high probability in the near future.

However, locality exhibited by flows in the backbone has been observed to be very poor [527]. This leads to a much lower cache hit ratio and degenerates to a linear search for every lookup. In summary, caching can be useful when used in conjunction with other algorithms, but that precludes the need for fast prefix lookups.

15.4 Binary Tries

The binary trie is the simplest of a class of algorithms that is tree-like. The term *trie* comes from "re*trie*val" and is pronounced "tree." However, most often to verbally distinguish a trie from a general tree, it is pronounced as "try." The trie data structure was first proposed in the context of file searching [169], [238].

A binary trie is a tree structure that allows a natural way to organize IP prefixes and uses the bits of prefixes to direct the branching. Each internal node in the tree can have zero, one, or two descendants. The left branch of a node is labeled 0 and the right branch is labeled 1. For instance, the binary trie for the prefixes in Table 15.2 is shown in Figure 15.3.

In a binary trie, a node l represents a prefix formed by concatenating the labels of all the branches in the path from the root node to l. For example, the concatenated label along the path to node P_2 is 00001, which is the same as prefix P_2. Note that some of the nodes are shaded in gray while the remaining nodes are not. The gray-shaded nodes correspond to actual prefixes. These nodes contain the next-hop information or a pointer to it. As can be seen, prefixes can be either in the internal nodes or at the leaf nodes. Such a situation arises if there are exception prefixes in the prefix aggregation. For instance, in Figure 15.3, the prefixes P_2 and P_3 represent exceptions to prefix P_1.

P_1 0*
P_2 00001*
P_3 001*
P_4 1*
P_5 1000*
P_6 1001*
P_7 1010*
P_8 1011*
P_9 111*

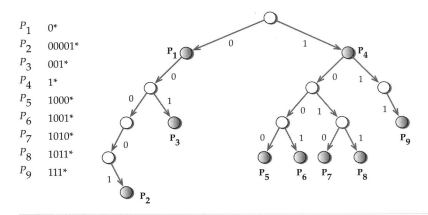

FIGURE 15.3   Binary trie data structure for the prefixes of Table 15.2.

Let us try to understand this better. A complete binary trie represents all the 5-bit address space. Each leaf represents one possible address. We can see that the address space covered by P_1 overlaps the addresses of P_2 and P_3. Thus, prefixes P_2 and P_3 represent exceptions to prefix P_1 and refer to specific subintervals of the address space of prefix P_1. Is it possible to identify such exception prefixes by simply looking at the trie in Figure 15.3? Indeed, yes. Exception prefixes are always descendants of an aggregation prefix. In Figure 15.3, P_2 and P_3 are exceptions of prefix P_1 since they are its descendants.

15.4.1 Search and Update Operations

We have seen so far how the prefixes are organized in a trie. The next question is given a destination address, how do the search/insert/delete operations work? The search in the trie is guided by the bits of the destination address starting from the root node of the trie. At each node, the left or right branch is taken depending upon the inspection of the appropriate bit. While traversing the trie, we may come across gray-shaded nodes that contain the prefix information. The search needs to remember this prefix information since it is the longest match found so far. Finally, the search terminates when there are no more branches to be taken and the best matching prefix will be the last prefix remembered.

Example 15.3   *Searching for the best matching prefix in a binary trie.*

Consider searching the binary trie shown in Figure 15.3 for an address that begins with 001. The search starts at the root node of the trie and the first bit is examined. Since it is a 0, the search proceeds toward the left branch and encounters the node with the prefix P_1. We remember P_1 as the best matching prefix found so far. Then, we move left as the second address bit is another 0; the node encountered this time does not have a prefix, so P_1 remains the best matching prefix so far. Next, we examine the third bit, which is a 1, and leads to prefix P_3. Since P_3 is a better matching prefix than P_1, it is remembered. Now, P_3 is a leaf node and, thus, the search terminates and the prefix P_3 is declared the best matching prefix.   ▲

The insert and delete operations are also straightforward to implement in binary tries. The insert operation proceeds by using the same bit-by-bit search. Once the search arrives at a node with no further branch to proceed, the necessary nodes are created to represent the prefix.

Example 15.4 *Inserting prefixes in a binary trie.*

Consider inserting prefixes 110 and 0110, referred to as P_{10} and P_{11}, respectively, in the binary trie shown in Figure 15.3. Figure 15.4 illustrates the insertion of both the prefixes. Since the first bit of P_{10} is 1, the search moves to the right and reaches the gray node P_4. Now the second bit is examined, which again guides the search right. As the third bit is 0, there is no left branch to take and, thus, a new node P_{10} is created and attached. The next-hop information for this prefix is stored in the node itself. Now consider inserting prefix P_{11}. After inspecting the bits, we find that there is no right branch to take on the node with prefix P_1. Thus, new nodes are added that create the path to prefix node P_{11}. ▲

Similar to the insert operation, the deletion of a prefix starts by a search to locate the prefix to be deleted. Once the node containing the prefix is found, different operations are performed depending on the node type. If it is an internal node (gray node), then the node is *unmarked*, indicating there is no more prefix information on it. For example, to delete prefix P_1, simply unmark it, which is equivalent to removing the next-hop information or nullifying it. If the node to be deleted is a leaf node, all the one-child nodes leading to the leaf node might have to be deleted as well.

A binary trie is implemented using two entries per node: one entry for bit 0 and the other for bit 1. Each entry contains two fields, *nhop* that stores the next-hop information and *ptr* that stores the pointer to the subtrie. If next-hop information is not present, the field is set to null and, similarly, if the subtrie is not present the *ptr* field is set to null. Note that the prefix information itself is not stored in each node. This is because it can be derived based on the current bit position being examined in the address that is being looked up. The implemen-

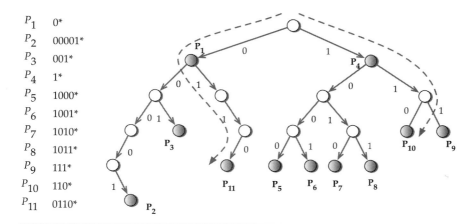

P_1	0*
P_2	00001*
P_3	001*
P_4	1*
P_5	1000*
P_6	1001*
P_7	1010*
P_8	1011*
P_9	111*
P_{10}	110*
P_{11}	0110*

FIGURE 15.4   Inserting new prefixes in a binary trie.

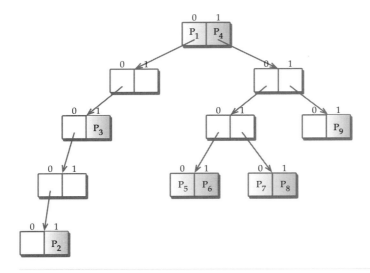

FIGURE 15.5 Implementation of the binary trie shown in Figure 15.3.

tation of the binary trie in Figure 15.3 is shown in Figure 15.5, in which prefixes indicate the presence of next-hop information and the arrows indicate the presence of pointers to subtries.

   Binary tries, in the worst case, during a search must traverse a number of trie nodes equal to the length of addresses; thus, the search complexity is $O(W)$. Inserting a prefix to a binary trie might require adding a sequence of W nodes, and the worst case update complexity is $O(W)$. Similarly, for deletion the worst-case time complexity is $O(W)$. In terms of memory consumption, the complexity is $O(NW)$ since each prefix at most can have W nodes. Note that some of the nodes are shared along the prefix paths and, thus, the upper bound might not be tight.

15.4.2 Path Compression

A binary trie can represent arbitrary-length prefixes but it has the characteristic that long sequences of one-child nodes may exist. Such long sequences are undesirable since the bits corresponding to those nodes would need to be examined even though no actual branching decision is made. This increases the search time more than necessary in some cases. Also, one-child nodes consume additional memory.

   Now, assume the objective is to reduce the search time and reduce the memory space; what can we do about it? One possibility is not to involve any of the bits corresponding to one-child nodes during inspection. If they do not need to be inspected, then we can eliminate them as well. This is exactly the idea behind *path compression*. By collapsing the one-way branch nodes, path compression improves search time and consumes less memory space. However, additional information needs to be maintained in other nodes so that a search operation can be performed correctly. Path compression is derived from a scheme called PATRICIA [502]; PATRICIA was meant primarily for storing strings of characters and it did not support longest prefix matching. It was later adapted for longest prefix matching [645].

P_1 0*
P_2 00001*
P_3 001*
P_4 1*
P_5 1000*
P_6 1001*
P_7 1010*
P_8 1011*
P_9 111*

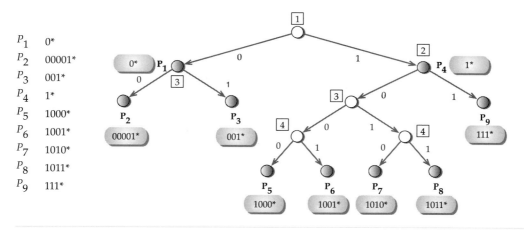

FIGURE 15.6   Path compressed trie data structure for the prefixes of Table 15.2.

Path compression applied to the binary trie in Figure 15.3 is shown in Figure 15.6. Observe that the nodes with prefixes P_2 and P_3 have moved up as immediate descendants of node P_1. The two nodes preceding P_2 in the binary trie have been removed since they are one-child nodes and redundant. Note that the node P_2, which was originally in the right branch of its parent, has moved as the left branch of P_1. This is because it is the only prefix in the entire left subtrie of node P_1. Note that prefix P_3, which was in the left branch of P_1 in the binary trie, has shifted to the right. The immediate descendant of P_1 in the binary trie can be eliminated and the decision of branching can be made at node P_1 itself. However, it requires extra information to be stored at node P_1—the bit number to be examined next in the prefix. Because one-child nodes are now removed, it is possible to jump directly to the bit where a significant decision needs to be made, thereby bypassing the inspection of some intermediate bits. In Figure 15.6, the bit numbers (or positions) that need to be examined are shown adjacent to the node enclosed in squares. Since one–child nodes are being removed, what will happen to the prefixes originally present in those nodes? They are simply moved to their nearest descendants. Hence, a list of prefixes needs to be maintained in some of the nodes.

A search in a path compressed trie proceeds in a manner similar to that in a binary trie by descending the tree under the guidance of the address bits. However, the search inspects only the bit positions indicated by the bit number in the nodes traversed. As a gray node (node with a prefix) is encountered, comparisons are performed with the actual prefix. These comparisons are necessary since the search can potentially skip some bits. If a match is found, the best matching prefix is remembered and the search proceeds further. The search ends when a leaf node has been reached or a mismatch is found.

Example 15.5 *Search for the best matching prefix in a path compressed trie.*
Consider the search for an address beginning with 001010 in the path compressed trie shown in Figure 15.6. The search starts with the root node. The node specifies that the bit number 1 needs to be examined. Since the first bit is 0, the search goes left and reaches the

prefix P_1 node. Now, we compare the prefix P_1 with the corresponding part of the address 0. Since they match, prefix P_1 is saved as the best matching prefix encountered so far. Now the bit number in node P_1 indicates that the third bit needs to be inspected, which guides the search to the right. Again, we check whether the prefix P_3 (001*) matches the corresponding part of the address (001010). Since they match and a leaf node is encountered, the search terminates and P_3 is the best matching prefix.   ▲

A path compressed trie has a search complexity of $O(W)$ in the worst case. Remember path compression is effective on a sparse binary trie. In the case of prefix distribution that results in a dense binary trie, height reduction is less effective. Using similar arguments, we can infer that the update complexity in the worst case is $O(W)$. Since path compressed tries are full binary tries, the total amount of memory required will be at most $2N - 1$, with N for the leaf nodes and $N - 1$ for the internal nodes. Hence the space complexity is $O(N)$. Thus, path compressed tries reduce the space requirements, but not the search complexity.

15.5 Multibit Tries

While binary tries can handle prefixes of arbitrary length easily, the search can be very slow since we examine one bit at a time. In the worst case, it requires 32 memory accesses for the 32-bit IPv4 address. If the cost of a memory access is 10 nanosec, the lookup will consume 320 nanosec. This translates to a maximum forwarding speed of 3.125 million packets per second (1/320 nanosec). At 40 bytes per packet, this can support an aggregate link speed of at most 1 Gbps. However, the increase in Internet traffic requires supporting aggregate link speeds as high as 40 Gbps. Clearly, sustaining such a high rate is not feasible with binary trie–based structures.

After closely examining the binary trie, we can ask: why restrict ourselves to only one bit at a time? Instead, examine multiple bits so that we can speed up the search by reducing the number of memory access. For instance, if we inspect 4 bits at a time, the search will finish in 8 memory accesses as compared to 32 memory accesses in a binary trie. This is the basic idea behind the multibit trie [661]. The number of bits to be inspected per step is called a *stride*. Strides can be either fixed-size or variable-size. A *multibit trie* is a trie structure that allows the inspection of bits in strides of several bits. Each node in the multibit trie has 2^k children where k is the stride. If all the nodes at the same level have the same stride size, we call it a *fixed* stride; otherwise, it is a *variable* stride.

As one can see, since multibit tries allow the data structure to be traversed in strides of several bits at a time, they cannot support prefixes of arbitrary lengths. To use a given multibit trie, a prefix must be transformed into an equivalent prefix of longer length to conform with the prefix lengths allowed by the structure. In the next section, we discuss some useful prefix transformation techniques that expand a prefix into an equivalent set of prefixes of longer lengths followed by a detailed discussion of various types of multibit tries.

15.5.1 Prefix Transformations

An IP prefix associated with the next-hop information can be expressed as an equivalent set of prefixes with the same next-hop information after a series of transformations. Various types

TABLE 15.3 Expansion of prefixes.

Prefix	Value	Expanded Prefixes
P_1	0*	000*, 010*, 011*
P_2	00001*	00001*
P_3	001*	001*
P_4	1*	100*, 101*, 110*
P_5	1000*	10000*, 10001*
P_6	1001*	10010*, 10011*
P_7	1010*	10100*, 10101*
P_8	1011*	10110*, 10111*
P_9	111*	111*

of transformation are possible but we restrict the discussion to the commonly used ones: *prefix expansion* and *disjoint prefixes*.

PREFIX EXPANSION

One of the most common prefix transformation techniques is *prefix expansion*. A prefix is said to be expanded if it is converted into several longer and more specific prefixes that cover the same range of addresses. For instance, consider the prefix 0*. The range of addresses covered by 0* can be also specified with the two prefixes 00* and 01*, or with the four prefixes 000*, 001*, 010*, and 011*.

Now the basic question is, how is this useful? If we do prefix expansion appropriately, a given set of prefixes of different lengths can be transformed into a set of prefixes that has fewer different lengths. Consider the set of prefixes in Table 15.3, which is the same set of prefixes shown in Table 15.2. These sets of prefixes have lengths ranging from 1 to 5 and have 4 distinct lengths. Now we want to transform these prefixes into an equivalent set with prefixes of lengths 3 and 2—two distinct lengths.

Prefix P_1 of length 1 cannot remain unchanged since the closest choice of length is 3. Hence we need to expand it into four equivalent prefixes of length 3. For the prefix 0*, some of the addresses will start with 000, 001, 010, and the rest will start with 011. Thus, the prefix 0* is equivalent to the union of four prefixes 000*, 001*, 010*, and 011*. Both of these prefixes will inherit the same next-hop information as the original prefix P_1. Similarly, we can expand prefix P_5 into two prefixes of length 5: 10000* and 10001*. Thus we can easily expand a prefix into multiple prefixes that are greater in length.

Now, by the same principle, if prefix P_4 is expanded into four prefixes 100*, 101*, 110*, and 111*, we find that prefix 111* already exists as prefix P_9. Since multiple copies of the same prefix are not desirable, we must break the tie somehow. In such cases, according to the longest matching rule, prefix P_9 is the correct choice during a search. In general, when a smaller length prefix is expanded in length and one of its expansions "collides" with an existing prefix, then we say the existing prefix *captures* the expansion prefix. In such cases, we simply get rid of the expansion prefix. In the example, we remove the expansion prefix 111* corresponding to P_4, since it has already been captured by the existing prefix P_9. The complete expanded prefixes are shown in the last column of Table 15.3.

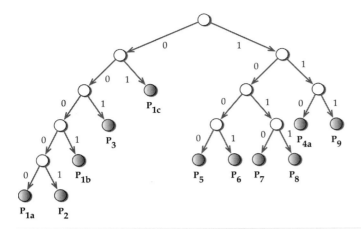

FIGURE 15.7   Disjoint prefix (leaf pushed) binary trie.

DISJOINT PREFIXES

As we have seen earlier, prefixes can overlap with each other. Furthermore, prefixes represent intervals of contiguous addresses. When two prefixes overlap, it means that one interval of addresses contains another interval of addresses. That is why an address lookup can match several prefixes. The longest prefix matching rule breaks the tie by choosing the prefix that matches as many bits as possible. Is it possible to avoid the longest prefix matching rule and still find the longest matching prefix? Indeed, it is.

The trick is to transform a given set of prefixes into a set of *disjoint prefixes*. In a disjoint set of prefixes, one prefix does not overlap with another. A trie used to represent disjoint prefixes will have prefixes at the leaf nodes and not at the internal nodes. Now, given a set of prefixes, how can you transform them into a set of disjoint prefixes? Construct a binary trie with the given set of prefixes. Now add leaf nodes to nodes that have only one child. These new leaf nodes represent new prefixes and they inherit forwarding information from the closest ancestor marked as a prefix. Then unmark the internal nodes containing the prefix.

The disjoint-prefix binary trie for the binary trie in Figure 15.3 is shown in Figure 15.7. Observe that new prefixes P_{1a}, P_{1b}, and P_{1c} have been added. They inherit the next-hop information from the original prefix P_1. Similarly, prefix P_{4a} inherits the next-hop information from prefix P_4. If an address lookup in the original binary trie ends up with prefix P_1 being the best match, then in the disjoint-prefix binary trie it will match P_{1a}, P_{1b}, or P_{1c}. Consider an example of looking up the prefix 01*. In the original binary trie, the best matching prefix is P_1. In the disjoint-prefix binary trie, the search will end with P_{1c}. Since P_{1c} has the same next-hop information as prefix P_1, the result will be equivalent. Since this transformation pushes all the prefixes in the internal nodes to the leaves, it is also known as *leaf pushing*.

15.5.2   Fixed Stride Multibit Trie

If all the nodes at the same level have the same stride size, then the multibit trie is called a *fixed stride multibit trie*. The fixed stride multibit trie, corresponding to the prefixes in Table 15.2, is

P_1	0*
P_2	00001*
P_3	001*
P_4	1*
P_5	1000*
P_6	1001*
P_7	1010*
P_8	1011*
P_9	111*

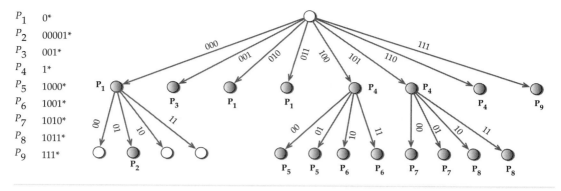

FIGURE 15.8   Fixed stride multibit trie data structure for the prefixes of Table 15.2.

shown in Figure 15.8. The example multibit trie uses a stride of 3 bits and 2 bits for all nodes in level 1 and level 2, respectively.

As noted earlier, some of the prefixes might have to be expanded to the lengths supported by the trie structure. Here prefixes of lengths other than 3 and 5 should be transformed into prefixes of lengths 3 and 5. Applying prefix expansion, prefix P_1 is expanded into four prefixes 000*, 001*, 010*, and 011* of length 3. One of the expanded prefixes 001* is the same as prefix P_3. What do we do about it? According to the prefix capture in Section 15.5.1, prefix P_3 captures the expanded prefix 001* of P_1 since it is more specific. In such cases, we have to retain the forwarding information of the existing prefix according to the longest matching rule. Now prefix P_5 of length 4 is expanded to two prefixes 10000* and 10001* of length 5. Similarly, prefixes P_6, P_7, and P_8 are expanded.

15.5.3 Search Algorithm

The search proceeds by breaking up the destination address into chunks that correspond to the strides at each level. Then these chunks are used to follow a path through the trie until there are no more branches to take. Each time a prefix is found at a node, it is remembered as the new best matching prefix seen so far. At the end, the last best matching prefix found is the correct one for the given address.

Example 15.6   *Search for the best matching prefix in a fixed stride trie.*

Consider searching for the best matching prefix for the address 100111 in the fixed stride trie shown in Figure 15.8. First, the address is divided into multiple chunks: a chunk made of the first 3 bits, 100, corresponds to level 1 of the trie; another chunk made of the next two bits, 11, corresponds to level 2 of the trie, and the last incomplete chunk consists of the remaining bits. Using the first chunk 100 at the root node leads to the prefix P_4 that is noted as the best matching prefix. Next, using the second chunk of 11 leads to the prefix P_6, which is updated to be the best matching prefix so far. Since the search cannot proceed further, the final answer is P_6. It can be seen that the number of memory accesses required is 2 as compared to 5 when using a binary trie for the same search.   ▲

Since a multibit trie is traversed in strides of k bits, the search time is bounded by $O(W/k)$.

15.5.4   Update Algorithm

Before examining how updates work, let us examine the concept of a local best matching prefix in multibit tries. A multibit trie can be viewed as a tree of one-level subtries. For instance, in Figure 15.8, there is one subtrie at the first level and three subtries at the second level. The prefix expansion in a subtrie is nothing but actually computing the local best matching prefix for each node. The best matching prefix is *local* because it is computed from a subset of prefixes of the entire prefix set.

Consider again the example in Figure 15.8. In the subtrie at the first level we are interested in finding the best matching prefix among prefixes P_1, P_3, P_4, and P_9. For the leftmost subtrie at the second level the best matching prefix will be selected from only prefix P_2. Similarly, in the second subtrie at the second level, the best matching prefix is selected from the prefix set P_5 and P_6 while for the rightmost subtrie it is selected from prefixes P_7 and P_8. Thus, multibit tries divide the problem of finding a best matching prefix into smaller subproblems in which the local best matching prefixes are selected from among a subset of prefixes. This works out to the advantage of prefix updates as illustrated by the following example.

Example 15.7   *Inserting prefixes to a fixed stride trie.*

Assume that we need to insert two prefixes, $1100*$ and $10111*$, to the fixed stride shown in Figure 15.8. These prefixes are referred to as P_{10} and P_{11}, respectively. Figure 15.9 illustrates the insertion of both prefixes P_{10} and P_{11}. Let us start with the insertion of prefix P_{10} by dividing it into chunks 110 and $0*$. We look up the root node using chunk 110, which leads to node P_4. Now we have the incomplete chunk $0*$ that needs to be expanded to $00*$ and $01*$ as the prefix length is 2 in the second level. Since P_4 does not have any children, we create four nodes as required for the second level and all of them are linked to P_4. The two nodes 00 and 01 are augmented with the prefix information P_{10} while the other two are not used. Note that only the subtrie rooted at P_4 has been affected, in this case creating the subtrie itself.

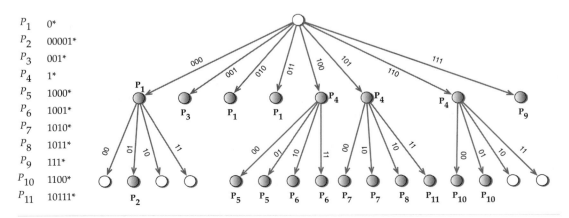

P_1 $0*$
P_2 $00001*$
P_3 $001*$
P_4 $1*$
P_5 $1000*$
P_6 $1001*$
P_7 $1010*$
P_8 $1011*$
P_9 $111*$
P_{10} $1100*$
P_{11} $10111*$

FIGURE 15.9   Inserting new prefixes in fixed stride multibit trie.

The insertion of prefix P_{11} proceeds by dividing P_{11} into chunks 101 and 11*. The search using these chunks leads to node P_8. Recall that prefix P_8 has been expanded to two prefixes, 10110* and 10111*. Now we compare the expanded prefix of P_8 and the new prefix P_{11}. We find that the best match is the new prefix P_{11}, which is of longer length. In other words, prefix P_{11} captures prefix P_8. Hence we update the node with the new prefix and new prefix length. To distinguish such cases, the length of the original prefixes needs to be stored in every node. Again note that the update is restricted to only a single subtrie rooted at P_4.   ▲

Since inserting new prefixes might require prefix expansion, deletion becomes more complex. Deletion involves removing expanded prefixes and, more importantly, updating the entries with the next best matching prefix. The problem is that original prefixes are not actually stored in the trie. Suppose we insert prefixes 100*, 101*, and 110* into the trie in Figure 15.8. This will overwrite the expanded prefixes for P_4 and the original prefix P_4 will not exist. Later if the prefix 110* gets deleted, the old best matching prefix of P_4 needs to be restored. Hence these operations require maintaining an additional structure for the original prefixes. Typically, these structures are maintained in the route control processor.

These examples show that the operation of inserting or deleting a prefix involves only an update of one of the subtries, since the best matching prefixes computed for each subtrie are independent of the other subtries. In other words, prefix update is completely local. The stride of a subtrie determines an upper bound on the time in which the actual updates will occur. If the stride of the subtrie has l bits, then at most 2^{l-1} nodes need to be modified. To illustrate this, consider the case of a prefix that has the last incomplete chunk of either 1* or 0*. If the stride on that subtrie is l bits, then half of the nodes will start with 0 and the other half will start with 1. Hence the prefix and the next-hop information corresponding to the incomplete chunk have to be inserted in half of the nodes.

The complexity of insertion and deletion includes the time for search, which is $O(W/k)$ where k is the size of the stride and the time to modify a maximum of 2^{k-1} entries. Thus, the update complexity is $O(W/k + 2^k)$. From the perspective of storage, a prefix might require the entire path of length W/k, and paths consists of one-level subtries of size 2^k. Hence, the memory complexity is $O(2^k NW/k)$ and increases exponentially with k.

15.5.5 Implementation

A fixed stride trie is typically implemented using arrays for each trie node and linking them using pointers. The trie nodes at different levels will have different array sizes as determined by the stride at that level. If the stride at a level is k, then the size of the array required will be 2^k. Each entry in the array consists of two fields: the field *nhop* contains the next-hop information and the field *ptr* contains the pointer to the subtrie, if any. The implementation of a fixed stride trie in Figure 15.8 is shown in Figure 15.10.

Since the stride at the first level is 3, we use an array containing $2^3 = 8$ elements for the first level. For the second level subtries we use arrays of size $2^2 = 4$ elements as the stride is 2. The prefix used to index into the array is shown adjacent to each element and note that this information is not stored. The presence of prefix information in an element indicates that the field *nhop* is not empty and stores the next-hop information associated with that prefix. The arrows indicate that the field *ptr* is not empty and point to the subtrie. Note the waste

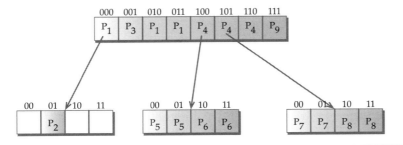

FIGURE 15.10   Implementation of the fixed stride multibit trie shown in Figure 15.8.

of space in the leftmost array in the second level that contains only prefix P_2; the rest of the three elements do not contain any information.

15.5.6 Choice of Strides

As we have seen earlier in the search algorithm, the number of memory accesses required, in the worst case, is bounded by the number of levels (alternatively referred to as the height) of the trie. The number of levels, in turn, is dependent on the size of the strides for each level. With large strides, the number of levels will be smaller as more bits are examined at each level. As a result, the number of memory accesses will be smaller. But at the same time, the amount of memory consumed will be larger. Hence the choice of strides represents a tradeoff between search speed and memory consumption. In the extreme case, using a single stride of size 32 bits, the search can be accomplished in one memory access, but the amount of memory consumed is rather large.

Generally, the number of levels of the trie is chosen depending on the desired lookup speed by the designer. For example, if the allowed budget time for a lookup is 30 nanosec and the speed of memory access is 10 nanosec, then the number of levels can be at most 3. It is clearly desirable that this constraint be satisfied with the least amount of memory possible. In other words, it is necessary to choose an optimal trie T that has at most three levels for a given prefix set but still occupies minimum storage. A space-optimized trie is heavily dependent on the size of the strides and, thus, choosing an optimal set of strides is important; see [658] for a dynamic programming-based approach.

15.5.7 Variable Stride Multibit Trie

If the nodes at the same level have different stride size, then the multibit trie is called a *variable stride multibit trie*. An example of a variable stride multibit trie is shown in Figure 15.11. We can see that the subtrie at level 1 has a stride of 2 bits. Some subtries at level 2 have strides of 2 bits and the rest 1 bit. As in a fixed stride multibit trie, each node will have the same information in addition to the stride length. This is needed since the search algorithms need to know at every subtrie how many bits need to be examined. Algorithms for search and incremental updates are very similar to a fixed stride multibit trie.

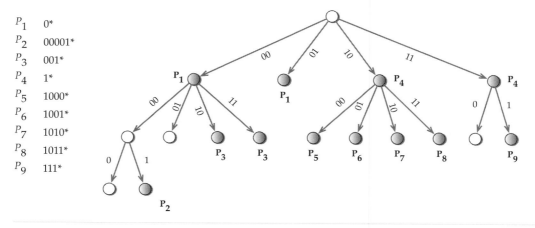

P_1	0*
P_2	00001*
P_3	001*
P_4	1*
P_5	1000*
P_6	1001*
P_7	1010*
P_8	1011*
P_9	111*

FIGURE 15.11   Variable stride multibit trie.

15.6 Compressing Multibit Tries

The aggressive use of prefix expansion in multibit tries introduces several new prefixes. These new prefixes inherit the same next-hop information as that of the original prefix. Furthermore, the use of large strides creates a greater number of contiguous nodes that have the same best matching prefix. For instance, take a look at the implementation of the fixed stride trie in Figure 15.10. It shows, for example, that the prefixes P_1 and P_4 in the first level are replicated, which means their next-hop information is repeated. Such redundant information can be compressed, saving memory and at the same time making the search faster because of the smaller height of the trie. After compression, the entire data structure can even fit into an L1 cache, which further speeds up the search as the access times are an order of magnitude faster than SRAM. In the next few sections, we discuss various compression schemes for multibit tries using bitmaps and compressed arrays.

15.6.1 Level Compressed Tries

Multibit tries, as we saw earlier, use prefix expansion to reduce the number of levels in a trie; however, this is at the expense of increased storage space. This can be viewed alternatively as compressing the levels of the trie, which sometimes is referred to as level compression. A closer examination of multibit tries shows that space is especially wasted in the sparsely populated regions of the trie. For instance, consider the binary trie in Figure 15.3. The trie region containing the prefix P_2 is sparse as no other prefixes are nearby. Now examine the fixed stride multibit trie variant of the binary trie in Figure 15.8. The leftmost subtrie contains only one prefix (P_2) and the rest of the the three locations are not used. Such sparse regions of a binary trie that contain long sequences of one-child nodes can be compressed by the technique called path compression discussed in Section 15.4.2. There is another trie-based scheme called level-compressed tries (LC-tries) that combines both path and level compression [529]. The main motivation behind this scheme is to reduce storage by ensuring that the resulting trie nodes do not contain empty locations.

The scheme starts with a binary trie and transforms it into an LC-trie in multiple steps. We illustrate this transformation using the binary trie shown in Figure 15.3. First, path compres-

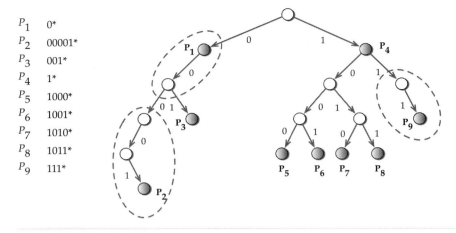

FIGURE 15.12   Identifying the paths to be compressed in the binary trie shown in Figure 15.3.

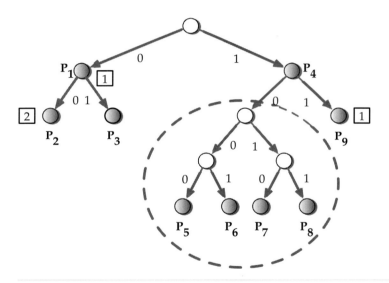

FIGURE 15.13   Identifying the levels to be compressed in the trie shown in Figure 15.12.

sion is applied as described in Section 15.4.2. This removes the sequences of internal nodes having one child. However, we need to keep track of the missing nodes somehow. A simple solution is to store a number called the *skip value* (s) in each node that indicates how many bits need to be skipped on the path. In Figure 15.12, we show that the sequences of nodes leading to P_2 and P_9 have only one child and hence are candidates for path compression. The path compressed trie is shown in Figure 15.13.

After path compression, level compression is used for compressing the parts of the binary trie that are densely populated. Instead of a node having two children, as in a binary trie, each internal node in a multibit trie is allowed to have 2^k children, where k is called the

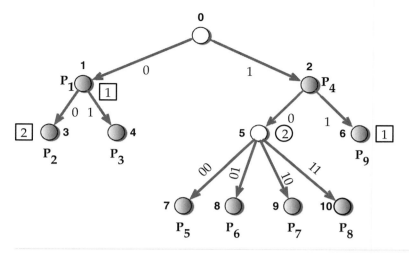

FIGURE 15.14   Level compressed trie resulting from the trie shown in Figure 15.13.

branching factor. The level compression is applied as follows. Every node *n* that is the root of a complete subtrie of maximum depth *d* is replaced by a corresponding one-level multibit subtrie. This process is repeated recursively on the children of the multibit trie thus obtained. Again, referring to Figure 15.13, we find that the trie rooted at the left child of node P_4 is a complete subtrie of depth 2. This trie can be replaced by a single-level multibit trie with four nodes as shown in Figure 15.14. The branching factor for the left child of node P_4 is 2, indicated by the number enclosed in a circle adjacent to the node. The rest of the internal nodes have a default branching factor of 1 and are not shown in Figure 15.13(a). The leaf nodes have a branching factor of 0.

Since at each step we replace several levels with a single-level multibit trie, this process can be viewed as the compression of levels of the original binary trie. Hence, the resultant trie is termed an LC-trie or simply an LC-trie. The main drawback of the scheme is that the structure of the binary trie determines the choice of strides at any given level without any regard for the height of the resulting multibit trie. This is because a subtrie of depth *d* can be substituted only if it contains all the 2^d leaves (a full binary subtrie). Hence, a few missing nodes might have a considerable negative influence on the efficacy of the level compression.

A simple optimization is to use a relaxed criterion where nearly complete binary tries are replaced with a multibit subtrie. In other words, if the nearly complete binary subtrie has a sufficient fraction of the 2^d tries at level *d*, then it is replaced with a multibit subtrie of stride *d* pointing to 2^d nodes. The parameter that controls the fraction is referred to as the *fill factor* $x, 0 < x \le 1$. Such a relaxed criterion will decrease the depth of the trie but will introduce empty nodes into the trie. However, in practice, this scheme results in substantial time improvements with only a moderate increase in space.

For optimizing storage, LC-tries do not the use the standard implementation technique that uses a set of child pointers at each internal node. Instead, an LC-trie is represented using a single array, and each trie node is an element in the array. An interested reader is encouraged to refer [529] for further details.

The LC-trie can be considered a special case of a variable stride trie. The algorithm for a variable stride trie using dynamic programming would indeed result in an LC-trie if it were the optimal solution for a given set of prefixes [658]. The worst-case time complexity for an LC-trie is $O(W/k)$ and the space complexity is $O(2^k NW/k)$, which are very similar to multibit tries.

15.6.2 Lulea Compressed Tries

Variable stride multibit tries and LC-tries attempt to reduce the height of the multibit tries by varying the stride. However, both schemes have problems. While variable stride multibit tries can be tuned for a shorter height to speed up the search, it is possible only at the expense of wasted memory due to the presence of empty locations in intermediate trie nodes. On the other hand, LC-tries choose strides such that the array locations in the trie node are completely filled without wasting any memory. However, it loses the flexibility to tune the height and consequently increases the height of the trie thereby making the search slower. The Lulea algorithm [173] uses fixed stride trie nodes of larger stride and employs bitmap compression to minimize storage. Using large strides in fixed stride multibit tries results in a greater number of contiguous nodes with the same best matching prefix and next-hop information. The Lulea algorithm takes advantage of this fact and compresses the redundant information using bitmaps, thereby reducing storage and still not incurring a high penalty in the search time. Before discussing the details of the algorithm, let us understand bitmap compression, using a simple example.

Example 15.8 *Compressing a prefix sequence.*
Consider the prefix sequence of $A_1A_1A_1A_1A_2A_2A_3A_3A_3A_1A_1A_1A_1A_4A_4A_4$. This can be represented using the bitmap 1000101001000100, where bit 1 indicates the start of a new prefix in the sequence and 0 indicates repetition. This bitmap is referred to as a *bitarr*. This bitmap alone is not sufficient to get back the original sequence since it does not capture any prefix information. Hence, a compressed sequence $A_1A_2A_3A_1A_4$ of the original sequence called *valarr* needs to accompany *bitarr*.
▲

To illustrate the concepts behind the Lulea algorithm, consider the fixed stride multibit trie in Figure 15.8 and its implementation in Figure 15.10. This trie has a stride of 3 for the first level and a stride of 2 for the second level. The implementation shows that some of the entries contain a prefix as well as a pointer to a subtrie. For instance, consider the first entry with prefix P_1. It contains the prefix information for P_1 and a pointer to the subtrie containing the prefix P_2. To minimize storage and make compression easier, each entry is allowed to contain either a prefix or a pointer but not both. Hence the prefixes in the intermediate nodes are pushed down to the trie leaves. The leaf nodes instead of a pointer store the next-hop information while the internal nodes just store pointers to children. Such a trie is referred to as a leaf pushed fixed stride multibit trie.

A leaf pushed fixed stride multibit trie for the trie shown in Figure 15.10 can be created as follows: prefix P_1 stored in the entry for 000 is pushed down to all the entries for 00, 10, and 11 in the subtrie while prefix P_2 in entry 01 is left intact. Prefix P_4 in entries 100 and 101

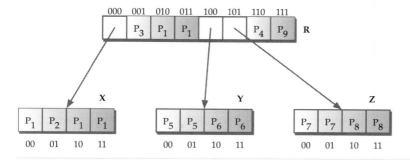

FIGURE 15.15 Implementation of leaf pushed fixed stride multibit trie for the trie in Figure 15.8.

P_1 0*
P_2 00001*
P_3 001*
P_4 1*
P_5 1000*
P_6 1001*
P_7 1010*
P_8 1011*
P_9 111*

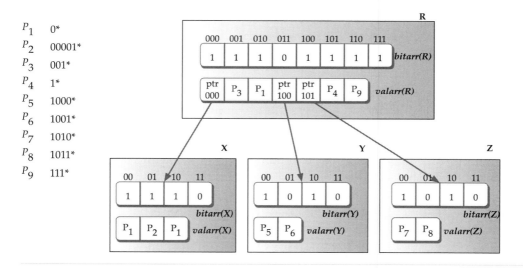

FIGURE 15.16 Lulea compressed trie data structure for the prefixes in Table 15.2.

are redundant and, thus, is eliminated. After these operations, each entry in a node contains either a stored prefix or a pointer, but not both. The final leaf pushed fixed stride multibit trie is shown in Figure 15.15.

Now let us look at how the bitmap compression can be applied to a leaf pushed trie. Consider the root node that has the sequence of $(ptr(000), P_3, P_1, P_1, ptr(100), ptr(101), P_4, P_9)$, where $ptr(xxx)$ represents the pointer stored in the entry xxx. Once the repeated values are replaced by the first value, we get the new sequence $(ptr(000), P_3, P_1, -, ptr(100), ptr(101), P_4, P_9)$. The resulting *bitarr* is 11101111, which indicates the repeated value positions by 0 and the *valarr* contains the sequence $(ptr(000), P_3, P_1, ptr(100), ptr(101), P_4, P_9)$. The same process is repeated for all the subtries and the final compressed trie is shown in Figure 15.16, which shows both the *bitarr* and *valarr* for each node.

What is the maximum number of entries that can be in *valarr*? Assume that a trie node contains N prefixes. Each prefix represents a range of addresses. These N prefixes partition the entire address space into no more than $2N + 1$ disjoint subranges. Since each subrange can be represented by at most one compressed entry, there can be at most $2N + 1$ entries in the *valarr*.

To analyze the space savings, assume that in a trie node array there are M elements, each of size W bits. Out of these M elements, if only S of them are distinct, then the total memory required is $M + SW$ bits. Directly storing the M elements will require MW bits. Now let us calculate the space savings between the leaf pushed trie and the Lulea compressed trie assuming the size of a prefix information and a pointer to the child is 32 bits each. Then the size of the root node of the leaf pushed trie in Figure 15.15 will be 256 bits (8×32) and each child will be 128 bits ($= 4 \times 32$) long. Hence, the entire trie will require 640 bits ($= 256 + 3 \times 128$) of storage space.

Now let us consider the Lulea compressed trie in Figure 15.16. The root node will need 8 bits for *bitarr* and 224 bits ($= 7 \times 32$) for *valarr*. All the child nodes will need 4 bits each for their *bitarr*. The leftmost child node will require 96 bits ($= 3 \times 32$) for *valarr* and the other two child nodes will need 64 bits ($= 2 \times 32$) each for their *valarr*. Hence the total space required is 468 bits ($= 8 + 224 + 4 \times 3 + 96 + 2 \times 64$), which represents a space savings of 172 bits. This space savings can be quite significant when very large strides are used.

SEARCH ALGORITHM

The search algorithm starts with the root node and uses the same number of bits as the stride at the first level to index into its *bitarr*. As the root trie node is compressed, the same index cannot be used to obtain the corresponding entry in *valarr*. Instead, the index to the uncompressed *bitarr* should be mapped to a different index in the compressed *valarr*. This mapping requires counting the number of 1 bits occurring in *bitarr* up to the indexed bit position. Then the count is used as the index into *valarr* to fetch the corresponding entry. If the entry is a pointer, the search fetches and follows it. This process is repeated for each subsequent level until either a leaf or a null pointer is reached. Let us walk through an example to understand this clearly.

Example 15.9 *Searching in a Lulea compressed trie.*

Consider searching for an address starting with 10011 (Figure 15.16). If the root node had not been compressed as in Figure 15.15, then using the first 3 bits (because the stride is 3 at the first level), the search would have directly fetched the pointer *ptr*(100), which could have taken into the second level. However, since the node is compressed, the first 3 bits 100 are used to index into *bitarr*(R) of the root node. To access the corresponding pointer entry, the number of 1 bits in *bitarr*(R) needs to be counted up to the indexed position 100. As the count is 4, the fourth element in *valarr*(R) of the root node containing the pointer *ptr*(100) needs to be fetched.

Using this pointer, the search proceeds to the second level. The stride at this level is 2 and hence the remaining two bits 11 of the address are used to index into *bitarr*(Y). Since the bit at this position is zero, the search terminates, meaning that it has reached a leaf node containing the best matching prefix. However, to access it, the search needs to find the position where the prefix occurs in *valarr*(Y). Hence the algorithm needs to count the number of 1 bits up to the bit position indexed by 11 in *bitarr*(Y) and use the count to index into *valarr*(Y) to find the best matching prefix. The number of 1 bits is 2 and hence the second element in *valarr*(Y) gives the correct result, P_6.   ▲

As can be seen, both steps require counting the number of 1 bits in *bitarr* until a given position. In the following section, we discuss how to efficiently count the number of 1 bits in a large bitmap.

COUNTING THE NUMBER OF 1 BITS

The original Lulea algorithm outlined in [173] used a fixed stride multibit trie of size 16, 8, and 8 for the first level, second level, and third level, respectively. As a result, for the first level, the algorithm requires a bitmap of size 64K. Obviously, naïve counting of such a large bitmap will be inefficient. Instead, the entire bitmap is divided into chunks of size 64 bits each and for each chunk the number of 1 bits is precomputed during trie construction. These precomputed counts are kept in a separate table called a *summary table*. An entry at position k in the summary table includes the cumulative count of 1 bits of the bit sequence up to chunk $k - 1$.

The size of the summary table is small compared to the actual bitmap size. For a rough idea of the space savings, we need to know the maximum value of the count that needs to be stored in an entry in the summary table. This occurs when all the bits in the bitmap are set to 1 and this needs to be stored in the last entry of the summary table. For a bitmap of size 64K divided into chunks of size 64 bits, the maximum value of the count will be 65472 ($= 64 \times 1024 - 64$). This will require 16 bits of storage for each entry and hence a total of 16K bits ($= 16 \times 1024$) is required, which is 25% of the memory used by the bitmap itself.

Now counting the number of bits up to a given position i in the bitmap consists of two steps. First, the chunk j into which i falls is computed as $j = \lfloor i/64 \rfloor$. Second, the entry at position j is accessed in the summary table. This gives the cumulative count of 1 bits in all chunks until $j - 1$. Finally, the bitmap chunk j itself is retrieved and the number of 1 bits is counted up to position $i - j \times 64$. The sum of the two values gives the final count.

Example 15.10 *Counting the number of 1 bits in a bitmap.*
    Let us try to count the number of 1 bits up to position P, which is the fourth bit located in chunk X as illustrated in Figure 15.17. For illustration, we use 8-bit chunks as opposed to

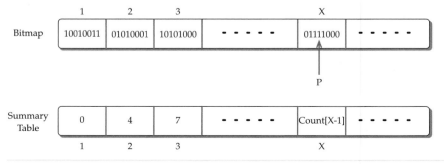

FIGURE 15.17   Counting the number of 1 bits up to position P.

the 64-bit chunks used in the algorithm. In Figure 15.17, the bitmap is laid out on top while the summary table is at the bottom. The first entry in the summary table is zero, the second entry has a count of 4 (because the bitmap in the first entry is 10010011, which has 4 bits set), and the third entry contains a count of 7 (the bitmap in the second entry 01010001 has 3-bit sets and this is added to the previous count of 4 for a cumulative count of 7). Note that the first entry is always zero.

To count the number of 1 bits up to position P, first the entry corresponding to chunk X in the summary table is accessed to retrieve the count of the number of 1 bits up to chunk $X - 1$. For the sake of discussion, let us assume that this value is $Count[X - 1]$. Then we retrieve the actual chunk X containing the bit sequence 01111000. Now the number of bits up to position 4 is counted. Since the first four bits of the sequence is 0111, the count is 3. Thus, the overall bit count until the desired position P is given by $Count[X - 1] + 3$.   ▲

The choice of the chunk size presents a tradeoff between memory and speed. If the chunk size is the entire bitmap, then the bits have to be counted every time. This could incur substantial CPU cost that decreases the lookup speed. However, if the chunk size is 1 bit then a simple retrieval of the count from the summary table gives the intended result. However, the downside is that the summary table will be much larger than the original bitmap. With a bitmap of 64K entries, the summary table will require 1M bits ($= 64 \times 1024 \times 16$).

Using the summary table during search requires at least three memory references: one for retrieving the appropriate entry from the summary table, another access for retrieving the actual chunk (depending on the chunk size and memory word size, it can be more than 1 memory accesses), and the final access for retrieving the element from *valarr*.

The Lulea algorithm applies the same optimization for next hops as well since the next-hops belonging to shorter prefixes will be at several consecutive locations. Because of such optimizations, the storage required by the Lulea algorithm is very compact. On a routing table with 38,816 prefixes, the compressed database size is 160 KB, which represents 4.2 bytes per prefix. Such compact storage can easily fit in the L1 cache of a conventional general-purpose processor or in an on-chip SRAM, a necessity for providing wire speed lookup at 40-Gbit speeds.

However, the optimizations made by the Lulea algorithm do introduce a few disadvantages. The counting of bits requires an additional memory access per node. The benefits of the optimization are heavily dependent on the structure of the forwarding table. Hence, predicting the worst-case storage requirements of the data structure as a function of prefixes is difficult. The implicit use of leaf pushing makes insertion inherently slow in the Lulea scheme. For instance, consider the insertion of a prefix to the root node entry that points to a subtrie containing thousands of leaf nodes. The next-hop information associated with the new prefix has to be pushed to thousands of leaf nodes and, furthermore, many entries in the summary table need to be updated, which could be slow. In the next section, we outline an algorithm that is as space efficient as Lulea but overcomes the incremental update problem by avoiding leaf pushing.

15.6.3 Tree Bitmap

The tree bitmap outlined in [198] is a fixed stride multibit trie algorithm that allows fast searches and uses storage comparable to the Lulea algorithm. However, the tree bitmap differs in one main aspect—the ability to allow fast insertions/updates. While fast lookups are needed for wire speed forwarding, from a commercial router perspective, fast insertions and modifications are clearly desirable. This is needed since the route updates/changes must modify a few hundred prefixes a second.

As observed in the previous section, inserting and deleting prefixes in the Lulea scheme require rebuilding the entire data structure. This is because the prefixes are expanded and pushed to trie leaves. If the packets are to be forwarded nonstop (which is now a requisite for a commercial router), we need to maintain two copies: one that is being used for lookups and the other for building the new data structure for prefix insertions and updating. This potentially doubles the memory requirements increasing the storage cost. In the case of implementations using a fixed size on-chip SRAM that stores the entire prefix database, the number of prefixes that can be stored is halved. The key to avoiding two copies and still allowing fast insertions and deletions is to eliminate leaf pushing. The tree bitmap algorithm achieves this by storing two bitmaps: one for pointers to the child and the other for prefixes. The next section outlines the principles and optimizations used in the design of a tree bitmap.

Design Rationale

The tree bitmap algorithm design considers that a multibit trie node is intended to serve two purposes—one to direct the search to its child nodes and the other to retrieve the forwarding information corresponding to the best matching prefix that exists in the node. It further emphasizes that these functions are distinct from each other. The design of the data structure reflects this observation by using two bitmaps per trie node instead of a single bitmap as in the Lulea algorithm. One bitmap used for storing internal prefixes belonging to that node is referred to as a *prefix bitmap* and the other bitmap used for storing the pointers to children is referred to as a *pointer bitmap*. Such a use of two bitmaps obviates leaf pushing, allowing fast insertions and updates.

Furthermore, the tree bitmap attempts to reduce the number of child node pointers by storing all the child nodes of a given trie node contiguously. As a result, only one pointer that points to the beginning of this child node block needs to be stored in the trie node. Such an optimization potentially reduces the number of required pointers by a factor of two compared to standard multibit tries. An additional advantage is that it reduces the size of the trie nodes. In such a scheme, the address for any child node can be computed efficiently using simple arithmetic, assuming a fixed size for each child node.

The tree bitmap algorithm attempts to keep the trie nodes as small as possible to reduce the size of a memory access for a given stride. A tree bitmap trie node is expected to contain the pointer bitmap, the prefix bitmap, the base pointer to the child block, and the next-hop information associated with the prefixes in the node. If the next-hop information is stored along with the trie node, it would make the size of the trie node much larger. Instead, the next-hop information is stored separately in an array associated with this node and a pointer to the first element is stored in the trie node. Would storing next-hop information in a separate array require two memory accesses per trie node: one for accessing the trie node and the other to fetch the next-hop information? The algorithm gets around the problem by a simple lazy evaluation strategy. It does not fetch the resulting next-hop information until the search is terminated. Once the search ends, the desired node is fetched. This node carries the next-hop information corresponding to a valid prefix present in the last trie node encountered in the path.

Finally, a tree bitmap uses only one memory access per node. With burst-based memory technologies, the size of a single random access can be large as 32 bytes. If the entire trie node has to be accessed in a single memory access, it cannot be larger than the optimal memory burst sizes. The size of the trie node greatly depends on the stride size. Consider a stride size of 8 bits. The pointer bitmap will require 256 bits (2^8). The prefix bitmap will require 511 bits ($2^9 - 1$) since there are $2^9 - 1$ possible prefixes of length 8 or less. In addition, we need another 4 bytes for storing the base pointers to the child and next-hop information. Hence the size of the trie node is approximately 100 bytes (($= 256 + 511 + 4 \times 8)/8$), which requires more than one memory access even with burst-based memory technologies. By using smaller strides, say 4 bits, the tree bitmap makes the bitmaps small enough that the entire node can be accessed in a single wide memory access. Use of small strides also keeps the update times bounded. Accessing the entire node includes both the bitmaps and the base pointers for the child and next-hop information. Since the bitmaps are smaller, special circuitry using combinatorial logic can be used to count the bits efficiently. This is unlike the Lulea algorithm, which requires at least two memory accesses because of large strides of 8 or 16 bits. Such large strides require a bitmap of larger size and necessitate the use of a separate summary table for counting, which must be accessed separately.

DATA STRUCTURE

Consider the root node of the fixed stride multibit trie implementation shown in Figure 15.10. The corresponding tree bitmap node is shown in Figure 15.18. As discussed earlier, a tree

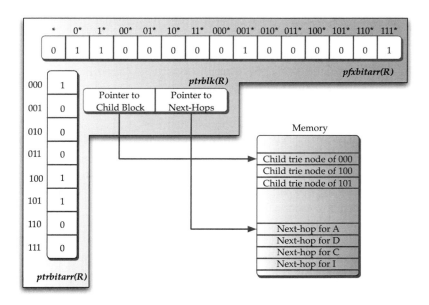

FIGURE 15.18   Structure of tree bitmap node corresponding to root node of trie in Figure 15.10.

bitmap node consists of two bitmaps. The first bitmap shown vertically is the pointer bitmap, which indicates the position where child pointers exist. In Figure 15.18, this bitmap is referred to as *ptrbitarr*. It shows that pointers exist for entries 000, 100, and 101. These pointers correspond to three subtries rooted at the entries 000, 100, and 101 in Figure 15.10. Now instead of storing explicit child pointers in a separate array as in the Lulea scheme, the tree bitmap node stores a pointer to the first child trie node as one of the fields in *ptrblk*.

The second bitmap shown horizontally is the prefix bitmap. It stores a list of all the prefixes within the first 3 bits that belong to the node. This bitmap is different from the Lulea scheme because it has a bit assigned for all possible prefixes of 3 bits or less. The bitmap positions are assigned to the prefixes in the order of 1-bit prefixes followed by 2-bit prefixes and so on. A bit in the prefix bitmap is set if that prefix occurs within the trie node. As you can see in Figure 15.10, the the root node contains the prefixes P_1, P_3, P_4 and P_9. Hence, the bit positions corresponding to prefixes 0*, 001*, 1*, and 111* are set in the prefix bitmap. The entire tree bitmap data structure for the fixed multibit trie in Figure 15.10 is shown in Figure 15.19.

SEARCH ALGORITHM

The search starts from the root trie node and using the same number of bits as the stride for the first level indexes into the pointer bitmap. Let us call this position P. If the bit in position P is set, it implies that there is a valid child pointer that points to a subtrie. To obtain the value of the child pointer, the number of 1 bits in the pointer bitmap is counted up to the indexed position P. Assuming the count is C and the base address to the child block in root trie node is A, the expression $A + (C - 1) \times S$ gives the value of the child pointer, where S refers to the size of each child trie node.

Before following the child pointer to the next level, the search examines the prefix bitmap to determine if one or more prefixes match. The bits used to match these prefixes are the same set of bits used to index the pointer bitmap. For the sake of discussion, let us assume that these bits are 111. First, the bit corresponding to prefix 111* is examined at position 15. The bit is set, which indicates that it is the best matching prefix. If the bit had not been set, the last bit would be dropped and the prefix bitmap would be again searched for prefix 11*. If there is no 11* prefix, the algorithm continues and checks for prefix 1*. The reader might note that the search algorithm has to perform multiple iterations proportional to the number of bits in the prefix. If all these iterations have to be performed in $O(1)$ time, they need to be executed in parallel. In custom ASICs (application-specific integrated circuits), this can be accomplished using dedicated combinatorial logic circuits and a priority encoder that returns the longest matching prefix. This could easily scale even for bitmaps as large as 256 or 512 bits.

If a matched prefix is found in the prefix bitmap, the next-hop information corresponding to the prefix is not fetched immediately. Instead, the matched prefix position and pointer to the trie node are remembered and the search continues to descend using the computed child pointer. As the search advances to lower levels, if better matched prefixes are found, the remembered matching prefix position and the pointer to the trie node are updated. The search terminates when it encounters a bit that is zero in the child pointer bitmap, meaning there are no children to continue further. At this point, the pointer to the next-hop information corresponding to the last remembered best matching prefix is computed using its trie

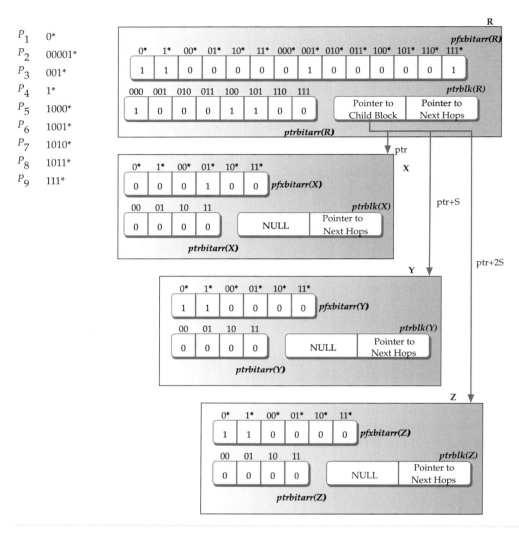

P_1 0*
P_2 00001*
P_3 001*
P_4 1*
P_5 1000*
P_6 1001*
P_7 1010*
P_8 1011*
P_9 111*

FIGURE 15.19   Tree bitmap data structure for the prefixes in Table 15.2.

node. This yields the desired result. For a better understanding of the search, let us look at an example.

Example 15.11   *Searching in a tree bitmap.*

   Consider the search for the address beginning with 10011 in the tree bitmap data structure shown in Figure 15.19. The search first examines the pointer bitmap *ptrbitarr(R)* of the root node using the first three bits 100. Since the bit is set, the search needs to examine the child subtrie. In the pointer bitmap, as it is the second bit set, the child pointer is computed as $ptr + (2 - 1) \times S = ptr + S$ where *ptr* is the base address and S is the size of the trie node. Before continuing the search to the child subtrie, the prefix bitmap *pfxbitarr(R)* is examined for matching prefixes. First, the entry corresponding to the first three bits of the address 100 is examined. Since the bit is not set, there is no matching prefix, the search then drops the last bit and examines the entry 10. This indicates there is no match and the search continues to the entry 1. Since the bit is set, a prefix match has been found, the pointer to the next-hop information is computed (similar to the computation of child pointer). This next-hop information is not fetched and instead remembered as the best matching prefix so far.

Now the computed child pointer is used to fetch the child node Y. Using the last two bits of the address 11, the child bitmap $ptrbitarr(Y)$ is examined. Since the bit corresponding to entry 11 is not set, there is no more child subtrie to examine. The prefix bitmap $pfxbitarr(Y)$ is examined for the entry 11. As the bit is not set, there is no matching prefix and the search continues to entry 1. The bit is set indicating the presence of matching prefix P_6. This prefix is updated as the best matching prefix; the pointer to its next-hop information is computed and fetched, terminating the search.   ▲

The algorithms for inserting and updating in a tree bitmap are very similar to the same set of operations in a multibit trie without leaf pushing. Inserting a prefix could change the entire trie node. In such cases, a new node is created and the original contents are copied, after which the necessary modifications are made and then atomically linked to the original trie. Since child nodes have to be contiguous, the entire child block may have to be copied even though only one child might require modifications. Performance results [198] show that the tree bitmap scheme is comparable to the Lulea scheme in terms of the size of the data structure and speed of lookup, but also provides the advantage of fast insertions. The algorithm lends itself to implementations ranging from software to architectures using an on-chip SRAM.

15.7 Search by Length Algorithms

A closer look at the algorithms discussed so far indicates that the main difficulty in finding the longest matching prefix is due to the presence of dual dimensions: prefix *length* and *value*; for instance, Table 15.2 lists several prefixes of different lengths and values. These dimensions present alternatives for designers in their effort to develop new speedy algorithms that consume less memory. Based on these dimensions, the search for the longest matching prefix can be based on either values or lengths of the prefixes.

The search using the length dimension can be either linear or binary. To facilitate the search, the prefixes can be organized in two different ways on the length dimension. As we have seen earlier, one approach is to arrange the prefixes in a trie. Searching of the trie can be viewed as a sequential search on length. Examining the binary trie discussed in Section 15.4 indicates that first it searches prefixes of length 1, then on prefixes of length 2, and so on. Furthermore, the prefixes are organized in such a way that the search space is reduced as the trie is descended. Another possible approach for organizing the prefixes is to use a hash table for each distinct length and employ either a linear search or binary search on these tables to locate the best matching prefix. To simplify the discussion, let L represent a sorted array in the increasing order of distinct prefix lengths. Each entry contains the length of the prefixes it represents and a pointer to the hash table that contains the actual prefixes. In the next few sections, we describe the linear and the binary search on L.

15.7.1 Linear Search on Prefix Lengths

Since we need to look for the longest matching prefix, the search begins with the table containing the longest prefixes. If the destination address is D and the longest prefix is l, the search extracts the first l bits and initiates a search in the hash table for length l entries. If an entry matches, then the search terminates since it is the best matching prefix. If not, the search moves to the first length smaller than l, say l', such that $L[l']$ is nonempty. Thus the search continues by examining the tables of decreasing prefix lengths until it either finds a match or runs out of lengths. In the worst case, this scheme needs to do a linear search among all distinct prefix lengths. Hence, it requires $O(W)$ time, where W is the address length in bits. Assuming the hash function is perfect, then an IPv4 lookup will require as much as 32 memory accesses while an IPv6 lookup will require as much as 128 memory accesses. If the hash function is not perfect, which usually is the case in practice, each hash probe for the matching prefix might require more than a single memory access. In such cases, an IPv4 lookup could cost more than 32 memory accesses.

15.7.2 Binary Search on Prefix Lengths

While a linear search requires $O(W)$ in the worst case, a better search strategy is to employ a binary search on the array L, which is described in [727]. The binary search starts at the median prefix length and divides the search space (in this case, prefix lengths) in each step by half. At every step, the hash table associated with that length is searched. Based on the results of searching the hash table, the choice of the half on which to continue the search is determined. The result can be one of the values: *found* or *not found*. If a match occurred at length l, then the search is directed to the half in which the lengths are strictly greater than l for a longer match. If no match was found at length l, then the search is continued on the half in which the lengths are strictly less than l.

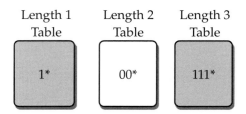

F I G U R E 15.20   Prefix length tables.

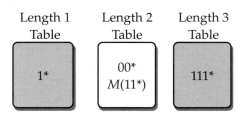

F I G U R E 15.21   Prefix length tables with markers.

Let us try to develop the concepts behind the algorithm by considering an example. Consider the set of prefixes $P_1 = 1*$, $P_2 = 00*$, and $P_3 = 111*$ shown in Figure 15.20. For the address that begins with 111, the search starts with the hash table of length 2. As can be seen, the hash will not match and the search will continue to length 1. However, there is a longer matching prefix P_3 in the length 3 table. To direct the search toward the right half, a *marker* needs to be introduced if there is a potential longer match. Hence, the marker 11* is added to the length 2 table as shown in Figure 15.21. Note that the marker is shown in italics as $M(11*)$. Now revisiting the search for the address starting with 111, the marker in the length 2 table will indicate a match and the search will move to the length 3 table and find the correct prefix P_3. While markers direct the search toward tables greater than the median length for specific matches, they also ensure that the probe failures at the median length table rule out the need to examine tables of length greater than the median.

How do we know where these markers need to be added and, even if we did, how do we know how many of them will be needed? For any prefix P, the marker needs to be added at all lengths where the binary search makes a decision about the half on which to continue the search. Since at most $\log_2 W$ length tables will be traversed on any search, the number of markers needed for any prefix P will be a maximum of $\log_2 W$.

The algorithm described so far is still not correct. This is because sometimes the markers can cause the search to follow false leads that may fail. Using the previous example, consider searching for an address starting with 110. The search starts at the length 2 table, which contains the prefix P_2 and the marker 11*. Since the marker matches the address, the search is directed toward the length 3 table. However, the search fails in this table since the prefix P_3 does not match. But the correct best matching prefix is in the length 1 table, prefix P_1. In this case, the marker 11* that was needed to find P_3 misleads the search. Hence, the search has to backtrack and examine the entire left half, resulting in a linear search.

To ensure the search is still logarithmic, each marker M in addition to its value is augmented to contain its best matching prefix, denoted by $bmp(M)$. This can be precomputed and stored during the time of the insertion of marker M into its hash table. When the algorithm uses marker M to continue the search for longer prefix lengths, it remembers $bmp(M)$. If the search on longer prefix lengths fails to produce anything interesting, then $bmp(M)$ is the answer, since it summarizes the results of backtracking. For the previous example, marker 11* is augmented with its *bmp* information as 1*, which is nothing but P_1. When the algorithm searches for the address starting with 110, it matches marker 11* in the length 2 table and remembers its *bmp*. Once the search fails in the length 3 table, the algorithm simply re-

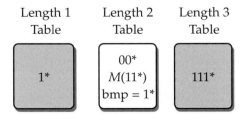

Length 1 Table	Length 2 Table	Length 3 Table
1*	00* *M(11*)* bmp = 1*	111*

FIGURE 15.22   Prefix length tables with markers and precomputed best matching prefix.

turns the *bmp* of the last matched marker, in this case, P_1. The final set of prefix length tables is shown in Figure 15.22. Here $bmp(11*)$ is shown as *bmp* adjacent to marker $M(11*)$.

The binary search requires $O(\log_2 W)$ time in the worst case. This assumes that the probes in a hash table for a given prefix consume only $O(1)$ time. Hence, an IPv4 address requires five hash lookups in the worst case. However, in practice, it is expected to take only two memory accesses since the majority of the prefixes are either 16 or 24 bits. For IPv6 addresses that are 128 bits long, seven hash lookups will be required. The use of markers makes the update more complex. The complexity arises from precomputing the best matching prefix for each marker, which itself is a function of the prefixes of the marker. The addition or deletion of a prefix may change the best matching prefix for many of the markers that are longer than the prefix entry being added or deleted. Since the added or deleted prefix can potentially be a prefix of $N-1$ longer entries, each of their $\log_2 W$ markers needs to be updated and hence the complexity is $O(N \log_2 W)$. The memory consumption is $O(N \log_2 W)$ as each prefix might need $\log_2 W$ markers.

The entire search data structure can be built using a simple algorithm. The algorithm takes as input a set of prefixes of different lengths and first determines the distinct prefix lengths. Using these distinct prefix lengths, the algorithm determines the sequence of length tables that the search needs to examine. Next, each prefix P is added into the corresponding table of length $length(P)$. While adding the prefix P, appropriate markers are also added into the tables of length $L < length(P)$ that the search will visit. Using a separate binary trie, the best matching prefix for each marker M is determined and stored along with it.

15.8   Search by Value Approaches

Sequentially searching all the prefixes is the simplest method to find the best matching prefix as outlined in Section 15.3. While this approach is not scalable as N becomes large, the exhaustive search does get rid of the length dimension. Alternatively, it is possible to use a binary search on the prefix values that could perform better than an exhaustive search. In the next section, we outline an algorithm described in [395] that uses a binary search on prefix values encoded as ranges.

15.8.1   Prefix Range Search

To find the longest matching prefix using a search on values requires the elimination of the length dimension. Since prefixes can be arbitrary lengths, one possible approach is to expand them such that all of them have a unique length. The addresses can be as long as 32 bits in the case of IPv4 and hence all the prefixes can be transformed into 32-bit length addresses. After transformation, the addresses are stored in a table in sorted order and a binary search on this table will find the longest prefix match. While this approach is simplistic, it requires a huge amount of memory as the number of entries in the table can be as much as 2^{32}. Fortunately, it is not necessary to store every address of a prefix since a great deal of information is redundant. Let us see how.

A prefix, as we have seen earlier, represents a well-defined contiguous range or interval of addresses. For example, the prefix 11* of a 5-bit length address represents the range of addresses in the interval [11000, 11111]. Hence, we can simply encode the prefix using the start and end of its interval. So the obvious question is why not just store these interval *endpoints*

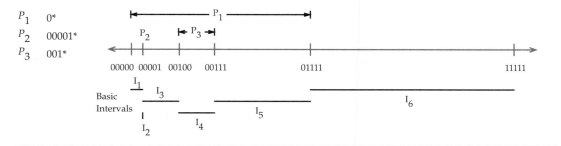

FIGURE 15.23   Prefix intervals identified by three prefixes 0*, 00001* and 001*.

instead of every single address in the table? After all, the best matching prefix is the same for all the addresses in the interval. Now the problem of finding the best matching prefix is reduced to locating the endpoints of the interval containing the destination address. In other words, this requires identifying the greatest starting endpoint smaller than or equal to a given address. For example, if we consider the three intervals [00100, 00111], [10000, 10001], and [11100, 11111], the address 00101 belongs to the interval [00100, 00111] since the endpoint 00100 is the greatest endpoint that is smaller than the address.

Unfortunately, the aforementioned approach will not work since one prefix interval might include other prefix intervals. For instance, consider only the prefixes P_1, P_2, and P_3 of the running example from Table 15.2. The interval corresponding to prefix P_1 is [00000, 01111], P_2 is [00001, 00001], and P_3 is [00100, 00111]. Prefix P_1 contains the intervals of prefix P_2 and P_3. For the address 01000, the greatest starting endpoint that is closer is 00100 and this endpoint belongs to interval [00100, 00111] of prefix P_3. However, as you can see, it is not the best matching prefix. The appropriate best matching prefix for address 01000 is associated with endpoint 00000 which belongs to interval [00000, 01111] of prefix P_1. The solution is to avoid overlapping intervals by partitioning the space of addresses into disjoint intervals between consecutive endpoints. For N prefixes, there will be a maximum of $2N + 1$ disjoint intervals, referred to as *basic intervals*. All the addresses that fall in a basic interval have the same best matching prefix. The basic intervals for prefixes P_1, P_2, and P_3 and their best matching prefixes are shown in Figure 15.23.

Using the endpoints of the basic intervals, the best matching prefix can be determined. However, a few of the basic intervals in Figure 15.23 do not have explicit endpoints (e.g., I_2 and I_4). This is due to the partitioning of a larger interval of a short prefix by smaller subintervals of longer prefixes. In such cases, the basic interval is associated with the closest endpoint to its left. Hence some endpoints are associated with two basic intervals, and thus endpoints are required to maintain two best matching prefixes denoted by $>$ and $=$: one for the interval they belong to and the other for the potential next basic interval.

The algorithm starts with the endpoints of the given prefixes, sorts them in ascending order, and builds a search table. Each entry in the table stores the endpoint and the next-hop information for the two best matching prefixes $>$ and $=$. The best matching prefixes for each entry are precomputed and stored in the table. The $>$ entry is used for addresses that are *strictly greater* than the endpoint and *strictly less* than the next endpoint in the sorted order. The $=$ entry is used for addresses that are *exactly equal* to the endpoint. The search table for

Prefix	>	=
00000	P_1	P_1
00001	P_1	P_2
00100	P_3	P_3
00111	P_1	P_3
01111	-	P_1
11111	-	-

FIGURE 15.24   Prefix range search table.

prefixes P_1, P_2, and P_3 and the ranges covered by the prefixes are shown in Figure 15.24, which actually shows the best matching prefixes instead of their next-hop information.

Consider the table entry for the endpoint 00001. This entry is associated with the basic intervals I_1 and I_2. This is because I_2 does not have an endpoint and hence it is associated with the endpoint to its left, i.e., 00001. For addresses strictly greater than 00001 but less than the next endpoint 00100, the best matching prefix is P_1, and hence it is stored in the > pointer. On the other hand, for addresses strictly equal to 00001, the best matching prefix is P_2, which is stored in the = pointer.

Now take a look at the entry for endpoint 00100. Unlike endpoint 00001, this entry is associated with a single basic interval I_3. Hence, for addresses strictly greater than 00100 but less than the next endpoint, the best matching prefix is P_3. Similarly, for addresses exactly equal to 00100, the best matching prefix is still P_3. Hence the pointers for > and = both contain prefix P_3.

Example 15.12   *Searching for the best matching prefix using prefix ranges.*
Let us try to find the best matching prefix for the address 00101 using the table shown in Figure 15.24. A binary search on the table for 00101 will terminate at the endpoint 00100 and since the address is greater than the endpoint 00100 and less than the next endpoint 00111, the > pointer will be used and hence the best matching prefix is P_3. However, if we were to find a matching prefix for the address 00111, the binary search will end at the endpoint 00111. Since the address is equal to the endpoint, the = pointer will be used to retrieve the best matching prefix P_3.
▲

The number of entries in the table can be at most $2N$ since each of the N prefixes can insert two endpoints. The table can be searched in $\log_2(2N)$ since at every step the binary search reduces the search space by half.

Another implementation can use a binary tree where each node contains the endpoint it represents and the next-hop information for the two best matching prefixes < and =. Of course, as in a binary tree, it needs to store pointers to the left subtree and right subtree as well. Again in this case, the worst-case time required is $\log_2(2N)$. A further reduction in search time is possible with the use of a multiway search tree of higher radix such as B-trees used in fast retrieval of records in a large database [151]. Such trees reduce the height of the tree and make the search faster. In some sense, this is analogous to multibit tries as compared to binary tries. In a multiway tree, if the radix is k, each internal node will contain $k - 1$ keys and k branches. An astute reader will observe that a search within a node is required for the appropriate branch to follow. If the entire node requires multiple memory access, then it could be potentially expensive. However, if the size of the entire node is such that it can fit in an L1 cache, then the search time might be negligible. Alternatively, use of wider memory can fetch the entire node in a single access.

Since the algorithm precomputes the best matching prefix for each basic interval, inserting or deleting a single prefix might result in having to update the best matching prefix for many basic intervals. The shorter the prefix, the higher the number of best matching prefixes that will require recomputation because it spans multiple basic intervals. In the worst case, we might have to update the best matching prefixes for N basic intervals. This is the case when all $2N$ endpoints are different and one prefix contains all the other prefixes.

Compared with trie-based algorithms, a binary search on prefix values is slower than the multibit tries. Further, the amount of storage it requires is higher than the multibit trie variants that employ compression. The hardware implementations of this scheme typically uses wider memory access and pipelining to make it faster.

15.9 Hardware Algorithms

The primary motivation for implementing lookups in hardware comes from the need for higher packet processing capacity driven by high Internet traffic growth. Such a growth leads to the development of faster interfaces that can support OC-192 and OC-768 speeds. At such high speeds, a software-based implementation using random access memory (RAM) is not fast enough. While the software implementations have the flexibility for later modifications, the need for such modifications is minimal. Since IPv4 is used so widely, disruptive modifications to addressing schemes or the best matching prefix mechanism seem unlikely in the near future.

15.9.1 RAM-Based Lookup

All the algorithms discussed in the previous sections use some form of RAM to store and retrieve their data structures. RAM supports the two major operations: writing a data item into a specific address and reading a data item from a given address. RAM can be used to perform the lookup in a single memory access if the destination address is used as a direct index (RAM address) into the memory. The data item stored in that address will be the next-hop information. For example, the IP address 172.12.180.20 can be directly used as an address to retrieve its next-hop information as shown in Figure 15.25.

Now the issue is, how large a memory will we need? The size of the RAM is determined by the number of bits in the destination address. In the case of IPv4, since the address contains 32 bits, 4 Gbytes ($2^{32} - 1$) of RAM is needed. The number of prefixes and their corresponding next-hop information do not have an effect on the required size of the RAM. Even if there are

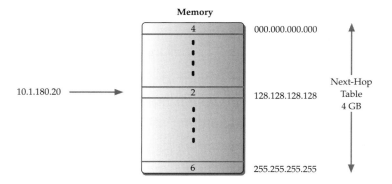

FIGURE 15.25   RAM based lookup.

fewer prefixes for IPv4, we need the entire 4 Gbytes. The size of the RAM grows exponentially with the number of bits in the address, when used as a direct index. For IPv6, which uses 128 bits for the destination address, the size of the memory required becomes impractical.

Another problem in using RAM is that updates require modifying half of the memory in the worst case. Consider the deletion of prefix 0* in IPv4 lookup. It requires modifying 2^{31} memory locations with new next-hop information. Because of such large updates, schemes based on RAM-based lookup are not used in practice.

15.9.2 Ternary CAM-Based Lookup

Content-addressable memories (CAMs) provide a performance advantage over conventional RAM-based memory search algorithms by comparing the desired information against the prestored entries simultaneously. This results in an order of magnitude reduction in search time. Since CAMs are an outgrowth of RAM technology, they employ conventional memory (usually SRAM) with the additional circuitry for comparisons that enable search operations to complete in a single clock cycle.

In RAM, data are stored at a particular location called address. A user supplies the address in order to retrieve the data. With CAM, the user supplies the data and gets the address back. The CAM searches through the memory in one clock cycle and returns the address where the data are found. An obvious question is, how to store the data in the first place? Data can be transferred to or from a CAM without knowing the memory address. Binary data are automatically written to the next free location.

With CAMs, a longest prefix matching operation on a 32-bit IP address can be performed using exact match search in 32 separate CAMs. The incoming IP address is given as input to all the CAMs. The output of the CAMs indicating the results of the match is fed through a priority encoder, which picks the CAM that has the longest matching prefix. Such a solution is expensive both in terms of cost and complexity.

Hence, a more flexible type of CAM that enables comparisons of the input key with variable length elements is desirable. Ternary CAMs were introduced to address this need. While a binary CAM stores one of two states 0 and 1 for each bit in a memory location, a ternary CAM (TCAM) stores one of the three states 0, 1, and X (don't care) for each bit in a memory location. The don't care state permits search operations beyond the exact match.

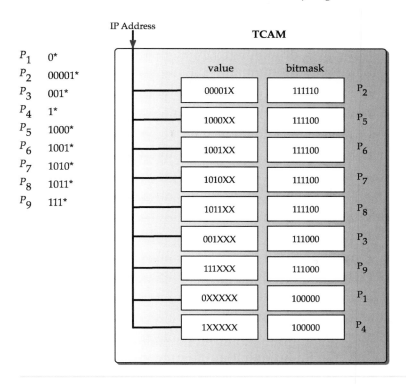

FIGURE 15.26 Storing the prefixes of Table 15.2 in TCAM.

A TCAM stores an element as a pair: a *value* and a *bitmask*, where each of them is the same length. The value field stores the actual value of the prefix, and the bitmask field is used to denote the length of the prefix. Let us see how this works. If a prefix is Y bits long, the most significant Y bits of the value field are assigned the same value as that of the prefix, and the remaining bits are set 0 or 1 or X. The most significant Y bits of the bitmask field are set to 1 and the remaining bits are set to 0. Thus, the bitmask indicate which bits in the value field are relevant. For example, a prefix of 110∗ will be stored as $(110XXX, 111000)$ assuming the elements are 6 bits long. The prefixes in Table 15.2 are stored in TCAM as shown in Figure 15.26. Note that the prefixes are stored in descending order of their lengths. An incoming key matches a stored element, if the bits of the value field for which the bitmask is 1 are identical to those in the incoming key.

Figure 15.27 illustrates how TCAM is used for longest prefix matching. The TCAM memory array stores the prefixes as value and bitmask pairs in decreasing order of prefix lengths. In Figure 15.27, P_i represent the prefixes of length i. The memory array matches the input key with all the elements in parallel. An element consisting of value and bitmask matches the key if and only if it is a prefix of that key. The memory array indicates the matched elements by setting the corresponding bits in a bit vector (in Figure 15.27, it is *matched_bitvector*). The length of the bit vector is equal to the number of TCAM memory locations. This bit vector is input to a priority encoder that outputs the location of the lowest bit that is 1 in the bit vector. This location corresponds to the longest matching prefix and can be used as an index or as an address to access the SRAM containing the next-hop information.

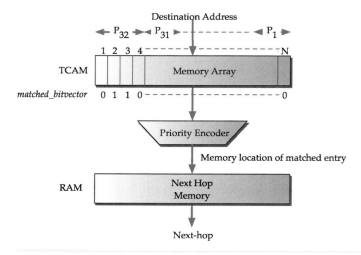

FIGURE 15.27 Ternary CAM-based lookup (adapted from [273]).

Having seen how TCAM can achieve high packet lookup throughput, it is logical to wonder what prevents their wide spread usage in high-end routers. The main disadvantages are as follows:

- *High Cost and Density:* Each bit of storage in TCAM requires two bits: one for the value and the other for the bitmask. In addition, extra circuitry is required for handling the logic of "don't cares." Hence the number of transistors required is two or three times higher than the regular SRAM. As a consequence, the number of bits stored in a given chip area is lower for TCAM, thus leading to lower density.

- *Power Efficiency:* Since all the elements are searched in parallel for an incoming key, the circuitry corresponding to each row that stores an unmatched element draws the same amount of electric current as the one that contains the matching key. An incoming address matches at most one prefix for each distinct prefix length. Since this is a small percentage of the number of entries stored, a TCAM consumes a lot of power even for a normal operation. However, an SRAM in a normal operation requires electric current only for the element being accessed by the address.

Because of these disadvantages, TCAM is not extensively used. While TCAM provides fast lookup in a single memory cycle, the next section outlines an algorithm that uses SRAM/DRAM memory but bounds each lookup to a maximum of two memory cycles.

15.9.3 Multibit Tries in Hardware

The basic scheme proposed in [274] is motivated by the need for a fast lookup solution in inexpensive pipelined hardware. It uses a two-level multibit trie with fixed strides for each level. The first level uses a stride of 24 bits while the second level uses a stride of 8 bits. This scheme is based on the following two key observations:

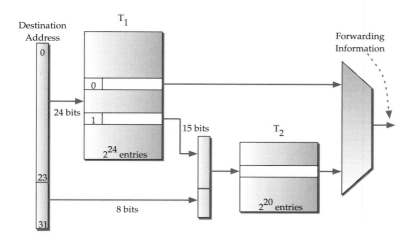

FIGURE 15.28   Implementation of multibit tries in hardware.

- Most of the prefix entries in routing tables in core routers are 24 bits or less. This is at-tributed to the aggressive route aggregation at intermediate routers.

- The cost of memory continues to decline while the density of memory doubles every year. As a result of these diverging trends, a large amount of memory is available at low cost. This observation leads to a situation in which large amounts of memory are traded off for lookup speed.

Based on the first observation, if a first stride of 24 bits is used, then for most cases the best matching lookups can be found in one memory access. For prefixes longer than 24 bits, they are expanded to the second level consisting of 32 bits using the prefix expansion described in Section 15.5.1.

This scheme is very similar to the fixed stride multibit trie with two levels: the first level with a stride of 24 and the second level with a stride of 8. However, internal nodes are not allowed to store prefixes. Hence, if a prefix corresponds to an internal node, it will be ex-panded to the second level. In other words, the internal prefixes are pushed to the leaves. This is similar to the leaf pushed fixed stride multibit trie outlined in Section 15.6.2.

The scheme, as realized in hardware, uses two tables as shown in Figure 15.28. The first level of the multibit trie is implemented using table T_1 and stores all the route prefixes that are 24 bits long or less. This table has a total of 2^{24} entries, addressed from 0 to $2^{24} - 1$. An entry in this table can be either of two types: one contains the next-hop information and the other contains a pointer to the corresponding subtrie at the second level. Each entry is 2 bytes wide and hence the table needs a memory of 32 Mbytes. The first bit is used to determine whether the entry stores the next-hop information or a pointer to the second level subtrie. Hence, only 15 bits are used for storing a pointer or next-hop information.

Table T_2 implements the second level of the multibit trie and stores all the route prefixes longer than 24 bits. The amount of memory required for table T_2 depends on the number of such prefixes. In the worst case, each of these prefixes will need a subtrie at the second level. Since the stride for the second level is 8 bits, each subtrie will occupy as many as $2^8 = 256$

entries in table T_2. These entries store only the next-hop information. If the number of next-hop routers does not exceed 255, then the size of each entry just needs to be a byte. If table T_2 contains 2^{20} entries of 1 byte each (1 Mbyte in total), then a total of $2^{20}/256 = 4096$ subtries at the second level can be supported.

Given an arbitrary prefix P, first it is examined to determine if it less than or equal to 24 bits long. If P is less than 24 bits, it is expanded into multiple prefixes of 24 bits each. For each entry in table T_1, addressed by these expanded prefixes, the first bit is set to zero to indicate that the rest of the bits contain the forwarding information. For example, prefix 172.25.0.0/16 will be expanded to a total of $2^{24-16} = 256$ prefixes ranging from 172.25.0.0 to 172.25.255.0. If P is longer than 24 bits, it is expanded to prefixes of size 32 bits. The first bit of the entry in table T_1 addressed by the first 24 bits is set to one to indicate that it contains a pointer. The pointer gives the start offset into table T_2 where the next-hop information is stored.

Now let us see how a route lookup occurs for an incoming packet. The first 24 bits of the destination address are used to index into table T_1 and read the corresponding entry. If the first bit in the entry equals zero, the remaining bits describe the forwarding information. If the first bit is set to one, the pointer stored in the remaining bits is multiplied by 256 and the result is added to the last 8 bits of the incoming destination address. This value is used to index into table T_2 to retrieve the next-hop information. These operations can be implemented in hardware using concatenation and shifting of bits.

The major advantage of this scheme is speed. It requires a maximum of two memory accesses. Since the implementation is in hardware, the memory accesses can be pipelined or paralleled. As a result, the lookup operation on average takes one memory access time. Updates can be time consuming. This algorithm is in contrast to most of the other algorithms that seek to minimize the storage requirements of the data structure.

15.10 Comparing Different Approaches

Now that we have covered many algorithms, it will be useful to gain some insights about how these algorithms compare with each other. While the worst-case algorithmic complexities of the various schemes are known, an understanding of how they perform in real-life conditions will be useful.

Simple binary tries and path compressed tries are primarily used for software implementations. For example, BSD and Linux implementations use these algorithms for routing packets that originate from the application. Since these implementations are typically used in servers and desktop machines where the number of packets to be routed and the number of prefixes are very small, the need for fast lookups is not felt.

An LC-trie is a compacted version of a path compressed trie in which complete subtries are level compressed from the top down. It uses an array-based storage and hence incremental updates are expensive. Since the internal nodes do not contain any next-hop information, a failed search at the leaf requires examination of separate tables, which is inefficient.

The Lulea scheme use multibit tries, but the wasted space in the trie nodes is compressed using bitmaps. Because of compression, the forwarding data structure is very compact, which fits the entire data structure in a cache, and hence lookup times are fast. But the updates are very slow due to compression and leaf pushing. However, the tree bitmap algorithm, by avoiding leaf pushing, provides guaranteed fast update times. Furthermore, the lookup speed and the size of the compressed forwarding data structure are still comparable to Lulea. Also the tree bitmap scheme provides sufficient flexibility for adapting the implementation to various memory architectures. Because of such attractive features, it is implemented by commercial router vendors.

The binary search on prefix lengths scales very well to IPv6 addresses that are 128 bits long. However, the main disadvantage is the use of hashing, which introduces nondeterminism, and hence there is no guarantee on worst case lookup times. Further implementing hash tables in hardware is tricky. Hence, it is better suited to implementation in software, and a few vendors have already done so. Incremental updates can be slow as they are more complex due to the use of markers.

The binary search on prefix ranges provides reasonably fast lookup performance and consumes only a reasonable amount of storage. Further improvement in lookup times can be made through the use of multiway branches, wider memory, and pipelined implementation. As the best matching prefixes are precomputed *a priori* for each interval, updates can be slow. In the worst case, the best matching prefix needs to be computed for all the intervals. However, the major advantage is that this algorithm is patent free and can be used without any encumbrance.

15.11 Summary

With the explosive growth of Internet traffic, routers, particularly in the core of the Internet, need to be efficient in the address lookup operation so that many packets they see can be handled extremely quickly. The forwarding capacity of a router is highly dependent on how quickly it can determine to which egress interface to transfer the packet, also known as *address lookup*. In this chapter, we presented various algorithms for fast IP address lookup. We first examined the requirements of these algorithms: lookup speed, memory usage, scalability, and updatability.

We started with naïve algorithms such as linear search and caching followed by trie-based algorithms with an emphasis on binary tries and multibit tries. We then studied different variants of multibit tries and their advantages and disadvantages. We briefly touched upon the taxonomy of IP lookup algorithms based on *length* and *value* dimensions. Based on this taxonomy, we described two algorithms that are variants of the binary search. We also looked in detail some of the hardware algorithms. Ultimately, the choice of the algorithm for fast lookup depends on the performance and cost constraints. Finally, note that essentially all algorithms discussed in this chapter are applicable to either IPv4 or IPv6 addressing; certainly, because of CIDR, many are tuned specifically for IPv4 addressing. Furthermore, some of the hardware-based approaches are bit specific and thus are typically designed for IPv4 addressing. Nevertheless, various impacts due to 32-bit or 128-bit addressing have been pointed out throughout the chapter.

Further Lookup

An excellent introduction to the IP address lookup problem can be found in [273], [658]. Various requirements and the metrics used to evaluate the algorithms are discussed in length in [273]. A detailed survey of the lookup algorithms and a comparative study of their performance in terms of lookup speed, scalability, and updatability can be found in [611]. Varghese

[712, Chapter 11] provides an excellent coverage of various algorithms and implementation insights.

The Patricia trie scheme [502] implemented in BSD unix is described in [645]. A dynamic prefix trie data structure, a variant of Patricia tries, that simplifies deletion and avoids recursive backtracking can be found in [180]. A lookup scheme based on the idea of caching is proposed in [133]. It uses a software scheme and integrates the lookup cache with the processor cache resulting in high speeds.

The idea of multi-ary tries has been discussed in [370] and [624] in a general context. However, detailed exposition of multi-ary tries under the context of prefixes and routing can be found in [661], who also describe the dynamic programming algorithms for fixed stride tries and variable stride tries that compute the optimal sequence of strides given a set of prefixes and the number of levels desired. Faster algorithms for the same purpose are proposed in [616] and [617] for fixed stride multibit tries and variable stride multibit tries, respectively. An algorithm that exploits the hierarchical structure of the memory in a system to speed up route lookups is described in [131]. Level-compressed tries that combine path compression and level compression were introduced in [529]. A modified version of an LC-trie that stores the internal node prefixes in the array storage representation is discussed in [586]. The Lulea algorithm, which attempts to minimize the storage requirements for the lookup data structure, is proposed in [173]. The tree bitmap algorithm is described in detail [198]. It also proposes quite a few optimizations that allows the memory access width of the algorithm at the cost of memory references. Another lookup algorithm that uses run length encoding instead of bitmap to efficiently compress the routing table is described in [159].

A binary search on prefix lengths is presented in [727] and various refinements to the basic scheme are proposed in [726]. In addition, [726] also presents algorithms that precompute the route lookup data structures used for searching. A binary search on intervals represented by prefixes is proposed in [395]. Several fast routing table lookup solutions based on binary and ternary CAMs are examined in detail [453]. Novel ideas for reducing power consumption in TCAMs are discussed in [760]. Hardware implementation of multibit tries that trades memory for fast lookups is described in [274].

Exercises

15.1. Review questions:

    (a) What are the main differences between classful addressing scheme and CIDR?

    (b) Explain why longest prefix match is important and define, in your words, the longest prefix matching problem.

    (c) What are the primary metrics of performance for evaluating a longest prefix matching algorithm?

    (d) What is the main difference between a binary trie and multibit trie?

    (e) What is prefix expansion and why is it required?

15.2. What is the maximum time allowed for a lookup in a router to sustain a data rate of 20 Gbps with an average packet size of 100 bytes? Assume that the router requires 15 ns per packet for other operations in the packet.

TABLE 15.4   Prefix table.

Prefix Label	Prefix
P_1	0*
P_2	10*
P_3	111*
P_4	10001*
P_5	1*
P_6	1001*
P_7	101000*
P_8	1010000*

15.3. For the prefixes in Table 15.4, construct a binary trie. Assuming each node in the binary trie requires a memory access, how many memory accesses will be required in the worst case during the search?

15.4. For the prefixes in Table 15.4, construct a path compressed trie. In this trie, how many memory accesses will be needed for looking up the 8-bit addresses 10011000 and 10100011? Do you see any improvement compared with the binary trie?

15.5. Draw a fixed stride multibit trie using the prefixes shown in Table 15.4. How many memory accesses will be required for the 8-bit addresses 10011000 and 10100011?

15.6. For the fixed stride multibit trie shown in Exercise 15.5, how much memory will be required to implement? Assume that the next hop and pointer require 4 bytes each.

15.7. Can you draw the implementation of leaf pushed fixed stride multibit trie for the trie in Exercise 15.5? Assuming that the next-hop and pointer require 4 bytes each, how much memory will it require? Does it require any extra information?

15.8. For the prefixes in Table 15.4, construct a Lulea compressed trie clearly indicating the *valarr* and *bitarr* for each node. Use a stride of 3 at the first level, and a stride of 2 for the second and the third levels.

15.9. Construct a tree bitmap for the prefixes shown in Table 15.4. Use a stride of 2 for the first level, a stride of 3 for the second level, and a stride of 2 for the third level.

15.10. Can you outline an efficient approach for counting the number of 1 bits in a bitmap of size 8? Can you extend it to bitmap of sizes 16, 24, 32, and so on? Can you observe what is the tradeoff?

16

IP Packet Filtering and Classification

Logic will get you from A to B. Imagination will take you everywhere.

Albert Einstein

Reading Guideline

A critical need in packet filtering and classification is consideration of algorithms that are efficient. However, different algorithms are appropriate for different scenarios. This chapter discusses algorithms of various complexity, pointing out their pros and cons, and explaining why certain algorithms may be more appropriate for one situation compared to another. The efficiency of an algorithm comes from trading off time, space, and sometimes data structures considered for implementation. Thus, some prior knowledge of computational complexity and data structure is helpful in understanding some of the details. The material starts with a discussion of naïve approaches, with increasingly complex approaches presented as the chapter progresses.

Because of the cost benefits of the Internet, it has been put to use for mission-critical functions executed by business organizations. Such users do not want their critical activities to be affected either by higher traffic sent by other organizations or by malicious intruders. As a result, some are willing to pay a premium price in return for better service from the network. At the same time, to differentiate themselves and maximize their revenue, commercial Internet service providers (ISPs) are looking for ways to share their IP network infrastructure to provide different levels of service, based on user requirements, at different prices.

To provide such differentiated services, service providers need mechanisms to identify and isolate packets from specific customers that provide customizable performance and bandwidth in accordance with customer requirements and pricing. Further, service providers must be able to route a packet not only based on the destination address or the shortest path to it, but also based on service level agreements (SLAs) between service providers or between a service provider and a customer. For a brief description of SLAs, see Section 23.10.

The forwarding engine needs to examine packet fields other than the destination address to identify the context of the packets and perform additional processing or actions for satisfying user requirements. Such actions might include dropping of the unauthorized packets, encrypting highly secure packets, prioritizing by special queueing, and scheduling. This process of differentiating packets is called *packet classification* and sometimes *packet filtering*. In this chapter, we focus on the algorithms for efficient packet classification.

16.1 Importance of Packet Classification

To illustrate the importance of packet classification, let us consider a few examples of how it can be used by an ISP to provide differential services. Consider the network shown in Figure 16.1. It shows an ISP (ISP1) serving three business customers, C1, C2, and C3, and another ISP (ISP2) that in turn serves the business customer C4 and a residential customer C5. Some of the following services provided by ISP1 require support for packet classification:

- *Providing preferential treatment for different types of traffic:* To provide different service guarantees for different types of traffic, an ISP might maintain different paths through the network for the same source and destination addresses, say, one for high-speed and real-time traffic and the other for the data traffic. For instance, all the video conference traffic originating from C2 and destined for C3 can be routed through a path that supports real-time traffic so that jitter is minimized. Hence, router R2 needs to distinguish video traffic from other types of traffic from C2.

- *Flexibility in accounting and billing:* An ISP needs flexible accounting and billing based on the traffic type so that different traffic can be charged at different prices. Voice over IP (VoIP) traffic is typically charged a higher price as compared to regular data traffic because of the delay and jitter guarantees required in delivering it. Such pricing models must distinguish one traffic type from the other as a basic primitive. For instance, in router R2, ISP1 needs to identify the VoIP traffic packets from C2 and collect packet and byte statistics so that customer C2 can be charged appropriately.

- *Managing customer expectations:* Consider an example in which ISP2 might be expanding its network to new residential areas. Initially, since the network is new and underloaded,

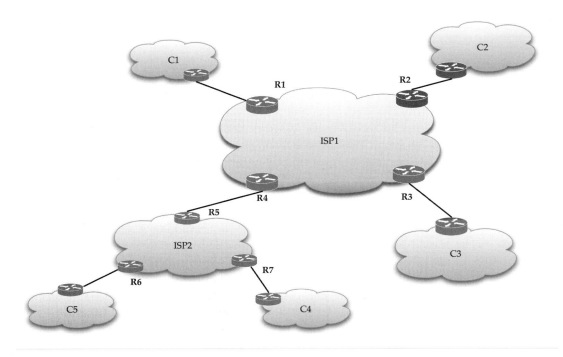

FIGURE 16.1   ISP1 and ISP2 networks connecting five customers C1, C2, C3, C4, and C5.

the customers in those areas could frequently obtain better bandwidth than the guaranteed minimum. As the spare capacity shrinks, the customer might receive a bandwidth closer to the guaranteed minimum, which could lead to the perception that the quality of the network has degraded. Hence, the ISP may wish to limit the maximum rate at which a customer can receive packets through the network to manage the expectations. By defining a specific rule at router R6 that distinguishes the traffic of one customer from another, ISP2 can ensure that no more than 2.5 Mbps of web traffic is delivered to residential customer C5.

- *Preventing malicious attacks:* Since malicious users can congest or overload the network affecting other customers, ISPs need to protect their network. Protective measures require the ability to identity malicious packets and drop them at the point of entry. Similar to the aforementioned examples, the ISP1 at router R3 needs to drop ICMP packets injected by C3, if they exceed the rate of 1 Kbps.

Based on the above examples, we can see that an ISP faces a variety of decision problems in regard to handling of traffic to provide differentiated services. A common requirement of all the above examples, from an ISP perspective, is the need for routers to have the ability to classify packets by examining the values of the header fields. Also, the routers should be capable of the additional processing needed to handle the packet, i.e., dropping it, encrypting it, making a copy, and so on.

The criteria for classification are often expressed in terms of *rules* or *policies* using the header fields of the packets. A collection of such rules or policies is called a *rule* or *policy database* or a *flow classifier* or simply a *classifier*. Each rule specifies a *flow* to which a packet may belong based on the conditions expressed in the rule. As a part of the rule definition, an *action* is specified for additional processing to be applied to the packet. An example of a policy is to encrypt the packets from a source address starting with prefix bits 1101∗, where "∗" refers to wildcard, bound to destination address 151.18.19.21 with a destination port of 80. Having seen the importance of packet classification, let us proceed to the next section, where we formally define the packet classification problem.

16.2 Packet Classification Problem

Consider a packet with d distinct header fields, denoted by $H_1, H_2, \ldots, H_d$, where each field is a sequence of bits. Such header fields for an IPv4 packet could represent, for example, the source address, destination address, source port, destination port, protocol field, and protocol flags. A packet from destination address 192.171.110.23, source address 192.142.55.45, a destination port of 80, a source port of 1800, and protocol field of *TCP* can be expressed by the combination (192.142.55.45, 192.171.110.23, 80, 1800, *TCP*). Valid header fields need not be restricted to the above set. They can even include layer 2 fields such as destination and source MAC addresses and application layer fields such as URL.

A classifier of a router consists of a set of rules denoted by $R_1, R_2, \ldots, R_N$. These rules are defined in a certain sequence, the importance of which will be examined at the end of this section. Each rule is a combination of d values, one for each header field. Each field in a rule is allowed four types of matches:

- *Exact match:* The exact match requires that the values on the rule field and the header field of the packet be identical. Such exact matches are useful for protocol and protocol flag fields.

- *Prefix match:* For a prefix match, the rule field should be a prefix of the header field; this could be useful for collecting statistics about packets originating from a subnetwork.

- *Range match:* In a range match, the packet header values should lie in the range specified by the rule; this is useful for specifying port number ranges.

- *Regular expression match:* In a regular expression match, the header field should match the string expression specified in the rule. Such regular expressions are useful for matching URLs for deep packet classification.

The use of wildcard ∗ in a field with preceding values represents a prefix match. On the other hand, the standalone occurrence of ∗ indicates that any arbitrary value can match.

In this chapter, we restrict ourselves to exact, prefix, and range matches. Each rule R_i ($0 < i \leq N$) has an associated directive or action act_i that specifies how to process the packet that matched this rule. This action represents the additional processing required for the packet—whether the packet should be discarded, encrypted, decrypted, and so forth.

A packet P *matches* a rule R if each field of P matches the corresponding field of rule R. The type of match for each field is implicit in its specification. Note a packet may match more than one rule in the classifier. Now let us consider an example of how a packet is classified.

Example 16.1 *Finding the matching rules for packets.*

Consider a classifier that has two rules specifying the four fields of an IP packet header: destination address, source address, destination port, source port, and protocol. Let the rules be $R_1 = (110*, 1*, < 1024, *, TCP)$ and $R_2 = (11*, 10*, < 1024, > 2048, TCP)$. A packet with header $(111111\ldots, 10010\ldots, 80, 3500, TCP)$ matches rule R_2 but not rule R_1. If R_2 has an action $act = discard$ associated with it, then the packet will be dropped. However, a packet with header $(110011\ldots, 10010\ldots, 80, 2500, TCP)$ matches both rules R_1 and R_2. Of course, now the question is which rule's action needs to be executed? Hence, a mechanism is needed to break the tie.   ▲

Since a packet may match more than one rule in the classifier, we associate a cost $cost(R_i)$ for each rule R_i to determine an unambiguous match. The goal is to find the rule with the least cost that matches a packet's header. However, what does this cost mean and how it is used? The cost is a measure that breaks the tie among multiple matching rules and it could represent almost anything—cost of the path the packet is going to take, cost of processing the packet, etc. The cost function is an attempt to generalize the ordering of rules in decreasing importance, which is often used in practice to choose between multiple matching rules. In a router, the rules are placed in a specific order such that a rule occurring earlier takes precedence over a rule occurring later. Thus, the goal of packet classification is to find the *earliest* matching rule. This is equivalent to assigning a $cost(R_i)$ equal to the position number of R_i in the classifier. The first rule assigned a cost of 1, the second rule assigned a cost of 2, and so on.

16.2.1 Expressing Rules

Now the obvious question is how these rules are specified in a router. Each router vendor provides different syntax for expressing rules, but the underlying primitives are pretty much the same. In this section, we focus on these basic primitives. Since rules contain fields of different types, each field is specified in a different fashion.

Source and destination address fields are expressed in prefix notation as described in Chapter 15. The prefix notation allows the expression of both exact and prefix matches. Source and destination port numbers can be either an exact specification or range specification or a wild card. Exact specifications are expressed by the clause "=." For example, an exact port specification of "= 1023" indicates that the port number field should be equal to 1023. The range specifications are indicated by various clauses "range," ">," "<," "≥," and "≤." The clause "> 1023" specifies any port number greater than 1023. Protocols are specified by an exact value or a wild card. Some valid protocol values are UDP, TCP, IGMP, and IGRP.

Typical actions specified are "Permit," "Deny," and "Count." However, much more sophisticated actions such as "Encrypt," "Decrypt," and "Rewrite" are possible, if the router supports the underlying functionality. A sample classifier that illustrates how the rules are specified is shown in Table 16.1.

16.2.2 Performance Metrics

Many algorithms exist for packet classification due to its importance and complexity. For comparison, we need some sort of metrics to analyze the strengths and deficiencies of each

TABLE 16.1   Expressing rules in a classifier.

Rule	Destination Address/Mask	Source Address/Mask	Destination Port	Protocol	Action
1	201.15.17.21/32	201.15.75.4/32	< 1024	*	Deny
2	201.18.20.25/24	201.15.100.10/32	= 80	TCP	Encrypt
3	201.15.20.25/24	201.15.100.10/32	< 1024	UDP	Permit
4	201.21.12.1/16	201.75.75.75/16	> 1023	TCP	Decrypt
5	201.21.12.1/24	201.75.75.75/24	*	UDP	Deny
6	*	*	*	*	Permit

of these algorithms. Similar to address lookup algorithms discussed in Chapter 15, the two widely used metrics for packet classification are the speed of search and the memory storage space occupied by the data structures of the algorithm. There are other metrics that are equally important as well. These are summarized below:

- *Speed:* As physical links are getting faster, the need for faster classification is greater than ever. This translates for every packet into making a classification decision in the time for handling a minimum-sized packet. This issue is far more pronounced for very-high-speed links. For example, at OC-768 rates (i.e., 40 Gbps) with a minimum packet size of 40 bytes, we need to handle 125 million packets per sec ($= 40 \times 10^9/(8 \times 40)$). Hence, a decision must be made in 8 nanosec ($= 8 \times 40/(40 \times 10^9)$). Speed is usually measured in terms of the number of memory accesses required. This is because memory accesses are expensive and constitute a dominant factor in worst-case execution time.

- *Memory space:* The smaller the amount of memory consumed by an algorithm, the greater the chances of using fast memory technologies like static random access memory (SRAM). On-chip SRAMs provide the fastest access time (around 1 nanosec) and they can be used as on-chip cache for software-based algorithm implementation. Hardware-based implementations embed SRAM on chips for faster access.

- *Faster updates:* As the classifier changes, because of the addition of new rules, deletion of existing rules, and changes in existing rules, the data structures maintained by an algorithm needs to be updated. The data structures can be categorized into those updated incrementally and those that need to be rebuilt from scratch each time the rule database changes. Such updates are generally not an issue for core routers where the rules are changed infrequently. However, edge routers that support dynamic stateful filtering or intrusion detection need to identify certain flows to be tracked; thus, faster updates are required.

- *Number of fields:* An ideal algorithm should be able to handle any number of header fields for classification.

- *Implementation flexibility:* For operation at wire speed, the algorithm should lend itself to hardware implementation. This does not mean that the software implementation is not desirable. An algorithm is attractive if it is implementable both in hardware and software.

16.3   Packet Classification Algorithms

In general, a packet classification algorithm consists of two stages: a *preprocessing stage* and a *classification stage*. The purpose of the preprocessing stage is to extract representative information from rules and build optimized data structures that capture the dependency among the rules. This data structure is consulted to find the least-cost matching rule for every incoming packet. The preprocessing stage is invoked only when new rules are added or deleted and existing rules are modified. Since these operations are infrequent, the preprocessing stage executes in the central CPU of the router.

In the classification stage, the actual packets are parsed and the headers are extracted. Using the values of the headers, the data structure built during the preprocessing stage is traversed to find the best matching rule. Since the classification stage runs in the data path, speed of classification is very important.

For the rest of chapter, let N be the number of rules in a classifier, W be the maximum length of each field in bits, and d be the number of dimensions or fields in the classifier. We will assume throughout this chapter that rules do not carry an explicit cost field as described in Section 16.2, and that the matching rule closest to the top of the list of rules in the classifier is the best matching rule. To illustrate various algorithms in the rest of the chapter, we will refer to the example classifier composed of eight rules as shown in Table 16.2, unless otherwise stated.

16.4   Naïve Solutions

The simplest algorithm is to store the rules in a linked list in the order of increasing cost. A packet is compared with each rule sequentially until a rule that matches all relevant fields is found. This approach is storage efficient since it requires only $O(N)$ memory locations. However, the time to classify a packet grows linearly with the number of rules N. If the number of rules is 10, we could require 10 memory accesses and if the number of rules increases to 100, the memory accesses also increase in proportion to 100. Thus, it has poor scaling properties.

Another approach to speed up classification is to use a cache in conjunction with an algorithm like linear search. The cache achieves low-latency classification requests by remembering previous classification results and searching these results first on the arrival of new

TABLE 16.2   An example classifier with eight rules.

Rule	F_1	F_2
R_1	00*	00*
R_2	0*	01*
R_3	0*	0*
R_4	10*	10*
R_5	11*	10*
R_6	11*	1*
R_7	0*	10*
R_8	*	11*

packets. When the result of a past classification request does not exist, a full classification is started using the associated algorithm and the result is cached.

For packet classification, the result of a full lookup using a header is cached along with the entire IP header. When a subsequent packet with the same header arrives, the results in the cache are directly used. Since cache access requires only a single access as compared to a full lookup, which requires several memory accesses in RAM, the lookup time is significantly reduced.

The efficiency of a caching scheme largely depends on the *temporal locality* of the traffic, which means the arrival of a packet implies a high probability of the arrival of another packet with the same IP header. Such a behavior can be explained by the fact that a file transfer or a web page is broken into a number of packets with the same IP header for transit. Hence, if we have saved a recently used classification result, there is a high probability that an incoming packet will hit the cache and will be classified without the need for a full lookup. Even though caching schemes might be attractive, they suffer from various problems. A significant issue is the requirement of more high-speed memory for caching full IP headers to achieve a hit rate of 80–90% (when compared to caching schemes for IP address lookup). However, caching is a general technique that can be combined with some of the fast algorithms discussed in later sections to improve the average case performance.

16.5 Two-Dimensional Solutions

Two-dimensional packet classification is a simpler version of the general packet classification problem. Studying it in detail could provide additional insights into developing solutions for the general case. Furthermore, two-dimensional classification is important on its own because of the variety of applications that use it. For example, applications like flow aggregation for MPLS and VPNs require handling of a larger number of rules that use source and destination network prefixes. Another use of two-dimensional classification is in firewalls where many rules contain distinct protocol ranges. Hence, it is possible to break up the classifier on more than two fields into multiple independent classifiers each with two fields and applying two-dimensional classification for each of them.

Since two-dimensional rules are the simplest generalization of the one-dimensional IP lookup problem, it is natural to think about extending those schemes for two-dimensional classification. In this section, we extend the binary trie–based schemes discussed in Chapter 15 to handle two fields.

16.5.1 Hierarchical Tries: Trading Time for Space

A binary trie, as described in detail in Section 15.4, organizes IP prefixes in a treelike fashion to identify the longest matching prefix. The internal nodes can be considered as decision points and the leaves represent the actual prefix. At each internal node, when the appropriate bit is examined, a 0 means the left branch needs to be traversed, while a 1 indicates that the right branch needs to be traversed. The path from the root to a leaf node gives the bit sequence of the prefix represented by the leaf node. Since IP prefixes can be represented as a range of addresses in a single dimension, the binary trie scheme is viewed as a solution for one-dimensional packet classification (also known as packet classification for a single field).

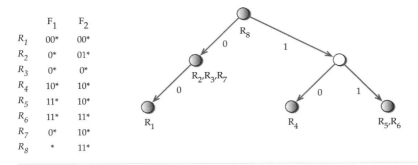

	F_1	F_2
R_1	00*	00*
R_2	0*	01*
R_3	0*	0*
R_4	10*	10*
R_5	11*	10*
R_6	11*	11*
R_7	0*	10*
R_8	*	11*

FIGURE 16.2   Constructing the F_1 trie needed for a hierarchical trie.

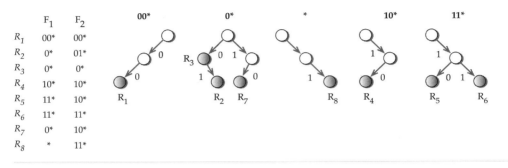

	F_1	F_2
R_1	00*	00*
R_2	0*	01*
R_3	0*	0*
R_4	10*	10*
R_5	11*	10*
R_6	11*	11*
R_7	0*	10*
R_8	*	11*

FIGURE 16.3   Constructing the F_2 tries needed for a hierarchical trie.

A hierarchical trie is a simple extension of the binary trie except that these tries can accommodate two fields. Let us illustrate hierarchical tries using the example classifier in Table 16.2 consisting of two fields, F_1 and F_2. First, construct a binary trie using the distinct prefixes of field F_1. This trie is shown in Figure 16.2 and is referred to as the F_1 trie. For the sake of clarity, the Figure 16.2 shows the nodes shaded if the prefix represented by each node appears in field F_1 of any one of the rules. Also, the rules containing the prefix of each node are indicated next to the node.

Now for each unique prefix in field F_1, construct a binary trie using the corresponding prefixes of field F_2. Consider the F_1 prefix 0*. It occurs in rules R_2, R_3, and R_7 and their F_2 prefixes are 01*, 0*, and 10*, respectively. We build a binary trie using these prefixes. Such a trie is called an F_2 trie. Each of the five distinct F_1 prefixes requires an F_2 trie and they are all shown separately in Figure 16.3. In Figure 16.3, above each trie, the corresponding F_1 prefixes are shown in boldface. As in the F_1 trie, the nodes corresponding to prefixes found in the rules are shaded.

The next logical problem is how to establish the relationship between the F_1 trie and the F_2 tries according to the rules in the classifier. An additional pointer is stored in each node of the F_1 trie that relates an F_1 prefix to its F_2 trie. This pointer, called the *next-trie* pointer, stores the pointer to the root of the F_2 trie corresponding to its F_1 prefix. The entire trie, known as the hierarchical trie, is shown in Figure 16.4 where the next-trie pointers are indicated by dashed arrows linking the F_1 trie to the F_2 tries. To briefly describe this, a hierarchical trie is

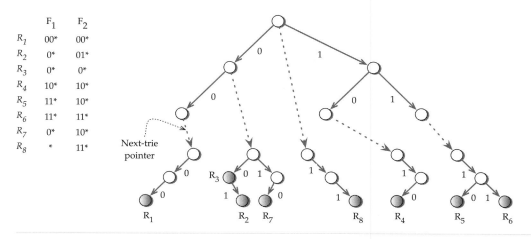

FIGURE 16.4 Combining the F_1 and F_2 tries for building a hierarchical trie.

constructed by first building a trie for all the F_1 prefixes, and for each F_1 prefix P an F_2 trie is associated that stores the rules whose F_1 prefix is exactly the same as P.

Once the hierarchical trie is constructed, the search proceeds as follows. Let us assume, for the sake of discussion, that the header fields F_1 and F_2 are extracted from the packet and denoted by $\grave{F}_1$ and $\grave{F}_2$, respectively. The search starts by traversing the F_1 trie to find the longest matching prefix for $\grave{F}_1$. The algorithm then searches the corresponding F_2 trie for the prefixes matching $\grave{F}_2$ and updates the best matching rule. The search then backtracks on the F_1 trie and traverses all the F_2 tries associated with all the ancestors of $\grave{F}_1$. During the traversal, it continues to keep updating the best matching rule, if an appropriate one is found. The algorithm terminates once the search backtracks to the root node and its corresponding F_2 tries are examined. The last recorded best matching rule during the search provides the final solution. To better understand this, let us go through an example of search.

Example 16.2 *Searching the hierarchical trie.*

Consider the classification of an incoming packet with fields $F_1 = 000$ and $F_2 = 000$ using the hierarchical trie shown in Figure 16.4. The search proceeds from the root of the F_1 trie to identify the longest prefix match for 000 and reaches the node corresponding to the prefix 00∗. The algorithm immediately fetches the next-trie pointer at this node and traverses its F_2 trie using 000. This leads to the matching rule R_1 and is recorded as the best matched rule so far. Now the search backtracks and moves up to the node with the prefix 0∗ in the F_1 trie and proceeds to its F_2 trie. During the traversal, the search encounters R_3, which matches the packet fields. Immediately, R_3 is compared with the best rule matched so far, R_1. According to our criteria, since rule R_1 occurs earlier in order in the classifier than R_3, R_1 is retained as the best matching rule. The search again backtracks on the F_1 trie reaching the root node. The next-trie pointer of the root node is fetched and its F_2 trie is traversed. It finds that none of the rules matches. Subsequently, the search terminates since there are no more nodes to backtrack and R_1 is declared to be the best matched rule. ▲

Why does the search need to backtrack? The reason that the search backtracks even after locating the first matching rule is to find the best matching rule. For example, assume the order of rules in the classifier is changed such that rule R_3 becomes the first entry and rule R_1 is moved to the third entry. In this case, the best matching rule for the same search in Example 16.2 will be R_3 since it occurs earlier than R_1 in the classifier. If the search has not backtracked, the best matching rule R_3 would be missed. Hence, the search algorithm needs to traverse all the F_2 tries in the path as it backtracks on the F_1 trie.

The lookup cost of this scheme is $O(W^2)$ for two fields. It follows from the observation that in the worst case, it is possible to end up searching W F_2 tries, one for each bit of the longest matching F_1 prefix. The cost of searching each F_2 trie is $O(W)$ and hence the overall search cost is $O(W^2)$. The amount of memory consumed is $O(NW)$. Observe that each prefix in field F_1 requires W nodes for the F_1 trie in the worst case. If all the N rules contain distinct F_1 prefixes, the worst-case memory required for F_1 trie is $O(NW)$. Each F_2 prefix also requires W nodes in the worst case, and hence the memory required for storing all the F_2 tries is $O(NW)$. Therefore, the overall memory required is $O(NW) + O(NW) = O(2NW)$, which is still $O(NW)$.

16.5.2   Set Pruning Tries: Trading Space for Time

Recall that hierarchical tries require traversals of many F_2 tries as the search backtracks on the F_1 trie. Can we do better by avoiding these traversals? The answer is yes. By replicating the rules of the prefixes of an F_1 prefix in its F_2 trie, such traversals can be eliminated. The resulting data structure is called a *set pruning trie* [171].

Constructing a set pruning trie is very similar to constructing a hierarchical trie. First, we build a trie using unique F_1 prefixes in the classifier. Each prefix in the F_1 trie points to an F_2 trie. Now the question is: which F_2 prefixes need to be stored? This is where the set pruning tries differ from hierarchical pruning tries.

For instance, consider the F_1 prefix of 00∗. Rule R_1 has this prefix and hence we need to store the corresponding F_2 prefix 00∗, as in hierarchical tries. However, if we need to eliminate the traversal of the F_2 trie at the prefix node 0∗ when the search backtracks on the F_1 trie, storing this rule alone is not sufficient. Since rules R_2, R_3, and R_7 also match whatever the prefix 00∗ matches, they also need to be included. Furthermore, the wildcard prefix ∗ of R_8 also matches whatever the prefix 00∗ matches. Thus, the F_2 trie for the prefix 00∗ contains the rules R_1, R_2, R_3, R_7, and R_8. The set pruning trie for the example classifier is illustrated in Figure 16.5. In other words, the F_2 trie for any prefix P should contain all the rules corresponding to the prefixes of P.

The search algorithm first matches the field F_1 of the incoming packet header in the F_1 trie. This yields the longest match on the F_1 prefix. Then the associated F_2 trie is traversed to find the longest match. As the search proceeds in the F_2 trie, we keep track of the best matching rule encountered so far. Since all the rules that have a matching F_1 prefix are stored in the F_2 trie, we are guaranteed to find the best matching rule.

Example 16.3   *Searching the set pruning trie.*

Consider searching for the best matching rule for a packet whose field F_1 value starts with 001 and F_2 value starts with 011 (Figure 16.5). The search begins by finding the longest

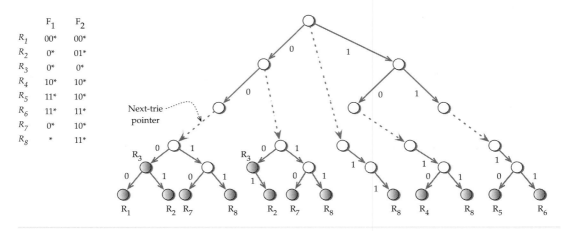

	F_1	F_2
R_1	00*	00*
R_2	0*	01*
R_3	0*	0*
R_4	10*	10*
R_5	11*	10*
R_6	11*	11*
R_7	0*	10*
R_8	*	11*

FIGURE 16.5   Set pruning trie data structure containing replicated rules.

matching prefix in the F_1 trie for 001. This yields the F_1 prefix 00* and using the next-trie pointer fetches the F_2 trie. This trie is traversed for the longest prefix matching 011. The search descends the F_2 trie and encounters rule R_3, which is remembered as the best matching rule. The search continues further and finds another rule R_2. After comparing this with R_3, the best matching rule so far is updated with R_2 since it occurs earlier in the classifier. Since the search cannot continue any further, R_2 is declared as the best matching rule.   ▲

Unfortunately, set pruning trees have a memory blowup problem. The problem arises because an F_2 prefix can occur in multiple tries. For instance, in Figure 16.5, the F_2 prefixes 0*, 01*, and 10* appear in trie associated with prefixes 00* and 0*. Hence, in the worst case for two fields, the set pruning trie consumes $O(N^2)$ memory. Since the algorithm attempts to find the longest matching prefix in the F_1 trie, followed by the F_2 trie, the number of memory accesses required is $2W$, which is $O(W)$.

To avoid the memory blowup, many optimizations have been presented [171], [572], and [734]. For instance, when two F_2 tries are identical, containing the same set of rules, only one copy is maintained. The corresponding F_1 next-trie pointers are updated to point to this single F_2 trie. This optimization greatly reduces the storage required and makes the set pruning trie attractive for implementing small classifiers in software.

16.5.3 Grid-of-Tries: Optimizing Both Space and Time

We saw from previous sections that hierarchical tries are on one end of the design spectrum and require large amounts of time while utilizing storage proportional to the number of rules and number of bits in the prefixes. On the other end of the design spectrum are the set pruning tries, which consume a large amount of storage but provide a time complexity proportional to the number of bits in the prefixes. Now the obvious question is, can we do the best of both? The grid-of-tries data structure, described in [662], is an attempt in this direction; it reduces the storage space by storing a rule only once as in hierarchical tries and still achieves the same time complexity as set pruning tries, i.e., $O(W)$.

A closer observation of the search algorithm for hierarchical tries indicates attempts to backtrack several times that are not necessary. To understand this better, let us revisit the search on hierarchical tries shown in Figure 16.4 with the following example.

Example 16.4 *Backtracking in hierarchical tries.*
Let us start the search using a packet header with values $F_1 = 000$ and $F_2 = 110$ (see Figure 16.4). We start by looking this up in the F_1 trie, which gives 00 as the best match. Using the next-trie pointer, we start the search for the matching F_2 prefix in the associated trie, containing the rule R_1. However, the search fails in the first bit 1. Hence, the search would have to back up on the F_1 trie and restart in the F_2 trie of 0*, which is the parent of 00*.   ▲

Such a backup in search is an attempt to find a matching rule that is shorter in the F_1 prefix, in this case 0*, and it should include all the bits in F_2 examined so far, including the failed bit. If the search algorithm could anticipate *a priori* such a sequence of bits, it could jump directly to the parent of R_7 from the failed point, the root of the F_2 trie associated with the F_1 prefix 00.

The grid-of-tries approach uses the key ideas of *precomputation* and *switch pointers* to speed up the search by jumping from the failed F_2 trie to another F_2 trie. The preprocessing algorithm that builds the trie identifies all the failure points in the F_2 tries and precomputes a switch pointer that allows the search to jump directly to the next possible ancestor F_2 trie that contains a matching rule. The jump occurs at the lowest point in the ancestor F_2 trie that has at least as good an F_2 prefix match as the current node. Furthermore, such a jump allows skipping over all rules in the next ancestor F_2 trie with shorter prefixes in the F_2 field.

A grid-of-tries data structure for the example classifier is shown in Figure 16.6. The switch pointers are illustrated by dashed arrows. Notice that the F_2 trie corresponding to F_1 prefix 00 contains two switch pointers: one from node A to node C and the other from node B to the

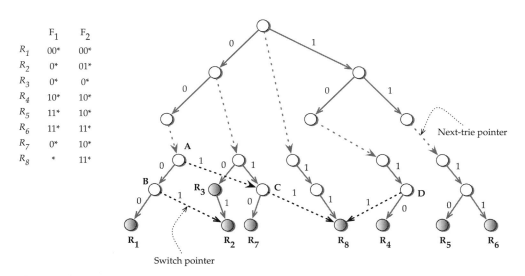

FIGURE 16.6   Using switch pointers to speed up search in grid-of-tries data structure.

node representing rule R_2. Note that node C and the node representing rule R_2 are in the F_2 trie corresponding to the F_1 prefix 0. Similarly, there is a switch pointer connecting node C to the node representing rule R_8, which is on the F_2 trie for the F_1 prefix $*$. The failed search for bit 1 in the F_2 trie containing rule R_4 is continued by introducing a switch pointer from node D to the node representing rule R_8.

Example 16.5 *Classifying a packet using grid-of-tries.*

Let us continue with the aforementioned example of classifying a packet with the header values $F_1 = 000$ and $F_2 = 110$ (Figure 16.6). The search when it fails while examining the first bit of F_2 uses the switch pointer to jump directly to node C in the F_2 trie containing rules R_2, R_3, and R_7. Similarly, when the search on the next bit fails again, we jump to the node containing rule R_8 in the F_2 trie associated with the F_1 prefix $*$. Hence the best matching rule for the packet is R_8.   ▲

As can be seen, the switch pointer eliminates the need for backtracking in a hierarchical trie without the storage of a set pruning trie. Essentially, it allows us to increase the length of the matching F_2 prefix without having to restart the search from the root of the next ancestor F_2 trie.

Now we can formally define a switch pointer as follows. Look at Figure 16.7. Let v be a node in the F_1 trie that represents an F_1 prefix, $P(v)$. The *lowest ancestor* of $P(v)$ is the longest prefix match for $P(v)$ in the F_1 trie. Let u denote the node corresponding to the lowest ancestor of $P(v)$ in the F_1 trie. Let $T(u)$ and $T(v)$ be the F_2 tries associated with nodes u and v, respectively. Assume that x is a node in trie $T(v)$ that represents the prefix s and there is no node representing the prefix $s0$, i.e., the search fails on a 0 bit at node x. If the F_2 trie corre-

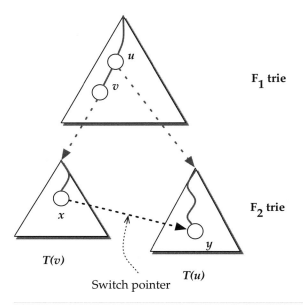

FIGURE 16.7   Formal definition of switch pointers in grid-of-tries (based on [273]).

sponding to the lowest ancestor $T(u)$ contains a node y that represents the prefix $s0$, then a switch pointer is placed at node x that points to y.

While the use of switch pointers speeds up the search, it has the disadvantage of sometimes missing best matches rules. We will give an example that illustrates this.

Example 16.6 *Missing the best matching rules in grid-of-tries.*

Consider classifying a packet with the header values of $F_1 = 000$ and $F_2 = 010$ (Figure 16.6). The search for 000 in the F_1 trie results in 00* as the best match. Using its next-trie pointer, the search continues on the F_2 trie for 010. However, it fails in the second bit 1 at node B. Hence, the switch pointer at node B is used to jump to the node representing rule R_2. The search terminates as it has reached the leaf node and R_2 is declared the best matching rule. Observe that the search has completely missed rule R_3, which also matches the packet header. While rule R_2 is the correct answer as it is lower in cost than rule R_3, according to our definition, in general the missed rule could have lower cost. ▲

To avoid missing the best matching least-cost rules, each node in the F_2 trie maintains an additional variable called *storedRule*. For a better explanation about what needs to be stored in the variable, let us consider a node x in the F_2 trie corresponding to the F_1 prefix P and the F_2 prefix Q. The variable *storedRule* in node x, denoted by *storedRule(x)*, is assigned the best matching rule whose F_1 prefix is a prefix of P and F_2 prefix is a prefix of Q. For instance, in the node representing rule R_2 this variable will store rule R_3 if rule R_3 is lower in cost than rule R_2.

Now let us analyze the worst-case time complexity and space complexity for the grid-of-tries. In the worst case, the number of memory accesses required is $2W$, which can be calculated as follows. First, traversing the F_1 trie to find the longest prefix match will consume at most W accesses. Next, the traversal of F_2 tries will consume at most W accesses. This is because in the F_2 tries we either traverse further down in the same trie or follow the switch pointer to jump to another trie. The maximum length of the F_2 field is also W bits, and hence the number of memory accesses required is bounded by W. Hence the total number of accesses is $W + W = 2W$, which is $O(W)$. To calculate the space complexity, observe that each rule is stored only once and each rule requires $O(W)$ of space. Hence, the amount of storage required is $O(NW)$.

16.6 Approaches for d Dimensions

In the previous sections, we examined various solutions for classifying packets with just two-dimensional classifiers. This section focuses on the different viewpoints and additional insights used by algorithms in subsequent sections. Section 16.6.1 takes a detailed look at the geometric view of the classification. Based on this view, the general packet classification problem is mapped onto a computational geometry problem. This mapping allows us to identify the theoretical lower bounds that illustrate the fundamental difficulty of the problem. Since a generalized solution is not practically feasible, Section 16.6.2 outlines many characteristics exhibited by real-life classifiers. These characteristics are effectively exploited by algorithms that work well in practice.

16.6.1 Geometric View of Classification: Thinking Differently

Recall from Chapter 15 that a prefix represents a contiguous interval on the number line. Similarly, a two-dimensional rule represents a rectangle in two-dimensional space, a three-dimensional rule represents a cube in three-dimensional space, and so on. A rule in d dimensions represents a d-dimensional hyper-rectangle in d-dimensional space. Therefore, a classifier can be viewed as a collection of hyper-rectangles, each labeled with a priority. A packet represents a point in this d-dimensional space with coordinates equal to the values of the header fields.

Given this geometric representation, classification of an arriving packet is based on finding the highest-priority rectangle among all the rectangles that encloses the point representing the packet. If higher-priority rectangles are drawn on top of lower-priority rectangles, classifying a packet is equivalent to finding the topmost visible rectangle containing a given point.

Consider the example classifier in Table 16.2 and its geometric representation in Figure 16.8. Field F_1 is represented as the x-axis and F_2 as the y-axis. In Figure 16.8, some prefix ranges are indicated by lines with ending arrows adjacent to each axis. The x-axis is divided into the three ranges, 00*, 01*, and 1*, while y-axis is divided into the four ranges, 00*, 01*, 10*, and 11*. Consider representing the rule R_1. Extend the range lines of the prefix 00* from the x-axis and similarly the range 00* from the y-axis. They form a box on the two-dimensional space marked R_1. Similarly, other rules can be represented in this two-dimensional space as indicated in Figure 16.8. Note that rule R_8 is overlapped by R_6. Now consider packet P with header fields 110 and 111. As can be seen from Figure 16.8, it is represented by the point P and overlapped by the regions representing the rules R_6 and R_8. Since the rectangle covered by R_6 is topmost, it is the best matching rule for packet P.

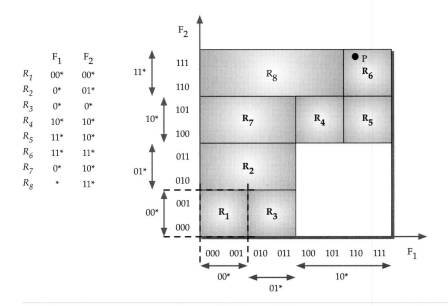

FIGURE 16.8 Geometric representation of rules. Each rule is shown as a shaded region.

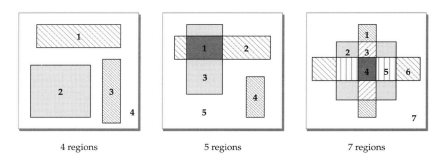

4 regions   5 regions   7 regions

FIGURE 16.9   Possible arrangements of 3 rules in two dimensions (adapted from [273]).

The advantages of taking a geometric view of packet classification are twofold. First, the packet classification problem can be viewed as a point location problem in a multidimensional space, a problem for which several algorithms have been reported [126], [127] and [145]. Mapping to the point location problem establishes lower bounds for space and time for packet classification. In the general case, for $d > 3$, the best bounds considering time or space are either an $O(\log^{d-1} N)$ time complexity with $O(N)$ space or $O(\log N)$ time complexity with $O(N^d)$ space. These algorithms are not practical for use in a high-speed router as illustrated in the following example.

Example 16.7 *Illustrating that lower-bound algorithms are not practical.*

Suppose a high-speed router needs to classify a packet within a time budget of 1 μsec, processing 1000 rules in five dimensions. Using an algorithm with time complexity of $O(\log^4 N)$ and space complexity of $O(N)$ will require 10,000 memory accesses. Even if we use the fastest memory with an access time of 10 nanosec, it will take 100 μsec for the classification operation to complete. Alternatively, using an algorithm of space complexity $O(N^d)$ and time complexity $O(\log N)$ will require 1000 Gbytes, which is prohibitively large.   ▲

The second advantage is that the geometric view provides more insight into the structure of a classifier by examining the number of distinct regions created by its rules. For a concrete understanding about distinct regions, let us walk through an example.

Example 16.8 *Number of distinct regions created by three two-dimensional rules.*

Consider a two-dimensional classifier with three rules. Figure 16.9 represents these rules geometrically as three rectangles in two-dimensional space and shows three possible arrangement of these rules. The leftmost figure illustrates one possible arrangement of these rules with four distinct regions. The three shaded regions represent the explicitly defined rules, and the fourth region corresponds to the default rule (implicitly defined) represented by white background. The figure in the middle shows the same rules with a different arrangement containing five distinct regions. Finally, the rightmost figure shows another arrangement with seven distinct regions.   ▲

As seen from the example, the number of distinct regions does not need be the same as the number of rules in the classifier since overlaps are possible. These distinct regions need to be kept track by any classification algorithm so that it can determine the region to which a newly arriving packet belongs. If the number of such regions in a classifier is large it will require more memory to represent the rules and as a result require a longer time to classify a packet.

Many classifiers containing the same number of rules can have different numbers of such distinct regions. If the number of regions is large, the classification algorithm needs to do more work to distinguish among these regions. It can be shown that the number of distinct regions created by N rules in d dimensions can be $O(N^d)$. Henceforth, these disjoint regions are referred to as *classification regions*. For our example classifier shown in Figure 16.8, there are nine classification regions.

16.6.2 Characteristics of Real-Life Classifiers: Thinking Practically

For packet classification, the algorithms must complete within a specified amount of time for N rules, which could range from a few thousands to tens of thousands. As described in the previous section, even algorithms with those lower bounds are not practical for use in a high-speed router. Fortunately, classifiers in real life exhibit many characteristics and a large amount of structures that can be taken advantage of when designing classification algorithms. These algorithms perform quite well on classifiers that satisfy one or more of these characteristics. However, in the worst case they perform very badly.

In this section, we outline many such characteristics identified by several independent analyses [47], [48], [275], and [687]. Even though we list many of the characteristics, not all of them are used by the algorithms described in subsequent sections. However, we believe that this will be beneficial for any designer of algorithms interested in coming up with new algorithms or adapting an existing algorithm for specific scenarios.

- *Number of fields:* While a typical implementation supports rules with eight fields, more than 50% of rules occurring in practice have only four fields specified.

- *Small set of values in a protocol field:* In most classifiers, the protocol field in the rules contains TCP or UDP or wildcard (∗) specifications. Some of the other specifications that do not occur often include ICMP, GRE, IGMP, (E)IGRP, and IP-in-IP.

- *Port field specifications are mostly ranges:* Many rules (approximately 10%) contain range specifications in destination and source port fields. These ranges are either the wild-card range (i.e., ∗) or the port ranges that distinguishe server ports (< 1024) from client ports (≥ 1024). Conversion of the range (≥ 1024) into prefixes requires splitting it into six intervals: [1024, 2047], [2048, 4095], [4096, 8191], [8192, 16383], [16384, 32767], and [32768, 65535]. As a consequence, any algorithm that converts ranges into prefix should have the ability to handle large numbers of rules in the classifier.

- *The number of disjoint classification regions is small:* This observation uses the geometric view described in Section 16.6.1 where rules form a number of distinct and possibly overlapping regions in d-dimensional space. While the number of such regions can be as much as $O(N^d)$ in the worst case, a survey of real-life classifiers indicates that such regions are linear in the number of rules N.

- *Source-destination matching:* Analysis of traffic traces with real-life classifiers indicates that most of the packets match at most five distinct source and destination address value combinations in the classifier. In the worst case, no packet matched more than 20 distinct source and destination address combinations.

- *Sharing of same field values:* It is common for different rules in the same classifier to share a number of the same field values. Such a sharing occurs when a network administrator wants to block every host in one group of IP addresses from communicating with any host in another group of IP addresses. A separate rule must be written for each pair of hosts in the two groups since the prefix notation is not sophisticated enough to allow such specifications in a single rule. Such a repetition of rules leads to sharing of source and destination address field values.

- *Redundant rules:* Observations indicate that about 15% of the rules are redundant. Some of the rules are backward redundant while others are forward redundant. A rule R is said to be *backward redundant* if there exists a rule T appearing earlier than R in the classifier and R is a subset of T. Hence, no packet will ever match R. For instance, if rule R_3 occurs earlier than R_2 in the example classifier shown in Table 16.2, then none of the packets will match R_2 since it is a subset of R_3. However, a rule is *forward redundant* if there exists a rule T that occurs after R in the classifier such that R is a subset of T, R and T have the same actions, and for each rule V occurring between R and T in the classifier either V is disjoint from R or V has the same action as R. Such forward redundancy rules can also be eliminated. A packet matching R will now match T yielding the same action.

16.7 Extending Two-Dimensional Solutions

When searching for a solution to a new problem, the natural tendency is to adapt or extend an existing solution to solve the problem. We extended the trie-based IP address lookup algorithms to classify rules in two dimensions. Similarly, can we extend the two-dimensional solutions to handle rules with d dimensions? In this section, we investigate such extensions by categorizing them into *naïve extensions* and *native extensions*.

16.7.1 Naïve Extensions

A naïve extension, as mentioned in [47], uses any efficient two-dimensional scheme and instead of a single rule at the leaf, a set of rules is stored. Even though the choice of the dimensions can be any two fields in the classifier, typically source and destination address fields are chosen. The primary motivation behind these choices is based on the source–destination matching observation outlined in Section 16.6.2. Recall that this observation indicates that when considering only the source and destination fields, almost all packets match at most five rules and no packet matches more than 20 rules. This will reduce the number of rules searched at the leaf between 5 and 20. A linear search on this reduced set of 20 rules will definitely perform better than the naïve linear search of the entire classifier, assuming the classifier contains more than 20 rules.

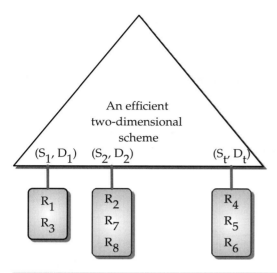

FIGURE 16.10   Extending a two-dimensional scheme for classification on arbitrary number of fields (adapted from [47]).

The naïve extension is illustrated in Figure 16.10. It uses an efficient two-dimensional scheme to find all distinct source–destination pairs $(S_1, D_1), (S_2, D_2), \ldots, (S_n, D_n)$ that match a packet header. Each distinct pair (S_i, D_i) is associated with a list of rules that contains (S_i, D_i) in source and destination fields. All the fields in the rule need not be stored in the list. Instead, only the fields other than the source and destination address need to be stored. The search first traverses all the rules associated with (S_1, D_1), then (S_2, D_2), and so on.

Use of such a data structure provides a few advantages. First, each rule is represented only once and hence the memory required is linear in the number of rules. However, sometimes replication of rules might be needed to reduce the number of source–destination pairs traversed during the search. Second, since only source and destination address fields are used for the search structure, the blowup of rules because of translating port range to prefixes is eliminated.

The grid-of-tries approach discussed in Section 16.5.3 is an efficient two-dimensional scheme for classifying address prefix pairs and it can be used as the two-dimensional scheme in Figure 16.10. However, this scheme cannot be generalized for $d > 2$. A solution proposed in [47] called *extended grid of tries* (EGT) uses the standard grid-of-tries scheme for two fields and extends the traversals to find all the matching rules.

16.7.2 Native Extensions

While naïve extensions use any two-dimensional scheme with a list of rules stored at the leaf, native extensions augment these schemes with additional levels of tries to accommodate multiple dimensions or fields. The general scheme is illustrated in Figure 16.11. In this section, we describe such extensions to hierarchical tries and set pruning tries.

Hierarchical tries for two fields can be extended recursively for rules with d dimensions as follows:

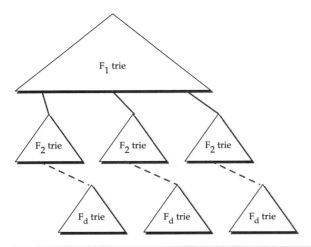

FIGURE 16.11 Extending a two-dimensional scheme natively for classification on arbitrary number of fields.

- If $d = 1$, the hierarchical trie is just a binary trie.

- If $d > 1$, first construct a binary trie corresponding to a field, say F_1, called the F_1 trie. This F_1 trie contains the distinct prefixes R_i that belong to field F_1 of all the rules in the classifier.

- For each prefix P in the F_1 trie, construct recursively a $(d - 1)$ hierarchical trie, say T_P, using the rules that specify P in the field F_1. The node representing the prefix P in the F_1 trie is linked to trie T_P using a pointer called the *next-trie* pointer.

A packet classification on an incoming packet header $(H_1, H_2, \ldots, H_d)$ proceeds recursively as follows. First, the F_1 trie is traversed based on the bits in H_1 and finds the node corresponding to the longest matching prefix of H_1. The next-trie pointer of the node is fetched and if it is not null, the search continues recursively on the $(d - 1)$ hierarchical trie. As the search proceeds, it keeps track of the best matching rule encountered so far. Once the search terminates on this $(d - 1)$ hierarchical trie, it backtracks on the F_1 trie. For each F_1 trie node encountered during backtracking, the search follows the next-trie pointer, if it is not null. Then it recursively traverses the $(d - 1)$ hierarchical trie stored at that node and updates the best matched rule, as more rules are encountered, if needed. This search algorithm is sometimes referred to as the backtracking search algorithms due to its recursive nature. It is left as an exercise to the reader to infer that the time complexity for d dimensions is $O(W^d)$.

Similar to hierarchical tries, set pruning tries can be extended to accommodate d dimensions. Let us start with a d-dimensional hierarchical trie consisting of an F_1 trie and its $(d - 1)$-dimensional hierarchical tries associated with the nodes. Now consider the set of nodes that represents prefixes shorter than a prefix P in the F_1 trie. Call this set of nodes T. In a d-dimensional set pruning trie, the rules in the $(d - 1)$ tries linked to all nodes in T are replicated to the $(d - 1)$-trie linked to prefix P. The replication of prefixes is carried out in a recursive fashion for the $(d - 2)$-tries and so on.

The search algorithm for a packet with the header $(H_1, H_2, \ldots, H_d)$ needs to traverse the F_1 trie only once to find the longest matching prefix node for H_1. If its next-trie pointer is not null, the search follows the F_2 trie and finds the longest matching prefix for H_2. The search continues in this fashion for F_3 tries and so on for all d-fields. Since the rules are replicated in a set pruning trie, the search algorithm ensures that all the matching rules will be encountered in its path. For d-fields, it can be shown that the memory required is $O(dWN^d)$. You can now see how the memory grows exponentially with the addition of each field. However, the time complexity for search is $O(dW)$, which is linear in the number of fields, unlike hierarchical tries.

16.8 Divide and Conquer Approaches

The main idea behind the divide and conquer approach is to partition the problem into multiple smaller subproblems and efficiently combine the results of these subproblems into the final answer. Recall that in Chapter 15 we outlined many efficient single-field search techniques under the context of longest prefix matching for IP address lookup. Hence, it is natural to consider whether these approaches can be effectively used; after all, the packet classification problem is nothing but a search on multiple fields.

A common theme of these divide and conquer algorithms is to decompose the packet classification problem into many longest prefix matching problems, one for each field, and combine the results of these longest prefix matches[1]. For decomposition, the classifier is sliced into multiple columns with the ith column containing all distinct prefixes of field i. Such columns are referred to as *field sets* and the field sets for the example classifier are shown in Figure 16.12. For each incoming packet, the longest prefix matching is determined separately for each of the fields. Now the key challenge is how efficiently the results of these prefix matches can be aggregated. The algorithms described in the next few sections differ mainly in two aspects:

- How the results are returned from the individual longest prefix matches, and

- How the individual results from these prefix matches are combined.

Dividing or decomposing the packet classification problem into many instances of single-field search problems offers several advantages. First, the search for each field can proceed independently, enabling the use of parallelism offered by modern hardware. Second, the search can be optimized by choosing different search strategies for each type of field. For example, source and destination address prefixes can employ longest prefix matches, while source and destination port ranges can use efficient range-matching schemes.

While there are compelling advantages, decomposing a multifield search problem raises subtle issues. Of course, the primary challenge is how to combine the result of the individual searches. Furthermore, it is not sufficient for a single-field search to return the longest matching prefix for a given field in the rule. This is because the best matching rule may contain a field that is not necessarily the longest matching prefix relative to other rules. Additionally,

---

[1]Note that ranges for source or destination ports can be converted into prefixes.

Rule	F_1	F_2
R_1	00*	00*
R_2	0*	01*
R_3	0*	0*
R_4	10*	10*
R_5	11*	10*
R_6	11*	11*
R_7	0*	10*
R_8	*	11*

F_1	F_2
0*	0*
00*	00*
10*	01*
11*	10*
*	11*

FIGURE 16.12   Field sets for the example classifier in Table 16.2.

the result of these single-field searches should be able to return more than one rule because packets may match more than one. In the next few sections, we discuss several algorithms that use the divide and conquer approach. We begin with the Lucent bit vector scheme.

16.8.1 Lucent Bit Vector

The Lucent bit vector scheme uses the divide and conquer approach [392]. It uses bit-level parallelism for accelerating the classification operation in any practical implementation. The basic idea is to first search for the matching rules of each relevant field F of a packet header and represent the result of each search as a bitmap. The final set of rules that matches the full packet header can be found by intersecting the bitmaps for all relevant fields F. Although this scheme is still linear in the number of rules, in practice searching through the bitmap is faster as a large number of bits can be accessed in a single memory access. While the original algorithm takes a geometric view and projects the rules to the corresponding dimensions, we describe a variant that uses tries.

The algorithm first partitions the classification problem in d-fields into d longest prefix matching problems, one for each field. Next, the unique prefixes for each field are identified and using these unique prefixes a separate data structure is constructed for finding the longest matching prefix. A bit vector of length N is associated with each prefix in the data structure and bit j in the bit vector is set if the prefix or its prefixes match rule R_j in the corresponding field of the classifier. In the bit vector, bit 1 refers to rule R_1, bit 2 refers to rule R_2, and so on. This process is repeated until all the bit vectors for each unique prefix of each field are constructed. Intuitively, bit vectors represent the matching rules corresponding to the prefix they represent.

The question now is what kind of data structures can be used and how they should be organized? Ideally, any data structure described in Chapter 15 can be used. However, for the sake of discussion, let us assume binary tries. We illustrate the construction of the data structure for the simple two-field classifier shown in Table 16.2. First, we identify the unique prefixes for fields F_1 and F_2, which are shown in Figure 16.12. Using these unique prefixes, we build two binary tries, one for field F_1 (F_1 trie) and the other for field F_2 (F_2 trie). Each

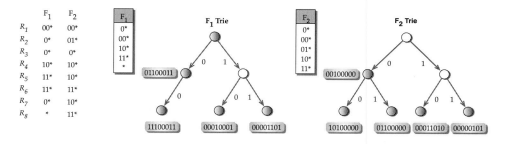

FIGURE 16.13 The F_1 and F_2 tries with bit vectors for Lucent scheme.

node containing a valid prefix is associated with a bit vector of size 8. The size of the bit vector, as noted earlier, is the same as the number of rules in the classifier.

The bit vector for each prefix is constructed by setting bit j if the prefix corresponding to rule R_j in the classifier matches the prefix corresponding to the node and its prefixes. Notice that in our example, the prefix 00* in field F_1 matches 00* and its prefixes 0* and *. These correspond to rules R_1, R_2, R_3, R_7, and R_8. Hence, the bit vector for the trie node corresponding to 00* has a value of 11100011 where the bits are numbered as 1 through 8 in increasing order from left to right. Similarly, the bit vector for each prefix is constructed. The binary tries for F_1 and F_2 along with the unique prefixes are shown in Figure 16.13.

Now, when a packet arrives with the header fields $H_1, \ldots, H_k$, the relevant headers that correspond to the fields in the classifier are extracted. Then for each field i, the corresponding trie is probed for the longest prefix match and the resulting bit vector B_i is read. Then the intersection of B_i is performed for all i using a bitwise AND operation. The resultant bit vector B_R contains all ones in bit positions that correspond to rules that matched. Since the rules are arranged in the order of cost, the position of the first bit set in bit vector B_R is the position of the rule in the classifier that best matches the packet header.

Example 16.9 *Classifying a packet using bit vectors.*

Let us determine how a packet with fields $F_1 = 001$ and $F_2 = 010$ gets classified (Figure 16.13). First, perform a longest prefix lookup in the F_1 trie that provides the bit vector 11100011 corresponding to prefix 00*. Next, probe the F_2 trie for the longest prefix match resulting in the bit vector 01100000 for the prefix 01*. Then, perform a bitwise AND operation that yields the result bit vector 01100000. Since the lowest bit position in the result bit vector is two, the best matching rule is R_2. ▲

Now that we know how the algorithm works, let us turn our attention to analyzing the memory access times and space requirements. Since the bit vectors are N bits in length, computing the bitwise AND requires $O(N)$ operations. It might be argued that in spite of using bitmaps the time complexity is still $O(N)$. If so, why is this approach any better than a linear search on the rules? This is because of the constant factor improvement possible using bitmaps. Since a group of bits is manipulated together in a single operation, constants are much lower compared to a naïve linear search. The size of this group is typically determined by the word size of the memory used. If w is the size of a word in memory, the total number of memory accesses required for these bit operations is $\lceil (N \times d)/w \rceil$ in the worst case. Notice that this worst case occurs when a packet does not match any rule in the classifier.

If commodity memory of 32 bits is used, the memory access is brought down by a factor of 32. A customized chip using wide memories, $w > 1000$, can even do better. As an example, consider a classifier containing 5000 rules with five dimensions and using a memory of $w = 500$ for classification. The number of memory accesses required is $5000 \times 5/500 = 50$. If the access time for the memory is 10 nanosec, the time to classify a packet is 500 nanosec. This implies that we can look up 2 million packets per sec, which is not achievable using naïve linear search.

Storage requirements can be calculated by observing that each field can have at most N distinct prefixes. As a consequence, each trie contains N bit vectors of size N bits each. Since there are d tries, one for each field, the total amount of memory required is $N \times N \times d$ bits, which translates to $\lceil N^2 \times d/w \rceil$ memory locations.

To conclude, while the cost of memory accesses is linear in the number of rules, i.e., $O(N)$, the constant factor of word size of the memory scales it down substantially. If the word size is 1000, the constant factor improvement could be a big gain in practice. However, the scheme suffers from the drawback of memory not being utilized efficiently. Practical observations indicate that the set bits in the bit vector are very sparse. Considerable savings in memory access could be achieved if we can selectively access portions of bit vectors that contain the set bits. In the next section, we outline an algorithm that uses aggregated bit vectors to identify the portions of actual bit vectors that need to be accessed.

16.8.2 Aggregated Bit Vector

The main motivation behind aggregated bit vector (ABV), described in [48], is to improve the performance of the Lucent bit vector scheme by leveraging the statistical properties of classifiers that occur in practice. In the Lucent bit vector scheme, in the case where the number of rules is large, the bit vector can be wider than the memory data bus. As a result, retrieving a bit vector requires several sequential memory accesses. To reduce the number of memory accesses, ABV takes advantage of the following observations: (1) the set bits in the bit vector are sparse, and (2) an incoming packet matches only a few rules. For example, in a 50,000 rule classifier, if only 6 bits are set in a bit vector of size 50,000, it is a waste to read the rest of the bits as a substantial number of memory accesses will be incurred. The algorithm uses two key ideas that takes advantage of these observations: *rule aggregation* and *rule rearrangement*.

The idea behind rule aggregation is to use a reduced-size bit vector that captures partial information from the original bit vector. This allows us to guess about the matching rules without comparing the bits in the original bit vectors. The reduced-size bit vector is called the *aggregate bit vector*.

For efficiently constructing an aggregated bit vector, an aggregation size A is selected. The original bit vectors are then partitioned into k blocks, each of size A bits, where $k = \lceil N/A \rceil$. If any of the bits in a block are set to 1, the corresponding bit in the aggregate bit vector is set to 1; otherwise, it remains 0. In other words, each group of A bits in the original bit vector is simply aggregated to a single bit in the aggregate bit vector. The aggregate size A can be tuned to optimize the performance of the entire scheme. A natural choice for A is the word size of the memory that makes it possible to fetch an aggregate bit vector in a single memory access.

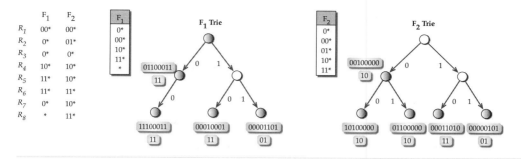

FIGURE 16.14   The F_1 and F_2 tries with original and aggregated bit vectors.

Conceptually, *ABV* uses the same data structures and bit vectors of size N constructed in the same manner as the Lucent bit vector scheme. In addition, each bit vector is associated with an aggregate bit vector that is built as described above. Figure 16.14 illustrates the *ABV* scheme using a trie along with the original bit vectors and aggregate bit vectors. The aggregate bit vectors are constructed from the original bit vectors using an aggregation size A of 4 bits. These aggregated bit vectors are shown below their original bit vector in the figure. Note that the original bit vector is stored in blocks of 4 bits so that each of them can be retrieved independently. For example, the original bit vector for the F_1 prefix $11*$ is 00001101. All the first four bits are 0 and hence the first aggregated bit is set to 0. The second aggregated bit is set to 1, since among the next four bits three are set to 1. The resulting aggregated vector is 01, which is shown below the original bit vector in the figure. Also the original bit vector is stored as two blocks, the first block containing the bits 0000 and the second block containing the bits 1101.

Now the search algorithm proceeds as follows. First, an independent search on d packet fields is performed to find the longest matching prefix on their respective tries. This search ends in returning the A bit aggregate bit vector associated with the longest matching prefix. Next, d aggregate bit vectors are intersected using a bitwise *AND* operation. For each bit set to 1 in the result aggregate bit vector, the corresponding d blocks of the original bit vectors are retrieved from memory and again a bit-wise *AND* operation is performed. In the resulting bit vector, the matching rules correspond to the bits set to 1.

Example 16.10   *Classifying a packet using ABV.*

Assume that we need to classify the same packet as in Example 16.9 with fields $F_1 = 001$ and $F_2 = 010$ (Figure 16.14). The search for the longest prefix match on the respective tries yields the prefix $00*$ for F_1 and $01*$ for F_2. The aggregate bit vectors 11 and 10 associated with these prefixes are retrieved. A bitwise *AND* operation on these aggregate bit vectors results in 10. This indicates that the first block of the original bit vectors for the matching prefix of F_1 and F_2 needs to be retrieved and intersected. The first block for the matching F_1 prefix is 1110 and for the F_2 prefix is 0110. The intersection of these two partial bit vectors results in 0110. The first bit set to 1 indicates the best matching rule, which is R_2. Hence, the number of memory accesses required for intersection of original bit vectors is two, assuming the word size is 4 bits. This presents a savings of 50% when compared with the the Lucent scheme, which requires four memory accesses. This is because the Lucent scheme requires accessing the second blocks of the original bit vectors for two fields.   ▲

While aggregation reduces the memory accesses in most cases, it also leads to *false matches* or *false positives*. This is due to the lack of information about which bit or bits in the original bit vector have led to a 1 in the aggregated bit vector. The worst case occurs when a false-positive occurs for every aggregated bit. For instance, consider the aggregate bit vectors corresponding to the F_1 prefix 00* and the F_2 prefix 10*. A bitwise *AND* of these aggregate bit vectors results in the bit vector 11. Hence, we need to retrieve both blocks of the original vectors and intersect them. The intersection of the first block (1110 for F_1 and 0001 for F_2) yields 0000, which might be a surprise even though the corresponding aggregate bit was a 1. This is what we call a false positive in which the intersection of an aggregate bit returns a 1 but there are no valid matching rules in the block identified by the aggregate. Hence, the packets containing the F_1 prefix 00* and the F_2 prefix 10* will incur extra memory access. To reduce the probability of false matches, a method for rearranging the rules in the classifier is proposed so that rules matching a specific prefix are placed close to each other. The details can be found in [48].

16.8.3 Cross-Producting

The cross-producting scheme outlined in [662] is motivated by the observation that the number of distinct prefixes for each field is significantly less than the number of rules in the classifier. For instance, our example classifier shown in Table 16.2 contains eight rules, but the number of distinct prefixes for both F_1 and F_2 is five, which is less than eight. While this reduction might not be much, in large classifiers it can be significant. This scheme, like any other divide and conquer approach, uses independent field searches and the results are combined to find the best matching rule.

Before examining the main idea, let us define what a crossproduct means. A *crossproduct* is defined as a *d*-tuple formed by drawing one distinct prefix from each field. For our example classifier, the crossproduct [0*, 11*] is formed by selecting the distinct prefixes 0* and 11* from fields F_1 and F_2, respectively. All the distinct prefixes for the example classifier are shown in Figure 16.12. Since F_1 and F_2 have five distinct prefixes each, a total of $5 \times 5 = 25$ crossproducts can be formed. The main idea behind utilizing crossproducts to find the best matching rule is based on the following proposition:

Proposition 16.1. *Given a packet P, if we do the longest matching prefix operation for each field P[i] and concatenate the results to form a crossproduct C, then the best matching rule for P is identical to the best matching rule for C.*

Proof. Let us prove this key observation by contradiction. Assume that the claim is not true. Since each field in C is a prefix of the corresponding field in P, every rule that matches C also matches P. Now the case in which P has a different matching rule implies that there is some other rule R that matches P but not C. This is possible only if there is some field i such that $R[i]$ is a prefix of $P[i]$ but not of $C[i]$ where $C[i]$ denotes the field i in crossproduct C. But since $C[i]$ is a prefix of $P[i]$, this can happen only if $R[i]$ is longer than $C[i]$. However, this contradicts our assumption that $C[i]$ is the longest matching prefix in field i.   ■

	F_1	F_2
R_1	00*	00*
R_2	0*	01*
R_3	0*	0*
R_4	10*	10*
R_5	11*	10*
R_6	11*	11*
R_7	0*	10*
R_8	*	11*

F_1
0*
00*
10*
11*
*

F_2
0*
00*
01*
10*
11*

Number	Cross Product	Best Matching Rule
1	(0*, 0*)	R_2
2	(0*, 00*)	R_2
3	(0*, 01*)	R_2
4	(0*, 1*)	R_8
.	.	.
.	.	.
.	.	.
29	(*, 10*)	-
30	(*, 11*)	R_8

FIGURE 16.15   Generation of crossproduct table for the rules of Table 16.2.

Thus the cross-producting algorithm begins by constructing independent data structures for d field sets, one for each field. These are used for the longest prefix matching operation of the corresponding packet field. To resolve the best matching rule, a table C_T is built consisting of all crossproducts. For each crossproduct in C_T, we precompute and store the best matching rule. The field sets and the crossproduct table for the example classifier are shown in Figure 16.15. For now let us not worry about how table C_T is organized as we will examine that later in the section.

For any incoming packet, the crossproduct C is constructed by performing a longest matching prefix on individual fields. Using C as a key, crossproduct table C_T is probed to locate the best matching rule. Hence, classifying a given packet header involves d longest matching prefix operations plus a probe of C_T for the best matching rule. These d prefix lookups can be carried out independently, thus lending to a parallel implementation in hardware.

Example 16.11   *Finding the best matching rule using cross-producting.*

Consider classifying the incoming packet, with values of $F_1 = 000$ and $F_2 = 100$ (Figure 16.15). Probing the independent data structures for the fields yields the longest prefix match for F_1 as 00 and for F_2 as 10. These prefixes yield the crossproduct (00, 10). The crossproduct is probed into table C_T which yields the best matching rule as R_7.   ▲

There are various ways in which the crossproduct table C_T can be organized. The simplest is the use of a direct lookup table such as an array. Using such a scheme requires labeling each prefix in the field set and that this label be returned as a result of longest prefix matching for each field. For example, the field sets of F_1 and F_2 can be labeled separately as $1, 2, 3, 4, 5$ since there are five distinct prefixes. Continuing with Example 16.11, the longest prefix matches for F_1 and F_2 will yield labels 2 and 4, respectively. From these labels, the index in the array can be determined as $2 \times 4 = 8$, which gives the best matching rule in a single memory access.

Use of a direct index table will require a large amount of memory. Can we reduce the memory consumption? A closer examination of the crossproducts shows that among the

25 entries, only 8 entries contain the original rules, which we call *original* crossproducts. The remaining ones are generated due to the crossproduct operation. Among the generated crossproducts some of them correspond to the original rule. To be precise, a match of the crossproduct implies a match for one or more of the original rules. For instance, the match for the crossproduct [00∗, 0∗] implies a match of the original rules R_1 and R_2. Such crossproducts are referred to as *pseudo*-crossproducts. Finally, some of the crossproducts do not map to any original rule such as [11∗, 00∗], which we call *empty* crossproducts.

Since there can be many empty crossproducts for classifiers containing hundreds and thousands of rules, the problem is mitigated by using a hash table instead of a direct lookup table. Using a hash table could save memory since it needs to store only the original crossproducts and the pseudo-crossproducts.

In spite of such optimizations, the naïve cross-producting algorithms suffer from exponential memory requirements. In the worst case, the number of entries in the crossproduct table can be as many as N^d. Even for smaller values, say, $N = 50$ and $d = 5$, the table size can reach as much as 50^5 entries! Assuming each entry requires 16 bits, the table needs 596 Mbytes of memory, which is prohibitively large. To reduce the memory, [662] suggest the use of *on-demand cross-producting*.

ON-DEMAND CROSS-PRODUCTING

The on-demand cross-producting scheme places a limit on the size of the crossproduct table and builds it on a need basis. Instead of building the entire crossproduct table a priori, the entries in the table are incrementally added. For each incoming packet, the longest prefix matching operations are performed on the individual fields and the crossproduct C is computed as in the naïve cross-producting scheme. Then the crossproduct table is probed using C. If the crossproduct table contains an entry for C, then the associated rule is returned. However, if no entry for C exists, the best matching rule for C is computed on the fly and an entry is inserted into the crossproduct table. Thus, it is expected that the first packet that adds such an entry will experience more latency. But subsequent packets with the same crossproduct C will benefit from fast lookups.

Thus, on-demand cross-producting can improve the building time of the data structure and the storage cost. In fact, the crossproduct table can be treated as a cache. To start with, the table will be empty. As entries are added with the arrival of packets, the table starts filling up. The subsequent addition of new entries may require the eviction of existing entries. Thus, a cache replacement policy that removes entries not recently used has to be implemented.

Since caching packet headers for classification is not considered effective, what suggests that caching based on cross-producting can be any better? It is because a single crossproduct can represent multiple headers. Hence the hit rates for the crossproduct cache can be expected to be much better than standard packet header caches.

16.8.4 Recursive Flow Classification

Recursive flow classification (*RFC*) attempts to reduce the memory requirements of the naïve cross-producting scheme by first creating smaller crossproducts and combining them in multiple steps to form larger crossproducts. Like the cross-producting scheme, *RFC* also performs independent parallel searches on the fields of the packet header. The results of these field searches are combined in multiple steps, unlike naïve cross-producting, which uses a single step. To do this efficiently, *RFC* uses two techniques:

- It uses equivalence classes (see Appendix B.4) for identifying the set of matched rules at each step. These equivalence classes represent concisely the rules matched by various header fields.

- For merging the results from different fields, *RFC* uses crossproduct tables to store the precomputed results.

In the next few sections, we develop these concepts by first examining packet classification in one dimension, followed by two dimensions, and then finally extending it to an arbitrary number of dimensions.

USE OF EQUIVALENCE CLASSES

Consider the example classifier shown in Table 16.2. To start with, let us consider only one field, say F_1, for classification. We can project the two-dimensional rules along the F_1 dimension that represents the domain of possible values for field F_1. This is shown in Figure 16.16. This dimension can be partitioned into intervals at the endpoints of each rule, and within each interval, a particular set of rules is matched. As can be seen from the Figure 16.16, there are four intervals [000, 001], [010, 011], [010, 101], and [110, 111] for F_1. Using these intervals, we partition the set of possible values 000–111 for this field into equivalence classes, where all values in a set match exactly the same rules. For example, F_1 values of 000 and 001 match rules R_1, R_2, R_7, and R_8 and hence they belong to the same equivalence class. In total, we can

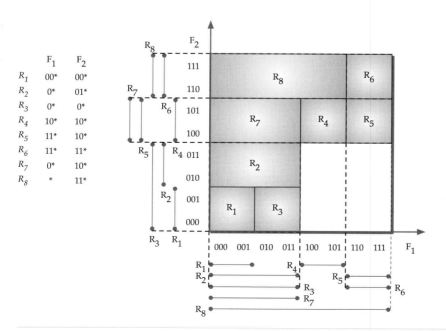

FIGURE 16.16   Geometric representation for identifying equivalence classes.

Equivalence class table for F$_1$

Equivalence Class	Rules Matched
EF$_1$-0	R_1, R_2, R_7, R_8
EF$_1$-1	R_2, R_3, R_7, R_8
EF$_1$-2	R_4, R_8
EF$_1$-3	R_5, R_6, R_8

Lookup table for F$_1$

F$_1$ Value	Equivalence Class
000	EF$_1$-0
001	EF$_1$-0
010	EF$_1$-1
011	EF$_1$-1
100	EF$_1$-2
101	EF$_1$-2
110	EF$_1$-3
111	EF$_1$-3

FIGURE 16.17   Equivalence classes and lookup table for field F_1.

Equivalence class table for F$_2$

Equivalence Class	Rules Matched
EF$_2$-0	R_1, R_3
EF$_2$-1	R_2
EF$_2$-2	R_4, R_5, R_7
EF$_2$-3	R_6, R_8

Lookup table for F$_2$

F$_2$ Value	Equivalence Class
000	EF$_2$-0
001	EF$_2$-0
010	EF$_2$-1
011	EF$_2$-1
100	EF$_2$-2
101	EF$_2$-2
110	EF$_2$-3
111	EF$_2$-3

FIGURE 16.18   Equivalence classes and lookup table for field F_2.

identify four such equivalence classes, corresponding to each interval. While in this example each interval corresponds to an equivalence class, in general this need not be the case.

Note that two points in the same interval always belong to the same equivalence class. Two intervals are in the same equivalence class if exactly the same rules project onto them. To find the best matching rule, we precompute two tables: one that maps each possible F_1 value to its equivalence class and the other that maps the equivalence class to the matched rules. The equivalence classes and the lookup tables for the F_1 dimension are shown in Figure 16.17. Similarly, we compute the equivalence classes for dimension F_2 and its lookup table, which are shown in Figure 16.18.

Even though symbols are used for equivalence classes in the figures, the actual implementation uses integers so that they can be used to index into another table. Hence they are sometimes referred to as *equivalence class identifiers* or simply *eqIDs*. Additionally, the equivalence class table does not store the rules explicitly as depicted in the figures. Instead, a bit vector as outlined in the Lucent bit vector scheme, is stored. The bit vector represents the matching rules by setting the appropriate bit position to 1. Now let us see an example of how a packet classification occurs in a single field.

Example 16.12 *Classifying a packet on a single field.*

Suppose we want to find the matching rules of a packet whose F_1 value is 001. Indexing into the F_1 lookup table using 001 shown in Figure 16.17 yields the equivalence class EF_1-0. Next, a lookup in the equivalence class table for EF_1-0 indicates the matched rules as R_1, R_2, R_7, and R_8. The rule R_1 is then declared the best matching rule since it occurs first in the order.   ▲

Of course, for the above one-dimensional classification example, there is no need for a separate equivalence class table. Instead of storing the *eqID* in the lookup table, the best matching rules can be directly stored, avoiding the lookup into equivalence class table. However, as we shall see later, the equivalence class tables provide a compact representation for intermediate results for classification on multiple fields.

USE OF CROSS-PRODUCT TABLES

Now let us extend the concept of equivalence classes to two-dimensional lookups involving fields F_1 and F_2. A packet's F_1 value can be used to lookup its one-dimensional tables in Figure 16.17 to obtain the *eqID*. This indicates the set of rules matched by F_1. Similarly, the F_2 value is used to lookup its one-dimensional tables in Figure 16.18. The resultant *eqID* represents the set of rules matched by F_2. However, what we are really interested in is the set of rules matched by both fields F_1 and F_2.

The intersection of the set of rules matched by F_1 and those matched by F_2 will provide the needed solution. However, computing such an intersection on the fly can be too expensive, especially if there are many rules (if there are N rules, an N-bitwise AND operation will be required, as we saw in the Lucent bit vector scheme). Hence, we compute the results of these intersections a priori and store them in a two-dimensional lookup table D, referred to as the crossproduct table.

Each entry in this two-dimensional crossproduct table D represents a set of rules matched by both fields F_1 and F_2. One dimension of the table D is indexed by the *eqIDs* of F_1 and the other by the *eqIDs* of F_2. Since the same set of matched rules may occur more than once in D, we assign a new set of *eqIDs* that represents these classes so that the table entries of D contain only *eqIDs*. If the matched rules by fields F_1 and F_2 are denoted by *eqIDs* m and n, respectively, then the entry $D[m][n]$ contains the *eqID* that represents the intersection of rules matched by both F_1 and F_2. Alternatively, these new equivalence classes represent distinct regions in the two-dimensional space of F_1 and F_2. Referring to Figure 16.16, it can be seen that there are nine distinct regions, each corresponding to an equivalence class.

Now let us see how we can precompute each entry in the two-dimensional crossproduct table. For the sake of discussion, let a represent the *eqID* for dimension F_1 and b for F_2. First, look up the set of rules matched by equivalence classes a and b. Second, compute the intersection of both the sets and identify the equivalence class to which the result belongs. Then store it as the entry for $D[a][b]$. The new equivalence classes and the resulting two-dimensional crossproduct table are shown in Figure 16.19. In Figure 16.19, the entry $D[EF_1$-0$][EF_2$-0$]$ contains the *eqID* EC-1. The entry corresponding to EC-1 in the equivalence class table indicates the rules matched for *eqIDs* EF_1-0 from F_1 and EF_2-0 from F_2. Note that the crossproduct table D is conceptually similar to the crossproduct table in the naïve cross-producting scheme

Equivalence class table

Cross product table

Equivalence Class	Rules Matched
EC-0	-
EC-1	R_1
EC-2	R_2
EC-3	R_3
EC-4	R_4
EC-5	R_5
EC-6	R_6, R_8
EC-7	R_7
EC-8	R_8

	EF_2-0	EF_2-1	EF_2-2	EF_2-3
EF_1-0	EC-1	EC-2	EC-7	EC-8
EF_1-1	EC-3	EC-2	EC-7	EC-8
EF_1-2	EC-0	EC-0	EC-4	EC-0
EF_1-3	EC-0	EC-0	EC-5	EC-8

FIGURE 16.19   Equivalence classes and the final crossproduct table.

outlined in Section 16.8.3, except that they are organized differently from an implementation perspective.

To perform classification, we need the one-dimensional lookup tables for field F_1 and F_2 and the two-dimensional crossproduct table and the final equivalence class table that maps the final result *eqID* to the matched rules. Now let us walk through an example of classifying a packet involving two fields using the tables shown in Figure 16.17, Figure 16.18, and Figure 16.19.

Example 16.13   *Classifying a packet on two fields.*

Consider classifying the packet with field values $F_1 = 000$ and $F_2 = 010$. We first search the F_1 lookup table for 000, which gives the result EF_1-0 (Figure 16.17). Next we search the F_2 lookup table for 010, which results in EF_2-1 (Figure 16.18). Using these *eqIDs*, EF_1-0 and EF_2-1, we index the two-dimensional crossproduct table to find the rules matched by both F_1 and F_2, which gives us EC_2. Finally, using the equivalence class table in Figure 16.19, we find that the rule R_2 matches the incoming packet, which is declared the best matching rule.   ▲

The final step can be eliminated by storing only the best matching rule directly in the crossproduct table. For instance, instead of storing EC_2, the rule R_2 could be stored. A pipelined implementation for classifying packets using two fields is shown in Figure 16.20.

EXTENDING TO d-DIMENSIONS

As we saw earlier, classification in two dimensions requires finding a pair of equivalence class identifiers and precomputing a two-dimensional crossproduct table to map those *eqIDs* to a single *eqID*. Extending it to three dimensions requires finding three equivalence class identifiers, say x, y, and z, one for each field, that indicate the matched rules by the corresponding field. Identifying rules that match all three dimensions requires computing the intersection of these three sets of rules.

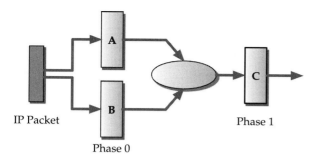

A - Lookup Table for field F_1 **B** - Lookup Table for field F_2

C - Crossproduct Table

FIGURE 16.20   Packet flow in recursive flow classification.

A straightforward approach is to use a three-dimensional crossproduct table where each entry $D[x][y][z]$ is precomputed by finding the intersection of the sets of rules matched in equivalence classes x, y, and z. This is similar to naïve cross-producting and, as we have seen earlier, does not scale very well because of large memory requirements.

An alternate approach is to use multiple two-dimensional crossproduct tables. To classify a packet in three dimensions, we can use two such two-dimensional crossproduct tables: one that merges the *eqID* x and *eqID* y to produce a single *eqID*, say a, which identifies the matching rules for both x and y; and the other that combines the *eqID* z and *eqID* a to another single *eqID*, say b, which identifies the intersection of the rules matched by a and z. The *eqID* b corresponds to the set of rules matched by all three header fields.

This idea can be extended to handle d dimensions using $d - 1$ separate two-dimensional tables. The purpose of each table is to merge two *eqIDs* into one *eqID*. Hence, with $d - 1$ two-dimensional crossproduct tables, d *eqIDs* can be reduced to just one. The order in which these *eqIDs* are combined will influence the contents of the tables that need to be precomputed.

A structure called a *reduction tree* is used to represent the order in which the *eqIDs* are merged. Each node in the tree represents a crossproduct table and the children represent the source of *eqIDs* used to index into the table. Note that many reduction trees are possible when the number of stages is greater than two. In such cases, *RFC* chooses the one that requires minimum memory.

In terms of performance, assuming the lookup proceeds from one stage to another in a sequential fashion, the number of memory accesses required is $O(P)$, where P is the number of stages. For a pipelined implementation, it will be $O(1)$. From a memory perspective, combining a pair of fields might require as much as N^2 memory since each field can have at most N distinct values and hence for d dimensions, the worst case is N^d. This corresponds to N^d classification regions, as we saw in Section 16.6.1. However, the real-life classifiers, as noted in [275], have only $O(N)$ regions instead of the worst case N^d, which requires $O(N^d)$ memory for both *RFC* and naïve cross-producting.

The simplicity and performance of *RFC* come at the cost of memory inefficiency. Memory usage for less than 1000 rules can range anywhere from a few hundred Kbytes to over 1 Gbytes of memory depending on the number of stages. The crossproduct tables used for aggregation also require significant precomputation for the proper assignment of an equivalence class identifier for the combination of the *eqIDs* of the previous steps. Such extensive precomputation precludes dynamic updates at high rates.

16.9 Tuple Space Approaches

Another unique approach to packet classification on multiple fields uses the notion of *tuple space*, which narrows the scope of a search by partitioning the rules using *tuples*. Before taking a look at the informal definition of a tuple, it is necessary to define a prefix length. Recall from Chapter 15, the number of bits specified in a prefix is referred to as the prefix length. A tuple represents a unique combination of prefix lengths of each field specified in the rules. There can be many tuples for a given classifier since the prefix lengths can be different for each rule and for each field. Such a resulting set of tuples defines the *tuple space*.

The main motivation behind this approach is that although classifiers contain many prefixes in the rules, the number of distinct prefix lengths tends to be small. As a result, the number of unique tuples is much less than the number of rules in the classifier. Hence, a linear search on the tuples might be more efficient than a linear search on the rules. In the worst case, the number of tuples can be as many as N, the number of rules in the classifier, which occurs rarely in practice.

For a concrete understanding of tuples, consider the example classifier shown in Table 16.2. The distinct prefix lengths for field F_1 are 0, 1, and 2. Similarly, the distinct prefix lengths for field F_2 are 1 and 2. Hence, the tuple space consists of $\{(0,1), (0,2), (1,1), (1,2), (2,1), (2,2)\}$ for a total of six tuples, one for each combination of prefix lengths from F_1 and F_2. Notice that this is less than the number of rules in the classifier, which is eight.

Formally, a tuple T is defined as a vector of d lengths, one for each field in the classifier. A rule in the classifier maps to tuple T, if for all i, the length of the prefix specified in field F_i consists of exactly $T[i]$ bits. For example, rule R_7 in the example classifier maps to tuple $(1,2)$

Rule	F_1	F_2	Tuple
R_1	00*	00*	(2, 2)
R_2	0*	01*	(1, 2)
R_3	0*	0*	(1, 1)
R_4	10*	10*	(2, 2)
R_5	11*	10*	(2, 2)
R_6	11*	11*	(2, 2)
R_7	0*	10*	(1, 2)
R_8	*	11*	(0, 2)

FIGURE 16.21   Mapping of rules in Table 16.2 to tuples.

since its prefix length for F_1 is 1 bit and for F_2 is 2 bits. Figure 16.21 shows the mapping of the example classifier rules to the tuples.

16.9.1 Tuple Space Search

The basic tuple space search approach performs an exhaustive search of the tuple space. Recall that a tuple specifies the valid bits of its constituent rules. Hence, we can probe the tuples for matching rules using a fast exact match technique such as hashing. For each tuple T, we associate a hash table $H(T)$ that contains all the rules mapped to T. To map a rule, we first identify the tuple T it belongs to, using the prefix lengths of its fields. Then we construct a key by concatenating the prefix bits of the rule specified by the tuple. Using this key, the rule is inserted into $H(T)$. For instance, the rule R_7 belongs to tuple $(1, 2)$ since its prefix lengths for fields F_1 and F_2 are 1 and 2, respectively. Now the key "010," generated by concatenating the prefix bits specified in the rule, is used to insert rule R_7 in the hash table for tuple $(1, 2)$. Figure 16.22 shows the contents of the hash table for each tuple.

The search algorithm decomposes a packet classification request into a number of exact match lookups into tuple hash tables. When a packet arrives, a search key is formed by concatenating the required number of bits from the packet header fields as specified by the first tuple T. Using this key, the hash table $H(T)$ is probed for an exact match. If there is a match for a rule, it is recorded as R_{best}, the best matching rule so far. The process is repeated for each T and if there is a match, say R, then it is compared with R_{best} and R_{best} is updated, if needed. After all the tuples have been examined, R_{best} contains the best matching rule.

Example 16.14   *Finding the best matching rule using tuple space search.*
    Consider classifying a packet with the values of 101 and 100 for fields F_1 and F_2, respectively. The search starts with the tuple $(0, 2)$ using the key "10," generated from the first two bits of field F_2 in the packet. Note that tuple $(0, 2)$ does not involve any bits from field F_1. This key "10" is used to probe the hash table for tuple $(0, 2)$ and we find that rule R_8 matches. Now R_8 is kept track of as R_{best}. For the next tuple $(1, 1)$, the key "11" is generated, which does not match any rule, and the search moves to the next tuple $(1, 2)$. Again, the generated key 110 does not match any rule, leaving R_8 still the best matching rule. Finally, tuple $(2, 2)$ is examined. The key "1010" matches rule R_4. Since R_4 is a better matching rule than R_8, R_{best} is updated with R_4. There are no more tuples to be examined and, therefore, R_4 is declared to be the best matching rule.   ▲

Tuple	Hash Table Entries
$(0, 2)$	R_8
$(1, 1)$	R_3
$(1, 2)$	R_2, R_7
$(2, 2)$	R_1, R_4, R_5, R_6

FIGURE 16.22   Contents of tuple hash tables.

While the search strategy might be straightforward, the search cost is proportional to the number of distinct tuples M. In practice, M tends to be much smaller than the number of rules N in the classifier. Hence, a linear search through the tuple set is likely to outperform the linear search of rules in the classifier. The algorithm lends itself to parallel implementation since the probes into separate tuples can be performed independently. However, the challenge lies in providing guaranteed worst-case lookup performance due to the unpredictability of the number of rules mapping to a single tuple.

Reference [660] indicates that using this scheme on a modest set of real-life classifiers reduces the number of searches by a factor of four to seven relative to an exhaustive search of all the rules in the classifier. The amount of storage required is $O(N)$ since each rule is stored only once in one of the hash tables. The main drawback of this algorithm is the use of hashing, which makes the time complexity of searches and updates nondeterministic.

16.9.2 Tuple Space Pruning

The key idea behind a *pruned tuple space search* [660] is to reduce the scope of an exhaustive search of tuples by performing independent searches on each field of the rules to find a subset of candidate tuples. It accomplishes this by first searching for an individual longest prefix match in each field and then probing only the tuples that are compatible with these individual matches.

The motivation behind tuple space pruning is that, in practice, classifiers seem to have very few prefixes for a given IPv4 address irrespective of whether it is a source or destination. Studies in [658] have shown that any IPv4 address in backbone routing tables has no more than six matching prefixes. In such cases, a naïve tuple search may require searching as many as $32 \times 32 = 1024$ tuples since each IPv4 address can be as long as 32 bits. However, using the observation as a heuristic, we might have to examine a total of only $6 \times 6 = 36$ tuples. This represents a cost savings of having to probe $1024 - 36 = 988$ tuples.

Let us illustrate tuple pruning using the example classifier in Table 16.2. We begin by constructing individual tries for both F_1 and F_2 prefixes as shown in Figure 16.23. Nodes that contain valid prefixes of rules store a list of tuples. These tuples contain the rules in the associated hash table that specify the prefix. In Figure 16.23, the shaded nodes represent the prefixes and the tuples they belong to are indicated adjacent to each node. Note that some nodes can have more than one tuple. A search on data structure begins by performing independent searches of F_1 and F_2 tries. The result of each search provides a list of candidate tuples for each field that corresponds to the longest prefix match. The final candidate list of tuples is constructed by the intersection of the tuple lists returned by the individual search.

Example 16.15 *Classifying a packet using tuple space pruning.*
Consider the classification of a packet with values $F_1 = 010$ and $F_2 = 100$ (Figure 16.23). Examining the F_1 trie, we find that the longest prefix matching $0*$ and the corresponding tuples are $(1, 1)$ and $(1, 2)$. Similarly, examining the F_2 trie provides the tuples $(1, 2)$ and $(2, 2)$. After intersecting these results, we get the tuple $(1, 2)$. Hence the hash table associated with tuple $(1, 2)$ is examined and we find that the best matching rule is R_7. ▲

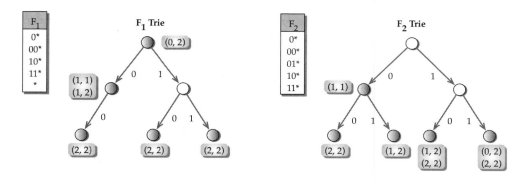

FIGURE 16.23   The F_1 and F_2 tries with tuples for rules in Table 16.2.

An attentive reader might notice that this approach is very similar to the parallel bit vector discussed in Section 16.8.1. While both schemes use independent matches in each field, the Lucent bit vector scheme searches through rules as opposed to tuples. Since the number of tuples grows much slower than the number of rules, the tuple pruning should scale better than the Lucent bit vector scheme.

16.10 Decision Tree Approaches

So far we have discussed two high-level approaches to packet classification on multiple fields. Our last approach examines the use of decision trees for multifield packet classification. A *decision tree* stores a single rule or a small list of possible matching rules at each leaf. The internal nodes contain enough information to guide the classification to a leaf node. During classification, a search key is constructed from the packet header fields. The tree is traversed using either individual bits or a subset of bits from the search key to make branching decisions at each node of the tree until a leaf node is reached. If the leaf node contains a single rule, then it is declared to be the best matching rule. In the other case, where a list of rules is stored in the leaf node, a linear search is required to identify the best matching rule.

A decision tree can be generalized by different parameters or degrees of freedom, as discussed below. Each degree of freedom offers a choice of various values from which to choose. A specific combination of choices for each degree of freedom generates a class of decision trees. Clearly, many such classes of decision trees are possible. Hence, it is worth taking a detailed look at these degrees of freedom to sharpen our understanding before delving into the details of the algorithms based on decision trees.

- *Sequential vs. interleaved search of the fields:* In trie-based schemes, described in Section 16.5, we saw that the search works only one field at a time. In these schemes, all the bits of field F_1 are tested before the search moves on to the bits of the next field F_2. Another interesting choice is to interleave the bit tests for all fields. For instance, in the root node, the fifth bit of a destination address field can be examined and if it is 0, the nineteenth bit in the source address field of the child node should be examined. Such interleaving allows the flexibility to locally optimize decisions at each node by choosing the next bit to test.

- *Bit test vs. range test for branching:* During search, at each node either a single bit or multiple bits in the search key can be examined to choose the next branch to follow. The use of a single bit limits the number of branches to two, while multiple bits can have many branches. Multiple branches have a tendency to reduce the height of the tree, which in turn could reduce the number of memory accesses. Multiple bits employ range checks such as $35 < S < 48$ for a given source address S. Range checks are slightly more general than bit tests.

- *Single rule vs. multiple rules in leaf nodes:* This degree of freedom allows the choice of either storing a single rule or multiple rules in the leaf nodes. Storing multiple rules in the leaf node reduces the overall storage required. However, a linear search will be needed when the search reaches the leaf node. As an example, consider a decision tree with 5000 leaves, each consisting of three rules. If we were to store only one rule per leaf node by increasing the height of the decision tree, this could require as much as $3 \times 5000 = 15{,}000$ extra nodes of storage. Hence, to balance storage with time, it might be advantageous to allow a small amount of linear searching at the end of tree search.

Based on these degrees of freedom, we can infer that trie-based schemes represent a class of decision trees with the choices of sequential search of fields, bit test for branching, and single rule in leaf node. In the next few sections, we examine a few decision tree algorithms based on geometrical cutting [276], [745]. Some of the decision tree approaches outlined are reminiscent of multi-attribute indexing methods in relational database systems such as grid file [528] and spatial indexing methods in geographic information systems such as *R*-tree [278], *R*∗-tree [65], *R*+-tree [628], and quad-tree [619].

16.10.1 **Hierarchical Intelligent Cuttings**

Hierarchical intelligent cuttings (*HiCuts*) [276] is another important approach. The notion of "cutting" comes from viewing the packet classification problem geometrically. Recall from Section 16.6.1 that each rule defines a *d*-dimensional rectangle in *d*-dimensional space. This *d*-dimensional space is cut or partitioned into equal-sized regions based on certain decision criteria. *HiCuts* uses heuristics that exploit the characteristics and structure of the classifier to arrive at the criteria of picking the dimension to partition and determining the number of cuts needed.

   HiCuts preprocesses the classifier and builds a decision tree. The root node of the decision tree covers the entire *d*-dimensional space, which is partitioned into smaller geometric subspaces by cutting across one of the *d* dimensions. These smaller geometric subspaces are represented by child nodes. Each subspace is recursively subdivided until no subspace has more than *binth* (bin threshold) number of rules.

   On receiving an incoming packet, its header fields are used to traverse a decision tree. The internal nodes contain information to guide the classification to a leaf node. The rules in the leaf node are then sequentially searched to determine the best matching rule. The tree is constructed such that the total number of rules stored in a leaf node is bounded by *binth* threshold. The characteristics of the decision tree such as its depth, degree of each node, and the local search decision to be made at each node are chosen during the preprocessing of the classifier.

Since the decision tree does not preclude an internal node containing more than two pointers, a fast mechanism is needed to access the correct child pointer. By using simple array indexing, we can choose the appropriate child pointer in one memory access regardless of the number of children at a node. To understand how the array indexing works, consider a single dimension. Now imagine a 5-bit address space that is partitioned into four equally spaced ranges [0, 7], [8, 15], [16, 23], and [24, 31]. Each range can be associated with a pointer that is stored as four consecutive elements in an array. To retrieve the pointer corresponding to a point, say 25, we can compute the quotient of 25 divided by the range width, which is 8. Since the quotient is 3, it retrieves the fourth element, assuming that the array indices start at 0. Hence the root node and internal nodes store a one-dimensional array the size of which is determined by the number of cuts, with each element containing a pointer to a child node.

A *HiCuts* partitioning of two-dimensional space for the example classifier is shown in Figure 16.24 and its decision tree is shown in Figure 16.25. In this example, we set the thresholds such that each leaf node contains at most two rules and an internal node may contain at most four children. As mentioned earlier, the root node represents the entire two-dimensional space. To keep the discussions simple, this two-dimensional space is divided into four equal partitions using the cuts A, B, and C along the F_1 dimension. These cuts yield the child nodes S, T, U, and V of the root node in the decision tree. The root node stores the pointers to these nodes in an array since the cuts yield equal-sized regions. The region of node U contains the rule R_4 and similarly node V contains the rules R_5 and R_6. If node S were to be a leaf node, then it should contain the four rules R_1, R_2, R_7, and R_8, which is not allowed by the threshold. Hence we need to further subdivide this partition across dimension F_2 into two equal partitions using cut D as shown in Figure 16.25. The new nodes W and X contain two rules each, which is allowed by the threshold, and hence there is no need for further partitioning. Similarly, node T is further subdivided to contain the leaf nodes Y and Z.

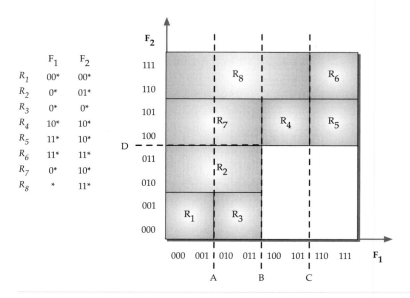

FIGURE 16.24   Geometric representation of *HiCuts*.

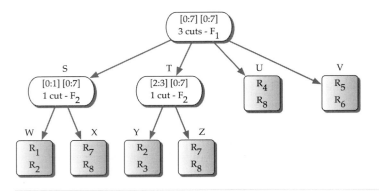

FIGURE 16.25   *HiCuts* decision tree for rules in Table 16.2.

Example 16.16   *Packet classification using HiCuts decision tree.*

An incoming packet with values for $F_1 = 000$ and $F_2 = 010$ is classified by first examining the root node (Figure 16.25). Based on the number of cuts across the F_1 dimension, we find that the value of $F_1 = 000$ falls on the range covered by the child stored at an array index of 0 and the search descends down to node S. At node S, the cut is across dimension F_2. We find that the $F_2 = 010$ value falls in the range of the child located at an array index of 0. It leads to node W and the rules R_1 and R_2. Each of these rules is examined one at a time and finally R_2 is declared to be the best matching rule. The classification of the packet required two memory accesses to reach the leaf node and another two memory accesses to linearly traverse the rules.   ▲

As can be seen, *HiCuts* decision trees are a class of generalized decision trees character-ized by an interleaved search of fields, range test for branching, and storing of multiple rules in the leaf node. The algorithm uses various heuristics to select the decision criteria at each node that minimize the height of the tree while controlling the amount of memory used.

- *Choice of number of cuts:* The number of cuts to be made is determined by a binary search algorithm suggested in [276]. The algorithm keeps doubling the number of cuts C until the storage space estimate for C becomes more than the predefined threshold. The storage cost for C cuts is computed by the sum of the rules assigned to each of the C cuts.

- *Choice of field:* Several approaches for selecting the field to cut have been suggested in [276]. One approach is to select a dimension that leads to the most uniform distribution of rules across nodes to create a balanced decision tree. Another simpler approach is to choose the field that has the largest number of distinct values. Hence the heuristic will select field F_2.

- *Maximize the reuse of child nodes:* With real-life classifiers, it is possible that many child nodes have identical sets of rules. In such cases storing each child separately leads to an increase in storage. Instead, a single child for each identical set is used, and the other child nodes with the same set of rules can point to this node.

The preprocessing algorithm uses the first two heuristics at every node in a repeated fashion until all the leaves have no more than *binth* rules. Due to the considerable preprocessing required, this scheme does not readily support incremental updates.

16.10.2 HyperCuts

In *HiCuts*, each internal node in the decision tree represents only the cuts in one dimension, irrespective of how many cuts are made in that dimension. Of course, the next logical question is why we cannot introduce cuts across multiple dimensions rather than a single dimension at each node. Intuitively, this can reduce the height of the decision tree, thereby reducing the number of memory accesses. A variant of *HiCuts*, called *HyperCuts*, outlined in [643], is a decision tree algorithm that takes advantage of this observation by introducing multiple cuts across multiple dimensions simultaneously. By forcing the cuts to create uniform regions, *HyperCuts* allows the efficient retrieval of pointers using array indexing.

Let us illustrate the *HyperCuts* algorithm using the example classifier in Table 16.2. For simplicity of discussion, consider the same set of cuts described in *HiCuts* and indicated in Figure 16.25, except that cut D is further extended along the F_1 dimension. The main difference is that cuts A, B, C, and D are applied simultaneously. This results in the new decision tree shown in Figure 16.26. As can be seen, the height of this tree is just one as compared to a height of three for the decision tree of *HiCuts*. However, the number of pointers from the root node has increased to six.

To speed up the search on an internal node, we can choose the same array indexing scheme as described in *HiCuts* except that it is extended to multiple dimensions as long as the width of the cuts is fixed in each dimension. In our example classifier, the root node is represented as a two-dimensional array of size $4 \times 2 = 8$ since the cuts across dimension F_1 divide it into four equal regions, and similarly a single cut across dimension F_2 divides it into two equal regions. However, only six of the pointers are used since two regions do not have any rules mapped.

Each internal node in the decision tree is associated with the information about the number of cuts across each dimension and the array of pointers for child nodes. Each time a packet arrives, the decision tree is traversed based on the information in the packet header to find a leaf node. Each leaf node stores a set of matching rules that is linearly traversed as in *HiCuts* to find the best matching rule that matches the packet.

A main drawback is that combining cuts in several dimensions can increase storage. For example, consider a *HiCuts* decision tree in which the root node has a single cut of F_1, which

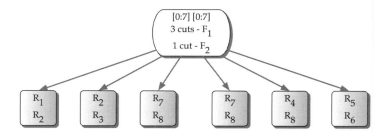

FIGURE 16.26   *HyperCuts decision tree for rules in Table 16.2.*

leads to two nodes A and B. Assume that A uses another five cuts on F_2 and B uses three cuts on F_2. The total amount of storage required for pointers is $2 + 6 + 4 = 12$. In *HyperCuts*, if we were to combine all the fields in a single node, the amount of storage required will be $2 \times 6 \times 4 = 48$ pointers. This is four times what is required for *HiCuts*.

In general, the decision tree approach provides a general framework characterized by various degrees of freedom. Using various choices for these degrees of freedom, there can be many potentially different algorithms with different performance characteristics and applicability in practice.

Performance experiments show that these decision tree approaches work well in practice except on classifiers that contain a large number of wildcards in one or more fields. Both *HiCuts* and *HyperCuts* do well for classifiers on edge routers since they have a simple structure using rules on source destination pairs. On classifiers in core routers, *HyperCuts* outperform *HiCuts* by a factor of 2 or 3 while utilizing an order of magnitude less memory than *HiCuts*.

16.11   Hardware-Based Solutions

So far, most of the solutions that we have discussed are algorithmic approaches. Due to the fast growth in Internet traffic, algorithms that can scale to millions of classification searches per second, millions of rules are needed. In the past few years, the industry has been increasingly employing TCAMs for performing packet classification since the searches can proceed in parallel irrespective of the number of rules. Unlike algorithmic approaches whose performance is highly dependent on the structure of the rules and classifiers, TCAMs perform an exhaustive search of rules in a memory cycle. The following section describes the use of TCAM in detail.

16.11.1   Ternary Content Addressable Memory (TCAM)

TCAM devices allow a parallel search over all rules in the classifiers. TCAMs differ from standard SRAM in their ability to store a "Don't Care" state in addition to a binary digit. Input keys are compared against every TCAM entry, thereby enabling them to retain single clock cycle lookups for arbitrary bit mask matches.

A TCAM stores each W-bit field as a $(value, bitmask)$ pair where *value* and *bitmask* are each W-bit numbers. For example, if $W = 4$, a prefix 01* is stored as the pair $(0100, 1100)$. An element matches a given input key k by checking if those bits of *value* for which the *bitmask* bit is 1 match those in the key k. To be precise, a key k matches a stored $(value, bitmask)$ pair if $k \ \& \ bitmask = value \ \& \ bitmask$, where & denotes a bitwise AND.

As we saw in Chapter 15, such a matching paradigm works well for prefix matching of IP addresses but is not well-suited to matching fields with ranges (e.g., port number range). The usual way to handle such a range specification is to replace each rule with several rules, each covering a portion of the desired range. This requires splitting the range into smaller ranges that can be expressed as $(value, bitmask)$ pairs. For example, the range 2–10 can be partitioned into a set of prefixes 001*, 01*, 100*, and 1010.

Use of TCAMs for classifying packets is similar to the use of TCAMs for address lookup, which is shown in Figure 15.27. A single TCAM stores all the rules. How can a single entry in the TCAM array store a rule consisting of multiple fields? Actually, a simple partitioning of the bits of TCAM entries can accommodate both fields. For instance, when TCAM entries are

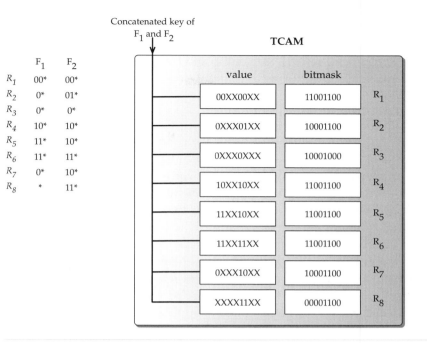

FIGURE 16.27   Representing the rules of Table 16.2 in TCAM.

8 bits wide, we can assign the first 4 bits for field F_1 and the remaining 4 bits for field F_2. For storing rule R_5, the F_1 prefix and the F_2 prefix are expanded to 10XX and 11XX, respectively, and their concatenation 11XX10XX is stored as the *value* field. Similarly, the bitmask for the F_1 prefix (1100) and the F_2 prefix (1100) are concatenated and stored as 11001100 in the *bitmask* field. The TCAM entries for all the rules of Table 16.2 are shown in Figure 16.27.

The order in which the prefixes are stored in TCAM corresponds to the priority ordering of rules. For an incoming packet, the interesting field values are extracted and passed as keys into the appropriate TCAM array. Each TCAM array compares the input key against every element in the array in parallel and outputs an N-bit bit vector. The N-bit bit vector indicates which rules match the input key and is passed through an N-bit priority encoder. The priority encoder indicates the address of the highest priority rule that matched. This address can be used to index into the RAM to find the action associated with the rule.

TCAMs are increasingly being deployed because of their simplicity and speed. They are becoming available in configurations with up to 18 Mbits, roughly half the size of the largest SRAMs. An 18-Mbit TCAM offers enough storage for up to 128K IPv4 rules, which is large enough to meet most near-term needs for general packet classification. Both faster and denser TCAMs can be expected in the near future. There are, however, some disadvantages to TCAMs—high cost per bit and high power consumption, as discussed earlier in Section 15.9.2. When used for packet classification, TCAMs have another disadvantage, *storage inefficiency*.

The storage inefficiency occurs due to the presence of ranges in some of the fields of the rules. These arbitrary ranges need to be converted into prefixes. In the worst case, a range covering a w-bit field may require $2(w - 1)$ prefixes. If a rule contains two fields with range specifications, it would require as many as $2(w-1)^2$ prefixes. Analysis of real-life classifiers indicates that the amount of TCAM storage required, in the worst case, is seven times the size of the classifier. Recent studies [181], [393], [654] and [760] seems to mitigate these issues.

16.12 Lessons Learned

In this chapter, so far, we have studied in detail algorithmic approaches and hardware-based approaches to packet classification. While theoretical worst-case complexities of these algorithms in terms of speed and memory consumption can be derived analytically, insights about how these algorithms perform in practice will be useful.

For the simpler case of classifying packets in two dimensions, trie-based approaches seem to work well. Among the trie-based approaches, the grid-of-tries approach provides fast search and scalable while consuming less memory.

As expected, the *RFC* algorithm outperforms other algorithmic approaches in search speed by effectively exploiting the characteristics of real-life classifiers, however, at the expense of memory. It does not support easy dynamic updates and its preprocessing time seems unpredictable. Hence, it might be better suited to applications in which memory is not an issue, dynamic changes are infrequent, and the search speed is more important.

The Lucent bit vector scheme provides comparable performance for medium-sized classifiers containing a few thousand rules and it consumes much less memory compared to *RFC*. Since it employs bitwise AND on all the bits representing the rules for identifying the matching ones, it does not scale well for larger classifiers. The aggregated bit vector scheme attempts to mitigate this by using summary bits; however, it suffers from false positives, leading to unpredictable average case search times. The dynamic updates for the Lucent bit vector scheme are slow as many bit vectors need to be reconstructed and typically the entire data structure is rebuilt.

For larger classifiers, the decision tree approaches seem to be attractive and provide a better trade-off between speed and memory. These approaches work well in practice with the exception of databases that contain large numbers of wildcards in one or more fields. However, the performance of decision tree approaches is governed by various parameters that are not characterized.

In general, the search time in the average case achieved by algorithmic approaches is based on exploiting certain assumptions and characteristics of the classifiers in practice. It is not clear whether these assumptions will continue to hold true in the future unless they are extensively validated. Hence, the worst-case performance of these algorithmic approaches might be much worse in reality.

However, solutions based on TCAM are independent of such assumptions because of an exhaustive search of the rules and at the same time provide the fastest search speed. Even though TCAMs have their own disadvantages of high power consumption and rule blowup due to port ranges, recent research in this direction seems to mitigate some of these issues.

Although this chapter discussed some of the principles employed in devising solutions for the packet classification algorithm, this is not the end. The reader is encouraged to think about other creative approaches that could lead to better algorithms.

16.13 Summary

The problem of packet classification is significant since it provides a way to discriminate packets and enables many differentiated services. We identified the key performance requirements of a packet classification algorithm as the number of memory accesses it needs and the amount of storage it occupies.

We started our discussion with naïve algorithms. We then studied algorithms for the simpler case of classifying packets using two fields. These algorithms are extensions of trie-based algorithms used for IP address lookup. After the discussion of these algorithms, we outlined various approaches for classifying a packet in arbitrary number of dimensions—divide and conquer, cross-producting, tuple space, and decision tree. We then described algorithms using each approach. Finally, we concluded the chapter with a discussion of hardware solutions using TCAMs.

Further Lookup

The survey papers by [277] and [686] provide a comprehensive coverage of the various algorithms for packet classification. In addition, two doctoral dissertations [273] and [658] summarize different classes of algorithms, and identify their asymptotic complexities and their pros and cons. Furthermore, an excellent overview of the various packet classification algorithms with sufficient insights about their performance and implementation can be found in [712, Chapter 12].

A study of the NSFNET backbone [141] indicates the possible use of caching to improve route lookup performance. Other studies [99], [455], [696] show that sufficient locality is present in the Internet traffic that can benefit by caching classification results. Li et al. [407] study how the caching architectures can be tuned, given limited silicon resources, to perform fast classifications. Use of approximate caches for packet classification using bloom filters is described in [120].

Set pruning tries and many optimizations to reduce its space occupancy are presented in [171], [572] and [734]. An *FIS*-tree data structure for packet classification in two fields is proposed in [220]. Another two-dimensional scheme that extends the quadtree for spatial data representation for packet classification is described in [101]. The grid-of-tries is described in [662].

There are a number of divide and conquer approaches; for example, the Lucent bit vector scheme [392], and the aggregate bit vector scheme [48]. A variation of the aggregation bit vector scheme that reduces the false positives while still keeping the benefit of bit vector aggregation is outlined in [406]. Cross-producting scheme is described in [662]. Gupta et al. [275] presented the *RFC* scheme that constructs partial crossproducts and combines them into the final crossproduct in multiple stages. Spitznagel [655] study a compressed representation for the tables used in *RFC* that trades memory accesses for space. Taylor et al. [687] presented the *Distributed Cross-Producting of Field Labels* that transforms the problem of aggregating the results from independent field searches into a distributed set membership query.

The tuple space approaches were presented in [660]. A detailed discussion about basic tuple search and pruned tuple search and an analysis of their performance relative to other approaches can be found in [658]. Srinivasan et al. [659] presented an *Entry-Pruned Tuple Search* algorithm that optimizes tuple space pruning by not maintaining separate lookup data structures for each field. Instead, the pruning information is maintained as a bitmap of tuples, which is then associated with the matches in the tuples.

The use of decision tree approaches for packet classification was described in [276], [745]. *Hypercuts*, presented in [643], allows the cuts to be on multiple dimensions and, in addition, proposes several storage-related optimizations.

TCAMs are used for packet classification. Spitznagel et al. [654] proposed the use of partitioned TCAMs to reduce power consumption and extension for storing arbitrary port ranges in TCAMs and also suggested how to partition TCAMs to dramatically reduce power consumption and propose extensions that eliminate the storage inefficiency for an arbitrary range of filters.

Exercises

16.1. Review questions:

(a) In your own words, define the packet classification problem.

(b) What are the different type of matches allowed in packet classification rules?

(c) Why is backtracking required in hierarchical tries?

(d) What is the main disadvantage of a set pruning trie?

(e) What is time complexity for classifying a packet in grid-of-tries?

(f) What is space complexity of grid-of-tries? Compare it with set pruning tries.

(g) How do the Lucent bit vector scheme and aggregated bit vector scheme differ?

(h) Can you explain the disadvantage of the cross-producting scheme?

(i) What are the disadvantages of using TCAM?

16.2. A router performs a route lookup followed by classification. If the route lookup operation takes 15 nanosec, how much time is available for packet classification to sustain a data rate of 40 Gbps with an average packet size of 100 bytes?

16.3. Can you give an example of a three-field classifier that shows the failure of grid-of-tries in finding the best matching rule?

TABLE 16.3 A two-field classifier.

Rule	F_1	F_2
R_1	0*	10*
R_2	0*	01*
R_3	0*	1*
R_4	00*	1*
R_5	00*	11*
R_6	10*	1*
R_7	11*	00*
R_8	*	00*

16.4. For the rules shown in Table 16.3, construct a hierarchical trie. What is the best matching rule for a packet with $F_1 = 0011$ and $F_2 = 0011$? How many memory accesses are required?

16.5. Draw a set pruning trie for the rules shown in Table 16.3. How many memory accesses are required to classify a packet with $F_1 = 0011$ and $F_2 = 0011$? Compare it with the hierarchical trie.

16.6. In the set pruning trie of Exercise 16.5, what is the maximum number of accesses required to classify a packet?

16.7. Construct a grid-of-tries using the rules in Table 16.3. Describe the steps involved in classifying the packet with $F_1 = 0011$ and $F_2 = 0011$?

16.8. Construct a Lucent bit vector data structure for the rules in Table 16.3. Use a bitmap of size 8 bits for representing the rules. For classifying a packet with $F_1 = 0011$ and $F_2 = 1111$, identify how many memory accesses will be required.

16.9. Calculate the overall memory required for the Lucent bit vector data structure in Exercise 16.8. Assume a size of 4 bytes for a trie pointer and 8 bits for each bitmap.

16.10. Construct a crossproduct table for the rules shown in Table 16.3.

16.11. Construct a table that maps rules to tuples, followed by a tuple space hash table for the rules shown in Table 16.3.

16.12. Draw the geometric view of the rules in Table 16.3 and identify the number of distinct regions.

16.13. Construct a *HiCuts* decision tree for the rules shown in Table 16.3.

16.14. Given an 8-bit wide TCAM, represent the rules specified in Table 16.3.

Part V: Toward Next Generation Routing

In this part, we bring together three routing paradigms: packet switching, circuit switching, and transport routing. Transport routing addresses the paradigm of routing in which a high-bandwidth entity such as OC-3 can be set up on a semi-permanent basis that can possibly have a considerable lag time for set up; with the new generation of switching equipment, it would be possible to set up transport routing and provisioning very quickly. In general, in the next generation routing, all three paradigms are likely to be juxtaposed together where all three operate on different time scales.

In Chapter 17, we first start with a discussion on request arrivals, quality of service, and the time unit on decision on routing for different service needs to show where different routing paradigms fit in. We then present quality of service routing for service classes with one or more attributes. We show the relation between QoS routing, dynamic call routing in PSTN, and widest path routing.

We present MPLS and GMPLS in Chapter 18. In both cases, the connection set up is typically for a virtual path that would serve as a bearer link to service networks (such as IP or PSTN) that requires such services. Along with this chapter, the reader may want to read Chapter 24 to understand the basic premise of transport routing as a general framework.

In Chapter 19, we discuss traffic engineering for MPLS networks. It may be noted that MPLS can be bearer for IP services where the unit of information is at the packet level; however, it is also possible for MPLS to be a bearer for virtual private networking, in which paths may be on a semi-permanent basis. Based on the deployment mode of an MPLS network, path set up and configuration can also be done on a short notice using RSVP-TE signaling.

In Chapter 20, we present VoIP routing in the IP-PSTN environment. This chapter brings different issues from packet and call routing into one place; in doing so, addressing issues from IP addressing, E.164 addressing, and SS7 addressing are considered together along with ISUP and SIP signaling issues and interworking.

17

Quality of Service Routing

The more precisely the position is determined, the less precisely the momentum is known in this instant, and vice versa.

Werner Heisenberg

Reading Guideline

Quality of Service routing includes aspects from both the circuit-switched world and the packet-switched world. Thus, some knowledge of circuit switching is helpful. In addition, understanding the material on dynamic call routing in the telephone network (Chapter 10) and its traffic engineering (Chapter 11), along with the link state routing protocol (Chapter 3) and the shortest and widest path routing algorithms (Chapter 2), is helpful in getting the most out of this chapter.

Quality of Service (QoS) is an important issue in any communication network; typically, this can be viewed from the perception of service quality. Eventually any service perception needs to be mapped to network routing, especially since QoS guarantee is required for a particular service class.

In this chapter, we discuss what QoS routing means and how different routing algorithms covered in this book may be extended to fit the QoS routing framework. Finally, we present a representative set of numerical studies with which we can understand the implications of different routing schemes and roles played by different network controls.

17.1 Background

We start with a brief background on QoS and QoS routing.

QUALITY OF SERVICE

To discuss *Quality of Service routing*, we first need to understand what *Quality of Service* means. Consider a generic request arrival to a network; if this request has certain resource requirements that it explicitly announces to the network at the time of arrival, then QoS refers to the network's ability to meet the resource guarantees for this request.

To understand QoS, we will first consider a network link; no routing is considered at this point. Assume that a request arrives at this network link for a 1-Mbps constant data rate. If the network link had bandwidth available that is more than 1 Mbps, then it can certainly accommodate this request. Thus, the arriving request received the specified QoS. Implicit in this is that the QoS will be continually met as long as this request is active; in other words, for the *duration* of the request, the QoS is met.

Suppose that the network link at the instant of the request arrival has less available bandwidth than the requested bandwidth. In this case, the request cannot be served. When there are many arriving requests requiring resource guarantees and the network link cannot accommodate them, another aspect related to QoS emerges. This aspect of QoS considers that arriving requests usually receive the service guarantee requested with an acceptable probability of not being turned away; in other words, blocking should not be high. That is, the blocking probability of arriving requests is another important consideration in regard to QoS. When we consider from this viewpoint, it is easy to see that traffic engineering and capacity expansion also play crucial parts in regard to QoS since if the network is not engineered with a reasonable capacity level, the likelihood of a request facing blocking would be high. Thus, blocking probability is an important factor in the perception of QoS and is traditionally known as *grade of service* (GoS).

In general, the term QoS is used much more broadly than its use in this chapter in the context of QoS routing. For example, "a network meets QoS" can be interpreted as meeting delay requirements through a network, not necessarily for a specific request.

QOS ROUTING

Consider now a network instead of just a link. Then for an arriving request that requires guaranteed resources, the network would need to decide what resources it has in its *different* links and paths so that the request can be accommodated. Thus, *QoS routing* refers to a network's ability to accommodate a QoS request by determining a *path* through the network that meets the QoS guarantee. Furthermore, an implicit understanding is that the network's performance is also optimized. In this sense, QoS routing cannot be completely decoupled from traffic engineering.

QoS ROUTING CLASSIFICATION

What are the types of resource guarantees an arriving request might be interested in? Typically, they are bandwidth guarantee, delay bound, delay jitter bound, and acceptable packet loss. We have already described a bandwidth guarantee. *Delay bound* refers to end-to-end delay being bounded. Jitter requires a bit of explanation. In a packet-based network, packets that are generated at equal spacing from one end may not arrive at the destination with the same spacing; this is because of factors such as delay due to scheduling and packet processing at intermediate routers, interaction of many flows, and so on. In real-time interactive applications such as voice or video, the interpacket arrival times for a call are equally spaced when generated, but may arrive at the destination at uneven time spacing; thus, interpacket delay is known as *jitter*. *Packet loss* refers to the probability of a packet being lost along the path from origin to destination.

Consideration of these four factors would, however, depend on whether the network is a circuit-based network or a packet-based network. To discuss this aspect and the critical elements related to QoS routing, we also need to consider time granularity in regard to an arriving request. By considering three time-related factors, arrival frequency, lead time for set up, and the duration of a session/connection, we broadly classify requests into three types as listed in Table 17.1. There are very specific outcomes of these classifications. In Type A, the network technology is either packet-switched or circuit-switched where circuit-switched networks require bandwidth guarantee while packet-switched networks may have one or all of the requirements: bandwidth guarantee, delay bound, jitter bound, and acceptable packet loss. However, Type B is generally circuit-oriented where a permanent or semipermanent bandwidth guarantee is the primary requirement; there is very little about on-demand switching. Routing for the Type B classification is traditionally referred to as *circuit routing* (for example, see [596, p. 136]); in recent literature, circuit routing is commonly known as *transport network routing* (covered in Chapter 24). Between Type A and Type B, there is another form of routing where some overlap of time granularity is possible. We classify this type that has overlapping regions as Type C; for example, routing for this type of service can be accomplished in MPLS networks and will be discussed later in Chapter 19.

Of these classifications, QoS routing arises for a Type A classification. It is thus helpful to consider a taxonomy for QoS routing to understand the relationship between networking paradigms and QoS factors (see Figure 17.1). In figure, we have included an identifier

TABLE 17.1 Service request type classification.

Type	Average arrival frequency	Lead time for setup	Duration of session
Type A	Subsecond/seconds time frame	A few seconds	Minutes
Type B	Day/week time frame	Weeks	Months to years
Type C	Multiple times a day	Minutes	Minutes to hours

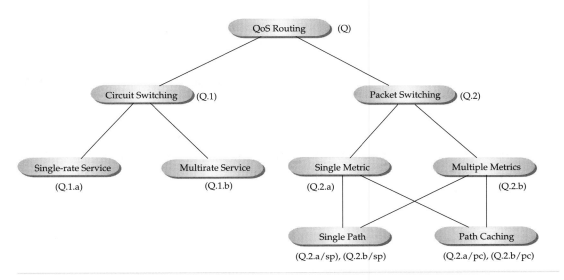

FIGURE 17.1   QoS routing taxonomy.

in parentheses for ease of illustration. First note that classification Q.1.a refers to routing in the current circuit-switched telephone network. You may note that we have already covered hierarchical and dynamic call routing in a telephone network in Chapter 10. An important point to note is that *both* hierarchical and all variations of dynamic call routing fall under Q.1.a in terms of meeting QoS. It may be noted that old hierarchical call routing meets the bandwidth guarantee of a new request if admitted; however, hierarchical call routing is not as flexible as dynamic call routing schemes and requires more bandwidth to provide the same level of service. This then helps in seeing that traffic engineering efficiency is an implicit requirement of QoS routing, a primary reason why dynamic call routing was pursued in the telephone network. A broader point is that QoS routing can be accomplished by different routing *schemes*—the drivers for QoS routing are developing routing schemes that address issues such as performance benefit, cost, routing stability, management, and so on. Thus, in general, dynamic or adaptive routing is preferred over fixed routing. Classification Q.1.a is a very important area in network routing. Besides circuit-switched voice, many problems in optical routing also fall under classification Q.1.a; this will be discussed later in Chapter 25.

Classification Q.1.b is an extension of Q.1.a. Multirate, multiservice circuit-switched QoS routing refers to the case in which there are more than one service classes and an arriving request for each service class has a different bandwidth requirement as opposed to Q.1.a, where all arriving requests have the same bandwidth requirement, for example, the current wired telephone network where per-request bandwidth is 64 Kbps. In case of Q.1.b, service classes are rigid and the bandwidth requirement of a request in each class is an integral multiple of the *base bandwidth rate*. If the base bandwidth rate in the circuit-switched voice network is 64 Kbps, then a switched video service can be defined as, say, 384 Kbps, which is then six times the base bandwidth rate. It may be noted that among the dynamic call routing schemes discussed earlier in Chapter 10, real-time network routing (RTNR), discussed in Section 10.6, has been deployed to handle multiple classes of services.

For the packet-switched branch of QoS routing, there are two aspects to consider: single attribute or multiple attributes. By single attribute, we mean only a single criterion, such as the bandwidth requirement, is used as a metric for a request that is considered for QoS routing. By multiple attributes, we mean that more than one factor, such as bandwidth and delay, is being considered for QoS routing. Note that we do not distinguish here by rates as we have done with Q.1.a and Q.1.b, although theoretically it is possible to discuss single rate and multiple rate. The reason this is grouped together is that packet-switched networks are usually not designed with a single bandwidth rate in mind—any arbitrary bandwidth rate is generally usable due to the packet switching nature. It is, however, indeed possible to deploy a private packet network, for example, a private voice over IP (VoIP) packet network, where all voice calls have the same data rate. Thus, Q.2.a has some aspects of Q.1.b, with the additional flexibility of arbitrary data rate of a request. For classification Q.2.b, multiple criteria are required to handle the decision-making process for an arriving request. This will be discussed in detail later in this chapter.

For both Q.2.a and Q.2.b, there are two possibilities in terms of path consideration: either a single path is considered, or paths are cached for alternate paths consideration. Note that for Q.1.a and Q.1.b, it has been common to consider path caching; in fact, a single path is rarely considered and is not shown in this classification.

DEFINING A REQUEST AND ITS REQUIREMENT

You may note that so far we have not defined a *request*. Typically, in a circuit-switching context, a request is labeled as a *call*; in a packet-switching context, especially in IP networking, a QoS request is labeled as a *flow*, while terms such as *SIP call* or *VoIP call* are also often used.[1] For a discussion on SIP, refer to Chapter 20. Note that the usage of the term *flow* here is not to be confused with *network flow* or *link flow* described earlier in Chapter 4.

When a call request arrives, there is a call setup phase that can typically perform functions such as route determination, signaling along the path to the destination, and QoS checking before the call is setup; in essence, a call request must always face a call setup time delay before it can be connected—this is also known as postdial delay; certainly, this should be minimized. For the services that require a QoS guarantee, the call setup phase needs to ensure that the QoS guarantee can be provided for the entire duration of the call; otherwise, the call request is denied by the network.

GENERAL OBSERVATIONS

It is quite possible that a network may not have the *functionality* to guarantee that it can meet QoS for an arriving request, but yet has the resources to meet the request. An IP network without integrated services functionality falls into this category. For example, a VoIP call can receive QoS in an IP network without the network explicitly having the ability to provide a guarantee at the time of request arrival. This can be possible, for example, if the network is engineered properly, or overprovisioned. In general, overprovisioning is not desirable since, after all, a network does cost real money in terms of capacity cost, switching cost, and so on.

---

[1]There is yet another terminology in the networking literature: *call flows*. This term refers to the flow or sequence diagram of messages in regard to establishing or tearing down a call over an SS7 network or in a SIP environment or when translation is required at a gateway going from SS7 to SIP, or vice versa.

Finally, much like best-effort traffic services, QoS routing can also have two components: intradomain and interdomain. Most of this chapter is focused on *intradomain* QoS routing. We will briefly discuss interdomain QoS routing at the end.

17.2 QoS Attributes

In the previous section, we mentioned the following factors in terms of attributes: *residual* bandwidth, delay, jitter, and packet loss. Note that any of these attributes are applicable under classification Q.2, while bandwidth is the only one applicable for classification Q.1. We will now discuss how to classify these attributes in terms of metrics.

Suppose that an arriving request has requirements for bandwidth, delay, jitter, and packet loss identified by $\bar{b}, \bar{\tau}, \bar{\zeta}$, and $\bar{L}$, respectively. The important question is how are measures for these factors accumulated along a path in terms of satisfying the guaranteed requirement of an arriving call? To understand this, we will consider a path that is made up of three links numbered 1, 2, and 3, and current residual bandwidth, delay, jitter, and packet loss measures for link i as b_i, τ_i, ζ_i, and L_i $(i = 1, 2, 3)$, respectively. We can then list the path measures as follows:

Type	Path measure	Requirement
Bandwidth	$\min\{b_1, b_2, b_3\}$	$\geq \bar{b}$
Delay	$\tau_1 + \tau_2 + \tau_3$	$\leq \bar{\tau}$
Jitter	$\zeta_1 + \zeta_2 + \zeta_3$	$\leq \bar{\zeta}$
Packet loss	$1 - (1 - L_1)(1 - L_2)(1 - L_3)$	$\leq \bar{L}$

You can see that the packet loss measure is a nonadditive multiplicative measure; however, it can be looked at from another angle. If L_i $(i = 1, 2, 3)$ is very close to zero, which is typically the case for packet loss, again due to traffic engineering requirements, the expression for path measure can be approximated as follows (see Appendix B.8):

$$1 - (1 - L_1)(1 - L_2)(1 - L_3) \approx L_1 + L_2 + L_3. \tag{17.2.1}$$

Thus, the packet loss measure becomes an additive measure. We can then classify the different attributes into two groups in terms of metric properties:

> *Additive:* Delay, jitter, packet loss
> *Nonadditive (concave):* Bandwidth

Broadly, this means that from a routing computation point of view, delay, jitter, and packet loss metrics can be classified under shortest path routing while the bandwidth requirement metric falls under widest path routing. It may be noted that a buffer requirement at routers along a path for an arriving request requiring a QoS guarantee is another possible metric that falls under the nonadditive concave property; however, unlike the rest of the metrics discussed so far, a buffer requirement is checked as the call setup signaling message is propagated along the path chosen, rather than being communicated through a link state advertisement.

To summarize, for classification Q.2, both additive and nonadditive concave metrics are possible, while for classification Q.1 only nonadditive concave is appropriate. In the next section, we will discuss adaptations of shortest path routing and widest path routing for a request requiring a QoS guarantee.

17.3 Adapting Shortest Path and Widest Path Routing: A Basic Framework

Out of different attributes classified into two categories, we will use one metric each from additive and nonadditive (concave) metric properties for our discussion here. Specifically, we will use delay for the additive property and bandwidth requirement for the nonadditive property. We assume here the reader is familiar with shortest path routing and widest path routing described earlier in Chapter 2. You may note that the discussion in this section is applicable only to classification Q.2.

From our discussion in Part I of this book, we know that applicability of a particular routing algorithm for a packet-switched network depends on whether the network is running a distance vector protocol or a link state protocol. While the basic idea of shortest path or widest path routing would work under both these protocol concepts, we will assume here that a link state protocol framework is used since most well-known intradomain routing protocol frameworks are link state based.

17.3.1 Single Attribute

We first consider that requests have a single additive metric requirement in terms of delay attribute. A simple way to adapt the shortest path routing algorithm paradigm here is by using delay as the link cost metric. Suppose a request arrives with the delay requirement no greater than $\bar{\tau}$.

> For an arriving request requiring a guaranteed delay requirement of $\bar{\tau}$, do the following:
> Compute the shortest delay using the shortest path first algorithm (Algorithm 2.4);
> if the result is less than $\bar{\tau}$, then admit the request; otherwise, deny the request.

Note that the request arrives for a particular destination. Thus, unlike the standard shortest path first (SPF) algorithm, here the shortest path computation must be computed only for the specific destination of a request. Consider again Algorithm 2.4 in Chapter 2; in step 3, once a new node k is identified with the minimum cost path, it can be checked whether this k is the destination of the request; if so, the algorithm can stop. At this point, this delay cost is then compared against the arriving request's delay requirement; if met, the request is accepted, otherwise it is denied.

What if the single metric is in terms of the bandwidth requirement of a request? This scenario is similar to the delay-based scenario. Suppose that an arriving request has a bandwidth requirement of $\bar{b}$. Then, we can use the following rule:

> For an arriving request with a guaranteed bandwidth requirement of $\bar{b}$, do the following:
> Compute the widest path using Algorithm 2.8 for the specific destination; if this value is higher than $\bar{b}$, then admit the request; otherwise, deny the request.

In many instances, it is desirable to obtain the widest path with the least number of hops for the path. Although this is sometimes referred to as the *shortest-widest* path, it is not a good name since shortest does not indicate the context in which this is meant. Thus, we will refer to it as the *least-hop-widest* path. How do we find the widest path with the least number of hops? Consider again Algorithm 2.8; In step 3 of this algorithm, k in S' with the maximum residual bandwidth is determined. Instead of storing just one k, the list of nodes where the maximum residual bandwidth is attained is determined. If this list happens to have more than one element, then k is chosen so that it is the least number of hops from source node i. In essence, this means that if there are multiple paths with maximum residual bandwidth, choose the one with the least number of hops; if there are still such multiple paths, one is randomly selected. In the same manner, a *least-hop-minimum delay* path can be determined when a delay metric is used.

17.3.2 Multiple Attributes

In this case, consider an arriving request specifying that both the delay as well as the bandwidth requirement must be satisfied. This can be addressed from the point of view of which factor is to be considered the dominant factor: delay or bandwidth; this, however, depends on which is found: a bandwidth feasible path while the delay is minimized, or a delay feasible path while maximizing available bandwidth.

Again, we can adapt the widest path and shortest path routing framework. To determine the minimum delay path that satisfies the bandwidth requirement of a request, we can initialize any link that does not meet the bandwidth requirement temporarily as a link with infinite delay; this method of considering a nonadditive metric requirement with an additive shortest path computation is generally known as *constrained shortest path routing*. Instead, if we were to determine a maximum residual bandwidth, i.e., the widest path while meeting the delay requirement, we can initialize any link that does not meet the delay requirement by temporarily setting the residual link bandwidth to zero; this form can be classified as *constrained widest path routing*. Note that for a constrained shortest path, the constraint is on bandwidth, while for a constrained widest path, the constraint is on delay. For source node i and destination node v, we present both routing algorithms in Algorithm 17.1 and Algorithm 17.2 for completeness. The notations are summarized in Table 17.2. Chapter 2 may also be consulted for comparison.

TABLE 17.2   Notation for QoS routing.

Notation	Remark
i	Source node
v	Destination node
$\mathcal{N}$	List of all nodes
$\mathcal{N}_k$	List of neighboring nodes of k
S	List of nodes considered so far
S'	List of nodes yet to be considered
τ_{ij}	Link delay on link i-j (set to ∞ if the link does exist, or not to be considered)
T_{ij}	Delay from node i to node j
b_{ij}	Residual bandwidth on link i-j (set to 0 if the link does exist, or not to be considered)
B_{ij}	Bandwidth available from node i to node j

ALGORITHM 17.1   QoS minimum delay path with bandwidth feasibility

$\mathcal{S} = \{i\}$ // permanent list; start with source node i
$\mathcal{S}' = \mathcal{N} \setminus \{i\}$ // tentative list (of the rest of the nodes)
for (j in $\mathcal{S}'$) do
    // check if i-j directly connected and link has required bandwidth $\bar{b}$
    if ($\tau_{ij} < \infty$ and $b_{ij} \geq \bar{b}$) then
        $T_{ij} = \tau_{ij}$ // note the delay cost
    else
        $\tau_{ij} = \infty; T_{ij} = \infty$ // mark temporarily as unavailable
    endif
endfor
while ($\mathcal{S}'$ is not empty) do   // while tentative list is not empty
    $Ttemp = \infty$ // find minimum-delay neighbor k
    for (m in $\mathcal{S}'$) do
        if ($T_{im} < Ttemp$) then
            $Ttemp = T_{im}; k = m$
        endif
    endfor
    if ($T_{ik} > \bar{\tau}$) then   // if minimum delay is higher than delay tolerance
        'No feasible path exists; request denied'
        exit
    endif
    if ($k == v$) then exit   // destination v found, done
    $\mathcal{S} = \mathcal{S} \cup \{k\}$ // add to permanent list
    $\mathcal{S}' = \mathcal{S}' \setminus \{k\}$ // delete from tentative list
    for (j in $\mathcal{N}_k \cap \mathcal{S}'$) do
        if ($T_{ij} > T_{ik} + \tau_{kj}$ and $b_{kj} > \bar{b}$) then   // if delay is less via k
            $T_{ij} = T_{ik} + \tau_{kj}$
        endif
    endfor
endwhile
if ($T_{iv} \leq \bar{\tau}$) then   // final check, if the path meets delay requirement
    'Request accepted'
else
    'No feasible path exists; Request denied'
endif

17.3.3 Additional Consideration

We next consider a general question: can we provide QoS routing in a packet environment where buffer guarantee at routers is also required? For this, assume that the packet network is an integrated services environment. For a request requiring bandwidth guarantee on demand, we need to consider also whether the router's scheduling algorithm can guarantee requests in terms of buffering, in addition to bandwidth guarantee on links. This brings in the issue of scheduling with routing. It has been shown that this combined problem can be addressed with a polynomial time algorithm that factors in capacity and constrained shortest path [766].

ALGORITHM 17.2   QoS widest path with delay feasibility

$\mathcal{S} = \{i\}$ // permanent list; start with source node i
$\mathcal{S}' = \mathcal{N} \setminus \{i\}$ // tentative list (of the rest of the nodes)
for (j in $\mathcal{S}'$) do
 // if i-j directly connected and link has required bandwidth $\overline{b}$
 if ($b_{ij} > \overline{b}$ and $\tau_{ij} < \infty$) then
  $B_{ij} = b_{ij}$; $T_{ij} = \tau_{ij}$
 else
  $b_{ij} = 0$; $B_{ij} = 0$; $\tau_{ij} = \infty$; $T_{ij} = \infty$ // mark temporarily as unavailable
 endif
endfor
while ($\mathcal{S}'$ is not empty) do   // while tentative list is not empty
 $Btemp = 0$ // find neighbor k with maximum bandwidth
 for (m in $\mathcal{S}'$) do
  if ($B_{im} > Btemp$) then
   $Btemp = B_{im}$; $k = m$
  endif
 endfor
 if ($B_{ik} < \overline{b}$) then   // bandwidth is higher than the request tolerance
  No feasible bandwidth path exists; request denied
  exit
 endif
 if ($k == v$) then exit   // destination v is found; done
 $\mathcal{S} = \mathcal{S} \cup \{k\}$ // add to permanent list
 $\mathcal{S}' = \mathcal{S}' \setminus \{k\}$ // drop from tentative list
 for (j in $\mathcal{N}_k \cap \mathcal{S}'$) do   // path has higher bandwidth
  if ($B_{ij} < \min\{B_{ik}, b_{kj}\}$) then
   $B_{ij} = \min\{B_{ik}, b_{kj}\}$
   $T_{ij} = T_{ik} + \tau_{kj}$
  endif
 endfor
endwhile
if ($B_{iv} \geq \overline{b}$) then   // final check; if path meets bandwidth requirement
 'Request accepted'
else
 'No feasible path exists; Request denied'
endif
end *procedure*

17.4 Update Frequency, Information Inaccuracy, and Impact on Routing

In the previous section, we have provided the computational framework for QoS routing for classification Q.2 by considering single or multiple attributes. What is missing is how often attribute information is obtained and/or when the computation is performed. To discuss these important aspects, we will again assume that a link state framework is used.

Ideally, it appears that if a node knows the state of each link in terms of the applicable attributes (either single or multiple) *instantaneously*, it can then invoke routing computation. There are, however, practical limitations on this utopian view:

- First, an instantaneous update is almost impossible in a real network; very frequent updates can lead to excessive information exchange, which can overload a network. In fact, it has now become a common practice in many routing protocols to include a hold-down time to assert that no updating of information is allowed that is more frequent than the hold-down time. Also note that if a particular link state is advertised too frequently due to a legitimate change in the link state status, some form of dampening is still applied by a receiving node to avoid having an undesirable performance consequence, and before flooding to its neighboring node; as an example of a similar situation, see Section 8.9 addressed for the border gateway protocol (BGP).

- Second, there are two possibilities in regard to routing computation: (1) perform the computation periodically, or (2) perform it on demand for every arriving request. The second option is usually avoided since an important requirement in QoS routing services is that the call setup time, also known as postdial delay, for an arriving request is as small as possible. There is an important lesson to be learned here from RTNR (refer to Section 10.6). In an almost fully mesh network environment with a separate signaling (SS7) network for link state message exchanges, RTNR was initially intended to be deployed with per call computation in mind; in actuality, the computation is based on the information queried for the *previous* call in order to avoid increasing postdial delay to an undesirable level. For a general packet network, performing routing computation on demand for each arriving request can be taxing on the CPU load of the node—thus, this is also not desirable.

- Finally, it is not hard to realize that if the link state information obtained at a node is delayed due to periodic/asynchronous update or dampening, i.e., the link state information is somewhat stale or inaccurate. Due to such inaccurate information, it is questionable if it is worth doing a per-call routing computation.

To summarize, for QoS routing, it is more appropriate to perform a routing computation periodically than on a per-call basis and build a routing table. Taking this entire scenario into account, the arrival of link state information and the timing of the routing computation are depicted in Figure 17.2. It may be noted that due to the periodic computation framework, instead of executing a constrained shortest path or constrained widest path on a per-pair basis, it can be performed on a source to all destination basis, albeit with the option that for a specific pair the computation can be triggered if needed. In any case, it is important to note that if there is a network link failure, usually link state flooding and routing computation are triggered immediately so that changes can be accommodated by each node.

There is, however, an important consequence of periodic/update and periodic routing table computation. Suppose that the routing is hop-by-hop and each node has only one entry for each destination identified by the next hop. When an actual request arrives, there may not be enough resources along the path (dictated by the routing table) to establish the call. Thus, this request is denied entry, which then affects the overall call-blocking probability. Note that just being locked into one path during two consecutive routing computations does not necessarily mean that all arrivals will be blocked during this window; it is important to note that during this time window, some exiting calls might be overreleasing resources that can be used by newly arrived calls (for example, refer to Figure 11.1). In any case, to maintain the GoS aspect of QoS, there are two possibilities: (1) updates must be done frequently enough so that the newly obtained path does not block too many calls, or (2) the network is engineered with enough capacity/resources so that the overall blocking effect is maintained at an accept-

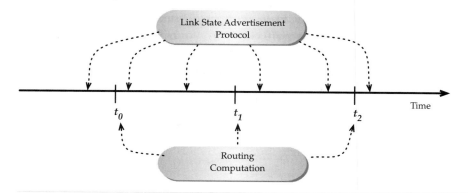

FIGURE 17.2   Time of link state advertisement and routing computing for QoS routing.

able level. The first option belongs to traffic engineering while the second option belongs to capacity expansion. Note that it is also important to maintain the GoS aspect of QoS to avoid excessive user-level retry in case calls are blocked.

Since an important goal of QoS routing is to provide good traffic engineering, we may ask the following question: can we consider more than one path from a source to a destination? This will partly depend on whether the network is capable of providing hop-by-hop routing and/or source routing. In the case of hop-by-hop routing, the only option for multiple paths is if there are two paths of equal cost, i.e., equal-cost multipath (ECMP). It is difficult to find multiple equal-cost paths in a constrained-based routing environment. In a source routing environment, multiple paths can be cached ahead of time, which then leads to the possibility of *alternate routing* options.

17.5 Lessons from Dynamic Call Routing in the Telephone Network

There has been extensive experience with alternate routing for dynamic call routing for the telephone network; see Chapter 10 for details and also Section 11.6. First, we summarize the typical network environment for dynamic call routing in telephone networks:

- The network is fully mesh, or nearly fully mesh.

- Calls attempt a direct link path first (if available); then an alternate path is attempted; alternate paths are made of at most two links.

- The path cost is nonadditive, concave.

- The alternate paths to be considered and the number of such paths to be cached depend on specific routing schemes.

- Call setup can be based on progress call control or originating call control.

- In the presence of originating call control, a call crankback can be performed to try another path, if such a path is listed in the routing table.

- Routing schemes use a link state framework.

- Link state update, setup messages, and crankback messages are carried through out-of-band signaling, for example, using an SS7 network.

- The main performance measure is minimization of call-blocking probability, which can dictate the choice of a routing scheme. However, factors such as messages generated due to call setup, crankback, and link state updates can also be deciding factors.

   There are key lessons learned from such a dynamic call routing environment:

- In general, a dynamic call routing scheme increases throughput, but has a metastability problem beyond a certain load to capacity ratio in the network.

- A trunk reservation feature is used to protect a direct link from being excessively used by alternate routed calls to avoid metastable behavior. In essence, trunk reservation works as a link-level admission control. An important consequence that sounds counterintuitive is that a network may not accept a call even if it has capacity under certain conditions.

- For effective traffic engineering, especially under overload conditions, several control measures such as dynamic overload control, call gapping, and hard to reach may need to be invoked.

   To reiterate, dynamic call routing in a telephone network is operating for *single-rate* homogeneous service—all are voice calls requiring the same amount of bandwidth for the duration of the call. The first question then is what changes in regard to a heterogeneous service environment where arriving calls require *differing* bandwidth. This is discussed in the next section.

17.6 Heterogeneous Service, Single-Link Case

To understand an important difference going from a single-rate service case to a multiple-rate service case, we illustrate a performance scenario that has important implications for QoS routing. Note that this analysis requires some understanding of offered load in Erlangs (Erls) and call blocking, discussed earlier in Section 11.2 using just a single link without the presence of routing. The results discussed below are summarized in Table 17.3.

   Consider a service that requires 1 Mbps of bandwidth during the duration of the call. Assume that the link capacity is 50 Mbps; thus, this link can accommodate at most 50 such calls simultaneously, that is, the effective capacity of the link is 50. Assume that the call arrival pattern is Poisson with an average call arrival rate at 0.38 per second, and that the average duration of a call is 100 seconds. Using Eq. (11.2.2), we can determine that the offered load is $0.38 \times 100 = 38$ *Erls*. Furthermore, using the Erlang-B loss formula Eq. (11.2.3), we can find that 38 *Erls* offered to a link with 50 units of capacity results in a call-blocking probability of 1%. Since most networking environments would like to maintain a QoS performance requirement for call blocking below 1% probability, we can see that users will receive acceptable QoS in this case. Note that to meet QoS, there are two issues that need to be addressed: (1) each call must receive a bandwidth guarantee of 1 Mbps, if admitted, and (2) the call acceptance probability is below 1% so that users perceive that they are almost always going to get a connection whenever they try.

TABLE 17.3   Call blocking for different services under various scenarios.

Link capacity (Mbps)	a_{low} (Erls)	a_{high} (Erls)	m_{low} (Mbps)	m_{high} (Mbps)	Reservation (Yes/No)	$\mathcal{B}_{low}$	$\mathcal{B}_{high}$	$\mathcal{W}_{composite}$
50	38.0	—	1	—		1.03%	—	1.03%
50	19.0	1.9	1	10	No	0.21%	25.11%	12.66%
85	19.0	1.9	1	10	No	0.05%	0.98%	0.52%
85	22.8	1.9	1	10	No	0.08%	1.56%	0.75%
85	22.8	1.9	1	10	Yes	1.41%	0.94%	1.20%
85	22.8	1.9	1	10	Yes, *Prob* = 0.27	1.11%	1.10%	1.11%

Next, consider the situation where we allow a new 10-Mbps traffic stream on the *same* 50-Mbps link to be shared with the basic 1-Mbps traffic stream. We start by splitting the 38 *Erls* of offered load equally, i.e., 19 *Erls* to the 1-Mbps traffic class and 19 *Erls* to the 10-Mbps traffic class. However, note that each 10-Mbps call requires 10 times the bandwidth of a 1-Mbps call. Thus, a more appropriate equitable load for a 10-Mbps traffic stream would be 1.9 *Erls* (= 19/10) when we consider traffic load level by accounting for per-call bandwidth impact. The calculation of blocking with different traffic streams and different bandwidth requirements is much more complicated than the Erlang-B loss formula; this is because the Erlang-B formula is for traffic streams where all requests have the *same* bandwidth requirement. The method to calculate blocking in the presence of two streams with differing bandwidth is known as the Kaufman–Roberts formula [356], [597]. Using this formula, we can find that the blocking probability for a 1-Mbps traffic class will be 0.21%, while for a 10-Mbps traffic class it is 25.11%.

We can see that for the same amount of load exerted, the higher-bandwidth traffic class suffers much higher call blocking than the lower-bandwidth service in a *shared* environment; not only that, the lower-bandwidth service in fact has much lower blocking than the acceptable 1% blocking. If we still want to keep the blocking below 1%, then there is no other option than to increase the capacity of the link to a higher-capacity link (unless the network is completely partitioned for each different service). After some testing with different numbers, we find that if the link capacity is 85 Mbps, then with 19 *Erls* load of 1-Mbps traffic class and 1.9 *Erls* load of 10-Mbps traffic class, the call blocking would be 0.05% and 0.98%, respectively. The important point to note here is that with the introduction of the higher-bandwidth traffic class, to maintain a 1% call-blocking probability for each class, the link capacity is required to be 70% (= (85 − 50)/50) more than the base capacity.

Now, consider a sudden overload scenario for the 1-Mbps traffic class in the shared environment while keeping the overall capacity at the new value: 85 Mbps. Increasing the 1-Mbps traffic class by a 20% load while keeping the higher bandwidth (10 Mbps) traffic class at the same offered load of 1.9 *Erls*, we find that the blocking changes to 0.08% and 1.56%, respectively. What is interesting to note is that although the traffic for the lower-bandwidth call has increased, its overall blocking is still below 1%, while that of the higher-bandwidth call has increased beyond the acceptable threshold level; yet there has been *no* increase in traffic load for this class. These are sometimes known as *mice and elephants* phenomena. Here mice are the lower-bandwidth service calls, while elephants are the higher-bandwidth service calls. However, unlike IP-based TCP flows (see [272]), the situation is quite different in a QoS-based environment—it is the mice that get through while elephants get unfair treatment.

This suggests that some form of admission control is needed so that higher-bandwidth services are not treated unfairly. One possibility is to extend the idea of trunk reservation to *service class* reservation so that some amount of the link bandwidth is logically reserved for the higher-bandwidth service class. Taking this into account, assume that out of 85 Mbps of capacity, 10 Mbps of capacity is reserved for the elephant (10-Mbps) service class; this means that any time the available bandwidth drops below 10 Mbps, no mice (1 Mbps) traffic calls are allowed to enter. With this change in policy, with 20% overload for mice traffic from 19 *Erls*, while elephant traffic class remains at 1.9 *Erls*, we find that the call blocking for mice traffic would be 1.41% and 0.94%, respectively—that is, the elephant traffic class is not affected much; this is then good news since through such a service class–based reservation concept, certain traffic classes may be protected from not getting their share of the resources. Now, if an equitable blocking is still desirable for both service classes, even though only the low-bandwidth stream is overloaded, then some mechanisms are needed to increase the blocking for the elephant service class. A way to accomplish this is to consider a probabilistic admission control; this rule can be expressed as follows:

- *An amount of bandwidth threshold may be reserved for higher-bandwidth calls, which is activated when the available bandwidth of the link falls below this threshold. As a broad mechanism, even when this threshold is invoked, lower-bandwidth calls may be admitted based on meeting the acceptable probabilistic admission value.*

To compute blocking for each traffic class with differing bandwidth *and* a probabilistic admission control and reservation, an approach presented in [480] is used. In Table 17.3 we list the probabilistic admission control case along with reservation and no reservation for the higher-bandwidth traffic class; you can see that equity in call blocking can be achieved when, with reservation, 27% of the time low-bandwidth calls are still permitted to be admitted.

We now consider the other extreme when only high-bandwidth 10-Mbps calls, still with 38 *Erls* of traffic, are offered. To keep call-blocking probability at 1%, with 38 *Erls* of offered load, a link would still need 50 units of *high-bandwidth* call-carrying capacity; this then translates to a raw bandwidth of 50×10 Mbps $= 500$ Mbps. Thus, we can see that depending on whether a network link faces low-bandwidth calls, or a mixture of low- and high-bandwidth calls, or just (or mostly) high-bandwidth calls, for the same offered load exerted, the link requires vastly different raw link bandwidth to maintain a QoS performance guarantee.

Finally, while we discuss call blocking for each individual traffic class, it is also good to have a network-wide performance objective in terms of bandwidth measure. Suppose that a_{low} is the offered load for the low-bandwidth traffic class that requires m_{low} bandwidth per call; similarly, a_{high} is the offered load for high-bandwidth traffic, and m_{high} is the bandwidth requirement per call of high-bandwidth calls, then a bandwidth blocking measure is given by

$$\mathcal{W}_{composite} = \frac{m_{low} a_{low} \mathcal{B}_{low} + m_{high} a_{high} \mathcal{B}_{high}}{m_{low} a_{low} + m_{high} a_{high}}. \qquad (17.6.1)$$

These composite performance measure values for the cases considered above are also listed in Table 17.3. We can see that while this composite measure is a good overall indicator, it can miss unfair treatment to high-bandwidth calls.

Generalizing from two service classes to the environment where *each* arriving call i has an arbitrary bandwidth requirement m_i, the composite bandwidth blocking measure, known as Bandwidth Denial Ratio (BDR), is given by

$$\mathcal{W}_{\text{composite}} = \frac{\sum_{i \in \text{Blocked Calls}} m_i}{\sum_{i \in \text{Attempted Calls}} m_i}. \tag{17.6.2}$$

However, we have learned an important point from our illustration of low- and high-bandwidth traffic classes that higher-bandwidth classes may suffer higher blocking. We can still consider a simple generalization determine if a similar occurrence is noticed when *each* call has a differing bandwidth. Based on profiles of calls received, they may be classified into two or more groups/buckets in terms of their per-call bandwidth requirements, and then apply the above measure to each such group. For example, suppose that a network receives calls varying from a 64-Kbps requirement to a 10-Mbps requirement; calls may be put into, say, three buckets: 0 to 3 Mbps, 3 Mbps to 7 Mbps, and higher than 7 Mbps. If higher-bandwidth groups have a significantly higher-bandwidth blocking rate than the average bandwidth blocking rate for all calls, then this is an indicator that some form of admission control policy is needed so that the higher-bandwidth call groups do not necessarily have a significantly higher-bandwidth blocking rate.

17.7 A General Framework for Source-Based QoS Routing with Path Caching

We now consider a general alternate call-routing framework where calls are heterogeneous. To consider a general framework, we first summarize several goals of QoS routing:

- Reduce the impact on the call setup time by keeping it as low as possible.

- Minimize user-level retry attempts, i.e., it is preferable to do retry *internally* to the network as long as the call setup time is not drastically affected. It is important to note that user-level retry attempts cannot be completely avoided, at least in a heavily loaded network, i.e., a network where the ratio of traffic to network bandwidth is at a level beyond the normally acceptable tolerance for service guarantee.

- Allow the capability for the source node to select a path from a number of possible routes very quickly for each arriving request. Also, allow *crankback* capability as an optional feature.

- Allow a call admission control feature that can be invoked.

To keep call setup time minimal and the need to minimize user-level retry along with the recognition that on-demand route determination can be taxing suggests that having multiple path choices can be beneficial in a QoS routing environment; this is often referred to as *alternate path routing*. Since path caching is necessary to be able to do alternate path routing, we refer to it as the *path caching option*. With multiple path choices, knowing that due to inaccurate/stale information blocking on a path selected cannot be completely ruled out, crankback is a nice optional feature to try another path quickly, thus avoiding user-level retry.

Finally, a framework should allow the ability to incorporate a number of routing schemes so that network providers can choose the appropriate one depending on their performance and systems configuration goal.

17.7.1 Routing Computation Framework

The basic idea behind this framework addresses the following: how is the selection of paths done, when are they selected, and how are they used by newly arrived requests? For calls requiring bandwidth guarantees, another important component that can complicate the matter is the definition of the cost of a path based on possibly both additive and nonadditive properties. Later, we will consider our framework using an extended link state protocol concept. Before we discuss this aspect, we describe a three-phase framework [469]: (1) Preliminary Path Caching (PPC) phase, (2) Updated Path Ordering (UPO) phase, and (3) Actual Route Attempt (ARA). Each of these phases operates at different time scales.

The first phase, PPC, does a preliminary determination of a set of possible paths from a source to destination node, and their storage (caching). A simple case for this phase is to determine this set at the time of major topological changes. PPC, in the simplest form, can be thought of as topology dependent, i.e., if there is a change in the major topological connectivity, then the PPC phase may be invoked. This can be accomplished by a topology update message sent across the network in a periodic manner. This process can be somewhat intelligent, i.e., if a link availability is expected to be less than a certain threshold for a prolonged duration or if the link is scheduled for some maintenance work, then PPC can also be used for pruning the link and a new topology update, thus letting nodes determine a new set of cached paths. Essentially, PPC uses a coarse-grain view of the network and determines a set of candidate paths to be cached. A simple mechanism to determine the set of paths for each source node to each destination node may be based on hop count or some administrative weight as the cost metric using the k-shortest paths algorithm (refer to Section 2.8). Thus, for this phase, we assume the link cost metric for determining a set of candidate paths to be additive.

The second phase, UPO, narrows the number of QoS acceptable paths; this module uses the most recent status of all links as available to each source node. Since the PPC phase has already cached a set of possible paths, this operation is more of a compare or filter to provide a set of QoS acceptable paths. Furthermore, for a specific service type or class, this phase may also *order* the routes from most acceptable to least acceptable (e.g., based on path residual bandwidth), and will, in general, have a subset of the routes "active" from the list obtained from the PPC phase. In this phase, the cost metric can be either additive, e.g., delay requirement, or nonadditive, i.e., bandwidth requirement, or a combination, where one is more dominant than the other. Another important factor to note about the UPO phase is that the value of the link state update interval may vary, with each node being able to select the interval value; for simplicity, we will refer to this as the routing link state update interval (RUI). This phase should be more traffic dependent (rather than on-demand per call) with a minimum and maximum time window on the frequency of invocation.

The third phase is ARA. From the UPO phase, we already have a reasonably good set of paths. The ARS phase selects a specific route on which to attempt a newly arrived flow. The exact rule for selecting the route is dependent on a specific route selection procedure. The main goal in this phase is to select the actual route as quickly as possible based on the pruned available paths from the UPO phase.

There are several advantages of the three-phase framework:

- Different routing schemes can be cast in this framework.

- It avoids on-demand routing computation; this reduces the impact on the call setup time significantly since paths are readily available; i.e., there is no "cost" incurred from needing to compute routes from scratch *after* a new flow arrives.

- The framework can be implemented using a link state routing protocol with some extension. For the PPC phase, some topology information, for example, needs to be exchanged at coarse-grain time windows. During the UPO phase, periodic update on the status of link usage is needed at a finer grain time window. Since different information about links is needed at different time granularity for use by the PPC and the UPO phase, we refer to this as the *extended* link state protocol concept.

- Each of the three phases can operate independently without affecting the other ones. For example, in the PPC phase, the k-shortest paths can be computed either based on pure hop count or other costs such as link speed–based interface cost. In some schemes the UPO phase may not be necessary.

A possible drawback of the framework is that path caching will typically require more memory at the routers to store multiple paths; this will certainly also depend on how many paths are stored. However, with the drop in memory price, a path caching concept is more viable than ever before. Additionally, there is some computational overhead due to k-shortest path computation on a coarse-scale time window. Our experience has been that k-shortest path computation takes only a few seconds to generate 5 to 10 paths in a 50-node network on an off-the-shelf computer. Thus, this overhead is not remarkable since it is done in the PPC phase. If needed, a router architecture can be designed to include a separate processor to do this type of computational work periodically.

17.7.2 Routing Computation

Consider the source destination node pair $[i, j]$. The set of cached paths for this pair determined at time t (the PPC phase time window) is denoted by $\mathcal{P}_{[i,j]}(t)$ and the total number of paths given by $\#(\mathcal{P}_{[i,j]}(t))$. For path $p \in \mathcal{P}_{[i,j]}(t)$, let $\mathcal{L}^p_{[i,j]}(t)$ denote the set of links used by this path.

Let $b_\ell(t)$ be the available capacity of link ℓ at time t (obtained using the link state protocol for the UPO phase). Then, from a bandwidth availability perspective, the cost of path p for $[i, j]$ is determined by the nonadditive concave property of the available capacity on the bottleneck link along the path:

$$z^p_{[i,j]}(t) = \min_{\ell \in \mathcal{L}^p_{[i,j]}(t)} \{b_\ell(t)\}. \tag{17.7.1}$$

Since the path is known from the PPC phase, this filter operation is quite simple. If the index p is now renumbered in order of the most available bandwidth to the least available bandwidth at time t, that is, from the widest path, the next widest path, and so on, then we have

$$z^1_{[i,j]}(t) \geq z^2_{[i,j]}(t) \geq \cdots \geq z^{\#(\mathcal{P}_{[i,j]}(t))}_{[i,j]}(t). \tag{17.7.2}$$

Similar to node i, all other source nodes can use the same principle to determine their own ordered path sets.

How is the available capacity of various links known to nodes i? This can be determined by receiving used capacity of various links through a link state protocol, either in a periodic or an asynchronous manner. Note the availability of the bandwidth on a link is dependent on whether trunk reservation is activated. Suppose the capacity of link ℓ is C_ℓ, and the currently occupied bandwidth as known at time t (based on link state update) is $u_\ell(t)$. In the absence of trunk reservation, the available bandwidth on link ℓ is given by

$$b_\ell(t) = C_\ell - u_\ell(t). \tag{17.7.3}$$

If, however, a part of the link bandwidth $r_\ell(t)$ for link ℓ is kept for trunk reservation at time t, then

$$a_\ell(t) = C_\ell - u_\ell(t) - r_\ell(t). \tag{17.7.4}$$

The former is sometimes also referred to as the residual bandwidth and the second as the available or allowable bandwidth.

There are two important observations to note. First, if the last update value of $u_\ell(t)$ changes dramatically, it can affect the decision process. Thus, in practice, an exponential weighted moving average value $\bar{u}_\ell(t)$ is more appropriate to use than the exact value from the most recently obtained measurement; a discussion on how the exponential weight moving average can be computed is given in Appendix B.6. Second, the reservation allocation for different service classes may be different; thus, it may be beneficial to keep different sets of alternate paths to consider for different service classes. This means that each service class is essentially sliced into a virtual topology.

17.7.3 Routing Schemes

The computation described above can be used in a number of ways. An obvious one is the maximum available capacity-based scheme (widest path); furthermore, the availability can be proportioned to different paths to select a weighted path, similar to dynamically controlled routing (DCR) (see Section 10.4).

The decision on computation routes may depend on whether the information is periodically updated. Finally, the crankback feature availability is a factor to consider; here we will assume that the crankback is activated only at the source node. This means that during the call setup phase, an intermediate node does not try to seek an alternate path; instead, it returns the call control to the originating node when the call does not find enough resources on the outgoing link for its destination.

Recall that a fundamental component of the QoS routing framework used here is path caching. With this, in the PPC phase, a k-shortest paths algorithm (refer to Section 2.8) is used to generate a set of paths, which is cached. At this phase, the cost metric used is additive. For the routing schemes, an extended link state protocol is used to disseminate the status of the link (different information) at the PPC phase and the UPO phase. Since paths are already cached, the UPO phase can use a simple filtering mechanism to order paths based on available bandwidth (for services that require bandwidth guarantee for QoS). If there are services that have other QoS requirements such as path delay, these requirements can be easily incorporated in the UPO phase as additional filters.

Recall that an important goal of reducing the impact on flow setup time is addressed by the framework through the notion of path caching. Due to the three-phase framework, the newly arrived flow attempts one of the paths already pruned by the UPO phase—so there is no on-demand route computation delay in this phase. Depending on the periodicity of the UPO phase and the arrival of the link state advertisement, the pruned path set can have outdated information. Thus, some newly arrived flows can be assigned to a path that may not have any available bandwidth at this instant. This cannot be completely avoided unless the frequency of the update interval is reduced; if this is done, then more frequent link state advertisement would be necessary, which leads to an increase in network traffic.

17.7.4 Results

For performance studies, we consider maximum available capacity routing with periodic update and crankback (MACRPC), as well as for no crankback (MACRPNC). Note that MACRPC uses the shortest widest path on residual bandwidth, but with trunk reservation turned on, and the computation is periodic. For comparison, the utopian scheme, maximum available capacity routing with instantaneous computation (MACRIC), is also considered. This is possible since we have used a simulation environment where the instantaneous feature can be invoked. Also, we consider a sticky random routing scheme that extends the dynamic alternate routing scheme (see Section 10.5) to the multiservice case, which is labeled as cached sticky random adaptive routing (CaSRAR). Results presented here are based on call-by-call routing simulation for randomly arriving calls that follow the Poisson process.

Revisit Homogeneous Traffic Case

We first start with results on call blocking for the homogeneous service case as the number of cached paths K changes from 2 to 15 (for a 10-node fully connected network); this is reported in Figure 17.3 for both the case of no reservation and with trunk reservation set at 40%; while a very high trunk reservation value such as 40% is rarely used in an operational network, the intent here is to show how results are influenced, with and without trunk reservation. It is interesting to note that for the no reservation case, the increase of cached paths does not necessarily result in improvement in performance for *all* routing schemes. We see improvement only for MACRPNC. However, with trunk reservation activated, performance can improve with the increase in K for *all* routing schemes. This substantiates the claim on performance degradation in the absence of trunk reservation as reported in [385]. Furthermore, our result shows that this behavior is not necessarily consistent for *all* routing schemes. For the utopian scheme, MACRIC, the performance degrades drastically as K increases when

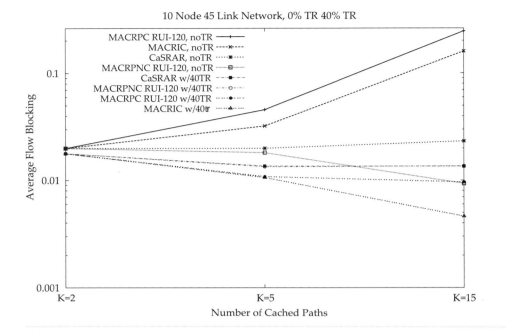

FIGURE 17.3   Homogeneous service fully-connected network (with and without trunk reservation).

there is no trunk reservation. Although this may sound surprising, this is possibly caused by overuse of multiple-link paths through instantaneous checking, which leads to local optimization and bistability. We observe the same problem with MACRPC when there is no trunk reservation. Overall, CaSRAR and MACRPNC are more robust in the absence of trunk reservation. However, in the presence of high trunk reservation, as K increases we found that MACRIC and MACRPC had better performances than CaSRAR and MACRPNC. Overall, these results show that path caching is indeed helpful; however, the actual routing schemes and factors such as trunk reservation do matter.

SERVICE CLASS PERFORMANCE

Next we discuss the case of heterogeneous services where three different service classes with differing bandwidth requirements for each service class are offered. We consider two cases: the network capacity in the first one is dimensioned[2] for low BDR (less than 1%) while the second one is dimensioned for moderately high BDR (over 5%). From the scenario where the network is dimensioned for low BDR (Figure 17.4(a)), we found that in the presence of trunk reservation, as K increases the BDR decreases for all schemes (similar to the homogeneous case). However, this is not true when the network is dimensioned for moderate BDR (Figure 17.4(b)), even in the presence of moderate trunk reservation. The pattern is somewhat closer to the homogeneous case with no trunk reservation. What we can infer is that even in

---

[2]Dimensioning or sizing refers to determining the capacity needed in a network to carry a given traffic offered load at a prespecified level of performance guarantee.

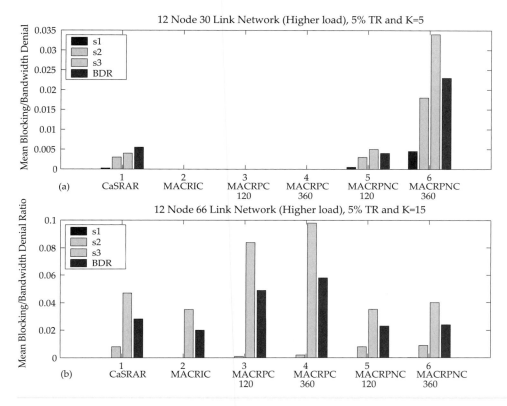

FIGURE 17.4   Performance of different routing schemes (and update periods),
(a) low-load, sparsely connected case, (b) higher-load case.

the presence of trunk reservation, the ability to hunt over multiple paths through crankback
is beneficial in a network designed for low BDR, but crankback can be detrimental when
the network is designed for moderately high BDR as it impacts network performance (and
also can lead to higher flow setup time due to frequent path hunting). See [695] for more
details.

Now we discuss briefly the role of the UPO phase. Recall that different routing update interval (RUI) parameter values can be used for the UPO phase. As one would guess, with more
frequent updates (i.e., for a smaller value of RUI), the inaccuracy in link state information decreases. It is observed that both schemes MACRPC and MACRPNC give better performance
with more frequent updates as would be intuitively guessed. However, it appears that inaccuracy in link state information can be well compensated by the availability of crankback
in a network designed for low BDR. Specifically, we note that MACRPC with an RUI of 360
seconds has much lower BDR than MACRPNC with an RUI of 120 seconds (Figure 17.4(a)).
However, the reverse relation holds when the load is moderately high (Figure 17.4(b)). We
also saw in an earlier example (through MACRIC) that instantaneous information update is
not always beneficial in terms of network performance (as well as negatively affecting flow
setup time considerably). Overall, we can infer that inaccuracy in link state information is not

necessarily bad, and in fact, can be well compensated through path caching; in any case, the specifics of the routing scheme do play a role here.

So far we have discussed performance using the network-wide indicator bandwidth blocking rate. We are next interested in understanding the effect on each service class. For this, we have considered three service classes in increasing order of bandwidth requirement, i.e., the first service (s1) class has the lowest bandwidth requirement per flow, while the third service class (s3) has the highest bandwidth requirement per flow. For a network dimensioned for low BDR, we found that with a moderate to large number of path caching, CaSRAR and MACRPNC tend to give poorer performances to the higher bandwidth service class (s3), whether the network is fully or sparsely connected (Figure 17.4(a) is shown here for the sparsely connected case). Furthermore, the inaccuracy of routing information due to the update interval of the UPO phase does not seem to affect MACRPC for different service classes but can noticeably affect MACRPNC (Figure 17.4(a)). To check whether the same behavior holds, we increased the load uniformly for all service classes. We made some interesting observations (Figure 17.4(b)): the lowest bandwidth service (s1) has uniformly low flow blocking for *all* routing schemes; however, the highest bandwidth service class (s3) is affected worst under MACRPC at the expense of the lower bandwidth classes; i.e., MACRPC is more unfair to higher-bandwidth services as the network load uniformly increases. In general, we found that CaSRAR works better than the other schemes in providing smaller variation in performance differences seen by different service classes.

CALL ADMISSION CONTROL

While it is known that higher-bandwidth, reservation-based services experience worse performance than lower-bandwidth, reservation-based services in a single-link system [380], these results indicate that this behavior holds as well in a network *with* dynamic routing and trunk reservation. In other words, routing and trunk reservation can*not* completely eliminate this unfairness. Thus, in a network, if fairness in terms of GoS to different service classes is desirable, then additional mechanisms are needed. In this context, a concept called *service reservation* beyond traditional trunk reservation has been proposed [471]. This concept can be manifested, for example, through source-based admission control at the time of flow arrival. While a good source-based admission control scheme for a general topology network in the *presence* of QoS routing operating in a link state protocol environment and trunk reservation remains a research problem, a probabilistic source-based admission control scheme for fully connected networks in the presence of routing and for two services case has been presented [471]. The ability to provide service fairness in terms of fair GoS using this source-based admission control scheme in the presence of routing and trunk reservation is shown in Figure 17.5. This is shown for three different values of network load with two service class scenarios (shown for normal load "lf-1.0," 5% s2 overload "lf-s2," and 5% network-wide overload "lf-1.05," all for MACRPC). The right-most entries (corresponding to $p = 1$) denote the *no* source-based admission control case. As we can see, with the increase in load, the higher-bandwidth service suffers the most in the absence of source-based admission control. As the admission control parameter is tuned (by changing p toward 0.8) to invoke different levels of source-based admission control, it can be seen that service-level fairness in terms of GoS can be achieved.

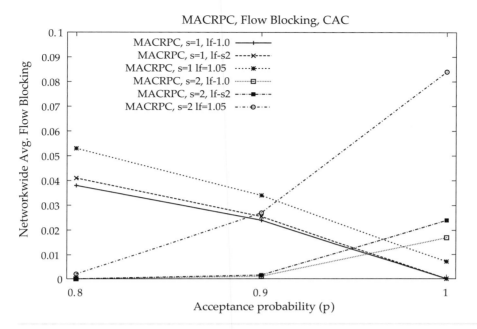

FIGURE 17.5   Performance impact in the presence of source-based admission control.

DYNAMIC TRAFFIC

Finally, we discuss network performance impact due to network traffic dynamics. To show this we consider a homogeneous service, fully connected network where one source-destination node pair has dynamic traffic while the rest of the traffic pairs have stationary traffic (no source-based admission control is included here). For our study, the dynamic traffic has been represented through a time-dependent, stationary process that follows a sinusoidal traffic pattern. For the case with no trunk reservation, we have found that MACRPC has much worse performance than both CaSRAR and MACRIC as traffic changes for the dynamic traffic class (pair); CaSRAR adapts very well with traffic changes, although it has no UPO phase. It is interesting to note that just the presence of dynamic traffic between a source-destination node pair can cause the rest of the (stationary) traffic to show dynamic performance behavior (Figure 17.6).

We also considered the case in which trunk reservation is imposed; purposefully, we set the reservation at an unusually high value of 40% to understand the performance implication—the result is shown in Figure 17.7; from this figure, we note two phenomena: (1) MACRPC performs better than CaSRAR for dynamic traffic, and (2) the imposition of dynamic performance on the stationary traffic (from the dynamic traffic class) is no longer there. Also, we found that the overall performance improves in the presence of trunk reservation in a dynamic traffic scenario (similar to the stationary traffic case). From these results an important question, although not directly within the purview of routing, arises: should a network allow a dynamic traffic stream/class to impose its behavior on a stationary traffic stream? In other words, should a stationary traffic stream suffer higher flow blocking just because the load for the dynamic traffic stream is increasing? This cannot be addressed alone

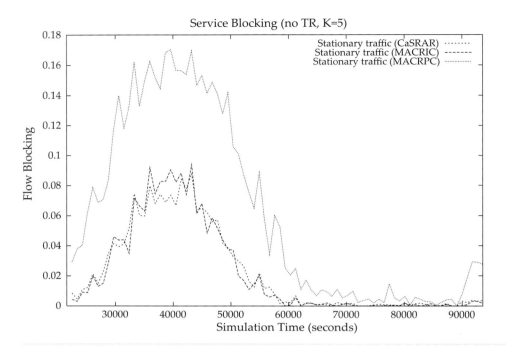

FIGURE 17.6   Dynamic performance behavior of *stationary* traffic due to the influence of dynamic traffic (no trunk reservation).

through the three-phase QoS routing framework or any other QoS routing framework. However, the impact can be controlled through the use of trunk reservation and under controls; this is where lessons on controls discussed earlier in Section 11.6 in Chapter 11 may be taken into consideration.

17.8 Routing Protocols for QoS Routing

17.8.1 QOSPF: Extension to OSPF for QoS Routing

The OSPF extension for QoS routing mechanisms, described in RFC 2676 [23], is commonly referred to as QOSPF. Earlier in Section 6.3, we discussed OSPF packet formats. You may note that every hello, database description, and LSA contains an options field that is 1 byte long. One of the bits in the options field, originally known as the T-bit to indicate if a originating router is capable of supporting Type of Service (TOS), was later removed [505]. Instead, the QOSPF specification proposed to reclaim this bit and renamed it as the Q-bit to indicate that the originating router is QoS routing capable. When this bit is set, two attributes are announced with a link state: bandwidth and delay.

An important aspect about the QOSPF protocol is that it specifies the path computation mechanism, which is divided into the pre-computed option and the on-demand option. For the pre-computed path option, a widest path version of the Bellman–Ford approach based on bandwidth was proposed (refer to Section 2.7.2 for a similar discussion). For the on-demand computation, a widest shortest path version of Dijkstra's algorithm that considered band-

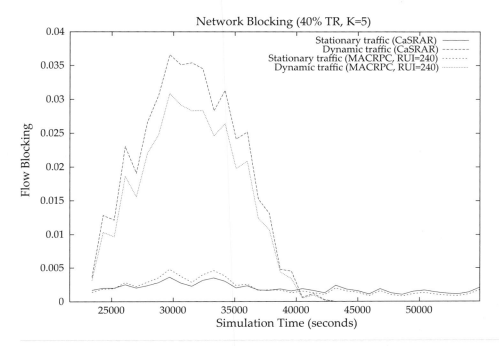

FIGURE 17.7   Performance of dynamic and stationary traffic (with trunk reservation).

width and hop count was proposed; this is essentially least-hop-widest path routing discussed earlier in this chapter.

Note that in QOSPF, as part of the protocol, both the path computation algorithm and the attributes to be exchanged were specified. It is important to distinguish this approach from traffic engineering extensions of OSPF (refer to Chapter 18) in which the extension on exchange of information has been standardized, while the actual path computation mechanism is left for the provider to decide.

17.8.2 ATM PNNI

Asynchronous Transfer Mode (ATM) technology is a packet-mode networking technology with its own protocol stack architecture and addressing. In ATM, all packets are of fixed 53-byte size, known as *cells*. The Private Network–Network Interface (PNNI) protocol [39], originally defined around the mid 1990s, is the standards for QoS routing in ATM Networks. PNNI is based on a link state routing protocol framework; it has the basic elements of a link state routing protocol such as the hello protocol, database synchronization, and flooding. However, PNNI is a *topology state* protocol since besides the status of links, the status of nodes can also be flooded; accordingly, the PNNI topology state element (PTSE) is equivalent of the link state advertisement.

Parameters associated with a topology state are divided into two categories: metrics and attributes; the main distinction is whether information is considered on an entire path basis ("metric"), or an individual node or link basis ("attribute"). Examples of metrics are: cell delay variation and maximum cell transfer delay. Attributes are either performance-related such as the cell loss ratio, the maximum cell rate, and the available cell rate, or policy-related such as the restricted transit. Since packet sizes are fixed, cells as units is used instead of the raw bandwidth rate; thus, effectively, the maximum cell rate refers to the total bandwidth of a link, and the available cell rate refers to the available bandwidth, both counted in cells as units. Although all information required for computing paths is provided, PNNI did not prescribe any specific way to do path computation; in this sense, PNNI is a visionary protocol and is one of the early routing protocols to decouple the routing information exchange part from the routing algorithm computation.

PNNI allows crankback and alternate routing, much like DNHR and RTNR for dynamic call routing in the telephone network. Crankback can be local; that is, the control for a connection need not be sent back to the ingress switch for performing crankback. By using addressing hierarchy, PNNI also handles scalability on information dissemination and storage. That is, through addressing hierarchy, aggregation of information about a group of nodes and links that are at the same hierarchy is performed—such a group is identified as a *peer group*; the aggregated information about a peer group is disseminated, instead of announcing the PTSE for each element within the group. Thus, a peer group can be thought of as a domain; in this sense, PNNI has both intra- and inter-domain flavors in the same protocol. Although PNNI routing is considered to be source-routing based, this is only true only within a peer group; to reach an address that is in a different group, a *designated transit list* is created that identifies the peer groups the connection control message is to visit during the connection set up; once such a request reaches a group identified in the designated transit list, the group is responsible for actual source route within its group to the appropriate egress node, which, then, hands off to the next peer group for further processing. For additional details, see [39].

17.9 Summary

In this chapter, we have discussed QoS routing. We started by discussing what QoS means, and the scope of QoS routing and its inherent relation to traffic engineering. Based on arrival and service frequency, we have also identified how different services may be classified into three types of classifications; this is summarized in Table 17.1. We have indicated that QoS routing falls under the Type A classification.

We then presented a taxonomy for QoS routing and showed how QoS routing can be divided based on different types of networks, and whether one or more attributes are to be considered in the QoS routing decision, especially for packet-switched networks.

We then discussed extendability of widest and shortest path routing to QoS routing. An important issue to consider here is that periodic updating of information induces inaccuracy on link state information—thus, to properly address service performance, a path caching mechanism that allows alternate path routing can be helpful; this is presented as a three-phase framework. Performance results are presented to understand the interrelation in the presence of heterogeneous guaranteed services, update frequency, traffic dynamism, and so on.

The importance of QoS routing goes beyond the telephone network. It is also applicable to MPLS, optical, and wavelength routing when service requests with guaranteed resource requirements are to be connected on demand and quickly. In subsequent chapters, the specific applicability of QoS routing will be discussed.

Before we conclude, we briefly comment on QoS guarantee in a generic best-effort network such as the Internet. This QoS guarantee issue should not be confused with QoS routing. In an intradomain environment running a best-effort model, QoS guarantee for services are quite possible if the network is engineered to meet QoS guarantee—this might require over-provisioning. A better environment is a differentiated services environment, where priority to certain packets can be given by using a router's scheduling algorithm (refer to Chapter 22) for services that require certain guarantee—in this case, the overprovisioning can be moderate since the routers have mechanisms to discriminate packets that require guarantee and those that do not. MPLS is also a mechanism to enable QoS quarantee [750], [751]. In an interdomain environment, it is much more difficult since each provider on a path for a request that requires QoS guarantee would need to have the proper mechanisms—this is difficult in practice since it might not be possible to enforce every provider to provide the same QoS guarantee. However, instead of stringent QoS guarantee, it is possible to provide certain quality through broad service-level agreements (SLAs) (see Section 23.1 for examples of SLAs). SLAs are possible to among different providers through which traffic may flow. Thus, meeting SLA agreements can be thought of as meeting "soft" QoS guarantee.

Further Lookup

QoS routing has been investigated by many researchers over the years, spanning a variety of issues. We have identified the key issues in this chapter. For further reading, you may consult works such as [2], [21], [22], [23], [24], [28], [58], [130], [248], [269], [270], [306], [307], [377], [432], [435], [436], [476], [477], [518], [524], [585], [644], [650], [651], [710], [766]. Through such work, you can, for example, obtain better understanding of issues such as inaccuracy of information in QoS routing, the relation between routing and scheduling, differences in performance between different routing schemes, and so on. For analysis of PNNI routing, [210], [437]. While most work consider intradomain, interdomain QoS routing is an important problem; see [622].

Exercises

17.1 How is QoS routing different from best-effort routing?

17.2 Explain constrained shortest path routing and its variations when you consider different attributes.

17.3 What is the relation between caching paths for alternate routing and inaccuracy in information available due to periodic link state update? For example, if there is no path caching, how would this inaccuracy effect routing performance?

17.4 How does bandwidth guarantee required by services affect the performance it receives in a heterogeneous bandwidth environment?

17.5 Compare QOSPF and PNNI.

18

MPLS and GMPLS

Are we there yet?

Any kid in the back seat of a car

Reading Guideline

Understanding the material presented in this chapter requires basic knowledge of packet switching, circuit switching, and routing protocols. In addition, the background presented in the chapter on quality-of-service routing is helpful.

In this chapter, we present emerging environments that are beneficial to routing and traffic engineering. In this regard, there are two key frameworks: MPLS and GMPLS. MPLS is designed for packet-switched arbitrary-rate services, whereas GMPLS is suited for circuit-oriented periodic or on-demand services with bidirectional requirements. We also discuss extensions required in routing protocols that can be useful in MPLS/GMPLS environments. It may be noted that both MPLS and GMPLS are evolving standards. This chapter captures the basic features.

18.1 Background

We start by reviewing the IP routing process: at a router, the routing table is determined, from which the forwarding table is built; when a packet arrives, the IP address of the destination is looked up and mapped against the forwarding table to identify the appropriate routing interface. The underlying principle is to operate on a packet-by-packet basis, while tracking a microflow and forwarding on the same path if possible. However, there are times when controlling the flow of packets for a class of traffic streams that is beyond a single packet or even a microflow is desirable. In an IP environment, IP traffic engineering that involves link weight determination in an OSPF/IS-IS environment can direct overall flow control; we have previously discussed IP traffic engineering in Chapter 7. However, in IP traffic engineering, all packets for a destination follow the same path due to a destination-oriented routing paradigm of IP routing; certainly, if equal cost paths are found, traffic can take two or more different paths to the destination. However, having a well-defined control mechanism that allows packets to take different paths to a particular destination, for example, depending on the type of packet or class of traffic, is desirable. This can be thought of as *traffic engineering with more knobs or controls*, to distinguish it from IP traffic engineering based on link weight settings as the primary control.

To attain the ability to define a path and force packets for a particular traffic stream or class or affinity to take this path, it is imperative that a mechanism is needed that allows the path to be defined/identified, independently of designated packets traversing this path. The question then is how to define such paths. It may be noted that the notion of defining a path is done in the voice telephone network where call setup is done first using, for example, ISUP signaling. Here, instead, we are interested in having a mechanism that can work for packet traffic, yet that is not necessarily meant to work on a per-microflow or per-call basis. There is an important point to be noted here—if we were to look for a mechanism to work on a per-call basis that can also work in an IP environment, a session initiation protocol (SIP) can be considered; SIP is, however, at the end-to-end basis. Rather, we are interested in a mechanism that is on a router-to-router basis. Thus, it should be clear that to set up a path between two routers, a mechanism for signaling such a setup is certainly required that is not necessarily on a per-microflow basis. Second, if part of the goal is to do traffic engineering with more knobs/controls, the state of the network still needs to be communicated among the routers.

As we know, one way to communicate the state of the network is to communicate link state information through a link state routing protocol. But standard link state protocols defined for IP networks carry only a single metric value for each link, typically to represent the cost of the link. For traffic engineering with more knobs/controls, additional information, such as the bandwidth of a link, must be communicated. This means that to learn about the state of a network, a link state protocol paradigm can still be used, however, with additional information about each link. Finally, to engineer a network, it should be possible to invoke different routing path computation framework, which is preferably decoupled from the link state update mechanism; such a path communication framework may also depend on the specific operational use and requirements of a network for services it provides (for example, refer to Table 17.1).

In addition to packet-based traffic, the need for a traffic engineering framework is also felt for circuit-mode connections for services provided in a modular switching environment such as optical wavelength switching or time-division multiplexed switching.

With the above background, we now discuss enabling environments that allow signaling setup and traffic engineering with more knobs/controls to be done.

18.2 Traffic Engineering Extension to Routing Protocols

We can see from the earlier discussions that at a minimum, the bandwidth of a link must be communicated for the purpose of traffic engineering. In addition, a link may allow higher reservable bandwidth than the announced bandwidth due to statistical multiplexing gain for certain types of traffic, meaning that oversubscription may be tolerable. Also, a link may announce currently unreserved bandwidth, which is useful for routing path computations, but which is not necessarily based on a shortest path computation. Since a network may provide more than one type of prioritized services, it would be useful to announce the unreserved bandwidth allowed for each priority class. Also, a network provider might want to use a different metric, other than the standard link metric; this link metric might have meaning that is internal only to the provider. To summarize, the following attributes of a link are desirable:

- Maximum link bandwidth that is usable

- Maximum reservation bandwidth in case oversubscription is allowed

- Unreserved bandwidth available at different priority levels

- Traffic engineering metric.

The question is: how is this information communicated? This is where two popularly deployed link state routing protocols—OSPF and IS-IS, presented in Chapter 6—come into the picture. These two protocols have been extended to allow for communication of the above attributes. The actual extensions are somewhat different in GMPLS compared to MPLS due to additional requirements in circuit-mode connections. We will cover them as we introduce MPLS and GMPLS in the following sections; specifically, refer to Sections 18.3.4 and 18.4.4, respectively.

18.3 Multiprotocol Label Switching

Multiprotocol Label Switching (MPLS) is a mechanism that addresses several issues discussed above; it is meant for packet-based services. Briefly, MPLS adds a *label* in front of a packet, i.e., as another header so that routers know how to act based on this label. To be able to act based on a label, routers must be *label-switched routers* (LSRs), and each LSR must maintain a valid

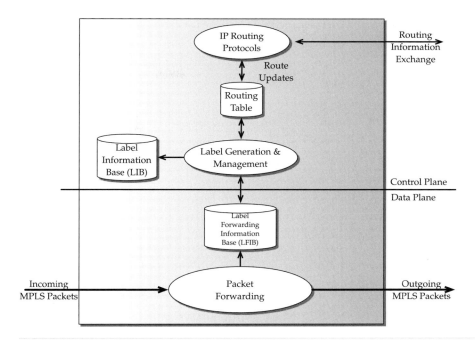

FIGURE 18.1   Conceptual architecture of an MPLS label-switched router.

mapping from the label of an incoming packet ("incoming label") to a label to be attached to the packet before being sent out ("output label"). This, in turn, means that LSRs maintain states in terms of input/output labels associated with a particular path, referred to as a *label-switched path* (LSP), which may be designated for a particular class of traffic flows. Note that an LSP must already be established between two routers so that packets can follow this path. To establish a path, a *label distribution protocol* is used. Certainly, the next question then is: how do we know this is the best path for the particular class of traffic flows? This will depend on the traffic engineering requirements of the network, and on the service requirements of the traffic flow to be carried by the LSP. In this section, we present the basic foundation for MPLS and the enabler for traffic engineering in MPLS; in Chapter 19, we will present examples of routing and traffic engineering design using MPLS networks.

In Figure 18.1, a conceptual architecture of an MPLS LSR with IP functionality is shown. As can be seen, there are two planes: a control plane and a data plane. At the control plane, IP routing protocols can exchange routing information, and another component manages label distribution and binding. It also maintains a *label information base (LIB)*, and creates a *label forwarding information base (LFIB)*. The packet arriving on the data plane consults the LFIB for proper forwarding in an outgoing direction.

It may be noted that the establishment of an LSP is a connection-oriented functionality— that is, a path must be set up before traffic can use this path. An established LSP may or may not have any packet traffic flow on it; furthermore, the packet traffic flow rate on an LSP can vary from one instant of time to another. Many traffic flows may be combined on a specific LSP; usually, such a flow aggregation is based on some affinity, such as a traffic class. The aggregated flow constructed on some affinity basis is referred to as a *traffic trunk*. Typically, a traffic trunk is defined on an ingress/egress LSR pair basis and is carried on an LSP. Note that all traffic between the same two ingress/egress LSRs may be split into multiple traffic trunks; each traffic trunk is then mapped into an LSP. Thus, a traffic trunk is a logical entity, while an LSP is a transport manifestation of this logical entity.

18.3.1 Labeled Packets and LSP

A label in MPLS is 20 bits long and is part of a 32-bit MPLS shim header. A packet with an MPLS shim header will be referred to as an *MPLS packet*. If an MPLS packet is to carry an IP packet, then we can think of MPLS as being placed in layer 2.5 (see Figure 18.2). Note that it still requires the help of layer 2 for packet delivery on a link-by-link basis between two adjacent LSRs; the main difference for MPLS from being at layer 2 is that it does provide a form of routing through labels and LSPs across multiple hops. Suppose that layer 2 is a Packet over SONET (PoS) technology between two LSRs. Since PPP is used for Packet over SONET, we have PPP as the layer 2 protocol to deliver an MPLS packet from one LSR to the other.

The 32-bit shim header also includes 3 experimental bits, a bit ("S" bit) to indicate this label is the last label (bottom of the stack) in the case of stacked labels, and 8 bits for the time-to-live (TTL) field (see Figure 18.3). The experimental bits are meant to describe services that require different priorities. The label values 0 to 15 are reserved. For example, label 0 (*explicit null label*) refers to a packet that is to be forwarded based on the IPv4 header and is allowed only at the bottom of the label stack. Similarly, label 2 serves as the explicit null label for IPv6. The explicit null label can be used by an egress router to signal its immediate upstream router ("penultimate hop router"). In turn, the egress router receives packets from the penultimate hop router with label value 0 and can also learn any priority information included in the experimental bits, which it can use for IP forwarding.

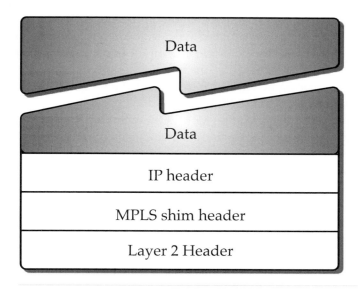

FIGURE 18.2   MPLS header as layer 2.5.

There is another terminology, *tunnel*, closely related to an LSP. A tunnel provides a transport service between two routers so that packets for a specific stream can flow without being label swapped at any intermediate switches or routers. A tunnel in MPLS can be realized by a well-defined LSP, often based on serving certain traffic engineering requirements. Thus, typically, such tunnels have longevity. Furthermore, note that tunnel is a generic name used in networking; it is not limited just to MPLS.

When an MPLS packet arrives at an LSR, the incoming label is swapped with an outgoing label; this assumes that an LSP is already defined and lookup tables at LSRs have appropriate entries. Before sending it out to a received MPLS packet, the TTL value is decreased by one. If the TTL value becomes zero, then the MPLS packet is to be dropped. Since the TTL field is 8 bits long, the likelihood of a path having more than 255 hops is zero. Consider Figure 18.4. Here an MPLS packet with label 16 arrives at LSR3 from LSR1 and is swapped with label 17 for transmittal to LSR4; similarly, the MPLS packet with label 17 that arrives from LSR2 at LSR3 is swapped with label 18 for transmittal to LSR4. In this case, the assumption is that two LSPs, LSR1-LSR3-LSR4 and LSR2-LSR3-LSR4, are already defined.

| 0 | 1 | 2 | 3 | 4 | 5 | 6 | 7 | 0 | 1 | 2 | 3 | 4 | 5 | 6 | 7 | 0 | 1 | 2 | 3 | 4 | 5 | 6 | 7 | 0 | 1 | 2 | 3 | 4 | 5 | 6 | 7 |

Label (20 bits)	Exp (3 bits)	S	TTL (8 bits)

FIGURE 18.3 MPLS shim header.

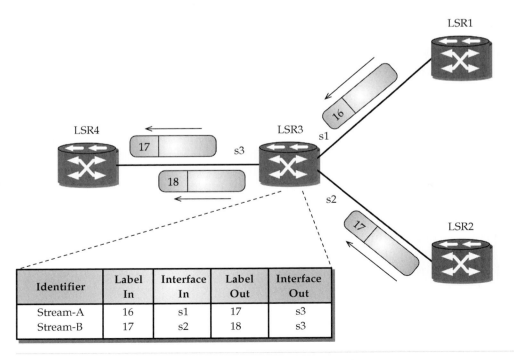

Identifier	Label In	Interface In	Label Out	Interface Out
Stream-A	16	s1	17	s3
Stream-B	17	s2	18	s3

FIGURE 18.4 Label swapping and label switched paths.

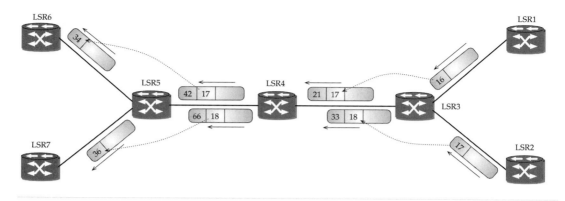

FIGURE 18.5   Label-switched paths using an LSP tunnel ("tunnel within a tunnel").

It may be noted that MPLS also allows stacked labels. A stack tag means that an MPLS shim header may appear more than once; each one is related to a particular LSP between certain points. Thus, a stacked MPLS packet has the following form:

| Layer 2 header | MPLS shim header | MPLS shim header | Data |

Consider Figure 18.5. Here, there is already an LSP set up from LSR3 to LSR5; this is referred to as an *LSP tunnel*. Now consider two LSPs, one from LSR1 to LSR6 and the other from LSR2 to LSR7, that use the LSR3-LSR5 tunnel as a logical link on their paths. For illustration, consider the LSP from LSR1 to LSR6. The packet with label 16 arrives at LSR3 and is swapped with label 17; due to the LSP tunnel LSR3-LSR5, an additional label 21 will be added—the "S" bit for this label would not be set. When this MPLS packet arrives at LSR4, the top label 21 will be swapped to 42 and the "S" bit is set—this means that it is as if LSR4 is thinking about the LSP tunnel LSR3-LSR4 and does not care about any labels in the MPLS packet. When this packet arrives at LSR5, the role of the label with value 42 will end here since this is the end of the tunnel; because of LSP path LSR1-LSR6, the second label 17 will be swapped to 34 and will be forwarded to LSR6. Similarly, the packet from LSR2 to LSR7 will be handled. There is an important point to note here: having an already established LSP tunnel between two routers does not imply that all LSPs that traverse this tunnel need to have the same label value. Now, if the LSP from LSR1 to LSR6 is established as long-lived, it will serve as a tunnel for other traffic; thus, we arrive at a scenario known as a *tunnel within a tunnel*. Note that we have described only the functionality—when and where to establish LSP tunnels in a network to encapsulate other LSPs that can be driven by traffic engineering decisions.

Note that examples discussed so far do not show how MPLS receives an IP packet. An MPLS network must have *edge* routers, which are the point where a native IP packet is prepended with an MPLS label; these routers are known as *label edge routers* (LERs). This is shown in Figure 18.6.

Another concept that is associated with an LSP is a *forwarding equivalence class* (FEC). In a network, an FEC streamlines packets based on, for example, traffic engineering requirements. Thus, an FEC must then have an association with at least one LSP in the MPLS network so that packets for this FEC have an actual path to follow along with any QoS considerations.

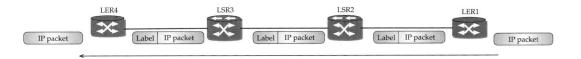

FIGURE 18.6   Label-edge routers and label-switched routers.

That is, an FEC does not define a path; one or more LSPs are used for carrying packets for an FEC.

18.3.2 Label Distribution

Between defining an FEC and establishing an LSP for this FEC, there is another very important phase known as *label distribution*. The basic idea of label distribution is analogous to route information exchange, somewhat similar to BGP. In BGP, two BGP speakers exchange IP prefix reachability information, while in MPLS two LSRs exchange label-related information. A key difference is that while a BGP speaker computes and determines an outgoing routing decision, in MPLS, an LSR essentially trusts that the label information is based on a valid LSP and does only a sanity check with its appropriate neighbor about label mapping; note that label mapping to an LSP may be generated by an outside entity, such as a traffic engineering manager. So that labels distributed can be associated with LSRs, each LSR must have an identifier (LSR ID) that must be unique in the MPLS domain. Typically, a router's IP address serves as the LSR ID.

In MPLS, *label binding* refers to directly associating a specific label to a specific FEC. Consider two LSRs, LSR-u and LSR-d, that have agreed to bind a specific label, for packets that will be sent from LSR-u to LSR-d; in regard to this label binding, LSR-u is referred to as the *upstream LSR* and LSR-d as the *downstream LSR*. The decision to bind a label is made by the LSR, which is downstream with respect to this binding. The downstream LSR then informs the upstream LSR of the binding. We can then say that labels are *assigned* from upstream to downstream, while label bindings are *communicated* from downstream to upstream. In MPLS, the distribution of labels for label binding can be accomplished through a *label distribution protocol* (LDP) [16]. Two approaches for label distribution are possible in the LDP paradigm in MPLS: in the *downstream on-demand* approach, a downstream LSR can distribute an FEC label binding when it receives a request explicitly from an upstream LSR on demand; in the *downstream unsolicited* approach, an LSR can distribute label bindings to LSRs without receiving an explicit request.

In the terminology of LDP, two LSRs that can exchange label/FEC mapping information are referred to as *LDP peers* using a bidirectional *LDP session*. TCP is used for setting an LDP session. The question is how. Do they need to set up an LSP first to exchange this information? That is, is there a chicken and egg problem? Fortunately, no. Recall that an MPLS router actually serves in dual-mode, IP for control plane and MPLS for data plane. Thus, two adjacent LSRs can use the IP only-mode to set up this TCP session, bypassing the MPLS plane.

18.3.3 RSVP-TE for MPLS

The MPLS framework originally defined the basic specification for a label distribution protocol [16]. Recently, the *Resource ReSerVation Protocol with Traffic Engineering extension* (RSVP-TE, in short) has become the *de facto* label distribution protocol for the purpose of traffic engineering. Thus, we will focus on RSVP-TE. We first start with a brief overview of RSVP.

RESOURCE RESERVATION PROTOCOL: OVERVIEW

RSVP is a connection setup protocol in a packet network. Originally, it was defined in the context of an *integrated services* (int-serv) framework. RSVP is considered a *soft-state* approach; this means that if a reservation refreshing message is not sent periodically for a session that has been setup, the session is teared down after a given interval. Such a soft-state measure is appealing in a networking environment where both best-effort and guaranteed bandwidth services are offered in a connection-less packet mode. RSVP is not scalable when there are many on-going end-to-end conversations on a network due to the number of messages that would be generated. However, for RSVP-TE, the setup messages are generated only for LSPs—this number is much smaller than if used for signaling of end-to-end sessions; furthermore, RSVP refresh reduction is possible in RSVP-TE. Here, we first highlight a few key elements about RSVP that are applicable to RSVP-TE.

All RSVP messages have a common header (see Figure 18.7). A key field in the header is the message type. Originally, seven main message types have been defined: Path, Resv, PathErr, ResvErr, PathTear, ResvTear, and ResvConf; they are used in regard to connection set-up and connection teardown. Additional types have been added for a variety of purposes, see [318] for an updated list. However, discussion of all of them is outside the scope of this book.

An RSVP message includes one or more RSVP objects, with each object having a certain significance in regard to a specific message; it may be noted that some objects can be optional. An RSVP object consists of four fields (see Figure 18.8): Length (16 bits), Class Number (8 bits), Class Type (8 bits), and Object Contents (variable). Length is the total object length in bytes, in multiples of 4 bytes. Class Number (Class-Num) identifies an object class, while Class Type (C-Type) identifies unique information within a class.

0 1 2 3 4 5 6 7	0 1 2 3 4 5 6 7	0 1 2 3 4 5 6 7 0 1 2 3 4 5 6 7
Version = 1 Flags	Message type	RSVP Checksum
(4 bits) (4 bits)	(1 byte)	(2 bytes)
Send-TTL	Reserved	RSVP Length
(1 byte)	(1 byte)	(2 bytes)

FIGURE 18.7   RSVP common header.

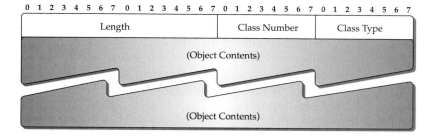

FIGURE 18.8   RSVP object format.

In Figure 18.9, an RSVP Path message is shown using a notation known as a *Backus–Naur Form* (BNF). Objects are specified using angle brackets, such as "<SESSION>"; an optional object is enclosed in square brackets, such as "[<INTEGRITY>]." as "[<POLICY-DATA>...]." A molecular object can be formed from atomic objects, which in turn can be part of a more complex object. The notation "::=" is used to separate the name of an object on the left side that is defined by the set of objects on the right side. It may be noted that the Path message example shown in Figure 18.9 is actually for RSVP-TE, not for the original version of RSVP; this is included for ease of our discussion in regard to traffic engineering.

RSVP-TE

RSVP-TE is the extension to RSVP, which has been developed for use in establishing LSPs, particularly geared to traffic engineering. The label distribution approach in RSVP-TE is based on the downstream on-demand approach. There are certain key differences between RSVP-TE and the original RSVP. For example, RSVP-TE is used for signaling between routers to set up LSP flows, unlike the original RSVP, which was used between hosts to set up microflows. Because of this RSVP-TE does not have the scalability problem that RSVP faces in regard to management of microflows. RSVP-TE is used to set up directional unicast LSPs, while RSVP allows multicast flow setup.

Traffic trunks, discussed earlier, can be carried in RSVP-TE–defined LSPs. A traffic trunk may be split on two LSPs established from an ingress LER to an egress LER, or multiple traffic trunks may be combined to be carried on a single LSP. Thus, such LSPs serve as traffic engineering tunnels. Note that RSVP-TE does not dictate how to decide or when to create different traffic trunks or when to split a traffic trunk into multiple LSPs; rather, the role of RSVP-TE is that of an *enabler* from a functional point of view, while the actual decision is left to operational network providers.

RSVP allows three service-type specifications, which can be used by RSVP-TE when setting up LSPs. The service types described within the purview of int-serv are Guaranteed Quality-of-Service [637], Controlled-load Service [747], and Null Service [76]. For example, if the tunnel requires a bandwidth guarantee, the Peak Data Rate parameter is specified and that it is a guaranteed service request. In case of Null Service, it need not specify resource requirements. However, controlled-load service can provide a full guarantee if the network is under little congestion; but in case of congestion, some delay would be experienced by packets.

```
⟨Path Message⟩ ::=        ⟨Common Header⟩ [ ⟨INTEGRITY⟩ ]
                          ⟨SESSION⟩ ⟨RSVP-HOP⟩
                          ⟨TIME-VALUES⟩
                          [ ⟨EXPLICIT-ROUTE⟩ ]
                          ⟨LABEL-REQUEST⟩
                          [ ⟨SESSION-ATTRIBUTE⟩ ]
                          [ ⟨POLICY-DATA⟩ ... ]
                          ⟨sender descriptor⟩

⟨sender descriptor⟩ ::=   ⟨SENDER-TEMPLATE⟩ ⟨SENDER-TSPEC⟩
                          [ ⟨ADSPEC⟩ ]
                          [ ⟨RECORD-ROUTE⟩ ]
```

FIGURE 18.9   RSVP path message in Backus-Naur form.

TABLE 18.1   RSVP object examples for MPLS (An up-to-date list is maintained at [318]).

Object Name	Used in	Class Number	Examples: Class type with value in parentheses (source RFC listed as [] from bibliography)
INTEGRITY	Path, Resv	4	RSVP Integrity (1) [51]
SESSION	Path, Resv	1	LSP Tunnel for IPv4 (7) [43]
RSVP-HOP	Path, Resv	3	IPv4 (1), IPv6 (2) [90]
TIME-VALUES	Path, Resv	5	Time Value (1) [90]
FILTER-SPEC	Resv	10	LSP Tunnel for IPv4 (7) [43]
SENDER-TEMPLATE	Path	11	LSP Tunnel for IP4 (7) [43]
RSVP-LABEL	Path	16	Type 1 Label (1) [43]; Generalized Label (2) [74]
LABEL-REQUEST	Path	19	No label range (1) [43]; generalized label request (4) [74]
EXPLICIT-ROUTE	Path	20	Type 1 Explicit Route (1) [43]
POLICY-DATA	Path	14	Type 1 (1) [90]
SENDER-TSPEC	Path	12	Integrated Services (1) [90]
RECORD-ROUTE	Path	21	IPv4 (1) [43]
SESSION-ATTRIBUTE	Path	207	LSP Tunnel (7) [43]
DETOUR	Path	63	IPv4 (7) [541]
FAST-REROUTE	Path	205	Type 1 (1) [541]
UPSTREAM-LABEL	Path	35	Same as in RSVP-LABEL

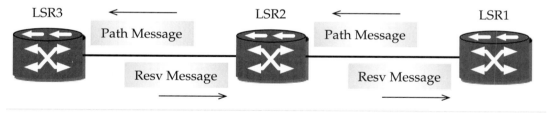

FIGURE 18.10   Example: label distribution using RSVP-TE.

Consider Figure 18.10, where we want to set up the traffic engineering tunnel LSR1-LSR2-LSR3. Here the Path message generated at LSR1 is renewed at LSR2 for destination LSR3. LSR3, being the destination, generates the Resv message in the reverse direction to LSR2 and which in turn forwards to LSR1. Thus, in RSVP-TE, the Resv message then helps accomplish the label-binding function.

What does the Path message include when it is initially generated by a router such as LSR1? And what specifics are key when it is used for RSVP-TE? In general, the Path message generated at LSR1 will include objects as shown in Figure 18.9; it can contain certain key objects:

- SESSION: It identifies the session type through a C-Type field set to LSP-TUNNEL-IPv4 and includes the IP address of the destination.

- LABEL-REQUEST: This indicates that a label binding is requested.

- RSVP-HOP: This indicates the IP address of the sending RSVP-capable router, and the outgoing interface.

- EXPLICIT-ROUTE: This is included to specify that this particular path LSR1-LSR2-LSR3 is to be followed; however, RSVP-TE allows an intermediate node to modify this EXPLICIT-ROUTE object to allow for any local rerouting; in this case, both the original and the modified EXPLICIT-ROUTE objects are stored. Certainly, any such local rerouting assumes that the intermediate router is somehow aware of this reroute. Note that such a local reroute is essentially a crankback feature.

- RECORD-ROUTE: This serves as a form of acknowledgment so that the sending node knows if the path specified was the actual route taken. The visited routers are added as a subobject to RECORD-ROUTE, as the Path message is being forwarded downstream.

- SESSION-ATTRIBUTE: This is included for the purpose of session identification as well as troubleshooting. Setup priorities and holding priorities are included here.

- SENDER-TEMPLATE: This is used primarily to announce the IP address of the sender along with an LSP identifier.

- ADSPEC: This is advertising information that may be carried in a Path. This information is passed to the local traffic control at a router, which returns an updated ADSPEC, which, in turn, is forwarded in Path messages sent downstream.

- SENDER-TSPEC: The traffic characteristics of the tunnel are defined through this object; it uses int-serv as C-Type. This field contains Token bucket rate (r in bytes/sec), Token bucket size (b in bytes), Peak data rate (p in bytes/sec), Minimum policed unit (m in bytes), and Maximum packet size (M in bytes). The service option can be specified as either Guaranteed QoS, Controlled-Load Service, or Null Service.

Similarly, the Resv message includes SESSION, RSVP-HOP, TIME VALUES, STYLE, FLOWSPEC, FILTERSPEC, and LABEL. There is an important connection between QoS, SENDER-TSPEC, and FLOWSPEC. Note that RSVP-TE is specifically designed to setup QoS-enabled LSPs, where QoS parameters may include diffserv parameters. QoS is specified using the SENDER-TSPEC object. The egress node, in return, creates and sends the FLOWSPEC object.

In addition to Path and Resv messages, there are other message types as well; for example, the PathErr message is generated if an LSP tunnel could not be established at any of the intermediate routers; the PathTear message is used to tear down an LSP session. Furthermore, a new optional message type, Hello, has been introduced to determine if an adjacent LSR is reachable.

Each RSVP object, shown in Figure 18.8, must contain a valid and unique class number and a class type. Relevant objects that are included in an RSVP Path message are shown in Figure 18.9. To address any future requirements, the ability to add new objects is possible by defining new class numbers. In Table 18.1, a representative set of objects with class number and class type is listed; an up-to-date list is maintained at [318].

You may note that RSVP-TE Path and Resv message can contain many parameters and sub-parameters. In fact, if we were to print a Path message with each field listed separately, it would run to several pages. Instead, you might want to consult [744], in which samples of RSVP-TE messages are available—they are very helpful in understanding RSVP-TE messages. Below, we briefly illustrate the important contents of the SENDER-TSPEC object.

Example 18.1 *Illustration of traffic characteristics of SENDER-TSPEC.*

As mentioned earlier, traffic characteristics have three key parameters: Token bucket rate (r in bytes/sec), Token bucket size (b in bytes), and Peak data rate (p in bytes/sec). For the concept of token bucket, refer to Chapter 23.

If the data are generated at a steady rate of 625,000 bps to provide guaranteed service, then $r = 625,000$ bps, and $b = 1$ byte. Note that $b = 1$ means that the token is spent immediately. In this case, the peak rate does not play a role. If, however, b is given to be 1000 bytes, then credits can be accumulated to use in a future time slot as long as it is allowed by the peak rate. For example, if p is also set at 625,000 bps, then the bucket size value is not meaningful. If, however, p is set at 630,000 bps, then even if the token is received at 625,000 bps, not all need to be used up; i.e., it can accumulate 1000 bytes credit for up to 5 sec, so that it can transmit at 630,000 bps at the end of the 5 sec.   ▲

An important issue during the label distribution phase is loop detection. This might give the impression that EXPLICIT-ROUTE, when created at the originating router, will check and provide a loopless path that would be sufficient; the difficulty is that local rerouting along the path may not be ruled out. Thus, RSVP-TE relies on the RECORD-ROUTE object; more specifically, when an intermediate router processes the RECORD-ROUTE object, it checks all subobjects inside this object to see if it is already listed as one of the nodes visited to detect looping.

Once a traffic engineering tunnel is set up, traffic trunks can use them. However, a tunnel might get broken, for example, if one of the links between intermediate routers goes down. In this case, the MPLS network faces the issue of generating a new tunnel in place of the original tunnel so that traffic trunks have a path to destination. However, there is a lag time from when a link fails to when a new path is established. This lag time, however, may not be acceptable to customers and services that rely on the network for reliable services. Consider, for example, TCP-based services that are using this link—in fact, in most actual networks, the bulk of the services are TCP-based. From the transport layer protocol perspective, a new TCP connection can be established on the new path after a failure (when the old connection times out). However, this also introduces delay; furthermore, if n TCP sessions are using such a link, $3n$ messages will be generated due to a TCP connection set phase. Such a delay and overhead also impact user perception on service reliability. To nullify such a delay and overhead, MPLS has introduced the concept of *fast reroute*. This means that when a TE tunnel (LSP) using Path message is established, a backup path is also established that is along routers and links that do not belong to the first path; this setup can be on a local basis or on an end-to-end basis between two LERs. To provide this functionality, RSVP-TE has also added two new objects, FAST-REROUTE and DETOUR. To control the backup for a protected LSP, the FAST-REROUTE object is used. The bandwidth to be guaranteed is the value in the bandwidth field of the FAST-REROUTE object. An LSP that is used to reroute traffic around a failure in one-to-one backup is referred to as a *detour LSP*; then, the DETOUR object is used in the one-to-one backup method to identify detour LSPs.

A question remains on how an RSVP-TE message is sent between two MPLS routers. In practice, MPLS routers are actually integrated IP/MPLS routers; thus, an RSVP-TE message is sent over the IP control plane, on a hop-by-hop basis.

To summarize, RSVP-TE supports the following key functionalities: downstream-on-demand label distribution, explicitly routed LSPs, and allocation of network resources. It also allows tracking of the actual route traversed by an LSP and loop detection. Rerouting is possible through FAST REROUTE.

18.3.4 Traffic Engineering Extensions to OSPF and IS-IS

Earlier in Section 18.2, we highlighted the additional attributes that need to be communicated about a link by the routing protocol. This information is then used by a routing computation module, whether centralized or decentralized, to determine LSPs for traffic trunks. Here, we summarize the extensions to protocols OSPFv2 and IS-IS for MPLS traffic engineering.

EXTENSIONS TO OSPFv2

In OSPFv2, several link state advertisement (LSA) types have been already defined. One of them is known as the *opaque LSA* [149], briefly described earlier in Section 6.2.8. The intended use of opaque LSA is to allow a general LSA feature so that it might be useful for any future extension. With regard to the opaque LSA, three link state types have been presented for the scope of flooding; they are known as type 9, type 10, and type 11 for local subnet flooding, intra-area flooding, and flooding in the entire autonomous systems, respectively.

For MPLS traffic engineering, opaque LSA type 10 is used; this limits flooding to an intra-area of an OSPF domain; it is known as a traffic engineering LSA (TE LSA). The TE LSA [355] contains a standard header that includes information such as link state age, advertising router, and link state sequence number; in addition, it uses nested TLV to contain information needed for TE LSA. At the top level, there are two TLVs: (1) a router address TLV and (2) a Link TLV. The Router Address TLV contains the IP address of the advertising router; this is preferably a stable and reachable address such as loopback addressing (refer to Section 8.4).

The link TLV contains several sub-TLVs. The key ones have already been mentioned in Section 18.2: maximum link bandwidth, maximum reservation bandwidth, unreserved bandwidth available at different priority levels, and traffic engineering metric. Note that IP differentiated services requirements can be mapped to the different priority levels. In addition, the link is identified as to whether it is a point-to-point or a multi-access link.

EXTENSIONS TO IS-IS

Recall that in IS-IS, each intermediate system, i.e., the router, advertises link state protocol data units (LSPs), which are analogous to LSAs in OSPF. The basic format of an LSP is a fixed header followed by a number of TLV-formatted information units. The important point to note is that LSPs already use TLV encoding. Thus, IS-IS extensions for traffic engineering define new TLVs. Most importantly, it replaces the extended IS reachability TLV that was originally defined in the IS-IS protocol. The new TLV types are 22, 134, and 135 (see Table 6.2 in Chapter 6).

The proposed extended IS reachability TLV specifies the following: system ID, default metric, and then through sub-TLVs the fields needed for traffic engineering such as maximum link and reservable link bandwidths are specified, the same ones as in an OSPF extension.

18.4 Generalized MPLS

Generalized MPLS (GMPLS), as the name suggests, is an extension of MPLS. Certainly, the question then is why such an extension is needed and what it is meant for. Recall that MPLS has been designed to *switch* packets using a labeling mechanism. Yet, there is the need for an MPLS control-type functionality for controls that is beyond just switching packets, such as wavelength switching, time division multiplexing, and fiber (port) switching. This mode of switching is traditionally referred to as *circuit switching* or *circuit routing* since a dedicated path and physical resources must be allocated for a service from one end to another. Note that the term *circuit switching* is commonly used with telephony voice circuit switching, and is not always the best terminology to use in this generalized context.

It is worth noting that originally GMPLS was intended for wavelength (lambda) switching where a dedicated path is required to be set up for a wavelength path end to end; however, it was soon realized that there is a need for a similar framework that can be used for other switching as well. GMPLS is thus intended for the following four switching capabilities:

- *Packet-Switch Capable (PSC)* for IP packets or ATM cell-level switching

- *Time-Division Multiplexing Capable (TDMC)*: for timeslot-based circuit switching

- *Lambda-Switch Capable (LSC):* for wavelength switching at optical cross-connects

- *Fiber-Switch Capable (FSC):* for fiber-level switching at optical cross-connects.

Collectively, for brevity, we will refer to them as *generalized modular switching* (GMS) instead of calling them circuit switching. Note that this is our terminology, used here for ease of discussion, to save us, each time, from discussing/mentioning switching for all of the above four technologies. GMS is thus distinguished from the umbrella suite GMPLS; *modular* is used here since in GMPLS, switching can only be on well-defined modular values that are tied to specific technology, for example, OC-3 rate in SONET or T1 for TDM. This certainly makes sense for TDM, LSC, and FSC. What about PSC? This requires some elaboration. Consider Packet over SONET. Here the data stream is coming as IP packets at an *arbitrary* data rate that is then mapped to a specific SONET frame rate through *asynchronous* mapping in an envelope mode; similarly, if IP traffic is to be carried over TDM switching, then PSC capability means that IP packets coming at an arbitrary data rate are mapped to, say, a T3 rate for TDM switching. Thus, for PSC, an incoming stream is mapped to modular value of a data rate, which may not be completely filled.

In essence, GMPLS is an umbrella suite that encompasses signaling and traffic engineering for generalized modular switching services. Note that GMPLS encompasses MPLS due to PSC, however, with a few twists. While in MPLS, a bandwidth request can be any quantity, in GMPLS the bandwidth request corresponds to one of the well-defined modular values such as asynchronous mapping of T1 or Packet over SONET at an OC-3 rate. That is, if PSC is used in a GMPLS environment, then the requirement must be mapped to one of the allowable packet switching types along with the associated bandwidth. In addition, due to the bidirectional nature of the services offered by GMPLS, LSPs must be set up in each direction for a path to be operational.

It is important to note that GMPLS is not necessarily for IP traffic, since it is meant for generalized modular switching. Thus, a GMPLS tunnel does not need to start or end on an

IP router; rather, it starts and ends on *similar* GMS nodes. To allow for different GMS types, a generalized label has been defined for GMPLS; this is discussed later.

GMPLS allows LSP setup for protection for any link-related failure. Thus, a primary and secondary (backup) LSP can be set up if and when needed. Thus, an LSP setup indicates whether it is primary or secondary. Furthermore, protection types such as dedicated $1 + 1$, dedicated 1:1, and shared 1:N can be announced. Furthermore, GMPLS allows control and data channels to be separate.

18.4.1 GMPLS Labels

GMPLS defines several types of labels. The most common one is known as a *generalized label request*. A generalized label request includes three pieces of information: LSP encoding type, switching type, and generalized payload identifer (G-PID) (see Figure 18.11). LSP encoding types are values such as Packet, Ethernet, SDH, and Lambda (see Table 18.2). Switching type refers to the generalized modular switching discussed earlier with additional details. For example, within PSC, variations in implementation such as asynchronous mapping, bit synchronous mapping, and byte synchronous mapping are possible; thus, they are identified separately while G-PID specifies the payload identifier (see Table 18.3 and Table 18.4). Note that since GMPLS is for generalized modular switching, bandwidth encoding for well-defined data rates such as DS0, DS1, and Ethernet has also been defined.

In addition to the standardized label request discussed so far, GMPLS allows a generalized label, port, and wavelength label by using a 32-bit field without specifying any details. For wavelength switching, there are three fields defined, each of 32 bits; they are identified wavelength ID, start of a label, and end of a label (Figure 18.12).

TABLE 18.2   GMPLS LSP encoding type.

Value	Type
1	Packet
2	Ethernet
3	ANSI PDH
5	SONET ANSI T1.105/ SDH ITU-T G.707
7	Digital Wrapper
8	Lambda (photonic)
9	Fiber
11	FiberChannel
12	ITU-T G.709 Optical Data Unit (ODU*k*)
13	ITU-T G.709 Optical Channel

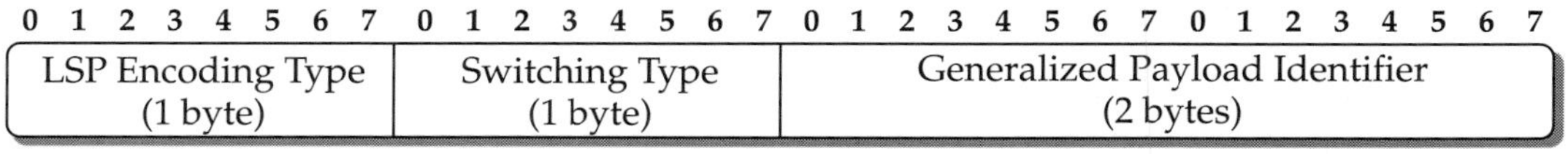

```
0 1 2 3 4 5 6 7 0 1 2 3 4 5 6 7 0 1 2 3 4 5 6 7 0 1 2 3 4 5 6 7
```

LSP Encoding Type (1 byte)	Switching Type (1 byte)	Generalized Payload Identifier (2 bytes)

FIGURE 18.11   Generalized label request.

TABLE 18.3   GMPLS switching types.

Value	Type
1	Packet-Switch Capable-1 (PSC-1)
2	Packet-Switch Capable-2 (PSC-2)
3	Packet-Switch Capable-3 (PSC-3)
4	Packet-Switch Capable-4 (PSC-4)
51	Layer-2 Switch Capable (L2SC)
100	Time-Division-Multiplex Capable (TDM)
150	Lambda-Switch Capable (LSC)
200	Fiber-Switch Capable (FSC)

```
0 1 2 3 4 5 6 7  0 1 2 3 4 5 6 7  0 1 2 3 4 5 6 7  0 1 2 3 4 5 6 7
```

Waveband Identifier (4 bytes)
Start Label (4 bytes)
End Label (4 bytes)

FIGURE 18.12   GMPLS label for wavelength switching.

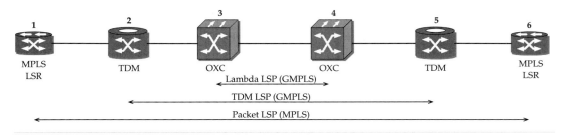

FIGURE 18.13   Label stacking and hierarchical LSPs: MPLS/GMPLS.

18.4.2 Label Stacking and Hierarchical LSPs: MPLS/GMPLS

It is possible to coordinate between MPLS and GMPLS through label stacking to create hierarchical LSPs. This can be useful when operating over multiple technologies such as in multilayer networking (refer to Section 25.3). In this section, we present a simple example to illustrate this nested label concept over multiple technologies.

In Figure 18.13, we consider a label from an MPLS router marked as node 1 to an MPLS router marked as node 6. This LSP is an MPLS packet level LSP. This LSP, however, is connected on the GMPLS LSP between node 2 and node 5, which are TDM switches. Note only that the TDM level GMPLS-based LSP is carried further over two optical switches, marked as node 3 and node 4. Thus, the original LSP between node 1 and node 6 is nested in GMPLS LSP between node 2 and node 5, which, in turn, is nested in another GMPLS LSP between node 3 and node 4.

TABLE 18.4   Examples of GMPLS Generalized Protocol identifier (G-PID).

Value	Type	Technology
0	Unknown	All
6	Asynchronous mapping of DS3/T3	SDH
7	Asynchronous mapping of E3	SDH
8	Bit synchronous mapping of E3	SDH
9	Byte synchronous mapping of E3	SDH
13	Asynchronous mapping of E1	SDH
14	Byte synchronous mapping of E1	SDH
15	Byte synchronous mapping of 31 * DS0	SDH
16	Asynchronous mapping of DS1/T1	SDH
17	Bit synchronous mapping of DS1/T1	SDH
18	Byte synchronous mapping of DS1/T1	SDH
19	VC-11 in VC-12	SDH
22	DS1 SF Asynchronous	SONET
23	DS1 ESF Asynchronous	SONET
24	DS3 M23 Asynchronous	SONET
25	DS3 C-Bit Parity Asynchronous	SONET
28	POS—No Scrambling, 16-bit CRC	SDH
29	POS—No Scrambling, 32-bit CRC	SDH
30	POS—Scrambling, 16-bit CRC	SDH
31	POS—Scrambling, 32-bit CRC	SDH
32	ATM mapping	SDH
33	Ethernet	SDH, Lambda, Fiber
34	SONET/SDH	Lambda, Fiber
36	Digital Wrapper	Lambda, Fiber
37	Lambda	Fiber
38	ANSI PDH	SDH
43	FiberChannel-3 (Services)	FiberChannel
44	HDLC	SDH
45	Ethernet V2/DIX (only)	SDH, Lambda, Fiber
46	Ethernet 802.3 (only)	SDH, Lambda, Fiber
47	G.709 ODUj	G.709 ODUk (with $k > j$)
48	G.709 OTUk(v)	G.709 OCh (ODUk mapped into OTUk(v))
53	IP/PPP (GFP)	G.709 ODUk (and SDH)
54	Ethernet MAC (framed GFP)	G.709 ODUk (and SDH)
55	Ethernet PHY (transparent GFP)	G.709 ODUk (and SDH)
58	Fiber Channel	G.709 ODUk, Lambda, Fiber

18.4.3 RSVP-TE for GMPLS

For GMPLS LSP setup, several setup and confirmation message types are used in RSVP-TE (see Table 18.6). Note that RSVP uses unreliable delivery since it is embedded in an UDP packet. However, GMPLS requires reliable delivery of setup messages; to accomplish this, a simple reliable delivery mechanism is used where the sender retransmits the setup message until it receives an acknowledgment message from the neighbor. Message retransmission can be frequent, such as every 10 millisec; however, an exponential decay mechanism can be used to reduce frequency if successive tries do not result in a response, along with a maximum timeout value to indicate failure to establish a tunnel.

To carry a GMPLS label, RSVP-TE creates an object by including its own 32-bit header to identify the length of the message, class number, and C-Type. For the generalized label request shown in Figure 18.11, the corresponding RSVP-TE object is shown in Figure 18.14.

0 1 2 3 4 5 6 7	0 1 2 3 4 5 6 7	0 1 2 3 4 5 6 7	0 1 2 3 4 5 6 7
Length (2 bytes)		Class Number (1 byte)	C-Type (1 byte)
LSP Encoding Type (1 byte)	Switching Type (1 byte)	Generalized Payload Identifier (2 bytes)	

FIGURE 18.14   RSVP-TE generalized label request object for GMPLS.

Similarly, the RSVP-TE header is included for other GMPLS labels. Note that class number and class type depend on the RSVP object (see Table 18.5).

Recall that label setup is bidirectional in GMPLS. To accomplish this in RSVP-TE, an Upstream-Label is included in the Path message. The class number for Upstream-Label is 35 and the C-Type used is as in RSVP-LABEL (see Table 18.5). The data rate of particular GM-PLS connections is also communicated through an RSVP-TE Path message using SENDER-TSPEC. Since the data rates for GMPLS generalized modular switching are well-defined, the actual rate does not need to be coded; instead, a mapping value is provided for well-known data rates; this is listed in Table 18.7.

18.4.4 Routing Protocols in GMPLS

The role and use of a routing protocol in GMPLS are primarily to enable traffic engineering of GMPLS-based connection-oriented networks. There are two important points to note:

- GMPLS-based networks require a link state–based framework for communicating status of links.

TABLE 18.5   RSVP object examples for GMPLS, in addition to Table 18.1

Object Name	Used in	Class Number	Examples: Class-Type with value in () (source RFC listed as [] from bibliography)
RSVP-LABEL	Path	16	Generalized Label (2) [74]; Wavelength switching (3) [74]
LABEL-REQUEST	Path	19	Generalized label request (4) [74]
SENDER-TSPEC	Path	12	Integrated Services (1) [90]; G.709 (5) [542]
Upstream-Label	Path	35	Same as in RSVP-LABEL
RECOVERY-LABEL	Path	34	Same as in RSVP-LABEL
PROTECTION	Path	37	Type 1 Protection (1) [74]
NOTIFY-REQUEST	Path, Resv	195	IPv4 request (1) [74]

TABLE 18.6   GMPLS message type and RSVP-TE protocol messages.

GMPLS Message type for setup	RSVP-TE Protocol message
LSP Setup	Path
LSP Accept	Resv
LSP Confirm	ResvConfirm
LSP Upstream Error	PathErr
LSP Downstream Error	ResvErr
LSP Downstream release	PathTear
LSP Upstream release	PathErr
LSP Notify	Notify

TABLE 18.7 Data rate for GMPLS.

Signal Type	Bit rate (in Mbps)	32-bit Encoding value (in hex)
DS0	0.064	0x45FA0000
DS1	1.544	0x483C7A00
E1	2.048	0x487A0000
DS2	6.312	0x4940A080
E2	8.448	0x4980E800
Ethernet	10.000	0x49989680
E3	34.368	0x4A831A80
DS3	44.736	0x4AAAA780
STS-1	51.840	0x4AC5C100
Fast Ethernet	100.000	0x4B3EBC20
E4	139.264	0x4B84D000
FC-0 133M	—	0x4B7DAD68
OC-3/STM-1	155.520	0x4B9450C0
FC-0 266M	—	0x4BFDAD68
FC-0 531M	—	0x4C7D3356
OC-12/STM-4	622.080	0x4C9450C0
GigE	1,000.000	0x4CEE6B28
FC-0 1062M	—	0x4CFD3356
OC-48/STM-16	2,488.320	0x4D9450C0
OC-192/STM-64	9,953.280	0x4E9450C0
10GigE-LAN	10,000.000	0x4E9502F9
OC-768/STM-256	39,813.120	0x4F9450C0

- The actual path computation algorithm need not be within the scope of these routing protocols.

You may recall that we have made the above points about link state routing protocols earlier in Chapter 3. Since GMPLS can be used for different technologies, the routing protocols need to be somewhat generic so that it can be used in any of these technologies. At the same time, there are existing routing protocol frameworks that can be applicable, for example, OSPF and IS-IS.

To satisfy the traffic engineering requirements, extensions to OSPF/IS-IS have been developed. However, first, we need to note that a "link" in a GMPLS network may not necessarily be a physical link. Depending on the technology, it may ride over another physical technology; this will be discussed later in the context of multilayer networking. Second, an LSP that has been set up can serve as a point-to-point link for other nodes. Third, a protection notion about a link may be helpful to communicate that might be useful to the routing path computation module. Thus, in general, it is safer to refer to a link in the context of GMPLS routing as a *TE link*. Since in GMPLS, control and data planes are completely separate, appropriate identifiers must be used so that the end of a link can be identified, which is discussed later in Section 18.4.5.

GMPLS extension of traffic engineering of a routing protocol relies on MPLS traffic engineering extensions. For example, sub-TLVs discussed earlier for MPLS such as maximum link bandwidth are still applicable. In addition, the following new sub-TLV types have been defined:

- Link Local/Remote Identifiers to provide support for unnumbered links.

- Support for link protection to announce the protection capability available for a link; this may be useful for path computation. Typical values are unprotected, shared, dedicated $1 + 1$, and dedicated 1:1.

- Interface Switching Capability Descriptor to identify generalized modular switching capabilities (see Table 18.3).

- Shared Risk Link Group (SRLG) identification: This issue arises from multilayer networking since multiple TE links at a particular layer may be using the same "link" at a lower layer, such as at a fiber level. Alternately, a TE link may belong to multiple SRLGs. This information may be used for path computation as well. We will describe SRLGs in detail later in the context of multilayer networking.

We now briefly highlight some of extensions to OSPF and IS-IS.

OSPF AND IS-IS EXTENSION

To allow for TE link information exchange, OSPF again uses the *opaque* LSA—this is the same as done for MPLS. That is, for TE usage, an opaque-type value of 10 is specifically assigned; this LSA is known as a TE LSA. From an information encoding point of view, a TE LSA uses one top-level TLV along with nested sub-TLVs, when needed, for an unnumbered link identifier, link protection type, SRLG, and so on. The sub-TLVs then include the extension information described above. Note that this is a continually evolving standard; for an up-to-date list of sub-TLVs for TE, refer to [317].

In IS-IS, all routing information encoding is always done using TLVs. Thus, for GM-PLS, new TE TLVs have been defined to capture information such as protection information, SRLGs, and so on, which we have listed earlier in this section (also, see Table 6.2 in Chapter 6). An up-to-date list of defined TLVs can be found at [316].

18.4.5 Control and Data Path Separation and Link Management Protocol

Control and data path separation in GMPLS is quite different from IP networks. Recall that, in IP networks, there is no separation of control and data traffic carried in terms of physical channel or partitioned bandwidth. An IP link carries both control and data traffic—the separation of control traffic is identified either at the IP protocol type field level (e.g., OSPF packet) or port level (e.g., RSVP packet). In an IP/MPLS environment, although there is separation of control traffic and MPLS packet forwarding, they both use the same logical link between two routers.

Now let us consider a different example: PSTN with SS7 for signaling. In PSTN, data traffic means the voice calls, which are carried on TDM trunks, while control traffic, for example ISUP messages for call setup, is sent over the SS7 network—that is, on a complete separate network. Thus, ISUP call setup messages on the SS7 network traverse on a completely different path or channel than the voice circuits (for example, refer to Figure 12.9).

In GMPLS, control and data path separation is similar to the PSTN/SS7 architecture. In GMPLS, they are separated through use of separate channels; the difference between the GMPLS approach and that of PSTN is that GMPLS does not define a completely separate

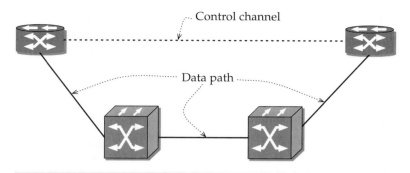

FIGURE 18.15 Control and data path separation in GMPLS.

network architecture like SS7 for PSTN. Instead, a separate channel may be dedicated for the delivery of control traffic from data traffic (Figure 18.15). More importantly, the channel that carries control traffic need not be on the same physical path as the data traffic. Recall that in PSTN, the voice circuit to be used is identified in an SS7 message through trunk ID (that is, TCIC code, refer to Section 12.8); similarly, in GMPLS the data link must be explicitly identified so that this can be communicated through exchange of control traffic.

Separation of control and data paths, however, necessitates the need for a link management control so that the two ends of a data link can communicate link-level information, which is completely decoupled from control information for traffic engineering. To accomplish this, the link management protocol (LMP) has been defined for use in a GMPLS networking environment [396].

LMP provides two core functions: (1) *control channel management*, and (2) *link property correlation*. The role of control channel management is to establish and maintain bidirectional control channels between adjacent nodes. The establishment step is accomplished using a configuration message; after establishment, a keep-alive message is generated frequently to ensure that the control channel is up. The link property correlation function of LMP allows aggregation of multiple, parallel data links into a TE link along with synchronization of the properties of the TE link. Multiple-link situation arises, for example, when there are multiple data links between ports of two adjacent nodes—they can be combined into a TE link through the link property correlation function.

From Figure 18.15, we can see that it is quite possible in GMPLS to have control channels on a completely different physical facility from the data channels. To check that both channels/paths are operating normally, additional functions have been defined. There are two additional LMP functions for this purpose: link connectivity verification and fault management. Link connectivity verification is accomplished by sending a test message over the data channel; in turn, a test status message is sent back over the control channel to complete such verification. That is, the verification is performed only at the end nodes of a TE link.

When the control channel uses a different path from the data channel, additional issues come up. Specifically, two scenarios are possible: (1) the channel that carries the control messages has a failure, while the data path is intact, (2) the control path is intact; however, a part on the data path is lost due to a failure. In the first case, the data path will continue to carry data traffic—an issue is that if an established data path is to be torn down, this cannot be communicated due to the control channel being down. The second scenario is more problematic. Here the control-ends think that the data path is working fine; however, it is not any more. To address this problem, LMP fault management procedure has been additionally defined. Briefly, the LMP fault management procedure is based on a ChannelStatus message exchange that uses the following messages: ChannelStatus, ChannelStatusAck, ChannelStatusRequest, and ChannelStatusResponse. The ChannelStatus message is sent unsolicited and is used to notify an LMP neighbor about the status of one or more data channels of a TE link. The ChannelStatusAck message is used to acknowledge receipt of the ChannelStatus message. The ChannelStatusRequest message is used to query an LMP neighbor about the status of one or more data channels of a TE Link. The ChannelStatusResponse message is used to acknowledge receipt of the ChannelStatusRequest message and to indicate the status of the queried data links.

Finally, it is worth noting that, to use LMP, the end nodes are first established with a control channel; then, addresses of the ends from the control channel are used to establish an LMP communication.

18.5 MPLS Virtual Private Networks

While MPLS was originally intended for controlled IP traffic engineering, it has been found useful in virtual private networking (VPN) as well. Suppose that a corporate customer has offices in different physical locations distributed geographically; such customers would like to lease a seamless "network" service through a provider that has geographic presence in these areas and would require this provider to carry their corporate intranet IP-based traffic. Since MPLS is a label-based concept, a VPN backbone provider can assign (provision) a unique label along with an LSP for traffic between any two sites, including any bandwidth guarantee using RSVP-TE label setup. To do that, the VPN backbone provider would need to have LERs, commonly referred to as *provider edge (PE) routers*, at each site to which customers can be connected through their *customer edge (CE) routers*. To the customer, it appears to be a point-to-point link with a guaranteed data rate. For the VPN backbone provider, the MPLS-based approach is appealing since it can combine different customers' traffic on the *same* network by doing software-based provisioning of LSPs for different customers. Such VPN service is also known as provider-provisioned VPN (PPVPN) service; generic requirements for PPVPN can be found in [517] and terminologies are described in [17].

While conceptually this approach is appealing, there is a key issue: address conflict. What does address conflict mean? Private IP addressing as defined in RFC 1918 [592] has become a popular mechanism for addressing hosts within almost any organization or corporate environment. Because of the prevalence of private IP addressing, two organizations may both choose to number hosts in the address block, say 10.0.0.0/8; certainly, within an organization the address allocation from this address block is unique for different hosts and subnets. If both these organizations are to be supported by a VPN provider, then address conflict will exist since both are using the same address block. Thus, an identifier is needed so that they can be distinguished and the traffic is properly routed through the VPN provider's network.

A possible approach for identifier tracking is that every router in the provider's network maintains routing information for all customers' networks. While this is a possible approach, it raises a second issue: scalability; this is because there would be limitations based on how many sites a router can support in terms of amount of routing information.

18.5.1 BGP/MPLS IP VPN

The BGP protocol has been extended for IP VPN to provide a mechanism for private IP traffic for different customers to be carried on the same network. The bearer network is maintained by a VPN network provider.

Consider a customer that has three locations separated by geographic distance. It uses private IP address block, 10.0.0.0/8, for numbering its internal networks at these three different locations, and numbers its private IP prefixes (routes) for its three geographically separate locations as 10.1.0.0/16, 10.2.0.0/16, and 10.3.0.0/16. Thus, traffic generated at 10.1.0.0/16 destined for other sites would need to be routed to either 10.2.0.0/16 or 10.3.0.0/16. In public domain Internet, if we were to consider three such public IP address blocks or prefixes (routes), they would belong to three different organizations; we would then consider the respective AS numbers they belong to and then do a BGP advertisement to announce the prefixes along with the AS number. However, when prefixes belong to a private IP address space, they cannot be advertised simply as in public BGP since another customer might number its private IP address space exactly in the same manner. If then a VPN provider were to carry traffic for both these customers, it would need to have physically partitioned channel in order to avoid address space conflict.

Instead of doing separate partitioned channels for different customers, another approach is possible using the BGP/MPLS approach. In this approach, to support different customers with addresses coming from the same private address space, the private IP prefix is prepended with a VPN *route distinguisher*; together this address family is known as a *VPN-IPv4 address family*. For clarity, a private IP prefix associated with this address family will be referred to as a *VPN-IPv4 route*. An advantage of BGP multiprotocol extension is that it allows multiple address families to coexist, thus allowing BGP to carry routes for multiple address families. To make the BGP/MPLS approach work, customers and IP VPN providers are separated in term of traffic hand-off. Each customer needs to have its traffic exit through a CE router that is then sent to the PE router. When a PE router receives a private IP-based route from a CE router, the PE router changes it to a unique VPN-IPv4 route for announcement on the BGP/MPLS VPN network.

In a nutshell, BGP/MPLS takes the following approach:

- Each VPN-IPv4 route is assigned an MPLS label.

- The PE router, when generating a BGP advertisement, announces the assigned MPLS label with the VPN-IPv4 route.

- For labels received for the VPN-IPv4 route, LSPs are established, as necessary, between PE routers using, for example, the RSVP-TE protocol.

- Actual data flow uses the already established appropriate LSPs.

In order to allow for association between route information and actual packet forwarding, each PE router needs to maintain *VPN routing and forwarding tables*, or *VRFs*, so that the association works at the time of actual packet forwarding. Also note that when the VPN-IPv4 route is advertised by a PE router, the next-hop attribute in the BGP message points to this PE router.

Type 0

Type Field (2 bytes)	Administrator Subfield (2 bytes)	Assigned Number Subfield (4 bytes)	IPv4 Address (4 bytes)

Type 1 and Type 2

Type Field (2 bytes)	Administrator Subfield (4 bytes)	Assigned Number Subfield (2 bytes)	IPv4 Address (4 bytes)

F I G U R E 18.16   Encoding of route distinguisher in BGP/MPLS.

The VPN Route Distinguisher in the VPN-IPv4 address family is an 8-byte address; thus, collectively, the VPN-IPv4 address looks like a 12-byte address. Since IPv4 can be from the same private address space used by different customers, the VPN-IPv4 then helps uniquely distinguish routes. The Route Distinguisher part is divided into a type and value field of 2 bytes and 6 bytes, respectively. Currently the type field takes on three values: 0, 1, and 2. The value field is made up of two subfields: an administrator subfield and an assigned number subfield; their length depends on the specific type field, as shown in Figure 18.16. In case of Type 0, the administrator subfield contains the 2-byte AS number; this can be a public or private AS number. The assigned number subfield serves as a degree of freedom allowing the customer that has that AS number to assign any internal numbering scheme. For example, a customer might internally want to do such a numbering scheme based on its different locations. Type 2 is similar to Type 0 where the 4-byte AS number extension takes the 4-byte administrator subfield, whereas the assigned number subfield is limited to 2 bytes. For Type 1, the administrator subfield is set to 4 bytes, but it must be an IPv4 address; thus, the assigned number subfield is limited to 2 bytes. The IPv4 address in Type 1 can be a public IP address. We can see that type and value fields together can uniquely identify different routes, when the administrator subfield uses public addressing for an AS number or IP address. However, a customer may choose to announce the route distinguisher using a private AS number or private IP address in the administrator subfield; if this is so, then the provider may choose to impose restrictions on the assigned number subfield so that the routes are distinct within the provider's network.

Example 18.2 *Illustration of labeling in BGP/MPLS IP VPN between two customer edge routers of a customer.*

In Figure 18.17, we show an IP packet being sent from CE router CE1 to CE router CE2 at another location; this packet is intended for a network/subnet served behind CE2 (not shown in the figure). Note that the CE1 is connected to the PE router, PE1; similarly, CE2 is connected to PE2. In between PE1 and PE2, the provider has LSRs; we show two in Figure 18.17—we assume that between PE1 and PE2, a label-switched path (LSP) is already established for carrying this customer's traffic.

First, CE1 forwards the packet to its default gateway, which happens to be PE1. On receiving this packet from CE1, PE1 consults its VRF for route lookup. For the VRF to know that there is a route to networks supported behind CE2, PE1 would need to have already received an advertisement from PE2 about networks supported by CE2. Suppose that the VPN-IPv4

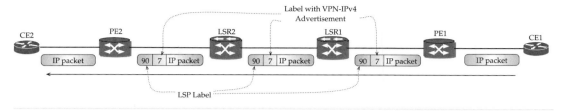

FIGURE 18.17   Prepending of two labels in BGP/MPLS packet forwarding.

address for the networks supported behind CE2 is advertised as MPLS label 7. The VRF at PE1 would then have an entry for MPLS label 7. Thus, the IP packet will first be prepended with the MPLS label 7. Second, PE1 would look up the LSP setup between PE1 and PE2 that happens to have label 90; thus, the packet will now be prepended by label 90 for using this LSP for packet forwarding.   ▲

We can see from the above example that every IP packet will have *two* labels prepended: the inner label is for route distinguisher, and the outer label is for packet forwarding through the MPLS VPN provider's network. It is important to note that the route distinguisher is advertised through a BGP advertisement; it is not included in the actual packet forwarding; instead an MPLS label is used. This MPLS label (inner label) is the information advertised with the route distinguisher.

While injection of two labels on an IP packet certainly incurs additional overhead, it provides the ability to clearly separate the route distinguishing part and the LSP part; furthermore, because of this advantage, a provider has the flexibility to use a single LSP between two sites to carry traffic between different customers. Alternately, for the same customer and for traffic between two sites, the VPN provider's network may choose to set up multiple LSPs through its network for its traffic engineering requirements while satisfying the customers' goals; note that use of multiple LSPs to split traffic between two sites is transparent to the customers. VPN traffic engineering will be discussed later in Section 19.2. Moreover, there is a scalability advantage since only one label for each pair of ingress/egress points is sufficient through the backbone, instead of having different labels for different customers for the same pair of ingress/egress points.

Now consider again Figure 18.17. If the customer edge routers are also MPLS LERs, then the LSP can possibly be set up from CE1 to CE2 where the receiving customer edge router is responsible for the outer label processing. However, to reduce load on customer edge routers, a function known as *Penultimate Hop Popping* is performed at the provider edge router ("penultimate hop router") in which the outermost label of an MPLS packet is removed before the packet is passed to the customer edge router. To activate this function, the edge router must indicate to the penultimate hop router to use implicit null label 3.

Example 18.3 *Illustration of multiple customers in a BGP/MPLS IP VPN provider's network.*
Consider the topology shown in Figure 18.18, where two customers are identified as Customer 1 and Customer 2. Each uses private address space 10.0.0.0/8, which is assigned to each site as indicated in Figure 18.18; their CE routers in a particular location connect through the same PE routers in the MPLS VPN provider's network.

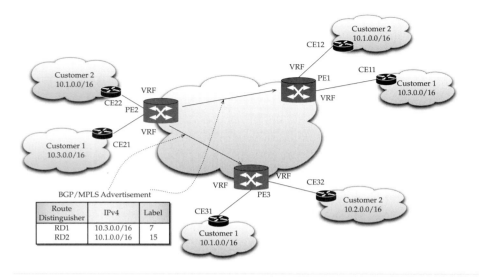

FIGURE 18.18   BGP/MPLS example with multiple customers.

At each PE router, two VRFs would be maintained, one for each customer. When initialized, the network will use an M-BGP protocol to exchange route information, rather than VPN-IPv4 route information, as shown in Figure 18.18 from PE2 to other PE routers—note that when advertising this route, the BGP announcement will include the next hop as that of this PE. Similarly, the other PE routers will advertise route information (not shown in the figure).   ▲

It may be noted that the BGP extended community attribute can also be used with distribution of VPN routing information. For example, this can be helpful to customers in controlling the distribution of routes.

Because of policy capabilities of BGP, two customers can choose to exchange traffic through MPLS VPN if they have any business relations. In the above example, Customer 1 might want to exchange data with Customer 2; once the policy determining which routes are allowed to which customer is finalized, the policies can be set up at the PE routers. It should be clear by now that PE routers are acting as BGP speakers.

In general, it may be noted that BGP/MPLS is not designed for all types and sizes of customers or VPN services. For example, a customer that has just two sites in two different locations might choose to tunnel any internal traffic through the public Internet by encrypting the data such as IPSec, especially if it does not require any guaranteed QoS; such a customer does not necessarily need to use the BGP/MPLS approach. This means that the BGP/MPLS approach is meant more for moderate to large customers that have multiple sites and required guaranteed QoS and that have an interest in sharing business traffic with other similar customers; for them, a BGP/MPLS provides a good alternative in an IP framework while bypassing the public Internet. It is worth pointing out that BGP/MPLS is meant for a single VPN provider to operate its network in order to carry traffic for different customers; it does not solve the need of a customer that wants a VPN tunnel between two locations that are not served by a single VPN provider's network—in such cases, often a VPN provider becomes the contract ("official") provider who, in turn, makes contractual arrangements, with another provider to carry the traffic for the customer to locations with which the official provider has no connectivity.

What about traffic engineering from the point of view of a VPN provider? This will be discussed later in Section 19.2. Finally, we point out that BGP/MPLS does not have multicasting capability as originally envisioned; extensions have been planned.

18.5.2 Layer 2 VPN

In the previous section, we discussed layer 3 VPN using the BGP/MPLS IP VPN approach. Many customers with offices located in geographically disparate sites are also interested in layer 2 VPN services; this means that they have a layer 2, e.g., Ethernet, subnet that they would like to tunnel across a wide-area network so that all sites for a particular customer look as if they are on the same LAN by, e.g., using standard Ethernet frames. Such a service is known as a *Virtual Private LAN Service (VPLS)*, *transparent LAN services (TLS)*, or "emulated" Ethernet service [449].

A key advantage of MPLS is its label-based forwarding mechanism to carry any type of packets. Thus, input to an MPLS router need not be only IP packets; it can be any other protocol packets as long as the ingress MPLS router has the functionality to accept them. When such a packet arrives, MPLS routers add the MPLS label header and forward it to a destination at another site. For example, such a received packet at an MPLS ingress router can be an Ethernet frame [449]. Thus, this frame will be carried as an MPLS packet through the wide-area network and is delivered to an egress MPLS router that will strip off the MPLS header and deliver the Ethernet frame on the LAN at the other site.

Example 18.4 *A layer 2 over MPLS example.*
Consider two geographically disparate sites that a customer needs to be on subnet 10.2.3.0/24. Now consider an IP packet generated at host 10.2.3.49 destined for 10.2.3.22 where the second host is at a different site.

As shown in Figure 18.19, the IP packet generated at host 10.2.3.49 will be encapsulated as an Ethernet frame that will arrive at the layer 2 CE device, which forwards it to a provider ingress PE router. The Ethernet frame is then encapsulated in an MPLS packet using an LSP to reach the egress PE router where the MPLS label is stripped, and the Ethernet frame is sent to the customer edge device for forwarding to the LAN for delivery to 10.2.3.22. As far as the two hosts are concerned, the Ethernet frame stayed on the same LAN.   ▲

Note that for such services, for each Ethernet frame that arrives at the ingress MPLS router, two labels are added: (1) at MPLS level to identify the input source, and (2) a label that serves as the LSP tunnel from the ingress to the egress router. Once the packet arrives at the egress router, it first removes the label associated with the LSP tunnel, and then for the second label, a lookup table is checked to find the destination interface to the customer edge device.

Finally, layer 2 VPN creates an interesting scenario in that more than two sites for the same customer might want to be connected in the same emulated LAN. Consider again Fig-

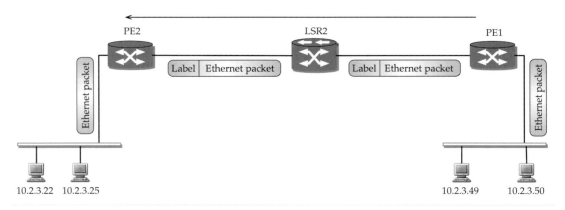

FIGURE 18.19   Layer 2 VPN over MPLS.

ure 18.19 and imagine a third site for the same customer that is also on 10.2.0.0/16 subnet. In this case, copies of the packet would need to be sent to each destination requiring multiple LSP setup. While this approach is doable, it requires many more LSPs to be set up as well as extra packet transmission, resulting in more bandwidth consumption than necessary. In fact, this approach can be thought of as the conference service on PSTN where more than two parties can join conference, while the underlying connections are all unicast based. The point-to-multipoint (P2MP) concept for use in MPLS is currently being developed to consider this scenario. Later in Section 19.2.4, we will discuss traffic engineering implications for point-to-point and P2MP virtual private LAN services form the point of view of a VPN provider.

18.6 Summary

In this chapter, we have presented two enabling environments for traffic engineering: MPLS and GMPLS. While MPLS is exclusively used for packet-based networks, GMPLS can be used for both packet-based as well as circuit-based networks on modular well-defined data rate levels. Furthermore, GMPLS addresses bidirectional requirements in circuit-mode connections, whereas MPLS is unidirectional. For traffic engineering information exchange in MPLS/GMPLS environment, OSPF and IS-IS protocols have been extended to contain traffic engineering information. It is important to note that the TCP/IP stack is used for exchanging control traffic information in GMPLS. This requires GMPLS nodes to have IP addresses; this aspect is also significant when you consider nested label stacking between MPLS and GMPLS. This is shown earlier in Figure 18.13. Such use of the TCP/IP stack allows label exchanges to be performed seamlessly.

In subsequent chapters, we will discuss how these enabling environments help traffic engineering in a variety of networks.

Further Lookup

MPLS architecture is described in RFC 3031 [604], while label stack encoding is discussed in RFC 3032 [603]. LDP specification is described in RFC 3036 [16]. For extensive coverage of MPLS, see books such as [167], [211], [263], [271]. Originally, MPLS has been defined for use within a provider's network ("intra-AS"). Recently, there have been works on extending MPLS for inter-AS; see RFC 4216 [765], RFC 4726 [214].

There have been two primary candidates for traffic engineering specifications in MPLS: RSVP-TE and *Constraint-Routing Label Distribution Protocol* (CR-LDP, in short). The MPLS working group within IETF recommended discontinuation of any new effort on CR-LDP in 2003 [18]. Extensions of RSVP to RSVP-TE for MPLS have been described in RFC 3209 [43] and RFC 4090 [541], and updates in RFC 4420 [215]; applicability of RSVP-TE to LSP is described in RFC 3210 [44]. We refer readers interested in CR-LDP to RFC 3212 [337], RFC 3213 [35], and RFC 3214 [37].

RFC 2764 [255] presents a framework for IP-based VPN. The concept of BGP/MPLS was presented in RFC 2547 [601], and has been updated in RFC 4364 [602]; for recent work, see RFC 4684 [448] and RFC 4797 [588]. Example 18.3 is adapted from [632]. Layer 2 VPN encapsulation of Ethernet over MPLS has recently been described in RFC 4448 [449]; see also RFC 4761 [376] and RFC 4762 [398].

GMPLS was originally described in RFC 3471 [73] and has been updated in RFC 4201 [372], and RFC 4328 [542]. For routing extensions to GMPLS, see RFC 4202 [375], RFC 4203 [374], and RFC 4205 [373]. GMPLS extension to G.709 is described in RFC 4328 [542]. LMP is described in RFC 4204 [396]. Details on GMPLS can be found in [212].

Exercises

18.1. What are the key differences between MPLS and GMLPS?

18.2. What is the FAST-REROUTE object used for?

18.3. Sample messages for RSVP-TE are available at [744]. Investigate how different values are specified in RSVP-TE messages.

18.4. Explore the current activities of the MPLS-related working groups at IETF.

18.5. Explain the main differences between BGP/MPLS IP VPN and layer 2 VPN.

19

Routing and Traffic Engineering with MPLS

If one sticks too rigidly to one's principles,
one would hardly see anybody.

Agatha Christie

Reading Guideline

Understanding of the material presented in this chapter requires a basic knowledge of MPLS as presented in Chapter 18. In addition, material on IP traffic engineering (Chapter 7), network flow modeling (Chapter 4), and quality-of-service routing (Chapter 17) for three classes of problems considered is helpful. Note that each problem class should be read independently with the background material from the respective chapters identified here.

In this chapter, we present the applicability of MPLS for routing and traffic engineering for a set of representative real-world problems. Specifically, three problem classes are considered: (1) an integrated IP/MPLS environment for IP traffic engineering, (2) VPN traffic engineering/routing using MPLS, and (3) a voice over MPLS network, that is, an MPLS network where voice or multimedia real-time interactive service is provided. Our discussion here is primarily limited to intradomain traffic engineering.

MPLS is usable in a variety of ways. It is, however, important to understand *how* it is used so that for any future network design or services deployed, it is possible to explore if and how MPLS can be used.

19.1 Traffic Engineering of IP/MPLS Networks

In this section, we discuss traffic engineering of IP networks, where IP/MPLS-integrated routers are deployed. Recall that IP traffic engineering was discussed earlier in Chapter 7.

19.1.1 A Brisk Walk Back in History

We will first provide a short historical context for the emergence of MPLS for IP traffic engineering. This is only a very brief overview focusing on a few key facts in regard to IP traffic engineering and is not focused on providing precise details of what happened when. For a detailed history of MPLS and its forerunners, refer to [167], [263].

By the mid to late 1990s, it was realized that some form of traffic engineering of IP networks was needed. At that time, large IP networks were either using OSPF or IS-IS protocol and, primarily, either simple hop-based link weight or the inverse of link capacity as link weights. As discussed in Chapter 7, there are many network situations where it is possible for some links to have very high utilization if link weights are not assigned properly.

Somewhat independent of the above development, there were concerns about the IP forwarding engine's ability to handle a large volume of traffic. Concepts such as IP switching, tag switching, and aggregate router-based IP switching (ARIS) emerged in 1996. It was soon recognized that a standard switching approach for packets is needed, which led to the MPLS workgroup being chartered by IETF in early 1997. By 1999, the role of MPLS in IP traffic engineering was well recognized [42], [45], citing the limitation of OSPF/IS-IS in being able to move traffic away from heavily utilized links due to lack of any control mechanism.

By 2000, however, it was reported that there was indeed a systematic way to determine OSPF/IS-IS link weights for IP traffic engineering [233], [565] (see also [234], [566]). Certainly, this was good news for many ISPs who wanted to continue to run IP-only routers in their network. Six years later, many large ISPs continue to successfully run IP-only networks with good traffic engineering through optimal link weight determination coupled with good traffic matrix determination.

Certainly, MPLS has its place in IP traffic engineering; in fact, many other large ISPs currently successfully run IP/MPLS networks for controlled IP traffic engineering. It is also important to note that MPLS has now found roles in many arenas, as discussed in Chapter 18.

Thus, whether IP-only is better or worse than IP/MPLS for IP traffic engineering is often a matter of opinion and preference; furthermore, this is also tied to customers that a provider is serving as well as personnel and expertise locally available. Next, we will present the essence of IP/MPLS traffic engineering in a provider's network in its own right.

19.1.2 MPLS-Based Approach for Traffic Engineering

The basic question is how to control traffic movement through a network if we do not like current traffic flows on different links. Ideally, it is desirable to somehow force traffic to a certain path. This is where one of the benefits of MPLS comes into the picture; that is, an LSP can be set up, where desired and when desired, and the bandwidth flow can be limited.

First note that once a tunnel is set up through MPLS, at the IP level, specifically to the routing protocol, it appears as a logical link. In an IP/MPLS network, the routing protocols such as OSPF and IS-IS are used in extension mode, i.e., OSPF-TE or IS-IS-TE, which provides bandwidth information. This information is then used by the traffic engineering component to determine LSPs.

WHEN TRAFFIC DEMAND IS FIXED

We first start with an example in which the traffic demand is fixed and given.

Example 19.1 *A simple example of IP/MPLS traffic engineering*
We first discuss the four-node example we presented earlier in Chapter 7 for IP traffic engineering; the topology with demand and link capacity is reproduced in Figure 19.1. We assume that the goal is to minimize maximum link utilization; then, the optimal solution is to send 54.54 Mbps of the total east-west traffic on the southern route and the rest of the 5.45 Mbps on the northern route.

In the case with IP traffic engineering, the best we can do is to allow all flow, 60 Mbps, to take the southern route, which is accomplished by choosing link weights so that the cost of the southern route is lower than the northern route. Now consider employing MPLS on this same network. We can now set up two LSPs, one with the guaranteed bandwidth set to 54.54 Mbps (southern path) and the other to 5.45 Mbps (northern path), while the MPLS router on the left provides proportional load balancing in terms of packet flow.   ▲

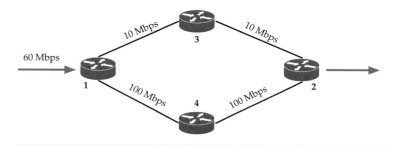

FIGURE 19.1   Four-node example with different link capacity.

ON DETERMINING OPTIMAL LSPS

To determine where and how many endpoint tunnels (from ingress to egress) are needed, the multicommodity network flow (MCNF) model presented in Eq. (4.4.10) can be used; note the subtle difference with the multicommodity shortest-path routing flow (MCSPRF) model Eq. (7.5.1). Recall that Eq. (4.4.10) is shown for minimizing maximum link utilization; as appropriate, the objective function can be replaced by either a piece-wise linear approximation of the delay function or by a composite function; refer to Eq. (7.6.25) and Eq. (7.6.20), respectively. Once the MCNF model is solved, we will obtain a solution where paths with positive flows will be identified; these paths are then prime candidates for becoming LSP tunnels in the IP/MPLS network. Note that the MCNF model may identify some paths with very small positive flow amounts; such paths may not need to be considered in the final optimal LSPs selected.

As an alternative to the MCNF approach, a constrained shortest-path first approach for LSP determination can be taken. Typically, this approach cannot provide optimal flows under certain situations. This will be illustrated later in the context of MPLS VPN traffic engineering.

TIME-DEPENDENT TRAFFIC VARIATION

Traffic does change quite significantly, for example, in a 24-hour time cycle. Instead of considering a single traffic matrix, multiple traffic matrices for different hours during the day may be estimated. Thus, the MCNF model can be solved on each of these matrices separately. The resulting paths with positive flows for each such traffic matrix are very likely to be different. Yet, some paths will be common. If so, such paths can be candidates for LSP tunnels to be set up as explicit LSP routes where the bandwidth allocation can be varied from one time period to another. For the ones not common, LSPs can be set up on a time-dependent basis. An important issue to keep in mind is that tearing down and setting up LSPs can affect the end-user's performance. Thus, minimizing such an impact is also important.

WHEN TRAFFIC DEMAND IS NOT FIXED

Next we consider the case in which the traffic demand is not fixed; this case is not to be confused with the case of time-dependent variation in traffic demand. In an IP network, demand is stochastic as it can vary instantaneously; this is sometimes referred to as *elastic demand*. Thus, from measurements, we can at best determine *projected* demand, not fixed demand. Going back to Example 19.1, assume that 60 Mbps is the projected demand. Thus, the traffic may fluctuate from this value at any instant, possibly going over 60 Mbps. Thus, if the network is set up with a guaranteed bandwidth on each LSP, then any traffic over 60 Mbps will be denied entry to the network. This is certainly not a good situation in an IP traffic environment. Thus, LSPs would need to be set up carefully to allow for fluctuations and also knowing that one path has much less bandwidth. For instance, when the RSVP-TE Path message is initiated, the LSP on the north path is set up with int-serv under a guaranteed service option to limit it from having to handle any load fluctuations. The LSP on the south path can be set up using RSVP-TE with int-serv under a controlled load service option to allow for flexibility for any overload. Note that there is no instantaneous service or delay guarantee; however, in a best-effort IP network or in a differentiated services IP network, this allows for service-level agreements to be met.

An alternate solution is to set up two LSPs on the south path, where one is set up at 54.54 Mbps with a guaranteed option, and the other one is set up with a null service option. Depending on the implementation, another option is to allow any traffic over 54.54 Mbps to be routed as IP traffic based on the shortest-path first routing decision made by the interior gateway protocols (IGPs). For this to happen, we need to ensure that the links on the north route are set with high weights, so that the IGP does not select this as a preferred route for overflow traffic. Another point to note is that the ingress node must be able to handle overflow of traffic from the first LSP to the second LSP. That is, at any time in IP/MPLS networks, both link weight setting and MPLS LSP setup is possible; while this provides flexibility, it also results in a certain amount of complexity in determining and managing link weights as well as MPLS LSPs so that the network is efficiently used.

JOINT LINK WEIGHTS AND MPLS LSP ENGINEERING

As stated above, the joint traffic engineering optimization problem determining link weights for the IGP *and* optimal MPLS LSPs is a complex problem for an integrated IP/MPLS network. We will consider a special case here that allows us to approach this problem through decoupling.

Suppose that a large ISP has a set of critical customers with web servers running directly off this ISP. Because of service-level agreements, it is decided that these customers would get specialized treatment. Thus, at the entry points to the network, traffic trunks for such customers can be defined that point to the address block of web servers; accordingly, LSP tunnels can be set up. Thus, when a real user's packet arrives at the ingress router, it will go on a "fast track" LSP to the destination, since such LSP tunnels have already been established for user traffic delivery. Alternately, the maximum rate that is allowed to be handled can also be limited using the same idea. In other words, controlled traffic engineering can be helpful in providing special treatment to large customers.

However, all the rest of the users can use standard IP-mode service. Given this, we can approach the joint optimization somewhat differently through a two-stage approach:

- For customers with SLAs, estimate traffic demands and determine the optimal LSPs using the MCNF approach (refer to Section 4.4.2).

- Determine residual link capacities after allocating bandwidth resources for the required LSP.

- For traffic estimated (that does not fall under SLAs provided through MPLS LSPs), consider the link weight optimization approach using an MCSPRF approach (refer to Section 7.5), in which case link capacity is considered to be the residual link capacity.

In addition, some customers might require failure protection as part of the SLA, which can be supported through FAST-REROUTE option in MPLS and by providing backup LSPs. In general, customers with varied levels of protection requirement might need to be accommodated through MPLS tunnels. To traffic engineer a network for this requirement, the transport modeling approach presented later in Section 24.4 can be used; see also [665], [750], [751].

TUNNEL IN THE MIDDLE

Finally, it may be noted that through label stacking, MPLS allows LSP tunnels to be set up in the "middle"; see Figure 18.5 for an example of tunnel in the middle. Based on traffic profile and knowledge of a specific network, it is quite possible to consider the option of creating tunnels in the middle. However, the general problem of selecting tunnels at the end and also determining *where* in the middle is a difficult combinatorial optimization problem.

GENERAL REMARK

We can see from the above discussion that in an IP/MPLS environment, the traffic engineering approach and decision depend on what types of customers a provider is serving, and the level of guarantee needed for meeting demand volume request.

19.2 VPN Traffic Engineering

It may be noted that VPN is a widely used terminology for a broad variety of VPN services, including accessing a corporate network from home through a VPN service. Here, the meaning of VPN is different in that a corporate customer may have offices in different physical locations distributed geographically; such customers would like to lease a seamless virtual link connectivity through another provider ("VPN provider") that has geographic presence in these areas. In this section, we present an MPLS VPN traffic engineering approach for such virtual services. Note that this usage of MPLS is quite different from the IP/MPLS traffic engineering issue for an ISP. Here, our focus is networks provided by MPLS/VPN providers, which is not to be confused with public ISPs. Such a VPN is also known as a provider-provisioned VPN (PPVPN); generic requirements and terminology for PPVPN can be found in [17], [517].

In Chapter 18, we discussed how PPVPN can be accomplished using BGP MPLS. We briefly review a few key points for the purpose of VPN traffic engineering. Here, we consider the case where the connectivity is provided at layer 3, i.e., a layer 3 VPN service. For illustration, we will assume that each customer has its own address block. From the point of view of the VPN provider, it will be necessary to have an LER where a customer is connected at layer 3, and then another LER at the other geographic location to connect back to the customer. Thus, within the VPN provider's networks, the LERs serve as ingress and egress points and the provider can have multiple LSRs for transiting traffic; such VPN providers are referred to as MPLS VPN providers, or more generally, as PPVPN providers. Note that the ingress and egress points serve as locations for LSP tunnels to originate and terminate to serve different customers. Furthermore, LERs are referred to as *provider edge (PE)* routers, while the routers at customer sites are referred to as *customer edge (CE)* routers.

Later, we will consider another VPN concept called *layer 2 VPN* (refer to Section 18.5.2). In this case, the customer edge is *not* a router. This is discussed later in Section 19.2.4.

19.2.1 Problem Illustration: Layer 3 VPN

We will illustrate a routing/traffic engineering problem from the perspective of an MPLS VPN provider who will be referred to as *ProviderStealth*. This provider has three customers:

TABLE 19.1   Customer demand matrix.

Customer ID	Locations between		Bandwidth Requirement
Customer A (27.27.0.0/16)	Kansas City (27.27.1.0/24)	San Francisco (27.27.128.0/24)	45 Mbps
	Kansas City (27.27.1.0/24)	New York (27.27.192.0/24)	60 Mbps
	San Francisco (27.27.128.0/24)	New York (27.27.192.0/24)	20 Mbps
Customer B (42.84.0.0/16)	San Francisco (42.84.0.0/20)	New York (42.84.128.0/20)	80 Mbps
Customer C (2.4.0.0/16)	San Francisco (2.4.0.0/20)	New York (2.4.128.0/20)	100 Mbps

TABLE 19.2   LSPs chosen as traffic engineering tunnels.

Customer ID	Origin-Destination	LSP for TE Tunnel
Customer A	SF-KC (for 27.27.1.0/24)	LER-SF1 ··· LSR-SF ··· LSR-KC ··· LER-KC1
	KC-SF (for 27.27.128.0/24)	LER-KC1 ··· LSR-KC ··· LSR-SF ··· LER-SF1
	KC-NY (for 27.27.192.0/24)	LER-KC1 ··· LSR-KC ··· LSR-NY ··· LER-NY1
	NY-KC (for 27.27.1.0/24)	LER-NY1 ··· LSR-NY ··· LSR-KC ··· LER-KC1
	SF-NY (for 27.27.192.0/24)	LER-SF1 ··· LSR-SF ··· LSR-NY ··· LER-NY1
	NY-SF (for 27.27.128.0/24)	LER-NY1 ··· LSR-NY ··· LSR-SF ··· LER-SF1
Customer B	SF-NY (for 42.84.128.0/20)	LER-SF1 ··· LSR-SF ··· LSR-KC ··· LSR-NY ··· LER-NY2
	NY-SF (for 42.84.0.0/20)	LER-NY2 ··· LSR-NY ··· LSR-KC ··· LSR-SF ··· LER-SF1
Customer C	SF-NY (for 2.4.128.0/20)	LER-SF2 ··· LSR-SF ··· LSR-NY ··· LER-NY2
	NY-SF (for 2.4.0.0/20)	LER-NY2 ··· LSR-NY ··· LSR-SF ··· LER-SF2

Customer A, Customer B, and Customer C. Customer A has locations in three cities: San Francisco (SF), Kansas City (KC), and New York (NY), while Customer B and Customer C have locations only in San Francisco and New York. We assume that each of these customers has already obtained an IP address block as follows:

         Customer A: 27.27.0.0/16
         Customer B: 42.84.0.0/16
         Customer C: 2.4.0.0/16

Customer A decides to activate only three subnets at a /24 level: 27.27.1.0/24 for KC, 27.27.128.0/24 for SF, and 27.27.192.0/24 for NY. Customer B has decided to equally divide its address space in its two locations using /20 and, thus, has allocated 42.84.0.0/20 to SF and 42.84.128.0/20 to NY. Customer C has also used the same address allocation rule for its address block, i.e., 2.4.0.0/20 to SF and 2.4.128.0/20 to NY. Each customer has a bandwidth requirement between its different sites as located in Table 19.1.

ProviderStealth has LERs and LSRs at a PoP in each city and customers would need to have connectivity to each PoP's LERs at respective locations; ProviderStealth's responsibility then is to meet the demand requirement of each customer in its MPLS VPN network. ProviderStealth's core network links are assumed to be OC-3 (155 Mbps), which provides an OC-3 rate in each direction. The entire network topological view is shown in Figure 19.2(a). From the bandwidth requirement, we can see the total bandwidth requirement between SF and NY is 200 Mbps; since ProviderStealth has only OC-3 capacity between SF and NY, it cannot meet the total bandwidth requirement using this direct link. By inspecting its capacity

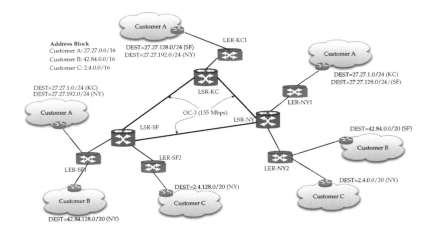

(a) Network connectivity view, including addressing

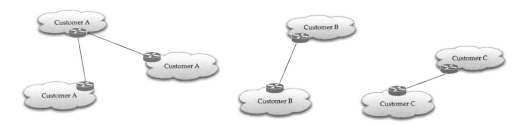

(b) Logical view to customers

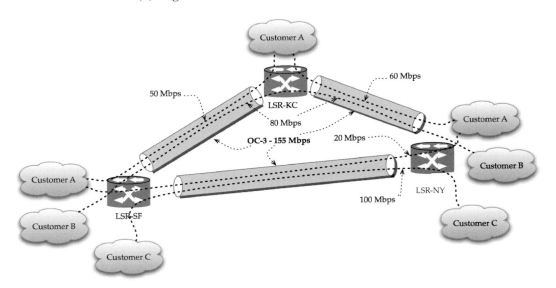

(c) Routing view to MPLS-VPN provider

FIGURE 19.2   MPLS-VPN routing/traffic engineering example.

in the entire network, it can route Customer B's requirement through KC taking the path SF to KC to NY using LSRs in each city. Accordingly, ProviderStealth will set up label-switched paths for traffic engineering tunneling for Customer B.

The LSPs in each direction are listed in Table 19.2 where FECs can be assigned based on the network destination for each customer. Note that LSPs are unidirectional; thus, two LSPs must be set up to meet the bidirectional requirement on bandwidth. The routes for the LSPs are shown in Figure 19.2(c), while the logical connectivity view to each customer would be made apparent of the MPLS VPN network by the MPLS VPN provider and is shown for each customer Figure 19.2(b).

Remark 19.1. *Customers' private addressing and MPLS VPN.*

In the above illustration, we used different IP address blocks for different customers. It is now common for organizations to use private IP address block such as 10.0.0.0/8 for numbering within their organizations with different subnets defined for different locations. Because of this, it is possible that two different customers have the same private address subnets, say 10.5.3.0/24 assigned for their own locations. This may look conflicting from the point of view of proper routing within the MPLS network. However, this is not an issue if BGP/MPLS IP VPN functionality [601], [602], presented earlier in Section 18.5.1, is used, which uses route distinguishers to distinguish between two customer's subnet addresses. Regardless of the numbering issue, the traffic engineering problem faced by the VPN provider is the same as if the address blocks were unique.   ◆

19.2.2 LSP Path Determination: Constrained Shortest Path Approach

Assume that MPLS routers are equipped with a constrained shortest path first (CSPF) algorithm, which is similar to shortest path algorithm, Algorithm 2.4, described in Chapter 2. There are two main differences/requirements: (1) a link is considered only if it has the bandwidth available to meet the request, and (2) a path must be computed only for a given destination, say, v. A simple way to address the first difference is to prune links that do not meet the bandwidth requirement by temporarily setting the link cost to infinity. For the second requirement, the algorithm needs to stop as soon as the path is found. For completeness, the basic idea of a CSPF algorithm is listed in Algorithm 19.1 using the same notation as used in Chapter 2; note that this algorithm is particularly stated for meeting *bandwidth* constraint. Other resource constraints can be considered as well by appropriately changing Step 2 of this algorithm.

To use CSPF for the problem illustrated, we first note that in our case, the link cost may be set to the hop count. The bandwidth availability can be determined at each router based on OSPF-TE or IS-IS-TE for traffic engineering. With this information, a sequence of steps would need to be performed that can be invoked at each router *independently* as follows:

1. Set up TE 100-Mbps tunnel for Customer C (at LSR-SF from SF to NY, and reverse)
    Available link bandwidth: SF-NY: 55 Mbps; SF-KC: 155 Mbps; KC-NY: 155 Mbps

2. Set up TE 80-Mbps tunnel for Customer B (at LSR-SF from SF to NY, and reverse)
    Available link bandwidth: SF-NY: 55 Mbps; SF-KC: 75 Mbps; KC-NY: 75 Mbps

---

ALGORITHM 19.1   Constrained shortest path first algorithm: from node i to node v, for bandwidth constraint, computed at time t

---

1. Network $\mathcal{N}$ and cost of link $d^i_{km}(t)$ and available bandwidth on $b^i_{km}(t)$ on link k–m, as known to node i at the time of computation, t.

2. For link k–m, if available bandwidth, $b^k_{km}(t)$, is smaller than bandwidth request $\bar{b}$, then set link cost temporarily to infinity, i.e., $d^i_{km}(t) = \infty$.

3. Initially, consider only source node i in the list of nodes considered ("permanent list"), i.e., $\mathcal{S} = \{i\}$; mark the list with all the rest of the nodes as $\mathcal{S}'$ ("tentative list"). Initialize

    $$\underline{D}_{ij}(t) = d^i_{ij}(t), \quad \text{for all} \quad j \in \mathcal{S}'.$$

4. Identify a neighboring node (intermediary) k not in the current list $\mathcal{S}$ with the minimum-cost path from node i, i.e., find $k \in \mathcal{S}'$ such that $\underline{D}_{ik}(t) = \min_{m \in \mathcal{S}'} \underline{D}_{im}(t)$

    if k is the same as destination v, *stop.*

    Add k to permanent list $\mathcal{S}$, i.e., $\mathcal{S} = \mathcal{S} \cup \{k\}$,

    Drop k from tentative list $\mathcal{S}'$, i.e., $\mathcal{S}' = \mathcal{S}' \backslash \{k\}$.

    If $\mathcal{S}'$ is empty, *stop.*

5. Consider neighboring nodes $\mathcal{N}_k$ of the intermediary k (but do not consider nodes already in permanent list $\mathcal{S}$) to check for improvement in the minimum-cost path, i.e.,

    for $j \in \mathcal{N}_k \cap \mathcal{S}'$

    $$\underline{D}_{ij}(t) = \min\{\underline{D}_{ij}(t), \underline{D}_{ik}(t) + d^i_{kj}(t)\} \tag{19.2.1}$$

    go to Step-4.

---

3. Set up TE 20-Mbps tunnel for Customer A (at LSR-SF from SF to NY, and reverse)
    Available link bandwidth: SF-NY: 35 Mbps; SF-KC: 75 Mbps; KC-NY: 75 Mbps

4. Set up TE 45-Mbps tunnel for Customer A (at LSR-SF from SF to KC, and reverse)
    Available link bandwidth: SF-NY: 35 Mbps; SF-KC: 30 Mbps; KC-NY: 75 Mbps

5. Set up TE 60-Mbps tunnel for Customer A (at LSR-KC from KC to NY, and reverse)
    Available link bandwidth: SF-NY: 35 Mbps; SF-KC: 30 Mbps; KC-NY: 15 Mbps

Since MPLS tunnel setup is undirectional, each direction must be set up separately. The change in *available capacity* at each link after each step is also noted above. Step 2 above requires further explanation. Since after Step 1, link SF-NY has only 55 Mbps left, CSPF will prune this link since it cannot meet the 80-Mbps requirement, which will result is choosing path SF-KC-NY.

Note again that CSPF is performed by each router indendently based on its current view of bandwidth availability. Suppose that requests were submitted and invoked in the following order in which the first two steps from the above are swapped:

1′. Set up TE 80-Mbps tunnel for Customer B (at LSR-SF from SF to NY, and reverse)
    Available link bandwidth: SF-NY: 75 Mbps; SF-KC: 155 Mbps; KC-NY: 155 Mbps

2'. Set up TE 100-Mbps tunnel for Customer C (at LSR-SF from SF to NY, and reverse)
Available link bandwidth: SF-NY: 75 Mbps; SF-KC: 55 Mbps; KC-NY: 55 Mbps

3. Set up TE 20-Mbps tunnel for Customer A (at LSR-SF from SF to NY, and reverse)
Available link bandwidth: SF-NY: 55 Mbps; SF-KC: 55 Mbps; KC-NY: 55 Mbps

4. Set up TE 45-Mbps tunnel for Customer A (at LSR-SF from SF to KC, and reverse)
Available link bandwidth: SF-NY: 55 Mbps; SF-KC: 10 Mbps; KC-NY: 55 Mbps

Now we can see that after the fourth step, there is not enough unsplit tunnel bandwidth left in the network to accommodate the final request of 60 Mbps. It may be noted that in the above case, you can go back and release the first LSP that was already set up in order to rearrange and fit them all. That is, in most cases, the CSPF approach works quite well; however, the order can matter and it is important to be careful. Otherwise, extra work/steps would be needed to reset some LSP tunnels. This is an issue, in particular, if network bandwidth is tight. If there is plenty of bandwidth, CSPF should not have trouble finding feasible paths. However, rearrangement can be time consuming for a large network, especially if it were to be done at the command line.

19.2.3 LSP Path Determination: Network Flow Modeling Approach

In this section, we discuss how to arrive at an optimal traffic engineering solution from the point of view of the MPLS VPN provider using a network flow optimization approach. For the small network example we have discussed, we can use functionalities such as OSPF-TE or IS-IS-TE to obtain bandwidth information about different links, and then issue a tunnel set-up command at the router's command line interface, which invokes the constrained shortest path approach. While this is a doable approach, it is not a scalable approach as the network size grows; in addition, the impact of the order of the CSPF invocation is difficult to predict in a large network.

Thus, in a large network environment, it would be necessary to do global optimization for the best traffic engineering solution. Here, for ease of illustration, we will still consider the same example as in the previous section and discuss how optimization is performed. In addition, the following discussion shows how network flow modeling presented earlier in Chapter 4 can be used for VPN traffic engineering.

The network has a total of eight LERs/LSRs in the ProviderStealth's network, of which five are LERs. Thus, a simple way to look at it is that we need to consider a 5×5 traffic demand matrix. However, this is often not necessary since instead of using an LER-level view, we can consider a PoP-level view. That is, there are three PoPs, one each in San Francisco, Kansas City, and New York. Thus, the core network routing is the key problem here rather than how an LER is connected to an LSR at a particular PoP. Second, the core network links are usually where the capacity is more constrained; here, we have used an OC-3 link. A link between an LER and an LSR in the same PoP may be on a gigabit Ethernet LAN—certainly, this bandwidth is not as tight of a constraint as the core network link. Thus, we can abstract the problem at the PoP-to-PoP level as a three- node in which a node represents a PoP.

Thus, we will consider the PoP-to-PoP network problem. We have three distinct customers that we need to track separately. However, we do not need to consider each direct path separately; for this model, bidirectionality can be used, which reduces the number of constraints to be considered. After the solution is obtained, the LSPs can be generated based on direction.

To see how to model the problem, consider Customer B, for which we need to choose from two possible *candidate* paths: either direct on SF to NY or the alternate one from SF to KC to NY. We can assign two unknowns for these two possible paths, and impose the binary requirement that only one of them must be chosen, i.e., the following decision requirement:

x_B_sf_ny + x_B_sf_kc_ny = 1

where x_B_sf_ny can take either the value 0 or 1; this is similar for x_B_sf_kc_ny. Certainly, both cannot be 1 in the final solution since that will then violate the above equation. In the same way, we can write for other demands. Since there are five demands (three for Customer A and one each for Customers B and C), we will have a total of five such equations. Note that if we were to consider each direction separately, we would have 10 equations—an unnecessary increase in the number of equations, which becomes very prominent in solving a large network problem. Next, we need to consider the bandwidth constraint on each core link. Consider the OC-3 link with a capacity of 155 Mbps between SF and KC. This link can be used by any of the paths for each of the customers, as long as the capacity is not exceeded. We need to consider the fact that if a path for a customer between two locations is chosen, then this path must be allocated the demand requirement. We will use the demand requirement as stated earlier in Table 19.1. If, for example, path x_A_sf_kc_ny is chosen, then on each link, SF-KC and KC-NY, 80 Mbps would need to be allocated. Since the unknowns are defined as binary variables, we can multiply such a variable by the demand amount. If we now consider all of the possible candidate paths for different customers and locations, we see that for the SF-KC link the following condition must be satisfied:

45 x_A_sf_kc + 60 x_A_kc_sf_ny + 20 x_A_sf_kc_ny
        + 80 x_B_sf_kc_ny + 100 x_C_sf_kc_ny <= 155

Since not all Xs can take a value of 1, these capacity constraints must work in concert with the decision requirements. Finally, an objective function may be considered that is appropriate for the provider. For simplicity, we will assume here that the "cost" of each possible path is one. We can write the entire optimization problem as follows:

```
Minimize    x_A_sf_kc + x_A_sf_ny_kc + x_A_kc_ny + x_A_kc_sf_ny
        +   x_A_sf_ny + x_A_sf_kc_ny + x_B_sf_ny + x_B_sf_kc_ny
        +   x_C_sf_ny + x_C_sf_kc_ny
subject to
d45_A_sf_kc:   x_A_sf_kc  + x_A_sf_ny_kc  = 1
d60_A_kc_ny:  x_A_kc_ny + x_A_kc_sf_ny  = 1
d20_A_sf_ny:  x_A_sf_ny  + x_A_sf_kc_ny  = 1
d80_B_sf_ny:  x_B_sf_ny  + x_B_sf_kc_ny  = 1
d100_C_sf_ny: x_C_sf_ny  + x_C_sf_kc_ny  = 1
l_sf_kc: 45 x_A_sf_kc +  60 x_A_kc_sf_ny + 20 x_A_sf_kc_ny
    + 80 x_B_sf_kc_ny + 100 x_C_sf_kc_ny <= 155
l_sf_ny: 45 x_A_sf_ny_kc + 60 x_A_kc_sf_ny + 20 x_A_sf_ny
    + 80 x_B_sf_ny + 100 x_C_sf_ny <= 155
l_kc_ny: 45 x_A_sf_ny_kc + 60 x_A_kc_ny +  20 x_A_sf_kc_ny
    + 80 x_B_sf_kc_ny + 100 x_C_sf_kc_ny <= 155
```

```
Integer
 x_A_sf_kc  x_A_sf_ny_kc  x_A_kc_ny  x_A_kc_sf_ny
 x_A_sf_ny  x_A_sf_kc_ny  x_B_sf_ny  x_B_sf_kc_ny
 x_C_sf_ny  x_C_sf_kc_ny
End
```

Note that the above is the format accepted by CPLEX, a linear optimization tool discussed earlier in Chapter 4. Note that each decision equation or constraint is identified at the beginning of the line with a name; for ease of tracking, we have embedded the demand value and location/customer information in such names for decision requirements, for example, d80_B_sf_ny, and link names, for example, l_sf_kc. Also, note that to indicate the binary nature of the path choice, the unknowns must be declared as "Integer," which means binary by default in CPLEX. On solving this model, we obtain the following solution:

```
 x_A_sf_kc = 1, x_A_kc_ny = 1, x_A_sf_ny = 1,
 x_B_sf_kc_ny = 1, x_C_sf_ny = 1.
```

All of the rest of the decision variables are zero. We can see that for Customer B, path SF-KC-NY is selected. Accordingly, this solution can be implemented by generating LSPs in each direction by taking into account the LER-LSR path; this is shown earlier in Table 19.2. It may be noted that this problem does not have a unique solution. For instance, Customer C could have routed on SF-KC-NY instead of Customer B; the capacity constraints will still be satisfied and the objective cost as defined here would be the same. Thus, sometimes additional factors need to be taken into account in defining the objective function such as whether any cost weight should be given to any customer, or on link utilization, or if twice the weight should be placed on two-link paths. Accordingly, the objective function can be adjusted in the above model. For example, if we were to give twice the weight to longer paths, then the objective function will take the following form:

```
Minimize    x_A_sf_kc + 2 x_A_sf_ny_kc + x_A_kc_ny + 2 x_A_kc_sf_ny
      +     x_A_sf_ny + 2 x_A_sf_kc_ny + x_B_sf_ny + 2 x_B_sf_kc_ny
      +     x_C_sf_ny + 2 x_C_sf_kc_ny
```

Note that for the above problem, the optimal solution would not change by using this modified objective. We have listed the above objective to illustrate another point. Suppose the fact that unknowns are to be binary is not declared, i.e., the part with "Integer" is left out. What does this mean? This means that decision equations must be satisfied, but each can take fractional values at the solution. In fact, for this problem by ignoring the binary requirement, with the modified objective, we find that the solution for Customer A remains the same. Customer B will be routed on the direct SF-NY route, while Customer C's requirement will be split over two paths: 55% on the direct SF-NY path and 45% on the SF-KC-NY path; that is, a 55-Mbps tunnel is created on the SF-NY path and another 45-Mbps tunnel on the SF-KC-NY path. Recall our discussion earlier about a traffic trunk being split on two LSPs; this is an example of how this can be generated through a network flow modeling approach; here, the customer requirement is a traffic trunk.

There are several additional points to note:

- For the same objective function considered, the total bandwidth required to accommodate all demands with nonsplit LSPs is more than with split LSPs. This is an important observation that is a result of linear programming theory: integer linear programming solution cost is either equal to or more than linear programming solution cost when the same objective function is minimized where the cost coefficient in the objective function is nonnegative. For instance, in the above example, with modified cost function, the split solution results in a total network bandwidth requirement of 350 Mbps as opposed to 385 Mbps for the nonsplit solution, out of a total bandwidth of $3 \times 155 = 465$ Mbps in the core network.

- The above observation can be used either to decide to split traffic trunks into multiple paths if the network capacity is tight, or to temporarily delay capacity expansion cost.

- The decision to split a traffic trunk for a customer into multiple paths could itself depends on the terms of the SLA with the customer. Accordingly, this requirement can be taken into account in the modeling phase. In particular, for the customers for which a traffic trunk split is allowed, the path variables can be defined using continuous variables, and for the customers where the traffic trunk split is not, the path variables are defined using binary variables, as presented earlier; the network flow modeling framework can handle this mixed-mode scenario, which results in a mixed integer linear programming problem.

- In addition to customer traffic, a network carries control traffic and management traffic. Thus, on each link, a certain amount of bandwidth can be set aside. This can also be incorporated in the network modeling approach. For example, if 10 Mbps is to be set aside on each link for control and management traffic, then the link capacity constraint requirement "<= 155" can be replaced by "<= 145." The same idea can also be used if no links are to be allocated to its fullest capacity in anticipation of future requests.

- The bandwidth requested by a customer may vary depending on time. This scenario occurs when customers have plants in different countries around the globe. When coupled with pricing for such service, a time-varying bandwidth requirement may be requested. If so, it would be necessary to do a network reoptimization periodically because of the time-dependent demands; the model discussed above can still be used except that the bandwidth demand value at the time of re-optimization will change while the network capacity will remain the same. The important issue to note is that if an LSP for an explicit route is to be released and a new one is to be established, some customers may be affected; therefore, minimizing this effect is important. However, bandwidth change on an already existing explicit route has little impact.

- Typically, the bandwidth requirement for customers is based on service-level agreements (SLAs). Often, at any particular instant, the tunnels established may not be fully utilized by the customer, and/or if one customer is using them, another customer may not use them at the same instant. Thus, a "bank"-style approach can also be taken. For example, a bank guarantees that it has the funds for your account; they do so in the hope that not all customers will withdraw all their money at the same time. A similar approach is possible in VPN networking. Suppose that we assume that each customer is likely to use about 80% of their bandwidth requirement on an average. Then, this can be taken into account in LSP generations since RSVP-TE includes a controlled services option. This can be taken into account in the network modeling approach; for example, the 45-Mbps requirement of Customer A can be replaced by 36 Mbps ($= 0.8 \times 45$), and similarly for others. Accordingly, the link capacity constraints can be adjusted.

If the network is large and a large number of customers are to be supported, then the number of tunnels to be set up will also grow. Thus, the use of an automated configuration management system to invoke tunnel setup would be required; such a management system can also check for label assignment and addresses mapping issues to ensure that different customers paths are assigned properly. Second, the number of candidate paths that is to be considered in the network flow modeling approach can be generated using the k-shortest path approach (refer to Section 2.8). The network modeling formulation for the general case is Model (4.5.3), presented in Chapter 4, and is thus not repeated here.

Finally, while CPLEX is efficient in solving linear programming problems, it is time consuming to solve large *integer* linear programming problems due to their combinatorial nature. Thus, other specialized algorithms may be developed. A detailed discussed about such approaches can be found, for example, in [564].

19.2.4 Layer 2 VPN Traffic Engineering

In layer 2 VPN, the CE device is a layer 2 device, not a router; we have discussed the basic concepts behind layer 2 VPN using MPLS in Section 18.5.2 to provide virtual private LAN service. We first briefly explain again why such a service is appealing. Consider again a customer that has corporate offices in two different locations where their layer 2 facility is Ethernet based. This customer wants a connectivity between these two sites instead of assigning separate IP address blocks so that it appears as if it is part of the same LAN. This way, it can have a common *supernetted* subnet that covers both sites, and the entity to be shipped between different site is Ethernet frames. This approach of using Ethernet as the bearer is also appealing for carrying any protocol other than IP.

The question from the point of view of a layer 2 VPN provider is how to route such a layer 2 request between the customer's sites. For the VPN provider, there might be many such requests from different customers to facilitate. In each case, the customers enter a VPN network from a CE device to an ingress edge router in which lookup tables for LSPs to the ingress node must be configured. In fact, conceptually this picture is not different than the view show in Figure 19.2(c). We can now assume Customer A to be the one wanting a layer 2 service between SF and KC; it is similar for other customers.

Thus, for the VPN provider to do traffic engineering based on many customers' requests, the basic network flow model is then the same as the one described in Section 19.2.3, which is a nonsplittable multicommodity network flow (MCNF) model, for determining optimal LSP paths through the provider's network. In this case, the demand volume request can be based on the customers' own estimates of how much they need, which becomes the bandwidth request to the VPN provider, for example, in terms of an SLA.

Consider again Figure 19.2(c). Note that both customers B and C have demand requests between SF and NY. In this specific example, the routes through the network were found to be different based on the optimization goal. Suppose that the route selected were found to be the same—they are both to use SF-KC-NY route. This poses an interesting question for the VPN provider—should the provider combine these demands into a single LSP on the route SF-KC-NY? From the traffic point of view, two labels can be used to differentiate traffic for different customers at the edge devices while using a common LSP for both. An advantage of combining such a request is that there are fewer numbers of tunnels to manage and track within the VPN provider's network; however, if each customer has a different bandwidth request, it must be ensured at the ingress point that no individual customer's agreed upon bandwidth receives more than its share.

Now consider the case of P2MP virtual private LAN service (refer to Section 18.5.2). From the service provisioning point of view, the P2MP scenario would then require a tree structure for delivery within the MPLS networks. Such a tree structure can be addressed in the following ways: (1) set up multiple point-to-point LSP tunnels as before; the ingress router generates multiple copies to be sent on each LSP tunnel destinations, or (2) the MPLS has multicast functionality. How do we handle these from the point of view of traffic engineering by the VPN provider?

If multiple tunnels are to be set up due to the lack of P2MP capability in the MPLS VPN network, then different requests for different sites need to be identified first and then the point-to-point model from Section 19.2.3 can again be used for determining optimal tunnels.

If the network is equipped with multicast functionality, then for the P2MP case, a candidate tree generation concept instead of a candidate path generation concept in the MCNF model is required. The generation of such candidate trees for use in an MCNF modeling framework is discussed in [564, § 4.6.2].

19.2.5 Observations and General Modeling Framework

From the illustration of different scenarios above for VPN traffic engineering, whether layer 2 or layer 3, we can say that the VPN routing/traffic engineering problem can be classified as a Type B classification according to our service classification tabulated earlier in Table 17.1. Thus, this use of MPLS is primarily a transport network service mode. If such requests are to be set up on semi-permanent basis, and different customer requests might arrive over a time horizon for tunneled services, then from the point of view of the VPN provider, the network traffic engineering problem can fall under the transport network routing framework as discussed in Chapter 24. This means that there are multi–time period VPN transport routing problems to consider for the provider, for which the model presented in Section 24.3 is applicable.

It is also possible that some customers might want protection and restoration of traffic engineering tunnels through a VPN provider's network. From MPLS functionality point of view, FAST-REROUTE can be used. From modeling the route selection for primary and backup path for many such requests, while some might have partial protection, the model presented later in Section 24.4 can be used.

Finally, under certain situations, dynamic transport and reconfigurability of LSP tunnels for customers are also permissible; if so, then a Type C classification, listed in Table 17.1, is also applicable.

19.3 Routing/Traffic Engineering for Voice Over MPLS

Real-time interactive applications such as voice and multimedia can also be carried over MPLS. This means that for the duration of the call, a connection is set up for a voice call and MPLS then provides a reserved path for the voice call through an MPLS network, much like circuit-switching for packet delivery. The connection setup aspect can be, for example, SIP based; this will be discussed later in Section 20.4.3. In general, voice over MPLS can mean either (a) voice over IP over MPLS, or (b) voice directly over MPLS. Sometimes, Voice over ATM over MPLS is also listed under this category.

To directly do voice over MPLS, the basic idea is to set up LSP tunnels as traffic trunks and then multiplex multiple calls on the same LSP. Such LSPs may be set up on an end-to-end basis with the MPLS network to carry a voice call, or a call may travel over multiple LSPs.

The MFA forum [487] has standardized the LSP structure for multiplexing voice calls, which is shown in Figure 19.3. An LSP has an outer label that identifies an LSP for two end-points, and one or more VoMPLS primary subframes. Between the outer label and the primary subframe, an optional inner label is also allowed. Each primary subframe carries four fields: channel ID, payload type, counter, and length. The channel ID field is to identify VoMPLS channels. Up to 248 channels can be multiplexed within a single LSP tunnel; however, using the inner label, the stacked label property of MPLS can be invoked to allow multiple different streams within an LSP.

You may note that VoMPLS falls under a Type A classification, listed earlier in Table 17.1. This means that a call is to be established as soon as the request arrives, bandwidth is to be reserved on a label switch path for each voice call, and, thus, link capacity resources are used on the link an LSP traverses. On average the duration of such a call is relatively short. Note that an LSP is *not* set for each call; rather, an LSP is set up to serve as trunkgroups between MPLS routers, on a periodic basis—thus, there is no scalability issue of setting up such LSPs

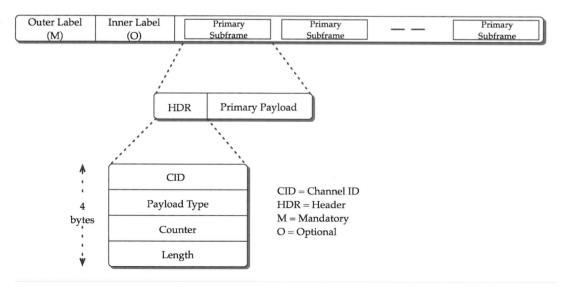

FIGURE 19.3   LSP Structure in VoMPLS (adapted from [487]).

using, say RSVP-TE. This is not to be confused with call setup signaling on a per-call basis that can be based on either ISUP messaging or SIP.

In a sense, this usage of MPLS for voice service is essentially the same as QoS routing discussed in greater depth in Chapter 17; thus, we refer the reader to this chapter for further details. Note that here MPLS LSPs serve as trunkgroups between two routers, thus forming a virtual topology in which calls are to be routed. The alternate call routing concept discussed in Chapter 17 means that in this MPLS environment, the MPLS routers for voice services would need to have alternate call routing capability.

If alternate call routing functionality is not available, then LSP tunnels set up for trunk-groups would serve as direct links. The capacity of these links would need to be engineered so that call-blocking probability is kept low at an acceptable Grade of Service (GoS). Voice traffic engineering was discussed earlier in Chapter 11 and will be discussed again in the context of VoIP in Chapter 20.

An important issue is that call traffic volume can vary over time. Either LSP can be set up statically with plenty of bandwidth that does not change over time, or it may frequently be set up along the way with the required capacity to meet GoS. Note that this may result in bandwidth allocation to an LSP, and then deallocation later since it is not needed due to a drop in call traffic volume.

In the presence of dynamic traffic, the dynamic allocation/deallocation problem can actually lead to network instability showing oscillatory behavior when there is no control. This oscillatory behavior is shown in Figure 19.4(a); here, to denote change in offered traffic over time, a sinusoidal traffic arrival curve is used that is subjected to allocation and deallocation of bandwidth on a tunnel if the blocking is below acceptable QoS tolerance or if it is above QoS tolerance, respectively. If simple controls such as hold-down timer between updates are used, it is possible to arrive at a stable environment (see Figure 19.4(b)). Thus, in a dynamical setup of LSPs, to meet service guarantee requirements, it is important to consider the LSP bandwidth update procedure in a way that avoids network instability. See [266] for further discussion.

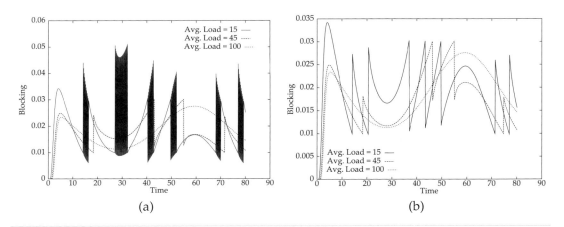

FIGURE 19.4   Transient performance due to an LSP bandwidth allocation/deallocation scheme: (a) instability; (b) corrected through control.

19.4 Summary

In this chapter, we have presented a set of routing and traffic engineering problems in which MPLS can be used. In general, an MPLS traffic engineering-based approach requires several issues to be considered, such as path management, traffic assignment, network information dissemination, and network management [42]. We have highlighted several approaches for traffic assignment and path management for different MPLS-based environments.

An important flexibility about MPLS is that depending on the service offered, it can fall in one of the three classifications identified earlier in Table 17.1. MPLS provides powerful capabilities if you know how to use it. In this chapter, we have presented a set of examples to illustrate the flexibility of MPLS. Furthermore, we have illustrated how various routing paradigms, including a network flow-based modeling approach, can be helpful in determining the optimal routing for a particular problem.

Further Lookup

For a historical treatment of the "birth" and development of MPLS, including an excellent organization of historical Internet drafts and RFCs on this subject, see [263].

For provider-provisioned VPN, see [517]. For discussions on issues related to traffic engineering an IP network, see [42], [45]. For dynamic two-layer reconfigurability, see [474]. For an early implementation of MPLS-enabled switch for routing, see [111].

For a dynamic MPLS environment with anticipation of future service request, a network would need to consider minimum interference routing [349], [757]. Such services can be served also using mechanisms that are variations of trunk reservation; the concept of trunk reservation is discussed elsewhere in this book.

Additional information about voice over MPLS can be found with the MFA forum; see [487].

Exercises

19.1. Discuss where and how MPLS-based IP/MPLS traffic engineering is different from "pure" IP traffic engineering.

19.2. Consider the network illustrated in Figure 19.2(a) and the network flow modeling approach described in Section 19.2.3.

   (a) Extend the model if the traffic trunk for Customer C is allowed to be split. Determine the optimal flows and tunnels if the objective function is the same as the one discussed in Section 19.2.3.

   (b) Assume that Customer B requires protection through FAST-REROUTE using a backup path. Extend the model to accommodate this change. Does the network have enough capacity to accommodate this request? If not, determine minimum additional capacity needed on each link if the objective is to load balance the network with no link having more than 70% utilization once this new capacity is added.

   (c) Each customer requests full protection back up tunnels with dedicated tunnels. Does the network have enough capacity to meet this request? If not, determine the minimum additional capacity needed in the network to serve this request, and determine the optimal LSP tunnel routing configuration.

(d) Suppose that Customer A wants full-protection backup path; Customer B wants partially-protection backup up path with 50% guarantee; and Customer C requests a basic MPLS tunnel service with guarantee bandwidth, but no protection for failure. Present a network flow optimization model. Determine if additional capacity is needed in the network. If so, determine the minimum additional capacity needed and the optimal LSP tunnel routing configuration.

19.3. Generalize the network flow model for traffic engineering for a network of L links and K demands in which K_1 customers require path protection, K_2 customers allow traffic trunk to be split, and K_3 require no-split traffic trunks.

20

VoIP Routing: Interoperability Through IP and PSTN

If I knew where the songs came from, I'd go there more often.

Leonard Cohen

Reading Guideline

This chapter assumes an understanding of many pieces covered in this book. In particular, the reader is expected to be familiar with Internet routing architectures (Chapter 9), call routing in PSTN (Chapter 10 and Chapter 13), including some understanding of SS7 signaling (Chapter 12), IP traffic engineering (Chapter 7) and, voice traffic engineering (Chapter 11).

Voice service over IP networks has received considerable attention in the past decade. This raises the issue of providing seamless voice service (and/or multimedia services) between an IP and PSTN. In this chapter, we present call routing in this hybrid IP-PSTN environment; for brevity, it will be referred to as voice over IP (VoIP) routing. As you will see in this chapter, there are several different scenarios possible. We consider a representative set of scenarios to show how call flow and routing would work when IP and PSTN are interconnected.

20.1 Background

Introduction of VoIP requires hybrid integration of the PSTN and the Internet. There are three key environments that we will discuss in detail in this chapter. They are broadly referred to as VoIP environments and can be briefly stated as follows:

- A call originating and terminating in PSTN that can use IP networks and protocols

- A call originating in a telephone network terminating in the Internet, and vice versa

- Call handling/routing work in an all-IP environment.

You may note that in some cases we have used the term "Internet" and in other cases as "IP network"; there is a subtle difference that we want to distinguish for VoIP environments. When we use "Internet," we refer to the public Internet, while "IP networks" is used to refer to the possibility that it is primarily a private IP network while not ruling out the public Internet.

To consider the above environments, there are several issues to consider. At the heart of it all is addressing. For example, the two key addressing schemes, E.164 for PSTN and IPv4 for Internet, must somehow interwork. Second, due to extensive deployment of SS7 for PSTN, SS7 point code addressing and protocols must also be interworked. Third, number portability is also another important factor to consider. All of these must work seamlessly without breaking the current service functionality and unduly affecting quality of service (QoS) requirements. Furthermore, there has been widescale deployment of private IP addressing both in residential and in enterprise settings that uses NAT/NAPT functionality; this then brings up the practical necessity of making VoIP work in such an environment where VoIP devices behind a private IP address interface with the public network and still can communicate.

This brief discussion indicates that there are many complexities involved. Thus, it is tempting to say "why don't we just move to an all-IP public network" for VoIP and Internet services with the hope that this resolves or minimizes all of the problems and issues mentioned above. While this may occur over time, this is not currently or in the near future completely possible for practical, real-life reasons, such as the huge installation base of PSTN and PSTN equipment, regulatory policies in different countries, widescale use of private IP addressing, and implementation costs. Thus, in this chapter, we consider the three environments listed earlier, along with the addressing factor for routing. Because of the mix of many components, there are many possible approaches; we present only a representative set of scenarios; by no means is this an exhaustive list. We have left enough room here for you to wonder and imagine other possible combinations.

20.2 PSTN Call Routing Using the Internet

In this section, we will discuss call routing for plain old telephone service (POTS) where the Internet is part of the call.

20.2.1 Conceptual Requirement

Chapter 13 covered PSTN architecture and routing in detail; you may note that our discussion on routing was focused primarily on a central office switch to another central office switch when a call is dialed by a user, in a traditional TDM environment. The segment from the user's residential telephone to the central office switch was assumed to be analog and directly connected through copper-wire technology. This segment is often referred to as the *user-to-network interface* (UNI).

Briefly, the UNI in the case of telephone service provides the following functionalities: when a user picks up the receiver at a residential phone, the central office recognizes that the receiver is off-hook and the central office provides the dial tone. On hearing the dial tone, the user dials a sequence of numbers to reach the desired party. If the central office is successful in setting up the call, the user hears the ring. A conversation takes place if the other party picks up the receiver, and the voice analog signal is carried by the UNI to the central office; this is then converted to 64 Kbps for carrying it, for example, through the PSTN using TDM switching. When the conversation ends, either party can hang up the receiver—at this point the phone goes to an on-hook state.

What if we want to replace the analog UNI segment with a different technological environment? There are two aspects to consider: (1) replacing the direct line with a nondirect networked connectivity mode, and (2) changing the communication mode to a packet-based communication environment. In addition, it is imperative that both the phone end and the central office end have the proper *adapters* so that the basic as well as any add-on functions, of the phone service for the UNI part is still possible. Interestingly, a number of technical environment has been proposed in the past two decades to address this; we limit our discussion to a representative set of scenarios in order to bring in the routing component.

For example, for the UNI part when a data protocol can be used, there are several possibilities. For our illustration, we will assume that the ITU-T Q.931 protocol is used; this protocol provides the basic call control for access signaling.[1] Call signaling within the network part in PSTN, however, would still use SS7. Then, the message flow for call establishment and call teardown, commonly referred to as *call flow*, would be as shown in Figure 20.1. As you can see, Q.931 has a set of messages that can essentially be considered to equivalent be to SS7 ISUP messages.

Suppose that the nondirect connectivity mode uses an IP-based environment. That is, what happens if the UNI is an IP network, and we want to use Q.931 for UNI call signaling? Then, the VoIP adapter must have the functionality to create Q.931 packets that are then car-

---

[1]ITU-T recommendation H.323 is an umbrella protocol suite for packet-based multimedia services in a local area network environment. The call signaling component of H.323 is described in H.225.0, which, in turn, is a subset of ITU-T recommendation Q.931 protocol. Thus, you will see the use of an expression such as "call setup is done using H.323," which then refers to relevant Q.931 messages used for call setup signaling. Thus, H.323 and Q.931 are often used interchangeably, although it is certainly important to understand the context of this usage.

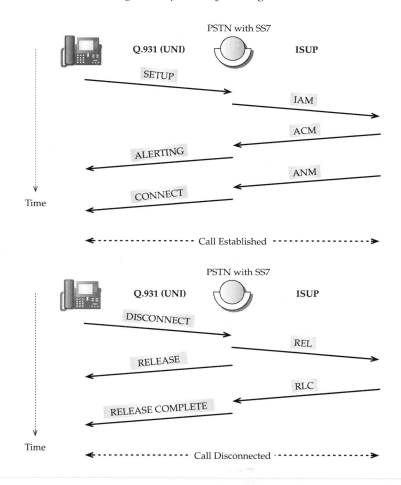

FIGURE 20.1 Call flow: setup and tear-down with Q.931 and SS7 ISUP.

ried over the IP networks. To accommodate the PSTN network side of UNI for the above situation, the central office switch must have an adapter as well. This adapter is a *media/signaling gateway*. This gateway function then has the ability to communicate with the VoIP adapters for exchange of call connection-related messages and packetized voice packets. This gateway function can either be physically integrated with the central office switch, or can be a separate server that also has circuit-mode functionality to communicate with a central office switch. In this environment, call management is accomplished by the central office switch through this gateway. Note that the gateway end would need to have an IP-based interface, so that it can receive and interpret packetized Q.931 messages; furthermore, it must also have an interface that can talk to the central office switch end for using SS7 messaging. In return, the gateway would need to generate a Q.931 message to send to the VoIP adapter over an IP network. This basic conceptual picture allows a regular phone to be used to connect to the PSTN, as shown in Figure 20.2.

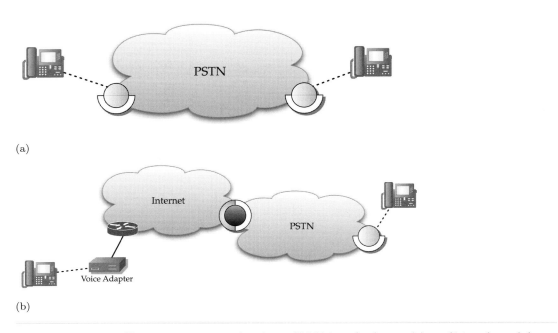

FIGURE 20.2   Change in access technology (UNI) in telephony: (a) traditional model, (b) incorporation of an IP network.

20.2.2 VoIP Adapter Functionality

Going beyond the conceptual picture, there are certain practical considerations to take into account. While the VoIP adapter would require a telephone jack such as an RJ-11 jack to connect a regular telephone to it, it would also need to have an Ethernet RJ-45 jack since this is the most commonly deployed physical interface for connectivity to an IP network through Ethernet. In addition, the VoIP adapter would require the following:

- It has IP-stack based software and can generate IP packets as dictated by the upper layer for transmission in Ethernet format.

- It has add-on software/functionality to generate Q.931 messages, voice packetization, and the ability to generate certain sounds such as the dial tone that can be heard through the phone receiver.

You can see that instead of a plain telephone relying on the central office switch for all its functionalities, it would use the VoIP adapter as a proxy for the central office switch.

20.2.3 Addressing and Routing

Next, we need to reconcile two addressing schemes: E.164 for PSTN and IPv4 for Internet. Note that IPv6 has similar issues when reconciling with E.164 and is not discussed separately. A basic requirement is that a telephone number is to be associated/homed with a specific central office switch; this number can be associated as a native number or through number porta-

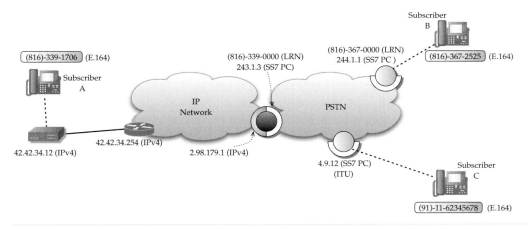

FIGURE 20.3 Addressing and call routing example.

bility. The second requirement is that both the VoIP adapter and the media/signaling gateway attached to the central office switch must have IP addresses associated with them. In addition, due to the use of SS7 signaling in PSTN, SS7 point code addressing also comes into the picture. Note that the VoIP adapter requires authorization to talk to the signaling/media gateway, which can be accomplished if the subscriber obtained the VoIP provider from its telephone service provider and the authorization process is, for instance, preconfigured, which may include hardware address information of the VoIP adapter. It should be noted that the provider is not required to statically assign the IP address for each VoIP adapter.

We will now illustrate call routing using Figure 20.3. Consider E.164 number +1-816-339-1706. This is located in country code +1 (North American zone, specifically in the United States), in area code 816. Thus, the central office to which this number is natively homed is identified by 816-339, and has the SS7 point code assigned as 243.1.3. For ease of discussion, the service provider that is assigned this address block and maintains this central office switch will be referred to as Provider X. We assume that the signaling/media gateway associated with this switch has the static IP address, 2.98.179.1. Allocation of the number +1-816-339-1706 to the subscriber ("Subscriber A") by Provider X requires a local postal address. For example, in the United States, this is required, for example, for billing purposes and also to determine the location for 911 emergency service. Using the VoIP adapter, Subscriber A would need to initialize the service. This involves the follows steps:

- Subscriber A connects a POTS telephone to the VoIP adapter.

- Subscriber A connects the VoIP adapter to the Internet.

- The VoIP adapter acquires an IP address, say, through the dynamic host configuration protocol; we assume the address is 42.42.34.12.

- The VoIP adapter with IP address 42.42.34.12 starts an initial configuration session with the gateway that has IP address 2.98.179.1. Since both these devices have Internet addresses, any packet generated by the VoIP adapter will be routed through the Internet to destination IP 2.98.179.1 (refer to Chapter 9). Through this process, the telephone number of the subscriber, +1-816-339-1706, is verified and stored in the VoIP adapter.

Now, Subscriber A is ready to make a call. Suppose that Subscriber A wants to call another subscriber ("Subscriber B") with the number +1-816-367-2525. Both of these are then local numbers. Assume that Subscriber B uses a POTS to connect to the central office switch through analog copper-wire technology to the central office switch identified as 816-367. The following steps occur:

- Subscriber A picks up the receiver. The phone then goes to the off-hook stage. The VoIP adapter with IP address 42.42.34.12 generates a notification packet that is routed to the gateway with IP address 2.98.179.1; the gateway checks with the central office switch, identified with location routing number (LRN) 816-339-0000, if dail tone service can be provided. If the switch allows the dial tone capability, then the gateway sends a response packet back to 42.42.34.12 that affirms dial tone service. The VoIP adapter then generates the sound for a dial tone so that the subscriber can hear the dial tone; note that the dial tone sound is *not* generated by the gateway; the gateway only generates a data packet that indicates the dial tone service while the VoIP adapter generates the actual sound.

- Since this is a local call, Subscriber A dials the numbers 367-2525 to reach Subscriber B. The called party number 367-2525 and calling party number 339-1706 are inserted in the payload of the packet generated by the VoIP adapter to send to the gateway 42.42.34.12; this packet is then the Q.931 SETUP message that is embedded in an IP packet.

- The central office that is home to Subscriber A will determine the outgoing trunkgroup and routing, consulting the SCP database, if needed, for delivery to 367-2525; refer to Chapter 9 for different scenarios that are possible and details for this call routing. The originating central office with its SS7 point code address (marked as 243.1.3) will identify the SS7 address for the next TDM switch, and so on, until the signal arrives at the destination central office (marked with SS7 point code address 244.1.1). Once Subscriber B's phone rings, an SS7 message will be communicated back to the original central office. This switch will be in charge of generating a Q.931 ALERTING message via its gateway to indicate call ringing for Subscriber A. On receiving this message, the VoIP adapter will generate the audio for the ring so that Subscriber A can hear it.

- If Subscriber B now picks up the receiver (called party off-hook state), then its home central office switch will generate an SS7 ANM message directed to the originating central office. Once it receives this message, the originating central office switch will generate a Q.931 CONNECT message through its gateway, which is transmitted to the VoIP adapter. Thus, as soon as the VoIP adapter receives the message, it will indicate the CONNECT state by stopping the ring tone and providing connectivity for voice communication.

Once the CONNECT state is established, the voice packetization will take place as soon as the conversation begins. This packetization will occur at the gateway for an audio stream directed to Subscriber A and at the VoIP adapter for an audio stream directed to Subscriber B; in the former case, the source IP address is 42.42.34.12 and the destination IP address is 2.98.179.1; in the latter case, the source IP address is 2.98.179.1 and the destination IP address is 42.42.34.12. Both ends would need to do translation of voice packets. At the VoIP adapter, it would need to convert to analog form in the 4-KHz range for the plain telephone, and while at the gateway, it would need to convert to 64-Kbps PCM coding known as G.711 standard for transmission in TDM in the core of the PSTN. Unless, another audio conversion standard is introduced at the gateway, G.711-based VoIP packets will be carried to the VoIP adapter; thus, the VoIP adapter must be G.711 aware so that it is compatible with the gateway.

How does either end of the IP network know whether IP packets received are for the call signaling function or for the voice media since this is not distinguishable at the IP address level? The distinction is accomplished at the transport layer protocol through different transport port numbers. Furthermore, for reliable delivery, the Q.931 packets can be carried in a TCP session; because some voice packet loss may be tolerable, voice packets, however, can be carried over UDP.

Suppose that Subscriber A would like to call a number located in a different country such as the number +91-11-62345678 ("Subscriber C"). The process would be the same as above. The subscriber will dial the number including the international access code. This will be transmitted as payload in a Q.931 SETUP message to the central office switch. The central office switch will recognize that it is an international call and will then do call routing as described earlier in Chapter 13. This would require SS7 point code address translation at an international gateway to the ITU point code address (not shown in figure); the SS7 messages can then be properly routed to the destination central office identified by ITU SS7 point code 4.9.12 so that the call communication to +91-11-62345678 can be established.

Next, how would a call originating with Subscriber B destined for Subscriber A work? In this case, the SS7 call establishment message, IAM, will start from SS7 point code 244.1.1 and will be routed to point code 243.1.3, the home central office for Subscriber A. The gateway will generate a Q.931 setup message with payload containing called and calling number that will be encapsulated in an IP packet that will be routed to 44.44.34.12. Once the phone rings at Subscriber A's end, an ALERTING message will be generated by the VoIP adapter to let the central office know that the phone is ringing so that this information can be communicated back to Subscriber A. Similarly, a call from Subscriber C will be routed to Subscriber A.

Finally, we consider an example in which Subscriber A calls another subscriber ("Subscriber D" with E.164 number +1-816-339-1605) who is also homed of the same central office switch and uses the service through the VoIP adapter (Figure 20.4). When Subscriber D registers, the gateway notes the IP address of Subscriber D's VoIP adapter, which is currently is at IP address 27.14.32.10. Now, Subscriber A calls Subscriber B; the Q.931 SETUP message encapsulated in an IP packet is still sent to the gateway, which then realizes that Subscriber B is also a VoIP customer. Thus, the gateway retransmits this Q.931 SETUP message to the VoIP adapter of Subscriber B through the Internet, by announcing its IP address as the source IP address; thus, IP address translation would take place at the gateway. Because of this address translation, on receiving this message, the VoIP Subscriber D responds back to the gateway, not directly to the VoIP adapter of Subscriber A. Once a connection is established, the encoded voice packets for the audio stream would still go through the gateway from Subscriber A to Subscriber D, and vice versa. It is important to note that once the call is established, it is conceptually possible and feasible to allow a direct audio stream from the VoIP adapter address 42.42.34.12 of Subscriber A to the VoIP adapter address 27.13.32.10 of Subscriber B and back without involving the gateway, a concept referred to as *antitromboning*. Despite this possibil-

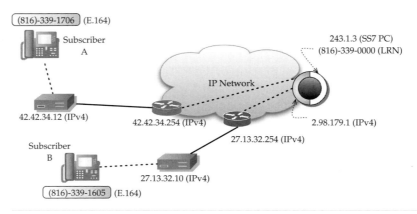

FIGURE 20.4   Call between two subscribers who are homed to Provider X's switch.

ity, this may not be used in practice, for example, due to (1) Provider X being required by regulatory reasons to track all call flows through a call detailed record, and (2) Provider X having no ability to monitor and troubleshoot if a subscriber complains about service quality when direct audio stream communication was allowed from one subscriber to another.

20.2.4 Service Observations

There are certain important observations to note:

- If there is a loss of electricity at the home where the VoIP adapter is located, then the subscriber cannot avail itself of the phone service; this is not to be confused with purposefully turning off the VoIP power much like turning off a mobile phone. That is, always-available service is not completely possible unlike a plain telephone connected to the central office switch where the electricity for the phone is directly supplied by the central office.

- Subscriber A's VoIP adapter is not tied to the IP address 44.44.34.12. The subscriber may decide to take the VoIP adapter and connect to the Internet from anywhere in the world. What gateway 2.98.179.1 really needs to be able to identify is that the VoIP adapter is authorized and that it has an assigned E.164 number. This is possible since this information is payload in a Q.931 message, which is then carried in an IP packet.

- The ability to take the VoIP adapter and connect to the Internet from anywhere is of great interest to the subscriber. This way any local call to which the telephone number was assigned is still a local call. In essence, this functionality provides *location portability*.

- The location portability, however, creates a problem in certain situations. For example, Subscriber A who has the number +1-816-339-1706 takes the VoIP adapter anywhere and connects to the Internet, and then dials for 911 emergency service. The 911 operator will understand it as coming from the physical postal address on record. Recently, there have been new discussions on how to to handle such issues; refer to [523] for recent developments.

Certainly, some of the above issues are beyond the purview of network routing. Yet, they provide a perspective on how location portability is possible while certain regulatory issues are to be completely addressed.

20.2.5 Traffic Engineering

From a service perspective, an important aspect is QoS. Since packets for call control that use Q.931 and for media are being transmitted over the Internet, a stringent end-to-end guarantee such as delay or jitter is not possible. Still it is important to take measures to provide good QoS. To understand this, we consider a few factors.

PER-CALL BANDWIDTH REQUIREMENT

First, we need to understand the bandwidth requirement of a call. Suppose that G.711 PCM coding is used for voice; its data rate in TDM is 64 Kbps. The sampling rate for G.711, as originally specified for use in the TDM environment, is every 10 millisec. However, for IP packetization a different time frequency can be used. A typical packetization frequency is 20 millisec; this results in 50 packets (=1 seconds/20 millisec) generated per second. Thus, for a 64-Kbps rate, the packet size for voice-coded information generated every 20 millisec will be

$$\frac{64 \text{Kbps}}{50} = 1280 \text{ bits} = 160 \text{ bytes}.$$

There is overhead involved in carrying a 160-byte packet in an IP network. The best-known mechanism for audio payload for IP is real-time transport protocol (RTP) that has a 12-byte overhead. RTP, in turn, is carried over UDP protocol that has an 8-byte overhead. IP default header is 20 bytes. Thus, each packet has 40 bytes of header overhead when counted at the IP level. Together, a G.711 codec voice packet generated every 20 millisec will be of size 200 bytes (=160 + 40). Note that this size packet is generated 50 times every second. Thus, the IP-level date rate is

200 bytes × 8 bits per byte × 50 packets per second = 80 Kbps.

Thus, a 64-Kbps voice circuit is translated to an 80-Kbps rate at the IP level.

Now consider layer 2 overhead. Suppose that each packet is carried in an Ethernet frame. Note that Ethernet has 26 bytes overhead per frame due to 8 bytes of preamble, 14 bytes for the header, and 4 bytes for the trailer. Thus, at the Ethernet level, every G.711 packet of 160 bytes, which is 200 bytes at the IP level, becomes 226 bytes, and the data rate becomes 90.4 Kbps. Thus, Ethernet induces another 13% overhead over the IP packet.

Another common technology in the core of the Internet is SONET technology such as OC-3 which uses Packet over SONET (PoS) for carrying IP packets over SONET. PoS uses PPP protocol with HDLC to frame an IP packet; in most common cases, this induces 5 bytes of header and 4 bytes of trailer for a total of 9 bytes. Thus, PoS overhead is 4.5% (9 bytes over 200 bytes) to carry an IP packet, and a G.711 voice call data rate at the SONET level is 83.2 Kbps.

ADDITIONAL FACTORS/OVERHEAD

In addition to header overhead, there are other overheads for each call. For example, call control is accomplished using Q.931. Typically, four minimum messages are generated when a call is established (SETUP, ALERTING, CONNECT, and CONNECT ACK); the ALERTING message is repeated frequently until the user at the other end picks up the phone. At connection tear-down, three messages are generated (DISCONNECT, RELEASE, and RELEASE COMPLETE). During the life of a call, some additional messages are generated that are periodic; for example, RTCP control messages are generated for RTP synchronization; typically, this is limited to 5% of the RTP data traffic. Now consider the duration of a voice call, which is often estimated to be on average 3 min. Thus, we can see that Q.931 one-time messages for each call are almost insignificant overhead; most overhead is due to RTCP. These various control packets are small, typically 50 to 100 bytes; this means that a 160-byte G.711 coded packet is still the most dominant packet size.

Thus, considering a 80-Kbps data rate for voice coding over IP, we see two types of overhead: one for the layer 2 technology and the other for call control/management overhead. It is safe to say that together, this overhead is not more than 20%. Thus, a simple rule of thumb for an equivalent data rate by taking this overhead into account is 80 Kbps × 20% overhead = 96 Kbps. Thus, a simple rule to consider is that the path of a call within the IP network can receives a 96-Kbps data rate. This is then primarily affected by any bottleneck link segment along the path. Assuming Internet service providers in the core are maintaining capacity with good traffic engineering objectives, the bottleneck then potentially falls into two places: (1) the place at which the user connects the VoIP adapter, or (2) the IP link from the service provider's gateway to the Internet. For example, if the user connects the VoIP adapter from behind a cable modem or DSL service, it is often possible to get the data rate estimated above.

CALL-CARRYING CAPACITY

Now consider the IP link from the gateway connected to the Internet. Note that this gateway will be required to handle traffic for all its customers with VoIP adapters. The gateway, being IP based, might not have the proper mechanism to block any calls; however, as we go through this analysis, you will see how the call rate can be monitored for QoS. Assume that this link is an OC-3 link; the SONET payload available to PoS for OC-3 is 149.760 Mbps. Assume that call bandwidth, including overhead as estimated above, is 96 Kbps at the PoS level. Thus, this link has a maximum call-carrying capacity of 1560 (= 149.76 Mbps/96 Kbps). Using the Erlang-B loss formula, we can find that to keep call blocking below 1%, it can handle around 1531.2 *Erlangs* of offered load. Thus, this load-to-link capacity ratio gives a utilization around 97%. This is where we need to understand another important fact. While packets for a particular call stream are essentially equispaced when generated, the call arrival is not. Call arrival is typically assumed to follow Poisson arrival, and each stream is then dictated by the start time of a particular call. Because of the nature of the call arrival and its impact, ensemble packet arrival will have Poisson arrival behavior. Now, from a discussion of IP traffic engineering (refer to Section 7.1.4 in Chapter 7), we know that link utilization should be kept at a level so that packet delay is not impacted. Suppose that we want to keep utilization at less than 60% to avoid unduly impacting packet delay; this is especially important for VoIP traffic. This means that we cannot let utilization build up to 97% as would be indicated by the call capacity–based computation, which is meant for pure circuit-mode assessment. In other words, we need to be restrictive to the more stringent requirement, i.e., the minimum of these two utilization values, which is at 60%. Note that 60% of the SONET payload capacity of 149.76 Mbps is 89.86 Mbps. At 96 Kbps of call bandwidth, effective call capacity becomes 936 voice call units. At this call capacity, we do not need to again impose the blocking requirement since this is already accounted for; in other words, 936 can also serve as the acceptable maximum offered load in Erlangs. With this offered load, call blocking will be essentially zero. If we now assume that the average call duration time is 3 min, then for an OC-3 link interface to the gateway, we can find that an acceptable call arrival rate is approximately

$$\frac{936 \, Erlangs}{180 \, sec} = 5.2 \, calls/sec.$$

Thus, the provider can monitor the call arrival rate at its gateway, for example, by monitoring Q.931 SETUP message arrivals. If the average call arrival rate is found to be noticeably more than 5.2, then this would mean that this OC-3 does not have sufficient capacity to provide good QoS; it is then time to do capacity expansion. The caveat here is that 5.2 is not a magic number for use with an OC-3 link. As you can see from the above discussion, it depends on the following key factors: (1) use of G.711 coding, (2) overhead estimation to determine call data rate, (3) link utilization and call blocking trade-off, and (4) average call duration. By following the above analysis, you can compute an acceptable arrival rate if any of these factors change.

20.2.6 VoIP Adapter: An Alternative Scenario

We now revisit the VoIP adapter issue. While we presented a specialized hardware–based adapter scenario earlier, this is not always necessary. In other words, a computer or a laptop can be loaded with a software-based adapter to mimic the same signaling or media generation functionality; this can then talk to the IP interface of a central office, and in essence, the phone number can be associated with the computer regardless of where it is located since the communication between the computer and the central office is over the public Internet. Certainly, it is possible to have the computer turned off; thus, for an incoming call, a central office with an answering machine service is desirable.

20.3 PSTN Call Routing: Managed IP Approach

The scenario we are about to describe is similar to the above scenario, with certain differences. In this instance, a subscriber receives the adapter from the provider ("Provider Y") but is required to physically locate it at the postal address provided at the time of sign-up. Certainly, this does not provide the flexibility of taking the phone anywhere as discussed above, but it addresses regulatory requirements and serves many customers who would like a nonmobile phone that provides POTS. Such a service paradigm is currently being deployed primarily in a cable modem environment, known as PacketCable, where the cable service provider wants to provide a home telephone service to its cable customers [456], [457]. A VoIP phone connectivity using an RJ-11 jack along with a RJ-45 jack for a computer connectivity is available through an embedded Multimedia Terminal Adapter (eMTA). The name implies that

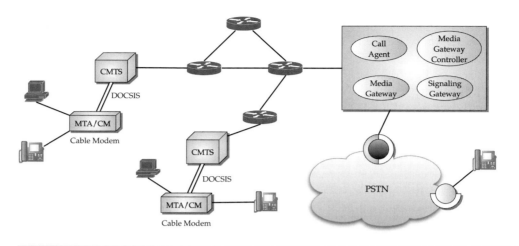

FIGURE 20.5   Cable IP networking for telephony services.

in addition to voice, such terminals are intended for multimedia services. The conceptual architecture for cable IP networking for telephony is shown in Figure 20.5.

In this case, the IP networking available at the customer's home utilizes a cable facility that already belongs to Provider Y through the cable modem service. For example, the customer through cable modem service has IP accessibility. Thus, the eMTA is connected to the cable line (refer to Figure 20.5). The cable modem is visible to the the Cable Modem Termination System (CMTS) using Data Over Cable Service Interface Specification (DOCSIS) protocol, for example, with DOCSIS v. 1.1, which provides QoS. It should be noted that CMTS is primarily a layer device that can be integrated with a layer 3 router for IP communication. Thus, from the eMTA at home to the call manager server, this entire IP network is in the jurisdiction of Provider Y; this forms a managed IP backbone for Provider Y (see Figure 20.5). PacketCable specification requires each endpoint to have a fully qualified domain name. Thus, from the call management server, DNS lookup can be performed and IP address allocation to different MTAs can be done. Since mobility is not involved here, we can say that each telephone number is then associated with an IP address that may likely remain the same for a long time, although for ease of tracking a fully qualified domain name (FQDN) is used with each IP address and it has the form such as billing-number.provider.NetworkRouting.net. Another way to look at it is that this network is intradomain and the provider might choose to run OSPF or integrated IS-IS for all its IP routers in this network. Thus, traffic engineering of this IP network would require optimal link weight determination in order to maintain minimum delay (refer to Chapter 7). Alternately, the provider might choose to deploy integrated IP/MPLS routers, in which case MPLS-based controlled traffic engineering can be used (refer to Chapter 19).

From a deployment point of view, the actual network architecture can be quite different depending on the cable network provider's physical domain. For example, if a cable network provider's service is limited to a geographic location, such as a metropolitan area, the exit point to PSTN can be in one location as shown in Figure 20.5; however, if the cable provider has multiple geographic locations, or a nationwide presence, it might consider the option of routing a call originating in one geographic location to another geographic location, typically considered as a toll call, *within* its network, rather than entering PSTN and exiting in another location back to the cable network. This might be desirable from a cost perspective and for business reasons. In this scenario, the provider will have call management functions in different geographic locations, and depending on the number dialed by a subscriber, the actual point of leaving the network might be different, resulting in early-exit routing or late-exit routing within its IP network. Accordingly, traffic engineering and routing issues are to be addressed.

It may be noted that the cable provider would typically need to lease high-bandwidth circuits such as OC-3 from a telecommunication provider for connections between different metropolitan areas in order to form its IP network. The telecommunication provider, in turn, would then need to address such requests through transport network routing and provisioning.

20.4 IP-PSTN Interworking for VoIP

From the above discussions, you can see that IP-PSTN internetworking in general is an important issue. In this section, we discuss the generic IP-PSTN interworking where a call starts in PSTN and ends in IP networks, or vice versa. This is different than the previous examples in that the actual end devices are not both analog telephones.

For use with telephony and multimedia real-time two-way applications, the session initiation protocol (SIP) has been developed for an IP environment. For ease of reference, a SIP-compliant phone will be referred to as a *SIP phone*. It may be noted that SIP handles only the session control aspect of a call; the actual media for a call is packetized and handled using RTP. An important point to note about SIP is that it can be used end-to-end from one end device to another; it does not require a separate protocol for the network part. You can contrast this with the situation in PSTN where, within a network, ISUP SS7 messaging is used while for the end device to central office, a different protocol such as Q.931 is used.

20.4.1 Gateway Function

In our discussion earlier, we introduced the role of a gateway for interfacing between IP and PSTN to use an analog POTS telephone through a VoIP adapter. We now discuss the general functionality of a gateway interfacing an IP network providing SIP phone services and PSTN.

A conceptual picture is presented in Figure 20.6. There are three components, media gateway controller, signaling gateway, and media gateway, in this conceptual architecture. The signaling gateway receives SS7 ISUP messages from PSTN and passes them to the media gateway controller; these messages are then translated to SIP-equivalent messages for transmittal over the IP network; similarly, the media gateway controller receives SIP messages, which are passed to the signaling gateway to generate equivalent SS7 ISUP messages for transmittal over the PSTN.

An important role of the media gateway controller is to control the audio streams, typically through a PCM-coded circuit-switched voice channel on the PSTN side and using RTP on the IP side. Note that the controller would need to act and inform the media gateway regarding media handling based on the status of a call, such as being established or released. The controller is the brain in this system and is required to maintain states and translation

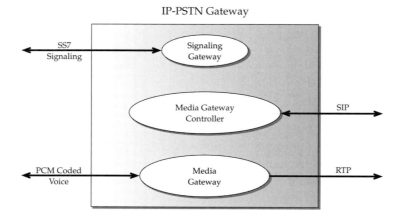

FIGURE 20.6   Gateway interfacing between PSTN and IP.

for different connections. As you can see, there are two interfaces involved in this architecture: one between the signaling gateway and the media gateway controller, and the other between the media gateway controller and the media gateway; for the former, a protocol known as the streaming control transmission protocol (SCTP) can be used, while for the latter, the MEGACO/H.248 protocol can be used. However, the signaling gateway, the media gateway, and the media gateway controller are often bundled together as an integrated gateway server since from an operational network point of view, such equipment is easier to manage; in this case, a vendor can do exchanges internally without using SCTP or MEGACO. Note that this equipment would handle two different addressing schemes: SS7 point code addressing and IP addressing. Furthermore, it will need to handle three different types of links: SS7 signaling link, circuit trunkgroup for PCM voice, and IP link.

20.4.2 SIP Addressing Basics

We now briefly describe SIP functions as related to call routing. Similar to SS7 ISUP messages, SIP defines a set of messages and actions for call establishment and release. SIP, however, has an important additional feature: it has a built-in portability concept. That is, it is not necessary to tie a physical address. To provide this mobility function, SIP has its own addressing mechanism that looks similar to an email address—this is known as a SIP Universal Resource Identifier, or *SIP URI* in short. For telephony use in E.164 addressing, the SIP standard also allows the use of another format called *tel URI*.

Suppose that a subscriber has a SIP phone with E.164 number +1-816-323-2208. Its tel URI for SIP-based service can be listed as

tel:+18163232208

or, in more human readable form as

tel:+1-816-323-2208

Both are acceptable usage since "-" is to be ignored during call processing. The equivalent SIP URI refers to the same SIP phone, but requires additional information, as shown below:

    sip:+18163232208@proxy1.NetworkRouting.Net;user=phone

There are three important points to note: (1) the telephone number part is listed in globally unique E.164 form in its entirety with prefix "+" so that the country code and, consequently, the numbering plan for a particular country can be easily identified (refer to Section 13.1); (2) while the SIP URI has some similarity to the standard email address, the at-sign "@" is used here to separate out and identify a domain name; in this case, the fully qualified domain name (FQDN) for the proxy (to be discussed soon) is listed as proxy1.NetworkRouting.Net; (3) the last part, user=phone, identifies it as a phone.

When such a SIP phone is initially activated in a network, it would need to contact a registration server to obtain a domain name and an IP address, and a proxy server through which all communication to this phone is accomplished. As you can see, the SIP URI described above includes the domain name information of the proxy in the URI. Most importantly, both the proxy server and the SIP phone must be associated with valid IP addresses.

20.4.3   SIP Phone to POTS Phone

Suppose that a subscriber with a SIP phone ("Subscriber A"), identified through tel URI tel:+1-816-323-2208, would like to talk to another subscriber ("Subscriber B") who has a POTS telephone with telephone number +1-816-367-2555. We next describe the call setup and call release between these two heterogeneous phones being served by different types of networks. In the process, we will identify the relevant SIP messages that are associated with SS7 ISUP messages.

Suppose that the IP address of the SIP phone with tel URI tel:+1816-323-2208 is identified as 27.5.16.22 and the proxy's IP address as 27.5.16.1. The gateway is identified for the IP network with IP address 27.0.1.2 and for the SS7 network with SS7 point code 241.1.7. On the PSTN side, the call terminates at the central office switch with SS7 point code 244.1.1; this switch is the home to the Subscriber B's analog POTS phone. In Figure 20.7, we indicate addresses of all entities along with the call flow and messages generated. Note that Subscriber A would dial 3442525 based on local dialing instructions. The SIP phone will generate an SIP INVITE message; the key components of this message are as follows:

    INVITE sip:+18163672525@proxy1.NetworkRouting.net;user=phone SIP/2.0
    From: <sip:+18163232208@proxy1.NetworkRouting.net;user=phone>
    To: <sip:+18163672525@proxy1.NetworkRouting.net;user=phone>
    Via: SIP/2.0/TCP host-a-client.NetworkRouting.net:5060

First note that SIP is a text-based protocol. The first word in the first line indicates that it is an INVITE message followed by SIP URI and the version number of the SIP protocol used. The SIP URI for Subscriber A's SIP phone is listed in the "From" field, and the "To" field contains the SIP URI of the destination. Although Subscriber A has dialed 3672525, the SIP phone is responsible for conversion to the correct SIP URI. Note that only at the IP header level (not shown) will the source IP be identified as 27.5.16.22 and the destination IP address as 27.5.16.1. The "Via" field indicates that TCP is used for reliable delivery of this message; in addition, the domain name for Subscriber A's SIP phone and the transport layer port number ("5060") are included. Since the SIP session is to be established using TCP, a TCP connection setup will need to be initiated to 27.5.16.1, thus requiring first exchange of TCP establishment

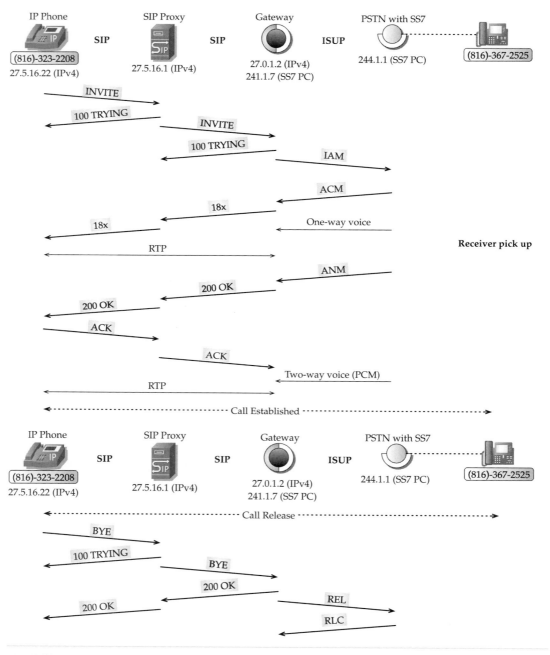

FIGURE 20.7   Call flow example from SIP phone to PSTN.

segments: SYN, SYN-ACK, and ACK (not shown in Figure 20.7). Once the TCP session is setup, the SIP INVITE message will be sent. Note that in this case, the device addresses are found to be from the same IP subnet (27.5.16.0/24), and thus the packet delivery will use any local mechanism such as Ethernet.

Once the SIP INVITE message is received at the proxy, it will generate a "100 TRYING" response message to the SIP phone to indicate that it is trying to establish a connection in the forward direction; this response message will be transmitted back through the TCP session already in place. The proxy will also initiate a SIP session with the gateway 27.0.1.2 and regenerate a SIP INVITE message as follows:

```
INVITE sip:+18163672525@gw1.NetworkRouting.net;user=phone SIP/2.0
From: <sip:+18163232208@proxy1.NetworkRouting.net;user=phone>
To: <sip:+18163672525@proxy1.NetworkRouting.net;user=phone>
Via: SIP/2.0/TCP proxy1.NetworkRouting.net:5060;branch=z9hG4bK2d4790.1
Via: SIP/2.0/TCP host-a-client.NetworkRouting.net:5060;branch=z9hG4bK74bf9
   ;received=27.5.16.22
```

There are a few subtle differences in this INVITE message compared to the first one: (1) a new "Via" field is appended to indicate the node visited, and to identify the IP address from which it has received the request, and (2) a "branch" parameter is included for both "Via" listings with a unique tag for the purpose of identifying this call. Note that another TCP connection is set up between the SIP proxy and the gateway that will generate TCP connection setup messages (not shown). The proxy is responsible for internally mapping the two TCP sessions to its two sides for this call, and the branch information it inserted can be helpful for this tracking purpose. The IP address for the second TCP connection may require packet forwarding through one or more routers in the SIP service provider's network, which is not shown here.

Next, the gateway would need to translate information from the SIP INVITE message to the ISUP IAM message. We assume that the gateway is directly connected to the SS7 interface of the terminating central office switch (identified with point code 244.1.1) using an SS7 F-link (refer to Chapter 12). In this message, the originating point code (OPC) will be 241.1.7 and the destination point code (DPC) will be 244.1.1. The called party number will include 816-367-2525, identifying that it is based on E.164 but using a national coding scheme, i.e., country code information is excluded. Thus, the initial address message (IAM) will contain the following key information:

```
Routing Label: OPC=244.1.7 DPC=244.1.1
TCIC: 45
Message Type: IAM
CalledPartyNumber:
  NatureofAddressIndicator: National
  NumberingPlan: E.164
  Digit: 816-367-2525
CallingPartyNumber:
  NatureofAddressIndicator: National
  NumberingPlan: E.164
  Digits: 816-328-2208
```

Note that a free trunk circuit, as identified through TCIC (refer to Section 12.6.2), must first be available before this IAM message will be generated. It may be noted that IAM is a bit-oriented protocol—information is represented using text for ease of understanding the content of the IAM message.

On receiving this IAM message, the destination central office will ring the phone of Subscriber B and generate an ACM message. When this is received at the gateway, a "183 Session Progress" message will be generated by the gateway to the proxy; this message will use the TCP connection already set up earlier. The proxy, in turn, will regenerate a "183 Session Progress" to the SIP phone. Thus, the entire call setup is complete only when this message is received at the originating SIP phone. As you can see, there is also the overhead of TCP connection setup (not shown) that adds to the call setup time.

Once the call is set up, the gateway sets up an RTP session with the SIP phone and seizes a TDM voice circuit with the destination central office. Note that the IP packets that contain RTP data usually would not go through the proxy. How does the gateway know not to do so, or, for that matter, which voice packetization scheme to use? So far, we have only mentioned G.711; there are other standardized schemes possible that might be available with the SIP phone. To determine these two pieces of information, the original SIP INVITE message from the SIP phone would actually contain more information than the headers we have shown earlier; this information is provided through a protocol format called session description protocol (SDP), which contains the IP address information of the SIP phone, the coding scheme it would use, and the transport port number where it will be expecting an RTP session from the gateway if the session is connected.

Similar to the call setup, the call release works and it is shown starting with an SIP BYE message generated by Subscriber A, which will be mapped to the REL message for the PSTN part.

There are certain important points to note: SIP ANM does not generate a response message while a SIP "200 OK" results in an ACK message in response. Furthermore, when the SIP BYE message reaches the gateway, it does not necessarily wait for the response RLC message to the REL message to arrive from the PSTN part; it goes ahead and issues a "200 OK" message. Thus, it is important to note that there is no exact, direct one-to-one mapping between SIP and SS7 ISUP messages.

20.4.4 POTS Phone to SIP Phone

Next, we consider call flow in the reverse direction. That is, the call starts from the POTS phone subscriber and is destined for the subscriber who has an SIP phone.

For the PSTN phone to SIP phone, the basic idea is similar to the detailed discussion given above for the other direction; thus, we present a pictorial view in Figure 20.8 and will not discuss this in detail. However, we want to address one issue. When an IAM message arrives at the gateway, it would need to identify the proxy to which the initial INVITE message is to be sent. This means that the gateway would need to be configured with information about the proxy server. If there is more than one proxy server, then the gateway needs to determine the proxy server to which the call setup message should be forwarded. To identify that quickly, through a separate mechanism, each proxy server informs the gateway of the current list of phones it serves, or the changes since the last announcement. Thus, the gateway can maintain a lookup table so that whenever a call arrives, it immediately knows to which proxy server to forward the SIP message.

20.4.5 PSTN-IP-PSTN

We next consider the interworking scenario where a call originates in PSTN, then goes to an IP network using a gateway, and then reenters PSTN. This involves SS7 ISUP signaling in two different segments: at the beginning and then again at the end. In between, SIP messaging is

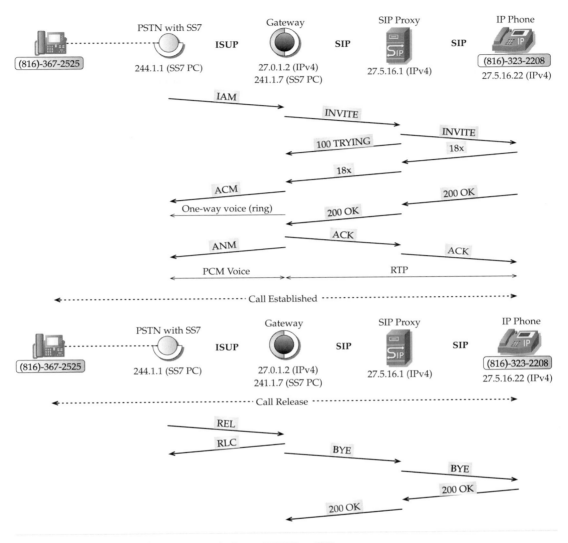

FIGURE 20.8 Call flow example from PSTN to SIP.

used to carry call control information. The call flow is similar to the ones discussed earlier. For completeness, it is depicted in Figure 20.9.

When a call goes from PSTN to IP back to PSTN, it implicitly implies that all PSTN features available to users are seamless. We next illustrate an example of what we mean by seamless. Suppose that a subscriber calls a number that uses PSTN-IP-PSTN architecture, and that the subscriber wants to make it a collect call, i.e., the recipient is suppose to pay for the call, also known as an operator-assisted call. Typically, the dialing rule for making a collect call is different. This information is then captured in the SS7 IAM message; however, SIP does not have a message type indicator to indicate this information. Thus, such information, which is to be carried back to the terminating PSTN segment, would get lost if address conversion is done only from PSTN to IP.

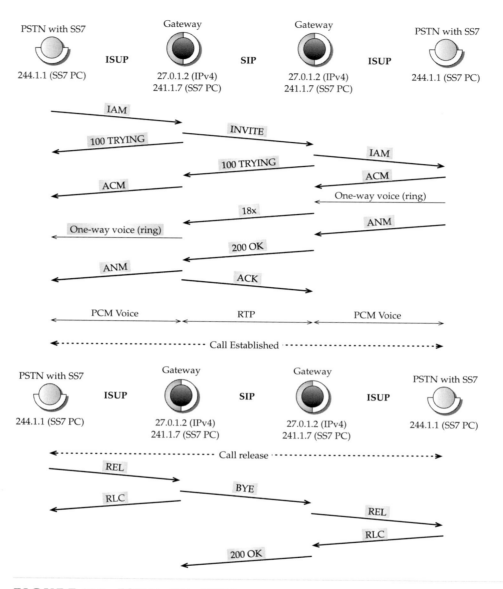

FIGURE 20.9 PSTN to IP to PSTN.

To avoid specific information loss such as the one described in the above illustration, SIP is invoked somewhat differently than the normal mode; this is referred to as *SIP for Telephones (SIP–T)* [715]. As before, a SIP INVITE message will be generated; in this case, however, the entire ISUP IAM message will be *encapsulated* in the SIP INVITE message. A difficulty with this is that an SIP INVITE message typically carries media information using the session description protocol—this is the payload. For SIP to include multiple payloads such as SDP, ISUP messages, and so on, SIP–T uses multipart MIME (multipurpose Internet mail extension) for separation of different payload contents. Note that MIME was originally standardized for simple mail transfer protocol (SMTP) to carry binary data of different types such as pictures, and is, in fact, a standard that can be thought independently of SMTP. MIME has been found to be useful in many other application layer protocols, such as HTTP. There is, however, yet another issue; there are several ISUP variations standardized and deployed throughout the world; thus, such MIME encoding is also required to identify the standard that is used so that the terminating PSTN segment would know how to handle it.

If SIP is used to serve the role of the bearer for ISUP messages for delivery to the terminating PSTN segment, can this be accomplished directly in a TCP session, bypassing the need for SIP? The answer is no. An advantage of SIP–T is that it allows call control that uses proxy servers for call routing; this feature would be lost if ISUP information is carried directly over TCP. Furthermore, the original PSTN segment does not necessarily know ahead of time whether there is a terminating PSTN segment after traversing through an IP network. Thus, SIP–T provides a richer functionality than carrying directly over TCP. From this discussion, you have probably realized by now that for PSTN-IP interworking discussed in the previous section, the gateway will generate SIP as SIP–T, since it does not know if the call necessarily ends in the IP network.

Finally, it may be noted that from an SS7 ISUP message, the routing label that contains the origination and destination point codes, and the TCIC information are excluded from being included in SIP–T. Note that OPC and DPC included in ISUP messages are on a trunkgroup basis connecting two TDM switches (refer to Figure 12.9 in Chapter 12); every TDP switch serves a service switching point (SSP) and changes OPC and DPC codes as the call control is being forwarded. Thus, we can imagine that the IP part in PSTN-IP-PSTN as the replacement for a trunkgroup by an IP network; because of this, TCIC information is not needed. Since we do not know ahead of time whether the call is PSTN-IP-PSTN or just PSTN-IP, there is no meaningful way the SS7 routing label information can be used. Thus, for PSTN-IP-PSTN, the gateway that starts the PSTN termination segment can just use its SS7 point code as the originating point code and generate the destination point code based on where the call is to be routed next.

20.4.6 Traffic Engineering

For an interworking environment, PSTN-IP, IP-PSTN, or PSTN-IP-PSTN, proper traffic engineering is required to provide good QoS. It may be noted that often the intersection point where PSTN meets IP is also the demarcation boundary between two different providers. Thus, traffic engineering issues center around offered load in a particular network although call flows in an interoperable environment. Thus, for the PSTN end, the voice traffic engineering and SS7 traffic engineering principles are applicable; refer to Chapter 11 and Section 12.12. For IP, the discussion presented in Section 20.2.5 is applicable, with one key difference. In a generic interoperable IP/PSTN environment, G.711 coding may not be the only voice coding scheme in use. In fact, the voice coding can be different. Thus, the per-call bandwidth calculation will be different depending on the coding scheme. In Table 20.1, we listed IP bandwidth requirements for a representative set of standardized voice coding schemes; the calculation is similar to the one discussed in Section 20.2.5. You will note that RTP and IP overhead, which is 40 bytes (including UDP overhead), together remains the same; the code schemes, the packetization time, and actual sample rate influence the IP call bandwidth data rate.

Once the coding information is known, through network monitoring and call flow logs we can then determine the percentage of calls for different coding schemes that is currently

TABLE 20.1   Call bandwidth requirement for different voice coding schemes (overhead = 40 bytes: 20 for IP, 8 for UDP, and 12 for RTP).

Codec	Raw Rate (Kbps)	Sampling Interval (ms)	Packet GenFreq (ms)	Packets per sec	Bits per frame	Bytes per frame	IP Level (bytes)	IP Rate (Kbps)
G.711	64	10	20	50	1280	160	200	80.0
G.726	32	5	20	50	640	80	120	48.0
G.726 (2)	24	5	20	50	480	60	100	40.0
G.728	16	5	30	33.33	480	60	100	26.7
G.728	10	10	30	33.33	300	37.5	77.5	20.7
G.729	8	10	20	50	160	20	60	24.0
G.723 (1)	5.3	30	30	33.33	159	19.875	59.875	16.0
G.723 (2)	6.4	30	30	33.33	192	24	64	17.1
iLBC	15.2	20	20	50	304	38	78	31.2
iLBC	13.3	30	30	33.33	399	49.875	89.875	24.0

using a network. Taking this information along with the IP data rate, cumulative bandwidth requirements and performance guarantees can be assessed.

An important point to note is that in a PSTN-IP-PSTN environment, IP is used as a bearer, and the endpoints are in PSTN networks with an E.164 telephone address. Since E.164 information is carried in the header of SIP-T and the entire ISUP information is carried as a payload, it is not necessary for the IP network part to use public IP address space—private IP addresses can be used; in other words, it is important to realize that the IP part can be carried as a private IP network instead of being transmitted over the public Internet. There is another advantage to it being a private IP network; the traffic carried is only for call signaling and packetized voice and is not influenced by IP packets due to many services in a public Internet. This means the performance variation for the IP segment deployed through a private IP network is much less and quality of service is easier to guarantee through traffic engineering.

20.4.7   Relation to Using MPLS

Note that once a call enters an IP provider's network, it would be routed based on its internal routing and technology environment. A provider may have a separate IP/MPLS network for this purpose. In this case, LSPs can be setup to serve traffic trunks between two LERs, and then call bandwidth is allocated based on traffic measurements. Thus, the QoS routing framework, discussed in Chapter 17, is also applicable here. In particular, the discussion presented in Section 19.3 in regard to routing with voice over MPLS is also applicable.

20.5   IP Multimedia Subsystem

In the managed IP section approach (refer to Section 20.3), we presented a managed IP approach for circumventing the copper wire–based UNI part highlighted in Figure 20.2(a). There is another emerging approach for replacing this segment for a mobile Internet, referred to as the IP multimedia subsystem (IMS). Interestingly, the basic IMS architecture is also applicable for the wired managed IP approach discussed in Section 20.3. We first start with a brief motivation for and overview of IMS.

The current cellular networks are primarily voice-circuit mode based. The underlying physical technology for cellular networks is either time-division multiple access (TDMA) or code-division multiple access (CDMA); both of these are voice optimized to provide dedicated channels to voice calls. As we move forward to providing multimedia services, it has become clear that the wireless mobile part would need to be an IP-based common platform so that a variety of services in addition to voice can be provided. Thus, IMS addresses the conceptual architectural framework for providing IP-based services in the mobile wireless world. It may be noted that IMS is constructed to address layer 3 requirements without dictating how layer 2 or layer 1 would work, thus allowing evolution of underlying wireless technology to take its own path for layer 1 and layer 2 for data-optimized applications. This is where two standardizations have emerged: third-generation partnership project (3GPP), and third-generation partnership project 2 (3GPP2). The essential difference between 3GGP and 3GPP2 is that 3GPP has evolved from GSM specifications, i.e., from a TDMA-based approach, while 3GPP2 has evolved from North American ANSI/TIA/EIA-41 standards and CDMA2000 (1x, 1xEV-DO), i.e., from a CDMA-based approach to allow for data mode applications. Discussions on TDMA/CDMA are beyond the scope of this book. An important point to note about IMS is that due to the clear separation of layer 3 from the physical technology, the basic architecture is not just limited to wireless; in fact, the managed IP approach discussed in Section 20.3 for a cable modem environment can implement the IMS architecture due to a well-defined separation of IP from the underlying technology for layer 2.

20.5.1 IMS Architecture

The goal of IMS is to make the mobile Internet a reality through the availability of a common platform for multimedia services. This provides support for mobility, for session management, and for QoS negotiation. Furthermore, it must be able to interwork with PSTN.

For session control, SIP has been chosen for IMS. It may be noted that SIP was originally designed for a *wired* Internet for session control of VoIP and multimedia services. Fundamentally, since SIP has no restriction to be on the wired part, it became the right choice for IMS as well. However, 3GPP has mandated several extensions to SIP for IMS to cope with radio access network characteristics. Note that for interworking between IP and PSTN, it is necessary to have signaling interoperability between SIP and SS7 ISUP; this is discussed separately in Section 20.4. Besides SIP, Diameter [104], [430], also specified by IETF, is used for authentication, authorization, and accounting in IMS.

In IMS, there are several different entities defined to serve different functions (see Figure 20.10). Some of these functions can be combined into one node, with the actual configuration or integration depending on the goal and the scope of the network provider that deploys IMS. For example, a regional provider's scope could be quite different from a national provider. Since IMS is IP based, all entities must have an IP address. Broadly, IMS defines the following functions:

- *Home Subscriber Server* (HSS): HSS is similar to home location register (HLR) in current cellular network; it contains various user information, especially identifying the home location of a user. In 3GPP, Subscriber Location Functions (SLF) is another entity that maps users' addresses to HSS, while in 3GPP2, this is not defined separately.

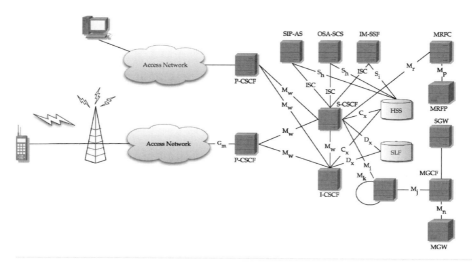

FIGURE 20.10   IMS architecture.

- *Call/Session Control Function* (CSCF): This serves the role of a SIP server for processing SIP signaling and is further divided into three categories: Proxy-CSCF (P-CSCF), Interrogating-CSCF (I-CSCF), and Serving-CSCF (S-CSCF). P-CSCF plays the same role as the proxy server discussed in Section 20.4. I-CSCF is listed in the DNS record; this is queried by a remote SIP entity, for example, during a session setup. S-CSCF is also a SIP server with its key responsibility as the SIP registration server; furthermore, it provides translation services for SIP routing when a telephone number is dialed.

- *Application Server* (AS): This has the function of executing IP multimedia services.

- *Media Resource Function* (MRF): This handles all media-related issues such as transcoding between different codes.

- *Breakout Gateway Control Function* (BGCF): This function is used to select the appropriate gateway/network based on a called number.

- *PSTN Gateway*: This provides the interface to PSTN as shown earlier in Figure 20.6.

It may be noted that multiple instances of the above functions may exist in a network. More details about IMS and description of functions can be found, for example, in [106].

20.5.2 Call Routing Scenarios

NONMOBILE CASE

We first consider the case where a user is using the service from its home network; in addition to 3GPP and 3GPP2, this scenario also works for wired IMS deployment in a technological environment such as cable modem. Since the user's phone is an IP-based SIP phone, for a call routing from IMS to PSTN, the call flow would be as illustrated in Figure 20.7; note that in this figure, registration is not shown. Thus, the S-CSCF function would also be required when a user activates the phone or turns it on after a power loss. Beyond registration, there are three additional possibilities:

1. The called number is in the same or is outside the geographic region but is provided by the same IMS provider.

2. The called number is provided by a different telephone service provider that is a traditional PSTN-based provider.

3. The called number is provided by a different telephone service provider that is also an IMS-based wired/wireless VoIP provider.

The first option can be handled by doing intradomain routing in the IP network assuming that the IMS provider has a wired IP network spanning both geographic regions; this network may run OSPF or IS-IS protocol. An internal lookup service is needed to identify the IP address of the dialed number.

In regard to either of the last two cases, an important function is to identify the receiving provider. To do that, there are two possible options: (1) the carrier identification code (CIC) of the long-distance provider of the caller, or (2) the CIC code of the receiving provider. In an IMS-based environment, the CIC code of the long-distance provider most likely will be the same as the user's local IMS-based provider, for instance, due to bundling of services at a competitive price. Thus, for a called number, the receiving provider's CIC code would need to be considered, and then, accordingly, used to identify the gateway node for routing. If it is going to a traditional PSTN, then the IP-PSTN internetworking discussed earlier will be applicable. If it is another IMS-based provider, then there are two possible options: (1) whether the entire transaction is carried through public IP-based network space, or (2) a session border controller is used. For the first approach, packets for the session and media stream will be routed by possibly traversing one or more autonomous systems in the public Internet. However, the second option, known as a *session border controller* (SBC)–based approach, has become an emerging option due to a service provider's interest in maintaining and providing stringent quality of service, as protection from divulging information such as IP address blocks about its network, and for full management of session and media control in a seamless manner. From the perspective of economics as well, SBCs are becoming popular. Interestingly, for either option, policy-based routing can play a role (refer to Section 9.5).

MOBILE USER CASE

When IMS is deployed in a mobile environment such as 3GPP or 3GPP2, a user may visit another network but still want to use the VoIP or multimedia service. A function similar to the visitor location register (VLR) function currently used in cellular networks (whether IS-95 based or GSM-based) is needed in the IMS architecture for its mobility function; this can be accomplished through deployment of P-CSCF in visited networks. However, in an IP-based world, this is not entirely necessary. When a user turns on an IMS-based phone in a visited network, it will generate a message to the P-CSCF for registration that is stored in the handset. Since both the home and visited network are IP-based, and although they may span a large geographic region, they may be connected by an intradomain wired IP network running, for example, OSPF or IS-IS protocol; thus, the actual message for registration can be routed through this network to the actual PC-CSCF in the home network.

There is another important issue to consider when a user activates services from a visited network in an IP-based environment. First, the user must be authenticated; second, the user must be assigned an IP address. Consider the following two approaches for IP address assignment:

- If the user is always assigned an IP address from its home network for service delivery, then an IP-over-IP tunnel would need to be set up from the user's device in the visited network to the home network to carry packets back and forth for this user. This option is especially necessary if P-CSCF is located only in the home network. In this case, the network has a tunneling overhead cost.

- If the user is assigned a new IP address for service delivery from the IP address block of the visited network upon registration, then the role of P-CSCF in a visited network becomes apparent. In this case, the IMS network would need to provide addressing mapping based on registration for routing an incoming call properly to this user.

It may be noted that we have touched on the above issues only briefly; this falls in the area of mobile IP, a detailed discussion is outside the scope of this book.

20.6 Multiple Heterogeneous Providers Environment

So far we have discussed issues and requirements for call routing in a mixed IP-PSTN environment so that interoperability is possible. We now briefly consider a multiple *heterogeneous* provider environment. After all, we live in a world that consists of network providers of various types and sizes, vendors who come up with solutions that hopefully make economic and business sense. In this section, we illustrate two real-world examples, which we label as "Via routing," and "Carrier Selection Alternative."

20.6.1 Via Routing

Our example involves a call that can be routed through a mix of networking technologies and providers to reach a certain destination, e.g., a 1-800 call routing in the United States. Here, the user dials an 800-number that is being answered in *another* country.

First, it must be noted that the 800-number must have a routable (telephone) number in the North American numbering plan so that the originating TDM knows where to route it. When a subscriber dials the 800-number, the call setup message is routed in the SS7 network to an SCP that provides the translation service to a routable number and identifies the CIC number of the provider. Based on this information, the setup message is routed in the SS7 network to a terminating central office; this switch then forwards the call to a location at which there is a logical trunkgroup from the destination central office. Note that this location is still in the North American region; once the call arrives at this location, there are several possible options that logically can be depicted as shown in Figure 20.11 (marked as "2"). There are three possible options from this location to another country (marked as "3"):

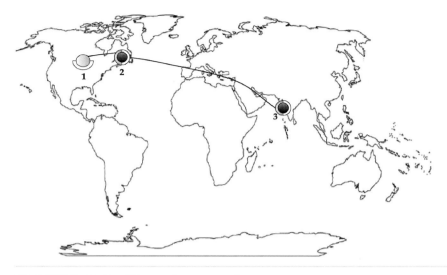

FIGURE 20.11 Via routing example.

1. There is yet another outgoing facility or mechanism used to forward this information/call further; this can be on private line–leased circuits that terminate in a different country.

2. The outgoing point serves as a gateway and is connected to a private IP network provided by a private IP network provider, which then carries the call through a shared private virtual network mechanism, including logical tunneling, to another country.

3. The outgoing point is a gateway connected to the public Internet; in this case, the call can be routed to any part of the world.

In the first case, leased circuit routing will fall under transport routing from one country to another country. In both the second and the third options, SIP-T can be used and the final termination may convert back to a regular TDM signal or be completed on SIP phones.

Note that the same principle can be used for indirectly dialing an international number as well. That is, a user wanting to make an international call first dials an 1-800 number. Once the user's call is connected to this 1-800 number location, the user now enters the actual international number. In this case, it is possible to set up an intermediate point mechanism like the site marked as "2" to which the 1-800 number for international access is first routed; from here one of the above options is used for termination at the site marked as "3"; from site 3, local or national routing within that country can be invoked to the final destination. Note that option 2 and/or option 3 can also be set for dialing an international number directly. This bypasses the need for the call to be carried over PSTN entirely, saving the carrier in the originating country from having to pay a fee to the terminating carrier in the destination country. Such models have become quite popular in recent years, again due to pricing advantage, the drop in transport cost, and the reduction in settlement charges.

20.6.2 Carrier Selection Alternative

We next consider an alternative to route selection where a local exchange carrier has a switch with an IP gateway interface for SIP-T. This then differs from the related discussion in Chapter 13, especially due to the availability of an IP interface along with the multiprovider environment.

When a subscriber dials a number, the central office switch first analyzes the dialed number. It then confers with a server connected to the SCP to determine which provider *owns* that number, or which provider is responsible for handling/delivering the call (especially for international calls). Thus, the address translation database contains identifiers such as the CIC number of the provider that is to handle delivery of the call. Along with that CIC number, the receiving IP address may be included so that a switch with an IP interface (SIP-T) can then route this call to the IP interface. Thus, in a hybrid IP-PSTN environment, conceptually a different carrier may be selected for routing if the economics make sense.

20.7 All-IP Environment of VoIP Services

Finally, we discuss the all-IP environment for delivering VoIP services. In a basic conceptual sense, a call delivery would not be any different than a standard client-server–type interaction in the current Internet for TCP-based services. The packets generated due to VoIP services would be routed at the IP level between any two points as is currently done in IP networks, hopping from one AS to another AS.

Note that in TCP-based services such as email, the originator *somehow* already knows the email address of the receiver or the web URL. Thus, this information can be used with DNS lookup to identify the receiving node's IP address; the packets are then routed. The issue in VoIP is how to *somehow* know the address or identifier of a user so that the call can be routed properly. Broadly, there are two options.

In the first option, a presence protocol-based approach can be used to identify when the user is connected to the Internet: this is transmitted to others who are on the user's "buddy list." This is generally the model of the instant messaging system, which is also applicable to recognizing the presence of users for VoIP services in an all-IP environment. Typically, these are proprietary protocol-based approaches in which users from one system may not be able to access users in another system.

In the second option, the user still uses an E.164 address as the telephone address. In this case, the issue of identifying where the user is currently connected arises. This can be handled in two different ways: (1) use an SS7-type all-call query approach to identify which provider handles the delivery of service for this user, or (2) use a standardized approach for a telephone number to IP address lookup.

You might immediately realize that this second approach is strikingly similar to the DNS-based approach for a domain name system for IP address lookup. In fact, for an E.164 address, a DNS-based approach called ENUM [208], [209], [369] has been proposed. In this approach, an E.164 address is transmitted to the DNS server using the extension, .e164.arpa, to find the IP address. In this case, being in a public protocol environment, the SIP-based approach for call setup can be used while the registration process in SIP will dictate what IP address will be listed for an E.164 for ENUM. Currently, this approach is being extensively investigated and testing for trial implementation is being pursued [108], [155], [200].

20.8 Addressing Revisited

To conclude, we revisit addressing issues. Originally, the role of the telephone numbering plan was to determine where to route a call geographically; to do that, PSTN was connected by trunkgroups between switches of various types in a hierarchical manner. The hierarchical mechanism was used to avoid looping.

With the introduction of SS7 networking, the switches have been identified by SS7 point code addressing; note that end users have continued to use the telephone number–based addressing scheme. Since each central office switch with SS7 capability serves a set of subscribers, we can say that all these subscriber telephone number addresses have direct mapping to the SS7 number for routing. Thus, to route a call to a number, a lookup needs to be performed to determine the next hop for SS7 call setup message forwarding; the decision depends on whether the call was a local call or a long-distance call. With the introduction of multiple providers for long-distance calls in the United States, a new addressing system was created, called the carrier identification code (CIC), so that for long-distance calls, mapping between the originating user's phone number and CIC is maintained; this CIC is further mapped to an SS7 point code for call forwarding and to identify outgoing trunkgroups.

With the introduction of a multiple local access provider, a telephone number address block is provided to each provider. With number portability, subscribers can take their telephone numbers from one service provider to another provider. An important issue is how do we know the *home* switch of a telephone number for call delivery? To determine this, the concept of location routing number (LRN) has been introduced. Essentially, a call is routed based on the LRN for final delivery to the actual user. This means that during the establishment of the call, one of the providers needs to do a lookup to determine to which provider/LRN to route the call.

Let us now discuss the IP addressing scheme. For IP addressing, an important difference from the telephone numbering scheme is that both end devices and routers share addresses from the same addressing scheme so that a packet can be routed based on this addressing scheme. From an address assignment point of view, the IP address is assigned on a block basis to an entity. For anyone to be able to route to this entity at its allocated IP address, the entity must sign up with an ISP that will take charge of announcing this address block to the rest of the Internet through BGP protocol. To do that, the IP address block must be associated with another addressing scheme called the autonomous system number (ASN). Thus, when an IP address block is advertised to the rest of the Internet, a "home" ASN must be associated with it so that routing tables can be formed at the right places so that they point backward toward this address block correctly.

An important point about addressing for the IP world is that additional addressing schemes have been defined for user-level addressing, for example, domain name, email address, web URL, and, more recently SIP URI (address). Since all of them fundamentally depend on the domain name, a lookup from the domain name to an IP address must be done; this function is provided through the domain name system that is globally available.

If we now compare these, we can say that a CIC code is to a telephone address as an ASN is to an IP address. This way of looking at different address relationships tells us that an address must have a home provider, where the provider must have a defined address as well. Furthermore, such homing to a provider is preferably changeable over time; i.e., the customer has the ability to change a provider without having to change their numbering.

With IP-PSTN interworking, we have arrived at a situation in which a telephone number is associated with an IP address. The question is how and where is this association taking place. Consider Figure 20.8 shown earlier. Here, the E.164 number +1-816-367-2525 has a direct association with IP address 27.0.1.2; certainly, this associate is through a masked PSTN network. For +1-816-323-2208, which is from a SIP phone, an IP address 27.5.16.22 is directly associated with the SIP phone. The larger question is how do we identify this association at the time of making a call. From the above discussion, it is clear that there are two possible models: (1) for each number dialed, identify a provider CIC number and find a route to this provider, and let the provider take care of final delivery, or (2) do a direct DNS lookup for a telephone number to an IP address through ENUM, and then route the call. Theoretically, and in practice, both of these are viable options and are currently being used. Although the second option looks more globally desirable, much like website domain name translation to an IP address, whether this will happen, and how quickly, remains to be seen. This is affected by factors such as regulatory policies, which may be different from one country to another, and the economics of delivery due to business relations between providers.

Finally, in general, addressing and routing remain a critical issue for any communication network, in particular for Internet and PSTN. We point out that while the address space exhaustion is an issue, simply increasing the address space, for example, by increasing the number of digits or bits, does not necessarily solve addressing and routing issues. In many ways, addressing and routing issues are intertwined. For example, we know that hierarchical addressing allows the route determination to be based on higher-level information, rather than more specific information. Alternately, a flat addressing scheme requires more specific information to be listed for routing table lookup, thus losing the benefit of hierarchy, and, thereby, route aggregation. However, a benefit of flat addressing is that it allows movability—that is, a customer does not need to renumber its network, yet can change a provider—this has come true both for IPv4 prefixes as well as for local number portability in PSTN. On the other hand, more specific information needed for routing for an addressing scheme, however, increases the size of the routing table. We have noticed that with routing table growth for IPv4 addressing, as discussed earlier in Chapter 9; a similar effect is starting to take place with local number portability in PSTN. The challenge then lies in building more efficient lookup algorithms that exploit every feature, along with improvement in hardware speed. Certainly, if the growth in routing table is slower than this progress in lookup algorithm and hardware speed, then even with a large routing table, we can say that the system has the ability to handle it; however, it is not apparent that this growth in the table will be slower. Thus, this brings a new challenge for networking researchers to think about innovative addressing schemes that are scalable from a routing point of view yet provide flexibility. Furthermore, any such addressing scheme must have the ability to be deployable in a live network. Thus, we anticipate many interesting and important issues to be tackled on addressing and routing in the coming years.

20.9 Summary

In this chapter, we covered VoIP routing. Because of IP-PSTN interactions and interworking, many different scenarios for service routing and delivery are possible. We present a representative set of approaches. Furthermore, we discuss IMS and how this can impact/influence routing.

The interworking environment also brings to light the issue of addressing for routing. In fact, similar to an email address the E.164 address is now taking the role of a service-based address, for service delivery for multimedia services. The issue of how E.164 is mapped to an IP address is currently a topic of considerable interest as this may entail regulatory issues that can be different in different countries. How this plays out remains to be seen.

As a final thought, in this book, we have attempted to provide a detailed coverage of network routing for a variety of communication networks. Thus, we pause for a moment and look back at the history of communication networking over the past 150 years or so. We can say that, broadly, we can classify the communication networking paradigm into three groups: telegraphy, telephony, and the Internet. If we look at time spacing, going from the birth of telegraphy (in 1844) to the birth of telephony (in 1876), it took thirty years; from the birth of telephony to the birth of the Internet (in 1969) took nearly a hundred years. Thus, we can ask a general question: is there any new communication paradigm that is different from the way have thought or done so far, i.e., from telegraphy, telephony, and the Internet? If so, what would be addressing and routing be? While we do not claim to have the answers to these questions, we hope that our coverage of routing in different networks, along with issues covered and lessons learned, will help in a small way toward such a vision.

Further Lookup

The basic concept for telephony for IP over cable, PacketCable, is discussed in [456], [457]; detailed specifications for PacketCable can be found at [103]. A discussion on IP-PSTN signal mapping between SIP INVITE and ISUP IAM messages can be found in [107]; also refer to [341] for SIP-PSTN call flows. Any hybrid environment brings new vulnerability; for example, see [12]. The emerging IMS architecture is discussed in detail in [106]. Both IMS architecture and cable providers are exploring/deploying IPv6 addresses; this is partly based on the issue of lack of IP address blocks (even private IP address) for large-scale deployment of services. In most such deployment, address need not be moved out of a provider since an IP address is potentially softly associated with a customer for the duration of service the customer signs up for. However, in many other situations, there is need for a customer to take the address block with them and have provided service by another provider; in this regard, there have been developments on provider-independent IPv6 addressing; see [281].

DNS mapping for E.164 addressing is discussed in [208], [209], which assigned certain aspects of ENUM to the Internet Architecture Board; a discussion on responsibility and the relation to ITU–T can be found in [369]. ENUM request for SIP is described in [561]. Enumservice that contains PSTN signaling is discussed in [415]. For recent trial implementation of ENUM, see [108], [155], [200].

You may note that there is more material available on the subject of routing than is possible to cover in a book. Two useful resources are Wikipedia [739] for a general description of any routing-related topic (and others as well) and Google [258] to search for things not found in Wikipedia or for detailed information through a series of complex searches. We can almost guarantee that someone has written about it, or, better, created a presentation that helps you learn. We also encourage you to read, participate, discuss, and learn from many routing- and addressing-related mailing lists, maintained by standards bodies and the networking research community.

Exercises

20.1 Describe the IP-PSTN-IP call routing scenario, along with the different protocol messaging involved in call routing setup.

20.2 Consider the call-carrying capacity discussed in Section 20.2.5 for G.711. Determine call-carrying capacity on a OC-3 link for other codec schemes listed in Table 20.1 due to packetization.

20.3 Discuss the key differences between ISUP signaling and SIP.

20.4 Identify what type of information may not be mappable when an ISUP IAM message is converted to a SIP INVITE message.

20.5 Discuss where and why SIP–T needs to use MIME.

Appendix A: Notations, Conventions, and Symbols

In most people's vocabularies, design means veneer. It's interior decorating. It's the fabric of the curtains of the sofa. But to me, nothing could be further from the meaning of design. Design is the fundamental soul of a human-made creation that ends up expressing itself in successive outer layers of the product or service.

Steve Jobs

A.1 On Notations and Conventions

In networking, we sometimes say that there are not enough bits or digits in a particular address space or a protocol field. We faced a somewhat similar situation with notations; we found that Roman and Greek scripts do not have enough alphabets to avoid repeats. Certainly, we were tempted to use Assamese and Tamil scripts to ensure uniqueness. In the end, we choose notations in this book so that the material is relatively easy to read and without being confusing due to repeats.

Typically, any two generic nodes are identified by i and j. If there is a link between them, then it is noted as i-j. You will note that k is used as a generic intermediate node between two nodes i and j. If we are referring to the pair of nodes i and j, for example, to denote traffic or to find a path between them, this pair is noted as i:j, so that it is not confused with link i-j. This set of notations is used in Chapter 2 and Chapter 3. A general comment is that historically k-shortest path is the common name in which the shortest, next best shortest, up to k paths are to be determined. Certainly, this k has no relation to the use of k to mean an intermediate node.

For notations related to network flow modeling and traffic engineering, we have used a somewhat different notation. Typically, for a link, we use single identifier ℓ. Similarly, to identify a pair of nodes for demand, we use a single identifier k. The demand volume, regardless of the unit of measure, is denoted by h_k for demand pair k. This basic notational guideline is followed in Chapter 4, Chapter 7, Chapter 19, Chapter 24, and Chapter 25. Since most of these chapters have quite a few notations, we have often used tables to list all notations that are relevant to a chapter. There are, however, commonality in notations; still, keeping similar looking summary tables on notations avoid the need to refer back to Chapter 4 when this notations were first introduced.

Other than that, the notations in the rest of the chapters are of primarily of local significance, i.e., they are unique within a chapter itself. In many occasions, we do reuse i, j, or k to denote an index or an iteration; however, all such usage should be clear within the context they are used.

When possible, we use a number range, rather than a set notation. For example, $k = 1, 2, \ldots, K$ is to denote the demand pairs from $k = 1$ to the maximum value of K. However, in certain situations, a set notation is helpful to use; in such cases, a set is denoted, say $\mathcal{K}$, and an element k in this set is indicated as $k \in \mathcal{K}$.

In general, we follow a single letter notation for describing a unit, for example, c for capacity. However, there are a few cases where we do use an abbreviation, due to such use being well known. For example, RTT is used to denote the round trip time; it does not mean $R \times T \times T$. Because of this possibility, on several occasions we have used the sign "$\times$" explicitly, primarily for clarity. We tend to use a descriptive name more commonly in a subscript, such as T_{average}. Otherwise, a subscript is used as an index, such as h_k to denote demand volume for demand identifier k. In a few places, we have also used a slightly different notation, $T[k]$, to convey the same meaning; typically, we use this notation when we are referring to memory location, such as location k for array T.

In this book, we use *byte* to indicate an 8-bit element, instead of using the traditional name *octet*.

Equation numbering has three parts: the chapter number, the section number, then the equation number within this section. For example, (7.6.12) refers to equation number 7 lo-

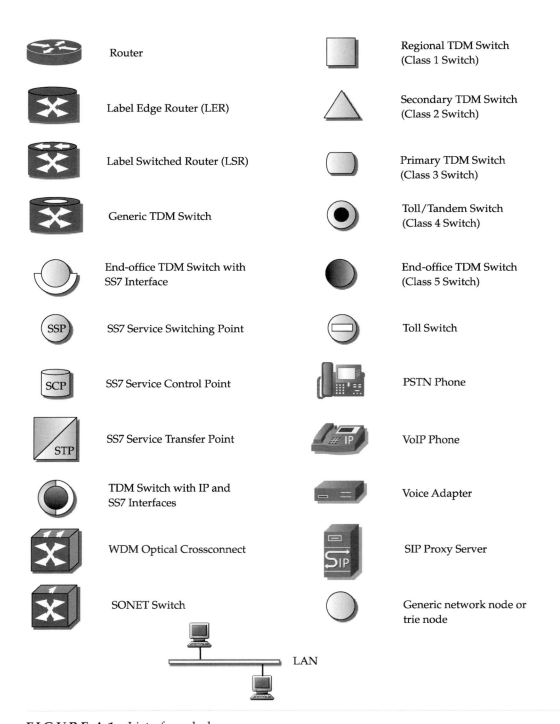

FIGURE A.1   List of symbols.

cated in Section 6 of Chapter 7. In general, we refer to an equation number as "Eq. (7.6.12)" regardless of whether it is an equation or an inequality or a formulation. In certain cases, we do qualify an equation number using a term to indicate what it reflects; for example, Formulation (7.6,8), or Constraints (7.4.8).

References cited are marked in square brackets (e.g., [21]).

In algorithmic description, we use double-slash ("//") to indicate the start of comments.

As a general convention, we use private addresses or unassigned addresses for illustrating examples.

A.2 Symbols

The book uses a variety of symbols to denote different types of nodes such as routers and switches. The entire list is summarized in Figure A.1.

Appendix B: Miscellaneous Topics

If you don't know what to do with many of the papers piled on your desk, stick a dozen colleagues' initials on 'em, and pass them along. When in doubt, route.

Malcolm Forbes

B.1 Functions: Logarithm and Modulo

The logarithm of n to the base b is written as $\log_b n$. If we write this quantity as z, then the equivalence between log and the exponent is as follows:

$$z = \log_b n \quad \Leftrightarrow \quad b^z = n.$$

There are two values of the base that are of interest to us, $b = 2$ and $b = e$, where e is the Euler number (e $= 2.718128$). The relation between different bases, shown below between bases e and 2, is as follows:

$$\log_2 n = \log_2 e \times \log_e n.$$

For base e, ln is often used instead of log; similarly, for base 2, lg is used. Thus, the above relation can also be written as: $\lg n = \lg e \times \ln n$.

You may note that we use the notation $\log^d N$ in several places in this book; this means $(\log N)^d$, which is not the same as $\log N^d (= d \log N)$.

The **modulo** function is written in short as mod. Thus, $a \bmod m$ means the non-negative remainder when a is divided by m, which results in a number in the range $[0, \ldots, m - 1)$. Thus, $1 = 10 \bmod 3$.

B.2 Fixed-Point Equation

A fixed point equation has the following form:

$$x = \mathcal{F}(x). \tag{B.2.1}$$

That is, when the continuous function $\mathcal{F}$ is applied to input x, then the output is x as well. It is important to note that the fixed point equation makes sense when the domain and the range of function $\mathcal{F}$ is the same. It is called a fixed point equation since there must be at least an x, i.e., a fixed point for which Eq. (B.2.1) holds; that is, the input is equal to the output.

An approach to solve a fixed point equation is to consider a simple iterative process where the input leads to output which, in turn, becomes input. In other word, if at iteration k, we have x_k, then using Eq. (B.2.1), we can compute a *new x* through the following recursion:

$$x_{k+1} = \mathcal{F}(x_k). \tag{B.2.2}$$

Then x_{k+1} becomes input in Eq. (B.2.1) to obtain x_{k+2}, and so on. The process stops when the difference between two consecutive x's, $|x_{k+1} - x_k|$, is below an acceptable threshold, ε. The general fixed-point algorithm is outlined in Algorithm B.1. Note that at the time of initialization, $x_{old} \neq x_{new}$ and as long as the difference is more than ε; a simple rule is to set x_{old} to be 2ε away from x_{new}. Finally, the actual value of ε depends on accuracy desired for a specific problem; for some problems, it might be acceptable to set $\varepsilon = 0.01$ while for others $\varepsilon = 10^{-6}$ might be required. Sometimes, the relative difference as percentage is more appropriate when the solution is far away from zero; in such cases the stopping rule $|x_{k+1} - x_k| \leq \varepsilon$ is replaced by $\frac{|x_{k+1} - x_k|}{|x_k|} \leq \varepsilon$; accordingly, the continuation condition in the while-loop of the algorithm is adjusted along with ensuring that no divide-by-zero situation occurs, i.e., as $|x_{old} - x_{new}| > \varepsilon |x_{old}|$.

ALGORITHM B.1   Fixed Point Algorithm.

procedure FixedPointSolution ($\mathcal{F}$)
Initialize x_{new}
Set $x_{old} = x_{new} \pm 2\varepsilon$
while ($|x_{old} - x_{new}| > \varepsilon$) do
   $x_{old} = x_{new}$
   $x_{new} = \mathcal{F}(x_{old})$
endwhile
return(x_{new})
end *procedure*

It is important to note that a fixed point equation does not necessarily always have a unique solution or stable point. Consider the function $\mathcal{F}(x) = 3x(1 - x)$ in the domain $[0, 1]$. In this case, the fixed point equation $x = \mathcal{F}(x)$ has two solutions: $x = 0$ and $x = \frac{2}{3}$. A fixed point equation is known to have a unique solution if the absolute value of the first derivative of $\mathcal{F}(x)$ is in the interval $(0,1)$, i.e., $|\mathcal{F}'(x)| \leq D$, where constant $D \in (0, 1)$. If $\mathcal{F}'(x) \neq 0$, then the rate convergence of the fixed point equation is linear, i.e., $\lim_{k \to \infty} = \frac{|x_{k+1} - x^*|}{|x_k - x^*|} = C$, where $C = |\mathcal{F}'(x)| < 1$ and x^* denoted the solution.

In general, while the fixed-point iterative scheme looks innocent, the convergence can be a problem depending on the starting point (x^0) and the function, $\mathcal{F}(x)$. This point is well described in [86] for $\mathcal{F}(x) = 4ax(1 - x)$ (where a is a parameter); it was shown (through animation) that the fixed point algorithms leads to convergence easily if $a \leq 0.75$, and alternates between two points for $0.75 < a \leq 0.86237$, and becomes very chaotic beyond $a = 0.86237$.

The fixed point equation problem arises also "in cube", i.e., in the n-dimensional space, instead of a single variable described so far. In this case, there are n unknowns, and there is a set of n equations connecting a set of functions $\mathcal{F}_1, \mathcal{F}_2, \ldots, \mathcal{F}_n$ as follows:

$$
\begin{aligned}
x_1 &= \mathcal{F}_1(x_1, x_2, \ldots, x_n) \\
x_2 &= \mathcal{F}_2(x_1, x_2, \ldots, x_n) \\
\cdots &= \cdots \\
x_n &= \mathcal{F}_n(x_1, x_2, \ldots, x_n).
\end{aligned}
\tag{B.2.3}
$$

If we denote n unknowns by vector $x = (x_1, x_2, \ldots, x_n)$, then we can write the above in the following compact form $x = \overrightarrow{\mathcal{F}}(x)$, where $\overrightarrow{\mathcal{F}}$ denotes the vector function, $\overrightarrow{\mathcal{F}} = (\mathcal{F}_1, \mathcal{F}_2, \ldots, \mathcal{F}_n)$. Certainly, the domain and the range need to be the same.

B.3 Computational Complexity

Many problems we have described in this book require algorithmic approaches. Algorithmic approaches can be broadly classified into the ones that can be described in terms of finite number of operations of the input, and the ones that are numeric precision oriented. An example of the former is Dijkstra's shortest path algorithm, and an example of the latter is the fixed-point algorithm. For the former, we usually consider the complexity of the algorithm in terms of the input length n and present with a notation known as *big-Oh*, such as $O(\log n)$ and

$O(n^2)$. For the latter, the performance of an algorithm is described in terms of linear, super-linear, or quadratic convergence. For example, Newton's method to find a root of an equation has quadratic convergence in general (there are exceptions, see [739]); we have already discussed earlier about the convergence of the fixed-point equation.

In this section, we consider the class of algorithms that can be determined in a finite number of steps, for input of size n. When we say finite number of operations or steps, we do not mean small. For instance, for input size n, a particular algorithm might require 2^n operations or steps to arrive at the solution; certainly, 2^n is finite, but can be very large. For example, suppose that each operation takes 1 microsec ($= 10^{-6}$ sec), then for $n = 50$, 2^n turns out to be 36 years! If someone comes up with a clever algorithm to solve the same problem in n^3 operations, then for $n = 50$, it will take only 0.13 sec. Thus, it is very helpful if we can count the number of operations of an algorithm. Often, however, we do not need to find the exact number of operations; an approximation that is off by a fixed multiplier and, thus, that can be written as a function of n is helpful enough. Suppose that the clever algorithm actually takes $5 \times n^3$ operations, then the total time will be 0.65 sec. We still know that this approach is much better than the algorithm that takes 2^n operations. The role of the big-oh notation is to give an approximate, yet clear idea of the time (in terms of the number of operations) it takes for an algorithm to solve a problem.

Formally, consider two functions f and g in the domain $\{1, 2, 3, 4, \ldots\}$. We say $f(n)$ is "order $g(n)$" or "big-Oh of $g(n)$" and write as

$$f(n) = O(g(n))$$

if there exists a positive constant C and a positive number k such that

$$|f(n)| \leq C|g(n)|$$

whenever $n > k$.

The meaning of k is best understood for algorithms in the sense that we are only interested in input size that is at least of size k. Going back to our previous example, if we were to consider the input size to be $n = 3$, then 2^n is smaller than n^3; this would have given us a false sense of security that the algorithm with 2^n operations is better than the algorithm with n^3 operations. In other words, we look at an algorithm from the growth point of view as n increases.

What are $f(n)$ and $g(n)$ then? The function, $f(n)$, represents the number of operations of an algorithm for input size n. Typically, the function, $g(n)$, is considered from a set of well-known functions that have clear demarcations, such as n, $\log n$ (here we assume base 2), $n \log n$, n^2. In Figure B.1, we plot a few such functions to illustrate their growth as n increases. You may note that we have also plotted $O(1)$—this means that regardless of the input size n, the number of operations remains constant.

With time complexity, it is also important to consider space complexity. Sometimes, an algorithm has a good time complexity, but requires a large storage space for operations to be computed. For example, an algorithm might have the time complexity $O(n)$ but requires $O(n^4)$ storage space—in this case, you have to determine if this storage is possible from an implementation of view. Such time-space trade-off is an important issue in determining appropriate lookup algorithms and classification algorithms as you will find when you read Chapter 15 and Chapter 16.

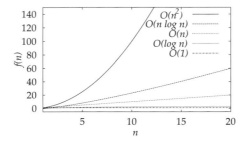

FIGURE B.1 Growth of some common functions.

B.4 Equivalence Classes

To understand equivalence classes, first consider a *relation*. A relation defines the notion of a connection between two elements, but in a broader sense than a function. For example, an element may be related to more than one elements. Consider a set X. Then, an element a in X may be related to two elements b and c, also in X. We can then use the ordered pair (a, b) and (a, c) to indicate that both belong to the same relation R. Using the set notation, we can write $(a, b) \in R$, $(a, c) \in R$.

There are three properties we need to define in order to define an equivalence relation. A relation R on a set X is reflexive if $(a, a) \in R$ for every $a \in X$; a relation R is symmetric if for all $a, b \in X$, if $(a, b) \in R$, then $(b, a) \in R$; a relation R is transitive if for $a, b, c \in X$, if $(a, b) \in R$ and $(b, c) \in R$, then $(a, c) \in R$. Then, a relation R on a set X is called an equivalence relation if it satisfies reflexive, symmetric, and transitive relations.

Now suppose that R is an equivalence relation on a set X. Then the set of all elements that are related to an element $a \in X$ is called an equivalence class of a. In other words, an equivalence class helps grouping or partitioning "like minded" elements.

As an example, consider the set $X = \{1, 2, \ldots, 10\}$. We define the relation R to mean that "3 divides $x - y$." You can verify that R is an equivalence relation. We can see that the set X can be partitioned to three equivalence classes consisting of elements that follow "3 divides $x - 1$," the elements that follow "3 divides $x - 2$," and the elements that follow "3 divides $x - 3$," which then correspond to the subsets $\{1, 4, 7, 10\}$, $\{2, 5, 8\}$, and $\{3, 6, 9\}$, respectively.

B.5 Using CPLEX

We will illustrate use of CPLEX [158] for a simple load balancing problem over two paths. First note that in CPLEX, a problem can be written almost similar to how an optimization problem formulation is written. There are, however, two main caveats: (1) variables should be all on the left-hand side, and (2) no "divide by" entry such as **x1/10** is allowed. A nice feature in CPLEX is that you can give a name to each constraint (dual multiplier). Thus, a formulation in CPLEX looks like

```
Minimize
  r
Subject to
  demandflow: x1 + x2 = 11
```

```
  link1utilization: x1  - 10 r <= 0
  link2utilization: x2  - 15 r <= 0
End
```

Note that in CPLEX, it is not necessary to indicate that the variables are non-negative; this is implicitly assumed. If the above model is saved in a file, say, load-balance.lp, then in CPLEX, it can be invoked at the command prompt CPLEX> as follows:

```
CPLEX> read load-balance.lp
CPLEX> optimize
CPLEX> display solution variables -
Variable Name         Solution Value
r                        0.440000
x1                       4.400000
x2                       6.600000
```

The dual solutions can be displayed as follows:

```
CPLEX> display solution dual -
Constraint Name          Dual Price
demandflow                0.040000
link1utilization         -0.040000
link2utilization         -0.040000
```

When some of the variables are integer-valued they must be explicitly noted. Furthermore, besides declaring the appropriate variables as integer, it is necessary to provide an upper bound. Without the bound, CPLEX assumes the integer variables to be binary variables. For instance, in the above problem, suppose we want the routing flow variable, x1 and x2, to be integer as well; then, they must be declared under Integer. Thus, by changing the variables to Integer and providing bounds, we can write the above problem as:

```
Minimize
  r
Subject to
  demandflow: x1 + x2 = 11
  link1utilization: x1  - 10 r <= 0
  link2utilization: x2  - 15 r <= 0
Bounds
  0 <= x1 <= 11
  0 <= x2 <= 11
Integer
  x1 x2
End
```

Note that Bounds for variables are set to 11, knowing that the total demand volume is 11. Also note that CPLEX requires Bounds to be listed before Integer. The solution this time will be:

```
CPLEX> display solution variables -
Variable Name         Solution Value
r                        0.466667
x1                       4.000000
x2                       7.000000
```

Finally, a nifty feature in CPLEX is that you can create a file with other comments/documentation before the declaration **minimize** and after the declaration **End**, which CPLEX ignores. Thus, to run the above model for continuous capacity variables, you can move up **End** and list it before **Bounds** declaration, without completely deleting **Bounds** and **Integer** declaration from the file. You might want to test and see what happens (and why) if you do *not* include the entire **Bounds** part, but do include **Integer** part.

While in this book, we have used CPLEX to illustrate how to solve linear programming or mixed integer linear programming problems, there are many other commercial and public-domain solvers available.

B.6 Exponential Weighted Moving Average

Consider the following data collected at time $t = 1, 2, 3, 4, 5$:

$t =$	1	2	3	4	5
$M_{measured} =$	5	10	7	8	11

We will assign 0.75 weight to the old value and 0.25 weight to the new value to compute a new weighted or smoothed value. For the initial step, we will use the value as given, that is, 5 in this case at time $t = 1$. Thus, the smoothed value at time $t = 2$ is $0.75 \times 5 + 0.25 \times 10 = 6.25$. Now, to compute the value at $t = 3$, the smoothed value from $t = 2$, i.e., 6.25 is weighted with the new value; thus, we calculate the smoothed value at time $t = 3$ as $0.75 \times 6.25 + 0.25 \times 7 = 6.43$ (rounded to two decimal places), and so on. Below, we show the results for different values of the weight, α (all values to two decimal places), both in tabular and graphical forms:

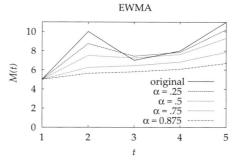

Event	Original	$\alpha =$			
t	Measurements	0.25	0.50	0.75	0.875
1.00	5.00	5.00	5.00	5.00	5.00
2.00	10.00	8.75	7.50	6.25	5.63
3.00	7.00	7.44	7.25	6.44	5.80
4.00	8.00	7.86	7.63	6.83	6.07
5.00	11.00	10.21	9.31	7.87	6.69

FIGURE B.2 Exponential weighted moving average.

From the computed values, we note that the numbers are being evened out to around 6 or 7 when $\alpha = 0.75$, taking away fluctuations such as 10 and 11 from the measured values. When the weight is reduced to, say 0.25, we see that the smoothed value is pulled toward the measured value; this is natural since higher weight is given to the new measurement. That is, the weight can make a big difference. However, whether to give a low or high weight to the newly measured value depends on a specific problem and is not generalizable. In any case, if we denote the weight by α (with $0 \leq \alpha \leq 1$), then the smoothed computation at time $t + 1$ can be written based on knowing the smoothed value at t and the newly measured value at time $t + 1$ as follows:

$$M_{\text{smoothed}}(t + 1) = \alpha M_{\text{smoothed}}(t) + (1 - \alpha)M_{\text{measured}}(t + 1).$$

The above approach is called the exponential weighted moving average (EWMA) method. It may be noted that if $\alpha = 0$, then this is nothing but considering the new measurement without giving any weight at all to the old measurement, and if $\alpha = 1$, then no weight is given to the new measurement. Most implementation of the EWMA method that require fast computation uses a weight value α that is a negative power of 2 so that it can be computed efficiently using the shift operation in registers.

B.7 Nonlinear Regression Fit

There are many forms of non-linear regression fits. We will focus here on a simple and effective model that fits a relation of the following form:

$$y = a + cx^m.$$

Note that it has "shift" factor a, which can be thought of as the degree of freedom; thus, we can focus on determining c and m. If we subtract a from both sides, we get $y - a = cx^m$. If we use $\tilde{y} = y - a$, we then have $\tilde{y} = cx^m$. Taking log of both sides (note: any base will do here), we get

$$\log \tilde{y} = \log c + m \log x.$$

Now, if we have a set of measurements (x_i, y_i), $i = 1, 2, \ldots, n$, we can write

$$\log \tilde{y}_i = \log c + m \log x_i, \quad i = 1, 2, \ldots n$$
$$\log x_i \log \tilde{y}_i = \log c \log x_i + m(\log x_i)^2, \quad i = 1, 2, \ldots n,$$

where the second set is multiplied throughout by $\log x_i$. Now, summing each set of relations over $i = 1, 2, \ldots n$, we have the following two equations:

$$\sum_{i=1}^{n} \log \tilde{y}_i = n \times \log c + m \sum_{i=1}^{n} \log x_i,$$

$$\sum_{i=1}^{n} (\log x_i \times \log \tilde{y}_i) = \log c \sum_{i=1}^{n} \log x_i + m \sum_{i=1}^{n} (\log x_i)^2.$$

This is a simultaneous equation of two unknowns, m and $\log c$. Thus, they can solved by substitution to arrive at the following solution:

$$\log c = \frac{\left(\sum_{i=1}^{n}\log\tilde{y}_i\right) \times \left(\sum_{i=1}^{n}(\log x_i)^2\right) - \left(\sum_{i=1}^{n}\log x_i\right)\left(\sum_{i=1}^{n}(\log x_i \times \log\tilde{y}_i)\right)}{n\times\sum_{i=1}^{n}(\log x_i)^2 - \left(\sum_{i=1}^{n}\log x_i\right)^2},$$

$$m = \frac{n\sum_{i=1}^{n}(\log x_i \times \log\tilde{y}_i) - \left(\sum_{i=1}^{n}\log x_i\right)\times\left(\sum_{i=1}^{n}\log\tilde{y}_i\right)}{n\times\sum_{i=1}^{n}(\log x_i)^2 - \left(\sum_{i=1}^{n}\log x_i\right)^2}.$$

While the above formula looks complicated, it is fairly simple to do using a tool such as Excel. Thus, m and c can be obtained for the regression fit. Recall that the shift parameter, a, is the degree of freedom and it must be chosen properly and carefully so that the fit is acceptable. Often, a regression fit is done to determine the future trend. Thus, a set of different values of a can be considered so that an upper and a lower bound on trend projection, i.e., an acceptable band, can be obtained.

The projection on IP prefix growth, shown in Eq. (9.3.1), is based the above approach. The measurement data used was for the first day of each quarter from the beginning of year 2002 to April 1, 2006, obtained from [303]; a quick side note is that the first quarter data for 2002 was for January 3, 2002, due to non-availability of data on January 1, 2002. For each data input day, the measurements during the day was averaged first, which then became the input to the regression model. Although we show an equation with specific values in Eq. (9.3.1), a band can also be determined to indicate over- and under-estimate by appropriately choosing the degree of of freedom, a.

B.8 Computing Probability of Path Blocking or Loss

Consider that probability of blocking on link 1 is b_1 and that on link 2 is b_2. Now consider the path that is composed of the links, 1 and 2. Assuming that the link blocking probability is independent, the probability of blocking for the path is given by

$$\mathcal{P}_{[2]} = 1 - (1 - b_1)(1 - b_2).$$

This result can be generalized to any number of links. Suppose that a path has L links and probability of blocking on link ℓ is given by b_ℓ, $\ell = 1, 2, \ldots, L$. Then the path blocking probability, b, is given by

$$\mathcal{P}_{[L]} = 1 - \prod_{\ell=1}^{L}(1 - b_\ell) \tag{B.8.1}$$

where $\prod_{\ell=1}^{L}(1 - b_\ell)$ denotes the product $(1 - b_1)(1 - b_2)\cdots(1 - b_L)$.

When $b_\ell \ll 1$, the path blocking probability can be approximated with a simpler result. Consider again the path with two links. On expanding, we get

$$\mathcal{P}_{[2]} = 1 - (1 - b_1 - b_2 + b_1 b_2) = b_1 + b_2 - b_1 b_2.$$

If b_1 and b_2 are very small, then the product $b_1 b_2$ is negligible compared to either b_1 or b_2 and can be ignored. For example, if $b_1 = b_2 = 0.01$, then $b_1 b_2 = 0.0001$. That is, we can write

$$\mathcal{P}_{[2]} \approx b_1 + b_2.$$

Generalizing for a path with L links, we have

$$\mathcal{P}_{[L]} \approx b_1 + b_2 + \cdots + b_L = \sum_{\ell=1}^{L} b_\ell. \tag{B.8.2}$$

This result is also useful if we replace the probability of blocking on a link by packet loss probability of a link buffer for data networks. That is the packet loss probability of a path is determined by knowing the packet loss probably of each link that is a constituent of this path; the additive property shown above holds since packet loss probability of a link is usually very small.

B.9 Four Factors in Packet Delay

There are four factors that cause delay for a packet starting from nodal processing to traversing a link. They are: (1) nodal processing delay, τ_{proc}, (2) transmission delay, τ_{trans}, (3) propagation delay, τ_{prop}, and (4) queueing delay, τ_{queueing}. Thus, we can write the overall link delay, τ_{delay}, as

$$\tau_{\text{delay}} = \tau_{\text{proc}} + \tau_{\text{trans}} + \tau_{\text{prop}} + \tau_{\text{queueing}}.$$

The nodal processing delay is due to packet checking at a node (e.g., a router) for bit error and determining which output link to use—usually, this factor needs to be the least dominant of all the four factors.

Transmission delay refers to the time to send bits into a link. If link speed is c bits per second, and the packet length is κ bits, then the transmission delay is

$$\tau_{\text{trans}} = \frac{\kappa \times 1000}{c} \text{ millisec.}$$

For example, a 40 byte packet (i.e., 320 bits) on a 56 Kbps links would have transmission delay of about 5.7 millisec ($= 320 \times 1000/56000$). The same packet size on a OC-3 link ($=155$ Mbps) would have a transmission delay of 0.002 millisec. In other words, the transmission delay is fairly significant on a low-speed link.

Propagation delay refers to the time it takes to send a bit from one end of the link to the other end, i.e., it depends on the length of the link and the propagation speed of the medium. If length of the link is specified as D kilometers (Km), and the propagation speed as s meters per second (m/sec), then propagation delay is

$$\tau_{\text{prop}} = \frac{D \times 1000 \times 1000}{s} \text{ millisec.}$$

For example, the propagation speed of coaxial cable and fiber cables is about 2×10^8 m/sec. Thus, the propagation delay on a 300 Km long link at this propagation speed is 1.5 millisec. If the distance is only 10 Km, then the propagation speed reduces to 0.05 millisec. Thus, on a long-distance link, the propagation delay can be a fairly signification factor.

Queueing delay depends on the stochastic arrival property of packets and the packet service rate of links. For simplicity, we assume here that it is an $M/M/1$ system, i.e., the packet arrival to a link follows Poisson process with rate λ and the packet service rate (μ) is exponential distributed (refer to Appendix B.12.2). If the link speed is c bits/sec, and the average packet size is κ (in bits), then the service rate is $\mu = c/\kappa$ packets/sec. Then, the $M/M/1$ average queueing delay is given by

$$\tau_{\text{queueing}} = \frac{1}{c/\kappa - \lambda} \sec = \frac{1000}{c/\kappa - \lambda} \text{ millisec.}$$

It is easy to see that the queueing delay is the most dominant factor when the arrival rate, λ, approaches the service rate, c/κ.

B.10   Exponential Distribution and Poisson Process

EXPONENTIAL DISTRIBUTION

The probability density function (PDF) and the cumulative distribution function (CDF) of the exponential distribution with rate parameter λ are given by

$$\begin{aligned} f(t) &= \lambda\,e^{-\lambda t}, \quad t \geq 0 \text{ (PDF)} \\ F(t) &= 1 - e^{-\lambda t}, \, t \geq 0 \text{ (CDF)}, \end{aligned} \qquad \text{(B.10.1)}$$

respectively. The mean and the variance of the exponential distribution are given by $E(X) = 1/\lambda$, $V(X) = 1/\lambda^2$. The PDF and CDF are plotted for $\lambda = 1, 5, 10$ (Figure B.3).

POISSON PROCESS

A Poisson process is a stochastic process. It is easy to understand a Poisson process through its relation to the exponential distribution. For a Poisson process with rate λ, the inter-arrival time between two events are independent and exponentially distributed with parameter $1/\lambda$.

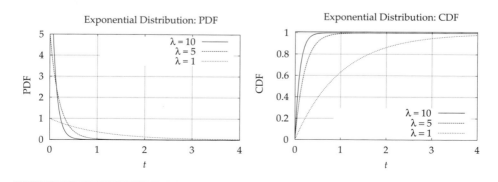

FIGURE B.3   PDF and CDF of the exponential distribution.

Thus, if we consider event $i-1$ occurring at time τ_{i-1} and the next event i occurring at time τ_i, we can look at the difference in the event time $s_i = \tau_i - \tau_{i-1}$. Then, we can say that the inter-arrival time, s_i, between $i-1$-th event and the i-th event is probabilistically related to the exponential distribution and can be written as:

$$Pr\{s_i \le t\} = 1 - e^{-\lambda t}.$$

Note that the right-hand side is the CDF of the exponential distribution. An important property of a Poisson process is that the sum of two independent Poisson processes is a Poisson process with rate being the sum of the two.

GENERATING EXPONENTIAL RANDOM VARIABLE

Generating exponential distribution involves a trick known as the inverse-transform technique [399]. Note that for the exponential distribution, the CDF is monotonically increasing and takes a value between 0 and 1, i.e., $0 \le F(t) \le 1$. Thus, if we consider from the point of view of the y-axis of the CDF function, we need to generate $F(t)$ uniformly between 0 and 1, to get back t, rather a random t. Let us write u for this uniform value, then $u = F(t)$, and thus, we can write $u = 1 - e^{-\lambda t}$. Rearranging, we get $e^{-\lambda t} = 1 - u$. Taking log (base e) of both sides, we get $-\lambda t = \ln(1 - u)$. This implies

$$t = -\frac{1}{\lambda} \ln(1 - u).$$

Thus, for a given rate λ, we can obtain a specific value t if we can generate a uniformly distributed number, u. You may note that we can make another minor change here by using u instead of $1 - u$, since if u is from a uniform distribution, so is $1 - u$. Thus, we can safely replace u in place of $1 - u$, i.e.,

$$t = -\frac{1}{\lambda} \ln(u).$$

Most programming environments provide a function called rand, which generates uniform random number between 0 and 1; in fact, each invocation gives a different random value. Thus, rand can be used in place of u for generating the exponentially distributed random number t. This is discussed next for Poisson process.

ON GENERATING POISSON ARRIVAL PROCESS

Recall that the Poisson Process with rate λ has the property that the inter-arrival time is exponentially distributed with mean $1/\lambda$. Then, we can generate τ_i for event $i = 1, 2, 3, \ldots$ (assume $t_0 = 0$) for a Poisson process in the following manner:

1. Generate $u \sim$ rand

2. Set $\tau_i = \tau_{i-1} - \frac{1}{\lambda} \ln u$.

This process is continued until you have generated the desired number of events, n, or τ_n reaches a desired value. In the following table, we show the result of the above approach for $\lambda = 10$ packets per sec, and the generated packet arrival events until around 1 sec. As you can see, due to the randomness of arrivals, you might not get exactly 10 arrivals in 1 sec.

TABLE B.1   Generated events for Poisson arrival ($\lambda = 10$ pps).

Event i	Inter-arrival time (millisec)	Event time, τ_i (millisec)
1	71.85	71.85
2	263.72	335.57
3	553.33	888.90
4	18.28	907.18
5	26.75	933.93
6	51.12	985.05
7	202.66	1187.71

B.11   Self-Similarity and Heavy-Tailed Distributions

SELF-SIMILARITY AND LONG-RANGE DEPENDENCY

A stochastic process $A(t)$ is self-similar if $A(\alpha t) = \alpha^H A(t)$ where H is the self-similarity parameter, known as the Hurst parameter.

Long-range dependency means that events that are distant in time are correlated, i.e., a process measured at time t and again at time $t + \Delta t$ are correlated; this correlation is measured through an auto-correlation function. Note that long-range dependency does not necessarily imply self-similarity. A certain type of self-similarity, known as second-order self-similarity that preserves auto-correlation independent of time aggregation, (but not all) typically implies long-range dependency.

HEAVY-TAILED DISTRIBUTIONS

There is a connection between self-similarity and heavy-tailed distributions, especially for Internet traffic. The presence of heavy tail in the distribution for transfer sizes is believed to be the cause of self-similarity in Internet traffic. Additionally, the Hurst parameter is observed to be $H > 1/2$, which reflects that burstiness at different time scales.

A distribution is considered to have a heavy-tail if

$$Pr[X > t] \approx t^{-\alpha}, \text{ as } t \to \infty, 0 < \alpha < 2.$$

A heavy-tailed distribution is also known as a power-law distribution.

There are many heavy-tailed distributions. A well-known heavy-tailed distribution is the general Pareto distribution; its probability density function is given by

$$f(t) = \alpha k^\alpha / t^{\alpha+1}, \quad \alpha, k > 0, \ t \geq k.$$

and the cumulative probability distribution is given by

$$F(t) = P[X \leq t] = 1 - (k/t)^\alpha,$$

where α is known as the shape parameter and k is known as the scale parameter. The PDF and CDF of the general Pareto distribution are plotted in Figure B.4, keeping scale parameter

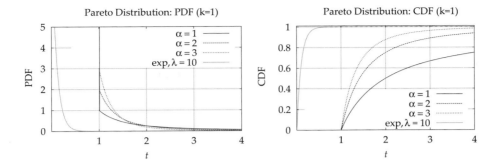

FIGURE B.4 PDF and CDF of Pareto distribution (including exponential with $\lambda = 10$).

k fixed at 1. For visual comparison, the exponential distribution with $\lambda = 10$ is also plotted alongside.

Visually, from the figure on the left side, we can see that the exponential distribution has a much shorter tail than heavy-tailed distributions.

The mean of the general Pareto distribution is given by $E(X) = k\alpha/(\alpha - 1)$ if $\alpha > 1$, and the variance by $V(X) = k^2\alpha/[(\alpha - 1)^2(\alpha - 2)]$ if $\alpha > 2$. For heavy-tailed distributions, the mean and the variance are essentially "infinite." Note that for the general Pareto distribution, the mean becomes infinite as $\alpha \to 1$.

There is an important behavior to understand about heavy-tailed distributions. Suppose we consider a probability distribution describing the transmission time of a packet to have a mean of 10 millisec. If the distribution were considered "light tailed," then if after 10 millisec the packet has not completed its transmission, the expected amount of time until it does complete transmission will be less than 10 millisec. If the distribution describing the transmission time were exponential, the expected remaining transmission time would equal 10 millisec. If the distribution describing the transmission time were heavy tailed, the expected remaining transmission time would be greater than 10 millisec. This can be explained by the fact that if a packet does not complete its transmission in 10 millisec, then it is much more likely to be one of the few that have an exceedingly long transmission time.

For additional material on self-similarity, long-range dependency, and heavy-tailed distributions, see [160], [404], [545], [551], [739], [740], [741].

B.12 Markov Chain and the Birth-and-Death Process

A Markov chain is a stochastic process in a countable state space (i.e., in discrete space, $S = \{0, 1, 2, \ldots\}$) with the special property that the probability of being at state j at a time instant depends only on the state i it was at in the previous time instant, not on any time instant before that; this is commonly known as the memoryless property. Time can be described either in discrete time or continuous time; these then correspond to the discrete time Markov chain, and the continuous time Markov chain, respectively. We will consider here the continuous time Markov chain, for which the memoryless property can be stated in terms of transition probability as follows:

$$p_{ij}(t) = Pr\{x(t + u) = j \mid x(u) = i\}, \quad t > 0$$

This is interpreted as follows: if we are at state i at time u, then for any small positive time movement t since u, the transition probability for time t, $p_{ij}(t)$, to be at state j depends only on being at state i at time u, nothing before that. Now, if we consider all the possible states to which transition is possible given that the system is at state i, we can write

$$\sum_{j \in S} p_{ij}(t) = 1, \quad \text{for each } i. \tag{B.12.1}$$

The initial system at start time, $t = 0$, is described by $p_{ii}(0) = 1$; otherwise, $p_{ij} = 0$, for $j \neq i$. If we now consider the steady-state system, i.e., to describe the system that is irrespective of time, then it is more meaningful to consider the steady-state transition *rate* q_{ij} for going from state i to j. It can be shown (by using limits) that Eq. (B.12.1) translates to the following in a steady-state:

$$\sum_{j \in S} q_{ij} = 0, \quad \text{for each } i. \tag{B.12.2}$$

Now, the steady state probability of being at particular state j, to be denoted by p_j (note: the difference from $p_{ij}(t)$), is related to the transition rate q_{ij} by the following set of equations [481]:

$$\begin{aligned}
p_0 q_{00} + p_1 q_{10} + p_2 q_{20} + \cdots &= 0 \\
p_0 q_{01} + p_1 q_{11} + p_2 q_{21} + \cdots &= 0 \\
p_0 q_{02} + p_1 q_{12} + p_2 q_{22} + \cdots &= 0 \\
\cdots
\end{aligned} \tag{B.12.3}$$

This set is known as the set of balance equations. It may also be noted that the sum of the steady-state probabilities, p_j, must add to 1, i.e.,

$$p_0 + p_1 + p_2 + \cdots = 1. \tag{B.12.4}$$

For many well-known systems, q_{ij} is known or can be derived; thus, by solving Eq. (B.12.3) along with Eq. (B.12.4), we can determine the steady-state probability p_j.

B.12.1 Birth-and-Death Process

The birth-and-death process is a specialized continuous-time Markov chain. This name has stuck on since this model was first used for population birth and death modeling; in a networking system, a better name would be to call it the arrival-and-service process. In this special Markov chain, q_{ij} is measurable only to its immediate one-step neighbors, i.e., going from i to $i+1$ or to $i-1$; the transition rate is zero, otherwise. Typically, we write going from i to $i+1$ as λ_i ("birth" or arrival rate) and from i to $i-1$ as μ_i ("death" or service rate). Thus, the transition rate can be summarized as follows:

$$q_{i,i+1} = \lambda_i, \quad i = 0, 1, 2, \ldots$$

$$q_{i,i-1} = \mu_i, \quad i = 1, 2, \ldots \tag{B.12.5}$$

$$q_{i,j} = 0, \quad \text{for } j \neq i+1, i-1, i.$$

Due to relation (B.12.2) and (B.12.5), we can see that for the birth-and-death process, we can write:

$$q_{ii} = -(\lambda_i + \mu_i), \quad \text{for } i \text{ not in a boundary state.}$$

Rewriting Eq. (B.12.3) with this knowledge, we have

$$
\begin{aligned}
-p_0\lambda_0 + p_1\mu_1 &= 0 \\
p_0\lambda_0 - p_1(\lambda_1 + \mu_1) + p_2\mu_2 &= 0 \\
p_1\lambda_1 - p_1(\lambda_2 + \mu_2) + p_3\mu_3 &= 0 \\
\cdots
\end{aligned}
\tag{B.12.6}
$$

The above balance equations for the birth-and-death process can be easily depicted pictorially by considering each state i and observing the balance maintained by rates going into a state and out of that state as shown below:

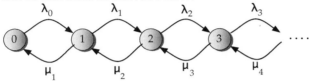

For example, consider state 1 to which from state 0 where steady-state probability is p_0, rate λ_0 is going in; similarly, from state 2 where steady-state probability is p_2, rate μ_2 is going into state 1; now, from state 1 where steady-state probability is p_1, two rates λ_1 and μ_1 are going out. Thus, $p_0\lambda_0 + p_2\mu_2 = p_1(\lambda_1 + \mu_1)$; this is the second equation in (B.12.6). Similarly, others can be determined. The set of equations given by Eq. (B.12.6) can be analytically solved by substitution and elimination to obtain:

$$p_1 = \frac{\lambda_0}{\mu_1}p_0, \quad p_2 = \frac{\lambda_1}{\mu_2}p_1 = \frac{\lambda_0\lambda_1}{\mu_1\mu_2}p_0, \ldots$$

Generalizing, we have

$$p_j = \frac{\lambda_0\lambda_1\cdots\lambda_{j-1}}{\mu_1\mu_2\cdots\mu_j}p_0, \quad j \geq 1.
\tag{B.12.7}$$

Now, plugging in p_j to Eq. (B.12.4), we can obtain p_0, and thus all p_js; this assumes that the summation converges, i.e., the following condition is satisfied:

$$\sum_j \frac{\lambda_0\lambda_1\cdots\lambda_{j-1}}{\mu_1\mu_2\cdots\mu_j} < \infty.$$

B.12.2 *M/M/1* System

An $M/M/1$ system is a special case of the birth-and-death process. In this system, transition rate for birth (arrival) is the same regardless of the state; similarly, the transition rate for death (service) is the same regardless of the state.

$$
\begin{aligned}
q_{i,i+1} = \lambda_i = \lambda, \quad & i = 0, 1, 2, \ldots \\
q_{i,i-1} = \mu_i = \mu, \quad & i = 1, 2, \ldots \\
q_{i,j} = 0, \quad & \text{for } j \neq i+1, i-1, i.
\end{aligned}
\tag{B.12.8}
$$

This system is stable if the arrival rate, λ, is smaller than the service rate, μ. Here, the steady-state probability p_j, given in Eq. (B.12.7), reduces to:

$$p_j = \left(\frac{\lambda}{\mu}\right)^j p_0, \quad j = 1, 2, \ldots$$

Accounting for Eq. (B.12.4), we can write

$$p_j = \left(1 - \frac{\lambda}{\mu}\right)\left(\frac{\lambda}{\mu}\right)^j, \quad j = 1, 2, \ldots$$

An advantage of knowing the steady-state probability is that the average number in a system can be determined as follows:

$$N = \sum_{j=0}^{\infty} j p_j = \frac{\lambda}{\mu - \lambda}. \tag{B.12.9}$$

(We have shown above the result of the summation, without showing the detailed derivation.) A well known result, known as Little's law, says that the average number in a system and the average delay is related as follows: $N = \lambda \times T$. Thus, we can write the average delay as:

$$T = \frac{1}{\mu - \lambda}. \tag{B.12.10}$$

Consider now a network link with bandwidth C and average packet size κ, the service rate becomes $\mu = C/\kappa$. If we denote utilization by $\rho = \lambda/\mu = \lambda \times \kappa/C$, then we can rewrite the average delay in a network link as

$$T = \frac{1}{C/\kappa - \lambda} = \frac{\kappa}{C(1 - \rho)}. \tag{B.12.11}$$

B.12.3 Trunk Reservation Model

The trunk reservation model presented in Section 11.8.3 is also a specialized case of the birth-and-death process. In this case, the total number of state is finite, which is the total number of circuits on a link, denoted by c circuits; thus, $S = \{0, 1, \ldots, c\}$. If r is the trunk reservation parameter, then the arrival rate up to state $c - r$ is to allow both direct and alternate routed arrival, i.e, $\lambda_{\text{direct}} + \lambda_{\text{alternate}}$, and after the state $c - r$ is reached, only direct routed calls are allowed, i.e. the arrival rate is λ_{direct}. Furthermore, in a circuit-switched environment, the service rate, $q_{i,i-1}$, is defined by the number circuits being currently occupied by calls assuming the average call duration to be $1/\mu$, i.e., the per-circuit service rate to be μ. Thus, we can write

$$\begin{aligned}
q_{i,i+1} &= \lambda_{\text{direct}} + \lambda_{\text{alternate}}, & i &= 0, 1, 2, \ldots, c - r - 1 \\
q_{i,i+1} &= \lambda_{\text{direct}}, & i &= c - r, \ldots, c - 1 \\
q_{i,i-1} &= i\mu, & i &= 1, 2, \ldots, c
\end{aligned} \tag{B.12.12}$$

Now consider the discussion in Section 11.8.3. In term of offered load, $\underline{A}$ and a described there, this means that

$$\begin{aligned}
\underline{A} &= (\lambda_{\text{direct}} + \lambda_{\text{alternate}})/\mu, & i &= 0, 1, \ldots, c - r - 1 \\
a &= \lambda_{\text{direct}}/\mu, & i &= c - r, \ldots, c - 1.
\end{aligned} \tag{B.12.13}$$

Thus, the result shown in Eq. (11.8.22a) can be derived from Eq. (B.12.7) using Eq. (B.12.12) and Eq. (B.12.13).

B.13 Average Network Delay

The results given by Eq. (B.12.9) is for the average number in a system. From a network link point of view, this result is then the average number of packets in a link assuming Poisson arrival and exponential service time. Now consider a network that consists of more than one link in which traffic from one link may feed to another; in other words, traffic interaction is possible. To obtain a simple network-wide result, an assumption is made that each link behaves as an $M/M/1$ system regardless of traffic interaction—this assumption is known as the *Kleinrock independence assumption*. Consider for link $\ell = 1, 2, \ldots, L$ in a network, the link arrival rate is y_ℓ and the link capacity c_ℓ; if we assume average packet size to be one, then the service rate can be represented by c_ℓ. Thus, using Eq. (B.12.9), the average number of packets on link ℓ can be written as

$$N_\ell = \frac{y_\ell}{c_\ell - y_\ell}. \tag{B.13.1}$$

Summing over all links, we can obtain the average number of packets in the network as

$$N = \sum_{\ell=1}^{L} N_\ell = \sum_{\ell=1}^{L} \frac{y_\ell}{c_\ell - y_\ell}. \tag{B.13.2}$$

Now, if there are $k = 1, 2, \ldots, K$ demand pairs in the network, and the arrival rate for demand k is h_k, then the total arrival rate is given by $H = \sum_{k=1}^{K} h_k$. Then using Little's law again, we can write the average network delay as

$$\mathcal{T} = \frac{N}{H} = \frac{1}{H} \sum_{\ell=1}^{L} \frac{y_\ell}{c_\ell - y_\ell}. \tag{B.13.3}$$

B.14 Packet Format: IPv4, IPv6, TCP, and UDP

In this section, we summarize packet formats for IPv4, IPv6, TCP and UDP. Recall that the pictorial views of packet formats with field locations were shown earlier in Figure 1.3 for IPv4 and IPv5, and in Figure 1.4 for TCP and UDP. We present below a brief explanation of the various fields associated with each of these packet types. Details are shown in Tables B.2, B.3, B.4, and B.5 for IPv4, IPv6, TCP, and UDP, respectively.

TABLE B.2 Explanation of IPv4 header fields.

IPv4 Header Field	Explanation
Version	Specifies IP version number, set to 4
Header Length	Length of the header (useful when options are included in header)
DiffServ/Type-of-Service	Six-bit Differentiated Service Code Point (two bits that follow are used for Explicit Congestion Notification) or 8-bit Type of Service Field
Total Length	Length of the IP packet in bytes
Identification	A unique identifier generated by the originator of the IP packet
Flags	The first bit is unused; if the second bit is set to 1, the packet is not to be fragmented; when a packet is fragmented, the third bit is set to 1 if a fragmented packet has more fragments to follow arising from the original IP packet
Fragment Offset	This field specifies the offset from the start of the header to the start of the fragment, counted in increments of eight bytes
Time to Live (TTL)	The life-time of packet as indicated by the originating host; the default value is set to 64, and a router that the packet visits decrements this value by 1, and discards if the value has become zero
Protocol Type	Indicates the protocol above IP for which this packet is for; this field indicates protocols such as TCP, UDP, ICMP, OSPF, IGRP
Header Checksum	Checksum computed on the header; see [89], [444]
Source Address / Destination Address	32-bit IPv4 addresses for source and destination
Options	Variable length optional field for extending IP header
Padding	bits added to follow Options field so that Options together with Padding is a multiple of 32 bits

TABLE B.3 Explanation of IPv6 header fields.

IPv6 Header Field	Explanation
Version	Specifies IP version number, set to 6
Traffic Class	Traffic class identifies Differentiated Services code points
Flow Label	An identifier to label packets from a source to a destination that require same treatment for the benefit of router processing
Payload Length	Length of the IPv6 packet, not counting header
Next Header	This field is used for indicating the protocol type field (as in IPv4), and also Options field
Hop Limit	This is the TTL field; This new name is given to indicate that it is counted in hops.
Source Address / Destination Address	128-bit IPv6 addresses for source and destination

TABLE B.4 Explanation of TCP header fields.

TCP Header Field	*Explanation*
Source Port / Destination Port	16-bit port address for applications
Sequence Number	The first byte identification counter of the current TCP segment
Acknowledgment Number	Informs the sequence number expected to receive to indicate bytes count received so far
Header Length	Length of header in multiples of 4 bytes
Reserved	Not used
ECN	For explicit congestion notification
Control Bits	Used for connection control functions; the well-known bits or Flags are SYN (synchronize for start of a TCP connection), ACK (for packet acknowledged), FIN (to end a TCP session), RST (to reset a TCP connection), PSH (to push data to upper layer immediately on arrival), URG (Urgent pointer indicator)
Advertised Window	Specifies the window size (in bytes) on how much outstanding data (in bytes) the sender is willing to receive from the receiver
Checksum	Checksum computed over header and payload
Urgent Pointer	Used with URG flag
Options	Optional field to extend the default header size; used for several features such as TCP window scaling, selective acknowledgment
Padding	bits added to follow Options field so that Options together with Padding is a multiple of 32 bits
	Note: TCP does not have a segment length field; the length of the payload data in TCP is derived from IP total length field, IP header length field, and TCP header length field

TABLE B.5 Explanation of UDP header fields.

UDP Header Field	*Explanation*
Source Port / Destination Port	16-bit port address for applications
Length	UDP Packet Length in bytes
Checksum	Checksum computed over the whole segment; optionally, this can be ignored

Solutions to Selected Exercises

▷ *Exercise 1.2:* The IP prefix is: 10.20.0.0/14.

▷ *Exercise 1.6(a):* Time-to-live field and header checksum.

▷ *Exercise 2.5(a):* The shortest path is 1-4-3-5.

▷ *Exercise 2.5(b):* The shortest path is 1-2-4-3-5.

▷ *Exercise 2.6:* The widest path is 2-4-6 with bandwidth 10.

▷ *Exercise 3.2:* Count-to-infinity and looping.

▷ *Exercise 3.6(a):* It will take $10 \times 4 = 40$ sec.

▷ *Exercise 4.2(b):* We name the flow variables based on clockwise or counter-clockwise directions. Thus, for demand between 1 and 3, there are two paths 1-2-3 and 1-4-3, resulting in paths variables x123 and x143; we can similarly name the flow variables for the other demand pairs. The linear programming formulation for the load balancing problem as an input to CPLEX is shown below:

```
Minimize r
subject to
      dem13: x123 + x143 = 25
      dem24: x234 + x214 = 30
      dem23: x23 + x2143 = 10
      link12: x123 + x214 + x2143 - 50 r <= 0
      link23: x123 + x234 + x23 - 50 r <= 0
      link34: x143 + x234 + x2143 - 50 r <= 0
      link41: x143 + x214 + x2143 - 50 r <= 0
End
```

▷ *Exercise 5.1(a):* The three timers are: Autoupdate Timer (default value: 30 sec), Expiration Timer (default value: 180 sec), and Garbage Collection Timer (60 sec longer than Expiration Timer).

▷ *Exercise 6.5:* This is a trick question. Try again.

▷ *Exercise 6.10:* Yes.

▷ *Exercise 7.2:* The linear programming formulation for the load balancing objective can be written as:

```
Minimize r subject to
    dem12: x142 + x1342 + x1432 + x132 = 15
    dem13: x13 + x143 + x1423 = 20
    dem23: x23 + x243 + x2413 = 10
    link13: - x1342 - x132 - x13 - x2413 + 30 r >= 0
    link14: - x142 - x1432 - x143 - x1423 - x2413 + 5 r >= 0
    link23: - x1432 - x132 - x1423 - x23 + 15 r >= 0
    link24: - x142 - x1342 - x1423 - x243 - x2413 + 10 r >= 0
    link34: - x1342 - x1432 - x143 - x243 + 10 r >= 0
End
```

On solving the above problem, we can obtain the dual solutions, which can possibly be used for link weights. Check if link weights so determined will work here.

▷ *Exercise 8.2:* By carrying the AS numbers visited in the path vector.

▷ *Exercise 9.4:* It is not necessary. However, to avoid the address spoofing problem, uRPF can be helpful.

▷ *Exercise 10.4:* Crankback is helpful because at the time of call setup, the call control can be cranked back to trying another path if the path first tried does not have any resources available. However, crankback is not beneficial if the network is overloaded since there might not be any alternate paths available—in this case, crankback causes unnecessary attempts without providing any benefit.

▷ *Exercise 11.2:* For the duration of a telephone call, bandwidth is required to be dedicated for the call; in other words, during this period, no other calls can use this bandwidth. A smaller duration means that another newly arriving call has the opportunity to use the released bandwidth for its call; otherwise, the call would be dropped.

▷ *Exercise 12.1(c):* The common fields are: Flag, BIB/BSN, FIB/FSN, S/LI, and FCS.

▷ *Exercise 12.2:* Two STPs are associated with an SSP for redundancy. Thus, if one fails, the other is still accessible to the SSP.

▷ *Exercise 13.2:* SS7 point code addressing is intended for the nodes in the network, not for end devices. The subscriber's telephone number addressing is completely decoupled from SS7 addressing.

▷ *Exercise 14.1(d):* The key elements of a router are: network interface, forwarding engine, queue manager, traffic manager, backplane, and route control processor.

▷ *Exercise 15.1(a):* A key difference is that the classful addressing scheme implicitly assumes that the netmask to be on specific bit boundaries while CIDR requires the network mask to explicitly announced.

▷ *Exercise 15.2:* 25 nanosec.

▷ *Exercise 15.4:* The memory accesses required would be 2 and 3, respectively.

▷ *Exercise 16.1(d):* It consumes more memory because of duplication.

▷ *Exercise 16.4:* The best matching rule is R_8; It would require 9 memory accesses.

▷ *Exercise 17.3:* The short answer is: having multiple paths cached compensates for inaccuracy in the information received since the last link state update.

▷ *Exercise 18.2:* The FAST-REROUTE object is used to protect an LSP so that in the event of a failure, traffic flow can use a backup LSP immediately, which is setup using DETOUR. The bandwidth guarantee for the backup path is specified in the FAST-REROUTE object.

▷ *Exercise 19.2(a):* The only change required is that the decision variables associated with customer C should be allowed to take fractional values. In CPLEX, this means that variables x_C_sf_ny and x_C_sf_kc_ny should be allowed to take continuous values and are not declared in the block for Integer. For completeness, the entire formulation is shown below:

```
Minimize    x_A_sf_kc + x_A_sf_ny_kc + x_A_kc_ny + x_A_kc_sf_ny
        + x_A_sf_ny + x_A_sf_kc_ny + x_B_sf_ny + x_B_sf_kc_ny
        + x_C_sf_ny + x_C_sf_kc_ny
subject to
d45_A_sf_kc:  x_A_sf_kc + x_A_sf_ny_kc = 1
d60_A_kc_ny:  x_A_kc_ny + x_A_kc_sf_ny = 1
d20_A_sf_ny:  x_A_sf_ny + x_A_sf_kc_ny = 1
d80_B_sf_ny:  x_B_sf_ny + x_B_sf_kc_ny = 1
d100_C_sf_ny: x_C_sf_ny + x_C_sf_kc_ny = 1
l_sf_kc: 45 x_A_sf_kc +  60 x_A_kc_sf_ny + 20 x_A_sf_kc_ny
    + 80 x_B_sf_kc_ny + 100 x_C_sf_kc_ny <= 155
l_sf_ny: 45 x_A_sf_ny_kc + 60 x_A_kc_sf_ny + 20 x_A_sf_ny
    + 80 x_B_sf_ny + 100 x_C_sf_ny <= 155
l_kc_ny: 45 x_A_sf_ny_kc + 60 x_A_kc_ny +  20 x_A_sf_kc_ny
    + 80 x_B_sf_kc_ny + 100 x_C_sf_kc_ny <= 155
Integer
x_A_sf_kc  x_A_sf_ny_kc  x_A_kc_ny  x_A_kc_sf_ny
x_A_sf_ny  x_A_sf_kc_ny  x_B_sf_ny  x_B_sf_kc_ny
End
```

▷ *Exercise 20.3:* There are many differences. We list here two differences: (1) ISUP messaging allows carrier identification code to be included while SIP does not have a similar field, (2) SIP allows different media to be communicated through session description protocol, which is not available with ISUP messaging.

▷ *Exercise 21.7:* The width of the bus should be 80 bits and approximately 134 bits, respectively.

▷ *Exercise 21.10:* 5 time slots.

▷ *Exercise 22.3:* The short answer is that it is better to avoid the network from reaching an unbearable congestion level by indicating (either implicitly or explicitly) to the TCP sources to reduce data transfer rate.

▷ *Exercise 23.2:* It allows to regulate traffic rate in an efficient manner in a packetized environment.

▷ *Exercise 24.1(a):* Since the flow variables are integer-valued, the only change needed is to declare $x_{kp\tau}$ as integer-valued.

▷ *Exercise 25.3:* Use the approach presented in Section 25.1.2.

Bibliography

[1] M. Adams, J. Coplien, R. Gamoke, R. Hanmer, F. Keeve, and K. Nicodemus, "Fault-tolerant telecommunication systems patterns," in *Pattern Languages of Program Design*, J. M. Vlissides, J. O. Coplien, and N. L. Kerth (Eds.), Addison-Wesley, pp. 549–562, 1996.

[2] S. Agarwal, C.-N. Chuah, S. Bhattacharyya, and C. Diot, "The impact of BGP dynamics on intradomain traffic," in *Proc. ACM SIGMETRICS'2004*, pp. 319–330, New York, NY, 2004.

[3] G. Agrawal and D. Medhi, "Lightpath topology configuration for wavelength-routed IP/MPLS networks for time-dependent traffic," in *Proc. of IEEE GLOBECOM'2006*, San Francisco, CA, November–December 2006.

[4] H. Ahmadi and W. E. Denzel, "A survey of modern high performance switching techniques," *IEEE Journal of Selected Areas in Communications*, vol. 7, pp. 1091–1103, 1989.

[5] A. V. Aho and D. Lee, "Hierarchical networks and the LSA N squared problem in OSPF routing," in *Proc. IEEE GLOBECOM'2000*, pp. 397–404, San Francisco, CA, November–December 2000.

[6] R. K. Ahuja, T. L. Magnanti, and J. B. Orlin, *Network Flows: Theory, Algorithms, and Applications*. Prentice-Hall, 1993.

[7] J. M. Akinpelu, "The overload performance of engineered networks with nonhierarchical and hierarchical routing," *AT&T Bell Labs Technical Journal*, vol. 63, pp. 1261–1281, 1984.

[8] C. Alaettinoglu, C. Villamizar, E. Gerich, D. Kessens, D. Meyer, T. Bates, D. Karrenberg, and M. Terpstra, "Routing policy specification language (RPSL)," *IETF RFC 2622*, June 1999. http://www.rfc-editor.org/rfc/rfc2622.txt

[9] M. Ali and H. Nguyen, "A neural network implementation of an input access scheme in a high speed packet switch," in *Proc. IEEE GLOBECOM'89*, pp. 1192–1196, 1989.

[10] Alliance for Telecommunications Industry Solutions, "555 Technical service interconnection arrangements (ICCF96-0411-014)," July 1998, ATIS Industry Numbering Committee. http://www.cnac.ca/numres/555/555_96041114.doc

[11] M. Allman, V. Paxson, and W. Stevens, "TCP congestion control," *IETF RFC 2581*, April 1999. http://www.rfc-editor.org/rfc/rfc2581.txt

[12] W. Allsopp, "VoIP—vulnerability over Internet protocol?" March 2004. http://www.continuitycentral.com/feature074.htm

[13] V. Alwayn, *Optical Network Design and Implementation*. Cisco Press, 2004.

[14] Amsterdam Internet Exchange, AMS-IX. http://www.ams-ix.net/

[15] T. Anderson, S. Owicki, J. Saxe, and C. Thacker, "High-speed switch scheduling for local area networks," *ACM Trans. on Computer Systems*, vol. 11, pp. 319–352, November 1993.

[16] L. Andersson, P. Doolan, N. Feldman, A. Fredette, and B. Thomas, "LDP specification," *IETF RFC 3036*, January 2001. http://www.rfc-editor.org/rfc/rfc3036.txt

[17] L. Andersson and T. Madsen, "Provider provisioned virtual private network (VPN) terminology," *IETF RFC 4026*, March 2005. http://www.rfc-editor.org/rfc/rfc4026.txt

[18] L. Andersson and G. Swallow, "The multiprotocol label switching (MPLS) working group decision on MPLS signaling protocols," *IETF RFC 3468*, February 2003. http://www.rfc-editor.org/rfc/rfc3468.txt

[19] T. Anjali, C. M. Scoglio, J. C. de Oliveira, L. C. Chen, I. F. Akyildiz, J. A. Smith, G. Uhl, and A. Sciuto, "A new path selection algorithm for MPLS networks based on available bandwidth estimation." in *Proc. International Workshop on Quality of Future Internet Services (QofIS'2002)*, Zurich, Switzerland, Lecture Notes in Computer Science, Springer, vol. 2511, pp. 205–214, October 2002.

[20] AOL Transit Data Network Settlement-Free Interconnection Policy. http://www.atdn.net/settlement_free_int.shtml

[21] G. Apostolopoulos, R. Guerin, and S. Kamat, "Implementation and performance measurements of QoS routing extensions to OSPF," in *Proc. IEEE INFOCOM'99*, pp. 680–688, New York, NY, March 1999.

[22] G. Apostolopoulos, R. Guerin, S. Kamat, and S. K. Tripathi, "QoS routing: A performance perspective," in *Proc. ACM SIGCOMM'98*, pp. 17–28, Vancouver, Canada, September 1998.

[23] G. Apostolopoulos, S. Kama, D. Williams, R. Guerin, A. Orda, and T. Przygienda, "QoS routing mechanisms and OSPF extensions," *IETF RFC 2676*, August 1999. http://www.rfc-editor.org/rfc/rfc2676.txt

[24] G. Apostolopoulos and S. K. Tripathi, "On reducing the processing cost of on-demand QoS path computation," in *Proc. ICNP'98*, pp. 80–89, Austin, TX, October 1998.

[25] G. Appenzeller, I. Keslassy, and N. McKeown, "Sizing router buffers," in *Proc. ACM SIGCOMM'2004*, Portland, OR, August–September 2004.

[26] G. Armitage, *Quality of Service in IP Networks*. Macmillan Technical Publishing, April 2000.

[27] G. R. Ash, *Dynamic Routing in Telecommunication Networks*. McGraw-Hill, 1997.

[28] G. R. Ash, "Performance evaluation of QoS-routing methods for IP-based multiservice networks," *Computer Communication*, vol. 26, pp. 817–833, 2003.

[29] G. R. Ash, *Traffic Engineering and QoS Optimization of Integrated Voice & Data Networks*. Morgan Kaufmann Publishers, 2006.

[30] G. R. Ash, R. H. Cardwell, and R. P. Murray, "Design and optimization of networks with dynamic routing," *Bell System Technical Journal*, vol. 60, pp. 1787–1820, 1981.

[31] G. R. Ash, K. K. Chan, and J.-F. Labourdette, "Analysis and design of fully shared networks," in *Proc. 14th International Teletraffic Congress (ITC14)*, pp. 1311–1320, Antibes, France, June 1994.

[32] G. R. Ash, F. Chang, and D. Medhi, "Robust traffic design for dynamic routing networks," in *Proc. IEEE INFOCOM'91*, pp. 508–514, Bal Harbour, Florida, April 1991.

[33] G. R. Ash and P. Chemouil, "20 years of dynamic routing in telephone networks: Looking backward to the future," *IEEE Global Communications Newsletter*, pp. 1–4, October 2004, note: appears as insert in the October 2005 issue of *IEEE Communications Magazine*.

[34] G. R. Ash, J. S. Chen, A. E. Frey, and B. D. Huang, "Real time network routing in a dynamic class-of-service network," in *Proc. 13th International Teletraffic Congress (ITC13)*, pp. 187–194, Copenhagen, Denmark, 1991.

[35] G. R. Ash, M. Girish, E. Gray, B. Jamoussi, and G. Wright, "Applicability statement for CR-LDP," *IETF RFC 3213*, January 2002. http://www.rfc-editor.org/rfc/rfc3213.txt

[36] G. R. Ash, A. H. Kafker, and K. R. Krishnan, "Servicing and real-time control of networks with dynamic routing," *Bell System Technical Journal*, vol. 60, pp. 1821–1845, 1981.

[37] G. R. Ash, Y. Lee, P. Ashwood-Smith, B. Jamoussi, D. Fedyk, D. Skalecki, and L. Li, "LSP modification using CR-LDP," *IETF RFC 3214*, January 2002. http://www.rfc-editor.org/rfc/rfc3214.txt

[38] S. Asthana, C. Delph, H. V. Jagadish, and P. Krzyzanowski, "Towards a gigabit IP router," *Journal of High Speed Networks*, vol. 1, no. 4, pp. 281–288, 1992.

[39] ATM Forum, "Private Network-Network Interface specification, version 1.1, af-pnni-0055.001," April 2002.

[40] AT&T Global IP Network Settlement-Free Peering Policy. http://www.att.com/peering/

[41] R. Y. Awdeh and H. T. Mouftah, "Survey of ATM switch architectures," *Computer Networks & ISDN Systems*, vol. 27, pp. 1567–1613, 1995.

[42] D. Awduche, "MPLS and traffic engineering in IP networks," *IEEE Communications Magazine*, vol. 37, no. 12, pp. 42–47, December 1999.

[43] D. Awduche, L. Berger, D. Gan, T. Li, V. Srinivasan, and G. Swallow, "RSVP-TE: Extensions to RSVP for LSP tunnels," *IETF RFC 3209*, December 2001. http://www.rfc-editor.org/rfc/rfc3209.txt

[44] D. Awduche, A. Hannan, and X. Xiao, "Applicability statement for extensions to RSVP for LSP-tunnels," *IETF RFC 3210*, December 2001. http://www.rfc-editor.org/rfc/rfc3210.txt

[45] D. Awduche, J. Malcolm, J. Agogbua, M. O'Dell, and J. McManus, "Requirements for traffic engineering over MPLS," *IETF RFC 2702*, September 1999. http://www.rfc-editor.org/rfc/rfc2702.txt

[46] J. Aweya, "IP router architectures: An overview," *International Journal of Communication Systems*, vol. 14, no. 5, pp. 447–475, May 2001.

[47] F. Baboescu, S. Singh, and G. Varghese, "Packet classification for core routers: Is there an alternative to CAMs," in *Proc. IEEE INFOCOM'2003*, pp. 53–63, April 2003.

[48] F. Baboescu and G. Varghese, "Scalable packet classification," in *Proc. ACM SIGCOMM'2001*, pp. 199–210, San Diego, CA, August–September 2001.

[49] S. Bahk and M. El Zarki, "Dynamic multi-path routing and how it compares with other dynamic routing algorithms for high speed wide area network," in *Proc. ACM SIGCOMM'1992*, pp. 53–64, Baltimore, MD, 1992.

[50] F. Baker, "Requirements for IP version 4 routers," *IETF RFC 1812*, June 1995, http://www.rfc-editor.org/rfc/rfc1812.txt.

[51] F. Baker, B. Lindell, and M. Talwar, "RSVP cryptographic authentication," *IETF RFC 2747*, January 2000. http://www.rfc-editor.org/rfc/rfc2747.txt

[52] K. Bala, I. Cidon, and K. Sohraby, "Congestion control for high speed packet switched networks," in *Proc. IEEE INFOCOM'90*, pp. 520–526, 1990.

[53] D. Banerjee and B. Mukherjee, "A practical approach for routing and wavelength assignment in large wavelength-routed optical networks," *IEEE Journal of Selected Areas in Communication*, vol. 14, no. 5, pp. 903–908, June 1996.

[54] D. Banerjee and B. Mukherjee, "Wavelength-routed optical networks: Linear formulation, resource budgeting tradeoffs, and a reconfiguration study," *IEEE/ACM Trans. on Networking*, vol. 8, no. 5, pp. 598–607, October 2000.

[55] S. Banerjee, J. Yoo, and C. Chen, "Design of wavelength-routed optical networks for packet switched traffic," *Journal of Lightwave Technology*, vol. 15, no. 9, pp. 1636–1646, September 1997.

[56] G. H. Barnes, R. M. Brown, M. Kato, D. J. Kuck, D. L. Slotnick, and R. A. Stokes, "The ILLIAC IV computer," *IEEE Trans. on Computers*, vol. 17, no. 8, pp. 746–757, August 1968.

[57] R. Barr and R. A. Patterson, "Grooming telecommunications networks," *Optical Networks Magazine*, vol. 2, no. 3, pp. 20–23, May/June 2001.

[58] C. Basso and P. Scotton, "Computing the widest shortest path in high-speed networks," U.S. Patent No. 6,370,119, April 9, 2002.

[59] T. Bates and R. Chandra, "BGP route reflection: An alternative to full mesh BGP," *IETF RFC 1966*, June 1996. http://www.rfc-editor.org/rfc/rfc1966.txt

[60] T. Bates, R. Chandra, and E. Chen, "BGP route reflection—an alternative to full mesh IBGP," *IETF RFC 2796*, April 2000, (Made obsolete by [62]). http://www.rfc-editor.org/rfc/rfc2796.txt

[61] T. Bates, R. Chandra, D. Katz, and Y. Rekhter, "Multiprotocol extensions for BGP-4," *IETF RFC 4760*, January 2007. http://www.rfc-editor.org/rfc/rfc4760.txt

[62] T. Bates, E. Chen, and R. Chandra, "BGP route reflection: An alternative to full mesh internal BGP (IBGP)," *IETF RFC 4456*, April 2006. http://www.rfc-editor.org/rfc/rfc4456.txt

[63] T. Bates, Y. Rekhter, R. Chandra, and D. Katz, "Multiprotocol extensions for BGP-4," *IETF RFC 2858*, June 2000. http://www.rfc-editor.org/rfc/rfc2858.txt

[64] A. Bavier, N. Feamster, M. Huang, L. Peterson, and J. Rexford, "In VINI Veritas: Realistic and controlled network experimentation," in *Proc. ACM SIGCOMM'06*, pp. 3–14, Pisa, Italy, September 2006.

[65] N. Beckmann, H. Kriegel, R. Schneider, and B. Seeger, "The R*-tree: An efficient and robust access method for points and rectangles," in *Proc. ACM SIGMOD'90*, pp. 322–331, 1990.

[66] F. Bedard, J. Regnier, and F. Caron, "Dynamically controlled routing using virtual nodes," U.S. Patent No. 5,526,414, June 11, 1996.

[67] M. Belaidouni and W. Ben-Ameur, "Super-additive approach to solve the minimum cost single path routing problem: Preliminary results," in *Proc. International Network Optimization Conference (INOC'2003)*, pp. 67–71, 2003.

[68] R. Bellman, "On a routing problem," *Quarterly of Applied Mathematics*, vol. 16, no. 1, pp. 87–90, 1958.

[69] W. Ben-Ameur, E. Gourdin, B. Liau, and N. Michel, "Optimizing administrative weights for efficient single path routing," in *Proc. Networks'2000*, Toronto, Canada, 2000.

[70] V. E. Beneš, "Rearrangeable three stage connecting networks," *Bell System Technical Journal*, vol. 41, pp. 1481–1492, 1962.

[71] V. E. Beneš, *Mathematical Theory of Connecting Networks and Telephone Traffic*. Academic Press, 1965.

[72] J. Bennett and H. Zhang, "Hierarchical packet fair queueing algorithm," in *Proc. ACM SIG-COMM'96*, pp. 143–156, Palo Alto, CA, August 1996.

[73] L. Berger (Ed.), "Generalized multi-protocol label switching (GMPLS): Signaling functional description," *IETF RFC 3471*, January 2003. http://www.rfc-editor.org/rfc/rfc3471.txt

[74] L. Berger (Ed.), "Generalized multi-protocol label switching (GMPLS) signaling resource reservation protocol-traffic engineering (RSVP-TE) extensions," *IETF RFC 3473*, January 2003. http://www.rfc-editor.org/rfc/rfc3473.txt

[75] L. Berger (Ed.), "GMPLS—communication of alarm information," *IETF RFC 4783*, December 2006. http://www.rfc-editor.org/rfc/rfc4783.txt

[76] Y. Bernet, A. Smith, and B. Davie, "Specification of the Null Service type," *IETF RFC 2997*, November 2000. http://www.rfc-editor.org/rfc/rfc2997.txt

[77] G. Bernstein, B. Rajagopalan, and D. Saha, *Optical Network Control: Architecture, Protocols, and Standards*. Addison-Wesley, 2003.

[78] D. Bertsekas, "Dynamic models of shortest path routing algorithms for communication networks with multiple destinations," in *Proc. 1979 IEEE Conference on Decision and Control*, pp. 127–133, Ft. Lauderdale, FL, 1979.

[79] D. Bertsekas, "Dynamic behavior of shortest path routing algorithms for communication networks," *IEEE Trans. on Automatic Control*, vol. AC-27, pp. 60–74, 1982.

[80] D. Bertsekas, *Network Optimization: Continuous and Discrete Models*. Athena Scientific, 1998.

[81] D. Bertsekas and R. Gallager, *Data Networks, 2nd Edition*. Prentice-Hall, 1992.

[82] R. Bhandari, *Survivable Networks—Algorithms for Diverse Routing*. Kluwer Academic Publishers, 1999.

[83] M. Bhatia, V. Manral, and Y. Ohara, "IS-IS and OSPF difference discussions," Internet draft, 2005.

[84] R. Bhatia, M. Kodialam, and T. Lakshman, "Fast network re-optimization schemes for MPLS and optical networks," *Computer Networks*, vol. 50, pp. 317–331, 2006.

[85] P. K. Bhatnagar, *Engineering Networks for Synchronization, CCS 7, and ISDN: Standards, Protocols, Planning and Testing.* Wiley–IEEE Press, 1997.

[86] A. Bogomolny, "Emergence of chaos (there is order in chaos)." http://www.cut-the-knot.org/blue/chaos.shtml

[87] E. Bouillet, J.-F. Labourdette, R. Ramamurthy, and S. Chaudhuri, "Lightpath re-optimization in mesh optical networks," *IEEE/ACM Trans. on Networking*, vol. 13, pp. 437–447, 2005.

[88] B. Braden, D. Clark, J. Crowcroft, B. Davie, S. Deering, D. Estrin, S. Floyd, V. Jacobson, G. Minshall, C. Partridge, L. Peterson, K. Ramakrishnan, S. Shenker, J. Wroclawski, and L. Zhang, "Recommendations on queue management and congestion avoidance in the Internet," *IETF RFC 2309*, April 1998. http://www.rfc-editor.org/rfc/rfc2309.txt

[89] R. Braden, D. Borman, and C. Partridge, "Computing the Internet checksum," *IETF RFC 1071*, September 1988, http://www.rfc-editor.org/rfc/rfc1071.txt.

[90] R. Braden Ed., L. Zhang, S. Berson, S. Herzog, and S. Jamin, "Resource ReSerVation Protocol (RSVP)—Version 1 functional specification," *IETF RFC 2205*, September 1997. http://www.rfc-editor.org/rfc/rfc2205.txt

[91] D. Braess, "Uber ein paradoxon der verkehrsplanung," *Unternehmensforschung*, vol. 12, pp. 258–268, 1968.

[92] W. S. Brainerd and L. H. Landweber, *Theory of computation.* John Wiley & Sons, 1974.

[93] H.-W. Braun, "Models of policy based routing," *IETF RFC 1104*, June 1989. http://www.rfc-editor.org/rfc/rfc1104.txt

[94] H.-W. Braun, "The NSFNET routing architecture," *IETF RFC 1093*, February 1989. http://www.rfc-editor.org/rfc/rfc1093.txt

[95] A. Bremler-Barr, Y. Afek, and S. Schwarz, "Improved BGP convergence via ghost flushing," in *Proc. IEEE INFOCOM'2003*, San Francisco, CA, 2003.

[96] J. Brewer and J. Sekel, "PCI express technology," *Dell Technology White Paper*, February 2004. http://www.dell.com/content/topics/global.aspx/vectors/en/2004_pciexpress?c=us&l=en&s=corp

[97] B. Briscoe, "Flow rate fairness: Dismantling a religion," Internet Draft, October 2006. http://www.cs.ucl.ac.uk/staff/bbriscoe/pubs.html#rateFairDis

[98] F. Brodersen and A. Klimetschek, "Anatomy of a high performance IP router." http://citeseer.ist.psu.edu/brodersen03anatomy.html

[99] N. Brownlee and M. Murray, "Streams, flows and torrents," in *Passive and Active Measurement Workshop*, April 2001.

[100] D. Brugman, "Call-routing analysis using SS7 data," *Bell Labs Technical Journal*, vol. 9, no. 4, pp. 133–138, 2005.

[101] M. M. Buddhikot, S. Suri, and M. Waldvogel, "Space decomposition techniques for fast layer-4 switching," in *Proc. Conference on Protocols for High Speed Networks*, pp. 25–41, August 1999.

[102] c7.com resource site. http://www.c7.com/

[103] CableLabs, "Packetcable^{TM}." http://www.packetcable.com/

[104] P. Calhoun, J. Loughney, E. Guttman, G. Zorn, and J. Arkko, "Diameter based protocol," *IETF RFC 3588*, September 2003. http://www.rfc-editor.org/rfc/rfc3588.txt

[105] R. Callon, "Use of OSI IS-IS for routing in TCP/IP and dual environments," *IETF RFC 1195*, December 1990. http://www.rfc-editor.org/rfc/rfc1195.txt

[106] G. Camarillo and M. A. García-Martín, *The 3G IP Multimedia Subsystem (IMS), 2nd Edition*. John Wiley & Sons, 2006.

[107] G. Camarillo, A. B. Roach, J. Peterson, and L. Ong, "Integrated Services Digital Network (ISDN) User Part (ISUP) to Session Initiation Protocol (SIP) Mapping," *IETF RFC 3398*, December 2002. http://www.rfc-editor.org/rfc/rfc3398.txt

[108] Canadian ENUM Working Group. http://www.enumorg.ca/

[109] Canadian Steering Committee on Numbering. http://www.cnac.ca/cscn/cscn.htm

[110] J. Carlson, P. Langner, E. Hernandez-Valencia, and J. Manchester, "PPP over simple data link (SDL) using SONET/SDH with ATM-like framing," *IETF RFC 2823*, May 2000. http://www.rfc-editor.org/rfc/rfc2823.txt

[111] M. Carson, J.-H. Hahm, S. Shah, and M. Zink, "MPLS-enabled routing algorithms," December 1999, DARPA Next-Generation Internet Principal Investigators' Meeting. http://www-x.antd.nist.gov/nistswitch/darpa.slides.ps.gz

[112] V. G. Cerf and R. E. Kahn, "A protocol for packet network interconnection," *IEEE Trans. on Communications*, vol. COM-22, pp. 627–641, May 1974.

[113] R. Chandra and J. Scudder, "Capabilities advertisement with BGP-4," *IETF RFC 2842*, May 2000, (Made obsolete by [114]). http://www.rfc-editor.org/rfc/rfc2842.txt

[114] R. Chandra and J. Scudder, "Capabilities advertisement with BGP-4," *IETF RFC 3392*, November 2002. http://www.rfc-editor.org/rfc/rfc3392.txt

[115] R. Chandra, P. Traina, and T. Li, "BGP communities attribute," *IETF RFC 1997*, August 1996. http://www.rfc-editor.org/rfc/rfc1997.txt

[116] J. Chandrashekar, Z. Duan, Z.-L. Zhang, and J. Krasky, "Limiting path exploration in path vector protocols," in *Proc. IEEE INFOCOM'2005*, pp. 2337–2348, Miami, FL, March 2005.

[117] T. J. Chaney, J. A. Fingerhut, M. Flucke, and J. S. Turner, "Design of a gigabit ATM switch," in *Proc. IEEE INFOCOM'97*, pp. 2–11, 1997.

[118] C.-S. Chang, *Performance Guarantees in Communication Networks*. Springer, 2000.

[119] F. Chang, "Routing-sequence optimization for circuit-switched networks," *AT&T Technical Journal*, vol. 68, no. 3, pp. 57–63, May/June 1989.

[120] F. Chang, W. Feng, and K. Li, "Approximate caches for packet classification," in *Proc. IEEE INFOCOM'2004*, pp. 2196–2207, Hong Kong, March 2004.

[121] H. J. Chao, "Next generation routers," *Proceedings of the IEEE*, vol. 90, pp. 1518–1558, 2002.

[122] H. J. Chao, K. L. Deng, and Z. Jing, "A petabit photonic packet switch (P3S)," in *Proc. IEEE INFO-COM'2003*, pp. 775–785, San Francisco, CA, April 2003.

[123] H. J. Chao, K. L. Deng, and Z. Jing, "Petastar: A petabit photonic packet switch," *IEEE Journal of Selected Areas in Communications*, vol. 21, no. 7, pp. 1096–1112, September 2003.

[124] H. J. Chao, C. H. Lam, and E. Oki, *Broadband Packet Switching Technologies*. John Wiley & Sons, 2001.

[125] H. J. Chao, S. Y. Liew, and Z. Jing, "A dual-level matching algorithm for 3-stage Clos-network packet switches," in *Proc. Hot Interconnects 11*, Stanford, CA, August 2003.

[126] B. Chazelle, "How to search in history," *Information and Control*, vol. 64, pp. 77–99, 1985.

[127] B. Chazelle and J. Friedman, "Point location hyperplanes and unidirectional ray shooting," *Computational Geometry: Theory and Applications*, vol. 4, pp. 53–62, 1994.

[128] P. Chemouil, J. Filipiak, and P. Gauthier, "Analysis and control of traffic routing in circuit-switched networks," *Computer Networks and ISDN System*, vol. 11, pp. 203–217, 1986.

[129] E. Chen, "Route refresh capability for BGP-4," *IETF RFC 2918*, September 2000. http://www.rfc-editor.org/rfc/rfc2918.txt

[130] S. Chen and K. Nahrstedt, "An overview of quality-of-service routing for the next generation high-speed networks: Problems and solutions," *IEEE Network*, vol. 12, no. 6, pp. 64–79, November/December 1998.

[131] G. Cheung and S. McCanne, "Optimal routing table design for IP address lookups under memory constraints," in *Proc. IEEE INFOCOM'99*, pp. 1437–1444, 1999.

[132] H. C. Chi and Y. Tamir, "Starvation prevention for arbiters of crossbars with multi-queue input buffers," in *Proc. COMPCON'94*, pp. 292–297, February 1994.

[133] T. Chiueh and P. Pradhan, "High performance IP routing table lookup using CPU caching," in *Proc. IEEE INFOCOM'99*, pp. 1421–1428, April 1999.

[134] F. M. Chiussi, J. G. Kneuer, and V. P. Kumar, "Low-cost scalable switching solutions for broadband networking: The ATLANTA architecture and chipset," *IEEE Communications Magazine*, vol. 35, no. 12, pp. 44–53, 1997.

[135] B.-Y. Choi, S. Moon, R. Cruz, Z.-L. Zhang, and C. Diot, "Practical delay monitoring for ISPs," in *Proc. ACM Conference on Emerging Network Experiment and Technology (CoNEXT'05)*, pp. 83–92, Toulouse, France, October 2005.

[136] J. S. Choi, N. Golmie, F. Lapeyrere, F. Mouveaux, and D. Su, "A functional classification of routing and wavelength assignment schemes in DWDM networks: Static case," in *Proc. of VII International Conference on Optical Communication and Networks*, New Jersey, 2000.

[137] X. Chu and B. Li, "Dynamic routing and wavelength assignment in the presence of wavelength conversion for all-optical networks," *IEEE/ACM Trans. on Networking*, vol. 13, no. 3, pp. 704–715, June 2005.

[138] Cisco Systems, "Policing and shaping overview," in *Tech Notes, Cisco Systems*, 2000.

[139] Cisco Systems, "Comparing traffic policing and traffic shaping for bandwidth limiting," in *Document ID: 19645, Tech Notes, Cisco Systems*, 2005. http://www.cisco.com/warp/public/105/policevsshape.html

[140] Cisco Systems, "Cisco CRS-1 carrier routing 8-slot line card chassis system description," in *Cisco Systems White Paper*, April 2006. http://www.cisco.com/en/US/products/ps5763/products_pre-installation_guide_chapter09186a008036e0d8.html

[141] K. Claffy, "Internet traffic characterization," Ph.D. dissertation, University of California, San Diego, CA, 1994.

[142] D. D. Clark, "What is "architecture"?" v 4.0 of 28 November 2005. http://find.isi.edu/presentation_files/Dave_Clark-What_is_architecture_4.pdf

[143] D. D. Clark, "The design philosophy of the DARPA Internet protocols," *ACM SIGCOMM Computer Communication Review*, vol. 18, no. 4, pp. 106–114, August 1988.

[144] D. D. Clark, "Policy routing in Internet protocols," *IETF RFC 1102*, May 1989. http://www.rfc-editor.org/rfc/rfc1102.txt

[145] K. L. Clarkson, "New applications of random sampling in computational geometry," *Discrete and Computational Geometry*, vol. 2, pp. 195–222, 1987.

[146] C. Clos, "A study of non-blocking switching networks," *Bell System Technical Journal*, vol. 32, no. 2, pp. 406–424, March 1953.

[147] R. Cole and J. Hopcroft, "On edge coloring bipartite graphs," *SIAM Journal on Computing*, vol. 11, pp. 540–546, 1982.

[148] R. Cole, K. Ost, and S. Schirra, "Edge-coloring bipartite multigraphs in $O(E \log D)$ time," *Combinatorica*, vol. 21, no. 1, pp. 5–12, 2001.

[149] R. Coltun, "The OSPF opaque LSA option," *IETF RFC 2370*, July 1998. http://www.rfc-editor.org/rfc/rfc2370.txt

[150] R. Coltun, D. Ferguson, and J. Moy, "OSPF for IPv6," *IETF RFC 2740*, December 1999. http://www.rfc-editor.org/rfc/rfc2740.txt

[151] D. Comer, "Ubiquitous B-tree," *ACM Computing Surveys*, vol. 11, no. 2, pp. 121–137, 1979.

[152] D. Comer, *Internetworking With TCP/IP, Volume 1: Principles Protocols, and Architecture, 4th Edition*. Prentice-Hall, 2000.

[153] M. Conte, *Dynamic Routing in Broadband Networks*. Kluwer Academic Publishers, 2003.

[154] J. P. Coudreuse and M. Servel, "PRELUDE: An asynchronous time-division switched network," in *Proc. IEEE International Conference on Communications (ICC'87)*, pp. 769–772, June 1987.

[155] Counry Code 1 ENUM LLC, "Provider ENUM trial." http://www.enumllc.com/

[156] J. Cowie and A. Ogielski, "Global routing instabilities during Code Red 2 and Nimda worm propagation," in *Presentation at NANOG23 Meeting*, Oakland, CA, October 2001. http://www.renesys.com/tech/presentations/pdf/Renesys-NANOG23.pdf

[157] J. Cowie, A. T. Ogielski, B. J. Premore, and Y. Yuan, "Internet warms and global routing instabilities," in *Proc. SPIE 2002 Conference*, July 2002. http://www.renesys.com/tech/presentations/pdf/renesys-spie2002.pdf

[158] CPLEX, *CPLEX User's Manual*. ILOG, 1999.

[159] P. Crescenzi, L. Dardini, and R. Grossi, "IP address lookup made fast and simple," in *Proc. 7th Annual European Symposium on Algorithms*, pp. 65–76, 1999.

[160] M. E. Crovella and A. Bestavros, "Self-similarity in world wide web traffic: Evidence and possible causes," *IEEE/ACM Trans. on Networking*, vol. 5, pp. 835–846, 1997.

[161] W. J. Dally, "Scalable switching fabrics for Internet routers," in *Avici Systems White Paper*, 2002. http://www.avici.com/technology/whitepapers/TSRfabric-WhitePaper.pdf

[162] W. J. Dally, M.-J. E. Lee, F.-T. An, J. Poulton, and S. Tell, "High performance electrical signaling," in *Proc. 5th International Conference on Massively Parallel Processing using Optical Interconnects*, pp. 11–16, 1998.

[163] W. J. Dally and B. Towles, *Principles and Practices of Interconnection Networks*. Morgan Kaufmann Publishers, 2004.

[164] G. B. Dantzig and M. N. Thapa, *Linear Programming 2: Theory and Extensions*. Springer, 1997.

[165] G. B. Dantzig, "On the shortest route through a network," *Management Science*, vol. 6, pp. 187–190, 1960.

[166] P. Datta and A. K. Somani, "Diverse routing for shared risk resource groups (SRRG) failures in WDM optical networks," in *Proc. BROADNETS'2004*, pp. 120–129, 2004.

[167] B. Davie and Y. Rekhter, *MPLS Technology and Applications*. Morgan Kaufmann Publishers, 2000.

[168] E. Davies and A. Doria, "Analysis of IDR requirements and history," IETF Internet-draft, 2006.

[169] R. De La Briandais, "File searching using variable length keys," in *Proc. Western Joint Computer Conference*, pp. 295–298. Spartan Books, 1959.

[170] R. de Rooij, "Bgphints." http://bgphints.ruud.org/

[171] D. Decasper, Z. Dittia, G. Parulkar, and B. Plattner, "Router plugins: A software architecture for next-generation routers," in *Proc. ACM SIGCOMM'98*, pp. 229–240, Vancouver, Canada, September 1998.

[172] J. J. Deegan, G. W. R. Luderer, and A. K. Vaidya, "Fast packet techniques for future switches," *AT & T Technical Journal*, pp. 36–50, March 1989.

[173] M. Degermark, A. Brodnik, S. Carlsson, and S. Pink, "Small forwarding tables for fast routing lookups," in *Proc. ACM SIGCOMM'97*, pp. 3–14, Cannes, France, 1997.

[174] P. Demeester, M. Gryseels, A. Autenrieth, C. Brianza, L. Castagna, G. Signorelli, R. Clemente, M. Ravaera, A. Jajszczyk, D. Janukowicz, K. V. Doorselaere, and Y. Harada, "Resilience in multilayer networks," *IEEE Communications Magazine*, vol. 37, no. 8, pp. 70–75, 1999.

[175] M. Devault, J. Y. Cochennec, and M. Servel, "The PRELUDE ATD experiment: Assessments and future prospects," *IEEE Journal on Selected Areas on Communications*, vol. 6, no. 9, pp. 1528–1537, December 1988.

[176] D. DeWitt and J. Gray, "Parallel database systems: the future of high performance database systems," *Communications of the ACM*, vol. 35, no. 6, pp. 85–98, 1992.

[177] A. Dhamdhere, H. Jiang, and C. Dovrolis, "Buffer sizing for congested Internet links," in *Proc. IEEE INFOCOM'2005*, pp. 1072–1083, Miami, FL, March 2005.

[178] E. W. Dijkstra, "A note on two problems in connection with graphs," *Numerische Mathematik*, vol. 1, pp. 269–271, 1959.

[179] E. W. Dijkstra and C. S. Scholten, "Termination detection for diffusing computations," *Information Processing Letters*, vol. 11, pp. 1–4, 1980.

[180] W. Doeringer, G. Karjoth, and M. Nassehi, "Routing on longest matching prefixes," *IEEE Trans. on Networking*, vol. 4, pp. 86–97, February 1996.

[181] Q. Dong, S. Bannerjee, J. Wang, D. Agarwal, and A. Shukla, "Packet classifiers in ternary cams can be smaller," in *Proc. ACM SIGMETRICS'2006*, St. Malo, France, June 2006.

[182] A. Doria, E. Davies, and F. Kastenholz, "Requirements for inter-domain routing," IETF Internet-draft, 2006.

[183] B. Douskalis, *IP Telephony: The Integration of Robust VoIP Services*. Prentice-Hall, 2000.

[184] R. D. Doverspike, "Algorithms for multiplex bundling in a telecommunications network," *Operations Research*, vol. 39, no. 6, pp. 925–944, 1991.

[185] R. D. Doverspike, "A multi-layered model for survivability in intra-LATA transport networks," in *Proc. IEEE GLOBECOMŠ91*, pp. 2025–2031, 1991.

[186] R. D. Doverspike, S. J. Phillips, and J. R. Westbrook, "Transport network architectures in an IP world," in *Proc. IEEE INFOCOM'2000*, pp. 305–314, Tel Aviv, Israel, 2000.

[187] R. D. Doverspike and J. Yates, "Challenges for MPLS in optical network restoration," *IEEE Communications Magazine*, vol. 39, no. 2, pp. 89–96, February 2001.

[188] J. Doyle and J. Carroll, *Routing TCP/IP, Volume II*. Cisco Press, 2001.

[189] J. Doyle and J. Carroll, *Routing TCP/IP, Volume I, 2nd Edition*. Cisco Press, 2006.

[190] L. Dryburgh and J. Hewett, *Signaling System No. 7 (SS7/C7): Protocol, Architecture, and Services*. Cisco Press, 2005.

[191] J. Duato, S. Yalamanchili, and L. Ni, *Interconnection Networks: An Engineering Approach*. Morgan Kaufmann Publishers, 2002.

[192] A. M. Duguid, "Structural properties of switching networks," Brown University, Tech. Rep. BTL-7, 1959.

[193] D. Dunn, W. Grover, and M. MacGregor, "Comparison of k-shortest paths and maximum flow routing for network facility restoration," *IEEE Journal on Selected Areas in Communications*, vol. 12, pp. 88–99, 1994.

[194] B. Dunsmore and T. Skandier, *Telecommunications Technologies Reference*. Cisco Press, 2003.

[195] A. Dutta and J.-I. Lim, "A multiperiod capacity planning model for backbone computer communication networks," *Operations Research*, vol. 40, pp. 689–705, 1992.

[196] R. Dutta and G. N. Rouskas, "A survey of virtual topology design algorithms for wavelength routed optical networks," *Optical Networks Magazine*, vol. 1, no. 1, pp. 73–89, January 2000.

[197] Z. Dziong, M. Pióro, and U. Körner, "State-dependent routing in circuit-switched networks: A maximum reward approach," in *Proc. 12th International Teletraffic Congress(ITC12)*, Turin, Italy, 1988.

[198] W. Eatherton, G. Varghese, and Z. Dittia, "Tree bitmap: hardware/sofware IP lookups with incremental updates," *ACM SIGCOMM Computer Communication Review*, vol. 34, pp. 97–122, April 2004.

[199] N. Endo, T. Kozaki, T. Ohuchi, H. Kuwahara, and S. Gohara, "Shared buffer memory switch for an ATM exchange," *IEEE Trans. on Communications*, vol. 41, no. 1, pp. 237–245, January 1993.

[200] ENUM Forum. http://www.enum-forum.org/

[201] D. Eppstein, "Finding the k shortest paths," in *Proc. 35th IEEE Symposium on Foundations of Computer Science*, pp. 154–165, 1994.

[202] D. Eppstein, "Bibliography on k shortest paths and other "k best solutions" problems." http://www.ics.uci.edu/~eppstein/bibs/kpath.bib

[203] M. Ericsson, M. G. C. Resende, and P. M. Pardalos, "A genetic algorithm for the weight setting problem in OSPF routing," *Journal of Combinatorial Optimization*, vol. 6, no. 3, pp. 229–333, 2002.

[204] A. K. Erlang, "The theory of probabilities and telephone conversations," *Nyt Tidsskrift for Matematik B*, vol. 20, 1909.

[205] A. K. Erlang, "Solution of some problems in the theory of probabilities of significance in automatic telephone exchanges," *Elektrotkeknikeren*, vol. 13, 1917.

[206] D. Estrin, "Policy requirements for inter administrative domain routing," *IETF RFC 1125*, November 1989. http://www.rfc-editor.org/rfc/rfc1125.txt

[207] European Radiocommunications Office. http://www.ero.dk/

[208] P. Faltstrom, "E.164 number and DNS," *IETF RFC 2916*, September 2000, (Made obsolete by RFC4632 [209]). http://www.rfc-editor.org/rfc/rfc2916.txt

[209] P. Faltstrom and M. Mealling, "The E.164 to Uniform Resource Identifiers (URI) Dynamic Delegation Discovery System (DDDS) Application (ENUM)," *IETF RFC 3761*, April 2004. http://www.rfc-editor.org/rfc/rfc3761.txt

[210] A. Farago, A. Szentesi, and B. Szviatovszki, "Allocation of administrative weights in PNNI," in *Proc. Networks'98*, pp. 621–625, Sorrento, Italy, 1998.

[211] A. Farrel, *The Internet and Its Protocols: A Comparative Approach*. Morgan Kaufmann Publishers, 2004.

[212] A. Farrel and I. Bryskin, *GMPLS: Architecture and Applications*. Morgan Kaufmann Publishers, 2006.

[213] A. Farrel, J.-P. Vasseur, and G. R. Ash, "A path computation element (PCE)-based architecture," *IETF RFC 4655*, August 2006. http://www.rfc-editor.org/rfc/rfc4655.txt

[214] A. Farrel, J.-P. Vasseur, and A. Ayyangar, "A framework for inter-domain multiprotocol label switching traffic engineering," *IETF RFC 4726*, November 2006. http://www.rfc-editor.org/rfc/rfc4726.txt

[215] A. Farrel (Ed.), D. Papadimitriou, J.-P. Vasseur, and A. Ayyangar, "Encoding of attributes for multiprotocol label switching (MPLS) label switched path (LSP) establishment using Resource Reservation Protocol-Traffic Engineering (RSVP-TE)," *IETF RFC 4420*, February 2006. http://www.rfc-editor.org/rfc/rfc4420.txt

[216] N. Feamster, H. Balakrishnan, and J. Rexford, "Some foundational problems in interdomain routing," in *Proc. 3rd ACM SIGCOMM Workshop on Hot Topics in Networks (HotNets-III)*, San Diego, CA, November 2004.

[217] N. Feamster, J. Borkenhagen, and J. Rexford, "Guidelines for interdomain traffic engineering," *ACM Computer Communication Review*, vol. 33, no. 5, pp. 19–30, 2003.

[218] A. Feldmann, A. Greenberg, C. Lund, N. Reingold, and J. Rexford, "NetScope: Traffic engineering in IP networks," *IEEE Network*, vol. 14, no. 2, pp. 11–19, March/April 2000.

[219] A. Feldmann, A. Greenberg, C. Lund, N. Reingold, J. Rexford, and F. True, "Deriving traffic demands for operational IP networks: Methodology and experience," *IEEE/ACM Trans. on Networking*, vol. 9, pp. 265–279, 2001, (an earlier version appeared in *Proc. ACM SIGCOMM'2000*).

[220] A. Feldmann and S. Muthukrishnan, "Tradeoffs for packet classification," in *Proc. IEEE INFOCOM'2000*, vol. 3, pp. 1193–1202, March 2000.

[221] A. Feldmann, O. Maennel, Z. M. Mao, A. Berger, and B. Maggs, "Locating Internet routing instabilities," in *Proc. ACM SIGCOMM'04*, pp. 205–218, Portland, OR, 2004.

[222] W. Feng, D. D. Kandlur, D. Saha, and K. G. Shin, "Stochastic fair blue: A queue management algorithm for enforcing fairness," in *Proc. IEEE INFOCOM'2001*, pp. 1520–1529, Anchorage, AK, 2001.

[223] J. Filipiak, *Modeling and Control of Dynamic Flows in Communication Networks*. Springer-Verlag, 1988.

[224] S. Floyd, "Connections with multiple congested gateways in packet-switched networks, Part 1: One-way traffic," *ACM SIGCOMM Computer Communication Review*, vol. 21, no. 5, pp. 30–47, 1991.

[225] S. Floyd and K. Fall, "Router mechanisms to support end-to-end congestion control," February 1997, Networking Research Group, LBL Labs.

[226] S. Floyd and K. Fall, "Promoting the use of end-to-end congestion control in the internet," *IEEE/ACM Trans. on Networking*, vol. 7, pp. 458–472, 1999.

[227] S. Floyd and V. Jacobson, "Random early detection gateways for congestion avoidance," *IEEE/ACM Trans. on Networking*, vol. 1, pp. 397–413, August 1993.

[228] S. Floyd and V. Jacobson, "The synchronization of periodic routing messages," *IEEE/ACM Trans. on Networking*, vol. 2, pp. 122–136, 1994, (an earlier version appeared in ACM SIGCOMM'93).

[229] S. Floyd and E. Kohler, "Profile for datagram congestion control protocol (DCCP) congestion control ID 2: TCP-like congestion control," *IETF RFC 4341*, March 2006. http://www.rfc-editor.org/rfc/rfc4341.txt

[230] S. Floyd, E. Kohler, and J. Padhye, "Profile for datagram congestion control protocol (DCCP) congestion control ID 3: TCP-friendly rate control (TFRC)," *IETF RFC 4342*, March 2006. http://www.rfc-editor.org/rfc/rfc4342.txt

[231] L. R. Ford, "Network flow theory," The Rand Corporation, Santa Monica, Tech. Rep. Paper P-923, 1956.

[232] L. R. Ford and D. R. Fulkerson, "A suggested computation for maximal multicommodity network flows," *Management Science*, vol. 5, pp. 97–101, 1958.

[233] B. Fortz and M. Thorup, "Internet traffic engineering by optimizing OSPF weights," in *Proc. IEEE INFOCOM'2000*, pp. 519–528, Tel Aviv, Israel, March 2000.

[234] B. Fortz and M. Thorup, "Optimizing OSPF/IS-IS weights in a changing world," *IEEE Journal on Selected Areas in Communications*, vol. 20, pp. 756–767, May 2002.

[235] B. Fortz, J. Rexford, and M. Thorup, "Traffic engineering with traditional IP routing protocols," *IEEE Communication Magazine*, vol. 40, no. 10, pp. 118–124, October 2002.

[236] M. Foster, T. McGarry, and J. Yu, "Number portability in the Global Switched Telephone Network (GSTN): An overview," *IETF RFC 3482*, February 2003. http://www.rfc-editor.org/rfc/rfc3482.txt

[237] R. Fourer, "2003 software survey: Linear programming," *ORMS Today*, vol. 30, no. 6, pp. 34–43, December 2003.

[238] E. Fredkin, "Trie memory," *Communications of the ACM*, vol. 3, pp. 490–499, September 1960.

[239] V. Fuller, T. Li, J. Yu, and K.Varadhan, "Classless inter-domain routing (CIDR): An address assignment and aggregation strategy," *IETF RFC 1519*, September 1993, http://www.rfc-editor.org/rfc/rfc1519.txt.

[240] V. Fuller, T. Li, J. Yu, and K. Varadhan, "Supernetting: An address assignment and aggregation strategy," *IETF RFC 1338*, June 1992. http://www.rfc-editor.org/rfc/rfc1338.txt

[241] A. Fumagalli and L. Valcarenghi, "IP restoration vs. WDM protection: Is there an optimal choice?" *IEEE Network*, vol. 14, no. 6, pp. 34–41, November/December 2000.

[242] P. Gajowniczek, M. Pióro, A. Szentesi, J. Harmatos, and A. Jüttner, "Solving an OSPF routing problem with simulated allocation," in *Proc. 1st Polish-German Teletraffic Symposium*, pp. 177–184, Dresden, Germany, 2000.

[243] L. Gao and J. Rexford, "Stable Internet routing without global coordination," *IEEE/ACM Trans. on Networking*, vol. 9, no. 6, pp. 681–692, 2001.

[244] J. J. Garcia-Luna-Aceves, "A distributed, loop-free, shortest-path routing algorithm," in *Proc. IEEE INFOCOM'1988*, pp. 1125–1137, 1988.

[245] J. J. Garcia-Luna-Aceves, "Loop-free routing using diffusing computation," *IEEE Trans. on Networking*, vol. 1, pp. 130–141, 1993.

[246] A. Gencata and B. Mukherjee, "Virtual-topology adaptaion for WDM mesh networks under dynamic traffic," *IEEE/ACM Trans. on Networking*, vol. 11, no. 2, pp. 236–247, April 2003.

[247] T. George, B. Bidulock, R. Dantu, H. Schwarzbauer, and K. Morneault, "Signaling System 7 (SS7) Message Transfer Part 2 (MTP2)—User Peer-to-Peer Adaptation Layer (M2PA)," *IETF RFC 4165*, September 2005. http://www.rfc-editor.org/rfc/rfc4165.txt

[248] D. Ghosh, V. Sarangan, and R. Acharya, "Quality of service routing in IP networks," *IEEE Trans. on Multimedia*, vol. 3, pp. 200–208, 2001.

[249] R. J. Gibbens, P. Hunt, and F. P. Kelly, "Bistability in communication networks," in *Disorder in Physical Systems: a Volume in Honour of John M. Hammersley*. G. Grimmett and D. Welsh (Eds.), Oxford University Press, pp. 113–127, 1990.

[250] R. J. Gibbens and F. P. Kelly, "Dynamic routing in fully connected networks," *IMA Journal of Mathematical Control and Information*, vol. 7, 1990.

[251] R. J. Gibbens, F. P. Kelly, and P. B. Key, "Dynamic Alternate Routing—modeling and behaviour," in *Proc. 12th International Teletraffic Congress (ITC12)*, pp. 3.4A3.1–3.4A3.7, Turin, Italy, 1988.

[252] V. Gill, J. Heasley, and D. Meyer, "The generalized TTL security mechanism (GTSM)," *IETF RFC 3682*, February 2004. http://www.rfc-editor.org/rfc/rfc3682.txt

[253] A. Girard, *Routing and Dimensioning in Circuit-Switched Networks*. Addison-Wesley, 1990.

[254] A. Girard and B. Sansò, "Multicommodity flow models, failure propagation and reliable network design," *IEEE/ACM Trans. on Networking*, vol. 6, pp. 82–93, 1998.

[255] B. Gleeson, A. Lin, J. Heinanen, G. Armitage, and A. Malis, "A framework for IP based virtual private networks," *IETF RFC 2764*, February 2000. http://www.rfc-editor.org/rfc/rfc2764.txt

[256] J. Goldsmith and T. Wu, *Who Controls the Internet?: Illusions of a Borderless World*. Oxford University Press, 2006.

[257] S. J. Golestani, "A self-clocked fair queueing scheme for broadband applications," in *Proc. IEEE INFOCOM'94*, pp. 636–646, Toronto, Canada, June 1994.

[258] Google. http://www.google.com/

[259] S. Gorinsky, A. Kantawala, and J. Turner, "Link buffer sizing: A new look at the old problem," in *Proc. IEEE Symposium on Computers and Communications (ISCC 2005)*, pp. 507–514, June 2005.

[260] J. Gottlieb, A. Greenberg, J. Rexford, and J. Wang, "Automated provisioning of BGP customers," *IEEE Network*, vol. 17, no. 6, pp. 44–55, November/December 2003.

[261] E. Gourdin, "Optimizing Internet networks," *ORMS Today*, vol. 28, no. 2, pp. 48–49, April 2001.

[262] L. Gouveia, P. Amaro, A. F. de Sousa, and R. Valadas, "MPLS over WDM network design with packet level QoS constraints based on ILP models," in *Proc. IEEE INFOCOM'2003*, pp. 576–586, San Francisco, CA, 2003.

[263] E. Gray, *MPLS: Implementing the Technology*. Addison-Wesley, 2001.

[264] T. Griffin and G. Huston, "BGP wedgies," *IETF RFC 4264*, November 2005. http://www.rfc-editor.org/rfc/rfc4264.txt

[265] T. G. Griffin, F. B. Shepherd, and G. Wilfong, "The stable paths problem and interdomain routing," *IEEE/ACM Trans. on Networking*, vol. 10, pp. 231–243, 2002.

[266] B. Groskinsky, D. Medhi, and D. Tipper, "An investigation of adaptive capacity control schemes in a dynamic traffic environment," *IEICE Tran. on Communications*, vol. E84-B, pp. 263–274, February 2001.

[267] W. D. Grover, *Mesh-based Survivable Networks: Options and Strategies for Optical, MPLS, SONET and ATM Networking*. Prentice-Hall, 2004.

[268] M. Gryseels, K. Struyve, M. Pickavet, and P. Demeester, "Survivability design in multi-layer transport networks," in *Proc. 6th ICTS*, 1998.

[269] R. Guerin and A. Orda, "QoS-based routing in networks with inaccurate information: Theory and algorithms," *IEEE/ACM Trans. on Networking*, vol. 7, pp. 350–364, 1999.

[270] R. Guerin, A. Orda, and D. Williams, "QoS routing mechanisms and OSPF extensions," in *Proc. 2nd IEEE Global Internet Mini-Conference*, pp. 1903–1908, Phoenix, AZ, November 1997.

[271] J. Guichard, F. Le Faucheur, and J.-P. Vasseur, *Definitive MPLS Network Designs*. Cisco Press, 2005.

[272] L. Guo and I. Matta, "The war between mice and elephants," in *Proc. 9th IEEE International Conference on Network Protocols (ICNP'2001)*, pp. 180–188, Riverside, CA, November 2001.

[273] P. Gupta, "Algorithms for routing lookups and packet classification," Ph.D. dissertation, Stanford University, Palo Alto, CA, 2000.

[274] P. Gupta, S. Lin, and N. McKeown, "Routing lookups in hardware at memory access speeds," in *Proc. IEEE INFOCOM'98*, pp. 1240–1247, April 1998.

[275] P. Gupta and N. McKeown, "Packet classification on multiple fields," in *Proc. ACM SIGCOMM'99*, pp. 147–160, Cambridge, MA, 1999.

[276] P. Gupta and N. McKeown, "Packet classification using hierarchical intelligent cuttings," in *Proc. Hot Interconnects'99*, 1999.

[277] P. Gupta and N. McKeown, "Algorithms for packet classification," *IEEE Network*, vol. 15, no. 2, pp. 24–32, March/April 2001.

[278] A. Guttman, "R-trees: A dynamic index structure for spatial searching," in *Proc. ACM SIGMOD'84*, pp. 47–57, 1984.

[279] D. Haenschke, D. A. Kettler, and E. Oberer, "Network management and congestion in the U.S. telecommunications network," *IEEE Trans. on Communications*, vol. COM-29, pp. 376–385, 1981.

[280] J. Hagstrom, "Braess's paradox—web-site." http://tigger.uic.edu/~hagstrom/Research/Braess/index.html

[281] T. Hain, "An IPv6 provider-independent global unicast address format," August 2006, Internet Draft. http://tools.ietf.org/wg/ipv6/draft-hain-ipv6-pi-addr-10.txt

[282] S. Halabi (with D. McPherson), *Internet Routing Architectures, 2nd Edition*. Cisco Press, 2000.

[283] R. Hanmer and M. Wu, "Traffic congestion patterns," in *Proc. 6th Annual Conference on Pattern Languages of Programs (PLoP99)*, Urbana, IL, August 1999.

[284] E. Hashem, "Analysis of random drop for gateway congestion control," Laboratory for Computer Science, Massachusetts Institute of Technology, Cambridge, MA, Tech. Rep. MIT-LCS-TR-465, 1989.

[285] B. M. Hauzeur, "A model for naming, addressing, and routing," *ACM Trans. Office Information Systems*, vol. 4, pp. 293–311, 1986.

[286] J. Hawkinson and T. Bates, "Guidelines for creation, selection, and registration of an autonomous system (AS)," *IETF RFC 1930*, March 1996. http://www.rfc-editor.org/rfc/rfc1930.txt

[287] J. F. Hayes, R. Breault, and M. K. Mehmet-Ali, "Performance analysis of a multicast switch," *IEEE Trans. on Communications*, vol. 39, no. 4, pp. 581–587, 1991.

[288] J. He, M. Bresler, M. Chiang, and J. Rexford, "Towards multi-layer traffic engineering: Optimization of congestion control and routing," to appear in *IEEE Journal on Selected Areas in Communications*.

[289] J. He, M. Chiang, and J. Rexford, "Can congestion control and traffic engineering be at odds?" in *Proc. IEEE GLOBECOM'2006*, San Francisco, CA, November/December 2006.

[290] C. L. Hedrick, "Routing Information Protocol," *IETF RFC 1058*, June 1988. http://www.rfc-editor.org/rfc/rfc1058.txt

[291] J. Heinanen and R. Guerin, "A single rate three color marker," *IETF RFC 2697*, September 1999, http://www.rfc-editor.org/rfc/rfc2697.txt.

[292] J. Heinanen and R. Guerin, "A two rate three color marker," *IETF RFC 2698*, September 1999, http://www.rfc-editor.org/rfc/rfc2698.txt.

[293] J. L. Hennessy and D. A. Patterson, *Computer Architecture: A Quantitative Approach, 4th Edition*. Morgan Kaufmann Publishers, 2006.

[294] M. Hidell, P. Sjodin, and O. Hagsand, "Router architectures," May 2004. http://www.imit.kth.se/mahidell/pubs/networking04_tutorial_final.pdf

[295] C. V. Hollot, Y. Liu, V. Misra, and D. Towsley, "Unresponsive flows and AQM performance," in *Proc. IEEE INFOCOM'2003*, San Francisco, CA, 2003.

[296] J. E. Hopcraft and R. M. Karp, "An $n^{5/2}$ algorithm for maximum matching in bipartite graphs," *SIAM Journal of Computing*, vol. 2, pp. 225–231, 1973.

[297] P. Hosein, "An improved congestion control algorithm for telecommunications signaling networks," *Telecommunication Systems*, vol. 16, pp. 379–398, 2001.

[298] D. J. Houck, K. S. Meier-Hellstern, F. Saheban, and R. A. Skoog, "Failure and congestion propagation through signaling controls," in *Proc. 14th International Teletraffic Congress (ITC14)*, pp. 367–376, Antibes, France, 1994.

[299] A. Huang and S. Knauer, "Starlite: A wideband digital switch," in *Proc. IEEE GLOBECOM'84*, pp. 121–125, 1984.

[300] J. Y. Hui and T. Renner, "Queueing analysis for multicast packet switching," *IEEE Trans. on Communications*, vol. 42, no. 2-4, pp. 723–731, Febraury 1994.

[301] C. Huitema, *Routing in the Internet, 2nd Edition*. Prentice-Hall, 2000.

[302] G. Huston, Personal Communication, May 2006.

[303] G. Huston, "BGP routing table analysis reports," 2006. http://bgp.potaroo.net/

[304] G. Huston, "CIDR report," 2006. http://www.cidr-report.org/

[305] G. Huston, "Where's the money?—Internet interconnection and financial settlements," January 2005, The ISP Column (Internet Society). http://ispcolumn.isoc.org/2005-01/interconns.html

[306] R.-H. Hwang, "LLR routing in homogeneous VP-based ATM networks," in *Proc. IEEE INFOCOM'95*, pp. 587–593, 1995.

[307] R.-H. Hwang, J. F. Kurose, and D. F. Towsley, "MDP routing in ATM networks using the virtual path concept," in *Proc. IEEE INFOCOM'94*, pp. 1509–1517, 1994.

[308] G. Iannaccone, C.-N. Chuah, S. Bhattacharyya, and C. Diot, "Feasibility of IP restoration in a tier-1 backbone," *IEEE Network*, vol. 18, no. 2, pp. 13–19, March/April 2004.

[309] Industrial Numbering Committee/Alliance for Telecommunications Industry Solution. http://www.atis.org/inc/

[310] Industrial Numbering Committee/Alliance for Telecommunications Industry Solutions, "Numbering and dialing plan within the United States," ATIS-0300076; last issue date: August 19, 2005. http://www.atis.org/inc/Docs/Finaldocs/US-Numbering-Dialing-Plan-08-19-05.doc

[311] Intel, "Public network signalling tutorial." http://resource.intel.com/telecom/support/ss7/SS7tutorial/tutorial.html

[312] International Dialing Codes. http://www.timeanddate.com/worldclock/dialing.html

[313] Internet Assigned Number Authority (IANA), "Address family numbers." http://www.iana.org/assignments/address-family-numbers

[314] Internet Assigned Number Authority (IANA), "BGP capability codes." http://www.iana.org/assignments/capability-codes

[315] Internet Assigned Number Authority (IANA), "BGP parameters." http://www.iana.org/assignments/bgp-parameters

[316] Internet Assigned Number Authority (IANA), "IS-IS TLV codepoints." http://www.iana.org/assignments/isis-tlv-codepoints

[317] Internet Assigned Number Authority (IANA), "OSPF traffic engineering TLVs." http://www.iana.org/assignments/ospf-traffic-eng-tlvs

[318] Internet Assigned Number Authority (IANA), "RSVP parameters." http://www.iana.org/assignments/rsvp-parameters

[319] Internet Assigned Number Authority (IANA), "Special-use IPv4 addresses," *IETF RFC 3330*, September 2002. http://www.rfc-editor.org/rfc/rfc3330.txt

[320] P. Iovanna, M. Settembre, and R. Sabella, "A traffic engineering system for multi-layer networks based on the GMPLS paradigm," *IEEE Network*, vol. 17, no. 2, pp. 28–37, March/April 2003.

[321] ISO, "Intermediate system to intermediate system routing information exchange protocol for use in conjunction with the protocol for providing the connectionless-mode network service (iso 8473)," *ISO/IEC 10589*, February 1990, (also see [537]).

[322] ITU-T, "List of Signalling Area/Network Codes (SANC): Position on 1 December 2004—complement to ITU-T recommendation Q.708 (03/99)." http://www.itu.int/itudoc/itu-t/ob-lists/icc/q708_767.html

[323] ITU-T Recommendation E.164, "The International Public Telecommunication Numbering Plan," February 2005, International Telecommunications Union–Telecommunication Standardization Sector.

[324] ITU-T Recommendation E.171, "International Telephone Routing Plan," 1993, International Telecommunications Union–Telecommunication Standardization Sector.

[325] ITU-T Recommendation Q.706, "Specifications of Signaling System No. 7: Message transfer part signalling performance," March 1993, International Telecommunications Union–Telecommunication Standardization Sector.

[326] ITU-T Recommendation Q.709, "Specifications of Signaling System No. 7: Hypothetical signalling reference connection," March 1993, International Telecommunications Union–Telecommunication Standardization Sector.

[327] ITU-T Recommendation Q.761, "Signalling System No. 7—ISDN User Part: Functional description," September 1997, International Telecommunications Union–Telecommunication Standardization Sector.

[328] ITU-T Recommendation Q.762, "Signalling System No. 7—ISDN User Part: General functions of messages and signals," December 1999, International Telecommunications Union–Telecommunication Standardization Sector.

[329] ITU-T Recommendation Q.763, "Signalling System No. 7—ISDN User Part: Formats and codes," December 1999, International Telecommunications Union–Telecommunication Standardization Sector.

[330] ITU-T Recommendation Q.769.1, "Signalling System No. 7—ISDN User Part: Enhancements for the support of number portability," December 1999, International Telecommunications Union–Telecommunication Standardization Sector.

[331] ITU-T Recommendation Q.850, "Usage of cause and location in the Digital Subscriber Signalling System No. 1 and the Signalling System No. 7 ISDN User Part," May 1998, International Telecommunications Union–Telecommunication Standardization Sector.

[332] S. Iyer and N. McKeown, "Techniques for fast shared memory switches," Stanford University, Stanford, CA, Tech. Rep. TR01-HPNG-081501, August 2001.

[333] A. R. Jacob, "A survey of fast packet switches," *ACM SIGCOMM Computer Communication Review*, vol. 20, no. 1, pp. 54–64, January 1990.

[334] V. Jacobson, "Congestion avoidance and control," in *Proc. ACM SIGCOMM'88*, pp. 314–329, 1988.

[335] V. Jacobson, "Modified TCP congestion control algorithm," April 1990, end2end-interest mailing list.

[336] J. Jaffe and F. Moss, "A responsive distributed routing algorithm for computer networks," *IEEE Trans. on Communications*, vol. 30, pp. 1758–1762, 1982.

[337] B. Jamoussi, Ed., L. Andersson, R. Callon, R. Dantu, L. Wu, P. Doolan, T. Worster, N. Feldman, A. Fredette, M. Girish, E. Gray, J. Heinanen, T. Kilty, and A. Malis, "Constraint-based LSP setup using LDP," *IETF RFC 3212*, January 2002. http://www.rfc-editor.org/rfc/rfc3212.txt

[338] Japan Internet Exchange: JPIX. http://www.jpix.ad.jp/

[339] L. Jereb, T. Jakab, and F. Unghváry, "Availability analysis of multi-layer optical networks," *Optical Networks*, vol. 3, pp. 84–95, 2002.

[340] L. Jereb, F. Unghváry, and T. Jakab, "A methodology for reliability analysis of multi-layer communication networks," *Optical Networks Magazine*, vol. 2, pp. 42–51, 2001.

[341] A. Johnston, S. Donovan, R. Sparks, C. Cunningham, and K. Summers, "Session initiation protocol (SIP) public switched telephone network (PSTN) call flows," *IETF RFC 3666*, December 2003. http://www.rfc-editor.org/rfc/rfc3666.txt

[342] Y. M. Joo and N. McKeown, "Doubling memory bandwidth for network buffers," in *Proc. IEEE INFOCOM'98*, pp. 808–815, April 1998.

[343] M. Joshi, A. Mansata, S. Talauliker, and C. Beard, "Design and analysis of multi-level active queue management mechanisms for emergency traffic," *Computer Communications*, vol. 28, pp. 162–173, February 2005.

[344] A. Jüttner, B. Szviatovszki, I. Mécs, and Z. Rajkó, "Lagrange relaxation based method for the QoS routing problem," in *Proc. IEEE INFOCOM'2001*, pp. 859–868, Anchorage, AK, April 2001.

[345] R. E. Kalaba and M. L. Juncosa, "Optimal design and utilization of communication networks," *Management Science*, vol. 3, no. 1, pp. 33–44, October 1956.

[346] H. Kanakia, "Datapath switch," AT&T Bell Labs Internal Memorandum, 1999.

[347] H. Kaplan, "Resilient IP network design," tutorial, 3rd IEEE International Workshop on IP Operations & Management, Kansas City, MO, October 2003.

[348] H. Kaplan, "NSR^{TM}: Non-stop routing technology," 2002. http://www.avici.com/technology/whitepapers/reliability_series/NSRTechnology.pdf

[349] K. Kar, M. Kodialam, and T. V. Lakshman, "Minimum interference routing of bandwidth guaranteed tunnels with MPLS traffic engineering applications," *IEEE Journal on Selected Areas in Communications*, vol. 18, pp. 2566–2579, 2000.

[350] T. Karagiannis, M. Molle, M. Faloutsos, and A. Broido, "A nonstationary Poisson view of Internet traffic," in *Proc. IEEE INFOCOM'2004*, Hong Kong, March 2004.

[351] M. Karol and M. Hluchyj, "Queueing in high performance packet switching," *IEEE Journal of Selected Areas in Communications*, vol. 6, pp. 1587–1597, Decemeber 1988.

[352] M. Karol, M. Hluchyj, and S. Morgan, "Input vs output queuing on a space division switch," *IEEE Trans. on Communications*, pp. 1347–1356, Decemeber 1987.

[353] R. Karp, U. Vazirani, and V. Vazirani, "An optimal algorithm for on-line bipartite matching," in *Proc. 22nd Annual ACM Symposium on Theory of Computing*, pp. 352–358, May 1990.

[354] D. Katz, "OSPF and IS-IS—a comparative anatomy," June 2000. http://www.nanog.org/mtg-0006/ppt/katz.ppt

[355] D. Katz, K. Kompella, and D. Yeung, "Traffic engineering (TE) extensions to OSPF version 2," *IETF RFC 3630*, September 2003. http://www.rfc-editor.org/rfc/rfc3630.txt

[356] J. S. Kaufman, "Blocking in a shared resource environment," *IEEE Trans. on Communications*, vol. COM-29, pp. 1474–1481, 1981.

[357] F. P. Kelly, "Blocking probabilities in large circuit-switched networks," *Advances in Applied Probability*, vol. 18, pp. 473–505, 1986.

[358] F. P. Kelly, "Routing in circuit-switched networks: Optimization, shadow prices and decentralization," *Advances in Applied Probability*, vol. 20, pp. 112–144, 1988.

[359] F. P. Kelly, "Routing and capacity allocation in networks with trunk reservation," *Mathematics of Operations Research*, vol. 15, pp. 771–793, 1990.

[360] F. P. Kelly, "Loss networks," *Annals of Applied Probability*, vol. 1, pp. 319–378, 1991.

[361] F. P. Kelly, "Network routing," *Philosophical Transactions of the Royal Society*, vol. A337, pp. 343–367, 1991.

[362] F. P. Kelly, "Charge and rate control for elastic traffic," *European Trans. on Telecommunications*, vol. 8, pp. 33–37, 1997.

[363] F. P. Kelly, "Fairness and stability of end-to-end congestion control," *European Journal of Control*, vol. 9, pp. 159–176, 2003.

[364] F. P. Kelly, A. K. Mauloo, and D. H. K. Tan, "Rate control for communication networks: Shadow prices, proportional fairness and stability," *Journal of the Operations Research Society*, vol. 49, pp. 2006–2017, 1997.

[365] S. Keshav, *An Engineering Approach to Computer Networking: ATM Networks, the Internet, and the Telephone Network*. Addison-Wesley, 1997.

[366] S. Keshav, "Naming, addressing, and forwarding reconsidered," August 2005. http://blizzard.cs.uwaterloo.ca/keshav/home/Papers/data/05/naming.pdf

[367] S. Keshav and R. Sharma, "Issues and trends in router design," *IEEE Communications Magazine*, vol. 36, no. 5, pp. 144–151, May 1998.

[368] A. Khanna and J. A. Zinky, "The revised ARPANET routing metric," in *Proc. ACM SIGCOMM'89*, pp. 45–56, September 1989.

[369] J. Klensin (Ed.), "The history and context of telephone number mapping (ENUM) operational decisions: Informational documents contributed to ITU-T Study Group 2 (SG2)," *IETF RFC 3245*, March 2002. http://www.rfc-editor.org/rfc/rfc3245.txt

[370] D. E. Knuth, *The Art of Computer Programming, Volume 3, Sorting and Searching, 3rd Edition*. Addison-Wesley, 1998.

[371] E. Kohler, M. Handley, and S. Floyd, "Datagram congestion control protocol (DCCP)," *IETF RFC 4340*, March 2006. http://www.rfc-editor.org/rfc/rfc4340.txt

[372] K. Kompella, Y. Rekhter, and L. Berger, "Link bundling in MPLS traffic engineering (TE)," *IETF RFC 4201*, October 2005. http://www.rfc-editor.org/rfc/rfc4201.txt

[373] K. Kompella and Y. Rekhter (Eds.), "Intermediate system to intermediate system (IS-IS) extensions in support of generalized multi-protocol label switching (GMPLS)," *IETF RFC 4205*, October 2005. http://www.rfc-editor.org/rfc/rfc4205.txt

[374] K. Kompella and Y. Rekhter (Eds.), "OSPF extensions in support of generalized multi-protocol label switching (GMPLS)," *IETF RFC 4203*, October 2005. http://www.rfc-editor.org/rfc/rfc4203.txt

[375] K. Kompella and Y. Rekhter (Eds.), "Routing extensions in support of generalized multi-protocol label switching (GMPLS)," *IETF RFC 4202*, October 2005. http://www.rfc-editor.org/rfc/rfc4202.txt

[376] K. Kompella and Y. Rekhter (Eds.), "Virtual private LAN service (VPLS) using BGP for auto-discovery and signaling," *IETF RFC 4761*, January 2007. http://www.rfc-editor.org/rfc/rfc4761.txt

[377] T. Korkmaz and M. Krunz, "Routing multimedia traffic with QoS guarantees," *IEEE Trans. on Multimedia*, vol. 5, pp. 429–443, 2003.

[378] A. Kortebi, L. Muscariello, S. Oueslati, and J. Roberts, "On the scalability of fair queueing," in *Proc. Third Workshop on Hot Topics in Networks (HotNets-III)*, San Diego, CA, November 2004.

[379] A. Kortebi, S. Oueslati, and J. Roberts, "Implicit service differentiation using deficit round robin," in *Proc. 19th International Teletraffic Congress (ITC19)*, Beijing, China, August 2005.

[380] B. Kraimeche and M. Schwartz, "Analysis of traffic access control strategies in integrated service networks," *IEEE Trans. on Communications*, vol. COM-33, pp. 1085–1093, 1985.

[381] D. Krioukov, "Routing → AS types." http://www.caida.org/analysis/routing/astypes/

[382] K. R. Krishnan, "Routing of telephone traffic to minimize network blocking," in *Proc. IEEE Conference on Decision and Control*, vol. 21, pp. 375–377, 1982.

[383] K. R. Krishnan, R. D. Doverspike, and C. D. Pack, "Improved survivability with multi-layer dynamic routing," *IEEE Communications Magazine*, vol. 33, no. 7, pp. 62–68, July 1995.

[384] B. Krithikaivasan, Y. Zeng, K. Deka, and D. Medhi, "ARCH-based traffic forecasting and dynamic bandwidth provisioning for periodically measured nonstationary traffic," *IEEE/ACM Trans. on Networking*, vol. 15, August 2007.

[385] R. S. Krupp, "Stabilization of alternate routing networks," in *Proc. IEEE ICC'82*, pp. 31.2.1–31.2.5, June 1982.

[386] J. Kurose and K. Ross, *Computer Networking: A Top-Down Approach Featuring the Internet: 3rd Edition*. Addison-Wesley, 2004.

[387] C. Labovitz, "Scalability of the Internet backbone routing infrastructure," Ph.D. dissertation, University of Michigan, Ann Arbor, MI, 1999.

[388] C. Labovitz, A. Ahuja, A. Bose, and F. Jahanian, "Delayed Internet routing convergence," *IEEE/ACM Trans. on Networking*, vol. 9, no. 3, pp. 293–306, 2001.

[389] C. Labovitz, A. Ahuja, and F. Jahanian, "Experimental study of Internet stability and wide-area network failures," in *Proc. Twenty-Ninth Annual International Symposium on Fault-Tolerant Computing (FTCS99)*, pp. 278–285, Madison, WI, June 1999.

[390] C. Labovitz, A. Ahuja, R. Wattenhofer, and S. Venkatachary, "The impact of Internet policy and topology on delayed routing convergence," in *Proc. IEEE INFOCOM'2001*, pp. 537–546, Anchorage, AK, 2001.

[391] C. Labovitz, G. R. Malan, and F. Jahanian, "Internet routing instability," *IEEE/ACM Trans. on Networking*, vol. 6, pp. 515–528, 1998.

[392] T. V. Lakshman and D. Stidialis, "High-speed policy based packet forwarding using efficient multidimensional range matching," in *Proc. ACM SIGCOMM'98*, pp. 203–214, Vancouver, Canada, September 1998.

[393] K. Lakshminarayanan, A. Rangarajan, and S. Venkatachary, "Algorithms for advanced packet classification with ternary CAMs," *ACM SIGCOMM Computer Communication Review*, vol. 35, no. 4, pp. 193–204, 2005.

[394] R. O. LaMaire and D. N. Serpanos, "Two dimensional round robin schedulers for packet switches with multiple input queues," *IEEE/ACM Trans. on Networking*, vol. 2, no. 5, pp. 471–482, October 1994.

[395] B. Lampson, V. Srinivasan, and G. Varghese, "IP lookups using multiway and multicolumn search," in *Proc. IEEE INFOCOM'98*, pp. 1248–1256, April 1998.

[396] J. Lang (Ed.), "Link management protocol (LMP)," *IETF RFC 4204*, October 2005. http://www.rfc-editor.org/rfc/rfc4204.txt

[397] L. Lasdon, *Optimization Theory for Large Systems*. Macmillan, 1970.

[398] M. Lasserre and V. Kompella (Eds.), "Virtual private LAN service (VPLS) using label distribution protocol (LDP) signaling," *IETF RFC 4762*, January 2007. http://www.rfc-editor.org/rfc/rfc4762.txt

[399] A. M. Law and W. D. Kelton, *Simulation Modeling and Analysis, 3rd Edition*. McGraw Hill, 2000.

[400] D. N. Lee, K. T. Medhi, J. L. Strand, R. G. Cox, and S. Chen, "Solving large telecommunications network loading problems," *AT&T Technical Journal*, vol. 68, no. 3, pp. 48–56, May/June 1989.

[401] W. Lee, M. Hluchyi, and P. Humblet, "Routing subject to quality of service constraints in integrated communication networks," *IEEE Network*, vol. 9, no. 4, pp. 46–55, July/August 1995.

[402] B. Leiner, "Policy issues in interconnecting networks," *IETF RFC 1124*, September 1989, (available only as postscript or PDF file). http://www.faqs.org/rfc/rfc1124.pdf

[403] B. M. Leiner, V. G. Cerf, D. D. Clark, R. E. Kahn, L. Kleinrock, D. C. Lynch, J. Postel, L. G. Roberts, and S. Wolff, "A brief history of the Internet," version 3.32, last revised 10 December, 2003. http://www.isoc.org/internet/history/brief.shtml

[404] W. E. Leland, M. S. Taqqu, W. Willinger, and D. V. Wilson, "On the self-similar nature of Ethernet traffic," in *Proc. ACM SIGCOMM'93*, pp. 183–193, September 1993.

[405] G. Li, D. Wang, C. Kalmanek, and R. Doverspike, "Efficient distributed restoration path selection for shared mesh restoration," *IEEE Trans. on Networking*, vol. 11, pp. 761–771, 2003.

[406] J. Li, H. Liu, and K. R. Sollins, "AFBV: a scalable packet classification algorithm." *ACM Computer Communication Review*, vol. 32, no. 3, p. 24, 2002.

[407] K. Li, F. Chang, D. Berger, and W. Fang, "Architectures for packet classification caching," in *Proc. IEEE ICON*, 2003.

[408] W. Liang and X. Shen, "Improved lightpath (wavelength) routing in large WDM networks," *IEEE Transactions on Communications*, vol. 48, pp. 1571–1579, 2000.

[409] W. Liang and X. Shen, "Finding multiple routing paths in wide-area WDM networks," *Computer Communications*, vol. 28, pp. 811–818, 2005.

[410] D. Lin and R. Morris, "Dynamics of random early detection," in *Proc. ACM SIGCOMM'97*, pp. 127–137, Cannes, France, September 1997.

[411] R. J. Lipton and J. F. Naughton, "Query size estimation by adaptive sampling," in *PODS 1990*, pp. 40–46, 1990.

[412] R. J. Lipton, J. F. Naughton, and D. A. Schneider, "Practical selectivity estimation through adaptive sampling," in *Proc. ACM SIGMOD'90*, pp. 1–11, 1990.

[413] Y. Liu, D. Tipper, and P. Sinpongwutikorn, "Approximating optimal spare capacity allocation by successive survivable routing," in *Proc. IEEE INFOCOM'2001*, pp. 699–798, Anchorage, AK, April 2001.

[414] Y. Liu, H. Zhang, W. Gong, and D. Towsley, "On the interaction between overlay routing and underlay routing," in *Proc. IEEE INFOCOM'2005*, pp. 2543–2553, Miami, FL, March 2005.

[415] J. Livingood and R. Shockey, "IANA registration for an enumservice containing public switched telephone network (PSTN) signaling information," *IETF RFC 4769*, November 2006. http://www.rfc-editor.org/rfc/rfc4769.txt

[416] Local Number Portability Administration Working Group. http://www.npac.com/cmas/

[417] Local Number Portability Administration Working Group, "LNPA WG interpretation of $N-1$ carrier architecture, Version 5.0," January 17, 2005. http://www.npac.com/cmas/co_docs/LNPA_WG_N-1_INTERPRETATION_v5.doc

[418] Local Number Portability Administration Working Group, "NP best practices matrix," 2005. http://www.npac.com/cmas/co_docs/LNPA_NP_Best_Practices_November_2005.doc

[419] London Internet Exchange, LINX. http://www.linx.net/

[420] K. Long, R. Tucker, S. Cheng, J. Ma, and R. Zhang, "A new approach to multi-layer network survivability: strategies, model and algorithm," *Journal of High Speed Networks*, vol. 10, no. 2, pp. 127–134, 2001.

[421] B. T. Loo, T. Condie, M. Garofalakis, D. E. Gay, J. M. Hellerstein, P. Maniatis, R. Ramakrishnan, T. Roscoe, and I. Stoica, "Declarative networking: Language, execution and optimization," in *Proc. ACM SIGMOD International Conference on Management of Data*, Chicago, June 2006.

[422] B. T. Loo, J. M. Hellerstein, and I. Stoica, "Customizable routing with declarative queries," in *Proc. 3rd Workshop on Hot Topics in Networks (ACM SIGCOMM HotNets-III)*, San Diego, CA, November 2004.

[423] B. T. Loo, J. M. Hellerstein, I. Stoica, and R. Ramakrishnan, "Declarative routing: Extensible routing with declarative queries," in *Proc. ACM SIGCOMM'05*, pp. 289–300, Philadelphia, PA, August 2005.

[424] E. Lorenz, "Predictability: Does the flap of a butterfly's wings in Brazil set off a tornado in Texas?" Washington, DC, December 1972, meeting of the American Association for the Advancement of Science.

[425] G. Lorenz, T. Moore, G. Manes, J. Hale, and S. Shenoi, "Securing SS7 telecommunications networks," in *Proc. 2001 IEEE Workshop on Information Assurance and Security*, pp. 273–278, West Point, NY, June 2001.

[426] P. Lothberg, "Sprintlink optical IP network," Joint NLANR Internet2 Techs Meeting, Las Cruces, NM, March, 1999. http://www.ncne.org/training/techs/1999/990307/Talks1/lothberg/sprint-i2/index.htm

[427] J. Lou and X. Shen, "Frame-based packet-mode scheduling for input-queued switches," *IEEE/ACM Trans. on Networking*, vol. 15, 2007.

[428] K. Lougheed and Y. Rekhter, "A border gateway protocol (BGP)," *IETF RFC 1105*, June 1989. http://www.rfc-editor.org/rfc/rfc1105.txt

[429] K. Lougheed and Y. Rekhter, "A border gateway protocol 3 (BGP-3)," *IETF RFC 1267*, October 1991. http://www.rfc-editor.org/rfc/rfc1267.txt

[430] J. Loughney, "Diameter command codes for third generation partnership project (3GPP) Release 5," *IETF RFC 3589*, September 2003. http://www.rfc-editor.org/rfc/rfc3589.txt

[431] J. Loughney, M. Tuexen (Ed.), and J. Pastor-Balbas, "Security considerations for signaling transport (SIGTRAN) protocols," *IETF RFC 3788*, June 2004. http://www.rfc-editor.org/rfc/rfc3788.txt

[432] K.-S. Lui and K. Nahrstedt, "Topology aggregation and routing in bandwidth-delay sensitive networks," in *Proc. IEEE GLOBECOM'2000*, pp. 410–414, San Francisco, CA, November–December 2000.

[433] C. Lund, S. Philips, and S. Reingold, "Fair prioritized scheduling in an input-buffered switch," in *Proc. International IFIP-IEEE Conference on Broadband Communications*, pp. 358–369, April 1996.

[434] J. Luo, J. Xie, R. Hao, and X. Li, "An approach to accelerate convergence for path vector protocol," in *Proc. IEEE GLOBECOM'2002*, pp. 2390–2394, Taipei, Taiwan, November 2002.

[435] Q. Ma and P. Steenkiste, "Quality-of-service routing for traffic with performance guarantees," in *Proc. IFIP Fifth International Workshop on Quality of Service (IWQoS'97)*, pp. 115–126, New York, May 1997.

[436] Q. Ma and P. Steenkiste, "Supporting dynamic inter-class resource sharing: A multi-class QoS routing algorithm," in *Proc. IEEE INFOCOM'99*, pp. 649–660, New York, March 1999.

[437] A. Magi, A. Szentesi, and B. Szviatovszki, "Analysis of link cost functions for PNNI routing," *Computer Networks*, vol. 34, no. 1, pp. 181–197, July 2000.

[438] P. Mahadevan, D. Krioukov, M. Fomenkov, B. Huffaker, X. Dimitropoulos, k.c. Claffy, and A. Vahdat, "The Internet AS-level topology: Three data sources and one definitive metric," *ACM SIGCOMM Computer Communication Review*, vol. 36, no. 1, pp. 17–26, January 2006.

[439] R. Mahajan, D. Wetherall, and T. Anderson, "Towards coordinated interdomain traffic engineering," in *Proc. 3rd ACM SIGCOMM Workshop on Hot Topics in Networks (HotNets-III)*, San Diego, CA, November 2004.

[440] A. Malis and W. Simpson, "PPP over SONET/SDH," *IETF RFC 2615*, June 1999. http://www.rfc-editor.org/rfc/rfc2615.txt

[441] G. Malkin, "RIP version 2 protocol applicability statement," *IETF RFC 1722*, November 1994. http://www.rfc-editor.org/rfc/rfc1722.txt

[442] G. Malkin, "RIP Version 2," *IETF RFC 2453*, November 1998. http://www.rfc-editor.org/rfc/rfc2453.txt

[443] G. Malkin and R. Minnear, "RIPng for IPv6," *IETF RFC 2080*, January 1997. http://www.rfc-editor.org/rfc/rfc2080.txt

[444] T. Mallory and A.Kullberg, "Incremental updating of the Internet checksum," *IETF RFC 1141*, January 1990, http://www.rfc-editor.org/rfc/rfc1141.txt.

[445] D. A. Maltz, G. Xie, J. Zhan, H. Zhang, G. Hjálmtýsson, and A. Greenberg, "Routing design in operational networks: A look from the inside," in *Proc. ACM SIGCOMM'2004*, pp. 27–40, Portland, OR, August 2004.

[446] P. Manohar, D. Manjunath, and R. K. Shevgaonkar, "Routing and wavelength assignment in optical networks from edge disjoint paths algorithms," *IEEE Communication Letters*, vol. 6, no. 5, pp. 211–213, May 2002.

[447] J. D. Marchland, "Braess's paradox of traffic flow," *Transportation Research*, vol. 4, pp. 391–394, 1970.

[448] P. Marques, R. Bonica, L. Fang, L. Martini, R. Raszuk, K. Patel, and J. Guichard, "Constrained route distribution for border gateway protocol/multiprotocol label switching (BGP/MPLS) internet protocol (IP) virtual private networks (VPNs)," *IETF RFC 4684*, November 2006. http://www.rfc-editor.org/rfc/rfc4684.txt

[449] L. Martini (Ed.), E. Rosen, N. El-Aawar, and G. Heron, "Encapsulation methods for transport of Ethernet over MPLS networks," *IETF RFC 4448*, April 2006. http://www.rfc-editor.org/rfc/rfc4448.txt

[450] M. Mathis, J. Semke, J. Mahdavi, and T. Ott, "The macroscopic behavior of the TCP congestion avoidance algorithm," *ACM SIGCOMM Computer Communication Review*, vol. 27, no. 3, pp. 67–82, July 1997.

[451] M. May, J. Bolot, C. Diot, and B. Lyles, "Reasons not to deploy RED," in *Proc. 7th International Workshop on Quality of Service (IWQoS'99)*, London, UK, 1999.

[452] R. Mazumdar, L. G. Mason, and C. Douligeris, "Fairness in network optimal flow control: optimality of product forms," *IEEE Trans. on Communications*, vol. 39, pp. 775–782, 1991.

[453] A. McAuley and P. Francis, "Fast routing table lookup using cams," in *Proc. IEEE INFOCOM'93*, pp. 1382–1391, 1993.

[454] C. J. McCallum, "An algorithm for finding the *k* shortest paths in a network," *Bell Laboratories Technical Memorandum*, 1973.

[455] S. McCreary and K. C. Claffy, "Trends in wide area IP traffic patterns," Cooperative Association for Internet Data Analysis - CAIDA, Tech. Rep., 2000.

[456] D. McIntosh, "Building a PacketCable network: A comprehensive design for the delivery of VoIP services," in *Society of Cable Telecommunications Engineers (SCTE) Cable Tec-Expo 2002*, 2002. http://www.packetcable.com/downloads/SCTE02_VOIP_Services.pdf

[457] D. McIntosh and M. Stachelek, "VoIP services: PacketCable delivers a comprehensive system," in *National Cable & Telecommunications Association—National Meeting (NCTA 2002)*, New Orleans, LA, 2002. http://www.packetcable.com/downloads/NCTA02_VOIP_Services.pdf

[458] N. McKeown, "Scheduling algorithms for input-queued cell switches," Ph.D. dissertation, University of California, Berkeley, CA, May 1995.

[459] N. McKeown, "A fast switched backplane for a gigabit switched router," *Business Communications Review*, vol. 27, no. 12, pp. 188–201, December 1997.

[460] N. McKeown, "The iSLIP scheduling algorithm for input-queued switches," *IEEE/ACM Trans. on Networking*, vol. 7, no. 2, pp. 188–201, April 1999.

[461] J. McQuillan, "Adaptive routing algorithms for distributed computer networks," BBN Report No. 2831, Bolk Beranek & Newman, May 1974.

[462] J. M. McQuillan, G. Falk, and I. Richer, "A review of the development and performance of the ARPANET routing algorithm," *IEEE Trans. on Communications*, vol. COM-26, pp. 1802–1811, 1978.

[463] J. M. McQuillan, I. Richer, and E. Rosen, "The new routing algorithm for the ARPANET," *IEEE Trans. on Communications*, vol. COM-28, pp. 711–719, 1980.

[464] J. M. McQuillan and D. C. Walden, "The ARPA network design decisions," *Computer Networks*, vol. 1, pp. 243–289, 1977.

[465] D. Medhi, "Traffic restoration design for self-healing networks," *AT&T Bell Laboratories—Technical Memorandum*, 1989.

[466] D. Medhi, "Diverse routing for survivability in a fiber-based sparse network," in *Proc. IEEE ICC'91*, pp. 672–676, Denver, Colorado, June 1991.

[467] D. Medhi, "A unified framework for survivable telecommunications network design," in *Proc. IEEE ICC'92*, pp. 411–415, Chicago, Illinois, June 1992.

[468] D. Medhi, "A unified approach to network survivability for teletraffic networks: Models, algorithms and analysis," *IEEE Trans. on Communications*, vol. 42, pp. 534–548, 1994.

[469] D. Medhi, "Quality of Service (QoS) routing computation with path caching: A framework and network performance," *IEEE Communications Magazine*, vol. 40, no. 12, pp. 106–113, December 2002.

[470] D. Medhi, "Network restoration," in *Handbook of Optimization in Telecommunications*, M. G. C. Resende and P. Pardalos (Eds.), Springer, pp. 801–836, 2006.

[471] D. Medhi and S. Guptan, "Network dimensioning and performance of multi-service, multi-rate loss networks with dynamic routing," *IEEE/ACM Trans. on Networking*, vol. 5, pp. 944–957, 1997.

[472] D. Medhi, S. Jain, D. Shenoy Ramam, S. R. Thirumalasetty, M. Saddi, and F. Summa, "A network management framework for multi-layered network survivability: An overview," in *Proc. IEEE/IFIP Conference on Integrated Network Management (IM'2001)*, pp. 293–296, Seattle, WA, May 2001.

[473] D. Medhi and R. Khurana, "Optimization and performance of network restoration schemes for wide-area teletraffic networks," *Journal of Network and Systems Management*, vol. 3, no. 3, pp. 265–294, 1995.

[474] D. Medhi and C.-T. Lu, "Dimensioning and computational results for wide-area broadband networks with two-level dynamic routing," *IEICE Trans. on Communications*, vol. E80-B, no. 2, pp. 273–281, 1997.

[475] D. Medhi and S. Sankarappan, "Impact of a transmission facility link failure on dynamic call routing circuit-switched networks under various circuit layout policies," *Journal of Network and Systems Management*, vol. 1, pp. 143–169, 1993.

[476] D. Medhi and I. Sukiman, "Admission control and dynamic routing schemes for wide-area broadband networks: Their interaction and network performance," in *Proc. IFIP-IEEE International Conference on Broadband Communications*, pp. 99–110, Montreal, Canada, April 1996.

[477] D. Medhi and I. Sukiman, "Multi-service dynamic QoS routing schemes with call admission control: A comparative study," *Journal of Network & Systems Management*, vol. 8, no. 2, pp. 157–190, June 2000.

[478] D. Medhi and D. Tipper, "Multi-layered network survivability—models, analysis, architecture, framework and implementation: An overview," in *Proc. DARPA Information Survivability Conference and Exposition (DISCEX'2000)*, vol. I, pp. 173–186, Hilton Head Island, South Carolina, USA, January 2000.

[479] D. Medhi and D. Tipper, "Some approaches to solving a multi-hour broadband network capacity design problem with single-path routing," *Telecommunication Systems*, vol. 13, pp. 269–291, 2000.

[480] D. Medhi, A. van de Liefvoort, and C. S. Reece, "Performance analysis of a digital link with heterogeneous multislot traffic," *IEEE Trans. on Communications*, vol. 43, pp. 968–976, March 1995.

[481] J. Medhi, *Stochastic Models in Queueing Theory, 2nd Edition*. Academic Press, 2003.

[482] A. Mekkittikul and N. McKeown, "A starvation-free algorithm for achieving 100% throughput in an input-queued switch," in *Proc. IEEE ICCCN'96*, pp. 226–231, October 1996.

[483] A. Mekkittikul and N. McKeown, "A practical scheduling algorithm to achieve 100% throughput in input-queued switches," in *Proc. IEEE INFOCOM'98*, pp. 792–799, March 1998.

[484] P. Merlin and A. Segall, "A failsafe distributed routing protocol," *IEEE Trans. on Communications*, vol. COM-27, pp. 1280–1288, 1979.

[485] D. Meyer and K. Patel, "BGP-4 protocol analysis," *IETF RFC 4274*, January 2006. http://www.rfc-editor.org/rfc/rfc4274.txt

[486] D. Meyer, J. Schmitz, C. Orange, M. Prior, and C. Alaettinoglu, "Using RPSL in practice," *IETF RFC 2650*, August 1999. http://www.rfc-editor.org/rfc/rfc2650.txt

[487] MFA Forum, "Voice over MPLS bearer transport implementation agreement," MPLS Forum 1.0, July 27, 2001. http://www.mfaforum.org/tech/MPLS1.0.pdf

[488] D. L. Mills, "A maze of twisty, turney passages–routing in the Internet swamp (and other adventures)," last updated May 18, 2005. http://www.cis.udel.edu/mills/database/brief/goat/goat.pdf

[489] D. L. Mills, "Exterior gateway protocol formal specification," *IETF RFC 904*, April 1984. http://www.rfc-editor.org/rfc/rfc904.txt

[490] D. L. Mills, "Autonomous confederations," *IETF RFC 975*, February 1986. http://www.rfc-editor.org/rfc/rfc975.txt

[491] Ministry of Communications and Information Technology, Department of Telecommunications, Government of India, "National numbering plan," April 2003. http://www.dot.gov.in/numbering_plan/nnp2003.pdf

[492] M. Minoux, "Optimal synthesis of a network with non-simultaneous multicommodity flow requirements," in *Studies in Graphs and Discrete Programming*, P. Hansen (Ed.), North-Holland, pp. 269–277, 1981.

[493] M. Minoux, "Discrete cost multicommodity network optimization problems and exact solution methods," *Annals of Operations Research*, vol. 106, pp. 19–46, 2001.

[494] N. F. Mir, *Computer and Communication Networks*. Prentice Hall, 2007.

[495] V. Misra, W.-B. Gong, and D. Towsley, "Fluid-based analysis of a network of AQM routers supporting TCP flows with an application to RED," in *Proc. ACM SIGCOMM'2000*, pp. 151–160, Stockholm, Sweden, August–September 2000.

[496] D. Mitra and J. B. Seery, "Comparative evaluations of randomized and dynamic routing strategies for circuit-switched networks," *IEEE Trans. on Communications*, vol. 39, pp. 102–116, 1990.

[497] P. V. Mockapetris, "Domain names—Concepts and facilities," *IETF RFC 1034*, November 1987, (originally as RFC882, published in November 1983). http://www.rfc-editor.org/rfc/rfc1034.txt

[498] G. Mohan, C. Siva Ram Murthy, and A. K. Somani, "Efficient algorithms for routing dependable connections in WDM optical networks," *IEEE/ACM Trans. on Networking*, vol. 9, pp. 553–566, 2001.

[499] E. F. Moore, "The shortest path through a maze," in *Proc. International Symposium on the Theory of Switching*, pp. 285–292, Cambridge, MA, April 1959.

[500] T. Moore, J. Kosloff, J. Keller, G. Manes, and S. Shenoi, "Signaling system 7 (SS7) network security," in *45th Midwest Symposium on Circuits and Systems (MWSCAS-2002)*, vol. 3, pp. 496–499, August 2002.

[501] K. Morneault, R. Dantu, G. Sidebottom, B. Bidulock, and J. Heitz, "Signaling System 7 (SS7) Message Transfer Part 2 (MTP2)—user adaptation layer," *IETF RFC 3331*, September 2002. http://www.rfc-editor.org/rfc/rfc3331.txt

[502] D. R. Morrison, "PATRICIA—practical algorithm to retrieve information coded in alphanumeric," *Journal of ACM*, vol. 15, pp. 514–534, October 1968.

[503] J. Moy, "The OSPF specification," *IETF RFC 1131*, (available only as postcript or PDF file). http://www.faqs.org/rfc/rfc1131.pdf

[504] J. Moy, *OSPF: Anatomy of An Internet Routing Protocol*. Addison-Wesley, 1998.

[505] J. Moy, "OSPF version 2," *IETF RFC 2328*, April 1998. http://www.rfc-editor.org/rfc/rfc2328.txt

[506] W. Mühlbauer, A. Feldmann, O. Maennel, M. Roughan, and S. Uhlig, "Building an AS-topology model that captures route diversity," in *Proc. ACM SIGCOMM'2006*, Pisa, Italy, September 2006.

[507] A. Mukherjee, "On the dynamics and significance of low frequency components of Internet load," *Journal of Internetworking: Research and Experience*, vol. 5, pp. 163–205, 1994.

[508] A. Mukherjee, L. H. Landweber, and J. C. Strikwerda, "Simultaneous analysis of flow and error control strategies with congestion-dependent errors," in *Proc. ACM SIGMETRICS'90*, pp. 86–95, Boulder, CO, 1990.

[509] B. Mukherjee, *Optical Communication Networks*. McGraw-Hill, 1997.

[510] B. Mukherjee, *Optical WDM Networks*. Springer, 2006.

[511] B. Mukherjee, D. Banerjee, S. Ramamurthy, and A. Mukherjee, "Some principles for designing a wide-area WDM optical network," *IEEE/ACM Trans. on Networking*, vol. 4, pp. 684–696, 1996.

[512] B. Mukherjee and H. Zang, "Survey of State of the Art," in *WDM Optical Networks: Principles and Practice*, K. M. Sivalingam and S. Subramaniam (Eds.), Springer, 2002.

[513] S. Murphy, "BGP security vulnerabilities analysis," *IETF RFC 4272*, January 2006. http://www.rfc-editor.org/rfc/rfc4272.txt

[514] S. Murphy, M. Badger, and B. Wellington, "OSPF with digital signatures," *IETF RFC 2154*, June 1997. http://www.rfc-editor.org/rfc/rfc2154.txt

[515] K. Murty, *Linear Programming*. Wiley, 1983.

[516] R. Nagarajan, "Threshold-based congestion control for SS7 signaling network in the GSM digital cellular network," *IEEE Trans. Vehicular Technology*, vol. 48, pp. 385–396, 1999.

[517] A. Nagarajan (Ed.), "Generic requirements for provider provisioned virtual private networks (PPVPN)," *IETF RFC 3809*, June 2004. http://www.rfc-editor.org/rfc/rfc3809.txt

[518] K. Nahrstedt and S. Chen, "Coexistence of QoS and best effort flows—routing and scheduling," in *Proc. 10th IEEE Tyrrhenian International Workshop on Digital Communications: Multimedia Communications*, Ischia, Italy, September 1998.

[519] Y. Nakagome and H. Mori, "Flexible routing in the global communication network," in *Proc. 7th International Teletraffic Congress (ITC7)*, pp. 426.1–426.8, Stockholm, Sweden, 1973.

[520] K. S. Narendra, "Recent developments in learning automata: Theory and applications," in *Proc. 3rd Yale Workshop on Applications of Adaptive Systems Theory*, vol. 3, pp. 90–99, 1983.

[521] K. S. Narendra, E. A. Wright, and L. G. Mason, "Application of learning automata to telephone traffic routing schemes," *IEEE Trans. on Systems, Man and Cybernatics*, vol. SMC-7, pp. 785–792, 1977.

[522] National Communication Systems, "SMS over SS7," Technical Information Bulletin 03-2, December 2003.

[523] National Emergency Number Association. http://www.nena.org/

[524] S. Nelakuditi, Z.-L. Zhang, R. P. Tsang, and D. H. C. Du, "Adaptive proportional routing: a localized QoS routing approach," *IEEE/ACM Trans. on Networking*, vol. 10, pp. 790–804, 2002.

[525] Network Simulator—ns-2. http://www.isi.edu/nsnam/ns/

[526] NeuStar, "ETNS register: European telephone numbering space." http://www.etns.org/

[527] P. Newman, G. Minshall, and L. Houston, "IP switching and gigabit routers," *IEEE Communications Magazine*, January 1997.

[528] J. Nievergelt, H. Hinterberger, and K. C. Sevcik, "The grid file: An adaptable, symmetric multikey file structure," *ACM Trans. Database Systems*, vol. 9, no. 1, pp. 38–71, 1984.

[529] S. Nilsson and G. Karlsson, "IP-address lookup using LC-tries," *IEEE Journal of Selected Areas in Communications*, vol. 17, pp. 1083–1092, June 1999.

[530] North American Network Operators' Group. http://www.nanog.org/

[531] North American Numbering Plan Administration (NANPA). http://www.nanpa.com/

[532] A. Nucci, B. Schroeder, S. Bhattacharyya, N. Taft, and C. Diot, "IGP link weight assignment for transient link failures," in *Proc. 18th International Teletraffic Congress (ITC18)*, pp. 321–330, Berlin, Germany, September 2003.

[533] T. Oetiker and D. Rand, "MRTG: Multi Router Traffic Grapher." http://www.mrtg.org/

[534] Y. Oie, T. Suda, M. Murata, D. Kolson, and H. Miyahara, "Survey of switching techniques in high speed networks and their performance," in *Proc. IEEE INFOCOM'90*, pp. 1242–1251, June 1990.

[535] Open Network Laboratory. http://onl.arl.wustl.edu/

[536] OPNET. http://www.opnet.com/

[537] D. Oran, "OSI IS-IS intra-domain routing protocol," *IETF RFC 1142*, February 1990, (re-publication of [321]). http://www.rfc-editor.org/rfc/rfc1142.txt

[538] S. Orlowski and R. Wessäly, "Comparing restoration concept using optimal network configurations with integrated hardware and routing decisions," in *Proc. Design of Reliable Communication Networks (DRCN'2003)*, pp. 15–22, Banff, Canada, 2003.

[539] T. J. Ott and K. R. Krishnan, "State dependent routing of telephone traffic and the use of separable routing schemes," in *Proc. 11th International Teletraffic Congress (ITC13)*, pp. 5.1.A.5.1–5.1A.5.6, Kyoto, Japan, 1985.

[540] A. Ozdaglar and D. Berteskas, "Routing and wavelength assignment in optical networks," *IEEE/ACM Trans. on Networking*, vol. 11, no. 2, pp. 259–272, April 2003.

[541] P. Pan, G. Swallow, and A. Atlas (Eds.), "Fast reroute extensions to RSVP-TE for LSP tunnels," *IETF RFC 4090*, May 2005. http://www.rfc-editor.org/rfc/rfc4090.txt

[542] L. Papadimitriou (Ed.), "Generalized multi-protocol label switching (GMPLS) signaling extensions for G.709 optical transport networks control," *IETF RFC 4328*, January 2006. http://www.rfc-editor.org/rfc/rfc4328.txt

[543] A. K. Parekh and R. G. Gallager, "A generalized processor sharing approach to flow control in integrated services networks: The single node case," *IEEE/ACM Trans. on Networking*, vol. 1, pp. 344–357, June 1993.

[544] A. K. Parekh and R. G. Gallager, "A generalized processor sharing approach to flow control in integrated services networks: The multiple node case," *IEEE/ACM Trans. on Networking*, vol. 2, pp. 137–150, April 1994.

[545] K. Park and W. Willinger (Ed.), *Self-Similar Network Traffic and Performance Evaluation*. Wiley-Interscience, 2000.

[546] W. R. Parkhurst, *Cisco BGP-4 Command & Configuration Handbook*. Cisco Press, 2001.

[547] C. Partridge, P. P. Carvey, E. Burgess, I. Castineyra, T. Clarke, L. Graham, M. Hathaway, P. Herman, A. K. S. Kohalmi, T. Ma, J. Mcallen, T. Mendez, W. C. Milliken, R. Pettyjohn, J. Rokosz, J. Seeger, M. Sollins, S. Storch, B. Tober, G. D. Troxel, D. Waitzman, and S. Winterble, "A 50-Gb/s IP router," *IEEE/ACM Trans. on Networking*, vol. 6, pp. 237–248, 1998.

[548] C. Partridge, *Gigabit Networking*. Addison-Wesley, 1994.

[549] M. M. B. Pascoal, V. Captivo, and J. C. N. Climaco, "An algorithm for ranking quickest simple paths," *Computers & Operations Research*, vol. 32, pp. 509–520, 2005.

[550] V. Paxson, "End-to-end routing behavior in the Internet," *IEEE/ACM Trans. on Networking*, vol. 5, pp. 601–615, 1997.

[551] V. Paxson and S. Floyd, "Wide-area traffic: The failure of Poisson modeling," *IEEE/ACM Trans. on Networking*, vol. 3, pp. 226–244, 1995.

[552] Peering Database. http://www.peeringdb.com/

[553] D. Pei, M. Azuma, D. Massey, and L. Zhang, "BGP-RCN: Improving BGP convergence through root cause notification," *Computer Networks*, vol. 48, no. 2, pp. 175–194, 2005.

[554] D. Pei, B. Zhang, D. Massey, and L. Zhang, "An analysis of convergence delay in path vector routing protocols," *Computer Networks*, vol. 50, no. 3, pp. 398–421, 2006.

[555] D. Pei, X. Zhao, L. Wang, D. Massey, A. Mankin, F. Wu, and L. Zhang, "Improving BGP convergence through assertions approach," in *Proc. IEEE INFOCOM'2002*, 2002.

[556] Performance Technologies, Inc., "SS7/IP Interworking Tutorial," 2001. http://www.pt.com/tutorials/iptelephony/

[557] Performance Technologies, Inc., "Tutorial on Signaling System 7 (SS7)," 2003. http://www.pt.com/tutorials/ss7/

[558] R. Perlman, "What is the fundamental difference between OSPF and IS-IS," August 31, 2002. http://archives.neohapsis.com/archives/microsoft/various/ospf/2002-q3/0303.html

[559] R. Perlman, *Interconnections, 2nd Edition*. Addison-Wesley, 2000.

[560] H. G. Perros, *Connection-Oriented Networks: SONET/SDH, ATM, MPLS and Optical Networks*. John Wiley & Sons, 2005.

[561] J. Peterson, "enumservice registration for session initiation protocol (SIP) addresses-of-record," *IETF RFC 3764*, April 2004. http://www.rfc-editor.org/rfc/rfc3764.txt

[562] L. Peterson and B. Davie, *Computer Networks—A Systems Approach, 4th Edition*. Morgan Kaufmann Publishers, 2007.

[563] M. Pickavet and P. Demeester, "Long-term planning of WDM networks: A comparison between single-period and multi-period techniques," *Photonic Network Communications*, vol. 1, no. 4, pp. 331–346, December 1999.

[564] M. Pióro and D. Medhi, *Routing, Flow, and Capacity Design in Communication and Computer Networks*. Morgan Kaufmann Publishers, 2004.

[565] M. Pióro, A. Szentesi, J. Harmatos, A. Jüttner, P. Gajowniczek, and S. Kozdrowski, "On open shortest path first related network optimization problems," in *Proc. IFIP ATM IP 2000*, Ilkley, England, July 2000, (see also [566]).

[566] M. Pióro, A. Szentesi, J. Harmatos, A. Jüttner, P. Gajowniczek, and S. Kozdrowski, "On open shortest path first related network optimization problems," *Performance Evaluation*, vol. 48, pp. 201–223, 2002, (see [565] for a preliminary version).

[567] D. M. Piscitello and A. L. Chapin, *Open Systems Networking: TCP/IP and OSI*. Addison-Wesley, 1993.

[568] Planetlab. http://www.planet-lab.org/

[569] PMC-Sierra, Inc., "A new architecture for switch and router design," v2r1, December 22, 1999. http://www.pmc-sierra.com/pressRoom/pdf/lcs_wp.pdf

[570] B. Prabhakar, N. McKeown, and R. Ahuja, "Multicast scheduling for input-queued switches," *IEEE Journal of Selected Areas in Communications*, vol. 15, no. 5, pp. 855–866, 1997.

[571] R. Pužmanová, *Routing and Switching*. Addison-Wesley, 2002.

[572] L. Qiu, G. Varghese, and S. Suri, "Fast firewall implementations for software-based and hardware-based routers," in *Proc. ACM SIGMETRICS'2001*, pp. 344–345, 2001.

[573] B. Rajagopalan, J. Luciani, and D. Awduche, "IP over optical networks: A framework," *IETF RFC 3717*, March 2004. http://www.rfc-editor.org/rfc/rfc3717.txt

[574] K. G. Ramakrishnan and M. A. Rodrigues, "Optimal routing in shortest-path data networks," *Bell Labs Technical Journal*, vol. 6, no. 1, pp. 117–138, 2001.

[575] K. K. Ramakrishnan and R. Jain, "A binary feedback scheme for congestion avoidance in computer networks with a connectionless network layer," in *Proc. ACM SIGCOMM'88*, pp. 303–313, August 1988.

[576] R. Ramamurthy and A. Ramakrishnan, "Virtual topology reconfiguration of wavelength-routed optical WDM networks," in *Proc. of IEEE GLOBECOM'2000*, pp. 1269–1275, San Francisco, CA, November 2000.

[577] K. Ramasamy, "Efficient storage and query processing of set-valued attributes," Ph.D. dissertation, University of Wisconsin, Madison, WI, 2001.

[578] K. Ramasamy, J. M. Patel, R. Kaushik, and J. F. Naughton, "Set containment joins: The good, the bad and the ugly," in *Proc. 26th International Conference on Very Large Databases (VLDB)*, September 2000.

[579] R. Ramaswami and K. N. Sivarajan, "Design of logical topologies for wavelength-routed networks," *IEEE Journal of Selected Areas in Communication*, vol. 14, no. 5, pp. 840–851, June 1996.

[580] R. Ramaswami and K. N. Sivarajan, *Optical Networks: A Practical Perspective, 2nd Edition*. Morgan Kaufmann Publishers, 2002.

[581] U. Ranadive and D. Medhi, "Some observations on the effect of route fluctuation and network link failure on TCP," in *Proc. 10th IEEE International Conference on Computer Communications and Networks (ICCCN'01)*, pp. 460–467, Scottsdale, AZ, October 2001.

[582] Y. Rapp, "Planning of a junction network in a multi-exchange area I," *General Principles Ericsson Tech.*, vol. 20, no. 1, pp. 77–130, 1964.

[583] Y. Rapp, "Planning of a junction network in a multi-exchange area II," *Extensions of the Principles and Applications. Ericsson Tech.*, vol. 21, no. 2, pp. 187–240, 1965.

[584] Y. Rapp, "Planning of a junction network in a multi-exchange area III," in *Proc. 5th International Teletraffic Congress (ITC5)*, New York, 1967.

[585] A. Rattanadilokochai, "QoS routing with inaccurate link-state information," M.S. Thesis, University of Missouri–Kansas City, May 2000.

[586] V. C. Ravikumar, R. Mahapatra, and J. C. Liu, "Modified LC-trie based efficient routing lookup," in *Proc. 10th IEEE International Symposium on Modeling, Analysis, and Simulation of Computer and Telecommunication Systems (MASCOTS'02)*, 2002.

[587] Y. Rekhter, "EGP and policy based routing in the new NSFNET backbone," *IETF RFC 1092*, February 1989. http://www.rfc-editor.org/rfc/rfc1092.txt

[588] Y. Rekhter, R. Bonica, and E. Rosen, "Use of provider edge to provider edge (PE-PE) generic routing encapsulation (GRE) or IP in BGP/MPLS IP virtual private networks," *IETF RFC 4797*, January 2007. http://www.rfc-editor.org/rfc/rfc4797.txt

[589] Y. Rekhter, S. Hotz, and D. Estrin, "Constraints on forming clusters with link-state hop-by-hop routing," University of Southern California, Tech. Rep. 93-536, August 1993.

[590] Y. Rekhter and T. Li, "A Border Gateway Protocol 4 (BGP-4)," *IETF RFC 1771*, March 1995, (Made obsolete by [591]). http://www.rfc-editor.org/rfc/rfc1771.txt

[591] Y. Rekhter, T. Li, and S. Hares, "A Border Gateway Protocol 4 (BGP-4)," *IETF RFC 4271*, January 2006. http://www.rfc-editor.org/rfc/rfc4271.txt

[592] Y. Rekhter, B. Moskowitz, D. Karrenberg, G. J. de Groot, and E. Lear, "Address allocation for private internets," *IETF RFC 1918*, February 1996. http://www.rfc-editor.org/rfc/rfc1918.txt

[593] Y. Rekhter (Ed.), "BGP protocol analysis," *IETF RFC 1265*, October 1991. http://www.rfc-editor.org/rfc/rfc1265.txt

[594] J. Rexford, Personal Communication, 2005.

[595] J. Rexford, "Route optimization in IP networks," in *Handbook of Optimization in Telecommunications*, M. G. C. Resende and P. Pardalos (Eds.), Springer, 2006.

[596] R. F. Rey (Ed.), *Engineering and Operations in the Bell System, 2nd Edition*. AT&T Bell Laboratories, 1983.

[597] J. W. Roberts, "A service system with heterogeneous user requirements: application to multi-services telecommunications systems," in *Performance of Data Communication Systems, and Their Applications*, G. Pujolle (Ed.), North-Holland, pp. 423–431, 1981.

[598] G. Rogers, D. Medhi, W.-J. Hsin, S. Muppala, and D. Tipper, "Performance analysis of multicast and priority-based routing under a failure in differentiated-services Internet," in *Proc. of IEEE MILCOM'99*, pp. 897–901, Atlantic City, NJ, October 1999.

[599] E. Rosen, "The updating protocol of ARPANET's new routing algorithm," *Computer Networks*, vol. 4, no. l, pp. l–19, 1980.

[600] E. Rosen, "Exterior gateway protocol (EGP)," *IETF RFC 827*, October 1982. http://www.rfc-editor.org/rfc/rfc827.txt

[601] E. Rosen and Y. Rekhter, "BGP/MPLS VPNs," *IETF RFC 2547*, March 1999, (Made obsolete by [602]). http://www.rfc-editor.org/rfc/rfc2547.txt

[602] E. Rosen and Y. Rekhter, "BGP/MPLS IP virtual private networks (VPNs)," *IETF RFC 4364*, February 2006. http://www.rfc-editor.org/rfc/rfc4364.txt

[603] E. Rosen, D. Tappan, G. Fedorkow, Y. Rekhter, D. Farinacci, T. Li, and A. Conta, "MPLS label stack encoding," *IETF RFC 3032*, January 2001. http://www.rfc-editor.org/rfc/rfc3032.txt

[604] E. Rosen, A. Viswanathan, and R. Callon, "Multiprotocol label switching architecture," *IETF RFC 3031*, January 2001. http://www.rfc-editor.org/rfc/rfc3031.txt

[605] E. Rosen, "Vulnerabilities of network control protocols: An example," *IETF RFC 789,* July 1981. http://www.rfc-editor.org/rfc/rfc789.txt

[606] G. Rosenbaum, C.-T. Chou, S. Jha, and D. Medhi, "Dynamic routing of restorable QoS connections in MPLS networks," in *Proc. 30th IEEE Conference on Local Computer Networks(LCN)*, Sydney, Australia, November 2005.

[607] M. Roughan, J. Li, R. Bush, Z. Mao, and T. Griffin, "Is BGP update storm a sign of trouble: Observing the Internet control and data planes during Internet worms," in *Proc. International Symposium on Performance Evaluation of Computer and Telecommunication Systems (SPECTS)*, Calgary, Canada, 2006.

[608] T. Roughgarden and E. Tardos, "How bad is selfish routing?" *Journal of the ACM*, vol. 49, no. 2, pp. 236–259, March 2002, (A preliminary version of this paper appeared in *Proc. 41st Annual IEEE Symposium on Foundations of Computer Science*, November 2000).

[609] G. Rouskas and M. Ammar, "Dynamic reconfiguration in multihop WDM networks," *Journal of High Speed Networks*, vol. 4, pp. 221–238, 1995.

[610] H. Rudin, "On routing and "delta routing": A taxonomy and performance comparison of techniques for packet-switched networks," *IEEE Trans. on Communications*, vol. 24, pp. 43–59, January 1976.

[611] M. A. Ruiz-Sanchez, E. W. Biersack, and W. Dabbous, "Survey and taxonomy of IP address lookup algorithms," *IEEE Network*, vol. 15, no. 2, pp. 8–23, March/April 2001.

[612] M. P. Rumsewicz and D. E. Smith, "A comparison of SS7 congestion control options during mass call-in situations," *IEEE/ACM Trans. on Networking*, vol. 3, pp. 1–9, 1995.

[613] T. Russell, *Signaling System # 7, 4th Edition*. McGraw-Hill, 2002.

[614] R. Sabella, P. Iovanna, G. Oriolo, and P. D'Aprile, "Routing and grooming of data flows into lightpaths in new generation network based on the GMPLS paradigm," *Photonic Network Communications*, vol. 7, no. 2, pp. 131–144, 2004.

[615] R. Sabella, M. Settembre, G. Oriolo, F. Razza, F. Ferlito, and G. Conte, "A multi-layer solution for path provisioning in new-generation optical/MPLS networks," *IEEE Journal on Lightwave Technology*, vol. 21, no. 5, pp. 1141–1155, 2003.

[616] S. Sahni and K. Kim, "Efficient construction of fixed-stride multibit tries for IP lookup," in *Proc. 8th IEEE Workshop on Future Trends of Distributed Computing Systems*, 2001.

[617] S. Sahni and K. Kim, "Efficient construction of variable-stride multibit tries for IP lookup," in *Proc. IEEE Symposium on Applications and the Internet (SAINT)*, pp. 220–227, 2002.

[618] J. Saltzer, "On the naming and binding of network destinations," *IETF RFC 1498*, August 1993. http://www.rfc-editor.org/rfc/rfc1498.txt

[619] H. Samet, "The Quadtree and related hierarchical data structures," *ACM Computing Surveys*, vol. 16, no. 2, pp. 187–260, 1984.

[620] S. Sangli, E. Chen, R. Fernando, J. Scudder, and Y. Rekhter, "Graceful restart mechanism for BGP," *IETF RFC 4724*, January 2007. http://www.rfc-editor.org/rfc/rfc4724.txt

[621] S. Sangli, D. Tappan, and Y. Rekhter, "BGP extended communities attribute," *IETF RFC 4360*, February 2006. http://www.rfc-editor.org/rfc/rfc4360.txt

[622] V. Sarangan, D. Ghosh, and R. Acharya, "Capacity-aware state aggregation for interdomain QoS routing," *IEEE Trans. on Multimedia*, vol. 8, pp. 792–808, 2006.

[623] L. J. Seamonson and E. Rosen, ""Stub" exterior gateway protocol," *IETF RFC 888*, January 1984. http://www.rfc-editor.org/rfc/rfc888.txt

[624] R. Sedgewick and R. Flajolet, *An Introduction to the Analysis of Algorithms*. Addison-Wesley, 1996.

[625] R. Sedgewick, *Algorithms in Java, Part 5: Graph Algorithms, 3rd Edition*. Addison-Wesley, 2004.

[626] S. Seetharaman and M. Ammar, "On the interaction between dynamic routing in the native and overlay layers," in *Proc. IEEE INFOCOM'2006*, Barcelona, Spain, April 2006.

[627] A. Segal, "Advances in verifiable fail-safe routing procedures," *IEEE Trans. on Communications*, vol. COM-29, pp. 491–497, 1981.

[628] T. K. Sellis, N. Roussopoulos, and C. Faloutsos, "The R+−tree: A dynamic index for multi-dimensional objects," in *Proc. 13th International Conference on Very Large Data Bases (VLDB)*, pp. 507–518, Brighton, England, 1987.

[629] C. Semeria, "Internet backbone routers and evolving internet design," in *Juniper Networks White Paper*, 1997.

[630] C. Semeria, "Internet processor II ASIC: Rate limiting and traffic-policing features," in *Juniper Networks White Paper, Part Number 200005-001*, 2000.

[631] C. Semeria, "T-series routing platforms: System and packet forwarding architecture," in *Juniper Networks White Paper, Part Number 200027-001*, 2002.

[632] C. Semeria, "RFC 2547bis: BGP/MPLS VPN fundamentals," in *Part Number 200012-001 03/01, Juniper network*, 2001. http://www.juniper.net/solutions/literature/white_papers/200012.pdf

[633] S. Sen, R. D. Doverspike, and M. S. Dunatunga, "Unified facilities optimizer," University of Arizona, Department of Systems and Industrial Engineering, Tech. Rep., January 1989.

[634] T. Seth, A. Broscius, C. Huitema, and H. P. Lin, "Performance requirements for signaling in Internet telephony," November 1998, Internet Draft. http://www.cs.columbia.edu/sip/drafts/draft-seth-sigtran-req-00.txt

[635] S. A. Shah and D. Medhi, "Performance under a failure of wide-area datagram networks with unicast and multicast traffic routing," in *Proc. IEEE MILCOM'98*, pp. 939–945, Bedford, MA, October 1998.

[636] A. Shaikh, A. Varma, L. Kalampoukas, and R. Dube, "Routing stability in congested networks: Experimentation and analysis," in *Proc. ACM SIGCOMM'2000*, pp. 163–174, Stockholm, Sweden, August–September 2000.

[637] S. Shenker, C. Partridge, and R. Guerin, "Specification of guaranteed quality of service," *IETF RFC 2212*, September 1997. http://www.rfc-editor.org/rfc/rfc2212.txt

[638] K. G. Shin and M. Chen, "Performance analysis of distributed routing strategies free of ping-pong-type looping," *IEEE Trans. on Computers*, vol. 36, pp. 129–137, 1987.

[639] J. Shoch, "Inter-network naming, addressing, and routing," in *Proc. IEEE Computer Conference (COMPCON)*, pp. 72–79, Washington, DC, 1978.

[640] M. Shreedhar and G. Varghese, "Efficient fair queueing using deficit round robin," in *Proc. ACM SIGCOMM'95*, pp. 231–242, Cambridge, MA, August–September 1995.

[641] S. Sibal and A. DeSimone, "Controlling alternate routing in general-mesh packet flow networks," in *Proc. ACM SIGCOMM'94*, pp. 168–179, London, United Kingdom, 1994.

[642] G. Sidebottom, K. Morneault, and J. Pastor-Balbas (Eds.), "Signaling System 7 (SS7) Message Transfer Part 3 (MTP3)—User Adaptation Layer (M3UA)," *IETF RFC 3332*, September 2002. http://www.rfc-editor.org/rfc/rfc3332.txt

[643] S. Singh, F. Baboescu, G. Varghese, and J. Wang, "Packet classification using multidimensional cutting," in *Proc. ACM SIGCOMM'2003*, pp. 213–224, Karlsruhe, Germany, August 2003.

[644] R. J. Sivasankar, S. Ramam, S. Subramaniam, T. S. Rao, and D. Medhi, "Some studies on the impact of dynamic traffic in QoS based dynamic routing environment," in *Proc. IEEE ICC'2000*, pp. 959–963, New Orleans, LA, June 2000.

[645] K. Sklower, "A tree-based packet routing table for berkeley unix," in *Proc. 1991 Usenix Winter Conference*, pp. 93–99, 1991.

[646] R. A. Skoog, "Engineering common channel signaling networks for ISDN," in *Proc. 12th International Teletraffic Congress (ITC12)*, pp. 915–921, Torin, Italy, 1988.

[647] D. Slepian, "Two theorems on a particular crossbar switching networks," 1952, unpublished manuscript.

[648] D. L. Slotnick, W. C. Borck, and R. C. McReynolds, "The soloman computer," in *Proc. AFIPS Sprint Joing Computer Conference*, vol. 22, pp. 97–107. Spartan Books, 1967.

[649] H. Smit and T. Li, "Intermediate System to Intermediate System (IS-IS) extensions for traffic engineering (TE)," *IETF RFC 3784*, June 2004. http://www.rfc-editor.org/rfc/rfc3784.txt

[650] B. R. Smith and J. J. Garcia-Luna-Aceves, "Efficient policy-based routing without virtual circuits," in *Proc. First International Conference on Quality of Service in Heterogeneous Wired/Wireless Networks (QSHINE'04)*, pp. 242–251, 2004.

[651] B. R. Smith and J. J. Garcia-Luna-Aceves, "A new approach to policy-based routing in the internet," in *Performance Evaluations and Planning Methods for the Next Generation Internet*, A. Girard, B. Sansò, and F. Vázquez-Abad (Eds.), Springer, pp. 99–124, 2005.

[652] SMS/800 Management Team, "Introduction to toll free services." http://www.sms800.com/

[653] R. Souza, P. Krishnakumar, C. Ozveren, R. Simcoe, B. Spinney, R. Thomas, and R.Walsh, "GIGAswitch: A high performance packet switching platform," *Digital Technical Journal*, vol. 27, no. 1, pp. 9–22, 1994.

[654] E. Spitznagel, D. Taylor, and J. Turner, "Packet classification using extended TCAMs," in *Proc. 11th IEEE International Conference on Network Protocols (ICNP)*, pp. 120–131, Washington, DC, 2003.

[655] E. W. Spitznagel, "Compressed data structures for recursive flow classification," Department of Computer Science and Engineering, Washington University, St. Louis, MO, Tech. Rep. WUCSE-2003-65, 2003.

[656] Sprintlink's BGP Policy. http://www.sprintlink.net/policy/bgp.html

[657] P. R. Srikantakumar, "Learning models and adaptive routing in telecommunication networks," Ph.D. dissertation, Yale University, New Haven, CT, 1980.

[658] V. Srinivasan, "Fast and efficient Internet lookups," Ph.D. dissertation, Washington University, Saint Louis, MO, August 1999.

[659] V. Srinivasan, "A packet classification and filter management system," in *Proc. IEEE INFOCOM'2001*, pp. 1464–1473, 2001.

[660] V. Srinivasan, S. Suri, and G. Varghese, "Packet classification using tuple space search," in *Proc. ACM SIGCOMM'99*, pp. 135–146, Cambridge, MA, 1999.

[661] V. Srinivasan and G. Varghese, "Fast address lookups using controlled prefix expansion," *ACM Trans. on Computer Systems*, vol. 17, pp. 1–40, February 1999.

[662] V. Srinivasan, G. Varghese, S. Suri, and M. Waldvogel, "Fast and scalable layer four switching," in *Proc. ACM SIGCOMM'98*, pp. 191–202, Vancouver, Canada, August 1998.

[663] S. Srivastava, G. Agrawal, and D. Medhi, "Dual-based link weight determination towards single shortest path solutions for OSPF networks," in *Proc. 19th International Teletraffic Congress (ITC19)*, pp. 829–838, Beijing, China, August–September 2005.

[664] S. Srivastava, G. Agrawal, M. Pióro, and D. Medhi, "Determining link weight system under various objectives for OSPF networks using a Lagrangian relaxation-based approach," *IEEE eTrans. on Network & Service Management*, vol. 2, no. 1, pp. 9–18, 2005.

[665] S. Srivastava, S. R. Thirumalasetty, and D. Medhi, "Network traffic engineering with varied levels of protection in the next generation internet," in *Performance Evaluations and Planning Methods for the Next Generation Internet*, A. Girard, B. Sansò, and F. Vázquez-Abad (Eds.), Springer, pp. 99–124, 2005.

[666] W. Stallings, *Data and Computer Communications, 8th Edition*. Prentice-Hall, 2007.

[667] M. Steenstrup (Ed.), *Routing in Communications Networks*. Prentice-Hall, 1995.

[668] W. R. Stevens, *TCP/IP Illustrated, Volume 1*. Addison-Wesley, 1994.

[669] J. W. Stewart III, *BGP4: Inter-Domain Routing in the Internet*. Addison-Wesley, 1999.

[670] D. Stiliadis and A. Varma, "Frame-based fair queueing: a new traffic scheduling algorithm for packet-switched networks," in *Proc. ACM SIGMETRICS'96*, pp. 104–115, 1996.

[671] J. Stokes, "PCI express: An overview," in *Arstechnica Web Site*, July 2004. http://arstechnica.com/articles/paedia/hardware/pcie.ars/1

[672] M. Stonebraker, "The case for shared nothing," *Database Engineering Bulletin*, vol. 9, no. 1, pp. 4–9, 1986.

[673] J. Strand, "Transport networks & technologies," 2001, Tutorial presentation, Optical Fiber Communication Conference (OFC 2001).

[674] J. Strand and A. Chiu, (Eds.), "Impairments and other constraints on optical layer routing," *IETF RFC 4054*, May 2005. http://www.rfc-editor.org/rfc/rfc4054.txt

[675] J. Strand, R. Doverspike, and G. Li, "Importance of wavelength conversion in an optical network," *Optical Networks Magazine*, vol. 2, no. 3, pp. 33–44, May/June 2001.

[676] H. Sullivan and T. R. Bashkow, "A large scale, homogeneous, fully distributed parallel machine," in *Proc. International Symposium on Computer Architecture (ISCA)*, pp. 105–124, March 1977.

[677] J. W. Suurballe, "Disjoint paths in a network," *Networks*, vol. 4, pp. 125–145, 1974.

[678] J. W. Suurballe and R. E. Tarjan, "A quick method for finding shortest pairs of disjoint paths," *Networks*, vol. 14, pp. 325–336, 1984.

[679] Syniverse Technologies, "A global perspective on number portability," May 2004. http://www.syniverse.com/pdfs/MNPReport.pdf

[680] E. Szybicki and M. Lavigne, "Alternate routing for a telephone system," U.S. Patent No. 4,284,852, August 18, 1981.

[681] Y. Tamir and H. C. Chi, "Symmetric crossbar arbiters for vlsi communication switches," *IEEE Trans. Parallel and Distributed Systems*, vol. 4, no. 1, pp. 13–27, January 1993.

[682] Y. Tamir and G. Frazier, "High performance multi-queue buffers for VLSI communication networks," in *Proc. 15th Annual Symposium on Computer Architecture*, pp. 343–354, June 1988.

[683] A. Tanenbaum, *Computer Networks, 4th Edition*. Prentice-Hall, 2003.

[684] A. Tang and S. Scoggins, *Open Networking with OSI*. Prentice-Hall, 1992.

[685] J. Tapolcai, T. Cinkler, and A. Recski, "On-line routing algorithms with shared protection in WDM networks," in *IFIP ONDM 2003, Budapest*, 2003.

[686] D. E. Taylor, "Survey and taxonomy of packet classification techniques," *ACM Computing Surveys*, vol. 37, no. 3, pp. 238–275, 2005.

[687] D. E. Taylor and J. S. Turner, "Scalable packet classification using distributed crossproducting of field labels," in *Proc. IEEE INFOCOM'2005*, pp. 269–280, March 2005.

[688] R. Teixeira, A. Shaikh, T. Griffin, and G. Voelker, "Network sensitivity to hot-potato disruptions," in *Proc. ACM SIGCOMM'2004*, pp. 231–244, Portland, OR, August–September 2004.

[689] Telcordia, "Telcordia Technologies specification of Signalling System number 7," GR-246-CORE, Issue 10, December 2005.

[690] Telecom Regulatory Authority of India, "Consultation paper on mobile number portability," July 2005. http://www.trai.gov.in/conpaper22jul05.pdf

[691] Telecom Regulatory Authority of India, "Recommendation on mobile number portability," March 2006. http://www.trai.gov.in/recomm8mar06.pdf

[692] TeleGeography, *Hubs and Spokes: A TeleGeography Internet Reader*. TeleGeography, Inc., 2000.

[693] TeleGeography, "Global communication submarine map," 2004. http://www.telegeography.com/products/map_cable/index.php

[694] S. Thiagarajan and A. K. Somani, "Optimal wavelength converter placement in arbitrary topology wavelength-routed networks," *Computer Communications*, vol. 26, pp. 975–985, 2003.

[695] S. R. Thirumalasetty and D. Medhi, "On the performance and behavior of QoS routing schemes," Technical Report, University of Missouri–Kansas City, 2000.

[696] K. Thompson, G. Miller, and R. Wilder, "Wide-area Internet traffic patterns and characteristics," *IEEE Network*, vol. 11, no. 6, pp. 10–23, November/December 1997.

[697] R. A. Thompson, *Telephone Switching Systems*. Artech House, 2000.

[698] M. Thorup and M. Roughan, "Avoiding ties in shortest path first routing," Technical Report, AT&T Labs-Research, 2003.

[699] F. Tobagi, "Fast packet switching architectures for broadband integrated services digital networks," *Proceedings of the IEEE*, vol. 78, pp. 133–167, 1990.

[700] D. M. Tow, "Network management—recent advances and future trends," *IEEE Journal on Selected Areas in Communications*, vol. 6, no. 4, pp. 732–741, 1988.

[701] P. Traina, "Autonomous system confederations for BGP," *IETF RFC 1965*, June 1996. http://www.rfc-editor.org/rfc/rfc1965.txt

[702] P. Traina, D. McPherson, and J. Scudder, "Autonomous system confederations for BGP," *IETF RFC 3065*, February 2001. http://www.rfc-editor.org/rfc/rfc3065.txt

[703] P. Traina (Ed.), "BGP-4 protocol analysis," *IETF RFC 1774*, March 1995. http://www.rfc-editor.org/rfc/rfc1774.txt

[704] J. Tünnissen, "BGP: the Border Gateway Protocol: Advanced Internet routing resources." http://www.bgp4.as/

[705] J. Turner and N. Yamanaka, "Architectural choices in large scale ATM switches," *IEICE Trans. on Communications*, vol. E81-B, no. 2, pp. 120–137, 1998.

[706] J. S. Turner, "New directions in communications (or which way to the information age)," *IEEE Communications Magazine*, vol. 24, no. 10, pp. 8–15, October 1986.

[707] University of Oregon Route Views Project, "Route views." http://www.routeviews.org/

[708] L. G. Valiant and G. J. Brebner, "Universal schemes for parallel communication," in *Proc. ACM Symposium of the Theory of Computing*, pp. 263–277, May 1981.

[709] I. van Beijnum, *BGP*. O'Reilly, 2002.

[710] P. van Mieghem, F. A. Kuipers, T. Korkmaz, M. Krunz, M. Curado, E. Monteiro, X. Masip-Bruin, J. Solé-Pareta, and S. Sánchez-López, "Quality of service routing," in *Quality of Future Internet Services: COST Action 263 Final Report* M. Smirnov et al. (Eds.), Springer, 2003.

[711] R. J. Vanderbei, *Linear Programming: Foundations and Extensions, 2nd Edition*. Kluwer Academic Publishers, 2001.

[712] G. Varghese, *Network Algorithmics*. Morgan Kaufmann Publishers, 2005.

[713] J.-P. Vasseur, M. Pickavet, and P. Demeester, *Network Recovery: Protection and Restoration of Optical, SONET-SDH, IP, and MPLS*. Morgan Kaufmann Publishers, 2004.

[714] J.-P. Vasseur (Ed.), Y. Ikejiri, and R. Zhang, "Reoptimization of multiprotocol label switching (MPLS) traffic engineering (TE) loosely routed label switched path (LSP)," *IETF RFC 4736*, November 2006. http://www.rfc-editor.org/rfc/rfc4736.txt

[715] A. Vemuri and J. Peterson, "Session initiation protocol for telephones (SIP–T): Context and architectures," *IETF RFC 3372*, September 2002. http://www.rfc-editor.org/rfc/rfc3372.txt

[716] VeriSign, "Intelligent database services." http://www.verisign.com/products-services/communications-services/intelligent-database-services/index.html

[717] VeriSign, "Future-proofing LNP architecture," 2004. http://www.verisign.com/stellent/groups/public/documents/white_paper/001949.pdf

[718] Verizon Business Policy for Settlement-Free Interconnection with Internet Networks. http://www.verizonbusiness.com/uunet/peering/

[719] D. Verma, *Supporting Service Level Agreements on IP Networks*. Macmillan Technical Publishing, 1999.

[720] C. Villamizar, R. Chandra, and R. Govindan, "BGP route flap damping," *IETF RFC 2439*, November 1998. http://www.rfc-editor.org/rfc/rfc2439.txt

[721] C. Villamizar and C. Song, "High performance TCP in ANSNET," *Computer Communications Review*, vol. 24, no. 5, pp. 45–60, October 1994.

[722] VINI—A Virtual Network Environment. http://www.vini-veritas.net/

[723] K. Vinodkrishnan, N. Chandhok, A. Durresi, R. Jain, R. Jagannathan, and S. Seetharaman, "Survivability in IP over WDM networks," *Journal of High Speed Networks*, vol. 10, no. 2, pp. 79–90, 2001.

[724] D. C. Walden, "Routing," BBN Memorandum, Bolk Beranek & Newman, June 1972.

[725] D. C. Walden, "The Bellman–Ford algorithm and "distributed Bellman–Ford algorithm"," 2003. http://www.walden-family.com/public/bf-history.pdf

[726] M. Waldvogel, "Fast longest prefix matching: Algorithms, analysis and applications," Ph.D. dissertation, Swiss Federal Institute of Technology, Zurich, 2002.

[727] M. Waldvogel, G. Varghese, J. Turner, and B. Plattner, "Scalable high-speed prefix matching," *ACM Trans. on Computer Systems*, vol. 19, pp. 400–482, November 2001.

[728] L. Wang, X. Zhao, D. Pei, R. Bush, D. Massey, A. Mankin, S. F. Wu, and L. Zhang, "Observation and analysis of BGP behavior under stress," in *Proc. ACM SIGCOMM Internet Measurement Workshop (IMW'2002)*, November 2002.

[729] W.-P. Wang, D. Tipper, B. Jæger, and D. Medhi, "Fault recovery routing in wide area packet networks," in *Proc. 15th International Teletraffic Congress (ITC15)*, pp. 1077–1086, Washington, DC, June 1997.

[730] Y. Wang, Z. Wang, and L. Zhang, "Internet traffic engineering without full mesh overlaying," in *Proc. IEEE INFOCOM'2001*, pp. 565–571, Anchorage, AK, 2001.

[731] Z. Wang and J. Crowcroft, "Quality-of-service routing for supporting multimedia applications," *IEEE Journal on Selected Areas in Communications*, vol. 14, pp. 1228–1234, 1996.

[732] J. G. Wardrop, "Some theoretical aspects of road traffic research," *Proc. Inst of Civil Engineers, Part-2*, vol. 1, no. 2, pp. 325–378, 1952.

[733] Y. Watanabe and T. Oda, "Dynamic routing schemes for international networks," *IEEE Communications Magazine*, vol. 38, no. 8, pp. 70–75, 1990.

[734] D. Watson, G. R. Malan, and F. Jahanian, "An extensible probe architecture for network protocol performance measurement," *Software Practice & Experience*, vol. 34, no. 1, pp. 47–67, 2004.

[735] J. H. Weber, "Some traffic characteristics of communication networks with automatic alternate routing," *Bell Systems Technical Journal*, vol. 41, pp. 769–792, 1962.

[736] J. H. Weber, "A simulation study of routing and control in communication networks," *Bell Systems Technical Journal*, vol. 43, pp. 2639–2676, 1964.

[737] M. Welzl, *Network Congestion Control: Managing Internet Traffic*. John Wiley & Sons, Ltd, 2005.

[738] R. White, D. McPherson, and S. Sangli, *Practical BGP*. Addison-Wesley, 2004.

[739] Wikipedia, the free encyclopedia. http://www.wikipedia.org/

[740] W. Willinger and V. Paxson, "Where mathematics meets the Internet," *Notices of the American Mathematical Society*, vol. 45, no. 8, pp. 961–970, August 1998.

[741] W. Willinger, M. S. Taqqu, and A. Erramilli, "A bibliographical guide to self-similar traffic and performance modeling for high-speed networks," in *Stochastic Networks: Theory and Applications*, F. P. Kelly, S. Zachary and I. Ziedins (Eds.), pp. 339–366. Clarendon Press, 1996.

[742] A. Winnicki and J. Paczynski, "An approach to design three-layer controlled telephone networks," *Large Scale Systems*, vol. 1, pp. 245–256, 1980.

[743] Wireshark. http://www.wireshark.org/

[744] Wireshark Sample Capture Files. http://wiki.wireshark.org/SampleCaptures

[745] T. C. Woo, "A modular approach to packet classification: Algorithms and results," in *Proc. IEEE INFOCOM'2000*, pp. 1213–1222, 2000.

[746] World Telephone Numbering Guide. http://www.wtng.info/

[747] J. Wroclawski, "Specification of the controlled-load network element service," *IETF RFC 2211*, September 1997. http://www.rfc-editor.org/rfc/rfc2211.txt

[748] T.-H. Wu, *Fiber Network Service Survivability*. Artech House, 1992.

[749] T. Wu, "Network neutrality, broadband discrimination," *Journal of Telecommunications and High Technology Law*, vol. 2, pp. 141–179, 2005.

[750] X. Xiao, A. Hannan, B. Bailey, and L. Ni, "Traffic engineering with MPLS in the Internet," *IEEE Network*, vol. 14, no. 2, pp. 28–33, March/April 2000.

[751] X. Xiao, T. Telkamp, V. Fineberg, C. Chen, and L. Ni, "A Practical Approach for Providing QoS in the Internet," *IEEE Communications Magazine*, vol. 40, no. 12, pp. 56–62, December 2002.

[752] B. Yaged, "Minimum cost routing for dynamic network models," *Networks*, vol. 3, pp. 193–224, 1973. (See also, B. Yaged, "Minimum cost routing for static network models," *Networks*, vol. 1, pp. 139–172, 1971.)

[753] X. Yang, "NIRA: A new Internet routing architecture," Ph.D. dissertation, Massachusetts Institute of Technology, September 2004.

[754] X. Yang, D. Clark, and A. Berger, "NIRA: A new inter-domain routing architecture," *IEEE/ACM Trans. on Networking*, vol. 15, December 2007.

[755] Y. Yeh, M. Hluchyj, and A. Acampora, "The Knockout Switch: A simple modular architecture for high performance packet switching," *IEEE Journal of Selected Areas in Communications*, pp. 1426–1435, October 1987.

[756] J. Y. Yen, "Finding the k shortest loopless paths in a network," *Management Science*, vol. 17, pp. 712–716, 1971.

[757] S. Yilmaz and I. Matta, "On the scalability-performance tradeoffs in MPLS and IP routing," in *Proc. SPIE ITCOM'2002: Scalability and Traffic Control in IP Networks*, Boston, MA, August 2002.

[758] N. Zadeh, "On building minimum cost communication networks," *Networks*, vol. 3, pp. 315–331, 1973.

[759] N. Zadeh, "On building minimum cost communication networks over time," *Networks*, vol. 4, pp. 19–34, 1974.

[760] F. Zane, G. Narlikar, and A. Basu, "CoolCAMs: Power–efficient TCAMs for forwarding engines," in *Proc. IEEE INFOCOM'2003*, pp. 42–52, April 2003.

[761] H. Zang, C. Ou, and B. Mukherjee, "Path-protection routing and wavelength assignment in WDM mesh networks under duct-layer constraints," *IEEE/ACM Trans. on Networking*, vol. 11, pp. 248–258, 2003.

[762] L. Zhang, "A new architecture for packet switching network protocols," Laboratory for Computer Science, Massachusetts Institute of Technology, Cambridge, MA, Tech. Rep. MIT-LCS-TR-455, 1989.

[763] L. Zhang, M. Andrews, W. Aiello, S. Bhatta, and K. R. Krishnan, "A performance comparison of competitive on-line routing and state-dependent routing," in *Proc. IEEE GLOBECOM'97*, pp. 1813–1819, December 1997.

[764] R. Zhang and M. Bartell, *BGP Design and Implementation*. Cisco Press, 2003.

[765] R. Zhang and J.-P. Vasseur, "MPLS inter-autonomous system (AS) traffic engineering (TE) requirements," *IETF RFC 4216*, November 2005. http://www.rfc-editor.org/rfc/rfc4216.txt

[766] W. Zhao and S. Tripathi, "Routing guaranteed quality of service connections in integrated services packet networks," in *Proc. of IEEE International Conference on Network Protocols (ICNP'97)*, pp. 175–182, Atlanta, GA, 1997.

[767] Y. Zhu, C. Dovrolis, and M. Ammar, "Dynamic overlay routing based on available bandwidth estimation: A simulation study," *Computer Networks*, vol. 50, pp. 742–762, 2006.

Index
